Nutraceutical Proteins and Peptides in Health and Disease

NUTRACEUTICAL SCIENCE AND TECHNOLOGY

Series Editor

FEREIDOON SHAHIDI, PH.D., FACS, FCIC, FCIFST, FIFT, FRSC
University Research Professor
Department of Biochemistry
Memorial University of Newfoundland
St. John's, Newfoundland, Canada

1. Phytosterols as Functional Food Components and Nutraceuticals, *edited by Paresh C. Dutta*
2. Bioprocesses and Biotechnology for Functional Foods and Nutraceuticals, *edited by Jean-Richard Neeser and Bruce J. German*
3. Asian Functional Foods, *John Shi, Chi-Tang Ho, and Fereidoon Shahidi*
4. Nutraceutical Proteins and Peptides in Health and Disease, *edited by Yoshinori Mine and Fereidoon Shahidi*

Nutraceutical Proteins and Peptides in Health and Disease

Edited by

Yoshinori Mine
Fereidoon Shahidi

Taylor & Francis
Taylor & Francis Group
Boca Raton London New York

A CRC title, part of the Taylor & Francis imprint, a member of the Taylor & Francis Group, the academic division of T&F Informa plc.

Published in 2006 by
CRC Press
Taylor & Francis Group
6000 Broken Sound Parkway NW, Suite 300
Boca Raton, FL 33487-2742

Printed in the United States of America on acid-free paper
10 9 8 7 6 5 4 3 2 1

International Standard Book Number-10: 0-8247-5354-2 (Hardcover)
International Standard Book Number-13: 978-0-8247-5354-2 (Hardcover)
Library of Congress Card Number 2005043917

Library of Congress Cataloging-in-Publication Data

Nutraceutical proteins and peptides in health and disease / Yoshinori Mine and Fereidoon Shahidi, editors.
p. cm. -- (Nutraceutical science and technology ; 4)
Includes bibliographical references and index.
ISBN 0-8247-5354-2
1. Functional foods. 2. Food--Protein conntent. 3. Protein drugs. 4. Proteins in human nutrition. I. Mine, Yoshinori. II. Shahidi, Fereidoon, 1951- III. Series.

QP144.F85N88 2005
612.3'98--dc22 2005043917

T&F informa

Taylor & Francis Group
is the Academic Division of T&F Informa plc.

Visit the Taylor & Francis Web site at
http://www.taylorandfrancis.com

and the CRC Press Web site at
http://www.crcpress.com

Preface

Recent advancement of science has demonstrated myriad biological activity associated with food proteins and peptides. Reports of the beneficial health effects of some peptides, short-chain and otherwise, have been appearing in the literature. A variety of such peptides may be prepared upon hydrolysis of proteins, using either chemical or enzymatic processes, from a number of source materials, including milk, soy protein, and other plant as well as animal sources. These peptides may have a positive influence on calcium absorption, regulation of serum cholesterol, and the action of proteins as immunomodulators. Furthermore, a number of peptides may possess antimicrobial properties and, hence, may enhance the body's defense mechanisms or be used for rendering microbial stability to foods. Hydrolysis of proteins may also lead to the production of hypoallergenic products, and some may have, in part, inhibitory effects for angiotensin-I-converting enzymes (ACE). Thus, they may have applications in the treatments for blood pressure, heart failure, and myocardinal infarction, as well as diabetic conditions. Bioactive peptides may also be influential in stress reduction through an opiate-like effect.

Quantitative structure–activity relationships have been developed for evaluating the biological role of peptides. Some of the effects are arising from the specific amino acid sequence of individual peptides. Meanwhile, the antioxidant activity of such products has been demonstrated, and this activity may be responsible, at least in part, for some of the biological properties of peptides. Finally, recent developments in proteomics may help in providing novel means for further clarifications regarding the activity and role of biopeptides.

We are grateful to all the authors for their state-of-the-art contributions that made the publication of this book possible. This is the first book covering discussions on bioactive proteins and peptides in the area of nutraceutical and functional foods and can serve as a compendium of information for biochemists, nutritionists, food scientists, and health professionals in universities as well as in government and industry research laboratories. The book could also be used as a reference by senior undergraduate and graduate students. We extend our appreciation to Jennifer Kovacs-Nolan for her help in editing the manuscript.

Yoshinori Mine
Fereidoon Shahidi

Editors

Fereidoon Shahidi, Ph.D., FACS, FCIC, FCIFST, FIFT, FRSC, has reached the highest academic level, university research professor, in the Department of Biochemistry at Memorial University of Newfoundland (MUN). He also is cross-appointed to the Department of Biology, Ocean Sciences Centre, and the aquaculture program at MUN. Dr. Shahidi is the author of over 500 research papers and book chapters, has authored or edited some 40 books, and has given over 300 presentations at scientific conferences. His research contributions have led to several industrial developments around the globe.

Dr. Shahidi's current research interests include different areas of nutraceuticals and functional foods as well as marine foods and natural antioxidants. Dr. Shahidi serves as the editor-in-chief of the *Journal of Food Lipids* and editor of *Food Chemistry* as well as an editorial board member of the *Journal of Food Science*; *Journal of Agricultural and Food Chemistry, Nutraceuticals and Food*; and the *International Journal of Food Properties*. He was the recipient of the 1996 William J. Eva Award from the Canadian Institute of Food Science and Technology in recognition of his outstanding contributions to food science in Canada through research and service. He also received the 1998 Earl P. McFee Award from the Atlantic Fisheries Technological Society in recognition of his exemplary contributions in the seafood area and their global impact.

He has been recognized as one of the most highly cited authors in the world in the discipline of agriculture, plant, and animal sciences and was the recipient of the 2002 ADM Award from the American Oil Chemists' Society. Dr. Shahidi is the recipient of the 2005 Stephen Chang Award from the Institute of Food Technologists. He is the past chairperson of Lipid Oxidation and Quality of the American Oil Chemists' Society as well as the agricultural and food chemistry division of the American Chemical Society. He serves as a member of the Expert Advisory Panel of Health Canada on Standards of Evidence for Health Claims for Foods, the Standards Council of Canada on Fats and Oils, the Advisory Group of Agriculture and Agri-Food Canada on Plant Products, and the Nutraceutical Network of Canada. He is a member of the Washington-based Council of Agricultural Science and Technology on Nutraceuticals.

Yoshinori Mine, Ph.D., received an MSc degree in 1987 from the Faculty of Agricultural Science (Food Science), Shinshu University, Japan and a Ph.D. degree in biochemistry from Tokyo University of Agriculture and Technology, Japan in 1993. He joined the University of Guelph, Canada in 1995 as a faculty member in the Department of Food Science. His research of over 15 years has been related to biologically active proteins and peptides in eggs and the molecular biology of egg allergens. Currently, he is professor and industrial research chair in Egg Material Science. He was the recipient of the PREA award (Premier's Research Excellence

Award) in 2000. He has published 11 book chapters, 10 review articles, and 86 peer-reviewed original papers in his field. Dr. Mine has also served as a member of the editorial boards of the *Journal of Agricultural Food Chemistry* and *International Journal of Food Science & Technology*. Dr. Mine is a member of several national and international egg material sciences, nutraceuticals, and functional foods research societies.

Contributors

N. Alizadeh-Pasdar
Food, Nutrition, and Health
Faculty of Agricultural Sciences
The University of British Columbia
Vancouver, B.C., Canada

Angelina M. Alvarez
Graduate School of Bioagricultural Sciences
Nagoya University
Nagoya, Japan

Seiichiro Aoe
Department of Home Economics
Otsuma Women's University
Otsuma, Japan

Hiba A. Bawadi
Department of Food Science
Louisiana State University
Baton Rouge, Louisiana

D. Bouglé
Service de Pédiatrie A
CHU de Caen
Caen, France

S. Bouhallab
Laboratoire de Recherches de Technologie Laitière
Rennes, France

Judy C.K. Chan
Food, Nutrition, and Health
Faculty of Agricultural Sciences
The University of British Columbia
Vancouver, B.C., Canada

Kazuhiro Chiba
Laboratory of Bio-organic Chemistry
Tokyo University of Agriculture and Technology
Tokyo, Japan

Paul D. Cotter
Department of Microbiology and National Food Biotechnology Centre
University College Cork
Cork, Ireland

K.J. Cross
School of Dental Science
The University of Melbourne
Melbourne, Australia

Jerzy Dziuba
Food Biochemistry
University of Warmia and Mazury
Olsztyn, Poland

Richard J. FitzGerald
Life Sciences Department
University of Limerick
Limerick, Ireland

Sylvie F. Gauthier
Institut des Nutraceutiques et des Aliments Fonctionnels (INAF)
Université Laval
Québec, Canada

Subrata Ghosh
Gastroenterology Section
Imperial College Faculty of Medicine
London, UK

Paule Emilie Groleau
Institut des Nutraceutiques et des Aliments Fonctionnels (INAF)
Université Laval
Québec, Canada

Benjamin Guesdon
Laboratoire de Physiologie de la Nutrition et du Comportement Alimentaire
Institut National Agronomique Paris-Grignon (INAP-G)
Paris, France

Hajime Hatta
Department of Food and Nutrition
Kyoto Women's University
Kyoto, Japan

Susan L. Hefle
Food Allergy Research and Resource Program
University of Nebraska
Lincoln, Nebraska

Colin Hill
Department of Microbiology and National Food Biotechnology Centre
University College Cork
Cork, Ireland

Chein-Soo Hong
Department of Internal Medicine
Yonsei University College of Medicine
Seoul, Korea

N.L. Huq
School of Dental Science
The University of Melbourne
Melbourne, Australia

Anna Iwaniak
Food Biochemistry
University of Warmia and Mazury
Olsztyn, Poland

Hidehiko Izumi
Nagoya University of Arts and Sciences
Nissin, Aichi, Japan

L.R. Juneja
Nutritional Foods Division
Taiyo Kgaku Co. Ltd
Yokkaichi, Japan

Akio Kato
Department of Biological Chemistry
Yamaguchi University
Yoshida, Japan

Takeo Kato
Food Research Center
Aichi Prefectural Government
Nagoya, Aichi, Japan

Kyu-Earn Kim
Department of Pediatrics
Yonsei University College of Medicine
Seoul, Korea

David D. Kitts
Food, Nutrition, and Health Faculty of Agricultural Sciences
The University of British Columbia
Vancouver, B.C., Canada

Hannu Korhonen
MTT Agrifood Research Finland
Jokioinen, Finland

Jennifer Kovacs-Nolan
Department of Food Science
University of Guelph
Ontario, Canada

Jun-ichi Kurisaki
Genetic Diversity Department
National Institute of Agrobiological Sciences
Tokyo, Japan

Eunice C.Y. Li-Chan
Food, Nutrition, and Health Faculty of Agricultural Sciences
The University of British Columbia
Vancouver, B.C., Canada

Jack N. Losso
Department of Food Science
Louisiana State University
Baton Rouge, Louisiana

Pertti Marnila
MTT Agrifood Research Finland
Jokioinen, Finland

Tsukasa Matsuda
Graduate School of Bioagricultural Sciences
Nagoya University
Nagoya, Japan

Hans Meisel
Federal Research Centre for Nutrition and Food
Institute for Dairy Chemistry and Technology
Kiel, Germany

Yoshinori Mine
Department of Food Science
University of Guelph
Ontario, Canada

Koko Mizumachi
Department of Animal Products Research
National Institute of Livestock and Grassland Science
Ibaraki, Japan

Akira Mori
Department of Applied Biological Chemistry
The University of Tokyo
Tokyo, Japan

Brian Murray
Department of Life Science
University of Limerick
Limerick, Ireland

Satoshi Nagaoka
Department of Food Science
Gifu University
Gifu, Japan

Shuryo Nakai
Food, Nutrition, and Health Faculty of Agricultural Sciences
The University of British Columbia
Vancouver, B.C., Canada

Soichiro Nakamura
Department of Bioscience and Biotechnology
Shimane University
Shimane, Japan

Masayuki Nakase
Department of Food Science for Health
Minami-kyushu University
Miyazaki, Japan

Toshihide Nishimura
Graduate School of Biosphere Science
Hiroshima University
Hiroshima, Japan

Satoshi Nagaoka
Department of Food Science
Gifu University
Gifu, Japan

Tadashi Ogawa
Graduate School of Agriculture
Kyoto University
Gokasho, Japan

M. Ozeki
Nutritional Foods Division
Taiyo Kagaku Co. Ltd.
Yokkaichi, Japan

Lisa Pichon
Laboratoire de Physiologie de la Nutrition et du Comportement Alimentaire
Institut National Agronomique Paris-Grignon (INAP-G)
Paris, France

Raymond J. Playford
Gastroenterology Section
Imperial College Faculty of Medicine
London, UK

Yves Pouliot
Institut des Nutraceutiques et des Aliments Fonctionnels (INAF)
Université Laval
Québec, Canada

T.P. Rao
Nutritional Foods Division
Taiyo Kagaku Co. Ltd.
Yokkaichi, Japan

E.C. Reynolds
School of Dental Science
The University of Melbourne
Australia

R. Paul Ross
Teagasc Dairy Products Research Centre
Moorepark
Fermoy, County Cork, Ireland

Prithy Rupa
Department of Food Science
University of Guelph
Ontario, Canada

Fereidoon Shahidi
Department of Biochemistry
Memorial University of Newfoundland
St. John's, NL, Canada

Makoto Shimizu
Department of Applied Biological Chemistry
The University of Tokyo
Tokyo, Japan

Yuichi Tada
Life Science Laboratory
Mitsui Chemicals, INC.
Mobara, Chiba, Japan

Yukihiro Takada
Technology and Research Institute
ITOCHU Corporation
Tokyo, Japan

Soichi Tanabe
Graduate School of Biosphere Science
Hiroshima University
Hiroshima, Japan

Steve L. Taylor
Food Allergy Research and Resource Program
University of Nebraska
Lincoln, Nebraska

Daniel Tomé
Laboratoire de Physiologie de la Nutrition et du Comportement Alimentaire
Institut National Agronomique Paris-Grignon (INAP-G)
Paris, France

Daniel J. Walsh
Limerick Institute of Technology
Limerick, Ireland

Jun Watanabe
Creative Research Initiative "Sousei" (CRIS)
Hokkaido University
Kita-ku, Sapporo, Japan

Ada H.-K. Wong
Department of Food Science
University of Guelph
Ontario, Canada

Table of Contents

SECTION I Nutrient Absorption System

SECTION II The Body's Defense System

SECTION V Hypoallergenic Foods

SECTION VI Modern Approaches to Bioactive Proteins and Peptides

Section I

Nutrient Absorption System

1 Nutraceutical Proteins and Peptides in Health and Disease: An Overview

Yoshinori Mine and Fereidoon Shahidi

CONTENTS

A growing body of scientific evidence in the past decade has revealed that many food proteins and peptides exhibit specific biological activities in addition to their established nutritional value as sources of protein (1–3). Bioactive peptides are specific protein fragments that positively impact body function or condition and ultimately may influence human health. These peptides are inactive within the sequence of the parent protein and can be released during proteolysis or fermentation. Gastrointestinal proteolytic enzymes release the peptides, and the fraction of these compounds absorbed may be sufficient for a person to be physiologically active (1,4). Recent developments of bioactive proteins and peptides in important areas related to nutraceutical proteins and peptides for the promotion of human health and prevention of chronic diseases have been presented in this volume. Leading authorities in the field have made contributions covering various important aspects of selected topics. Depending on the nature of the proteins, peptides, and their amino acid sequences, these proteins/peptides may exert a number of different biological activities *in vivo*, affecting the body's function. Six main areas related to bioactive proteins and peptides in which there is increasing scientific and commercial interest have been covered (5–7).

1.1 THE NUTRIENT ABSORPTION SYSTEM

Bioactive peptides are widely distributed in milk proteins; the class of caseinophosphopeptides (CPP) derived from casein hydrolysis is one of the well-known groups of milk peptides. CPP is an effective peptide for stabilizing amorphous calcium phosphate and for rendering calcium ions soluble, thus enhancing bioavailability by passive mechanisms. CPP also improves absorption of other trace elements (8,9). Coronary heart disease, caused by high cholesterol, is the leading cause of morbidity and mortality in Western societies. Dietary protein is considered useful as a regulator of serum cholesterol, and various food proteins and peptides have been found to possess biological activity for lowering cholesterol absorption. Metal-binding proteins and peptides are involved in several disease processes and are capable of acting as immunomodulators, suggesting a possible therapeutic role for them in human medicine. The mechanism of absorption of dietary substances, including the physiological role of bioactive peptides, remains poorly understood. The recent discovery of various pathways for the intestinal absorption of dietary substances and the importance of tight junctions opens a new window in this area. Tight-junction modulatory food factors improve the efficient absorption of functional substances.

1.2 THE BODY'S DEFENSE SYSTEM

The system involved in the body's defense is varied and complex (10,11). The investigation of the role of functional proteins and peptides in this area holds significant promise. One important group of bioactive peptides comprises those that have antimicrobial activity. New antimicrobial peptides are of great importance in agriculture, as natural preservatives, and in medical applications. Many resources such as mammalian, plant, and marine organisms produce antimicrobial peptides. The oral administration of specific antibodies is an attractive approach to passive immunization against various pathogens, in both humans and animals. Milk and egg are considered to be ideal sources for such applications. The biological mechanisms associated with orally administered specific bovine milk and colostral antibodies in fighting microbial infectious diseases are noteworthy. Egg yolk antibodies offer several advantages over conventional antibodies produced in mammals, making them excellent candidates for many potential medical and nutraceutical applications. The gastrointestinal tract and colostrum possess various growth factors that are associated with the body's defense mechanism. These factors from natural food resources are of great significance in bioactive peptides research. Antiangiogenic proteins, peptides, and amino acids have been found in various food sources, and the concept of bioactive amino acids, peptides, and polypeptides as antiangiogenic functional foods with clear nutritional and clinical potential has been documented. The cystatin superfamily includes a number of chicken cystatin-like cysteine proteinase inhibitors that are widely distributed in mammalian tissues and body fluids. The cysteine proteinase inhibitors have been suggested to be important in protecting the host from invading organisms and tissue destruction caused by endogenous cysteine proteinases in inflammation, tumor growth, and metastasis. Biological activities of recombinant cystatin have been investigated and these include anti-rotavirus, anti-*Salmonella*, and anticancer effects.

1.3 THE BODY'S REGULATORY SYSTEM

Angiotensin-I-converting enzyme (ACE) is a key enzyme in the regulation of peripheral blood pressure and electrolytes (12,13). Inhibitors of ACE are widely used in therapy for hypertension, heart failure, myocardial infarction, and diabetic nephropathy. The ACE-inhibiting effect has been observed *in vivo* with various peptides from food proteins. The blood pressure–lowering food-derived peptides are now one of the most important bioactive peptide groups for clinical application. Milk proteins are highly functional substances, and it has become clear during the last two decades that they are a source of biologically active peptides. Milk basic protein (MBP), an ingredient that strongly stimulates both bone formation and bone resorption *in vitro* and *in vivo*, has been shown to increase the number of osteoblastic cells and the amount of bone proteins such as collagen. MBP promotes bone formation and suppresses bone resorption in humans. CPP-amorphous calcium phosphate (ACP) has been demonstrated to not only help prevent the development of dental caries but also to repair enamel subsurface demineralized lesions representing early stages of the caries process. The CPP-ACP technology has been commercialized, and functional foods and oral care products are commercially available. Fibrinolytic enzymes can be found in a variety of foods, particularly in the traditional fermented foods in Asia. Like other potent fibrinolytic enzymes, these novel food-derived enzymes are useful for thrombolytic therapy. These enzymes have significant potential for food fortification and nutraceutical application in the effective prevention of cardiovascular disease.

1.4 THE BODY'S NERVOUS SYSTEM

Opioid peptides are defined as peptide-like encephalins that have both affinity for opiate receptors and opiate-like effects, which are inhibited by naloxone (14). Opioid peptides can be obtained by *in vitro* enzymatic hydrolysis of food proteins. Most of them are derived from bovine milk or human β-casein and therefore are known as β-casomorphins, but some are also found in α-casein, α-lactalbumin, and β-lactoglobulin. β-Casomorphins might fit into the category of "food hormones." Stress is the root cause of many in mental and physical abnormalities in humans. Research on green tea has revealed that L-theanine can induce both psychological and physiological relaxation. L-Theanine induces a strong α-wave in the brain, indicating its clear psychological relaxation effects in humans. Green tea L-theanine is widely recommended for various food applications.

1.5 HYPOALLERGENIC FOODS

Food allergies occur in some individuals as the result of abnormal immunoglobulin responses to a particular food or food component, usually naturally occurring proteins (15,16). Food allergies affect only a small percentage of the population; however, food allergic reactions can be quite severe and even life threatening in some of individuals. The avoidance of a specific food can be difficult, and complete avoidance is unlikely. Wheat is one of the world's most important grains, and a variety of flour and wheat products are consumed throughout the world. Hypersensitive responses to allergenic proteins in wheat flour have long been a public health problem. Glutenin and α-amylase inhibitors have been identified as the major allergens. Clinical evaluation has been showed that enzymatically treated hypoallergenic wheat flour could act as an anti-allergen via allergen-specific immunotolerance. Cow's milk allergy is also a serious health problem, since milk is widely used for infant formula. The αs1-casein and β-lactoglobulin, major casein and whey proteins in bovine milk, are absent in human milk and are believed to be the potent allergens in cow's milk. Caseins (αs2-, β- and κ-) have also been reported as milk allergens. A variety of hypoallergenic formulas have been produced and are widely available to milk-allergic patients. Although the hydrolysis of milk proteins is the only practical method to reduce the allergenicity at present, epitope-modified milk proteins, tolerogenic milk peptides, and combinations with probiotics are good antiallergic candidates. The prevalence of egg allergy is about 35% among food allergic reactions in children. The four major allergenic proteins in egg white are ovomucoid, ovalbumin, ovotransferrin, and lysozyme. Ovomucoid is the abundant allergen in egg white. A complete immunoglobulin E (IgE) epitope mapping of ovomucoid and ovalbumin has been completed, and this will enable us to tailor egg allergens to reduce their allergenicity. Soybean and soy products, which are an important protein source, are known to be one of the major allergenic foods. Three major soybean allergens — Gly m Bd60K, Gly m 30K, and Gly m Bd 28K — have been identified. The α-subunit of β-conglycinin and Gly m Bd 28K was eliminated from soybean seeds by molecular breeding. The strongest allergen, Gly m Bd 28K, was almost completely removed from defatted soymilk under the condition of limited pH and high ionic strength. Meat products, which are important for their high nutritional value and palatability, are generally considered less allergenic than other common food allergens; however, there is increasing evidence that even meat products can provoke allergic reactions in sensitized patients. The major beef allergens are bovine serum albumin (BSA) and bovine gamma globulin (BGG). In addition, actin, myoglobin, and tropomyosin sometimes cause allergic reactions. Heat treatment and enzymatic treatment reduce or abolish the allergenicity of meat. Clinical cases of allergic disorders caused by rice, which are less frequent than those caused by other common allergens, have been reported in Japan and in some European countries. Three strong IgE-binding proteins with molecular masses of 16 kDa and 33 to 35 kDa have been identified. A 16-kDa protein was first isolated from the salt-soluble fraction of rice seed proteins. The 26-kDa protein, which is the major component of the rice endosperm globulin fraction, was identified as an allergen. The

33-kDa protein was strongly stained with IgE antibodies from rice-allergic patients. Hypoallergenic rice was developed by means of enzymatically digesting the allergens, followed by extracting or washing out. The combination of pressurization and protease treatment was suggested to be effective in the removal of allergenic proteins from rice grains. The genetic suppression of a target rice allergen, a 16-kDa protein, has also been conducted. Buckwheat is widely consumed in Japan, Korea, and China because of its high productivity and outstanding nutritive value. Buckwheat products are increasing in popularity as health foods based on their high protein content, abundance of vitamin B_1, rutin, and dietary fiber, and low fat content. Although buckwheat allergy is not common, buckwheat is considered to be a very potent allergen. Even small amounts of the proteins in buckwheat can sometimes provoke serious hypersensitive reactions. Various proteins in buckwheat have been found to be allergenic; however, α-22-kDa protein is considered to be the major allergenic protein in buckwheat.

1.6 MODERN APPROACHES TO BIOACTIVE PROTEINS AND PEPTIDES

Computer databases are employed for biomacromolecule classification, taking into account the similarity between the sequence motifs; and proteins are identified based on the sequences or the fragments by the use of mass spectrometry (17,18). Such databases enable identification of bioactive peptides in the protein chain and the hierarchical classification of proteins. The information obtained can be useful for modeling the biomacromolecule structure. BIOPEP, established in 1999, consists of two major databases: proteins and bioactive peptides. A new quantitative structure-activity relationships (QSAR) system was proposed by combining principal component similarity (PCS) analysis, homology similarity analysis (HAS), and artificial neural networks (ANA). Random centroid optimization (RCO) is another technology for peptide designing in food research and development.

The modern molecular approach is becoming a powerful tool in the field of food science. The recent development of recombinant techniques has enabled not only the improvement of the functional properties and antimicrobial action of lysozyme but also the reduction of its allergenicity. Genetic engineering of lysozyme has been highly effective in elucidating the molecular mechanism of its structural and functional properties. Novel functional lysozymes can be designed and constructed using the post-translational modifications in a yeast expression system. This technique will contribute to the elucidation of the molecular mechanisms of diseases such as allergy and amyloidosis and will aid in the development of preventive approaches. Recently, the systematic synthesis of peptides and peptidomimetics has played a very important role in the structure elucidation and estimation of their active conformations in target receptors. In order to achieve the final goal of the structure analysis of peptides, synthetic methodology based on combinatorial, high-throughput peptide synthesis is required to explore the adequate peptide sequences among the enormous amount of oligopeptides now known.

Bacteriocins are antimicrobial peptides/proteins produced by bacteria. Lacticin 3147 is a two-peptide lantibiotic containing seven lanthionine groups in total. Lacticin 3147 was isolated by a strategy designed to screen kefir grains for a bacteriocin producer with potential use in dairy fermentation. A combination of classic techniques and modern bioinformatics, through the comparison of closely related bacteriocins, is a useful approach. Two novel approaches for such purposes have been proposed: (1) a new methodology for liquid-phase peptide synthesis using thermomorphic biphasic organic solutions, and (2) a new anodic method for the introduction of functional groups for the conformation control of proline moieties that enable the regulation of secondary structures of several kinds of peptides. There is increasing commercial interest in the production of bioactive peptides from various sources such as milk, egg, cereal, and fish proteins. Industrial-scale production of such peptides is, however, hampered by the lack of suitable technologies. Membrane-based separation techniques as unit operations in a general fractionation/purification process are powerful tools for peptide separation. Although some processes have reached commercial scale, a large number of processes are too complex and expensive to be commercialized. High-resolution purification techniques that could perform peptide separation in a minimum of processing steps are still needed. The recent development of proteomics may provide useful information to formulate new tools that could specifically bind bioactive peptides in one single step.

REFERENCES

1. H. Korhonen, A. Pihlanto. Food-derived bioactive peptides-opportunities for designing future foods. *Curr. Pharm. Des.*, 9: 1297–1308, 2003.
2. C.K. Ferrari, E.A. Torres. Biochemical pharmacology of functional foods and prevention of chronic diseases of aging. *Biomed. Pharmacother.*, 57: 251–260, 2003.
3. G.P. Zaloga, R.A. Siddiqui. Biologically active dietary peptides. *Mini. Rev. Med. Chem.*, 4: 815–821, 2004.
4. M. Yoshikawa, H. Fujita, N. Matoba, Y. Takenaka, T. Yamamoto, R. Yamauchi, H. Tsuruki H. K. Takahara. Bioactive peptides derived from food proteins preventing lifestyle-related diseases. *Biofactors*, 12: 143–146, 2000.
5. S. Arai, T. Osawa, H. Ohigashi, M. Yoshikawa, S. Kaminogawa, M. Watanabe, T. Ogawa, K. Okubo, S. Watanabe, H. Nishino, K. Shinohara, T. Esashi, T. Hirahara. A Mainstay of functional food science in Japan — history, present status, and future outlook. *Biosci. Biotechnol. Biochem.*, 65: 1–13, 2001.
6. Japan Health Food & Nutrition Food Association. Foods for specified health uses (FOSHU) — A guideline. http/www.health-station.com/jhnfa/ 2004.
7. T. Hirahara. Key factors for the success of functional foods. *Biofactors*, 22: 289–293, 2004.
8. D.D. Kitts, K. Weiler. Bioactive proteins and peptides from food sources. Application of bioprocesses used in isolation and recovery. *Curr. Pharm. Des.*, 9: 1309–1323, 2003.
9. M. Shimizu. Food-derived peptides and intestinal functions. *Biofactors*, 21: 43–47, 2004.

10. H. Meisel. Multifunctional peptides encrypted in milk proteins. *Biofactors*, 21: 55–61, 2004.
11. Y. Mine, J. Kovacs-Nolan. Biologically active hen egg components in human health and disease. *J. Poultry Sci.*, 41: 1–29, 2004.
12. N. Yamamoto, M. Ejiri, S. Mizuno. Biogenic peptides and their potential use. *Curr. Pharm. Des.*, 16: 1345–1355, 2003.
13. A. Kilara, D. Panyam. Peptides from milk proteins and their properties. *Crit. Rev. Food Sci. Nutr.*, 43: 607–633, 2003.
14. H. Meisel, R.J. FitzGerald. Opioid peptides encrypted in intact milk proteins. *Br. J. Nutr.*, 84: 27–31, 2000.
15. S.L. Hefle, S.L. Taylor. Food allergy and the food industry. *Curr. Allergy Asthma Rep.*, 4: 55–59, 2004.
16. H.A. Sampson. Update on food allergy. *J Allergy Clin Immunol.*, 113: 805–819, 2004.
17. J. Dziuba, P. Minkiewicz, D. Nalecz, A. Iwaniak. Database of biologically active peptide sequence. *Nahrung*, 43: 190–195, 1999.
18. J. Senorans, E. Ibanez, A. Clifuentes. New trends in food processing. *Crit. Rev. Food Sci. Nutr.*, 43: 507–526, 2003.

2 Calcium Binding Peptides

David D. Kitts

CONTENTS

2.1 INTRODUCTION

Calcium homeostasis in mammalian species is acutely regulated by the concentration of soluble calcium ions, which reach both intracellular and extracellular body compartments following absorption from the gastrointestinal tract. The intestine represents the principal organ where calcium ions enter the body compartmental pools by two primary intestinal transport mechanisms present in the intestinal lumen. In general, calcium transport at this level regulates calcium bioavailability from different food matrices, as the intestinal mucosal layer is the only biological barrier for calcium to cross in order to reach the portal and, subsequently, systemic circulations. Intestinal calcium transport involves translocation of soluble calcium ions from the intestinal lumen to the lateral space occupied by the lamina propria, where two distinct pathways, namely, cellular and paracelluar mechanisms, exist. Calcium ions are transported via the cytosolic compartment in the cellular pathway, facilitated by vitamin D, and are moved by the tight junctions between adjacent cells by passive absorption in the paracellular pathway. The rate of calcium transport in the cellular mechanism is saturable and attains a constant value at a certain intralumenal calcium concentration. Alternatively, the rate of calcium transport in the nonsaturable, paracellular pathway increases as a linear function of soluble concentration of the ion. Prominent cofactors that may enhance paracellular calcium transport include lactose, some amino acids, and some bioactive peptides, such as caseinphosphopeptides (CPP) from milk proteins.

The relative importance of milk proteins in health and well being has been traditionally associated with nutritional factors related to protein digestibility and availability of important amino acids required for the growth of the neonate. Casein has also been associated with calcium, as one third of the calcium in bovine milk is present in the serum phase as free Ca^{2+} or complexed with phosphate, and two thirds are partly incorporated in micellar calcium phosphate and partly bound to casein (1–3).

Individual casein fractions have recently been recognized to have potential physiological roles and possess various bioactive properties (4,5). Bioactive peptides are peptides that are usually inactive within the protein sequence but become active upon release during digestive or hydrolysis processes (5–7). There are instances when liberated peptides, in spite of having little nutritional significance, can adequately exert a physiological function (8). To complement this further, bioactive peptides obtained from enzymatic hydrolysis *in vitro* or from digestion products *in vivo* have a natural resistance to further enzymatic hydrolysis because of covalently bound phosphate groups and proline residues that exist in large amounts within the peptides (7). Sites of action for bioactive peptides are dependent on the specific sites to which they are exposed to after release from the native protein by proteolytic action (9). Thus, some prominent effects can be localized in the gastrointestinal tract after mucosal absorption.

Mellander (10–11) was the first to report the bioactivity of isolated peptide fractions derived from enzymatic hydrolysis of phosphate-rich caseinates. These peptides were associated with an affinity to form strong complexes with calcium, thereby increasing calcium ion solubility and thus enhancing calcium absorption. As a consequence of this property, skeletal mineralization was improved in raichitic infants administered phosphate-rich casein peptides (10). Since the discovery of CPP, numerous studies have confirmed the ability of CPP to inhibit calcium phosphate precipitation *in vitro* (12) and thus enhance the solubility and uptake of calcium *in vivo* from the distal intestine during luminal digestion of casein-containing meals (13–16). The yield of CPP generated from the digestion of casein has been estimated to be approximately 5 mg for 200 g casein per kg diet (17), a sufficient quantity that can improve paracellular calcium absorption *in situ* in the distal intestine of rats fed casein diets (18) and soy diets supplemented with CPP (19). Heaney et al. (20) have also shown improved calcium absorption in postmenopausal women when CPP was administered with calcium. Extensive research on CPP has focused mainly on its role of enhancing solubility and hence bioavailability of calcium during milk or casein meal digestion (17,21–28). Other workers, however, have reported no effect of CPP supplementation or casein feeding on improving calcium bioavailability (29,30). Since the distal small intestine is a primary location for the passive transport of calcium, the ability of CPP to prevent insolubilization of calcium ions would theoretically lead to greater calcium absorption from the small intestine. This chapter will discuss the calcium-binding activity of CPP, with emphasis on the structure–function aspects that underlie potential enhancement of calcium bioavailability.

2.2 SOURCES OF CALCIUM-BINDING PEPTIDES FROM MILK PROTEINS

Bovine milk proteins are classified into two primary fractions: (1) the many whey proteins, which consist of α-lactalbumin, β-lactoglobulin, immunoglobulins, lactoferrin, proteose-peptone fractions, transferrin, and serum albumin, and (2) the caseinates (31,32). Bovine caseins constitute 78 to 80% of total milk proteins and represent a major source of amino acids, calcium, and inorganic phosphate to the neonate (33). There is a relatively small proportion of bovine whey proteins that interacts with calcium ions (34), although this may be different in the case of human milk (35). Milk caseinates associate together as highly aggregated micelles, with characteristically high contents of proline and ester-bound phosphate, with low solubility at pH 4 to 5 (36). There are four casein primary fractions in bovine milk — α_{s1}, α_{s2}, β, and κ, at approximately 38%, 10%, 36%, and 12% of whole casein, respectively (37). The primary amino acid sequence of the polypeptide chain for these peptides is known, with a phosphate content ranging from 1 mole of phosphate per mole protein for κ-casein, to 5, 8, and 11 phosphate groups, respectively, for β-casein, α_{s1}-casein, and α_{s2}-casein (Table 2.1). Casein phosphate groups represent esterified serine monoesters that have usually highly anionic clusters available for binding to divalane catins such as calcium, zinc, and iron (31,37,38). With a complex network of calcium phosphate bridges, bovine caseins contain a proportion of calcium and inorganic phosphate in milk that is greater than would be expected from the physicochemical solubility of calcium phosphate in milk (36). Despite the strong interaction between calcium and the inorganic residues in milk, the CPP–Ca^{2+} binding constant will facilitate the release of Ca^{2+} during intestinal absorption.

Casein micelles consist of a network of submicelles that are enriched with κ-casein on the micelle surface and the presence of α_{s1}-, α_{s2}-, and β-caseinates in the core. The submicelles are linked together with calcium phosphate and, to some extent, both hydrogen and hydrophobic bonding (36). The micelle charge is neutralized when the calcium content exceeds 6 mM and the α_{s1}-, α_{s2}-, and β-caseins precipitate to form a curd. Bovine α- and β-caseins are characterized by a high frequency of phosphoseryl and proline residues, having a distinct amphipathic nature, with both the polar and hydrophobic domains organized in a tertiary structure (1). All caseins, especially β-casein, have a high content of proline, preventing the formation of secondary structures (e.g., α-helices and β-sheets/turns), thus allowing caseins to exist as flexible structures. As a result of the flexible structure, individual caseins are especially stable to denaturing agents such as heat or urea but are more susceptible to proteolysis than are typical globular proteins through enhanced penetration by exo- and endopeptidases. Predicted secondary structures around phosphorylation sites on Ca^{2+}-sensitive caseins often comprise an α-helix loop and an α-helix motif, with the sites of phosphorylation located in the loop region (39). The motif consisting principally of phosphoserines is essential for binding calcium, but possible conformation changes in the backbone of casein-binding peptides allow the residues remote from the phosphorylated motif to also interact with calcium (40).

TABLE 2.1
The Positioning of Phosphoseryl Residues on κ-, α_{s1}-, α_{s2}-, and β- Caseinates

κ-Casein

```
125                                                 138
Ile-Ala-Ser-Gly-Pro-Thr-Ser-Thr-Pro-Thr-Ile-Glu-Ala-Val

143                                     153
Val-Ala-Thr-Leu-Glu-Ala-Ser-Pro-Glu-Val-Ile
                        P
```

α_{s1}-Casein

```
39                                                  52
Glu-Leu-Ser-Lys-Asp-Ile-Gly-Ser-Gly-Ser-Thr-Glu-Asp-Gln
                            P       P

61      63                                                              79
Glu-Ala-Glu-Ser-Ile-Ser-Ser-Ser-Glu-Glu-Ile-Val-Pro-Asn-Ser-Val-Glu-Gln-Lys
            P       P   P   P                           P

112                         119
Val-Pro-Asn-Ser-Ala-Glu-Glu-Arg
            P
```

α_{s2}-Casein

```
1                                                                           20
Lys-Asn-Thr-Met-Glu-His-Val-Ser-Ser-Ser-Glu-Glu-Ser-Ile-Ile-Ser-Gln-Glu-Thr-Tyr
                            P   P                           P

29                  34
Asn-Pro-Ser-Lys-Glu-Asn
        P

55                                                                              75
Gly-Ser-Ser-Ser-Glu-Glu-Ser-Ala-Glu-Val-Ala-Thr-Glu-Glu-Val-Lys-Ile-Thr-Val-Asp-Asp
    P   P   P           P

127                                                                             147
Gln-Leu-Ser-Thr-Ser-Glu-Glu-Asn-Ser-Lys-Lys-Thr-Val-Asp-Met-Glu-Ser-Thr-Glu-Val-Phe
        P                       P                               P
```

β-Casein

```
10                                                  23
Gly-Glu-Ile-Val-Glu-Ser-Leu-Ser-Ser-Ser-Glu-Glu-Ser-Ile
                    P       P   P   P

33                                          44
Phe-Gln-Ser-Glu-Glu-Gln-Gln-Gln-Thr-Glu-Asp-Glu
        P
```

Caseinophosphopeptides (CPP) have been prepared from a variety of enzymatic hydrolysis procedures that have employed animal, bacterial, and fungal proteolytic enzyme sources (41–44). Soluble calcium derived from these CPP sources range from 7.4 to 24 mg Ca^{2+} mg^{-1} CPP. Peptides prepared by successive digestion treatment with pepsin and trypsin have a higher capacity to bind calcium, as evidenced by the ability to chelate 248 mg Ca^{2+} g^{-1} peptide, compared with

168 mg Ca^{2+} g^{-1} peptide for tryptic peptides (45). Significantly different ^{45}Ca diaysability results, reflecting relative yield of CPP have been shown using *in vitro* by stepwise addition of digestive enzymes and adjustments of pH simulated. These experiments digestion in humans and produced CPP from porcine pancreatic trypsin, bioprotease N100L, and porcine pancreatin (46).

Nuclear magnetic resonance (NMR) studies have indicated that the majority of highly phosphorylated sites are present in regions likely to be β-turns or loops in the peptide structure. This feature corresponds to different fractions of CPP. In fact, although different fractions of CPP share analogous sequence and possible function, they have distinctly different conformations. For example, the phosphoseryl motif present in the α_{s1}-casein (*f*59–79) peptide has a Glu^{61}–$SerP^{67}$ segment, which is a loop-type structure and is followed by a Pro^{73}–Val^{76} in a β-turn conformation (47). The same motif found in β-casein (*f*1–25) $SerP^{17}$–Glu^{20} is embedded in a β-turn along with sequences Val^{8}–Glu^{11}, Glu^{21}–Thr^{24}, and Arg^{1}–Glu^{4}, all of which are located in a loop structure (40). Dephosphorylated O-phosphoseryl peptide groups are more susceptible to enzymatic hydrolysis than are organic phosphate groups, which are present as monoesters, diesters, or triesters (11). Phosphopeptide α_{s1}-casein (*f*59–79) found in the ileum of rats consuming casein-containing diets is dephosphorylated at a slower rate than other peptide fragments in the intestinal lumen, implicating a high *in vivo* stability (21).

The density of phosphoseryl groups present in CPP in close vicinity varies and is a quality that can be used to distinguish specific forms of CPP derived from the same protein. Tryptic hydrolyzates of bovine casein micelles contain mineral-rich peptide fractions that are composed of 72% colloidal calcium, 49% inorganic phosphate, 27% nitrogen, and 82% micellar phosphoseryl residues in the native micelle. Distribution of phosphoserine moieties vary in caseins (α_s, β, and κ), and the extent of phosphorylation directly affects CPP mineral chelating affinity, in the order of $\alpha_{s2} > \alpha_{s1} > \beta > \kappa$ (48). α_{s2}-Casein, which represents only 10% of the total casein, contains the greatest amount of phosphate (e.g., 10 to 13 $molecule^{-1}$) and the greatest sensitivity to calcium (49). β-Casein phosphopeptides, with a molecular weight of about 3 kDa, are sufficient in micromolar concentration to bind and solubilize up to 40-fold the molar excess of phosphate and calcium (50). The anionic triplet (SerP-SerP-SerP-Glu-Glu) is a distinctive feature of all of the major phosphopeptide fractions characterized, whether the CPP are generated from *in vitro* or *in vivo* origin. The most common CPP derived from *in vitro* tryptic digests of whole bovine casein include β-casein-4P (*f*1–25), α_{s1}-casein-5P (*f*59–79), α_{s2}-casein-4P (*f*1–21), and α_{s2}-casein-4P (*f*46–70). A wide variety of phosphopeptide sequences have been characterized using different analytical techniques or hydrolytic enzymes (Table 2.2). Mineral binding to CPP has been associated with specific α_{s1}-casein (*f*45–55), α_{s1}-casein (*f*56–74), α_{s2}-casein (*f*55–75), and β-casein (*f*13–35) segments that contain the embedded anionic triplet. Phosphopeptides prepared using different *in vitro* or *in vivo* procedures share general amino acid profiles, analogous chemical natures, and physiological functions (2,17,28,50,51). The high negative charge of the phosphopeptide results in relatively higher resistance to further proteolysis (7). Several of the CPP derived

TABLE 2.2
Examples of CPPs Derived from Enzymatic Hydrolysis of β- and α_{s1}–, α_{s2}–Caseins

Caseinate	Mol Wt	Peptide	Enzyme	Source	Reference
β	3.1kDa	(*f*1–25)4P	Trypsin	*In vitro*	33,40,47,57,91–93
β		(*f*1–25)4P	Pepsin & Trypsin	*In vitro*	44
β	3.5kDa	(*f*1–28)4P	Trypsin	*In vitro*	22,47,53
β	3.3 kDa	(*f*2–28)4P	Trypsin	*In vitro*	41
β	3.6 kDa	(*f*1–29)4P	Trypsin	*In vitro*	41
α_{s1}	5.6 kDa	(*f*35-79)8P	Trypsin	*In vitro*	41
α_{s1}		(*f*59–79)5P	Gastric and Intestinal enzymes	*In vivo* (rats' ileum)	21
α_{s2}	2.7 kDa	(*f*1–21)4P	Trypsin	*In vitro*	47,91

from β-casein have been recovered from the intestinal digests of mini-pigs (52,53) and rats (13) only hours following ingestion of casein diets and in the stomach and duodenum of adult humans after milk digestion (54). CPP found in the cecum and colon contents of rats fed diets containing purified CPP was not distinguishable from that of rats fed casein that contained a comparable amount of phosphoserines (55). Moreover, CPP remnants have been recovered in the feces of casein-fed rats, suggesting that at least a part of the CPP formed in the small intestine is not hydrolyzed in the digestive tract but excreted intact (55). These results strongly indicate that CPP is not only resistant to hydrolysis by digestive enzymes but also can escape hydrolysis by enteric bacteria in the distal intestine.

2.3 PHYSICOCHEMICAL PROPERTIES OF CPP

Phosphopeptide fractions isolated from tryptic casein hydrolysate will sequester calcium phosphate over a range of neutral (e.g., 7) to alkaline (e.g., 10.5) pH values (1,12) and by doing so inhibit amorphous calcium phosphate precipitation (56–58). By linking the seryl-phosphate group to a calcium phosphate moiety within a nanometer-sized CPP particle (e.g., β-CPP *f*1–25), the amorphous dicalcium phosphate is stabilized. As a result, formation of hydroxyapatite crystals at an early stage is prevented (39). The binding kinetics of phosphoseryl groups will vary depending on the concentration of inorganic calcium. For example, concentrations of calcium, below 1 mM, will mainly result in exothermic binding of calcium ions by phosphoseryl residues. With an increase in calcium concentration to 1 to 3 mM, a second exothermic binding phase is followed by an endothermic reaction involving the phosphoseryl residues binding to carboxylate residues, thus producing increased self-association. The reaction becomes highly endothermic, and phosphoserine residues interact predominantly with carboxylate residues above a 3-mM calcium threshold concentration. When calcium concentrations reach 5 to 6 mM, binding to carboxylate residues minimizes intermolecular electrostatic repulsion and hydrophobic interaction of the hydrophobic domains. This will lead to the formation of large aggregates and eventual precipitation.

Inhibition of calcium phosphate precipitation can be achieved at a concentration of 10 mg L^{-1} of CPP, or higher at a pH of 6.5 with complete emulsion stability maintained with 100 mg L^{-1} (57). The strength of calcium sequestering is, in turn, highly correlated with the extent of phosphorylation in different fractions of CPP and the conditions (e.g., pH, ionic strength, temperature, and enzyme selection) of the peptide–metal interaction. Meisel (59) reported an apparent calcium-binding constant (K_{app} l mol^{-1}), for α_{s1}-casein (*f*43–48), with two phosphate groups to be 328 l mol^{-1}, compared with 629 l mol^{-1} for β-casein (*f*1–25), and 841 l mol^{-1} for α_{s1}-casein (*f*59–79) that contained four and five phosphate groups, respectively. Baumy et al. (60) identified maximum calcium binding (pK 6.57 to 7.10) on residues 17,18, and 19 on β-casein (*f*1–25) and Nagasawa et al. (18) reported optimal binding of CPP to ^{45}Ca at pH 8, with a 50% binding capacity to ^{45}Ca at 8 μg of CPP and 100% binding capacity at 25 μg CPP. The complete dephosphorylation of CPP results in loss of affinity for divalent metals (18) and mineral solubilization properties (22). Dephosphorylated CPP is not effective at sequestering divalent ions at concentrations higher than 100 mg L^{-1} (57), indicating that the binding sites for calcium and the functional units of CPP bioactivity are essentially associated with the phosphoserine residues. In contrast, although both glutamic and aspartic acids contribute a free carboxyl site for potential metal binding, this activity is relatively of minor importance (61).

CPP binding affinity (pK) to cations is greatly affected by changes in temperature, ionic strength, and the presence of calcium. Protonation of phosphate groups at low pH will reduce calcium binding and calcium phosphate solubilization (62). Increasing temperature from 20 to 40°C will catalyze endothermic binding of calcium, thereby greatly enhancing CPP–metal association (3). At a greater ionic strength, the presence of other cations with very similar binding kinetics as calcium (e.g., sodium) will result in competition with calcium for binding to phosphoseryl residues, thus lowering the calcium-binding ability of CPP. Excess calcium may also inhibit calcium chelation, as shown by a linear decrease in binding between 1 and 4 moles of calcium per mole of phosphopeptide. Only a minor effect on binding affinity was observed following saturation of β-casein (*f*1–25) with excess calcium (60).

Binding one calcium ion to ionized phosphoserine will decrease the bond strength between the oxygen and hydrogen atoms on the second acidic group. As a result, the functional binding sequence (e.g., SerP-SerP-SerP-Glu-Glu) of CPP is only partially important for the mineral–peptide interaction along with the neighboring residues that also afford an important function for effective binding (63). Amino acid residues located both upstream and downstream from this region can participate in binding. It is noteworthy that mimicking the amino acid sequence that spanned the phosphoseryl residue, (SerP-SerP-SerP-Glu-Glu) will result in loss of calcium-binding activity (64,65). Synthetic peptides corresponding to the phosphoserine-rich region of α_{s1}-casein bind less calcium than the entire tryptic peptide (*f*59–79), whereas the synthetic N-terminal (*f*59–63) and the corresponding C-terminal (*f*71–79) do not bind calcium (66). Park and Allen (67) have shown that altering the amino acid sequence by cleavage at the glutamic acid residue will result in a charge redistribution in the anionic cluster and significantly lower calcium-binding activity.

2.4 PROPOSED BIOACTIVE PROPERTIES OF CPP

2.4.1 Bioavailability of Minerals

Evidence for a physiological role of CPP in intestinal calcium absorption points more to an indirect luminal inhibition of precipitation by phosphate salts (23) than to a direct effect on the intestinal mucosal membrane (68). Calcium transport measured using a perfused rat ileum segment was decreased by over 90% with the addition of inorganic phosphate; however, in the presence of CPP, the inhibiting effect of phosphate on calcium absorption was only 46 to 60% (68). The addition of CPP to calcium induces a transient rise in free intracellular calcium ions in cultured intestinal tumor HT-29 and Caco-2 cells (69,70). These workers have speculated that CPP does not influence membrane-bound receptors or ion channels but rather acts as a calcium carrier for calcium-selective channels on the plasma membrane or becomes internalized as a calcium–peptide complex via endocytosis (71).

The level of dietary calcium, in particular, the calcium to CPP ratio, is an important factor for determining the effect of CPP on calcium absorption in animal studies. With an adequate dietary calcium intake, the resulting luminal calcium concentration will reduce the contribution of CPP required to enhance calcium binding to small or negligible levels (26,72). The minimum effective concentration of CPP to enhance calcium absorption, under marginal dietary calcium levels (0 to 0.35%) has been shown to be 0.7 g kg^{-1}. This translates to a CPP/Ca weight ratio of 0.2 in rats fed CPP-supplemented soy protein diets (72). A similar result has been reported by Lee et al. (16) who showed a concentration-dependent effect of CPP/Ca ratios that ranged from 0 to 0.35. Calcium balance is not altered by CPP when the CPP/Ca ratio exceeds 0.35 in rats fed soy protein diets. These findings support earlier studies that showed CPP to be ineffective at enhancing calcium utilization in rats fed soy diets supplemented with 3% CPP and containing adequate (e.g., 0.5%) calcium (73). Kopra et al. (29) demonstrated a similar result in rats fed whey protein diets containing 0.7% calcium and supplemented with CPP. In both studies, plasma calcium, apparent calcium absorption, or retention and femur mineralization in normal and vitamin D–deficient rats were not altered by the feeding of CPP. Supplementing whey protein diets containing 0.53% calcium with CPP at levels of 3.8 or 7.6 g 100 g^{-1} also had no effect on fractional calcium absorption in normal female rats (21). However, calcium absorption in the same female rats was elevated when lower-calcium-containing whey protein diets (e.g., 0.2%) were supplemented with 9 to 27% casein. There is evidence that the level of calcium in the diet, or the demands for calcium by the host, will influence the bioactive effect of CPP. For example, inclusion of CPP in low (e.g., 1.5%)-calcium diets successfully improved the utilization of calcium and recovered the specific gravity of egg shells in laying hens. This result contradicted other findings, in which at normal (e.g., 3.4%) calcium diets where calcium requirements for eggshell formation were satisfied, the addition of CPP had no effect on egg shell parameters (74).

2.4.2 CPP AND BONE HEALTH

As noted earlier, Mellander et al. (10) first reported that in rachitic patients, bone calcification increased and a potential for enhanced uptake of calcium in bone was realized with the administration of crude CPP formulations. Cell culture studies conducted many years later confirmed this observation with the finding that increases in the calcification of the diaphyseal area of explanted and cultured embryonic rat femora, tibiae, and metatarsal rudiments were attributed to the presence of CPP (22). More recently, Tsuchita et al. (75) reported a similar finding *in vivo* of calcified CPP in ovariectomized rats. Other bone calcification indices and different biomechanical end-point measurements have indicated that reducing the level of dietary protein intake from casein or lowering casein digestibility, thereby reducing CPP generation, was associated with both reduced bone mineralization and biomechanical strength (27). The low quality of phosphopeptides in casein foods has also been suggested to be a factor, since no effect of CPP was observed on the calcium content and bending movement of the femurs of weaning piglets and on the apparent absorption or retention of calcium in vitaminD–depleted rats (25). Bennett et al. (30) reported that the uptake of calcium into bone at low levels was not attributed to a direct acute effect of CPP. This conclusion supported previous studies that came to the same conclusion, that CPP had no effect on calcium absorption (65) or femur calcification (73) and that bone mineralization was influenced more by dietary calcium intake than by variation in protein source (76). Of considerable interest was the finding that decalcification of CPP prior to supplementation may contribute to a negative effect on calcium absorption in rats (30). However, other studies have shown that CPP, along with the isoflavone genistein, can bring about a synergistic effect at increasing the femoral dry weight, calcium content, alkaline phosphatase activity, and DNA content of both diaphyseal and metaphyseal tissues of young and elderly rats (77).

Common to these studies has been the lack of attempt to control for the presence of coexisting phosphatases, mineral bioavailability inhibitory agents, and other dietary factors when interpreting the CPP-enhancement results with different bivalent minerals. It is important to note that CPP-Ca complexes interact with intestinal phosphatases, thus influencing subsequent utilization of the released calcium. Intestinal phosphatases, which cleave calcium ions free from the CPP-Ca complex, have the highest activity in the ileal segment of the small intestine in rats (21), where passive absorption of calcium takes place. Notwithstanding this, CPP can interact with alkaline phosphatase to enhance bone mineralization and with tartrate-resistant acid phosphate to reduce bone resorption (78).

The presence of antinutrients such as phytates in soy products should also be considered in assessing the efficacy of CPP to enhance calcium bioavailability.

The interaction between phytates and calcium could lead to the formation of insoluble calcium salts, and significantly reduce a potential CPP-calcium interaction by adversely affecting the sequestering activity of CPP. This has been shown in studies where CPP was added for the purpose of inhibiting the phytate-induced effect responsible for reduced calcium absorption and had significantly better results than when bovine serum albumin, casein, or whey proteins were added to improve calcium absorption in rat pups fed oat-based and soy-based high phytate infant diets (79). In a human trial, CPP was associated with an improvement of calcium bioavailability from phytate-containing rice-based infant cereal by 26 to 27%, when the ratio of phytate to calcium was relatively low (phytatel/Ca μmol ratio of 30:12). CPP had no effect on calcium absorption from whole-grain-based infant meals and bread meals when phytate:calcium ratios were increased to 300:13.5 and 528:7.8 (80,81).

We should not, however, overlook the fact that other milk proteins such as whey proteins may also be involved in bone health. For example, greater calcium deposits into femoral tissue, compared with calcium supplements such as calcium carbonate, calcium lactate, or even calcium citrate, have been reported in rats fed whey supplemented with calcium (35). Moreover, milk whey protein has also been shown to be effective in the proliferation and differentiation of cultured osteoblast cells (82).

2.5 OTHER POTENTIAL APPLICATIONS FOR CPP

Being a natural digestion product of milk (54) and cheese (83), with resistance to further proteolytic hydrolysis (55,84) and possessing a high ability to form organophosphate salts with calcium, CPP have other potential uses such as in the making of anticariogenic toothpaste and mouthwash products (85). It has been proposed that the potential sequestering and localization of amorphous calcium on the tooth surface by CPP adequately maintain a localized state of calcium supersaturation with respect to its close association with hydroxyapatite, which, in turn, depresses enamel demineralization and enhances the potential for remineralization at the tooth surface (86). The inclusion of α_{s1}-casein into plaque has been shown to act as a calcium phosphate reservoir and a buffer against bacterial acid production via a proton-accepting affinity at pH 7. Along this line of thinking, Harper et al. (87) reported the anticariogenic activity of different cheeses, and they attributed this to the casein and calcium phosphate contents in the cheeses. Studies have shown that a 0.1% CPP-complex significantly reduces caries by 14%, compared with a 1% CPP-complex that produces 55% and 46% reductions in smooth surface and fissure caries activity, respectively (88). The mechanism underlying this result has been attributed to the binding character of CPP and S*treptococcus mutans* for calcium (89). Grenby et al. (90) also reported other minor proteins or peptides with phosphoseryl residues that were derived from the proteose-peptone fraction of β-casein with calcium ion sequestering activity.

Phosphopeptides, α_{s1}-casein (*f*59–79), and β-casein (*f*1–25), derived from raw and pasteurized bovine milk may also influence immune status. CPP derived

TABLE 2.3
Summary of Casein and CPP Effects on Calcium Bioavailability

Procedure	Result	Reference
In vivo balance studies	No effect of dietary protein source. No effect of synthetic or natural CPP.	25,26,65,66,76
Extrinsic or intrinsic labeling of calcium and collection of label in fecal excreta collection	Casein > Soya CPP no effect	15, 30,81
^{45}Ca or ^{47}Ca fractional absorption	Casein induced effect on Ca absorption. No effect of CPP	21, 30
Bone uptake of calcium radioactivity	Casein > Soya	27,73
	CPP no effect	30
Ligated duodenal loop	Positive CPP effect	24
Ligated ileal loop	Positive CPP and casein effect	13, 14–16 18,26,27,73
Mineralization composition	Casein > soya	73
	CPP – no effect	27, 25
Bone loss	CPP – positive effect	75,94

from α_{s1}-casein and β-casein inhibit concanavalin A (Con A)–induced proliferation of mouse spleen cells and rabbit Peyer's patch cells, while enhancing lipopolysaccharide (LPS)- and phytohemagglutinin (PHA)-induced proliferation in these cells (91,92). Other CPP fractions, consisting mainly of α_{s2}-casein (1–32) and β-casein (1–28), produced a similar effect on Con A-, LPS-, and PHA-induced proliferation of nude mouse spleen cells (93). Other findings include the enhancement of IgG, IgM, and IgA levels by CPP in cultured cells (93). Further studies have indicated that feeding CPP to mice could enhance mucosal immunity by stimulating a greater intestinal-specific IgA response toward antigens such as ovalbumin and β-lactoglobulin (94) (Table 2.3).

2.6 CONCLUSIONS

Caseinophosphopeptides (CPP) derived from tryptic digestion of bovine milk casein are effective macropeptides for stabilizing amorphous calcium phosphate and rendering calcium ions in a relatively soluble form for potential enhanced bioavailability by paracellular (passive) mechanisms. The phosphoserine residues are important for the inhibition of insoluble calcium phosphate formation. The generation of bioactive CPP from α_{s1}-casein, α_{s2}-casein, and β-casein are characterized by phosphoseryl residue groupings, which differ in affinity to modulate the precipitation of calcium phosphate by forming amorphous dicalcium phosphate nanoclusters. The anionic hydrophilic domain of CPP and the specific amino acid residues located both upstream and downstream in the region are also important for mineral binding affinity. Bioactive properties of CPP that include divalent mineral binding have been proposed by many, although considerable controversy exists as to the relative significance of these activities. It is certain that CPP can enhance the solubility of calcium; however, the solubility of calcium alone may have little to do with the absolute absorbability of the mineral. Various factors relating to the generation of CPP from different caseinates in the intestine and the association of these peptides with calcium, with considerations made to the chemical makeup of the intestinal milieu and intestinal transit time, need to be studied further to determine the significance of CPP-induced calcium bioavailability.

REFERENCES

1. H.E. Swaisgood. Chemistry of milk proteins. In: PF Fox (Ed.) *Developments in Dairy Chemistry*, Volume 1: Proteins. Essex, U.K.: Applied Science Publishers Ltd, 1982, 1–59.
2. D.W. West. Structure and function of the phosphorylated residues of casein. *J. Dairy Sci.*, 53: 333–352, 1986.
3. H.E. Swaisgood. Review and update of casein chemistry. *J. Dairy Sci.*, 76: 3054–3061, 1993.
4. N.P. Shah. Effects of milk-derived bioactivities: an overview. *Br. J. Nutr.*, 84: S3–S10, 2000.
5. D.D. Kitts, K. Weiler. Bioactive proteins and peptides from food sources. Applications of bioprocesses used in isolation and recovery. *Curr. Pharm. Des.*, 9: 1309–1323, 2003.
6. H. Meisel. Overview on milk protein-derived peptides. *Int. Dairy J.*, 8: 363–373, 1998.
7. H. Meisel, E. Schlimme. Milk proteins: precursors of bioactive peptides. *Trends Food Sci. Technol.*, 1: 41–43, 2000.
8. D.D. Kitts. Bioactive substances in food. *Can. J. Physiol. Pharmacol.*, 72: 423–434, 1994.
9. F.L. Schanbacher, R.S. Talhouk, F.A. Murray, L.I. Gherman, L.B. Willett. Milk-borne bioactive peptides. *Int. Dairy J.*, 8: 393–403, 1998.
10. O. Mellander. The physiological importance of casein phosphopeptide calcium salts. II. Peroral calcium dosage of infants. *Acta Soc. Med. Ups.*, 55: 247–255, 1950.
11. O. Mellander. Phosphopeptides: chemical properties and their possible role in the intestinal absorption of metals. In: RH Wasserman (Ed.) *The Transport of Calcium and Strontium across Biological Membranes.* New York: Academic Press, 1963, 265–276.
12. R.E. Reeves, N.G. Latour. Calcium phosphate sequestering phosphopeptide from casein. *Science*, 128: 472–472, 1958.
13. H. Naito, A. Kawakami, T. Imamura. *In vivo* formation of phosphopeptide with calcium-binding property in the small intestinal tract of the rat fed on casein. *Agric. Biol. Chem.*, 36: 409–415, 1972.
14. R. Sato, T. Noguchi, H. Naito. The necessity for the phosphate portion of casein molecules to enhance calcium absorption from the small intestine. *Agric. Biol. Chem.*, 47: 2415–2417, 1983.
15. R. Sato, T. Noguchi, H. Naito. Casein phosphopeptide (CPP) enhances calcium absorption from the ligated segment of rat small intestine. *J. Nutr. Sci. Vitaminol.*, 32: 67–76, 1986.

16. Y.S. Lee, G. Park, H. Naito. Supplemental effect of casein phosphopeptides (CPP) on the calcium balance of growing rats. *J. Jpn. Soc. Nutr. Food Sci.*, 45: 333–338, 1992.
17. Y.S. Lee, T. Noguchi, H. Naito. Phosphopeptides and soluble calcium in the small intestine of rats given a casein diet. *Br. J. Nutr.*, 43: 467, 1980.
18. T. Nagasawa, Y.V. Yuan, D.D. Kitts. Casein phosphopeptides enhance paracellular calcium absorption but do not alter temporal blood pressure in normotensive rats. *Nutr. Res.*, 11: 819–830, 1991.
19. Y.V. Yuan, D.D. Kitts. Estimation of dietary calcium utilization in rats using a biochemical functional test. *Food Chem.*, 44: 1–7, 1992.
20. R.P. Heaney, Y Saito, H Orimo. Effect of caseinphosphopeptide on absorbability of co-ingested calcium in normal postmenopausal women. *J. Bone Mineral Methabol.*, 12: 77–81, 1994.
21. R. Brommage, M.A. Juillerat, R. Jost. Influence of casein phosphopeptides and lactulose on intestinal calcium absorption in adult female rats. *Lait*, 71: 173–180, 1991.
22. H.W. Gerber, R. Jost. Casein phosphopeptides. Their effect on calcification of *in vitro* cultured embryonic rat bone. *Calcif. Tissue Intern.*, 38: 350–357, 1986.
23. Y. Li, D. Tomé, J.F. Desjeux. Indirect effect of casein phosphopeptides on calcium absorption in rat ileum *in vitro*. *Reprod. Nutr. Dev.*, 29: 227–233, 1989.
24. H.M. Mykkänen, R.H. Wasserman. Enhanced absorption of calcium by casein phosphopeptides in rachitic and normal chicks. *J. Nutr.*, 110: 2141–2148, 1980.
25. K.E. Scholz-Ahrens, N. Kopra, C.A. Barth. Effect of casein phosphopeptides on utilization of calcium in minipigs and vitamin-D-deficient rats. *Z. Ernährungswiss.*, 29: 295–298, 1990.
26. D.D. Kitts, Y.V. Yuan, T. Nagasawa, Y. Moriyama. Effect of casein, caseinophosphopeptides and calcium intake on ilea ^{45}Ca disappearance and temporal systolic blood pressure in spontaneously hypertensive rats. *Br. J. Nutr.*, 68: 765–781, 1992.
27. Y.V. Yuan, D.D. Kitts. Calcium absorption and bone utilization in spontaneously hypertensive rats fed on native and heat-damaged casein and soybean protein. *Br. J. Nutr.*, 71: 583–603, 1994.
28. H. Tsuchita, H. Suzuki, T. Kuwata. The effect of casein phosphopeptides on calcium absorption from calcium-fortified milk in growing rats. *Br. J. Nutr.*, 85: 5–10, 2001.
29. N. Kopra, K.E. Scholz-Ahrens, C.A. Barth. Effect of casein phosphopeptides on utilization of calcium in vitamin D-replete and vitamin D-deficient rats. *Milchwissenschaft*, 47: 488–492, 1992.
30. T. Bennett, A. Desmond, M. Harrington, D. McDonagh, R. FitzGerald, A. Flynn, K.D. Cashman. The effect of high intakes of casein and casein phosphopeptide on calcium absorption in the rat. *Br. J. Nutr.*, 83: 673–680, 2000.
31. A.M. Fiat, D. Migliore-Samour, P. Jolles, L. Drouet, C.B.D. Sollier, J. Caen. Biologically active peptides from milk proteins with emphasis on two examples concerning antithrombotic and immunomodulating activities. *J. Dairy Sci.*, 76: 301–310, 1993.
32. D.A. Clare, H.E. Swaisgood. Bioactive milk peptides: a prospectus. *J. Dairy Sci.*, 83: 1187–1195, 2000.
33. E. Schlimme, H. Meisel. Bioactive peptides derived from milk proteins. Structural, physiological and analytical aspects. *Die Nahrung*, 39: 1–20, 1995.
34. B. Lonnerdal, C. Glazier. Calcium binding by a-lactalbumin in human milk and bovine milk. *J. Nutr.*, 115: 1209–1216, 1985.

35. G.S. Ranhotra, J.A. Gelrotyh, S.D. Leinen, A. Rao. Bioavailabililty of calcium in a high calcium whey fraction. *Nutr. Res.*, 17: 1663–1670, 1997.
36. L. Hambraeus. Importance of milk proteins in human nutrition: physiological aspects. In: TE Galesloot, BJ Tinbergen (Eds.). *Milk Proteins.* 1984 Proceedings of the International Congress on Milk Proteins. Pudoc: Wageningen, The Netherlands, 1985.
37. P.F. Fox. Milk proteins as food ingredients. *Int. J. Dairy Technol.*, 54: 41–55, 2001.
38. N. Aît-Oukhatar, S. Bouhallab, F. Bureau, P. Arhan, J.-L. Maubois, D.L. Bouglé. In *vitro* digestion of caseinophosphopeptide-iron complex. *J. Dairy Res.*, 67: 125–129, 2000.
39. C. Holt, M.J. Van Kemenade, L.S. Nelson Jr, L. Sawyer, J.E. Harries, R.T. Bailey, D.W.L. Hukins. Composition and structure of micellar calcium phosphate. *J. Dairy Res.*, 56: 411–416, 1989.
40. K.J. Cross, N.L. Huq, W. Bicknell, E.C. Reynolds. Cation-dependent structural features of β-casein-(1-25). *Biochem. J.*, 356: 277–286, 2001.
41. M.A. Juillerat, R. Baechler, R. Berrocal, S. Chanton, J.C. Scherz, R. Jost. Tryptic phosphopeptides from whole casein. I. Preparation and analysis by fast protein liquid chromatography. *J. Dairy Sci.*, 56: 603–611, 1989.
42. O. Park, H.E. Swaisgood, J.C. Allen. Calcium binding of phosphopeptides derived from hydrolysis of αs-casein or β-casein using immobilized trypsin. *J. Dairy Sci.*, 81: 2850–2857, 1998.
43. D. McDonagh, R. FitzGerald. Production of caseinophosphopeptides (CPPs) from sodium caseinate using a range of commercial protease preparations. *Int. Dairy J.*, 8: 39–45, 1998.
44. Ono, Y. Takagi, I. Kunishi. Casein phosphopeptides from casein micelles by successive digestion with pepsin and trypsin. *Biosci. Biotech. Biochem.*, 62: 16–21, 1998.
45. S. Kennefick, K.D. Cashman. Investigation of an *in vitro* model for predicting the effect of food components on calcium availability from meals. *Int. J. Food Sci. Nutr.*, 51: 45–54, 2000.
46. N.L. Huq, K.J. Cross, E.C. Reynolds. A ^{1}H-NMR study of the casein phosphopeptide α_{s1}-casein (59–79). *Biochim. Biophys. Acta*, 1247: 201–208, 1995.
47. V. Gagnaire, A. Pierre, D. Molle, J. Leonil. Phosphopeptides interacting with colloidal calcium phosphate isolated by tryptic hydrolysis of bovine casein micelles. *J. Dairy Res.*, 63: 405–422, 1996.
48. D.D. Kitts, Y.V. Yuan. Caseinophosphopeptides and calcium bioavailability. *Trends Food Sci. Technol.*, 3: 31–35, 1992.
49. T. Aoki, K. Toyooka, Y. Kako. Role of phosphate groups in the calcium sensitivity of α_{s2}-casein. *J. Dairy Sci.*, 68: 1624–1629, 1985.
50. W. Manson, J. Cannon. The reaction of α_{s1}- and β-casein with ferrous ions in the presence of oxygen. *J. Dairy Res.*, 45: 59–67, 1978.
51. H. Naito, H. Suzuki. Further evidence for the formation *in vivo* of phosphopeptide in the intestinal lumen from dietary β-casein. *Agric. Biol. Chem.*, 38: 1543–1545, 1974.
52. H. Meisel, H. Frister. Isolation and chemical characterization of a phosphopeptide from *in vivo* digests of casein. In: CA Barth, E Schlimme (Eds.). *Milk Proteins: Nutritional, Clinical, Functional and Technological Aspects.* New York: Steinkopff, Darmstadt and Springer, 1986, 150–154.
53. M. Hirayama, K. Toyota, H. Hidaka, H. Naito. Phosphopeptides in rat intestinal digests after ingesting casein phosphopeptides. *Biosci. Biotech. Biochem.*, 56: 1128–1129, 1992.

54. B. Chabance, P. Marteau, J.C. Rambaud, D. Migliore-Samour, M. Boynard, P. Perrotin, R. Guillet, P. Jolles, A.M. Fiat. Casein peptide release and passage to the blood in humans during digestion of milk or yogurt. *Biochimie*, 80: 155–165, 1998.
55. T. Kasai, R. Iwasaki, M. Tanaka, S. Kiriyama. Caseinphosphopeptides (CPP) in feces and contents in digestive tract of rats fed casein and CPP preparations. *Biosci. Biotech. Biochem.*, 59: 26–30, 1995.
56. R. Sato, M. Shindo, H. Gunshin, T. Noguchi, H. Naito. Characterization of phosphopeptide derived from bovine β-casein: an inhibitor to intra-intestinal precipitation of calcium phosphate. *Biochim. et Biophysic. Acta*, 1077: 413–415, 1991.
57. R. Berrocal, S. Chanton, M.A. Juillerat, B. Pavillard, J.C. Scherz, R. Jost. Tryptic phosphopeptides from whole casein. II. Physiochemical properties related to the solubilization of calcium. *J. Dairy Res.*, 56: 335–341, 1989.
58. K.H. Ellegård, C. Gammelgard-Larsen, E.S. Sorensen, S. Fedosov. Process scale chromatographic isolation, characterization and identification of tryptic bioactive casein phosphopeptides. *Int. Dairy J.*, 9: 639–652, 1999.
59. H. Meisel. Biochemical properties of regulatory peptides derived from milk proteins. *Biopolymers*, 43: 119–128, 1997.
60. J.-J. Baumy, P. Guenot, S. Sinbandhit, G. Brulé. Study of calcium binding to phosphoserine residues of β-casein and its phosphopeptide (1–25) by ^{31}P NMR. *J. Dairy Res.*, 56: 403–409, 1989.
61. I.R. Dickson, D.J. Perkins. Studies of the interactions between purified bovine caseins and alkaline earth metal ions. *Biochem. J.*, 124: 235–240, 1971.
62. H. Meisel, C. Olieman. Estimation of calcium-binding constants of casein phosphopeptides by capillary zone electrophoresis. *Anal. Chim. Acta*, 372: 291–297, 1998.
63. D.W. West. Structure and function of the phosphorylated residues of casein. *J. Dairy Res.*, 53: 333–352, 1986.
64. H. Meisel, H. Frister. Chemical characterization of bioactive peptides from *in vivo* digests of casein. *J. Dairy Res.*, 56: 343–349, 1989.
65. B.G. Shah, B. Belonje, A. Paquet. The lack of effect of synthetic phosphoseryl peptide on calcium absorption by the rat. *Nutr. Res.*, 10: 1331–1336, 1990.
66. R.J. FitzGerald. Potential uses of caseinophosphopeptides. *Int. Dairy J.*, 8: 451–457, 1998.
67. O. Park, J.C. Allen. Preparation of phosphopeptides derived alpha s-casein and beta-casein using immobilized glutamic acid-specific endopeptidase and characterization of their calcium binding. *J. Dairy Sci.*, 81: 2858–2865, 1998.
68. D. Erba, S. Ciappellano, G. Testolin. Effect of caseinphosphopeptides on inhibition of calcium intestinal absorption due to phosphate. *Nutr. Res.*, 21: 649–656, 2001.
69. A. Ferraretto, A. Signorile, A. Fiorilli, G. Tettamanti. CPP influence on calcium uptake by the tumoral cells of intestinal origin HT-29 and Caco-2 in culture. In: *44th Congress of the Italian Society for Biochemistry and Molecular Biology*. Volume 13, 1989, 95.
70. A. Ferraretto, A. Signorile, C. Gravaghi, A. Fiorilli, G. Tettamanti. Casein phosphopeptides-induced $[Ca^{+2}]_i$ changes in individual tumor cells of intestinal origin HT-29 cultured in vitro. In: *45th Congress of The Italian Society for Biochemistry and Molecular Biology*. Volume 15, 2000, 149.
71. A. Ferraretto, A. Signorile, C. Gravaghi, A. Fiorilli, G. Tettamanti. Casein phosphopeptides influence calcium uptake by cultured human intestinal HT-29 tumor cells. *J. Nutr.*, 131: 1655–1661, 2001.

72. Y. Saito, Y.S. Lee, S. Kimura. Minimum effective dose of casein phosphopeptides (CPP) for enhancement of calcium absorption in growing rats. *Int. J. Vitaminol. Nutr. Res.*, 68: 335–340, 1998.
73. Y.V. Yuan, D.D. Kitts. Confirmation of calcium absorption and femoral utilization in spontaneously hypertensive rats fed casein phosphopeptide supplemented diets. *Nutr. Res.*, 11: 1257–1272, 1991.
74. K. Ashida, T. Nakajima, M. Hirabayashi, Y. Saito, T. Matsui, H. Yano. Effects of dietary casein \phosphopeptides and calcium levels on eggshell quality and bone status in laying hens. *Anim. Sci. Technol. (Japan)*, 67: 967–974, 1996.
75. H. Tsuchita, T. Goto, T. Shimizu, Y. Yonehara, T. Kuwata. Dietary casein phosphopeptides prevent bone loss in aged ovariectomized rats. *J. Nutr.*, 126: 86–93, 1996.
76. Y.V. Yuan, D.D. Kitts. Effects of dietary calcium intake and protein source on calcium utilization and bone biomechanics in the spontaneously hypertensive rat. *J. Nutr. Biochem.*, 3: 452–460, 1992.
77. Z.J. Ma, M. Yamaguchi. Synergistic effect of genistein and casein phospho-peptides on bone components in young and elderly rats. *J. Health. Sci.*, 46: 474–479, 2000.
78. T. Matsui, H. Yano, T. Awano, T. Harumoto, Y. Saito. The influences of casein phosphopeptides on metabolism of ectopic bone induced by decalcified bone matrix implantation in rats. *J. Nutr. Sci. Vitaminol.*, 40: 137–145, 1994.
79. M. Hansen, B. Sandström, B. Lönnerdal. The effect of casein phosphopeptides on zinc and calcium absorption from high phytate infant diets assessed in rat pups and Caco-2 cells. *Ped. Res.*, 40: 547–552, 1996.
80. M. Hansen, B. Sandström, M. Jensen, S.S. Sørensen. Casein phosphopeptides improve zinc and calcium absorption from rice-based but not from whole-grain infant cereal. *J. Pediatr. Gastroenterol. Nutr.*, 24: 56–62, 1997.
81. M. Hansen, B. Sandström, M. Jensen, S.S. Sørensen. Effect of casein phosphopeptides on zinc and calcium absorption from bread meals. *J. Trace Elem. Med. Biol.*, 11: 143–149, 1997.
82. Y. Takada, S. Aoe, M. Kumegawa. Whey protein stimulates cell proliferation and differentiation in osteoblastic MC3T3-E1 cells. *Biochem. Biophys. Res. Comm.*, 223: 445–449, 1986.
83. F. Roudot-Algaron, D. LeBars, L. Kerhoas, J. Einhorn, J.C. Gripon. Phosphopeptides from Comte cheeses: nature and origin. *J. Food Sci.*, 59: 544–547, 1994.
84. T. Kasai, T. Honda, S. Kiriyama. Caseinphosphopeptides (CPP) in feces of rats fed casein diet. *Biosci. Biotechnol. Biochem.*, 56: 1150–1151, 1992.
85. E.C. Reynolds. Anticariogenic complexes of amorphous calcium phosphate stabilized by casein phosphopeptides. *Spec. Care Dentist.*, 18: 8–16, 1998.
86. R.K. Rose. Effects of an anticariogenic casein phosphopeptide on calcium diffusion in streptococcal model dental plaques. *Arch. Oral Biol.*, 45: 569–575, 2000.
87. D.S. Harper, J.C. Osborn, R. Clayton, J.J. Hefferren. Cariostatic evaluation of cheeses with diverse physical and compositional characteristics. *Caries Res.*, 20: 123–130, 1986.
88. E.C. Reynolds, C.J. Cain, F.L. Webber, C.L. Black, P.F. Riley, I.H. Johnson, J.W. Perich. Anticariogenicity of calcium phosphate complexes of tryptic casein phosphopeptides in the rat. *J. Dent. Res.*, 74: 1272–1279, 1995.
89. R.K. Rose. Binding characteristics of *Streptococcus mutans* for calcium and casein phosphopeptide. *Caries Res.*, 34: 427–431, 2000.

90. T.H. Grenby, A.T. Andrews, M. Mistry, R.J.H. Williams. Dental caries-protective agents in milk and milk products: investigations *in vitro*. *J. Dent.*, 29: 83–92, 2001.
91. W. Manson, W.D. Annan. The structure of a phosphopeptide derived from α-casein. *Arch. Biochem. Biophys.*, 145: 16–26, 1971.
92. D.D. Kitts, R. Leung, S. Nakai. Extrinsic labeling of caseinophosphopeptides with 45calcium and recovery following thermal treatment. *Can. Inst. Food Sci. Technol. J.*, 24: 278–282, 1992.
93. T. Ono, T. Ohotawa, Y. Takagi. Complexes of casein phosphopeptide and calcium phosphate prepared from casein micelles by tryptic digestion. *Biosci. Biotech. Biochem.*, 58: 1376–1380, 1994.
94. M. Yamaguchi, M. Tezuka, S. Shimanuki, M. Kishi, Y. Tukada. Casein phosphopeptides in dietary calcium tofu enhance calcium availability in ovariectomized rats: prevention of bone loss. *Food Sci. Technol. Int. Tokyo*, 4: 209–212, 1998.

3 Mineral-Binding Proteins and Peptides and Bioavailability of Trace Elements

D. Bouglé and S. Bouhallab

CONTENTS

3.1 INTRODUCTION

Proteins have nutritional as well as functional properties such as peptide sequences produced by the digestion of dietary proteins, display opioid, and immunoregulating activities; and whole proteins and derived peptides also bind minerals (1,2).

Deficiencies in micronutrients, including trace elements (TE), affect nearly half of the world's population. Approximately 35% of children and 42% of women are deficient in iron (Fe), zinc (Zn), or both. Fe and Zn deficiencies impair immunity, growth, and cognitive development (3–5) and cause increased morbidity during pregnancy and delivery (6). Deficiencies usually result from diets providing low amounts of minerals and TE with a low bioavailability. This bioavailability depends on interactions with other components of the diet — phytates, vitamin C, reducing agents, minerals, TE themselves, and proteins (7) — and on competition for common absorption pathways (8). In many cases, these interactions decrease TE bioavailability, but they could also be used to stabilize TE and keep them soluble and available for absorption. We can infer from the complex interactions between minerals and proteins that these nutrients influence, with great efficiency, the metabolism of minerals.

3.2 BIOCHEMICAL INTERACTIONS AMONG MINERALS, TE, AND PROTEINS OR PEPTIDES

3.2.1 PROTEINS

Within a protein, the side chains of the amino acid residues (tryptophan, tyrosine, serine, cysteine) can interact with metals. On a free amino acid, the α-carboxyl and, to a lesser extent, the α-amino groups can also be involved in the interaction (9–11). α- and β-Caseins strongly bind bivalent and trivalent cations, calcium (Ca), Zn, Fe, manganese (Mn), and copper (Cu) (12–14); the affinity order is $Ca^{2+} < Mn^{2+} < Zn^{2+} < Cu^{2+} < Fe^{3+}$. This affinity is mostly attributable to sequence clusters — Ser(P)-Ser(P)-Ser(P)-Glu-Glu — and depends on the number of phosphoseryl residues (11 for α_{S2}-, 8 for α_{S1}, and 5 for β-caseins); the binding capacities of casein decrease with the decrease of pH for Ca, Mn, Zn, and Cu ions and with the increase of ionic strength for Ca and Mn. During the binding of Fe to caseins, which involves coordination bonds, Fe is oxidized from Fe^{2+} to Fe^{3+}, and casein is oxidized as well (15,16). The affinity of caseins for Fe is ~100 × stronger than for Ca and other cations, and the Fe^{3+} ions remain bound to caseins whatever the pH or ionic strength. In addition, the conformational changes of caseins and the cross-linking that occur differ between α- and β-caseins (9,17–19). These biochemical differences between cations and between caseins explain the results of *in vivo* studies.

Whey proteins, α-lactalbumin, and lactoferrin also bind minerals[17] through sites other than caseins. Ca binding is a main factor in α-lactalbumin spatial conformation and stability; this protein binds also Zn through a second specific site; lactoferrin binds Fe^{3+} through two specific sites (20).

Heme-Fe, hemoglobin (MW = 64500), contains 0.335% Fe; it contains 65% of body iron. It is made of globin (96% molecular weight) and heme. Fe is in a reduced state (Fe^{2+}) and has strong interactions with the surrounding protein structures: it is bound by four coordination links to the heme porphyrin and by a fifth link to the globin chain; the sixth valence is free and available for oxygen binding and transport (21).

3.2.2 PEPTIDES

Peptides with mineral-binding properties are derived from enzyme proteolysis of whole proteins. Several phosphopeptides (CPP, caseinphosphopeptides) with

TABLE 3.1
Primary Sequences and Phosphorylation of Main Caseinophosphopeptides (CPP) of Bovine Caseins

Caseinophosphopeptide	No. of Phosphate Residues	Native Casein and Localization
- QMEAESpISpSpSpEEIVPNSpVEQK	5	αs_1 (59–79)
- NTME HVSpSpSpEESIISpQETYK	4	αs_2 (2–21)
- NANEEEYSIGSpSpSpEESpAEVATEEVK		αs_2 (46–70)
- ELEELNVPGEIVESpLSpSpSpEESITR		β (2–25)
- KDIGSpESpTEDQAMEDIK	2	αs_1 (43–58)
- EQLSpTSpEENSK		αs_2 (126–136)
- TEIPTINTIASpGEPTSTPTTEAVESTVATL EDSpPEVIESPPEINTVQVTSTAV		α (117–169)
- VPQLEIVPNSpAEER	1	αs_1 (106–119)
- FQSpEEQQQTEDELQDK		β (33–48)

Sp: phosphoserine.

specific sequences are produced by the digestion of α_{S1}-, α_{S2}-, and β-caseins (22,23); these CPP vary in size and content in the phosphoseryl groups. The sequences of the main CPP (see Table 3.1) are resistant to further hydrolysis. The ability to bind minerals depends on the number of phosphoseryl groups and therefore on the charge of peptides and cross-linking between chains (17,22,23) and thus on the production process. Using pancreatin, Alcalase, or proteinase L660 for casein hydrolysis yields CPP, which can bind 24, 15, or 7 g Ca 100 g^{-1}, respectively (24).

High purity CPP can been produced by a combination of enzyme hydrolysis and membrane filtration; they can bind and keep soluble 6 g of Ca, 10 g of Cu, 12 g of Zn, or 12 g of Fe 100 g^{-1} of CPP. Fewer peptides produced from whey proteins, and these peptides are less specific. Peptides obtained by hydrolysis of β-lactoglobulin and α-lactalbumin seem to have a higher affinity for Fe than do native proteins. The role of peptides obtained from bovine serum albumin, immunoglobulins, or lysozyme is currently unknown (23).

Binding to minerals can cause a change in the conformation or in the bridges between peptide chains and thereby modify (inhibit or accelerate) their enzyme hydrolysis (25).

3.3 PROTEINS AND BIOAVAILABILITY OF TE

3.3.1 Calcium

Dietary proteins and phosphorus do not usually contribute to the variability in calcium absorption efficiency (26). Bone accretion of calcium also depends on urinary calcium excretion, which is strongly related to net renal acid excretion. Increasing intake of purified proteins from either animal or plant sources similarly increases urinary calcium. "Excess" dietary protein from animal or plant source may be detrimental to bone health, but its effect will be modified by other nutrients in the diet: the high amount of calcium in milk compensates for urinary calcium losses generated by milk protein, while the high potassium levels of plant protein foods decreases urinary calcium. The hypocalciuric effect of the high phosphate associated with the amino acids of meat offsets, at least partially, the hypercalciuric effect of the protein (27).

In addition, the enhancing effect of milk protein on insulin-like growth factor 1 (IGF-1) synthesis favors bone mineralization, even without changes in Ca absorption (28,29).

3.3.2 Zinc

A significant positive correlation is found between zinc absorption and the protein content in meals containing milk, cheese, beef, and egg. Animal proteins counteract the inhibiting effect of phytic acid (30,31); casein could be slightly inhibitory compared with other proteins, including cow's milk whey proteins (32,33).

3.3.3 Iron

Most of the dietary Fe is released from food in the stomach and enters a common pool of ferric ion (Fe^{3+}), from which point absorption is strongly dependent on pH and other food components (34).

Fe^{3+} is insoluble at pH values above 4 and creates large-weight hydroxides. Therefore, keeping Fe in a reduced ferrous form (Fe^{2+}) or bound to weak ligands enhances it absorption at the alkaline pH of the duodenum. Fe can be absorbed as Fe^{2+} or Fe^{3+} by two different pathways. Fe^{3+} is usually reduced by a ferrireductase located at the brush border membrane before being absorbed as Fe^{2+} (8).

The digestion of meat and fish proteins produces low-molecular-weight peptides and free amino acids such as cysteine, which enhance Fe absorption (35). Vegetal proteins such as soya, lupin, or pea have an inhibitory effect that is not fully reversed by dephytinization (36–38). Although phosphorylated proteins of cow's milk (caseins) and egg keep Fe soluble, they are claimed to inhibit its absorption (39,40) and to bind Fe so strongly that it cannot be released in a free form available for absorption in the proximal bowel. On the other hand, some other studies have shown that the bioavailability of iron-milk-protein–complexes (Fe-casein or Fe-whey proteins) is similar to iron salts (41) and that defatted milk, in fact, enhances Fe absorption (42).

Other components of milk such as fat, the high calcium concentration of cow's milk, or technological processes can play a role in mineral bioavailability (43,44).

Lactoferrin has a strong affinity for Fe. Human milk has a high concentration of lactoferrin, and human milk iron is well absorbed, supporting the hypothesis that this protein is involved in the absorption of iron. However, studies have shown that neither bovine nor human lactoferrin enhances absorption of the Fe in infant formulas (45).

Blood hemoglobin and meat myoglobin provide Fe bound to heme. Heme Fe is quite insoluble at low pH and soluble at alkaline pH. The heme molecule is taken up intact by the mucosal cell, by a pathway distinct from nonheme iron; it is released in the enterocyte and metabolized like nonheme Fe. Dietary factors which react with nonheme iron in the gut do not influence heme-Fe uptake (8), and therefore the bioavailability of heme-Fe is higher than that of nonheme-Fe (~15 to 35%) (8,46).

3.4 PEPTIDES AND BIOAVAILABILITY OF TE

The peptides produced by protein digestion can have different effects on mineral and TE bioavailability from native proteins, even if the binding properties are kept similar. Therefore, hydrolysis of cow's milk, breast milk, and egg proteins improves the bioavailability of Fe and Zn (39,40,47). The properties of phosphopeptides released by casein degradation will be reviewed later in detail.

While glycinin (11S) and conglycinin (7S) fractions released by the digestion of soybean protein are inhibitory for Fe absorption (37), fermentation of soy products enhances it, perhaps by the solubilizing effect of volatile acids (48,49).

Heme-Fe is soluble at the alkaline pH of the duodenum and is absorbed by a specific pathway. Due to the low solubility of pure heme, attempts to increase the concentration of hemoglobin iron by hydrolyzing its globin moiety gave products with a low bioavailability because peptides produced by this hydrolysis are required to prevent the polymerization of heme (50,51). When the process allows a balance between hydrophobic peptides and the strength of peptide–heme interactions, heme iron can be concentrated and its high absorption rate preserved (51).

Peptides released by meat digestion also prevent the polymerization of heme and enhance heme-Fe absorption (34). Cysteine-containing peptides and free cysteine enhance heme and nonheme-Fe absorption (35,51). So far, the mechanisms involved have not been fully defined.

Amino acids such as histidine and methionine have a positive effect on zinc absorption; however, histidine is a strong promoter of Zn loss through urine (32) that could lead to a negative metabolic balance. Some studies that investigated Fe and Zn stabilized with glycine found that amino acid chelates keep TE soluble in the gut and are not reactive with food ingredients, leaving more TE to be potentially available for absorption. TE are absorbed into mucosal cells as chelates, and hydrolysis of the complex occurs into the mucosal cell before delivery of TE to plasma (52–54).

3.5 ROLES OF CPP

CPP released during the digestion of caseins (55–57) are partly resistant to further digestion and can be found in gut fluid and in feces (56–59); this supports the argument for their functional properties.

The resistance to enzymatic digestion depends on the phosphorylation of CPP (60), and this resistance disappears when CPP are dephosphorylated (61,62). Binding minerals to CPP reinforces this resistance to hydrolysis (63,64). The ability of brush-border alkaline phosphatases to release minerals depends on CPP (60,65).

Studies on the influence of CPP on mineral metabolism have given conflicting results. These discrepancies were not unexpected, however, owing to the specificity of the interactions between the different minerals and the multiplicity of CPP.

3.5.1 Calcium

Most of the studies have been devoted to Ca, even though its affinity for CPP is low. The good bioavailability of cow's milk Ca could be attributed to the presence of phosphorylated caseins that maintain it in a soluble state (66).

In vitro studies generally have reported an enhancing effect of CPP on Ca absorption and bone accretion in relation with an increased Ca solubility (61, 66–68). This effect was better observed at low CPP concentrations, while a high CPP/Ca ratio could be inhibitory (69). These observations support the biphasic and dose-dependent binding of Ca on CPP (70). A greater effect is observed when Ca is bound to CPP (71): CPP could act as a carrier for an endocytosis-mediated absorption pathway (72). Finally, the studies have shown that the different fractions of casein hydrolyzates give uneven effects on Ca absorption, suggesting that specific interactions occur between Ca and the various CPP (67).

On the other hand, *in vivo* studies have given rather disappointing results, even though CPP seem to improve Ca balance when Ca needs exceed the dietary supply (68,73). CPP-bound Ca has been shown to prevent the inhibitory interaction of phosphates (74).

Some hypotheses to explain the differences between *in vitro* and *in vivo* data are that (1) the ratio between Ca and CPP must remain optimal in the digestive fluid to keep Ca soluble and to enhance its absorption, and (2) this ratio depends on the affinity of Ca for the different CPP; it must also be kept in mind that the absorption of Ca is highly regulated by active transport mechanisms such that colonic absorption compensates for variations in small bowel ones (75).

3.5.2 Zinc

The binding of Zn to CPP depends on the pH and ionic strength of the milieu (17,19); therefore, it can be released from CPP in the gut. In rat studies, CPP have given results similar to those of Ca. These results are more obvious when minerals are bound to CPP. The Ca–CPP complex lessens the inhibitory effect of phytates on Zn absorption; CPP enhance Zn absorption from an oat gruel and from soya-based infant formula (71); the absorption of Zn bound to a purified CPP of β-casein (β-CN 1–25) is also increased and protected from interactions with Fe (76,77); and in humans, CPP enhance Zn absorption from oat or wheat gruel (78).

3.5.3 Iron

Fe seems to be the TE which could at best benefit from the presence of CPP, owing to its low absorption rate, its susceptibility to digestive interactions, and its strong affinity for CPP. Studies on Fe have allowed the determination of the absorption mechanisms of minerals and the influence of CPP according to their native caseins. Fe that binds to CPP is resistant to pH variations and to luminal digestion; in turn, CPP themselves are less susceptible to enzyme degradation (25,64).

In rats, Fe binding to the β-CN 1–25 enhances its absorption (79) and bioavailability. In iron-deficiency anemia, it improves both Fe absorption (79) and red blood cell count and body stores (80). A human study with the same complex gave similar results (81); Fe bound to β-CN 1–25 is protected from Ca and Zn interactions (76,77,79).

The mucosal uptake of Fe-β CN 1–25 is distinct from ionic Fe; it could use an endocytosis pathway, maybe in a bound form (82). Similar results were shown with a non purified casein hydrolysate; they disappear when the peptides are dephosphorylated (83). However, results displayed with the specific peptide β-CN 1–25 remain to be confirmed for other CPP. It was shown *in vitro* that CPP from whole casein inhibit Fe absorption; this inhibition was prevented by the dephosphorylation of CPP (84). In fact, as previously shown for Ca (67), all CPP are not equal: some of them, mainly those issued from α-caseins, have an inhibitory effect, while β-casein CPP have an enhancing effect, such as the pure β-CN 1–25 (85). It is known that CPP from α-caseins are poorly absorbed and remain in the digestive fluid (65).

3.6 CONCLUSIONS

Although the interactions with minerals are fundamental for the functional properties of proteins, their role in nutrition is not fully understood. CPP can improve the absorption and bioavailability of TE and lessen their digestive interactions; the influence of CPP on TE nutrition is promising, especially in populations that are at risk of intakes below the recommended levels, but further studies are needed to determine the most efficient molecules.

REFERENCES

1. C. Bos, C. Gaudichon, D. Tomé. Nutritional and physiological criteria in the assessment of milk protein quality for humans. *J. Am. Coll. Nutr.*, 19: 191S–205S, 2000.
2. D.A. Clare, H.E. Swaisgood. Bioactive milk peptides: a prospectus. *J. Dairy Sci.*, 83: 1187–1195, 2000.
3. H.H. Sandstead, J.G. Penland, N.W. Alcock, H.H. Dayal, X.C. Chen, J.S. Li, F. Zhao, J.J. Yang. Effects of repletion with zinc and other micronutrients on neuropsychological performance and growth of Chinese children. *Am. J. Clin. Nutr.*, 68 (Suppl.): 470S–475S, 1998.

4. C. Castillo-Duran, F. Cassoria. Trace minerals in human growth and development. *J. Pediatr. Endocrinol. Metabol.*, 12: 589–601, 1999.
5. D. Bouglé, D. Laroche, F. Bureau. Zinc and iron status and growth in healthy infants. *Eur. J. Clin. Nutr.*, 54: 1–4, 2000.
6. N.F. Krebs, C.J. Reidinger, L.V. Miller, M.W. Borschel. Zinc homeostasis in healthy infants fed a casein hydrolysate formula. *J. Pediatr. Gastroenterol. Nutr.*, 30: 29–33, 2000.
7. S.J. Fairweather-Tait. Bioavailability of dietary minerals. *Biochem. Soc. Transac.*, 24: 775–780, 1996.
8. M.E. Conrad, J.N. Umbreit. Iron absorption and transport — an update. *Am. J. Hematol.*, 64: 287–298, 2000.
9. D.M. Byler, H.M. Farrell. Infrared spectroscopic evidence for calcium ion interaction with carboxylate groups of casein. *J. Dairy Sci.*, 72: 1719–1723, 1989
10. M.I. Reddy, A.W. Mahoney. *J. Agric. Food Chem.*, 43: 1436–1443, 1995.
11. E. Bernos, J.M. Girardet, G. Humbert, G. Linden. Role of the O-phosphoserine clusters in the interaction of the bovine milk α_{s1}-, β-, κ-caseins and the PP3 component with immobilized iron (III) ions. *Biochim. Biophys. Acta*, 1337: 149–159, 1997.
12. G. Harzer, H. Kauer. Binding of zinc to casein. *Am. J. Clin. Nutr.*, 35: 981–987, 1982.
13. D.W. West. Structure and function of the phosphorylated residues of casein. *J. Dairy Res.*, 53: 333–352, 1986.
14. H. Singh, A. Flynn, P.F. Fon. Zinc binding in bovine milk. *J. Dairy Res.*, 56: 249–263, 1989.
15. J. Hegenauer, P. Saltman, D. Ludwig, L. Ripley, A. Ley. Iron-supplemented cow milk. Identification and spectral properties of iron bound to casein micelles. *J. Agric. Food Chem.*, 27: 1294–1301, 1979.
16. T. Emery. Iron oxidation by casein. *Biochem. Biophys. Res. Commun.*, 182: 1047–1052, 1992.
17. G. Brulé, J. Fauquant. Interactions des protéines du lait. *Lait*, 62 :323–331, 1982.
18. JJ Baumy, P Guenot, S Sinbandhit, G Brulé. Study of calcium binding to phosphorine residues of ß-casein and its phosphopeptide (1-25) by ^{31}P. *J. Dairy Res.*, 56: 403–409, 1989.
19. –F. Gaucheron, Y. Le Graet, E. Boyaval, M. Piot. Binding of cations to casein molecules: importance of physicochemical conditions. *Milchwissenschaft*, 52: 322–327, 1997.
20. P. Caillot, D. Lorient. *Structures et technofonctions des protéines du lait.* Paris: Lavoisier Tec & Doc, 1998, 31–50.
21. H. Wajcman. *Hématologie, Précis des maladies du sang.* Paris: Ellipses, 1994, 97–113.
22. V. Gagnaire, A. Pierre, D. Mollé, J. Léonil. Phosphopeptides interacting with colloidal calcium phosphate isolated by tryptic hydrolysis of bovine casein micelles. *J. Dairy Res.*, 63: 405–422, 1996.
23. G.E. Vegarud, T. Langsrud, C. Svenning. Mineral-binding milk proteins and peptides; occurrence, biochemical and technological characteristics. *Br. J. Nutr.*, 84(Suppl. 1): S91–S98, 2000.
24. D. McDonagh, R.J. FitzGerald. Production of caseinophosphopeptides (CPP) from sodium caseinate using a range of commercial protease preparations. *Int. Dairy J.*, 8: 39–45, 1998.

25. S. Bouhallab, J. Léonil, J.L. Maubois. Complexation du fer par le phosphopeptide (1-25) de la caséine ß: action de l'alcalase et de la phosphatase acide. *Lait*, 71: 435–443, 1991.
26. R.P. Heaney. Dietary protein and phosphorus do not affect calcium absorption. *Am. J. Clin. Nutr.*, 72: 751–761, 2000.
27. L.K. Massey. Dietary animal and plant protein and human bone health: a whole foods approach. *J. Nutr.*, 133: 862S–865S, 2003.
28. G.M. Chan, K. Hoffman, M. McMurry. Effects of dairy products on bone and body composition in pubertal girls. *J. Pediatr.*, 126: 551–556, 1995.
29. P. James, J.P. Sabatier, F. Bureau, D. Laroche, P. Jauzac, P. Arhan, D. Bouglé. Influence of dietary protein and phyto-oestrogens on bone mineralization in the young rat. *Nutr. Res.*, 22: 385–392, 2002.
30. B. Sandström. Zinc absorption from composite meals. I. The significance of wheat extraction rate, zinc, calcium, and protein content in meals based on bread. *Am. J. Clin. Nutr.*, 33: 739–745, 1980.
31. B. Sandström. Bioavailability of zinc. *Eur. J. Clin. Nutr.*, 51(Suppl. 1): S17–S19, 1997.
32. B. Lönnerdal. Dietary factors influencing zinc absorption. *J. Nutr.*, 130: 1378S–1383S, 2000.
33. M.L. Pabon, B. Lönnerdal. Bioavailability of zinc and its binding to casein in milks and formulas. *J. Trace Elem. Med. Biol.*, 14: 146–153, 2000.
34. R.F. Hurrell.Bioavailability of iron. *Eur. J. Clin. Nutr.*, 51(Suppl. 1): S4–S8, 1997.
35. R.P. Glahn, D.R. Van campen. Iron uptake is enhanced in Caco-2 cell monolayers by cysteine and cysteinyl glycine. *J. Nutr.*, 127: 642–647, 1997.
36. B.J. Macfarlane, R.D. Baynes, T.H. Bothwell, U. Schmidt, F. Mayet, B.M. Friedman. Effect of lupines, a protein-rich legume, on iron absorption. *Eur. J. Clin. Nutr.*, 42: 683–687, 1988.
37. S.R. Lynch, S.A. Dassenko, J.D. Cook, M.A. Juillerat, R.F. Hurrell. Inhibitory effect of a soybean-protein-related moiety on iron absorption in humans. *Am. J. Clin. Nutr.*, 60: 567–572, 1994.
38. L. Davidsson, T. Dimitriou, T. Walczyk, R.F. Hurrell. Iron absorption from experimental infant formulas based on pea (*Pisum sativum*)-protein isolate: the effect of phytic acid and ascorbic acid. *Br. J. Nutr.*, 85: 59–63, 2001.
39. R.F. Hurrell, S.R. Lynch, T.P. Trinidad, S.A. Dassenko, J.D. Cook. Iron absorption in humans: bovine serum albumin compared with beef muscle and egg white. *Am. J. Clin. Nutr.*, 47: 102–107, 1988.
40. R.F. Hurrell, S.R. Lynch, T.P. Trinidad, S.A. Dassenko, J.D. Cook. Iron absorption in humans as influenced by bovine milk proteins. *Am. J. Clin. Nutr.*, 49: 546–552, 1989.
41. D. Zhang, A.W. Mahoney. Bioavailability of iron-milk-protein complexes and fortified cheddar cheese. *J. Dairy Sci.*, 72: 2845–2855, 1989.
42. D. Carmichael, J. Christopher, J. Hegenauer, P. Saltman. Effect of milk and casein on the absorption of supplemental iron in the mouse and chick. *Am. J. Clin. Nutr.*, 28: 487–493, 1975.
43. J.C. Barton, M.E. Conrad, R.T. Parmley. Calcium inhibition of inorganic iron absorption in rats. *Gastroenterology*, 84: 90–101, 1983.
44. M.L. Pabon, B. Lönnerdal. Effects of type of fat in the diet on iron bioavailability assessed in suckling and weanling rats. *J. Trace Elem. Med. Biol.*, 15: 18–23, 2001.
45. B. Lönnerdal. Effects of milk and milk components on calcium, magnesium, and trace element absorption during infancy. *Physiol. Rev.*, 77: 643–669, 1997.

46. C. Martinez, T. Fox, J. Eagles, S. Fairweather-Tait. Evaluation of iron bioavailability in infant weaning foods fortified with haem concentrate. *J. Pediatr. Gastroenterol. Nutr.*, 27: 419–424, 1998.
47. R.E. Serfass, M.B. Reddy. Breast milk fractions solubilize Fe(III) and enhance iron flux across Caco-2 cells. *J. Nutr.*, 133: 449–455, 2003.
48. B.J. Macfarlane, W.B. van der Riet, T.H. Bothwell, R.B. Baynes, D. Siegenberg, U. Schmidt, A. Tal, J.R.N. Taylor, F. Mayet. Effect of traditional oriental soy products on iron absorption. *Am. J. Clin. Nutr.*, 51: 873–880, 1990.
49. M. Hirabayashi, T. Matsui, H. Yano. Fermentation of soybean flour with *Aspergillus usamii* improves availabilities of zinc and iron in rats. *J. Nutr. Sci. Vitaminol.*, 44: 877–886, 1998.
50. P.A. Seligman, G.M. Moore, R.B. Schleisher. Clinical studies of hip: an oral heme-iron product. *Nutr. Res.*, 20: 1279–1286, 2000.
51. N. Vaghefi, F. Nedjaoum, D. Guillochon, F. Bureau, P. Arhan, D. Bouglé. Influence of the extent of hemoglobin hydrolysis on the digestive absorption of heme iron. An *in vitro* study. *J. Agric. Food Chem.*, 50: 4969–4973, 2002.
52. M.J. Salgueiro, M.B. Zubillaga, A.E. Lysionek, M.I. Sarabia, R.A. Caro, T. De Paoli, A. Hager, E. Ettlin, R. Weill, J.R. Boccio. Bioavailability, biodistribution, and toxicity of BioZn-AAS: a new zinc source. Comparative studies in rats. *Nutrition*, 16: 762–766, 2000.
53. O. Pineda, H.D. Ashmead. Effectiveness of treatment of iron-deficiency anemia in infants and young children with ferrous bis-glycinate chelate. *Nutrition*, 17: 381–834, 2001.
54. L.H. Allen. Advantages and limitations of iron amino acid chelates as iron fortificants. *Nutr. Rev.*, 60: S18–S21, 2002.
55. H. Naito, H. Suzuki. Further evidence for the formation *in vivo* of phosphopeptide in the intestinal lumen from dietary β-casein. *Agric. Biol. Chem.*, 38: 1543–1545, 1974.
56. Y.S. Lee, T. Noguchi, H. Naito. Phosphopeptides and soluble calcium in the small intestine of rats given a casein diet. *Br. J. Nutr.*, 43: 457–467, 1980.
57. H. Meisel, H. Frister. Chemical characterization of bioactive peptides from *in vivo* digest of casein. *J. Dairy Res.*, 56: 343–349, 1989.
58. T. Kasai, R. Iwasaki, M. Tanaka, S. Kiriyama. Caseinphosphopeptides (CPP) in faeces and contents in faeces and contents in digestive tract of rats fed casein and CPP preparations. *Biosci. Biotechnol. Biochem.*, 59: 26-30, 1995.
59. H. Meisel, H. Bernard, S. Fairweather-Tait, R.J. FitzGerald, R. Hartmann, C.N. Lane, D. McDonagh, B. Teucher, J.M. Wal. Detection of caseinophosphopeptides in the distal ileostomy fluid of human subjects. *Br. J. Nutr.*, 89: 351–359, 2003.
60. O. Mellander, G. Fölsch. Enzyme resistance and metal binding of phosphorylated peptides. In: EJ Bigwood (Ed.). *Protein and Amino Acids Functions*. Oxford: Pergamon Press, 1972, 569–579.
61. R. Sato, T Noguchi, H. Naito. The necessity for the phosphate portion of casein molecules to enhance Ca absorption from small intestine. *Agric. Biol. Chem.*, 10: 2415–2417, 1983.
62. R. Berrocal, S. Chanton, M.A. Juillerat, B. Pavillard, J.C. Scherz, R. Jost. Tryptic phosphopeptides from whole casein. II. Physicochemical properties related to the solubilization of calcium. *J. Dairy Res.*, 56: 335–341, 1989.
63. S. Bouhallab, N. Aït-Oukhatar, D. Mollé, G. Henry, J.L. Maubois, P. Arhan, D. Bouglé. Sensitivity of β-casein phosphopeptide-iron complex to digestive enzymes in ligated segment of rat duodenum. *J. Nutr. Biochem.*, 10: 723–727, 1999.

64. N. Aït-Oukhatar, S. Bouhallab, F. Bureau, P. Arhan, J.L. Maubois, D. Bouglé. *In vitro* digestion of caseinophosphopeptide — iron complex. *J. Dairy Res.*, 67: 125–129, 2000.
65. R. Brommage, M.A. Juillerat, R. Jost. Influence of casein phosphopeptides and lactulose on intestinal calcium absorption in adult female rats. *Lait*, 71: 173–180, 1991.
66. D.D. Kitts, Y.V. Yuan, T. Nagasawa, Y. Moriyama. Effect of casein, caseinphosphopeptides and calcium intake on ileal ^{45}Ca disappearance and temporal systolic blood pressure in spontaneously hypertensive rats. *Br. J. Nutr.*, 68: 765–781, 1992.
67. H.M. Mykkänen, R.H. Wasserman. Enhanced absorption of calcium by casein phosphopeptides in rachitic and normal chicks. *J. Nutr.*, 110: 2141–2148, 1980.
68. H. Tsuchita, T. Suzuki, T. Kuwata. The effect of casein phospshopeptides on calcium absorption from calcium-fortified milk in growing rats. *Br. J. Nutr.*, 85: 5–10, 2001.
69. Y. Li, D. Tomé, J.F. Desjeux. Indirect effect of casein phosphopeptides on calcium absorption in rat ileum *in vitro*. *Reprod. Nutr. Dev.*, 29: 227–233, 1989.
70. T. Nagasawa, Y.V. Yuan, D.D. Kitts. Casein phosphopeptides enhance paracellular calcium absorption but do not alter temporal blood pressure in normotensive rats. *Nutr. Res.*, 11: 819–830, 1991.
71. M. Hansen, B. Sandström, B. Lönnerdal. The effect of casein phosphopeptides on zinc and calcium absorption from high phytate infants diets assessed in rat pups and Caco-2 cells. *Pediatr. Res.*, 40: 547–552, 1996.
72. A. Ferraretto, A. Signorile, C. Gravaghi, A. Fiorilli, G. Tettamanti. Casein phosphopeptides influence calcium uptake by cultured human intestinal HT-29 tumor cells. *J. Nutr.*, 131: 1655–1661, 2001.
73. R.P. Heaney, Y. Saito, H. Orimo. Effect of caseinphosphopeptide on absorbability of co-ingested calcium in normal postmenopausal women. *J. Bone Miner. Res.*, 12: 77–81, 1994.
74. D. Erba, S. Ciappellano, G. Testolin. Effect of caseinphosphoptides on inhibition of calcium intestinal absorption due to phosphate. *Nutr. Res.*, 21: 649–656, 2001.
75. K. Shiga, H. Hara, T. Kasai. The large intestine compensates for insufficient calcium absorption in the small intestine in rats. *J. Nutr. Sci. Vitaminol.*, 44: 737–744, 1998.
76. J.M. Pérès, S. Bouhallab, C. Petit, F. Bureau, J.L. Maubois, P. Arhan, D. Bouglé. Improvement of zinc intestinal absorption and reduction of zinc/iron interaction using metal bound to the caseinophosphopeptide 1–25 of ß-casein. *Reprod. Nutr. Dev.*, 38: 465–472, 1998.
77. J.M. Pérès, S. Bouhallab, F. Bureau, J.L. Maubois, P. Arhan, D. Bouglé. Reduction of iron/zinc interactions using metal bound to the caseinophosphopeptide 1–25 of ß-casein. *Nutr. Res.*, 19: 1655–1663, 1999.
78. M. Hansen, B. Sandström, M. Jensen, S.S. Sørensen. Casein phosphopeptides improve zinc and calcium absorption from rice-based but not from whole-grain infant cereal. *J. Pediatr. Gastroenterol. Nutr.*, 24: 56–62, 1997.
79. J.M. Pérès, S. Bouhallab, F. Bureau, J.L. Maubois, P. Arhan, D. Bouglé. Absorption digestive du fer lié au caséinophosphopeptide 1–25 de la β-caséine. *Lait*, 77: 433–440, 1997.
80. N. Aït-Oukhatar, S. Bouhallab, F. Bureau, M. Drosdowsky, J.L. Maubois, P. Arhan, D. Bouglé. Bioavailability of caseinophoshopeptide bound iron in the young rat. *J. Nutr. Biochem.* 8: 190–194, 1997.
81. N. Aït-Oukhatar, J.M. Pérès, S. Bouhallab, D. Neuville, F Bureau, G. Bouvard, P. Arhan, D. Bouglé. Bioavailability of caseinophosphopeptide bound iron. *J. Lab. Clin. Med.*, 140: 290–294, 2002.

82. J.M. Pérès, S. Bouhallab, F. Bureau, D. Neuville, J.L. Maubois, P. Arhan, D. Bouglé. Mechanisms of absorption of caseinophosphopeptide bound iron. *J. Nutr. Biochem.*, 10: 215–222, 1999.
83. M.V. Chaud, C. Izumi, Z. Nahaal, T. Shuhama, M.L. Bianchi, O. Freitas. Iron derivatives from casein hydrolyzates as a potential source in the treatment of iron deficiency. *J. Agric. Food Chem.*, 50: 871–877, 2002.
84. A.C. Yeung, R.P. Glahn, D.D. Miller. Dephosphorylation of sodium caseinate, enzymatically hydrolyzed casein and casein phosphopeptides by intestinal alkaline phosphatase: implications for iron bioavailability. *J. Nutr. Biochem.*, 12: 292–299, 2001.
85. S. Bouhallab, V. Cinga, N. Aït-Oukhatar, F. Bureau, D. Neuville, P. Arhan, J.L. Maubois, D. Bouglé. Influence of various phosphopeptides of caseins on iron absorption . *J. Agric. Food Chem.*, 50: 7127–7130, 2002.

4 Cholesterol-Lowering Proteins and Peptides

Satoshi Nagaoka

CONTENTS

4.1 BACKGROUND

Hyperlipidemia, especially hypercholesterolemia, is one of the most important risk factors associated with heart disease, including ischemic heart disease (1–3). The increase in death rate due to ischemic heart disease such as angina pectoris based on coronary arteriosclerosis and myocardial infarction is remarkable even in countries such as Japan. The prevention and treatment of hypercholesterolemia and its association with food is considered important (4,5). Dietary protein is useful as a regulator of serum cholesterol concentration (6). In fact, diets low in saturated fat and cholesterol that include 25 g soy protein per day may reduce the risk of heart disease: A health claim made by the U.S. Food and Drug Administration suggested that intake of 25 g of soybean protein per day would lower serum cholesterol concentration by 5 to 10%, and this was approved in 1999 (7). "Food for specified health use" have been produced for the prevention and treatment of lifestyle-related diseases such as hypercholesterolemia in Japan. Very recently, such a food, using soy peptides with bound soy phospholipids, called CSPHP (or c-SPHP) was created (8–10). In this chapter, the process for the development of CSPHP and related research in the global arena are discussed.

It is often believed that dietary animal proteins produce a higher serum cholesterol concentration in rats than do vegetable proteins (11,12). However, many studies on the effects of dietary proteins on serum cholesterol in animal experiments have tended to focus on a comparison between soybean protein and casein (11–17). Only limited data are available concerning the effects of animal protein other than casein on serum cholesterol level. Thus, the effects of animal protein such as milk whey proteins (12, 18–25) and their peptides (24,25), egg protein (26–28) and its peptides (26), beef protein and its peptides (16,22,29,30), pork peptides (31), and fish protein (32,33) have been studied. In this chapter, we also introduce studies about the hypocholesterolemic action of animal proteins (19,20,24,25,28) and their peptides (24, 25) and describe the discovery of novel hypocholesterolemic peptides (25).

4.2 HYPOCHOLESTEROLEMIC MILK PROTEINS AND NOVEL HYPOCHOLESTEROLEMIC PEPTIDES

4.2.1 Serum Cholesterol and Bovine Milk Whey Protein

Milk whey is formed as a by-product of cheese and casein manufacturing and has been utilized marginally as a food until now. However, research on proteins in milk whey, lactose, and other minor constituents and on their physiological significance has advanced, and this may lead to the commercialization and use of milk whey in many fields. It is also expected that isolated biologically active components derived from milk whey would appear in great numbers and could be utilized as a food for specified health purposes in the future.

Dietary proteins have been shown to influence serum cholesterol level in a number of studies (11–17). As mentioned earlier, most research on the effects of dietary proteins on serum cholesterol levels in animal and human studies have focused on the comparison of soybean protein and milk casein. Only limited data are available concerning the effects of milk whey protein on cholesterol metabolism. In rabbits, whey protein feeding increases the plasma level of cholesterol compared with soybean protein feeding (11,12). In 1983, Sautier et al. (18) first reported the hypocholesterolemic effect of whey protein compared with casein in rats fed a cholesterol-free diet. Choi et al. (21), however, reported that the hypocholesterolemic action of whey protein compared with casein was only evident in rats fed a cholesterol-containing diet. Moreover, Lapre et al. (22) suggested that this effect of whey protein was not observed in rats fed a 0.15% cholesterol-containing diet. We previously reported that whey protein exhibited a greater hypocholesterolemic effect in comparison with casein or soybean protein in rats (19,20). Thus, the hypocholesterolemic action of whey protein as observed in Sautier's research in 1991 was first confirmed (19) and, with some exception, reconfirmed by later research (23).

4.2.2 The Hypocholesterolemic Action of Major Constituents of Bovine Milk Whey Protein

The effects of the major constituents of bovine milk whey protein such as β-lactoglobulin and α-lactalbumin have been investigated. A murine study found for the first time that β-lactoglobulin, α-lactalbumin, and their peptic hydrolyzates had cholesterol-lowering action (24). A soy protein hydrolysate (SPH) had a stronger serum-cholesterol-lowering effect than that of intact soy protein (34). Sugano et al. (34) have shown that SPH both decreased the blood cholesterol level and promoted fecal excretion of steroids, as compared with the effects of casein. They also reported that a high-molecular-weight fraction (HMF) with higher bile-acid-binding capacity was very potent in lowering cholesterol and produced significant reductions in cholesterol compared with a low-molecular-weight fraction (LMF) and intact isolated soy protein. Thus, in the case of soy protein, there is a relationship between the hypocholesterolemic action *in vivo* and bile-acid-binding capacity or hydrophobicity of proteins *in vitro*. By this concept, the relationship between the hypocholesterolemic action of several kinds of proteins and their hydrolyzates and bile-acid-binding capacity or hydrophobicity of the protein *in vitro*, including whey protein, was evaluated, as shown in Figures 4.1 and 4.2 (24). As soy HMF with higher bile-acid-binding capacity has a higher hypocholesterolemic activity than that of intact soy protein or soy LMF, whey HMF, β-lactoglobulin HMF, or α-lactalbumin HMF was expected to express a higher activity. However, the degree of hypocholesterolemic action of whey HMF, β-lactoglobulin HMF, or α-lactalbumin HMF with both higher bile-acid-binding capacity and higher hydrophobicity was the same as that of intact whey protein, β-lactoglobulin, or α-lactalbumin (see Figure 4.1). Thus, in the case of whey protein or its constituent proteins, no relationship exists between the hypocholesterolemic action *in vivo* and bile acid binding capacity or hydrophobicity of the protein *in vitro* (see Figure 4.2).

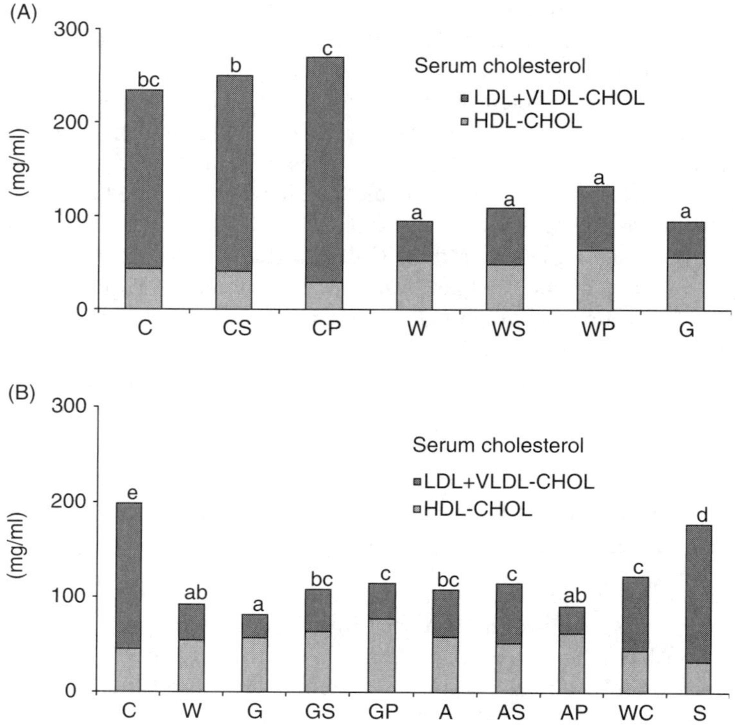

FIGURE 4.1 A: Effects of dietary casein (C), low molecular fraction (LMF) of casein peptic hydrolysate (CS), high molecular fraction (HMF) of casein peptic hydrolysate (CP), whey protein isolate (W), LMF of whey protein isolate-peptic hydrolysate (WS), HMF of whey protein isolate-peptic hydrolysate (WP), β-lactoglobulin (G). Means not followed by the same letter are significantly different. (*Source*: From S. Nagaoka. *J. Jpn. Soc. Nutr. Food Sci.*, 49, 303–313, 1996. With permission.)
B: Effects of dietary casein (C), whey protein isolate (W), β-lactoglobulin (G), LMF of β-lactoglobulin peptic hydrolysate (GS), HMF of β-lactoglobulin peptic hydrolysate (GP), α-lactalbumin (A), LMF of α-lactalbumin peptic hydrolysate (AS), HMF of α-lactalbumin peptic hydrolysate (AP) or whey protein concentrate (WC), soybean protein (S). Means not followed by the same letter are significantly different. (*Source*: From S. Nagaoka. *J. Jpn. Soc. Nutr. Food Sci.*, 49, 303–313, 1996. With permission.)

4.2.3 The Hypocholesterolemic Action of Bovine Milk Whey Protein, its Major Constituent Protein β-Lactoglobulin, and its Tryptic Hydrolysate

The first direct evidence that β-lactoglobulin tryptic hydrolysate (LTH) has hypocholesterolemic activity in comparison with casein tryptic hydrolysate (CTH) originated from the work in rats carried out in our laboratories (Table 4.1) (25). It has been postulated that the degree of serum-cholesterol-lowering

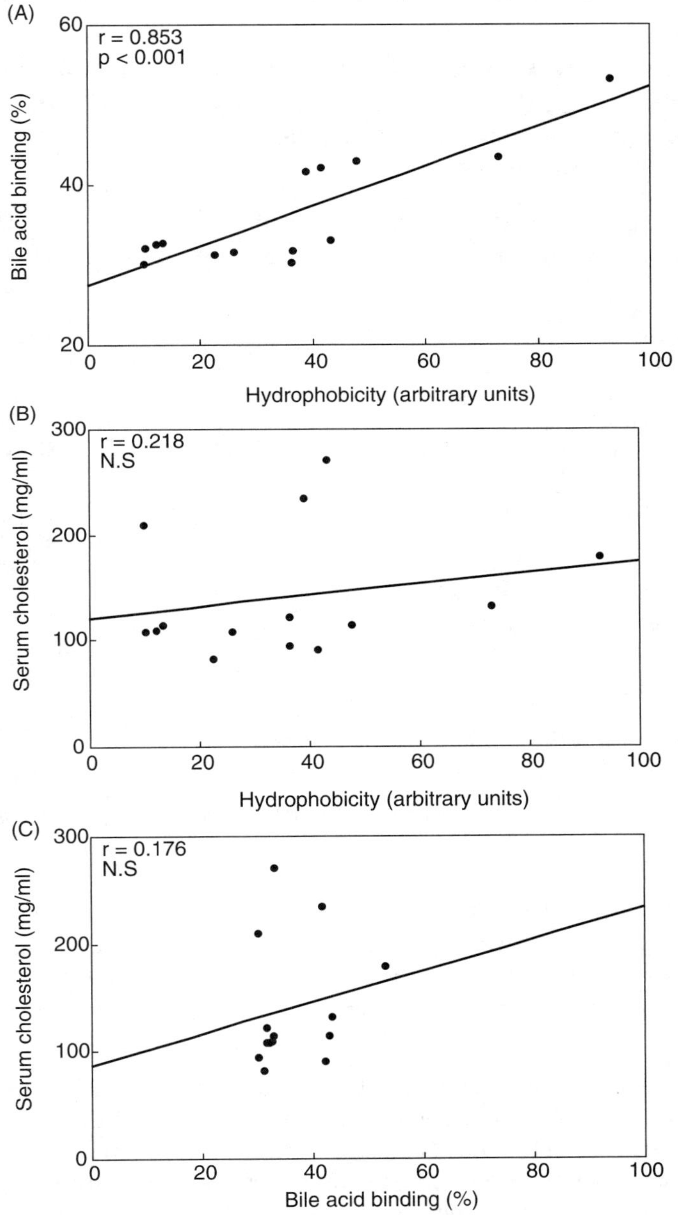

FIGURE 4.2 A: Relationship between the bile-acid-binding capacity and hydrophobicity. (*Source*: From S. Nagaoka. *J. Jpn. Soc. Nutr. Food Sci.*, 49, 303–313, 1996. With permission.) B: Relationship between the serum cholesterol level and hydrophobicity. (*Source*: From S. Nagaoka. *J. Jpn. Soc. Nutr. Food Sci.*, 49, 303–313, 1996. With permission.) C: Relationship between the serum cholesterol level and bile-acid-binding capacity. (*Source*: From S. Nagaoka. *J. Jpn. Soc. Nutr. Food Sci.*, 49, 303–313, 1996. With permission.)

TABLE 4.1
Effects of Dietary Casein Tryptic Hydrolysate (CTH), β-Lactoglobulin Tryptic Hydrolysate (LTH) on Body and Liver Weights, Food Intake, Serum and Liver Lipids and Fecal Steroid Excretion in Rats

	Diet Group	
	CTH	LTH
Body weight gain (g 14 days^{-1})	71.2 ± 4.7	68.8 ± 2.5
Liver weight (g 100 g^{-1} body weight)	5.46 ± 0.22	5.30 ± 0.20
Food intake (day 6, g day^{-1})	17.0 ± 0.8	16.0 ± 1.0
Serum (mgdl^{-1})		
Total cholesterol (a)	136.6 ± 6.5	97.0 ± 3.9 ***
HDL cholesterol (b)	46.8 ± 1.9	64.3 ± 3.8 **
LDL + VLDL cholesterol	89.8 ± 7.9	32.7 ± 4.5 ***
Atherogenic Index (b)/(a)	0.34 ± 0.03	0.66 ± 0.04 ***
Liver		
Total lipids (mg g^{-1} liver)	102.9 ± 5.8	93.1 ± 5.5
Cholesterol (mg g^{-1} liver)	17.6 ± 1.0	8.4 ± 0.6 ***
Fecal		
Dry weight (g 3 days^{-1})	1.73 ± 0.10	2.36 ± 0.04 ***
Neutral steroids (μmol 3 days^{-1})		
Cholesterol	506.1 ± 50.6	580.3 ± 10.8
Coprostanol	1.4 ± 0.5	53.1 ± 8.7 ***
Total	507.5 ± 50.3	633.4 ± 11.3 *
Acidic steroids (μmol 3 days^{-1})	65.1 ± 4.6	76.4 ± 2.7 *
Total steroids (μmol 3 days^{-1})	572.6 ± 54.1	709.8 ± 10.8 *

The data are means ± SEM (n = 6). Statistical significance compared to CTH group by Student's t-test (p<0.05, p<0.01, p<0.001).

Values were calculated as follows: LDL + VLDL cholesterol = Total cholesterol − HDL cholesterol.

Total steroids = neural steroids + acidic steroids.

Source: From S. Nagoka, Y. Futamura, K. Miwa, T. Awano, K. Yamauchi, Y. Kanamaru, T. Kojima, and T. Kuwata. *Biochem. Biophys. Res. Commn.*, 281, 11–17, 2001. With permission.

activity depends on the degree of fecal steroid excretion (acidic steroids + neutral steroids) (25). We also demonstrated that a higher fecal excretion of total steroids in rats fed LTH, indicating that the effect was, at least in part, due to an enhancement of fecal steroid excretion (25). There have been many subsequent studies on the hypocholesterolemic effects of proteins, most of which support the hypothesis that a peptide with high bile-acid-binding capacity could inhibit the reabsorption of bile acid in the ileum and decrease the blood cholesterol level (35). These possibilities may be applicable to the case of LTH on the basis of the evidence of fecal bile acid excretion and taurocholate-binding capacity presented in our study (25).

In the case of SPH, the hypocholesterolemic action was induced by the-inhibition of both cholesterol absorption accompanying the suppression of micellar solubility of cholesterol and ileal reabsorption of bile acids (8,36). A prior communication (34) demonstrated the hypothesis that the hypocholesterolemic peptides derived from soybean protein might influence the serum cholesterol level. However, until now, no one could find the hypocholesterolemic peptide in soybean protein nor in any other protein. We have hypothesized that the peptide derived from bovine milk, β-lactoglobulin, might induce a hypocholesterolemic action.

In recent studies, monolayers of Caco-2 cell cultures were used as a model system to examine the process of lipid metabolism (37–39). For example, Field et al. (37) reported that Caco-2 cells, like the small intestine, had the ability to absorb micellar cholesterol and to express marker enzymes such as alkaline phosphatase as small intestinal epithelial cells. The cholesterol micelles containing SPH significantly suppressed cholesterol uptake by Caco-2 cells compared with the cholesterol micelles containing CTH (36). Furthermore, LTH also directly inhibited the absorption of micellar cholesterol in Caco-2 cells *in vitro* (25). Cholesterol is rendered soluble in bile-salt-mixed micelles and then absorbed (40). The micellar solubility of cholesterol in the presence of LTH was significantly lower than that with CTH (25). Very interestingly, the micellar solubility of cholesterol in the presence of SPH was significantly lower than that with CTH (8). Sitosterol (41), sesamine (42), and catechin (43) also lowered the micellar solubility of cholesterol in conjunction with serum-cholesterol-lowering effects in rats. These findings, including LTH, suggest that suppression of the micellar solubility of cholesterol induces the inhibition of cholesterol absorption in the jejunum, and this may be closely related to the lowering of serum cholesterol.

To confirm the concept that suppression of the micellar solubility of cholesterol induces the inhibition of cholesterol absorption in the jejunum by whey protein feeding, jejunal micelles of rats fed a high-cholesterol diet containing whey protein or casein were examined by electron microscopy (24). A jejunal emulsion or micelle was prepared by ultracentrifugation before observation by electron microscopy. As shown in Figure 4.3, the jejunal emulsion of casein feeding was aggregated, while whey feeding dispersed the jejunal emulsion. The size of the casein emulsion in the intestine was larger than that of the whey emulsion, and the number of casein micelles was greater than that of the whey emulsion. Thus, we suggested that the intestinal nature of the emulsion and micelle can affect cholesterol absorption and serum cholesterol level; as well, there was a relationship between the number of jejunal micelles and hypocholesterolemic activity.

4.2.4 Identification of Novel Hypocholesterolemic Peptides Derived from Bovine Milk β-Lactoglobulin

It was suggested that the hypocholesterolemic action of LTH was induced by the suppression of cholesterol absorption, as evidenced by an *in vivo* cholesterol

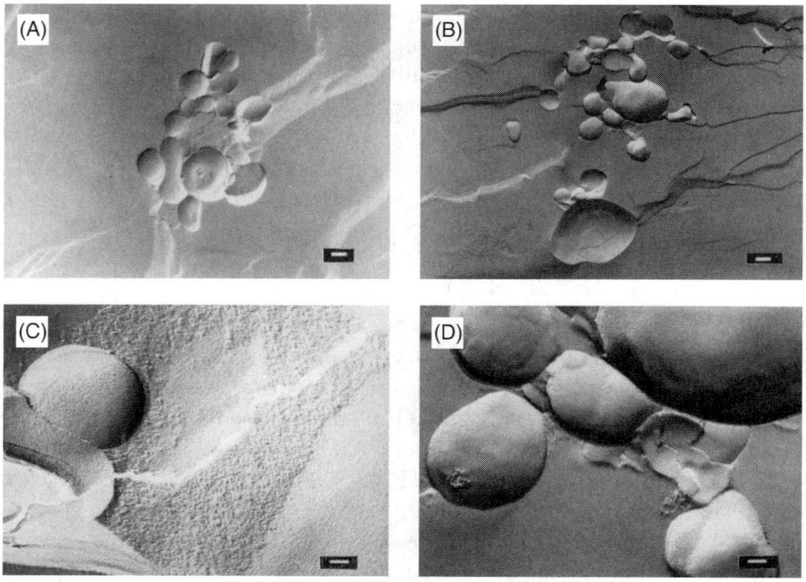

FIGURE 4.3 Freeze fracture transmission electron microscopy photograph of intestinal emulsion and micelle in rats fed the diets containing casein (A, C) or whey protein (B, D). The bars represents 200 nm (A, B) or 50 nm (C, D). (*Source*: From S. Nagaoka. *J. Jpn. Soc. Nutr. Food Sci.*, 49, 303–313, 1996. With permission.)

absorption study and an *in vitro* Caco-2 cell study (25). In an attempt to identify the active component related to the hypocholesterolemic action of LTH, we hypothesized that the peptide derived from bovine milk β-lactoglobulin might induce hypocholesterolemic action. The experimental system to evaluate cholesterol uptake in Caco-2 cells, called "Caco-2 cell screening," was useful for clarifying the active component underlying the inhibitory effect of peptide on cholesterol absorption from the small intestine (8,36). Thus, (1) the peptide sequences which suppress cholesterol absorption was first identified using Caco-2 cell screening, as shown in Figures 4.4 and 4.5. (2) After Caco-2 cell screening *in vitro*, the hypocholesterolemic activity of the new peptides was evaluated in animal studies. By using Caco-2 cell screening, four kinds of novel peptide sequences which inhibit cholesterol absorption *in vitro* were identified for the first time as shown in Table 4.2.

The hypothesis that hypocholesterolemic peptides derived from soybean protein might influence the serum cholesterol level had been advanced earlier (34). However, the hypocholesterolemic peptide in soybean protein or in any other protein source had not been examined. Sugano et al. (34) suggested that the hypocholesterolemic peptides derived from soybean protein consisted of peptides with molecular weights between 1 and 10 kDa. However, the hypocholesterolemic peptide of soybean protein remained unidentified. Although a previous study (31)

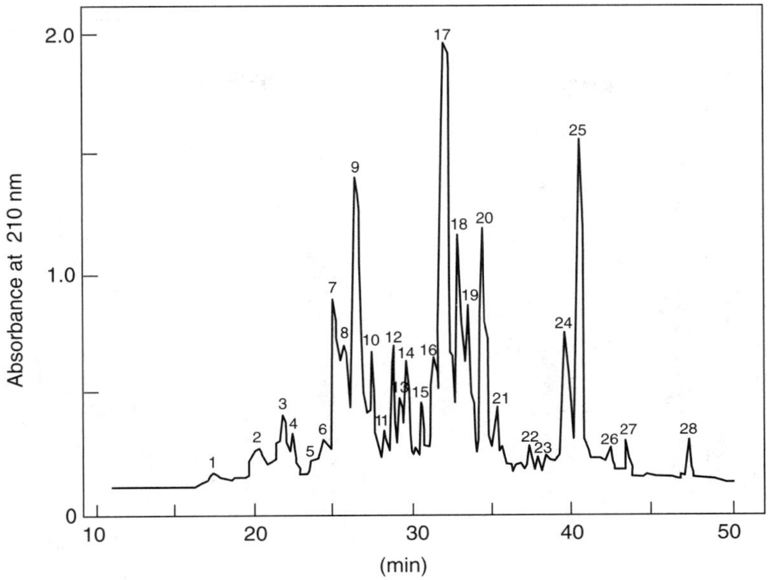

FIGURE 4.4 Purification of the peptides from β-lactoglobulin tryptic hydrolysate (LTH) by reversed-phase high-performance liquid chromatography (HPLC) on an ODS column. (*Source*: From S. Nagaoka, Y. Futamura, K. Miwa, T. Awano, K. Yamauchi, Y. Kanamaru, T. Kojima,T. Kuwata. *Biochem. Biophys. Res. Commn.*, 281, 11–17, 2001. With permission.)

demonstrated that the hypocholesterolemic effects of papain-hydrolyzed pork meat contained peptides with molecular weights of 3 kDa or less, the hypocholesterolemic peptide of pork protein was not identified.

After four kinds of novel peptide sequences that inhibited cholesterol absorption in Caco-2 cell screening *in vitro* were identified, we evaluated the hypocholesterolemic activity of these new peptides in animal studies. The newly identified IIAEK was chosen for the animal study because of its lack of peptic or tryptic digestive site and it was expected to elicit the hypocholesterolemic action *in vivo*. This novel peptide, derived from β-lactoglobulin and can influence the serum cholesterol level effectively, exhibited a greater hypocholesterolemic activity in comparison with β-sitosterol in rats (Table 4.3) and induced suppression of cholesterol absorption, as evidenced by the Caco-2 cell study. We speculate that IIAEK may decrease the micellar solubility of cholesterol and inhibited cholesterol absorption as in the case of LTH or SPH in our unpublished results. Myoglobin (44) and haptoglobin (45) are known to decrease the cholesterol solubility in model bile *in vitro*. Ahmed et al. (44) suggested that myoglobin decreased cholesterol solubility through acceleration of the rate of cholesterol migration from the micellar phase to the vesicular phase *in vitro*. Thus, IIAEK or LTH may decrease cholesterol solubility by increasing the amount of vesicular cholesterol rather than that of micellar cholesterol.

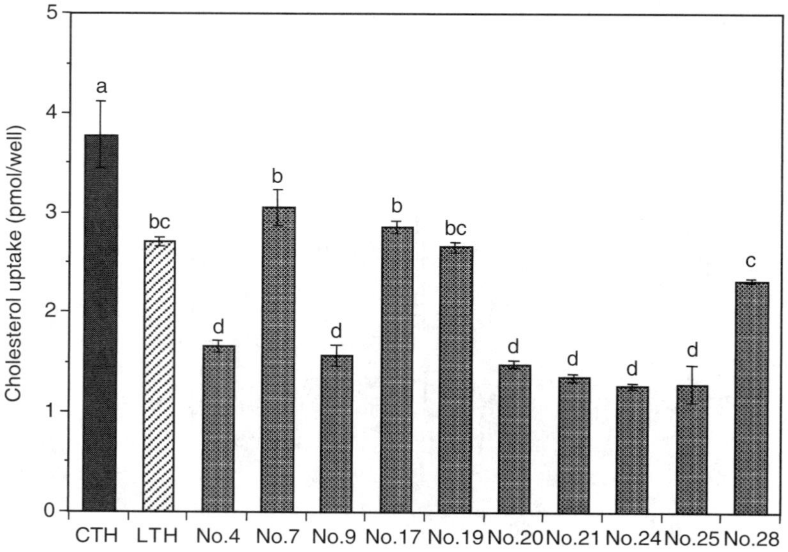

FIGURE 4.5 Effects of fractionated peptides derived from β-lactoglobulin on cholesterol absorption in Caco-2 cells *in vitro*. The digest (LTH) was fractionated by reversed-phase high-performance liquid chromatography (HPLC) as shown in Figure 4.4. After fractionation, cholesterol absorption from micelles containing the peptides from LTH was measured by the method shown in the text. (*Source*: From S. Nagaoka, Y. Futamura, K. Miwa, T. Awano, K. Yamauchi, Y. Kanamaru, T. Kojima,T. Kuwata. *Biochem. Biophys. Res. Commn.*, 281, 11–17, 2001. With permission.)

TABLE 4.2
New Hypocholesterolemic Peptides Derived from Bovine Milk β-Lactoglobulin by "Caco-2 Cell Screening"

Primary Structure	Residue Number
Gly-Leu-Asp-Ile-Gln-Lys	9~14
Val-Tyr-Val-Glu-Glu-Leu-Lys-Pro-Thr-Pro-Glu-Gly-Asp-Leu-Glu-Ile-Leu-Leu-Gln-Lys	41~61
Ile-Ile-Ala-Glu-Lys	71~75
Ala-Leu-Pro-Met-His	142~146

The finding of IIAEK should play a crucial role in clarifying the mechanisms by which dietary proteins induce differential effects on serum cholesterol level as well as in developing a functional food or an antiatherogenic drug. So far, the effect of the protein on cholesterol metabolism has mainly been studied on the

TABLE 4.3
Effects of Oral Administration of Casein Tryptic Hydrolysate (CTH), β-Lactoglobulin Tryptic Hydrolysate (LTH), β-Sitosterol or IIAEK on Body and Liver Weights, Food Intake, Serum HDL, and LDL + VLDL-Cholesterol in Rats[1]

	Group			
	CTH	**LTH**	**β-sitosterol**	**IIAEK**
Body weight gain ($g\ 4\ days^{-1}$)	10.4 ± 0.4[a]	10.5 ± 0.7[a]	9.5 ± 0.8[a]	8.7 ± 0.4[a]
Liver weight ($g\ 100\ g^{-1}$ body weight)	5.22 ± 0.07[b]	5.37 ± 0.05[ab]	5.17 ± 0.10[b]	5.54 ± 0.11[a]
Food intake (day 3, $g\ day^{-1}$)	8.5 ± 0.2[a]	8.7 ± 0.2[a]	8.3 ± 0.3[a]	8.4 ± 0.3[a]
Serum ($mg\ dl^{-1}$)				
Total cholesterol (a)	529.3 ± 14.4[a]	441.3 ± 12.0[b]	435.4 ± 12.8[b]	328.7 ± 11.3[c]
HDL cholesterol (b)	42.6 ± 1.6[b]	50.5 ± 2.4[a]	52.1 ± 1.7[a]	47.5 ± 0.8[ab]
LDL + VLDL-cholesterol[2]	486.7 ± 26.7[a]	390.8 ± 22.9[b]	383.3 ± 24.2[b]	281.2 ± 20.6[c]
Atherogenic Index (b)/(a)	0.08 ± 0.01[c]	0.11 ± 0.01[b]	0.12 ± 0.01[b]	0.15 ± 0.01[a]

[1]The data are means ± SEM ($n = 7$). Within a row, means with different superscript letters are significantly different ($p < 0.05$) by Duncan's multiple range test.

[2]Values were calculated as follows: LDL + VLDL cholesterol = total cholesterol − HDL

Source: From S. Nagoka, Y. Futamura, K. Miwa, T. Awano, K. Yamauchi, Y. Kanamaru, T. Kojima, and T. Kuwata. *Biochem. Biophys. Res. Commn.*, 281, 11–17, 2001. With permission.

amino acid composition, hydrophobicity, and bile-acid-binding capacity of the protein. Since the peptidic amino acid sequence that lowered serum cholesterol had been unknown, research on this hypocholesterolemic peptide posed difficulties. However, the discovery of IIAEK makes elucidation of the property of the hypocholesterolemic peptide and the search and evaluation of it possible. We now suggest that hypocholesterolemic peptides — until now only a hypothesis or the "unknown signal for cholesterol metabolic regulation" — are latent in the amino acid sequence of proteins.

4.3 CHOLESTEROL-LOWERING EGG PROTEINS AND PEPTIDES

4.3.1 Egg White Protein and Cholesterol Metabolism

Egg protein has a well-balanced amino acid composition with high biological value. However, egg is a cholesterol-rich food, and its use should always be strictly or carefully controlled in order to avoid hypercholesterolemia and related diseases. Cholesterol exists exclusively in the yolk, and egg white is cholesterol-free. Egg intake is thought to increase serum cholesterol concentrations in experimental animals (11) and in humans (46). Thus, extensive efforts are now globally underway for screening the hypocholesterolemic factors from eggs. Egg white contains a wide variety of proteins such as ovalbumin, ovomucin, ovotransferrin, and lysozyme, among others (47,48). Several reports have indicated that the quality and quantity of dietary proteins affect serum cholesterol level (11–17). Only a few reports have dealt with the effect of egg white protein (EW) on serum cholesterol level in rats (26) and humans (27). Moreover, the mechanism by which the hypocholesterolemic effect of EW is exerted in rats is not well understood. Yamamoto et al. (26) reported a hypocholesterolemic effect of EW or egg white protein hydrolyzates (EWP) in rats or mice fed a high-cholesterol diet. Recent studies of EW (27) suggested favorable effects on serum cholesterol levels in healthy young women. However, further studies are needed to establish EW's hypocholesterolemic action in human subjects.

4.3.2 Ovomucin and Serum Cholesterol

Though ovomucin (OV) is one of the EW proteins, there is little information available on its hypocholesterolemic action. The hen egg ovomucin was found to serve as a novel hypocholesterolemic protein (Table 4.4) (28). OV clearly demonstrated serum cholesterol-lowering effects as compared with casein. The major difference in amino acid compositions between casein and OV is in the levels of glycine and cysteine. The relationship between the serum-cholesterol-lowering activity of dietary protein and the amino acid composition has previously been reported (16,49,50). Sugiyama et al. (50) reported a significant negative correlation between blood cholesterol levels and the level of cysteine in intact dietary proteins. Thus, as OV contains a higher level of cysteine than casein, the differences in amino acid content may be related to the differences in serum cholesterol level seen in our study.

The amount of OV is equivalent to 3.5% (w/w) of the EW. OV is a highly glycosylated glycoprotein and a polymer of two subunits of protein-rich α-subunit (MW 220 kDa) and carbohydrate-rich β-subunit (MW 400 kDa) (51,52). Recent studies have suggested that OV exhibits both antiviral (53,54) and antitumor activities (55). All the sialic acid present in OV was described as *N*-acetylneuraminic acid (56). Our preliminary observations suggest that cholesterol absorption in Caco-2 cells or micellar solubility of cholesterol is unaffected by the *N*-acetylneuraminic acid *in vitro*.

A higher fecal excretion of cholesterol and acidic steroids by OV feeding has been observed, indicating that the effect is, at least in part, due to an enhancement of fecal steroid excretion (jejunal effects). Smith (57) reported that human gallbladder mucin binds to cholesterol in model bile. Thus, we speculate that an increased fecal excretion of cholesterol may be induced by binding of cholesterol to OV in the intestine. These possibilities may be applicable to the case of OV on the basis of the evidence of fecal bile acid excretion and bile-acid-binding capacity in our study (ileal effects). In fact, OV has a significantly greater bile-acid-binding capacity, accompanied by a greater increase in fecal bile acid excretion, than does casein. Thus, the hypocholesterolemic action of OV may involve both jejunal and ileal effects.

TABLE 4.4
Effects of Dietary Casein, Ovumucin on Body and Liver Weights, Food Intake, Serum and Liver Lipids, and Fecal Steroid Excretion in Rats[a]

	Diet Group	
	Casein	OV
Body weight gain, g 10 d^{-1}	25.9 ± 1.1	22.4 ± 2.5
Liver weight, g 100 g^{-1} body wt	3.81 ± 0.06	3.61 ± 0.07
Food intake, d 6, g d^{-1}	14.2 ± 0.7	13.6 ± 0.6
Serum, mmol l^{-1}		
Total cholesterol (a)	2.83 ± 0.21	1.96 ± 0.07[b]
HDL cholesterol (b)	0.53 ± 0.04	0.66 ± 0.06
LDL + VLDL cholesterol	2.30 ± 0.22	1.30 ± 0.08[b]
Atherogenic Index (b)/(a), mol n^{-1}	0.19 ± 0.02	0.34 ± 0.03
Triglyceride	0.38 ± 0.04	0.37 ± 0.03
Phospholipids	1.13 ± 0.08	1.04 ± 0.05
Liver		
Total lipids, mg g^{-1} liver	142.8 ± 4.41	124.0 ± 2.47[b]
Cholesterol, μmol g^{-1} liver	70.8 ± 2.4	66.9 ± 3.2
Triglyceride, μmol g^{-1} liver	26.8 ± 2.9	24.8 ± 2.1
Phospholipid, μmol g^{-1} liver	118.8 ± 3.2	98.7 ± 1.6
Fecal		
Dry weight, g $3d^{-1}$	2.42 ± 0.06	2.58 ± 0.10
Cholesterol, μmol $3d^{-1}$	252.2 ± 6.7	281.3 ± 9.5[c]
Bile acids, μmol $3d^{-1}$	126.5 ± 5.8	151.9 ± 7.2[c]

[a]Mean ± SEM of six rats.
[b]Significantly different from Casein group at $p < 0.01$.
[c]Significantly different from Casein group at $p < 0.05$.
Abbreviations: OV, ovumucin; HDL, high density lipoprotein; LDL, low density lipoprotein; VLDL, very low density lipoprotein.
Source: S. Nagaoka, M. Masaoka, Q. Zhang, M. Hasegawa, and K. Wanatabe. *Lipids*, 37, 267–272, 2002. With permission.

It was found that OV lowers serum cholesterol levels in rats and inhibits cholesterol absorption in Caco-2 cells (28). These results clearly suggest that suppression of cholesterol absorption by direct interaction between cholesterol-mixed micelles and OV in jejunal epithelia is involved in the mechanism of hypocholesterolemic action induced by OV (jejunal effects). This clearly demonstrates the hypocholesterolemic action of OV compared with that of casein in the animal model; hence, OV may be a more promising material compared with EW itself for the prevention and treatment of hypercholesterolemia. Present findings concerning the hypocholesterolemic action of OV may enhance the utilization of eggs, egg proteins, and their products.

4.4 CHOLESTEROL-LOWERING MEAT PROTEINS AND PEPTIDES

Very few investigations have been reported on meat proteins (11,16,22,29–31). Carroll and Hamilton (11) reported that dietary animal proteins, including those of pork and beef, exhibited higher plasma cholesterol levels in rabbits than those of plant proteins, including soybean protein. Moreover, Lapre et al. (22) suggested that beef proteins did not have a hypocholesterolemic action in rats. However, Jacques et al. (16) first reported that dietary beef protein had a hypocholesterolemic action in rats fed a cholesterol-enriched diet. Morimatsu et al. (31) studied the hypocholesterolemic action of pork meat protein in rats and suggested that rats fed the low-molecular-weight fraction (MW 3000 Da or less) of papain-hydrolyzed pork meat had a significantly lower plasma cholesterol than those fed untreated pork meat or soybean protein. Their study also suggested that peptides produced by papain hydrolysis of pork meat had a hypocholesterolemic activity through their interference with the steroid absorption process. The effects of cattle round muscle hydrolysate (CMH), cattle heart hydrolysate (CHH), or cattle red blood cell hydrolysate (CBH) on liver and serum lipid components in hypercholesterolemic rats have also been reported (29). Serum total cholesterol and serum low-density lipoprotein (LDL) + very-low-density lipoprotein (VLDL) cholesterol were significantly decreased and the serum high-density lipoprotein (HDL) cholesterol to serum total cholesterol ratio was significantly increased in the CBH group compared with those of the control group. CHH feeding significantly decreased liver total lipids and liver cholesterol concentration compared with casein feeding. Dietary CHH or CBH exhibited a hypocholesterolemic action in rats (29). Interestingly, in human studies, Holmes et al. (30) reported that the hypocholesterolemic action of dietary meat (beef) protein was the same as that of a soybean protein diet containing low fat and low cholesterol. Further studies are necessary to establish the favorable effect of beef and pork protein and peptides on serum cholesterol in humans.

4.5 CHOLESTEROL-LOWERING FISH PROTEINS

Kritchevsky et al. (32) first reported that dietary fish protein exhibited a hypocholesterolemic action in rats. Jacques et al. (16) reported that rats fed fish (cod) protein had a lower serum cholesterol than those fed casein on a cholesterol-enriched diet. In rabbits, cod-fish protein has been shown to increase serum cholesterol concentrations when compared with casein (12). Zhang and Beynen (33) also reported that plasma and liver cholesterol concentrations were significantly decreased in rats fed a fish (cod)-protein diet containing high cholesterol compared with casein feeding. However, there are some conflicting results from rabbits and rats fed fish protein (12). As there is no information about the hypocholesterolemic effect of fish protein in humans, further studies are needed to establish this effect in animal and human subjects.

4.6 CHOLESTEROL-LOWERING BUCKWHEAT PROTEINS AND PEPTIDES

Buckwheat protein (BWP) is an excellent supplement to cereal grains, although its digestibility is relatively low. Kayashita et al. (58) reported that BWP has a hypocholesterolemic action in rats fed a cholesterol-enriched diet. In their experiment, plasma cholesterol level was lower in the BWP group than in the LMF (low-molecular-weight fraction) group, whereas that in the HMF (high- molecular-weight fraction) group was intermediate. It was also suggested that consumption of BWP lowered plasma cholesterol and raised fecal neutral sterols in a cholesterol-enriched diet because of its low digestibility in rats. As there is no information available on the hypocholesterolemic effect of BWP in humans, further studies are needed in this area.

4.7 NOVEL CHOLESTEROL-LOWERING OLIGOPEPTIDES

As discussed earlier, a new hypocholesterolemic pentapeptide IIAEK was derived from bovine milk β-lactoglobulin (25). Enterostatin (VPDPR) is released from the amino terminus of procolipase during its conversion to colipase (59,60). Enterostatin inhibits food intake after central or peripheral administration in animals fed a high-fat diet (60–62). Very recently, Takenaka et al. (63) found that enterostatin (VPDPR) and its fragmented peptide DPR have a hypocholesterolemic effect in mice fed a high-cholesterol diet. They showed that fecal excretion of cholesterol and bile acids was increased significantly by both VPDPR and DPR. They also showed that DPR induced hypocholesterolemic action just 2 hours after a single oral administration. The oligopeptide-induced hypocholesterolemic action is an uncultivated area in the field of hypocholesterolemic proteins and peptides research, and we predict major progress in this area in the future.

4.8 CHOLESTEROL-LOWERING SOYBEAN PROTEINS, PEPTIDES, AND SOYBEAN PEPTIDES WITH BOUND PHOSPHOLIPIDS

4.8.1 Effects of Amino Acid Sequence of Soybean Protein on Cholesterol Metabolism (Especially, Effects on Cholesterol Absorption and Re-Absorption of Bile Acid in the Intestine)

Soy protein, a vegetable protein, is well known to reduce serum cholesterol in comparison with casein, an animal protein (11–17). A soy protein peptic hydrolysate (SPH) has been reported to have a stronger serum cholesterol lowering effect than that of intact soy protein (34). Sugano et al. (34) showed that SPH decreased both blood cholesterol level and promoted fecal excretion of steroids, as compared with the effects of casein. They found that the high-molecular-weight fraction (HMF) was very potent in lowering cholesterol and produced significant reductions compared with the soluble fraction and intact isolated soy protein. They also suggested that the hypocholesterolemic peptides derived from soybean protein consisted of those peptides with molecular weights between 1 and 10 kDa. However, the hypocholesterolemic peptide of soybean protein remains unidentified. Based on the results of fecal steroid excretion, they suggested that SPH might have inhibited cholesterol absorption more than casein. Although the soy protein effect on cholesterol absorption had previously been examined and compared with that of casein (13), the *in vivo* experimental system did not allow determination of its direct effect on cholesterol absorption. Saeki et al. (14) suggested that the inhibition of cholesterol absorption was not the major factor involved in the differential effects of dietary proteins on serum cholesterol. Moreover, Lovati et al. (64) suggested that the activation of LDL receptor activity in liver cells induced by soybean globulin may be related to the serum-cholesterol-lowering action of soy protein; hence, it was necessary to provide direct evidence for suppression of cholesterol absorption in the intestine by soybean protein. Cholesterol micelles containing SPH were found to significantly suppress cholesterol uptake by Caco-2 cells compared with cholesterol micelles containing casein tryptic hydrolysate (CTH) (36). In relation to this cell culture study, using radiolabeled cholesterol in rats, we reported that SPH inhibited the intestinal absorption of cholesterol *in vivo*, accompanied by a greater increase in fecal cholesterol excretion (8,36). Interestingly, the micellar solubility of cholesterol in the presence of SPH was significantly lower than with casein CTH (8). We have demonstrated clearly the inhibition of the micellar solubility of cholesterol causing suppression of cholesterol absorption by direct interaction between cholesterol-mixed micelles and SPH in the jejunal epithelia and as part of the mechanism of hypocholesterolemic action induced by SPH (jejunal events) (8). A significantly higher bile-acid-binding capacity of SPH compared with that of CTH, concomitant with a greater increase in fecal bile acid excretion was also observed. There have been many studies on the hypocholesterolemic effects of proteins, most of which have emphasized the hypothesis that a bile-acid-binding peptide could inhibit the reabsorption of bile acid in the ileum and could decrease the blood cholesterol level (ileal events). These possibilities may be applicable to the case of SPH based on the evidence of fecal bile acid excretion and bile-acid-binding capacity (8,34). Thus, the hypocholesterolemic action of SPH may be related to both jejunal and ileal events.

It is of interest to explore the kinds of amino acid sequences of soybean protein that are concerned with the suppression of cholesterol absorption or its hypocholesterolemic action. Lovati et al. (65) reported that the peptides (2271 Da) derived from soybean 7S globulin activated the LDL-receptor in Hep G2 cells. However, no evidence was provided to indicate the activity of this peptide *in vivo*. Very recently, Manzoni et al. (66) found that the soybean 7S globulin α' constituent subunit could induce LDL-receptor upregulation in Hep G2 cells and that this subunit of soybean interacted with thioredoxin 1 and cyclophilin B, that are involved in cell protection against oxidative stress in Hep G2 cells.

4.8.2 Effects of Amino Acid Composition of Soybean Protein on Cholesterol Metabolism

Several excellent reviews are available on the effects of the amino acid composition of soy protein on cholesterol metabolism (67–70). The focus has been centered on the difference in amino acid composition in various proteins. Huff and Carroll (49) speculated that the relative proportion of the specific amino acids was concerned with this effect. In the meantime, there are some reports that assert the involvement of large number of amino acids (16). The ratio of arginine and lysine might be important (15), and in this theory, there are some negative results (16,24,71). In the research literature on proteins, it is indicated that the serum cholesterol is lower in milk whey protein because the cystine content is higher (50). However, no correlation has been seen to exist among the lysine and arginine ratio, cysteine content in the protein, and serum cholesterol value (24). In addition, extensive analysis using the manifold protein is necessary on amino acid content in the protein and relevance to amino acid content ratio and serum cholesterol.

4.8.3 The Relationship between the Hypocholesterolemic Effect of Soybean Protein and the Endocrine System

Forsythe (72) has summarized the animal data on the effect of soy protein on blood thyroxine concentration and has suggested that the elevation in blood thyroxine concentrations preceding the decline in blood cholesterol concentrations is consistent with a potential mechanism for cholesterol-lowering effect. However, no relationship has been found to exist between the effect of soy protein and blood cholesterol concentrations via thyroid status in rats and hamsters (73). Results in humans are also inconsistent. Ham et al. (74) found that 50 g of soy protein per day significantly increased thyroxine levels to a very modest degree in men. However, with 25 g of soy protein in a subsequent study, there was no modification in thyroid hormone status, even though the hypercholesterolemic subjects did experience a significant decrease in blood cholesterol level (75).

Sugano et al. (76) reported that soybean protein intake produced an increase in serum glucagon levels in rats. McArthur et al. (77) suggested that soy protein meals stimulated less gastric acid secretion and gastrin release than did beef meals. Further studies on molecular and gene levels are needed to clarify the relationship between the endocrine system and cholesterol metabolism regulated by dietary protein.

4.8.4 Soy Isoflavonoids and Cholesterol Metabolism

Evidence for an independent effect of isoflavonoids on blood cholesterol concentrations has been demonstrated in rats, hamsters, monkeys, and humans (17,78–81). Cassidy et al. (79) have reported that 45 mg of isoflavonoids, but not 23 mg, resulted in a significant reduction in total and LDL cholesterol levels in young women. Similar findings were reported by Potter et al. (82) and Bakhit et al. (75).

In contrast, Gooderham et al. (83) showed that plasma total and HDL cholesterol concentrations did not change in normocholesterolemic men who consumed 60 g of an isolated soy protein supplement providing 131 mg day^{-1} of isoflavones for 28 days. Nestel et al. (84) reported no significant effect on blood cholesterol concentrations with the administration of 45 mg of genistein over a 5- to 10-week period. Another trial by Simons et al. (85) reported that daily intake of 80 mg of soy isoflavones for 8 weeks did not affect plasma cholesterol in postmenopausal women. Owing to the inconsistent findings about the effects of isoflavones on blood cholesterol concentrations, it remains uncertain whether soy isoflavones are responsible for the proposed hypocholesterolemic effects of soy.

Lovati et al. (64,65) observed in their studies with cell culture that a 7S globulin fraction or a peptide derived from the 7S globulin of soybeans enhanced the activity of the LDL receptor, suggesting that this component may be a cholesterol-lowering agent. Fukui et al. (86) suggested that the isoflavone-free soybean protein can exhibit a cholesterol-lowering action in rats. Very interestingly, our products SPHP (soy peptides with bound phospholipids) and the crude type of SPHP (CSPHP) containing negligible soy isoflavones have been shown to clearly affect blood cholesterol concentrations in rats and humans (8–10). These results demonstrated that soy isoflavones did not play a main role or contribute to the hypocholesterolemic action induced by soy peptides.

4.8.5 The Health Claim and Industrial Utilization of Soy Protein and Peptides

The ameliorative action of soybean protein and soy peptides on cholesterol metabolism has been extensively reported. The U.S. Food and Drug Administration suggested that a daily intake of 25 g of soy protein might reduce the risk of heart disease (7). Also, in Japan, "Foods for specified health use" for preventing and treating hypercholesterolemia are being created. As mentioned earlier, we developed CSPHP, a "Food for specified health use," using soy peptides with bound soy phospholipids, for the prevention and treatment of hypercholesterolemia; the process for its development is described below.

Though the U.S. Food and Drug Administration suggested that a daily intake of 25 g of soy protein might reduce the risk of heart disease, this is an unusual amount for application in daily life. Thus, it is important to carry out research which specifies a more active ingredient from the soybean protein. In this regard, a soybean protein hydrolysate (SPH) was originally developed by Sugano et al. (34). However, this product could not be approved by the Ministry of Health as "Food for specified health use" in Japan. Therefore, a powerful and active component had to be developed using SPH or soybean protein. An attempt was made to combine soybean phospholipids with soy peptides for the reinforcement of hypocholesterolemic activity because phospholipids have also been reported to possess a cholesterol-lowering effect (87,88). Sirtori et al. (89) studied the effects of textured soy protein containing 6% lecithin in type II hyperlipidemic patients, among whom some positive effects on increasing HDL cholesterol concentration were noted. However, there is little information about textured soy protein containing 6% lecithin, and no studies have yet been reported on the effects of proteins that bind phospholipids in large quantities (> 20%) and, in particular, their hydrolyzates. As mentioned earlier, SPHP (soy protein hydrolysate with bound phospholipids) developed by us (8–10) has a higher phospholipid content (~20%) than those reported by Sirtori et al. (89). We also prepared a crude type of soy protein hydrolysate with bound phospholipids (CSPHP or c-SPHP), which did not require centrifugation in its preparation.

The effects of CSPHP (c-SPHP) on cholesterol metabolism were compared with the effects of SPHP, casein, SPH, and chitosan. Serum and liver cholesterol levels were significantly decreased with CSPHP, SPHP, SPH, and chitosan feeding compared with casein feeding (8,9,90–92). SPHP was the most effective in lowering cholesterol (Table 4.5). *In vivo* radioisotope studies suggested that SPHP or chitosan feeding inhibited the absorption of cholesterol and the reabsorption of taurocholate in rats (8,9,90–92). A higher fecal excretion of total steroids (acidic steroids + neutral steroids) in rats fed SPHP was demonstrated, indicating that the effect was, at least in part, due to an enhancement of fecal steroid excretion (8). A Caco-2 cell culture study indicated that cholesterol uptake from micelles containing SPHP or CSPHP was significantly lower than that from micelles containing casein tryptic hydrolysate (CTH) or casein (8,9,90–92). Thus, the suppression of cholesterol absorption by direct interaction between cholesterol-mixed micelles and SPHP in intestinal epithelia was part of the mechanism of hypocholesterolemic action induced by SPHP. The micellar solubility of cholesterol *in vitro* was significantly lower in the presence of SPHP or CSPHP than with CTH or casein (8,9,90–92). The taurocholate-binding capacity of SPHP was significantly higher than that of CTH or SPH (8). Our previous studies (8,90–92) clearly suggest that inhibition of the micellar solubility of cholesterol causes the suppression of cholesterol absorption by direct interaction between cholesterol-mixed micelles and SPHP in jejunal epithelia and is part of the mechanism of hypocholesterolemic action induced by SPHP (jejunal events).

Clinical trials with CSPHP have been carried out, and human studies suggest that serum total cholesterol and LDL cholesterol levels are significantly decreased by CSPHP feeding. CSPHP is useful for both normalizing high cholesterol levels in hypercholesterolemic patients whose serum total cholesterol levels are > 220 mg dl^{-1} (5.70 mmol l^{-1}) and maintaining the serum cholesterol level in healthy individuals consuming high cholesterol meals (two egg yolks per day) (93). Daily intake of only 3 g of CSPHP for 3 months significantly reduced both serum total cholesterol and LDL cholesterol levels in hypercholesterolemic patients (Figure 4.6) (10). The U.S. Food and Drug Administration suggested that a daily intake of 25 g soy protein might reduce the risk of heart disease. However, it has been shown that a smaller quantity of CSPHP can achieve the same desirable outcome.

TABLE 4.5
Effects of Dietary Casein, Soyprotein, Soyprotein with Bound Phospholipids (SP), Soyprotein Peptic Hydrolysate (SPH), Soyprotein Peptic Hyderolysate with Bound Phospholipids (SPHP) on Body and Liver Weights, Food Intake, Serum and Liver Lipids and Fecal Steroid Excretion in Rats[1]

	Diet Group					
	Casein	Soyprotein	SP[4]	SPH[5]	SPHP[6]	Pooled SEM
Body weight gain, g 14 d^{-1}	25.9	27.4	24.7	29.9	25.3	1.8
Liver weight, g 100 g^{-1}						
body weight	4.14a	3.80b	3.73b	3.83b	3.51c	0.08
Food intake, d 6, g d^{-1}	14.8	15.1	14.9	15.7	14.5	0.4
Serum, mmol l^{-1}						
Total cholesterol (a)	3.62a	2.69ab	2.36b	2.21b	1.95b	0.32
HDL cholesterol (b)	0.70d	0.90c	1.09b	1.18b	1.38a	0.06
LDL + VLDL cholesterol[2]	2.92a	1.79b	1.27bc	1.03bc	0.57c	0.31
(b)/(a), mol/mol	0.22d	0.33c	0.47b	0.54b	0.71a	0.03
Triglyceride	0.58	0.66	0.72	0.70	0.67	0.10
Phospholipids	1.37	1.36	1.38	1.39	1.38	0.08
Liver						
Total lipids, mg g^{-1} liver	158.6a	127.2b	94.6c	88.8c	59.9d	3.2
Cholesterol, μmol g^{-1} liver	110.0a	74.3b	47.2c	37.2d	11.1e	2.8
Triglyceride, μmol g^{-1} liver	32.1a	30.1a	20.0b	21.8b	10.2c	1.9
Phospholipids, μmol g^{-1} liver	113.7a	93.2b	76.0c	71.5c	60.2d	2.1
Fecal						
Dry weight, g $3d^{-1}$	3.23c	3.57c	3.66bc	4.20b	4.91a	0.19
Neutral steroids, μmol $3d^{-1}$						
Cholesterol	150.6c	214.4ab	246.7a	243.4a	189.5bc	14.3
Coprostanol	15.5c	36.7c	58.5bc	85.9b	287.4a	15.7
Total	166.1d	251.1c	305.2bc	329.3b	476.9a	19.9
Acidic steroids, μmol $3d^{-1}$	191.6b	198.4b	167.7b	196.3b	242.3a	13.8
Total steroids[3], μmol $3d^{-1}$	357.7c	449.5b	472.9b	525.6b	719.2a	26.6

[1] Values are means (n = 6) and pooled SEM Within a row, means different superscript letters are significantly different ($p < 0.05$) by Duncan's multiple range test.

[2] Values were calculated as follows: LDL + VLDL cholesterol = total cholesterol − HDL cholesterol.

[3] Total steroids = neutral steroids + acidic steroids.

[4] SP: soyprotein with bound phospholipids.

[5] SPH: soyprotein peptic hydrolysate.

[6] SPHP: soyprotein peptic hydrolysate with bound phosphlipids.

Source: S. Nagaoka, K. Miwa, M. Eto, Y. Kuzuya, G. Hori, and K. Yamamoto. *J. Nutr.*, 129, 1725–1730, 1999. With permission.

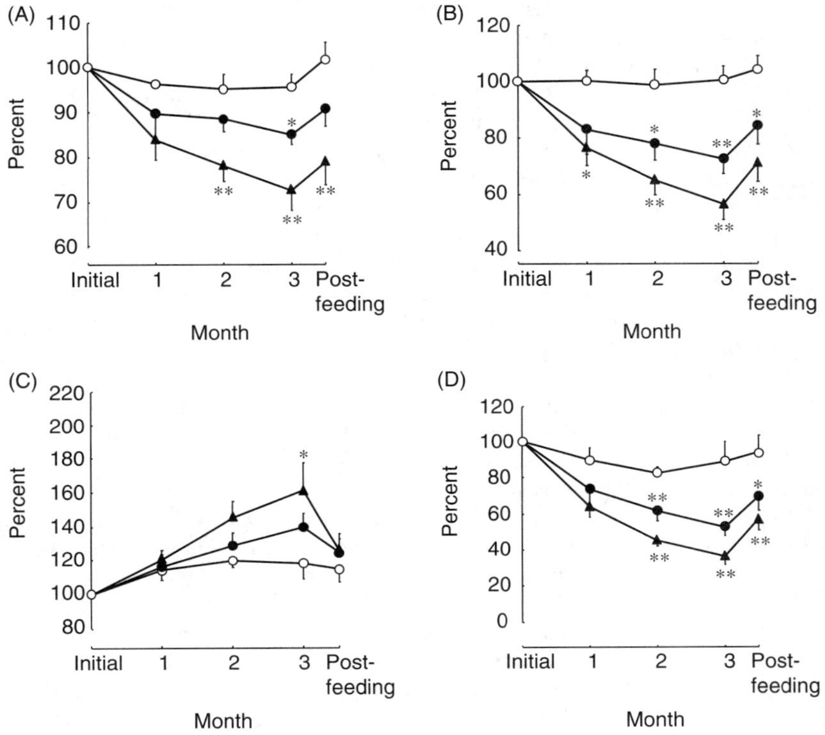

FIGURE 4.6 Relative changes of serum cholesterol levels in hypercholesterolemic volunteers. A: Total cholesterol; B: LDL-cholesterol; C: HDL-cholesterol; D: LDL/HDL ratio. c-SPHP, soy protein hydrolyzate with bound phospholipids; ○, placebo group; •, c-SPHP3g group; ▲, c-SPHP6g group. Each point shows the mean of the data from seven men per group and vertical bars represent SEM. Asterisks represent the significant difference from placebo group using Dunnett's test (*, $p < .05$; **, $p < .01$). (*Source*: From G. Hori, M.F. Wang, Y.C. Chan, T. Komatsu, Y. Wong, T.H. Chen, K. Yamamoto, S. Nagaoka, S. Yamamoto. *Biosci. Biotechnol. Biochem.*, 65, 72–78, 2001. With permission.)

REFERENCES

1. W.B. Kannel, W.P. Castelli, T. Gordon, P.M. McNamara. Serum cholesterol, lipoproteins, and the risk of coronary heart disease. The Framingham Study. *Ann. Intern. Med.*, 74: 1–12, 1971.
2. M.J. Martin, S.B. Hully, W.S. Browner, L.H. Kuller, D. Wentworth. Serum cholesterol, blood pressure, and mortality: implications from a cohort of 361,662 men. *Lancet*, 2: 933–936, 1986.
3. M.R. Law. Lowering heart disease risk with cholesterol reduction: evidence from observational studies and clinical trials. *Eur. Heart J.*, (Suppl. 1): S3–S8, 1999.
4. S.M. Grundy, M.A. Denke. Dietary influences on serum lipids and lipoproteins. *J. Lipid Res.*, 31: 1149–1172, 1990.

5. H.N. Ginsberg, S.L. Barr, A. Gilbert. Reduction of plasma cholesterol levels in normal men on an American Heart Association Step I diet or a Step I diet with added monounsaturated fat. *N. Engl. J. Med.*, 322: 574–579, 1990.
6. J.W. Anderson, B.M. Johnstone, M.E. Cook-Newell. Meta-analysis of the effects of soy protein intake on serum lipids. *N. Engl. J. Med.*, 333: 276–282, 1995.
7. Food and Drug Administration. Soy proteins and coronary heart disease. *Fed. Regist.*, 64: 57688–57733, 1999.
8. S. Nagaoka, K. Miwa, M. Eto, Y. Kuzuya, G. Hori, K. Yamamoto. Soyprotein peptic hydrolyzate with bound phospholipids decrease micellar solubility and cholesterol absorption in rats and Caco-2 cells. *J. Nutr.*, 129: 1725–1730, 1999.
9. S. Nagaoka, G. Hori, K. Yamamoto, K. Yamamoto, S. Yamamoto. Improvements in cholesterol metabolism induced by soypeptides with bound phospholipids. *J. Nutr.*, 132: 604S, 2002.
10. G. Hori, M.F. Wang, Y.C. Chan, T. Komatsu, Y. Wong, T.H. Chen, K. Yamamoto, S. Nagaoka, S. Yamamoto. Soy protein hydrolyzate with bound phospholipids reduces serum cholesterol levels in hypercholesterolemic adult male volunteers. *Biosci. Biotechnol. Biochem.*, 65: 72–78, 2001.
11. K.K. Carrol, R.M.G. Hamilton. Effects of dietary protein and carbohydrate on plasma cholesterol levels in relation to atherosclerosis. *J. Food Sci.*, 40: 18–23, 1975.
12. M.R. Lovati, C.E. West, C. Sirtori, A.C. Beynen. Dietary animal proteins and cholesterol metabolism in rabbits. *Br. J. Nutr.*, 64: 473–485, 1990.
13. Y. Nagata, N. Ishiwaki, M. Sugano. Studies on the mechanisms of antihypercholesterolemic action of soy protein and soy protein-type amino mixtures in relation to the casein counterparts in rats. *J. Nutr.*, 112: 1614–1625, 1982.
14. S. Saeki, H. Nishikawa, S. Kiriyama. Effects of casein or soybean protein on plasma cholesterol level in jejunectomized or ileectomized rats. *J. Nutr.*, 117: 1527–1531, 1987.
15. D. Kritchevsky. Vegetable protein and atherosclerosis. *J. Am. Oil Chem. Soc.*, 56: 135–140, 1979.
16. H. Jacques, Y. Deshaies, L. Savoie. Relationship between dietary proteins, their in vitro digestion products, and serum cholesterol in rats. *Atheroscelosis*, 61: 89–98, 1986.
17. F. Balmir, R. Staack, E. Jeffrey, M.B.B. Jimenez, L. Wang, S.M. Potter. An extract of soy flour influences serum cholesterol and thyroid hormones in rats and hamsters. *J. Nutr.*, 126: 3046–3053, 1996.
18. C. Sautier, K. Dieng, C. Flament, C. Doucet, J.P. Suquet, D. Lemonnier. Effects of whey protein, casein, soya-bean and sunflower proteins on serum, tissue and faecal steroids in rats. *Br. J. Nutr.*, 49: 313–319, 1983.
19. S. Nagaoka, Y. Kanamaru, Y. Kuzuya. Effects of whey protein and casein on the plasma and liver lipids in rats. *Agric. Biol. Chem.*, 55: 813–818, 1991.
20. S. Nagaoka, Y. Kanamaru, Y. Kuzuya, T. Kojima, T. Kuwata. Comparative studies on the serum cholesterol lowering action of whey protein and soybean protein in rats. *Biosci. Biotechnol. Biochem.*, 56: 1484–1485, 1992.
21. Y.S. Choi, S. Goto, I. Ikeda, M. Sugano. Interaction of dietary protein, cholesterol and age on lipid metabolism of the rat. *Br. J. Nutr.*, 61, 531–543, 1989.
22. J.A. Lapre, C.E. West, M.R. Lovati, C.R. Sirtori, A.C. Beynen. Dietary animal proteins and cholesterol metabolism in rats. *Int. J. Vitam. Nutr. Res.*, 59: 93–100, 1989.
23. X. Zhang, A.C. Beynen. Lowering effect of dietary milk-whey protein v. casein on plasma and liver cholesterol concentrations in rats. *Br. J. Nutr.*, 70: 139–146, 1993.

24. S. Nagaoka. Studies on regulation of cholesterol metabolism induced by dietary food constituents or xenobiotics. *J. Jpn. Soc. Nutr. Food Sci.*, 49: 303–313, 1996.
25. S. Nagaoka, Y. Futamura, K. Miwa, T. Awano, K. Yamauchi, Y. Kanamaru, T. Kojima,T. Kuwata. Identification of novel hypocholesterolemic peptides derived from bovine milk β-lactoglobulin. *Biochem. Biophys. Res. Commn.*, 281: 11–17, 2001.
26. S. Yamamoto, T. Kina, N. Yamagata, T. Kokubu, S. Shinjo, L. Asato. Favorable effects of egg white protein on lipid metabolism in rats and mice. *Nutr. Res.*, 13: 1453–1457, 1993.
27. L. Asato, M.F. Wang, Y.C. Chan, S.H. Yeh, H.M. Chung, S.Y. Chung, S. Chida, T. Uezato, I. Suzuki, N. Yamagata, T. Kokubu, S. Yamamoto. Effect of egg white on serum cholesterol concentration in young women. *J. Nutr. Sci. Vitaminol.*, 42: 87–96, 1996.
28. S. Nagaoka, M. Masaoka, Q. Zhang, M. Hasegawa, K. Watanabe. Egg ovomucin attenuates hypercholesterolemia in rats and inhibits cholesterol absorption in Caco-2 cells. *Lipids*, 37: 267–272, 2002.
29. K. Honda, M. Suzuki, M. Numata, T. Nakamura, Y. Futamura, Y. Kanamaru, S. Nagaoka. Effect of dietary protein hydrolyzate in livestock products on the plasma and liver lipid components in rats. *Congress proceedings of 45th International Congress of Meat Science and Technology*, Volume II, Yokohama, 1999, 698–699.
30. W.L. Holmes, G.B. Rubel, S.S. Hood. Comparison of the effect of dietary meat versus dietary soybean protein on plasma lipids of hyperlipidemic individuals. *Atherosclerosis*, 36: 379–387, 1980.
31. F. Morimatsu, M. Ito, S. Budijanto, I. Watanabe, Y. Furukawa, S. Kimura. Plasma cholesterol-supressing effect of papain-hydrolyzed pork meat in rats fed hypercholesterolemic diet. *J. Nutr. Sci. Vitaminol.*, 42: 145–153, 1996.
32. D. Kritchevsky, S.A. Tepper, S.K. Czarnecki, D.M. Klurfeld. Atherogenicity of animal and vegetable protein — influence of the lysine to arginine ratio. *Atherosclerosis*, 41: 429–431, 1982.
33. X. Zhang, A.C. Beynen. Influence of dietary fish proteins on plasma and liver cholesterol concentrations in rats. *Br. J. Nutr.*, 69: 767–777, 1993.
34. M. Sugano, S. Goto, Y. Yamada, K. Yoshida, Y. Hashimoto, T. Matsuo, M. Kimoto. Cholesterol-lowering activity of various undigested fractions of soybean protein in rats. *J. Nutr.*, 120: 977–985, 1990.
35. K. Iwami, K. Sakakibara, F. Ibuki. Involvement of post-digestion 'hydrophobic' peptides in plasma cholesterol-lowering effect of dietary plant proteins. *Agric. Biol. Chem.*, 50: 1217–1222, 1986.
36. S. Nagaoka, T. Awano, N. Nagata, M. Masaoka, G. Hori, K. Hashimoto. Soyprotein peptic hydrolyzate lowers serum cholesterol level and inhibits cholesterol absorption in CaCo-2 cells. *Biosci. Biotechnol. Biochem.*, 61: 354–356, 1997.
37. F.J. Field, E. Albright, S. Mathur. Regulation of cholesterol esterification by micellar cholesterol in CaCo-2 cells. *J. Lipid Res.*, 28: 1057–1066, 1987.
38. T.E. Hughes, W.V. Sasak, J.M. Ordovas, T.M. Forte, S. Lamon-Fava, E.J. Schaefer. A novel cell line (Caco-2) for the study of intestinal lipoprotein synthesis. *J. Biol. Chem.*, 262: 3762–3767, 1987.
39. T. Ranheim, A. Gedde-Dahl, A.C. Rustan, C.A. Drevon. Influence of eicosapentaenoic acid (20:5, n-3) on secretion of lipoproteins in CaCo-2 cells. *J. Lipid Res.*, 33: 1281–1293, 1992.

40. M.D. Wilson, L.L. Rudel. Review of cholesterol absorption with emphasis on dietary and biliary cholesterol. *J. Lipid Res.*, 28: 1057–1066, 1994.
41. I. Ikeda, K. Tanaka, M. Sugano, G.V. Vahouny, L.L. Gallo. Inhibition of cholesterol absorption in rats by plant sterols. *J. Lipid Res.*, 29: 1573–1582, 1988.
42. N. Hirose, T. Inoue, K. Nishihara, M. Sugano, K. Akimoto, S. Shimizu, H. Yamada. Inhibition of cholesterol absorption and synthesis in rats by sesamin. *J. Lipid Res.*, 32: 629–638, 1991.
43. I. Ikeda, Y. Imasato, E. Sasaki, M. Nakayama, H. Nagao, T. Takeo, F. Yayabe, M. Sugano. Tea catechins decrease micellar solubility and intestinal absorption of cholesterol in rats. *Biochim. Biophys. Acta*, 1127: 141–146, 1992.
44. H.A. Ahmed, M.L. Petroni, M. Abu-Hamdiyyah, R.P. Jazrawi, T.C. Northfield. Hydrophobic/hydrophilic balance of proteins: a major determinant of cholesterol crystal formation in model bile. *J. Lipid Res.*, 35: 211–219, 1994.
45. G. Yamashita, R. Secknus, A. Chernosky, K.A. Krivacic, R.T. Holzbach. Comparison of heptoglobin and apolipoprotein A-I on biliary lipid particles involved in cholesterol crystallization. *J. Gastroenterol. Hepatol.*, 11: 738–745, 1996.
46. S.L. Roberts, M.P. McMurry, W.E. Conner. Does egg feeding (i.e., dietary cholesterol) affect plasma cholesterol levels in humans? Results of double-blind study.*Am. J. Clin. Nutr.*, 34: 2092–2099, 1981.
47. R. Nakamura, M. Takayama, K. Nakamura, O. Umemura. Constituent proteins of globulin fraction obtained from egg white. *Agric. Biol. Chem.*, 44: 2357–2362, 1980.
48. K. Watanabe, Y. Tsuge, M. Shimoyamada, N. Ogama, T. Ebina. Antitumor effects of pronase-treated a double grafted tumor system. *J. Agric. Food Chem.*, 46: 3033–3038, 1998.
49. M.W. Huff, K.K. Carroll. Effects of dietary proteins and amino acid mixtures on plasma cholesterol levels in rabbits. *J. Nutr.*, 110: 1676–1685, 1980.
50. K. Sugiyama, S. Ohkawa, K. Muramatsu. Relationship between amino acid composition of diet and plasma cholesterol level in growing rats fed a high cholesterol diet. *J. Nutr. Sci. Vitaminol.*, 32: 413–423, 1986.
51. A. Kato, R. Nakamura, Y. Sato. Studies on changes in stored shell eggs. Part IV. changes in the chemical composition of ovomucin during storage. *Agric. Biol. Chem.*, 34: 1009–1013, 1970.
52. T. Itoh, J. Miyazaki, H. Sugawara, S. Adachi. Studies on the characterization of ovomucin and chalaza of the hen's egg. *J. Food Sci.*, 52: 1518–1521, 1987.
53. Y. Tsuge, M. Shimoyamada, K. Watanabe. Structural features of newcastle disease virus and anti-ovomucin antibody-binding glycopeptides from pronase-treated ovomucin. *J. Agric. Food Chem.*, 45: 2393–2398, 1997.
54. Y. Tsuge, M. Shimoyamada, K. Watanabe. Binding of ovomucin to newcastle disease virus and anti-ovomucin antibodies and its heat stability based on binding abilities. *J. Agric. Food Chem.*, 45: 4629–4634, 1997.
55. K. Watanabe, Y. Tsuge, M. Shimoyamada, N. Ogama, T. Ebina. Antitumor effects of pronase-treated a double grafted tumor system. *J. Agric. Food Chem.*, 46: 3033–3038, 1998.
56. D.S. Robinson, J.B. Monsey. Studies on the composition of egg-white ovomucin. *Biochem. J.* 121: 537–547, 1971.
57. B.F. Smith. Human gallbadder mucin binds biliary lipids and promotes cholesterol crystal nucleation in model bile. *J. Lipid Res.*, 28: 1088–1097, 1987.

58. J. Kayashita, I. Shimaoka, M. Nakajoh, M. Yamazaki, N. Kato. Consumption of buckwheat protein lowers plasma cholesterol and raises fecal neutral sterols in cholesterol-fed rats because of its low digestibility. *J. Nutr.*, 127: 1395–1400, 1997.
59. C. Erlanson-Albertsson. Pancreatic colipase. Structure and physiological aspects. *Biochim. Biophys. Acta*, 1125: 1–7, 1992.
60. C. Erlanson-Albertsson, A. Larsson. The activation peptide of pancreatic procolipase decreases food intake in rats. *Regul. Pept.*, 22: 325–331, 1988.
61. L. Lin, J. Chen, D.A. York. Chronic ICV entrostatin preferentially reduced fat intake and lowered body weight. *Peptides*, 18: 657–661, 1997.
62. S. Okada, L. Lin, D.A. York, G.A. Bray. Chronic effects of intracerebral ventricular enterostatin in Osborne-Mendel rats fed a high-fat diet.*Physiol. Behav.*, 54: 325–329, 1993.
63. Y. Takenaka, F. Nakamura, T. Yamamoto, M. Yoshikawa. Enterostatin (VPDPR) and its peptide fragment DPR reduce serum cholesterol levels after oral administration in mice. *Biosci. Biotechnol. Biochem.*, 67: 1620–1622, 2003.
64. M.R. Lovati, C. Manzoni, A. Corsini, A. Granata, R. Frattini, R. Fumagalli, C.R. Sirtori. Low density lipoprotein receptor activity is modulated by soybean globulins in cell culture. *J. Nutr.*, 122: 1971–1978, 1992.
65. M.R. Lovati, G.E. Manzoni, A. Arnoldi, E. Kurowska, K.K. Carroll, CR Sirtori. Soy protein peptides regulate cholesterol homeostasis in Hep G2 cells. *J. Nutr.*,130: 2543–2549, 2000.
66. C. Manzoni, M. Duranti, I. Eberi, H. Scharnag, W. Marz, S. Castiglioni, M.R. Lovati. Subcellular localization of soybean 7S globulin in Hep G2 cells and LDL receptor up-regulation by its α'- constituent subunit. *J. Nutr.*, 133: 2149–2155, 2003.
67. M. Messina, J.W. Jr Erdman. First international symposiun on the role of soy in preventing and treating chronic disease. *J. Nutr.*, 125: 567S–808S, 1995.
68. M. Sugano. Nutritional studies on the regulation of cholesterol meabolism. *J. Jpn. Soc. Nutr. Food Sci.*, 40: 93–102, 1987.
69. K. Sugiyama. Importance of sulfur-containing amino acids in cholesterol metabolism. *J. Jpn. Soc. Nutr. Food Sci.*, 42: 353–363, 1989.
70. S.M. Potter. Soy protein and cardiovascular disease: the impact of bioactive components in soy. *Nutr. Rev.*, 56: 231–235, 1998.
71. M.J. Gibney. The effects of dietary lysine to arginine ratio on cholesterol kinetics in rabbits. *Atherosclerosis*, 47: 263–270, 1983.
72. W.A.III Forsythe. Soy protein, thyroid regulation and cholesterol metabolism. *J. Nutr.*, 125: 619S–623S, 1995.
73. S.M. Potter, M.D. Berber-Jimenez, J. Pertile. Protein from soy concentrate and isolated soy protein alters blood lipids and hormones differently. *J. Nutr.*, 126: 2007–2011, 1996.
74. J.O. Ham, K.M. Chapman, D. Essex-Sorlie, R.M. Bakhit, M. Prabhudesai, L. Winter, J.W.Jr Erdman, S.M. Potter. Endocrinological responses to soy protein and fiber in mildly hypercholesterolemic men. *Nutr. Res.*, 13: 873–884, 1993.
75. R.M. Bakhit, B.P. Klein, D. Essex-Sorlie, J.O. Ham, J.W. Erdman, S.M. Potter. Intake of 25 g soy protein reduces plasma cholesterol in men with elevated cholesterol concentrations. *J. Nutr.*, 124: 213–222, 1994.
76. M. Sugano, N. Ishiwaki, Y. Nagata, K. Imaizumi. Effects of arginine and lysine addition to casein and soya-bean protein on serum lipids, apolipoproteins, insulin and glucagons in rats. *Br. J. Nutr.*, 48: 211–221, 1982.

77. K.E. McArthur, J.H. Walsh, C.T. Richardson. Soy protein meals stimulate less gastric acid secretion and gastrin release than beef meals. *Gastroenterology*, 95: 920–926, 1988.
78. M.S. Anthony, T.B. Clarkson, C.L. Hughes, T.M. Morgan, G.L. Burke. Soybean isoflavones improve cardiovascular risk factors without affecting the reproductive system of peripubertal rhesus monkeys. *J. Nutr.*, 126: 43–50, 1996.
79. A. Cassidy, S. Bigham, K. Setchell. Biological effects of isoflavonoids in young women: importance of the chemical composition of soybean products. *Br. J. Nutr.*, 74: 587–601, 1995.
80. T.B. Clarkson, M.S. Anthony, J.K. Williams, E.K. Honore, J.M. Cline. The potential of soybean phytoestrogens for postmenopausal hormone replacement therapy. *Proc. Soc. Exp. Biol. Med.*, 217: 365–368, 1998.
81. X. Pelletier, S. Belbraout, D. Mirabel, F. Mordret, J.L. Perrin, X. Pages, G. Debry. A diet moderately enriched in phytosterols lowers plasma cholesterol concentrations in normocholesterolemic humans. *Ann. Nutr. Metabol.*, 39: 291–295, 1995
82. S.M. Potter, R.M. Bakhit, D.L. Essex-Sorlie, K.E. Weingartner, K.M. Chapman, R.A. Nelson, M. Prabhudesai, W.D. Savage, A.I. Nelson, L.W. Winter, J.W. Erdman. Depression of plasma cholesterol in men by consumption of baked products containing soy protein. *Am. J. Clin. Nutr.*, 58: 501–506, 1993.
83. M.H. Gooderham, H. Adlercreutz, S.T. Ojala, K. Wahala, B.J. Holub. A soy protein isolate rich in genistein and daidzein and its effects on plasma isoflavone concentrations, platelet aggregation, blood lipids and fatty acid composition of plasma phospholipid in normal men. *J. Nutr.*, 126: 2000–2006, 1996.
84. P.J. Nestel, T. Yamashita, T. Sasahara, S. Pomeroy, A. Dart, P. Komesaroff, A. Owen, M. Abbey. Soy isoflavones improve systemic arterial compliance but not plasma lipids in menopausal and perimenopausal women. *Arterioscler. Thromb. Vasc. Biol.*, 17: 3392–3398, 1997.
85. L.A. Simons, M. Von Konigsmark, J. Simons, D.S. Celermajer. Phytoestrogens do not influence lipoprotein levels or endothelial function in healthy, postmenopausal women. *Am. J. Cardiol.*, 85: 1297–1301, 2000.
86. K. Fukui, N. Tachibana, S. Wanezaki, S. Tsuzaki, K. Takamatsu, T. Yamamoto, Y. Hashimoto, T. Shimoda. Isoflavone-free soy protein prepared by column chromatography reduces plasma cholesterol in rats. *J. Agric. Food Chem.*, 50: 5717–5721, 2002.
87. J.E. O'Mullane, J.N. Hawthorne. A comparison of the effects of feeding linoleic acid-rich or corn oil on cholesterol absorption and metabolism in rat. *Atherosclerosis*, 45: 81–90, 1982.
88. K. Imaizumi, M. Sakono, M. Sugano, Y. Shigematsu, M. Hasegawa. Influence of saturated and polyunsaturated egg yolk phospholipids on hyperlipidemia in rats. *Agric. Biol. Chem.*, 53: 2469–2474, 1989.
89. C.R. Sirtori, C. Zucchi-Dentone, M. Sirtori, E. Gatti, G.C. Descovich, A. Gaddi, L. Cattin, P.G. Da Col, U. Senin, E. Mannanino, G. Avellone, L. Colombo, C. Fragiacomo, G. Noseda, S. Lenzi. Cholesterol-lowering and HDL-raising properties of lecithinated soy proteins in type II hyperlipidemic patients. *Ann. Nutr. Metabol.*, 29: 348–357, 1985.
90. S. Nagaoka, H. Ishikawa, F. Shibayama, G. Hori, T. Hara, K. Yamamoto. Comparative studies on the improving effects of cholesterol metabolism induced by soyprotein sumizyme hydrolysate with bound phospholipids or chitosan. *Proceedings of the Third International Soybean Processing and Utilization Conference*, Tukuba, 2000, 207–208.

91. K. Morishita, K. Yamamoto, G. Hori, M. Tanaka, S. Kamiya, S. Nagaoka. Cholesterol-lowering effects of soy protein peptic hydrolyzate with bound phospholipids in rats; cross-over test, dose-response test, and comparison with materials which have cholesterol-lowering effect. *J. Jpn. Soc. Nutr. Food Sci.*, 52: 183–191, 1999.
92. G. Hori, K. Yamamoto, K. Morishita, S. Mukawa, S. Kamiya, S. Nagaoka. Cholesterol-lowering effects of isolated soybean protein hydrolyzate with bound phospholipidsin rats. *J. Jpn. Soc. Nutr. Food Sci.*, 52: 135–145, 1999.
93. G. Hori, K. Yamamoto,T. Kamiya, T. Hara, S. Nagaoka, N. Motoya, S. Yamamoto. The effects of soybean protein hydrolyzate with bound phospholipids on serum cholesterol levels in adult male subjects receiving high cholesterol diet. *J. Jpn. Soc. Clin. Nutr.*, 22: 21–27, 2000.

5 Heavy-Metal-Binding Proteins

Jennifer Kovacs-Nolan and
Yoshinori Mine

CONTENTS

5.1 INTRODUCTION

Heavy metals such as zinc (Zn) and copper (Cu) are ubiquitous components of the biosphere and are essential elements that participate in a variety of enzymatic reactions (1). Essential heavy metals act in catalytic sites of metalloenzymes and enable specific folding of many other metalloproteins (2). Cadmium (Cd), lead (Pb), and other nonessential metals (as well as the essential metal ions in excess) can exert toxic effects due to their high affinity for amino acid residues involved in important biological functions and their potential to displace essential metals required for biological processes (2). Living organisms have therefore evolved mechanisms that control and respond to the uptake and accumulation of both essential and nonessential heavy metals, including the chelation and sequestering of heavy metals by intracellular ligands (2,3). Some of the best characterized heavy-metal-binding ligands include metallothioneins (MT), which were first described by Margoshes and Vallee (4) in equine kidney. Since then, several MTs and MT-like proteins, often described as heavy-metal-inducible, heavy-metal-binding, cysteine-rich polypeptides, as well as other heavy-metal-binding proteins have been described and characterized from a number of different organisms (3).

Besides their role in heavy-metal regulation, metal-binding proteins have also been shown to be involved in several disease processes by demonstrating a direct therapeutic effect or being induced by disease conditions. Their importance has been underscored by research linking a number of human diseases, including Alzheimer's disease and prion diseases, to metal-binding proteins (5,6). The use of metal-binding proteins has also been suggested for micronutrient supplementation and environmental remediation. These applications as well as an overview of metal-binding proteins and peptides, their physiological significance, and their isolation and characterization will be discussed here.

5.2 METAL-BINDING PROTEINS AND PEPTIDES

5.2.1 Metallothioneins and Phytochelatins

MTs are a family of low-molecular-weight, heavy-metal-binding proteins, characterized by a high cysteine content and lack of aromatic amino acids and histidine, which bind 7 to 12 heavy metal atoms per molecule of protein (7,8). Since their initial discovery in equine kidney, MTs have been found to be broadly distributed among mammals, eukaryotic microorganisms, certain prokaryotes, and plants (9). In mammals, MTs have been suggested to play a role in metal metabolism and homeostasis; to regulate synthesis, assembly, or activity of zinc metalloproteins; and to protect against reactive oxygen species. MTs may be induced by not only metal ions but also by certain hormones, cytokines, and chemical and physical stress (1,2).

MTs subsequently isolated from other sources have been subdivided into three classes. Class I MTs (MT-I) contain 20 highly conserved Cys residues based on mammalian MTs and are widespread in vertebrates as well as in certain fishes, crabs, oysters, and mussels (2,3,9). Class II MTs (MT-II) lack obvious sequence homology to the archetypal mammalian MT and among each other and are found in *Drosophila*, sea urchins, fungi, cyanobacteria, and plants (2,9). A third group, MT-III, is the major metal-binding protein in plants and consists of nontranslationally synthesized, low-molecular-weight peptides, consisting of repeating units of γ-glutamate (Glu)-cysteine (Cys) units, which are involved in metal homeostasis (2,8,10). Widely distributed throughout the plant kingdom, class III MTs are actually aggregates of a heterogeneous population of polypeptides (9,11) that are believed to combine to form metal-binding complexes rather than occurring as individual metal-binding proteins (9). Five families of MT-III peptides are known, including phytochelatins (PC), and which of these peptides occur in particular plants depends on the species and extent of metal exposure (9). As well, other metallothionein-like and nonmetallothionein proteins have been reported in plants (12,13).

5.2.2 Metallochaperones

More recently, the study of complex trafficking pathways involving specific delivery molecules, or metallochaperones — proteins that bind and protect metal ions and deliver them to their target enzymes — has been of particular interest (14–16). While metallochaperones are not structurally defined, characteristics common to all include a conserved cysteine-containing metal-binding motif as well as apparent similarities between the metallochaperones and their target proteins (16).

Copper chaperones, in particular, from yeast and humans have been well characterized and can be divided into three functional groups: (1) the Atx1-like chaperones, (2) the Cu chaperones for superoxide dismutase, and (3) the Cu chaperones for cytochrome *c* oxidase (15). Homologs of those studied in yeast and humans have also been identified in bacteria, plants, and other animals (17) as having nickel (Ni) metallochaperones (18).

5.2.3 Animal Fibrous Proteins

Although not metal binding by design like MTs, PCs, and metallochaperones, several animal fibrous proteins (AFPs) have also demonstrated heavy-metal-binding capabilities. Studies of various industrial wastes, agricultural byproducts and biological materials indicated that AFPs, which are stable and water-insoluble fibers with high surface area (e.g., egg shell membrane, chicken feather, wool and silk), were capable of binding metals such as gold (Au), silver (Ag), Cu, Cd, Pb, Ni, and Zn (19–21).

Egg shell membrane (ESM) proteins, which are available in large quantities as a byproduct of the egg industry, are of particular interest in metal biosorption applications (21). ESM, which consists of protein fibers, is found between the egg white and the inner surface of the egg shell and has a high content of arginine, glutamic acid, methionine, histidine, cysteine, and proline (22). While the binding selectivity of ESM appears to be pH dependent, it has been suggested that not only electrostatic but also ligand-change reactions may be involved in the metal sorption by ESM (21).

5.3 PHYSIOLOGICAL FUNCTION OF METAL-BINDING PROTEINS AND PEPTIDES

It is widely believed that many metal-binding proteins and peptides play a role in metal homeostasis, and aid in the detoxification of toxic heavy metals (2). One of the main functions of MT is the maintenance of free metal Zn and Cu ions in the cells by acting as a homeostatic regulator and reservoir of ions and as a donor of Zn for Zn-dependent biological processes (23,24). Cadmium (Cd), a metal ion similar to Zn, is one of the most toxic heavy metals. The toxicity of Cd is due to its negative influence on the enzymatic systems of cells, resulting from substitutions of other metal ions (mainly Zn, Cu, and Ca) in metalloenzymes, and its very strong affinity for biological structures containing the thiol groups, including proteins and nucleic acids (23). The toxicity of Cd has been demonstrated in several target organs, including kidneys, lungs, liver, and bone, and inhalation in high amounts has been found to cause lethal pulmonary edema (25). By displacing Zn, Cd interferes with Zn absorption, distribution into tissues, and transport into cells and may inhibit its metabolic processes such as the cellular production of deoxyribonucleic acid (DNA), ribonucleic acid (RNA), and proteins (23).

Much research has been carried out to investigate the protective effect of metal-binding proteins and peptides against Cd toxicity. It has been suggested that MT may prevent toxicity from heavy metals such as Cd and mercury (Hg) by the formation of metal–MT complexes (23). The protective effect of Zn– and Cd–PC complexes on metal-sensitive enzymes has been investigated. Concentrations of Zn or Cd that would normally inhibit enzyme activity by 50% could be exceeded by 10- to 150-fold when the Zn or Cd was supplied as a metal–PC complex (26). Takagi et al. (27) found that when added to cell culture medium, heavy-metal-binding plant peptides (PC) enhanced the detoxification of Cd. Similarly, when the PC peptides were expressed by mammalian cells, they exhibited *in vitro* resistance to Cd toxicity.

Copper plays a key role in all living organisms, acting as a cofactor for many protein and enzymes; however, high Cu concentrations can be dangerous, leading to oxidative damage of proteins, lipids, and nucleic acids. Therefore, Cu concentrations must be carefully controlled so that it is available for essential enzymes but does not accumulate to toxic levels (17). Like MTs, metallochaperones help achieve metal ion homeostasis and, in the case of Cu chaperones, prevent neurological diseases due to metal ion imbalance (28). The importance of Cu chaperones can be emphasized by the fact that total Cu concentrations in living cells are in the micromolar range and that metalloenzymes therefore require an auxiliary method of acquiring Cu (14).

Alternative roles of metal-binding peptides in plants have also been proposed, including roles in iron (Fe) or sulfur metabolism (3).

5.4 ISOLATION OF METAL-BINDING PROTEINS AND PEPTIDES

The isolation and purification of metal-binding compounds generally involve the extraction from tissues using Tris or phosphate buffers, in the pH range of 7.2 to 8.6, often containing additives, such as sodium chloride (NaCl), or reducing agents (8). Concentration by lyophilization, ultrafiltration, dialysis against polyethylene glycol, or ammonium sulfate precipitation is often carried out, followed by sequential separation steps using gel filtration and ion exchange chromatography (8). The use of hydrophobic interaction chromatography has also been reported (29). More recently, the use of capillary electrophoresis for separation and high-performance liquid chromatography (HPLC) coupled with element-specific and molecule-specific detectors, such as inductively coupled plasma mass spectrometry (ICP-MS) and electrospray ionization–tandem mass spectrometry (ESI-MS/MS), has been described for the isolation and characterization of metal-binding proteins (30).

The isolation of animal fibrous proteins requires simply (1) the mechanical separation, in the case of ESM protein, (2) the removal of the membrane from the egg shell following immersion in acidic and then basic solutions, and (3) finally drying (31).

The isolation of metal-binding compounds from several plant species, including flax (10,32), horseradish (33), wheat germ (34), and Asian periwinkle *Littorina brevicula* (35), from several marine species, including shellfish (36), clam (37,38), crab (29), and rainbow trout (39), and from mushroom (12) has also been described.

5.5 APPLICATIONS OF HEAVY-METAL-BINDING PROTEINS

5.5.1 IMPLICATIONS IN HUMAN HEALTH AND DISEASE

Experimental autoimmune encephalomyelitis (EAE) is an animal model for the human autoimmune disease multiple sclerosis (MS). Proinflammatory cytokines such as interleukin-6 (IL-6) and tumor necrosis factor-alpha (TNF-α) are considered important in the induction and pathogenesis of EAE and of MS, which is characterized by significant inflammation, demyelination, neurological damage, and cell death (40). MT-I and MT-II are anti-inflammatory and neuroprotective proteins that are expressed during the EAE and MS disease processes (41). Penkowa and Hidalgo (40) found that the administration of exogenous zinc-metallothionein-II (Zn-MT-II) significantly decreased the expression of IL-6 and TNF-α in EAE and decreased the clinical symptoms, mortality, and leukocyte infiltration of the central nervous system in a rat model of EAE. They also found that the systemic administration of Zn-MT-II during EAE prevented demyelination and axonal damage and stimulated oligodendroglial regeneration as well as the expression of basic fibroblast growth factor (bFGF), transforming growth factor (TGF)-α, neurotrophin-3 (NT-3), NT-4/5, and nerve growth factor (NGF) (41). These results suggest that the administration of MT-II may be useful in the treatment of MS in humans. Using a mouse model of collagen-induced arthritis, Youn et al. (42) found that the administration of MT-I and MT-II reduced the incidence and severity of disease symptoms and significantly reduced the levels of proinflammatory mediators such as TNF-α.

MT also possesses antioxidant properties (43,44) that have demonstrated significant medical benefits. Islet transplantation, a promising therapy for Type 1 diabetes, often fails due to graft hypoxia or immune rejection, which generates reactive oxygen species. MT overexpression in pancreatic beta cells was found to provide broad resistance to oxidative stress by scavenging most kinds of reactive oxygen species and to reduce nitric oxide-induced beta cell death (45). The antioxidant properties of MT were also found to protect human central nervous system cells from radiation-induced cell death *in vitro* (46). Decreased levels of MT have also been found in the brains of patients with Alzheimer's disease (47).

The biological importance of MTs has been demonstrated in several mouse models, using transgenic mice devoid of MT-I and MT-II. Mice that did not express functional MT-I or MT-II were found to be more susceptible to pathological damage and apoptosis induced by compounds such as Cd, arsenic (As), and anticancer agents (e.g., cytosine arabinoside, bleomycin, melphalan, and cis-dichlorodiammineplatinum [II]) (48–51), as well as displaying depressed wound healing (52). Mice lacking MT were also found to be more susceptible to *Helicobacter pylori* colonization and gastric inflammation, indicating that MT may be protective against *H. pylori*-induced gastritis (53).

Hamza et al. (54) found that mice lacking the metallochaperone Atox1 exhibited growth retardation, skin laxity, hypopigmentation and seizures, due to related Cu deficiency, indicating the importance of metallochaperones in Cu homeostasis. Many neurological disorders have also been associated with the accumulation of toxic amounts of metal ions, in particular Cu, in the nervous system (28). The accumulation of toxic amounts of Cu in the brain can result in neuronal degeneration (28), resulting in diseases such as Menkes syndrome and Wilson disease, and may be linked directly to Cu chaperone proteins (55).

Finally, PCs and MT fragments have been reported to exhibit angiotensin converting enzyme (ACE)-inhibitory activity. ACE inhibitors have been suggested for the treatment of hypertension, coronary heart disease, heart arrhythmia, and myocardial infarction (10).

5.5.2 Nutritional Supplementation

More than one third of the world's population suffers from micronutrient deficiencies, including Fe and Zn deficiencies (56). Therefore, increasing the amount of bioavailable micronutrients in plant foods for human consumption is of particular significance (57).

The metal-binding capabilities of the metal-binding proteins and peptides in plants have been exploited to enhance the mineral content and bioavailability of minerals in various plant species. For example, rice is the staple food for half of the world's population, who consume it on a daily basis; however, it is a poor source of many essential micronutrients (56). Lucca et al. (58) overexpressed an endogenous MT-like protein in rice, creating rice with a higher Fe content and thus the potential to substantially improve Fe nutrition in Fe-deficient populations.

Elless et al. (59) found that mineral-enhanced edible plants could be used as a novel source of mineral dietary supplements, which would provide essential minerals in a more available form than do the current, inorganically based supplements. The Indian mustard plant *Brassica juncea* was grown hydroponically and was supplemented with various minerals, including Fe, Zi, manganese (Mn), and selenium (Se), resulting in increases of up to 500-fold in mineral concentrations. Ellis et al. found that the accumulated trace elements achieved greater soluble concentrations than those provided in supplements, perhaps owing to the biochemical complexation between the metals and metal-binding compounds, which maintains the metal in a soluble form that is available for metabolism. These authors also proposed that the consistent high concentrations of minerals in the edible plant tissue would allow small quantities of these plants to be processed into supplements supplying 100% of the recommended daily intake of these elements in a soluble form from a natural, vegetative source (59).

5.5.3 Biosorption and Environmental Remediation

The removal of heavy metals and radionuclides from the environment often requires physical or chemical means, including chemical oxidation and reduction, ion exchange, filtration, electrochemical treatment, evaporative recovery, and solvent extraction, and can be costly and inefficient, with low metal recovery (60).

The observation that some plant species are capable of surviving in contaminated areas has led to the concept of phytoremediation. Phytoremediation employs the use of plants to degrade, contain, or stabilize various environmental contaminants in soil, water, and air (61). It is a relatively new approach to removing contaminants from the environment (62), drawing on the ability of some plants to hyperaccumulate metals, due to the presence of their endogenous metal-binding proteins (63,64). It is an environmentally sound strategy for environmental cleanup, offering a low-cost method for soil and water remediation, with the possibility that some of the extracted metals may be recycled for value (65,66). Plants have been shown to successfully extract lead and cadmium from both soil (67) and water (68). However, even plants that are relatively resistant to environmental contaminants often exhibit growth retardation in the presence of contaminants (62). To this end, novel plant-bacterial remediation systems have been developed to encourage growth and increase metal accumulation in plants (62,69). The production of transgenic plants, genetically engineered to overexpress MTs and other metal-binding proteins in order to enhance metal uptake, has also been described (70–72).

It has been reported that various types of biomaterials are also able to accumulate inorganic ions in aqueous solutions, a process termed "biosorption" (31). ESM was found to accumulate significant amounts of uranium (U) and thorium (Th) from dilute aqueous solution (31,60,73), indicating potential industrial applications for the removal of actinides from wastewater (60). Ishikawa et al. (73) also found that this process could be optimized to selectively accumulate U or Th from a heterogeneous solution of heavy metals.

The binding of gold ions to ESM has also been examined and has been suggested as a method of facilitating gold recovery from sources such as electronic scrap and electroporating and leach solutions (21,74). ESM was found to accumulate gold ions in high yield, which were then easily desorbed from the ESM, with no apparent change in its gold uptake capacity (21,74).

5.6 Genetic Engineering to Modify Metal-Binding and Sequestering Characteristics

Significant research has been carried out to study the genes encoding metal-binding proteins and peptides, with regard to their regulation, biochemistry, and biotechnological potential (63). Candidate genes from mice (75), humans (69), yeast (72), and metal-tolerant plant species (63) have been examined and suggested for the production of transgenic plants to increase metal-binding protein/peptide expression and thereby enhance heavy-metal-binding capabilities and improve phytoremediation possibilities.

Unfortunately, the metal-binding properties of plants can also lead to the introduction of toxic metals into the food system. In most crops, Cd accumulation in the generative organs is minor compared with the accumulation in the roots. In flax, however, a high concentration of Cd is known to accumulate in the seeds, even at low soil Cd levels (32). The production of transgenic plants to modify the metal-binding properties and sequester Cd in the roots of plants, where it will not be consumed, has been examined. Yeargan et al. (76) found that tobacco plants expressing mouse MT-I contained less Cd in the shoots and retained more Cd in the roots, as compared with control tobacco plants.

Sauge-Merle et al. (77) described the insertion of a PC-encoding gene into *Escherichia coli*, to increase the ability of bacterial cells to accumulate heavy metals and observed a marked increase in cell PCs as well as up to a 50-fold increase in cellular metal contents when the bacteria were grown in the presence of heavy metals.

A further understanding of the molecular and genetic bases of the mechanisms of heavy metal binding will be an important aspect in the development of transgenic metal-binding organisms (78) and in the extension of their role in human health and environmental applications.

REFERENCES

1. D.H. Hamer. Metallothionein. Ann. Rev. Biochem., 55: 913–951, 1986.
2. P. Kotrba, T. Macek, T. Ruml. Heavy metal-binding peptides and proteins in plants. A review. *Collect. Czech. Chem. Commn.*, 64: 1057–1086, 1999.
3. C. Cobbett, P. Goldsbrough. Phytochelatins and metallothioneins: roles in heavy metal detoxification and homeostasis. *Annu. Rev. Plant. Biol.*, 53: 159–182, 2002.
4. M. Margoshes, B.L. Vallee. A cadmium protein from equine kidney cortex. *J. Am. Chem. Soc.*, 79: 4813–4814, 1957.
5. A.I. Bush. Metal complexing agents as therapies for Alzheimer's disease. *Neurobiol. Aging*, 23: 1031–1038, 2002.
6. B.S. Wong, D.R. Brown, M.S. Sy. A yin-yang role for metals in prion disease. *Panminerva Med.*, 43: 283–287, 2001.
7. B.A. Masters, E.J. Kelly, C.J. Quaife, R.L. Brinster, R.D. Palmiter. Targeted disruption of metallothionein I and II genes increases sensitivity to cadmium. *Proc. Natl. Acad. Sci. USA*, 91: 584–588, 1994.
8. W.E. Rauser. Phytochelatins. *Annu. Rev. Biochem.*, 59: 61–86, 1990.
9. W.E. Rauser. Structure and function of metal chelators produced by plants: the case for organic acids, amino acids, phytin, and metallothioneins. *Cell Biochem. Biophys.*, 31: 19–46, 1999.
10. E.C.Y. Li-Chan, F. Sultanbawa, J.N. Losso, B.D. Oomah, G. Mazza. Characterization of phytochelatin-like complexes from flax (*Linum usitatissimum*) seed. *J. Food Biochem.*, 26: 271–293, 2002.

11. E. Grill, E.-L. Winnacker, M.H. Zenk. Phytochelatins, a class of heavy-metal-binding peptides from plants, are functionally analogous to metallothioneins. *Proc. Natl. Acad. Sci. USA*, 84: 439–443, 1987.
12. H.-U. Meisch, I. Beckmann, J.A. Schmitt. A new cadmium-binding phosphoglycoprotein, cadmium-mycophosphatin, from the mushroom, *Agaricus macrosporus*. *Biochim. Biophys. Acta*, 745: 259–266, 1983.
13. H. Stone, J. Overnell. Non-metallothionein cadmium binding proteins. *Comp. Biochem. Physiol.*, 80c: 9–14, 1985.
14. T.V. O'Halloran, V.C. Culotta. Metallochaperones, an intracellular shuttle service for metal ions. *J. Biol. Chem.*, 275: 25057–25060, 2000.
15. A.C. Rosenzweig. Metallochaperones: bind and deliver. *Chem. Biol.*, 9: 673–677, 2002.
16. S. Clemens, C. Simm, T. Maier. Heavy metal-binding proteins and peptides. In: S.R. Fahnestock, A. Steinbüchel (Eds.). *Biopolymers*, Volume 8: Polyamides and Complex Proteinaceous Materials II. Weinheim, Germany: Wiley-VCH GmbH & Co. KGaA, 2003, 255–288.
17. A.C. Rosenzweig. Copper delivery by metallochaperone proteins. *Acc. Chem. Res.*, 34: 119–128, 2001.
18. H.K. Song, S.B. Mulrooney, R. Huber, R.P. Hausinger. Crystal structure of *Klebsiella aerogenes* UreE, a nickel-binding metallochaperone for urease activation. *J. Biol. Chem.*, 276: 49359–49364, 2001.
19. K. Suyama, Y. Fukazawa, Y. Umetsu. A new biomaterial, hen egg shell membrane, to eliminate heavy metal ion from their dilute waste solution. *Appl. Biochem. Biotechnol.*, 45–46: 871–879, 1994.
20. M. Goto, K. Suyama. Occlusion of transition metal ions by new adsorbents synthesized from plant polyphenols and animal fibrous proteins. *Appl. Biochem. Biotechnol.*, 84–86: 1021–1038, 2000.
21. S.-I. Ishikawa, K. Suyama, K. Arihara, M. Itoh. Uptake and recovery of gold ions from electroplating wastes using eggshell membrane. *Bioresour. Technol.*, 81: 201–206, 2002.
22. E.C.Y. Li-Chan, W.D. Powrie, S. Nakai. The chemistry of eggs and egg products. In: W.J. Stadelman, O.J. Cotterill (Eds.). *Egg Science and Technology*, Fourth Edition. New York: The Haworth Press, Inc., 1995, 105–175.
23. M.M. Brzóska, J. Moniuszko-Jakoniuk. Interactions between cadmium and zinc in the organism. *Food Chem. Toxicol.*, 39: 967–980, 2001.
24. L. Tapia, M. Gonzalez-Aguero, M.F. Cisternas, M. Suazo, V. Cambiazo, R. Uauy, M. Gonzalez. Metallothionein is crucial for safe intracellular copper storage and cell survival at normal and supra-physiological exposure levels. *Biochem. J.*, 378: 614–624, 2003.
25. S. Oishi, J.I. Nakagawa, M. Ando. Effects of cadmium administration on the endogenous metal balance in rats. *Biol. Trace Elem. Res.*, 76: 257–278, 2000.
26. R. Kneer, M.H. Zenk. Phytochelatins protect plant enzymes from heavy metal poisoning. *Phytochemistry*, 31: 2663–2667, 1992.
27. M. Takagi, H. Satofuka, S. Amano, H. Mizuno, Y. Eguchi, K. Hirata, K. Miyamoto, K. Fukui, T. Imanaka. Cellular specificity of cadmium ions and their detoxification by heavy metal-specific plant peptides, phytochelatins, expressed in mammalian cells. *J. Biochem. (Tokyo)*, 131: 233–239, 2002.
28. S.C. Burdette, S.J. Lippard. Meeting of the minds: metalloneurochemistry. *Proc. Natl. Acad. Sci. USA*, 100: 3605–3610, 2003.

29. S.G. Ang, P.S. Chong. Purification of metallothionein proteins from the crab, *Portunus pelagicus* — selectivity of hydrophobic interaction chromatography. *Talanta*, 45: 693–701, 1998.
30. A. Prange, D. Schaumloffel. Hyphenated techniques for the characterization and quantification of metallothionein isoforms. *Anal. Bioanal. Chem.*, 373: 441–453, 2002.
31. S.-I. Ishikawa, K. Suyama. Removal and recovery of heavy metal ions by animal fibrous proteins. *World Res. Rev.*, 13: 406–412, 2001.
32. B. Lei, E.C.Y. Li-Chan, B.D. Oomah, G .Mazza. Distribution of cadmium-binding components in flax (*Linum usitatissimum* L.) seed. *J. Agric. Food Chem.*, 51: 814–821, 2003.
33. H. Kubota, K. Sato, T. Yamada, T. Maitani. Phytochelatin homologs induced in hairy roots of horseradish. *Phytochemistry*, 53: 239–245, 2000.
34. T. Hofmann, D.I.C. Kells, B.G. Lane. Partial amino acid sequence of the wheat germ E_c protein. Comparison with another protein very rich in half-cystine and glycine: wheat germ agglutinin. *Can. J. Biochem. Cell Biol.*, 62: 908–913, 1984.
35. S.K. Ryu, J.S. Park, I.S. Lee. Purification and characterization of a copper-binding protein from Asian periwinkle *Littorna brevicula*. *Comp. Biochem. Physiol. Toxicol. Pharmacol.*, 134: 101–107, 2003.
36. Y. Dohi, K. Ohba, Y. Yoneyama. Purification and molecular properties of two cadmium-binding glycoproteins from the hepatopancreas of a whelk, *Buccinum tenuissimum*. *Biochim. Biophys. Acta*, 745: 50-60, 1983.
37. M.J. Bebianno, M.A. Serafim, D. Simes. Metallothioneins in the clam *Ruditapes decussates*: an overview. *Analysis*, 28: 386–390, 2000.
38. D.C. Simes, M.J. Bebianno, J.J. Moura. Isolation and characterization of metallothionein from the clam *Ruditapes decussates*. *Aquat. Toxicol.*, 63: 307–318, 2003.
39. L. Vergani, M. Grattarola, F. Dondero, A. Viarengo. Expression, purification, and characterization of metallothionein-A from rainbow trout. *Prot. Expr. Purif.*, 27: 338–345, 2003.
40. M. Penkowa, H. Hidalgo. Metallothionein treatment reduces proinflammatory cytokines IL-6 and TNF-alpha and apoptotic cell death during experimental autoimmune encephalomyelitis (EAE). *Exp. Neurol.*, 170: 1–14, 2001.
41. M. Penkowa, H. Hidalgo. Treatment with metallothionein prevents demyelination and axonal damage and increases oligodendrocyte precursors and tissue repair during experimental autoimmune encephalomyelitis. *J. Neurosci. Res.*, 72: 574–586, 2003.
42. J. Youn, S.H. Hwang, Z.Y. Ryoo, M.A. Lynes, D.J. Paik, H.S. Chung, H.Y. Kim. Metallothionein suppresses collagen-induced arthritis via induction of TGF-beta and down-regulation of proinflammatory mediators. *Clin. Exp. Immunol.*, 129: 232–239, 2002.
43. J.S. Lazo, S.M. Kuo, E.S. Woo, B.R. Pitt. The protein thiol metallothionein as an antioxidant and protectant against antineoplastic drugs. *Chem. Biol. Interact.*, 112: 255–262, 1998.
44. A. Viarengo, B. Burlando, N. Ceratto, I. Panfoli. Antioxidant role of metallothioneins: a comparative overview. *Cell Mol. Biol.*, 46: 407–417, 2000.
45. X. Li, H. Chen, P.N. Epstein. Metallothionein protects islets from hypoxia and extends islet graft survival by scavenging most kinds of reactive oxygen species. *J. Biol. Chem.*, 279: 765–771, 2004.

46. L. Cai, S. Iskander, M.G. Cherian, R.R. Hammond. Zinc- or cadmium-pre-induced metallothionein protects human central nervous system cells and astrocytes from radiation-induced apoptosis. *Toxicol. Lett.*, 146: 217–226, 2004.
47. W.H. Yu, W.J. Lukiw, C. Bergeron, H.B. Niznik, P.E. Fraser. Metallothionein III is reduced in Alzheimer's disease. *Brain Res.*, 894: 37–45, 2001.
48. J.S. Lazo, Y. Kondo, D. Dellapiazza, A.E. Michalska, K.H. Choo, B.R. Pitt. Enhanced sensitivity to oxidative stress in cultured embryonic cells from transgenic mice deficient in metallothionein I and II genes. *J. Biol. Chem.*, 270: 5506–5510, 1995.
49. Y. Kondo, J.M. Rusnak, D.G. Hoyt, C.E. Settineri, B.R. Pitt, J.S. Lazo. Enhanced apoptosis in metallothionein null cells. *Mol. Pharmacol.*, 52: 195–201, 1997.
50. Y. Liu, J. Liu, S.M. Habeebu, M.P. Waalkes, C.D. Klaassen. Metallothionein-I/II null mice are sensitive to chronic oral cadmium-induced nephrotoxicity. *Toxicol. Sci.*, 57: 167–176, 2000.
51. G. Jia, Y.Q. Gu, K.T. Chen, Y.Y. Lu, L. Yan, J.L. Wang, Y.P. Su, J.C.G. Wu. Protective role of metallothionein (I/II) against pathological damage and apoptosis induced by dimethylarsinic acid. *World J. Gastroenterol.*, 10: 91–95, 2004.
52. M. Penkowa, J. Carrasco, M. Giralt, T. Moos, J. Hidalgo. CNS wound healing is severely depressed in metallothionein I- and II-deficient mice. *J. Neurosci.*, 19: 2535–2545, 1999.
53. C.D. Tran, H. Huynh, M. van den Berg, M. van der Pas, M.A. Campbell, J.C. Philcox, P. Coyle, A.M. Rofe, R.N. Butler. *Helicobacter*-induced gastritis in mice not expressing metallothionein-I and II. *Helicobacter*, 8: 533–541, 2003.
54. I. Hamza, A. Faisst, J. Prohaska, J. Chen, P. Gruss, J.D. Gitlin. The metallochaperone Atox1 plays a critical role in perinatal copper homeostasis. *Proc. Natl. Acad. Sci. USA*, 98: 6848–6852, 2001.
55. A.K. Wernimont, D.L. Huffman, A.L. Lamb, T.V. O'Halloran, A.C. Rosenzweig. Structural basis for copper transfer by the metallochaperone for the Menkes/Wilson disease proteins. *Nat. Struc. Biol.*, 7: 766–771, 2000.
56. M.B. Zimmermann, R.F. Hurrell. Improving iron, zinc and vitamin A nutrition through plant biotechnology. *Curr. Opin. Biotechnol.*, 13: 142–145, 2002.
57. E. Frossard, M. Bucher, F. Mächler, A. Mozafar, R. Hurrell. Potential for increasing the content and bioavailability of Fe, Zn and Ca in plants for human nutrition. *J. Sci. Food Agric.*, 80: 861–879, 2000.
58. P. Lucca, R. Hurrell, I. Potrykus. Fighting iron deficiency anemia with iron-rich rice. *J. Am. Coll. Nutr.*, 21: 184S–190S, 2002.
59. M.P. Elless, M.J. Blaylock, J.W. Huang, C.D. Gussman. Plants as a natural source of concentrated mineral nutritional supplements. *Food Chem.*, 71: 181–188, 2000.
60. S.-I. Ishikawa, K. Suyama, I. Satoh. Biosorption of actinides from dilute waste actinide solution by egg shell membrane. *Appl. Biochem. Biotechnol.*, 77–79: 521–533, 1999.
61. H. Morikawa, O.C. Erkin. Basic processes in phytoremediation and some applications to air pollution control. *Chemosphere*, 52: 1553–1558, 2003.
62. B.R. Glick. Phytoremediation: synergistic use of plants and bacteria to clean up the environment. *Biotechnol. Adv.*, 21: 383–393, 2003.
63. S. Clemens. Developing tools for phytoremediation: towards a molecular understanding of plant metal tolerance and accumulation. *Int. J. Occup. Med. Environ. Health.*, 14: 235–239, 2001.
64. S. Cheng. Heavy metals in plants and phytoremediation. *Environ. Sci. Pollut. Res. Int.*, 10: 335–340, 2003.

65. R.L. Chaney, M. Malik, Y.M. Li, S.L. Brown, E.P. Brewer, J.S. Angle, A.J. Baker. Phytoremediation of soil metals. *Curr. Opin. Biotechnol.*, 8: 279–284, 1997.
66. M.S. Kambhampati, G.B. Begonia, M.F.T. Begonia, Y. Bufford. Phytoremediation of a lead-contaminated soil using morning glory (*Ipomoea lacunoas* L.): effects of a synthetic chelate. *Bull. Environ. Contam. Toxicol.*, 71: 379–386, 2003.
67. A.L. Salido, K.L. Hasty, J.M. Lim, D.J. Butcher. Phytoremediation of arsenic and lead in contaminated soil using Chinese brake ferns (*Pteris vittata*) and Indian mustard (*Brassica juncea*). *Int. J. Phytoremediation*, 5: 89–103, 2003.
68. T. Schor-Fumbarov, Z. Keilin, E. Tel-Or. Characterizaton of cadmium uptake by the water lily *Nymphae aurora*. *Int. J. Phytoremediation*, 5: 169–179, 2003.
69. R. Sriprang, M. Hayashi, M. Yamashita, H. Ono, K. Saeki, Y. Murooka. A novel bioremediation system for heavy metals using the symbiosis between leguminous plant and genetically engineered rhizobia. *J. Biotechnol.*, 99: 279–293, 2002.
70. L.E. Bennett, J.L. Burkhead, K.L. Hale, N. Terry, M. Pilon, E.A. Pilon-Smits. Analysis of transgenic Indian mustard plants for phytoremediation of metal-contaminated mine tailings. *J. Environ. Qual.*, 32: 432–440, 2003.
71. C. Gisbert, R. Ros, A. De Haro, D.J. Walker, M. Pilar Bernal, R. Serrano, J. Navarro-Avino. A plant genetically modified that accumulates Pb is especially promising for phytoremediation. *Biochem. Biophys. Res. Commn.*, 303: 440–445, 2003.
72. J.C. Thomas, E.C. Davies, F.K. Malick, C. Endreszl, C.R. Willians, M. Abbas, S. Petrella, K. Swisher, M. Perron, R. Edwards, P. Ostenkowski, N. Urbanczyl, W.N. Wiesend, K.S. Murray. Yeast metallothionein in transgenic tobacco promotes copper uptake from contaminated soils. *Biotechnol. Prog.*, 19: 273–280, 2003.
73. S.-I. Ishikawa, K. Suyama, K. Arihara, M. Itoh. Selective recovery of uranium and thorium ions from dilute aqueous solutions by animal biopolymers. *Biol. Trace Elem. Res.*, 86: 227–236, 2002.
74. S.-I. Ishikawa, K. Suyama. Recovery and refining of Au by gold-cyanide ion biosorption using animal fibrous proteins. *Appl. Biochem. Biotechnol.*, 70–72: 719–728, 1998.
75. A. Pan, M. Yang, F. Tie, L. Li, Z. Chen, B. Ru. Expression of mouse metallothionein-I gene confers cadmium resistance in transgenic tobacco plants. *Plant Mol. Biol.*, 24: 341–351, 1994.
76. R. Yeargan, I.B. Maiti, M.T. Nielsen, A.G. Hunt, G.J. Wagner. Tissue partitioning of cadmium in transgenic tobacco seedlings and field grown plants expressing the mouse metallothionein I gene. *Transgenic Res.*, 1: 261–267, 1992.
77. S. Sauge-Merle, S. Cuine, P. Carrier, C. Lecomte-Pradines, D.T. Luu, G. Peltie. Enhanced toxic metal accumulation in engineered bacterial cells expressing *Arabidopsis thaliana* phytochelatin synthase. *Appl. Environ. Microbiol.*, 69: 490–494, 2003.
78. C.S. Cobbett. Phytochelatins and their roles in heavy metal detoxification. *Plant Physiol.*, 123: 825–832, 2000.

6 Tight-Junction-Modulatory Factors in Food

Makoto Shimizu and Akira Mori

CONTENTS

6.1 MAJOR MECHANISMS FOR INTESTINAL ABSORPTION

The physiological functions of nutrients and other food-derived factors strongly depend on their absorbability at the intestinal epithelium. Various pathways for the intestinal absorption of dietary substances have been discovered during the past several decades, and the mechanisms for the intestinal absorption of major nutrients have been revealed. Four major pathways are currently known (Figure 6.1). Two of them are active transport pathways, one involving transporter-mediated transport and the other transcytosis. The other two involve passive transport via the paracellular and transcellular routes.

1. Transporter-mediated transport: Fundamentally important nutrients such as glucose, amino acids, di(tri)peptides, monocarboxylic acid, vitamin C, and calcium are known to be transported from the apical side of the intestinal epithelial cells to the basal side by plasma membrane transporters. These transporters specifically recognize their substrates and efficiently transport them in an energy-dependent manner.
2. Transcytosis: This pathway involves the process of internalization of extracellular substances and fluids by cell-membrane-derived intracellular vesicles. Intestinal epithelial cells may have certain receptor proteins on the apical membrane surface. When a ligand binds to such a receptor, this endocytotic process and subsequent transcellular migration of the

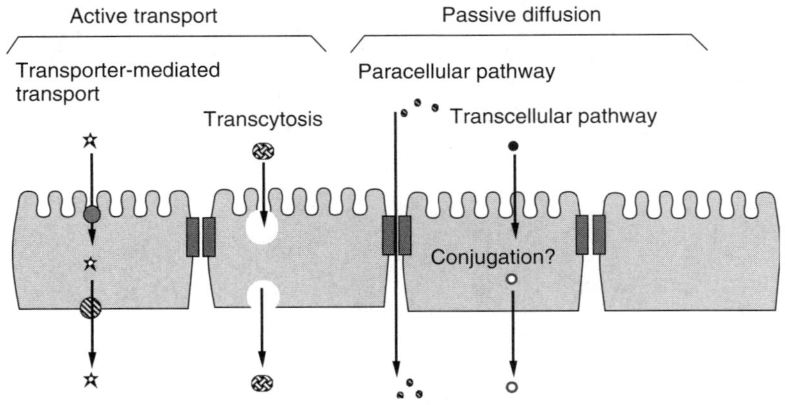

FIGURE 6.1 Four major pathways for intestinal absorption.

vesicles is accelerated (receptor-mediated transcytosis). The uptake of extracellular substances and fluids can also occur without the involvement of receptors (nonreceptor-mediated transcytosis). Transcytosis is known to be an absorption route for such macromolecules as proteins.

3. Paracellular diffusion: Epithelial cells in the monolayers are bound to each other and involve intracellular adhesive molecules such as tight junction proteins. The tight junction is a complex device assembled with many proteins and will be described later. It forms a channel-like structure between the cells and allows small water-soluble molecules to pass across the epithelial cell monolayers. Small molecular substances are transported via this route by passive diffusion.
4. Transcellular diffusion: This pathway is mostly for such hydrophobic molecules as cholesterol, carotenoids, long-chain fatty acids, and oil-soluble vitamins. These substances pass through the cell membrane because of their hydrophobic properties and are transcellularly transported via certain carrier proteins (for example, retinal is transported by the intestinal retinal-binding protein, IRBP). Calcium ions are also transported via this route, where the calcium-binding protein (CaBP) is involved in their transcellular migration.

This chapter focuses on the paracellular transport regulated by the tight junction. Paracellular transport is not a specific transport system, but it is likely to be an important pathway for the absorption of small water-soluble molecular compounds. Recent studies have revealed that the absorption of calcium ions, particularly in the lower small intestinal tract, occurs via the paracellular route (1). Oligopeptides are also likely to be transported paracellularly (2,3). On the one hand, since such transcellular transport systems as transporter-mediated transport and transcytosis are basically degradative pathways (4), bioactive oligopeptides will be readily hydrolyzed during their transcellular absorption, their bioactivity being eliminated. On the other hand, paracellular absorption can bypass the degradation enzymes present in the epithelial cells. This pathway is therefore thought to play an important role in the absorption of bioactive substances that are susceptible to the intracellular degradation enzymes.

The paracellular permeability of the epithelial cell monolayers depends on the tight junction. As described later, tight junction permeability is affected by many internal and external factors. Under certain conditions, the paracellular permeability is increased to enable the absorption of larger-molecular-weight substances or more efficient absorption of functional substances.

6.2 STRUCTURE OF THE TIGHT JUNCTION

The tight junction comprises many components. Although the key components remained unidentified for a long time, recent work during the past decade has found new proteins that are involved in the construction of tight junctions. Occludin, a component of tight junctions, was reported in 1993 by Furuse et al. (5) and has been recognized as a major component of the tight junction. Occludin is a 65-kDa protein with four transmembrane domains and two extracellular loops. On the one hand, the first extracellular loop of occludin, having glycine- and tyrosine-rich sequences, probably plays an important role in cell–cell adhesion. On the other hand, the second extracellular loop domain of occludin contains the amino acid sequence Leu-Tyr-His-Tyr, that is present in all mammalian occludins and is thought to enable cell adhesion recognition (6). The C-terminal intracellular domain of occludin is highly conserved among various spices. Occludin associates with ZO-1, an intracellular 220-kDa protein located in the tight junction, via this domain. This association is important for maintaining the structure and functions of the tight junction (7,8).

To date, three proteins, ZO-1, -2, and -3, have been identified as members of the ZO protein family (9). They have three PDZ domains, the SH-3, GuK, and proline-rich domains. The sequences of these domains have about 50 to 70% homology among the three ZO proteins. ZO-1 is located in the tight junctions of epithelial cells. It is also localized in the adherence junctions of fibroblasts and myocytes that do not have tight junctions. The GuK and proline-rich domains, respectively, are thought to play important roles in association with the C-terminal region of occludin and the actin filament. ZO-2 is a 160-kDa protein, its N-terminal domain having affinity for occludin or α-catenin, one of the adherence junction proteins. ZO-1 and ZO-2 associate with each other through the second PDZ domain. ZO-3 is a 130-kDa protein and can associate with the C-terminal region of occludin and ZO-1.

Since the discovery of occludin, it had been thought that occludin was the main molecule for cell–cell adhesion in tight junctions. However, the finding that even occludin-deficient mice were able to form a tight junction strand suggested the existence of other adhesion molecules. Another adhesion molecule was then discovered and named claudin (10). Subsequent studies have revealed different types of claudin (to date, 24 claudins have been reported) that form a multigene family. Claudins have four transmembrane domains like occludin, although little homology is apparent between occludin and claudins. The expression of claudins is tissue specific (11); for example, claudin-1 is expressed in various tissues, whereas claudin-3 is expressed only in the lung and liver. Claudin-6 is expressed in fetal tissues but not in adult tissues. Claudin-16 (paracellin-1) is located in the tight junction of the thick ascending limb of Henle and is related to the renal resorption of Mg^{2+} ions. Mutation of this gene causes massive renal magnesium wasting with hypomagnesemia. Different types of claudin are able to bind both homogeneously and heterogeneously. Since the adhesion intensity of the extracellular domains differs among the claudins, small spaces or pores may be formed among the claudins on cell surfaces. This enables small molecular substances to pass through the tight junction.

Cingulin and 7H6 are component proteins of the tight junction that were discovered in the early stages of the study of tight junction (12,13). Cingulin is a 140- to 160-kDa protein that is located in the tight junction submembranous region of epithelial and endothelial cells. This protein does not have a PDZ domain but has globular head and tail domains. Cingulin interacts with ZO-2, ZO-3, JAM, F-actin, and myosin that intervene in the globular head domain. 7H6 is a 155-kDa protein homologous to those proteins in the SMC family. The appearance of 7H6 at a tight junction is correlated with maintenance of the paracellular barrier function. Phosphorylation of 7H6 leads to its association with the tight junction, whereas a depletion of cellular adenosine triphophate (ATP) causes the dissociation of 7H6 from the tight junction. It has also been reported that many other proteins are involved in the formation of tight junctions, and so their complete structure is understandably complicated (14) (Figure 6.2).

6.3 INTRACELLULAR SIGNALING PATHWAY MODULATING THE TIGHT-JUNCTION PERMEABILITY

A change in paracellular permeability is sometimes accompanied by the activation of certain cellular signaling pathways. Ca^{2+}, protein kinase C (PKC), tyrosine kinase, and small G protein are involved in this process. PKC is a family of Ca^{2+}- and phospholipid-dependent serine/threonine protein kinases. PKC has many isoforms, at least 10 having been identified in mammals.

On the one hand, it is well known that such phorbol esters as TPA (12-O-tetradecanoyl phorbol-13-acetate) activate PKC, which increases paracellular permeability (15). On the other hand, the specific PKC inhibitor, bisindolylmaleimide, can suppress this increase in paracellular permeability induced by phorbol esters. These results indicate that the PKC pathway is involved in the regulation of tight junctions. However, the change in tight-junction permeability by phorbol ester stimulus varies with the strain of cell line, probably because multiple PKC isoforms are expressed in cells and tissues in specific manners. The lack of specific activators and inhibitors for each isoform, however, makes the relationship between PKC isoforms and tight-junction regulation unclear. Mullin

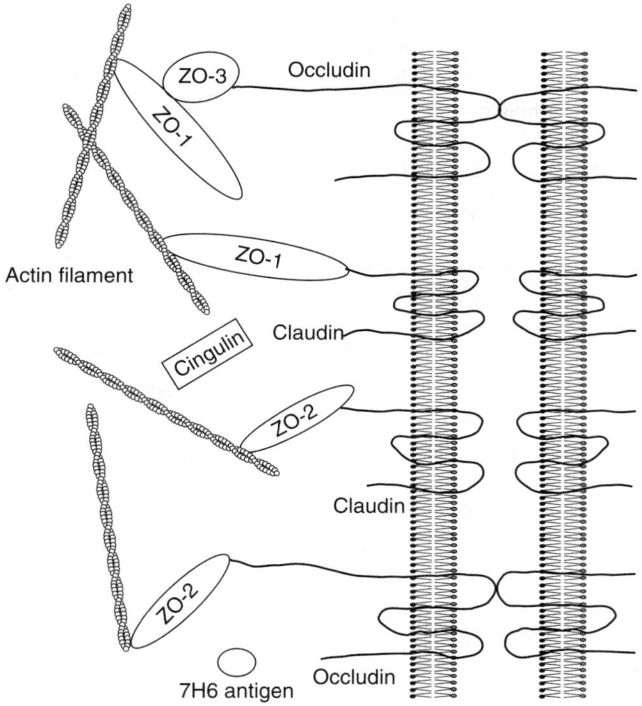

FIGURE 6.2 Schematic model of the molecular structure of the tight junction. (*Source*: Modified from S. Tsukita , M. Furuse , M. Itoh. *Curr. Opin. Cell Biol.*, 11, 628–633, 1999. With permission.)

et al. (16) however, have discovered a strong correlation between PKC-α associated with the membrane and the permeability of the tight junction, suggesting that transmigration of PKC-α from the cytosol to the cell membrane is involved in tight-junction regulation. The relationship between PKC isoforms and tight-junction permeability has also been studied by using cells that have been stably transfected with human PKC-α and -δ, the results showing that the PKC-α and -δ isoforms caused the leakiness of tight junctions (17).

Phosphorylation of serine/threonine residues is known to be the main function of PKC in regulating the cellular functions. Investigating the change in the phosphorylation state of tight-junction proteins would provide important information on the signaling pathways that regulate tight-junction permeability. The protein kinase, ZAK (ZO-1-associated kinase), associates with the SH-3 domain of ZO-1, phosphorylating the C-terminal residues of ZO-1. On the one hand, phosphorylation of the serine/threonine residues of occludin has been suggested to have a strong correlation with tight-junction formation (18). On the other hand, activation of PKC with 10^{-7} M TPA has been shown to cause dephosphorylation of the threonine residue of occludin in a time-dependent manner. This effect has been correlated with an abrupt decrease in the transepithelial ion permeability. PKC may be positioned upstream of the signal to regulate the permeability of epithelial cells and has the possibility of inactivating serine/threonine kinase or activating the phosphatases associated with occludin.

Small GTP-binding proteins are also known to be a factor regulating the tight junction (19). ROK-1, included in the Rho protein family, plays a redistribution role in the cytoskeleton by inactivating myosin phosphatase. ROK activation also causes direct phosphorylation of the serine residues of myosin, with subsequent contraction of actin/myosin. Thus, ROK alters the structure of the cytoskeleton, playing an important role in the regulation of tight junctions. Ras protein has also been suggested to regulate the tight junction. The Raf/MEK/ERK cascade is downstream of Ras. Transfection of Raf-1 in the rat parotid gland epithelial cell line Pa-4 caused the disappearance of tight junctions. The expression of occludin and claudin-1 was downregulated, with alteration of the localization of ZO-1 and E-cadherin. Transfection of Pa-4 with the active form of Ras, Raf-1, and MEK-1 decreased occludin mRNA (messenger ribonucleic acid), suggesting that the Ras/Raf/MEK/ERK cascade alters the epithelial tight junction by regulating the expression of occludin.

6.4 REGULATION OF TIGHT-JUNCTION PERMEABILITY BY INTERNAL FACTORS

Tight-junction permeability can be evaluated by measuring the transepithelial electrical resistance (TER) of the intestinal epithelium. TER measurement of the intestinal epithelial cell monolayers that have been cultured on a semipermeable filter (Figure 6.3) is frequently used to evaluate the integrity of the cell monolayer or to look for substances that modulate the paracellular permeability of the monolayer because TER is a highly sensitive parameter of tight-junction permeability. Measuring the transepithelial flux of paracellularly transported marker substances such as mannitol, inulin, and Lucifer Yellow is also useful to detect tight-junction-modulatory substances.

The permeability or barrier function of the tight junction is altered by various factors (20–23). Table 6.1 lists some of the factors that perturb the barrier function. Changing the physical conditions (e.g., osmotic pressure) or chemical conditions (e.g., extracellular calcium concentration) alters the tight-junction structure (23). A number of endogenous factors are also known to influence paracellular permeability, cytokines being typical examples of such endogenous factors (20,24).

1. Interferon-γ (IFN-γ): IFN-γ is a 25-kDa protein secreted from activated T cells, natural killer cells, and others. Polarized T84 intestinal epithelial cells showed a decrease in TER by a treatment for 72 hours with recombinant IFN-γ (20,24,25). This change was accompanied by an increase in mannitol flux, suggesting that IFN-γ altered tight junction permeability. It is known that the production of IFN-γ in colonic mucosa

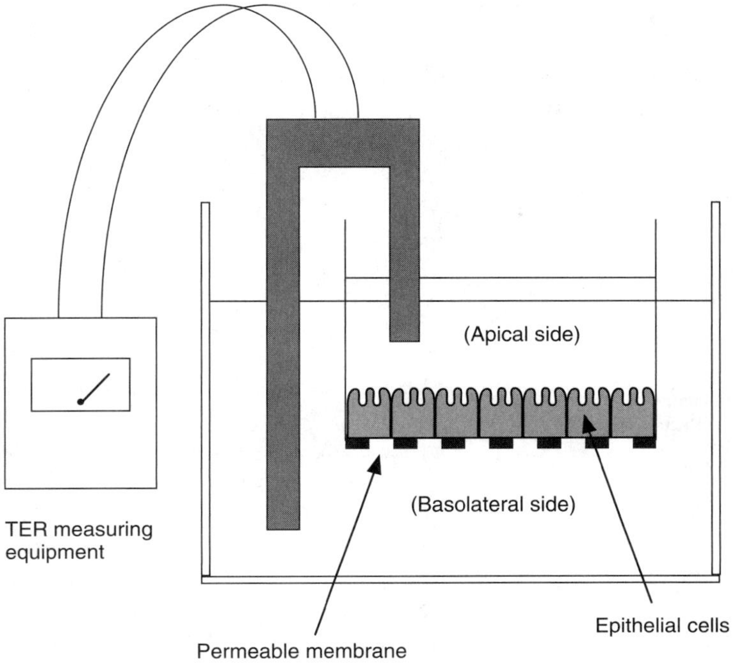

FIGURE 6.3 System for measuring the transepithelial electrical resistance (TER) of the intestinal epithelial cell monolayers.

is upregulated in patients with inflammatory bowel disease, thereby increasing paracellular permeability. The effect of IFN on intestinal epithelial permeability seems to be specific to the IFN-γ isoform, IFN-α, for example, having no effect on T84 cells. Details of the molecular mechanism for the effect of IFN-γ on the tight junction are still unknown, but IFN-γ-induced changes in the tight junction structure have been reported. For example, IFN-γ caused the displacement of ZO-1, ZO-2, and occludin in T84 cell monolayers, although the change in the phosphorylation state of each protein has not been observed.

2. Hepatocyte growth factor (HGF): HGF is also known to influence the permeability of the tight junction (20). The TER of T84 cells decreased when treated with HGF, the value being lowest after 48 hours of incubation. Treating undifferentiated MDCK cells with HGF inhibited the formation of adherence and tight junctions. The amount of cytoplasmic E-cadherin in MDCK cells stimulated with HGF is likely to be upregulated and the phosphorylation pattern of E-cadherin changed. Changes in the phosphorylation state of E-cadherin may alter the affinity of this adhesive protein to other proteins, thereby inhibiting the adherence junction formation.

TABLE 6.1
Various Factors that Influence Epithelial Barrier Function

Groups	Factors	Effect on TER
Cytokines and growth factors	IL-1	Decrease
	IL-4	Decrease
	IL-10	Decrease
	IL-13	Decrease
	HGF	Decrease
	IFNγ	Decrease
	TNFα	Decrease
	TGFα	Decrease
	TGFβ	Increase
Bacterium toxins	*Clostridium difficile* toxin	Decrease
	Zonula occludens toxin	Decrease
	Bacteroides fragilis toxin	Decrease
Physicochemical factors	High osmotic pressure	Decrease
	Depletion of oxygen	Decrease
Dietary substances	Protamine	Increase
	β-Lactoglobulin	Increase
	β-Casein	Increase
	L-tryptophan	Decrease
	Glucose	Decrease
	Nucreotide	Decrease
	Sodium caprate	Decrease
	Linoleic acid	Decrease
	γ-linolenic acid	Increase
	Eicosapentaenoic acid	Increase
	Chitosan	Decrease
	Cyclodextrin	Decrease
	Surfactant	Decrease
	Capsianoside	Decrease
	Paprila	Decrease
	Cayenne papper	Decrease
	Capsaicin	Decrease
	Green/black oeooer	Increase
	Nutmeg	Increase
	Piperine	Increase

Source: From J.M. Mullin, K.V. Laughlin, N. Ginanni, C.W. Marano, H.M. Clarke, A.P. Soler. *Ann. NY Acad. Sci.*, 915, 231–236, 2000; D. Rosson, T. O'Brien, J.A. Kampherstein, Z. Szallasi, K. Bogi, P.M. Blumberg, J.M. Mullin. *J. Biol. Chem.*, 272, 14950–14953, 1997; A. Sakakibara, M. Furuse, M. Saitou, Y. Ando-Akatsuka, S. Tsukita. *J. Cell Biol.*, 137, 1393–1401, 1997; A.M. Hopkins, D. Li, R.J. Mrsny, S.V. Walsh, A. Nusrat. *Adv. Drug Deliv. Rev.*, 41(3), 329–340, 2000. With permission.

3. Tumor necrosis factor-α (TNF-α): TNF-α has been reported to decrease TER of the monolayers of some intestinal epithelial cell lines (20,24). This decrease in TER was accompanied by increased mannitol flux in HT-29 cl.19A cells. The number of tight-junction strands and the depth of the tight junctions were also decreased by treatment with TNF-α. The TNF-α-induced decrease in barrier function has been suggested to occur via some type of kinase. Genistein, a tyrosine kinase inhibitor, blocked the TNF-α-induced decrease in the barrier function of LLC—PK-1 cells. In HT-29/B-6 cells, genistein and protein kinase A inhibitor H-8 blocked the effect of TNF-α, suggesting that tyrosine kinase and protein kinase A are involved in the effect of TNF-α on the epithelial cells (26).
4. TGF-β: TGF-β_1 is a 25-kDa protein with various bioactivities. It prevents cell proliferation/differentiation, and stimulates apoptosis. The epithelial barrier functions of intestinal T84 cells were enhanced by a TGF-β_1 treatment (20). On the one hand, TGF-β not only enhances the epithelial barrier function but also reduces the effects of other cytokines, IFN-γ, IL-4, and IL-10. On the other hand, it inhibits the formation of tight junctions stimulated by glucocorticoid in the murine mammary epithelial cell line 31EG4 (27). The effects of TGF-β_1 may therefore be different in different epithelial cells.

6.5 DIETARY SUBSTANCES THAT MODULATE PARACELLULAR PERMEABILITY

A variety of dietary substances have been reported to modulate tight-junction permeability (see Table 6.1). Some examples are given below.

1. Proteins and peptides: Protamine, a naturally occurring arginine-rich polycationic protein, has been reported to increase TER and reduce the paracellular permeability (28). Protamine was found to not directly interact with tight junctions but was likely to affect tight junctions through a secondary action via a cellular regulatory mechanism (29). Protamine therefore can be used as a tight-junction blocker, reducing the paracellular absorption of water and such carbohydrates as lactulose. We have observed that a less stable tight junction of intestinal Caco-2 cells cultured in a serum-free culture medium is stabilized by β-lactoglobulin from bovine milk (30). β-Lactoglobulin is likely to have modulated the cytoskeletal structure by stimulating the receptor tyrosine kinase, with subsequent activation of phospholipase C and PKC. In the same model experiment, the tight junction-stabilizing effect of a β-casein tryptic digest was also observed. The stabilizing activity was higher in the more hydrophobic and more acidic peptide fractions (31). Although the physiological significance of these phenomena is not known, these effects of milk proteins may be involved in the development of intestinal epithelial barrier functions during suckling.

2. Amino acids: Some amino acids have been reported to affect tight-junction permeability. Madara and Carlson (32) have found that L-tryptophan induces disruption of the tight junction and transepithelial macromolecular leakages in the hamster small intestinal epithelium. This effect was energy and Na^+ dependent, suggesting that the activation of the Na^+-nutrient co-transporter in the cell membrane of intestinal epithelial cells was involved in this altered junctional permeability. It is possible that dietary supplementation of amino acids such as tryptophan may result in loss of the intestinal epithelial barrier function to food-derived antigens.
3. Glucose: Intestinal glucose absorption in humans is known to be nearly proportional to the luminal concentration at up to 500 mM (33). This suggests that the intestinal absorption of glucose is not only via active Na^+-glucose co-transport (the reported Km value is 0.11 mM) but also via the paracellular pathway. Interestingly, the paracellular flux of glucose as well as that of other small molecular nutrients increases during active Na^+-glucose co-transport. According to the proposed model, active Na^+-glucose co-transport leads to accumulation of Na^+ and glucose within the subjunctional basolateral space, which draws water across the tight junction, resulting in greater absorption of water and small nutrients (33). The effect is referred to as the solvent drag mechanism. This increased paracellular permeability is accompanied by localized disruption of tight-junction structure. Recent studies have suggested that Na^+-glucose co-transport led to activation of the Na^+-H^+ exchanger in the apical cell membrane, leading to increased myosin light chain kinase (MLCK) activity (34). The resulting myosin light chain phosphorylation may contract the perijunctional actomyosin ring, triggering structural modification of the tight junction. The physiological role of this greater tight-junction permeability is thought to increase the mass transport of such nutrients as glucose by the paracellular pathway as rapidly as possible.
4. Nucleosides: Kishibuchi et al. (35) have reported that a nucleoside/nucleotide mixture (OG-VI consisting of inosine, guanosine monophosphate [GMP], cytidine, uridine, and thymidine) improved the gut mucosal barrier function. In their experiment, deterioration of the mucosal barrier function was induced by total parenteral nutrition (TPN). The increased macromolecular transmission in the intestine by the TPN treatment was accompanied by widening of the intercellular spaces. The addition of OG-VI reduced the width of these intercellular spaces, decreasing the mucosal permeability of larger-molecular-weight substances.
5. Fatty acids: Tight-junction permeability is modulated by various fatty acids. Sodium caprate, a dairy product constituent, is known to open tight junctions and increase the paracellular permeation of water-soluble nonelectrolytes (36) and macromolecules (37) without affecting epithelial viability. The reassembly of such tight-junction proteins as occludin and ZO-1 by caprate was observed by Lindmark et al. (38), and this may be

one of the reasons for the increased tight-junction permeability. On the one hand, Roche et al. (39) have reported that conjugated linoleic acids (CLA) increased paracellular epithelial permeability. CLA delayed the development of functional Caco-2 cell monolayers and altered the distribution of occludin and ZO-1, thereby increasing tight-junction permeability. On the other hand, Jiang et al. (40) have reported that polyunsaturated fatty acids such as γ-linolenic acid and eicosapentaenoic acid increased the TER value of endothelial cell monolayers by changing the expression level of occludin in the cells, although their effects on intestinal epithelial cells have not yet been studied.

6. Polysaccharides: Among the polysaccharides, chitosan (a deacetylated chitin) and its derivatives are known to enhance oral drug absorption by widening the tight junctions (41). Chitosans appear to bind tightly to the epithelium, inducing the redistribution of F-actin and the tight-junction protein ZO-1. It has been reported that the effects were mediated by cationic charges of chitosan (42). Although the chitosan treatment slightly perturbed the cell membrane, this effect was reversible, and the viability of the cells was not affected. These characteristics indicate that chitosan can be used as a generally safe absorption enhancer of hydrophilic macromolecular drugs.
7. Cyclodextrins: Cyclodextrins are known to bind hydrophobic substances such as fatty acids, cholesterol,and polyphenols. Incubation of epithelial cell monolayers with methyl β-cyclodextrin (MBCD) has been reported to alter TER and the transepithelial flux of mannitol (43). TER of the monolayer was increased by incubating with MBCD for 30 minutes but decreased by incubating for 2 hours or longer. The interaction between the tight-junction particles and the underlying cytoskeletal elements would change by incubating with MBCD because MBCD induces cholesterol efflux from the cell membrane. On the one hand, the increase in TER after a 30-minute incubation is accompanied by an increase in the number of tight-junction particles associated with the inner surface of the plasma membrane. On the other hand, the cytoskeletal organization during a long incubation with MBCD may physically disrupt the tight-junction network.
8. Surfactants: Surfactants are extensively used as absorption enhancers in oral drug delivery. This absorption enhancing effect is, at least partly, due to the solubilization of phospholipids and membrane proteins in the epithelial cells by a surfactant. In addition to the increased permeability via the transcellular route, an alteration of the tight junction permeability by surfactants has been suggested (44). Food-grade surfactants also affect the absorption of substances from the gastrointestinal tract. Mine and Zhang (45) have found that the TER of Caco-2 cell monolayers was decreased upon exposure to sucrose monoester fatty acids. By treating the cells with a surfactant, shortening in microvilli, disorganization of actin filaments, and structural separation of tight junctions were observed,

which may lead to opening the paracellular pathway and increasing the uptake of food allergens.

9. Other substances: We have observed that capsianoside, a diterpene glycoside from sweet pepper, decreased the TER of Caco-2 cell monolayers and increased the paracellular permeability (46). Capsianoside is incorporated into the cell membrane, resulting in the dysfunction of tight junctions by changing the F-actin/G-actin ratio. An increase in tight-junction permeability by such spices as paprika and cayenne pepper has been reported (47). Capsaicin is though to be involved in this phenomenon. However, extracts from green/black peppers or nutmeg increased TER, thus reducing the macromolecular permeability. Piperine, a major component of peppers, seems to play a role in reducing the permeability.

As described previously, cytokines play an important role in the regulation of tight-junction permeability. Since the production of cytokines by immune and epithelial cells in the intestine is affected by dietary substances, some foods could change tight-junction permeability by regulating the cytokine production. The paracellular permeability is therefore regulated or affected by many factors. The tight-junction-modulatory substances may be useful for inducing the efficient absorption of nutrients, functional food substances, and drugs and also for maintaining intestinal barrier functions against harmful substances such as allergens and toxicants. More information on tight-junction-modulatory substances and their mechanisms is needed.

REFERENCES

1. U. Karbach. Paracellular calcium transport across the small intestine. *J. Nutr.*, 122: 672–677, 1992.
2. J.R. Pappenhimer, C.E. Dahl, M.L. Karnovsky, J.E. Maggio. Intestinal absorption and excretion of octapeptides composed of D-amino acids. *Proc. Natl. Acad. Sci. USA*, 91: 1942–1945, 1994.
3. M. Satake, M. Enjoh. Y. Nakamura, T. Toshiaki, Y. Kawamura, S. Arai, M. Shimizu. Transepithelial transport of the bioactive tripeptide, Val-Pro-Pro, in human intestinal Caco-2 cell monolayers. *Biosci. Biotechnol. Biochem.*, 66: 378–384, 2002.
4. M. Heyman, J.F. Desjeux. Significance of intestinal food protein transport. *J. Pediatr. Gastroenterol. Nutr.*, 15: 48–57, 1992.
5. M. Furuse, T. Hirase, M. Itoh, A. Nagafuchi, S. Yoneyama, Sa. Tsukita, Sh. Tsukita. Occludin: a novel integral membrane protein localizing at tight junctions. *J. Cell Biol.*, 123: 1777–1788, 1993.
6. O.W. Blaschuk, T. Oshima, B.J. Gour, J.M. Symonds, J.H. Park, C.G. Kevil, S.D. Trocha, S. Michaud, N. Okayama, J.W. Elrod, J.S. Alexander. Identification of an occludin cell adhesion recognition sequence. *Inflammation*, 26 (4): 193–198, 2002.
7. S. Balada, J.A. Whitney, C. Folres, S. Gonzalez, M. Cereijido, K. Matter. Functional dissociation of paracellular permeability and transepithelial electrical resistance and disruption of the apical-basolateral intramembrane diffusion barrier by expression of a mutant tight junction membrane protein. *J. Cell Biol.*, 134: 1031–1049, 1996.

8. V. Wong, B.M. Gumbiner. A synthetic peptide corresponding to the extracellular domain of occludin perturbs the tight junction permeability barrier. *J. Cell Biol.*, 136: 399–409, 1997.
9. J. Haskins, L. Gu, E.S. Wittchen, J. Hibbard, B.R. Stevenson. ZO-3, a novel member of the MAGUK protein family found at the tight junction, interacts with ZO-1 and occludin. *J. Cell Biol.*, 141: 199–208, 1998.
10. M. Furuse, K. Fujita, T. Hiiragi, K. Fujimoto, S. Tsukita. Claudin-1 and -2: novel integral membrane proteins localizing at tight junctions with no sequence similarity to occludin. *J. Cell Biol.*, 141: 1539–1550, 1998.
11. K. Morita, M. Furuse, K. Fujimoto, S. Tsukita. Claudin multigene family encoding four-transmembrane domain protein components of tight junction strands. *Proc. Natl. Acad. Sci. USA.*, 96: 511–516, 1999.
12. S. Citi, H. Sabanay, R. Jakes, B. Geiger, J. Kendrick-Jones. Cingulin, a new peri-junctional component of tight junctions. *Nature*, 333: 272–276, 1988.
13. Y. Zhong, T. Saitoh, T. Minase, N. Sawada, K. Enomoto, M. Mori. Monoclonal antibody 7H6 reacts with a novel tight junction-associated protein distinct from ZO-1, cingulin and ZO-2. *J. Cell Biol.*, 120: 477–483, 1993.
14. S. Tsukita , M. Furuse , M. Itoh. Structural and signaling molecules come together at tight junctions. *Curr. Opin. Cell Biol.*, 11: 628–633, 1999.
15. G. Hecht, B. Robinson, A. Koutsouris. Reversible disassembly of an intestinal epithelial monolayer by prolonged exposure to phorbol ester. *Am. J. Physiol.*, 266: G214–G221, 1994.
16. J.M. Mullin, K.V. Laughlin, N. Ginanni, C.W. Marano, H.M. Clarke, A.P. Soler. Increased tight junction permeability can result from protein kinase C activation/translocation and act as a tumor promotional event in epithelial cancers. *Ann. NY Acad. Sci.*, 915: 231–236, 2000.
17. D. Rosson, T. O'Brien, J.A. Kampherstein, Z. Szallasi, K. Bogi, P.M. Blumberg, J.M. Mullin. Protein kinase C activity modulates transepithelial permeability and cell junctions in the LLC-PK1 epithelial cell line. *J. Biol. Chem.*, 272: 14950–14953, 1997.
18. A. Sakakibara, M. Furuse, M. Saitou, Y. Ando-Akatsuka, S. Tsukita. Possible involvement of phosphorylation of occludin in tight junction. *J. Cell Biol.*, 137: 1393–1401, 1997.
19. A.M. Hopkins, D. Li, R.J. Mrsny, S.V. Walsh, A. Nusrat. Modulation of tight junction function by G protein-coupled events. *Adv. Drug Deliv. Rev.*, 41: 329–340, 2000.
20. S.V. Walsh, A.M. Hopkins, A. Nusrat. Modulation of tight junction structure and function by cytokines. *Adv. Drug Del. Rev.*, 41: 303–313. 2000.
21. A. Nusrat, C. von-Eichel-Streiber, J.R. Turner, P. Verkade, J.L. Madara, C.A. Parkos. Clostridium difficile toxins disrupt epithelial barrier function by altering membrane microdomain localization of tight junction proteins. *Infect. Immun.*, 69: 1329–1336, 2001.
22. S.S. Koshy, M.H. Montrose, C.L. Sears. Human intestinal epithelial cells swell and demonstrate actin rearrangement in response to the metalloprotease toxin of Bacteroides fragilis. *Infect. Immun.*, 64: 5022–5028, 1996.
23. J.L. Madara. Increases in guinea pig small intestinal transepithelial resistance induced by osmotic loads are accompanied by rapid alterations in absorptive-cell tight-junction structure. *J. Cell Biol.*, 97: 125–136, 1983.
24. M. Heyman, J.F. Desjeux. Cytokine-induced alteration of the epithelial barrier to food antigens in disease. *Ann. NY Acad. Sci.*, 915: 304–311, 2000.

25. J.L. Madara, J. Stafford. Interferon-gamma directly affects barrier function of cultured intestinal epithelial monolayers. *J. Clin. Invest.*, 83: 724–727, 1989.
26. H. Schmitz, M. Fromm, C.J. Bentzel, P. Scholz, K. Detjen, J. Mankertz, H. Bode, H.J. Epple, E.O. Riecken, J.D. Schulzke. Tumor necrosis factor-alpha (TNF-alpha) regulates the epithelial barrier in the human intestinal cell line HT29/B6. *J. Cell Sci.*, 112: 137–146, 1999.
27. P.L. Woo, H.H. Cha, K.L. Singer, G.L. Firestone. Antagonistic regulation of tight junction dynamics by glucocorticoids and transforming growth factor-beta in mouse mammary epithelial cells. *J. Biol. Chem.*, 27: 404–412, 1996.
28. X. Shi, C.V. Gisolfi. Paracellular transport of water and carbohydrates during intestinal perfusion of protamine in the rat. *Am. J. Med. Sci.*, 311(3): 107–112, 1996.
29. C.G. Bentzel, M. Fromm, C.E. Palant, U. Hegal. Protamine alters structure and conductance of necturus gallbladder tight junctions without major electrical effects on the apical cell membrane. *J. Membr. Biol.*, 95: 9–20, 1987.
30. K. Hashimoto, T. Nakayama, M. Shimizu. Effects of β-lactoglobulin on the tight-junctional stability of Caco-2-SF monolayer. *Biosci. Biotechnol. Biochem.*, 62(9): 1819–1821, 1998.
31. M. Shimizu. Milk and intestinal functions. *Milk Sci. (Jpn.)*, 49: 203–206, 2000.
32. J.L. Madara, S. Carlson. Supraphysiologic L-tryptophan elicits cytoskeletal and macromolecular permeability alterations in hamster small intestinal epithelium *in vitro*. *J. Clin. Invest.*, 87: 454–462, 1991.
33. J.R. Turner. Show me the pathway! Regulation of paracellular permeability by Na^+-glucose contransport. *Adv. Drug Deliv. Rev.*, 41: 265–281, 2000.
34. J.J. Berglund, M. Riegler, Y. Zolotarevsky, E. Wenzl, J.R. Turner. Regulation of human jejunal transmucosal resistance and MLC phosphorylation by Na^+-glucose cotransport. *Am. J. Physiol.*, 281: G1487–G1493, 2001.
35. M. Kishibuchi, T. Tsujinaka, M. Yano, T. Morimoto, S. Iijima, A. Ogawa, H. Shiozaki, M. Monden. Effects of nucleosides and a nucleotide mixture on gut mucosal barrier function on parenteral nutrition in rats. *J. Parenter. Enter. Nutr.*, 21: 104–111, 1997.
36. T. Sawada, T. Ogawa, M. Tomita, M. Hayashi, S. Awazu. Role of paracellular pathway in nonelectrolyte permeation across rat colon epithelium enhances by sodium caprate and sodium caprylate. *Pharm. Res.*, 8: 1365–1371, 1991.
37. J.D. Soderholm, H. Oman, L. Blomquist, J. Veen, T. Lindmark, G. Olaison. Reversible increase in tight junction permeability to macromolecules in rat ileal mucosa *in vitro* by sodium caprate, a constituent of milk fat. *Dig. Dis.*, 43: 1547–1552, 1998.
38. T. Lindmark, Y. Kimura, P. Artrusson. Absorption enhancement through intracellular regulation of tight junction permeability by sodium chain fatty acids in Caco-2 cells. *J. Pharm. Exp. Ther.*, 284: 362–369, 1998.
39. H.M. Roche, A.M. Terres, I.B. Black, M.J. Gibney, D. Kelleher. Fatty acids and epithelial permeability: effect of conjugated linoleic acids in Caco-2 cells. *Gut*, 48: 797–802, 2001.
40. W.G. Jiang, R.P. Bryce, D.F. Horrobin, R.E. Mansel. Regulation of tight junction permeability and occluding expression by polyunsaturated fatty acids. *Biochem. Biophys. Res. Commn.*, 244: 414–420, 1998.
41. M. Thanou, J.C. Verhoef, H.E. Junginger. Oral drug absorption enhancement by chitosan and its derivatives. *Adv. Drug Deliv. Rev.*, 52: 117–126, 2001.

42. N.G.M. Schipper, S. Olsson, J.A. Hoogstraate, A.G. deBoer, K.M. Varum, P. Artursson. Chitosans as absorption enhancers for poorly absorbable drugs 2: mechanism of absorption enhancement. *Pharm. Res.*, 14: 923–929, 1997.
43. S.A. Francis, J.M. Kelly, J. McCormack, R.A. Rogers, J. Lai, E.E. Schneeberger, R.D. Lynch. Rapid reduction of MDCK cell cholesterol by methyl-β-cyclodextrin alters steady state transepithelial electrical resistance. *Eur. J. Cell Biol.*, 78: 473–484, 1999.
44. H.E. Junginger, J.C. Verhoef. Macromolecules as safe penetration enhancers for hydrophilic drugs — a fiction? *PSTT*, 1: 370–376, 1998.
45. Y. Mine, J.W. Zhang. Surfactants enhance the tight-junction permeability of food allergens in human intestinal epithelial Caco-2 cells. *Int. Arch. Allergy Immunol.*, 130: 135–142, 2003.
46. K. Hashimoto, H. Kawagishi, T. Nakayama, M. Shimizu. Effect of capsianoside, a diterpene glycoside, on tight junctional permeability. *Biochim. Biophys. Acta*, 323: 281–290, 1997.
47. E. J. Jarolim, L. Gajdzik, I. Haberl, D. Kraft, O. Scheiner, J. Graf. Hot spices influence permeability of human intestinal epithelial monolayers. *J. Nutr.*, 128: 577–581, 1998.

Section II

The Body's Defense System

7 Antimicrobial Peptides

Judy C.K. Chan and
Eunice C.Y. Li-Chan

CONTENTS

7.1 INTRODUCTION

Since the identification of the first antibacterial protein family, thionins, during the early 1970s, over 700 antimicrobial proteins and peptides have been identified in plants and animals (1). The widespread distribution of potent, broad-spectrum antimicrobial peptides in multicellular organisms suggests that they could be used to fend off a wide range of microbes, including bacteria, fungi, viruses, and protozoa. An increased expression of antimicrobial peptides in an organism could enhance its innate immunity and increase its resistance against microorganisms and diseases in the natural environment. Furthermore, recent studies have shown that these antimicrobial peptides perform multiple roles in host defense, inflammation, and tissue regeneration. A better understanding of the basic biological properties and actions of antimicrobial peptides may allow the engineering of pathogen-resistant crops and stimulate the development of novel therapeutic approaches to the treatment of diseases. It is therefore our intent to provide an overview of the current status of some of the more commonly found antimicrobial peptides that exist as peptides at the source or that can be derived by proteolysis, in terms of their widespread occurrence, their basic structural biology, mechanisms of their antimicrobial activity, and their applications in agriculture, foods, and clinical treatments.

7.2 SOURCES OF ANTIMICROBIAL PEPTIDES

7.2.1 Mammalian Sources

Most mammalian antimicrobial peptides share common features such as small size, cationic charge, and an amphipathic nature. Based on such features, antimicrobial peptides in mammals have been classified into two major families, defensins and cathelicidins.

7.2.1.1 Mammalian Defensins

Small cationic antimicrobial proteins were first detected in rabbit and guinea pig granulocytes in the mid-1960s (2,3) and were later identified to be the principal constituents of these granulocytes and human neutrophils. Structurally, these peptides belong to the superfamily defensin. Three types of defensins have been found among vertebrates: α-, β-, and θ-defensins (4). α-Defensins are small peptides with 29 to 35 residues and a six-cysteine motif, whose C1–C6, C2–C4, and C3–C5 pairings form three intramolecular disulfide bonds. All known α-defensins are found in humans, expressed by granulocytes, lymphocytes, and intestinal Paneth cells (5).

β-Defensins were discovered in bovine granulocytes and tracheal epithelial cells about 10 years ago. They differ from α-defensins by having up to ~45 residues, a different cysteine pairing (C1–C5, C2–C4, C3–C6), and relatively more lysine than arginine residues. A triple-stranded, antiparallel β-sheet with strands 2 and 3 involved in an β-hairpin has emerged as a common structural feature in both

α- and β-defensins (6,7). Unusual circular θ-minidefensins have recently been identified in rhesus monkeys and humans. This cyclic defensin has only 18 amino acids, including six cysteines that form three disulfide bridges. It was also demonstrated that the native cyclic form had threefold greater antimicrobial activity than its open chain analogue and had the ability to protect humans from HIV-1 infection (8–10).

7.2.1.2 Mammalian Cathelicidins

Cathelicidins are a group of structurally diverse antimicrobial peptides found, to date, only in mammalian species (11–13). Examples include PR-39 and protegrins from pigs (14–16), SMAP-29 from sheep (17,18), CAP-18 from rabbits (19,20), and bactenecin and indolicidin from cows (21). Unlike defensins, cathelicidins are stored as inactive propeptide precursors and processed upon stimulation. Members of this family have a conserved N-terminal cathelin domain (propiece) of about 100 residues and a C-terminal cationic antimicrobial domain of varied length (12 to 100 residues). Based on their structures, cathelicidins can be categorized into four classes: α-helical, β-sheet, proline-rich, and tryptophan-rich (Table 7.1; 17,19,22–30). Cathelicidins have a diverse range of antimicrobial activity. They are microbicidal against Gram-negative or Gram-positive bacteria, fungi, parasites, and enveloped viruses (13). The most interesting ability of some cathelicidins such as CAP-18 and SMAP-29 is their activity against *Pseudomonas aeruginosa,* irrespective of salt concentration. The salt-resistant, antimicrobial properties of CAP-18 and SMAP-29 suggest that these peptides of congeneric structures have potential for the treatment of bacterial infections in normal and immunocompromised persons and those with cystic fibrosis (31,32).

7.2.2 Plant Sources

Several families of antimicrobial peptides have been identified in plants. Ranging in size from 2 to 9 kDa, all plant antimicrobial peptides are globular, compact, and cysteine-rich peptides. Thionins were the first plant peptides to be described. Subsequently, the antimicrobial activites of various defensins, lipid transfer proteins, hevein- and knottin-like peptides, including MBP-1 from maize, IbAMP from the seeds of impatiens, snakins from potatoes, and shepherdins from the roots of shepherd's purse have been identified (33,34).

7.2.2.1 Plant Defensins

Formerly known as γ-thionins, plant defensins (45 to 54 residues in length, 5 kDa) are cysteine-rich cationic peptides that consist of triple-stranded antiparallel β-sheets and one α-helix that are stabilized by disulfide bonds (33). Members of the plant defensin family are highly homologous and share the conservation of 12 amino acid residues including the positions of eight cysteines, two glycines, one glutamic acid, and one aromatic amino acid (Table 7.2) (35). The remaining amino acids in these peptides are variable. This feature likely contributes to the divergence

TABLE 7.1
Primary Sequences of Representative Antimicrobial Peptides Classified as Cathelicidins[a]

Structure and Peptide ID	Origin	Amino Acid Sequences									References
α-Helical											
SMAP-29	Sheep	RGLRR	LGRKI	AHGVK	KYFPT	VLRRI	RIAG				17
$CAP18_{106-137}$	Rabbit	GLRKR	LRKFR	NKIKE	KLKKI	GQKIQ	GLLPK	LA			19
LL-37	Human	LLGDF	FRKSK	EKIGK	EFKRI	VQRIK	DFLRN	LVPRT	ES		22, 23
CRAMP	Mouse	GLLRK	GGEKI	GEKLK	KIGQK	IKNFF	QKLVP	QPE			24, 25
Proline-rich											
Bac-5	Cow	RFRPP	IRRPP	IRRPPF	YPPFR	PPIRP	PIFPP	IRPPF	RPPLG	PFPG	26
PR-39	Pig	RRRPR	PPYLP	RPRPP	PFFPP	RLPPR	IPPGF	PPRFP	PRFP		27
Disulfide bonds and/or β-sheet											
Bactenecin	Cow	RLCRI	VVIRV	CR							28
Protegrin	Pig	RGGRL	CYCRR	RFCVC	VGR						29
Tryptophan-rich											
Indolicidin	Cow	ILPWK	WPWWP	WRR							30

[a] Amino acids are represented in conventional single letter code. All antimicrobial peptides are amphipathic and contain more positively charged residues (arginine and lysine) than negatively charged ones (aspartic and glutamic acids). Cysteine residues, when present, form intramolecular disulfide bonds.

of their biological activities and has been assessed for defensins from various plant species. Two sulfur-rich cationic antimicrobial polypeptides, $\gamma 1$-purothionin and $\gamma 2$-purothionin, were first isolated from wheat endosperm (36), and the majority of all defensins identified, to date, have originated from plant seeds such as barley (37), broad beans (38), and radish (39,40). Nevertheless, members of this family have been characterized at the deoxyribonucleic acid (DNA) or protein level from various plant tissues in several plant species, including the leaves of spinach (41), pea pods (42,43), potato tubers (44,45), the fruits of bell pepper (46) and chili (47), and the flowers of tobacco (48,49) and petunia (49,50).

The antimicrobial activity of defensins was first reported from two isoforms isolated from radish seeds and now has been extensively studied in defensins from different species. Based on these antimicrobial properties, Broekaert et al. (35) classified defensins into three groups. Group I defensins can inhibit Gram-positive bacteria and fungi; those of group II are active against fungi but inactive against bacteria; and those of group III are active against both Gram-positive and Gram-negative bacteria but inactive against fungi. Segura et al. (41) later identified a fourth group from spinach tissues. Group IV defensins are active against Gram-positive and Gram-negative bacteria as well as against fungi. The structural basis for the different groups of defensins is not yet fully understood. It is postulated that there are multiple mechanisms of action among the different defensin groups (35). Segura et al. (41) suggested that the structural and functional divergence in plant defensins could have been driven by different environmental challenges of the different plant species. Defensins from different groups could be found in the same plant tissue as a way to achieve a broader antimicrobial barrier (41).

7.2.2.2 Lipid Transfer Proteins

Plant lipid transfer proteins (LTPs) have been isolated from several members of mono- and dicotyledonous species (51). LTPs are divided into two subfamilies with molecular masses of 9 kDa (LTP-1s) and 7 kDa (LTP-2s). All plant LTPs have some general characteristics in common: basic, high contents of proline, glycine, alanine and serine; low content of aromatic residues; and four disulfide bridges. The unexpected antimicrobial properties of LTPs were discovered by screening plant proteins for their ability to inhibit the growth of plant fungal and bacterial pathogens *in vitro* (52,53). Examples include LTPs isolated from maize seeds (54), maize leaves, and barley leaves (55), *Arabidopsis* and spinach leaves (56), and the wax of broccoli leaves (57). Although the actual antimicrobial mechanism of LTPs has not been fully elucidated yet, it is proposed that LTPs form a globular structure that consists of a bundle of four α-helices linked by flexible loops, with a hydrophobic cavity that may accommodate a variety of lipids. Hence, LTPs could be involved in the formation of hydrophobic cutin and suberin layers, which prevent water diffusion into the grain and consequently provide a defense against bacterial and fungal attacks (51).

TABLE 7.2
Primary Sequences of Selected Plant Defensins[a]

Peptide ID	Origin					Amino Acid Sequences								References
γ1-Purothionin	Wheat	—	KI**C**RR	RSAG**F**	**KGPC**M	SNKN**C**	AQY**C**Q	-Q**E**GW	GG**GNC**	DGPFR	R—**CK**	**C**IRQ**C**	-	36
γ2-Purothionin	Wheat	—	KV**C**RQ	RSAG**F**	**KGPC**V	SDKN**C**	AQV**C**Q	-Q**E**GW	GG**GNC**	DGPFR	R—**CK**	**C**IRQ**C**	-	36
γ-Hordothionin	Barley	—	RI**C**RR	RSAG**F**	**KGPC**V	SNKN**C**	AQV**C**M	-Q**E**GW	GG**GNC**	DGPLR	R—**CK**	**C**MRR**C**	-	37
Fabatin	Bean	LL	GR**C**KV	KSNR**F**	H**GPC**L	TDTH**C**	STV**C**R	-G**E**GY	KG**GDC**	HGLRR	R—**CM**	**C**L—**C**	-	38
Rs-AFP1	Radish	-U	KL**C**ER	PSGT**W**	SGV**C**G	NNNA**C**	KNQ**C**I	NL**E**KA	RH**GSC**	NYVFP	AHK**C**I	**C**YFP**C**	-	40
Rs-AFP2	Radish	-U	KL**C**QR	PSGT**W**	SGV**C**G	NNNA**C**	KNQ**C**I	RL**E**KA	RH**GSC**	NYVFP	AHK**C**I	**C**YFP**C**	-	40
P*sd*1	Pea	—	KT**C**EH	LADT**Y**	R**GPC**F	TNAS**C**	DDH**C**K	NK**E**HL	IS**GTC**	HNWK-	——**C**F	**C**XQN**C**	-	42
P*sd*2	Pea	—	KT**C**EN	LSGT**F**	**KGPC**I	PDGN**C**	NKH**C**R	NN**E**HI	LS**GRC**	RDDFX	——**C**W	**C**TRN**C**	-	42
p322	Potato	-A	RH**C**ES	LSHR**F**	**KGPC**T	RDSN**C**	ASV**C**E	-T**E**RF	SG**GNC**	HK—F	RRR**C**F	**C**TKP**C**	-	45
FST	Tobacco	-A	RE**C**KT	ESNT**F**	P**G**I**C**I	IKPP**C**	RKA**C**I	-S**E**KF	TD**GHC**	SKLL-	-RR**C**L	**C**TKP**C**	V	48
PPT	Petunia	—	RT**C**ES	QSHR**F**	H**G**T**C**V	RESN**C**	ASV**C**Q	-T**E**GF	IG**GNC**	RA—F	RRR**C**F	**C**TRN**C**	-	50

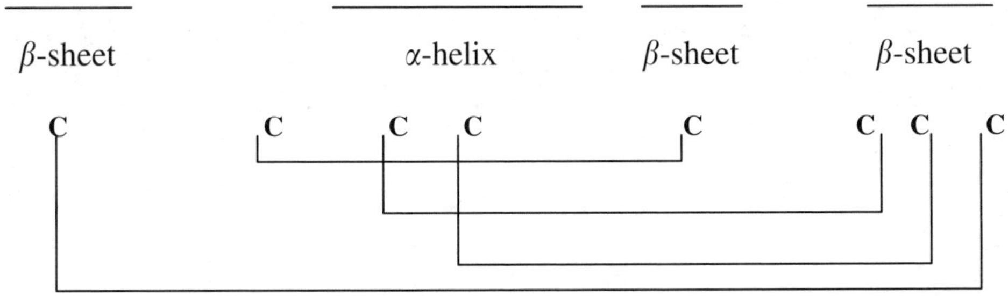

[a] Bold letters indicate conserved residues of the antimicrobial peptides. The common patterns of secondary structures and disulfide linkages are shown below the sequences.

7.2.2.3 Hevein- and Knottin-Type Peptides

Hevein is a small cysteine- and glycine-rich chitin-binding peptide of 43 amino acid residues first discovered in the latex of rubber trees (58). Since then, a number of hevein-like antimicrobial peptides such as Pn-AMPs from morning glory (*Pharbitis nil*) (59), GAFP from *Ginkgo biloba* (60), avesin A from oat seeds (61), Ee-CBP from the bark of spindle tree (*Euonymus europaeus* L.) (62), and Wj-AMP-1from the leaves of *Wasabia japonica* L. (63) have been reported. Hevein-type peptides are all highly basic and have pI values above 10. They consist of three folded β-strands, the second and third strands being linked by an α-helix. Hevein-type peptides have shown a wide range of *in vitro* antifungal activity potencies. Their activities, which were found to be strongly dependent on the ionic composition of the growth medium, are essentially abolished in high-salt media. However, it is not yet known how these chitin-binding proteins exhibit antifungal and antibacterial activities at the molecular level. Knottins are a group of structurally related small molecules with six disulfide-linked cysteines with a consensus pairing pattern (C1-C4, C2-C5, C3-C6).

Knottins fold into a triple-stranded β-sheet and form a "knot-like" feature. Few knottin-like antimicrobial peptides have been reported so far. Examples include two highly homologous antimicrobial peptides Mj-AMP-1 and Mj-AMP-2 (36 and 37 residues, respectively) from *Mirabilis jalapa* L. seeds (64,65) and PAFP-s (38 residues) from the seeds of *Phytolacca americana* (pokeweed) (66,67). These plant knottins show a broad antifungal activity with limited activity against bacteria.

7.2.2.4 Other Novel Plant Antimicrobial Peptides

Ib-AMP-1, Ib-AMP-2, Ib-AMP-3, and Ib-AMP-4 are four closely related peptides isolated from seeds of *Impatiens balsamina*, which were shown to be inhibitory to the growth of a range of fungi and bacteria (68). These 20 amino acid-long peptides are highly basic, with five arginine residues and four cysteine residues that form two intramolecular disulfide bonds. This family of peptides is the smallest of the antimicrobial peptides isolated, to date, from plants. A sequence search of the protein sequence databases for Ib-AMP-1 failed to identify any protein or peptide with statistically significant homology (69). Snakin-1 and snakin-2 are two novel 12-Cys antimicrobial peptides from the tubers, stems, flowers, shoot apex, and leaves of potato (70,71). They are active against fungal and bacterial plant pathogens and cause rapid aggregation of both Gram-positive and Gram-negative bacteria.

7.2.3 Marine Sources

Compared with the terrestrial sources, relatively few antimicrobial peptides have been identified with a marine origin. Although fishes and invertebrates rely heavily on their innate immune defenses for protection against pathogen insults, only a few antimicrobial peptides have been described from marine organisms. The following section will introduce a few antimicrobial peptides from marine sources. The primary sequences of some of these peptides are shown in Table 7.3.

7.2.3.1 Antimicrobial Peptides from Fishes

Of all antimicrobial peptides known in fishes, protamines probably represent the oldest known family. Found in fish milt, clupeines from *Clupea pallasii* (Pacific herring), salmine from chum salmon (*Oncorhynchus keta*), and iridine from rainbow trout (*Salmo irideus*) were first reported in 1957 (72). Protamines are a family of sperm nuclear basic proteins in fish milt. They consist of about 30 amino acid residues, including about 66% arginine on a mole basis; the arginine residues usually exist in clusters of four or five, and other residues follow a simple consensus amino acid composition (Table 7.3) (73–75). Protamines exhibit antimicrobial properties against a wide range of Gram-positive and Gram-negative bacteria, yeasts, and moulds (76–79). Probably due to the presence of many positively charged residues, protamines interact with the negatively charged cell surface and cause the permeabilization of cell envelopes (80,81).

Pardaxins are, in fact, neurotoxins used by fishes as a natural weapon against shark predation. Pardaxins have been isolated and characterized from the western Pacific Peacock sole (Pardachirus pavonius) (82) and the Red Sea Moses sole (Pardachirus marmoratus) (83,84). All known pardaxins contain a single peptide chain composed of 33 amino acids and are acidic, amphipathic, and hydrophobic peptides with a mass of about 3500 daltons. Their sequences are homologous, differing only at positions 5, 14, or 31 (see Table 7.3) (85). Oren and Shai (86) showed that pardaxins possess a wide range of antimicrobial activity.

Pleurocidin is a potent and wide-spectrum antibacterial peptide from the skin secretions of the winter flounder (*Pleuronectes americanus*) (87,88). This 25-residue peptide is quite basic and histidine rich. Pleurocidin forms a single, well-defined amphipathic helix and kills both Gram-positive and Gram-negative bacteria (89). However, its exact antimicrobial mechanism has not been fully elucidated (90,91). Small cysteine-rich antimicrobial peptides of the defensin family, known as hepcidin or LEAP-1 (liver-expressed antimicrobial peptide), have been predicted from expressed sequence tag databases from various fish species, including medaka, rainbow trout, Japanese flounder (92), winter flounder (93,94), Atlantic salmon (94), and long-jawed mudsucker (95). Nevertheless, only one fish hepcidin has been successfully isolated from the gill of hybrid striped bass (96). This 21-residue peptide is stabilized by four intramolecular disulfide bonds and has been shown to have antimicrobial activity against *Escherichia coli* (96).

New peptides are being discovered continually, and many of them carry with them certain unique features. For instance, a highly polar peptide, misgurin, rich in both basic and acidic residues, has been isolated from a mudfish. This peptide is unlike any other reviewed here and, despite an imperfect amphipathicity, is quite potent in killing bacteria and yeasts (97). A polar 19-residue α-helical peptide, parasin, has been isolated from the epithelial mucosa of the catfish (*Parasilurus asotus*) (98). Parasin is a highly cationic peptide, has a rod-like helical structure, and appears to penetrate and inactivate bacteria without causing membrane permeabilization. Hagfish intestinal antimicrobial peptides (HFIAPs) are a family of polycationic peptides containing brominated tryptophan residues in their sequence with activity against both Gram-positive and Gram-negative bacteria (99). Moronecidin is a novel 23-residue, C-terminally amidated antimicrobial peptide, isolated from the skin and gill of hybrid striped bass. Two isoforms, differing by only one amino acid, are derived from each parental species white bass (*Morone chrysops*) and striped bass (*Morone saxatilis*). A synthetic, amidated white bass moronecidin exhibited broad-spectrum antimicrobial activity that was retained at high salt concentration. An α-helical structure was confirmed (100).

7.2.3.2 Antimicrobial Peptides from Crustaceans

Since invertebrates lack the specific antibody immunity, a pre-existing presence of antimicrobial peptides is of great importance to their innate immune defense system. Broad antimicrobial activity has been detected in tissues of crustaceans such as shrimp (101,102), crayfish (102), squat lobster and Norway lobster (103), and a great variety of crabs (102,104,105).

However, few antimicrobial peptides have been isolated from crustaceans. The antibacterial peptide in shore crab (Carainus maenas) was the first one isolated from crustaceans (106). This 6.5 kDa proline-rich peptide has been shown to have activity against both Gram-negative and Gram-positive bacteria and contains the triplet Pro-Arg-Pro motif commonly seen in other proline-rich antimicrobial peptides. In addition to the 6.5 kDa peptide, an 11.5 kDa antibacterial peptide has been isolated from the shore crab (107). This 84 amino acid residue peptide has been shown to be cationic and hydrophobic and have specific activity for only Gram-positive marine or salt-tolerant bacteria. Callinectin is another proline-rich, partially sequenced 3.7 kDa basic antimicrobial peptide isolated from blue crab (C. sapidus) (108). This peptide has demonstrated activity against E. coli and possesses a proline-rich amino-terminal region. However, the proline residues are not arranged in the typical motifs seen in other proline-rich antimicrobial peptides, and the peptide does not show significant homology with any other known peptides (108).

Four structurally related antimicrobial peptides, named tachyplesin, big defensin, tachycitin, and tachystatin, have been identified from Japanese horseshoe crab (*Tachypleus tridentatus*). Tachyplesin is a 17-residue β-sheet forming cationic peptide isolated from the acid extracts of horseshoe crab, which has been shown to inhibit both Gram-negative and Gram-positive bacteria (109). In contrast, the "big defensin" is a 79-residue peptide with structure resembling that of mammalian β-defensins. "Big defensin" has demonstrated strong inhibitory activity against Gram-negative and Gram-positive bacteria as well as fungi (110). Tachycitin is a chitin-binding peptide of 73 amino acid residues containing five disulfide bonds. It showed inhibitory effect against the growth of both Gram-negative and Gram-positive bacteria as well as fungi and also

TABLE 7.3
Primary Sequences of Selected Antimicrobial Peptides from Marine Sources

Peptide ID	Origin	Amino Acid Sequences							References
Fish									
Clupeine Z	Herring	ARRRR	SRRAS	RPVRR	RRPRR	VSRRR	RARRR	R	73
Clupeine YII	Herring	PRRRT	RRASR	PVRRR	RPRRV	SRRRR	ARRRR		74
Clupeine YI	Herring	ARRRR	SSSRP	IRRRR	PRRRT	TRRRA	GRRRR		75
Pardaxin-1	Pacific sole	GFFAL	IPKII	SSPLF	KTLLS	AVGSA	LSSSG	EQE	82
Pardaxin-2	Pacific sole	GFFAL	IPKII	SSPIF	KTLLS	AVGSA	LSSSG	GQE	82
Pardaxin-3	Pacific sole	GFFAF	IPKII	SSPLF	KTLLS	AVGSA	LSSSG	EQE	82
Pardaxin-4	Red Sea sole	GFFAL	IPKII	SSPLF	KTLLS	AVGSA	LSSSG	GQE	84
Pardaxin-5	Red Sea sole	GFFAL	IPKII	SSPLF	KTLLS	AVGSA	LSSSG	DQE	84
Pleurocidin	Winter flounder	GWGSF	FKKAA	HVGKH	VGKAA	LTHYL			87
Hepcidin	Striped bass	GCRFC	CNCCP	NMSGC	GVCCR	F			96
Parasin	Catfish	KGRGK	QGGKV	RAKAK	TRSS				98
Moronecidin	White bass	FFHHI	FRGIV	HVGKT	IHKLV	TGT			100
Moronecidin	Striped bass	FFHHI	FRGIV	HVGKT	IHKLV	TGT			100

Crustaceans										
Tachyplesin	Horseshoe crab	KWCFR	VCYRG	ICYRR	CR					109
Tachystatin A2	Horseshoe crab	YSRCQ	LQGFN	CVVRS	YGLPT	IPCCR	GLTCR	SYFPG	STYGR	112
		CQRY								
Penaeidin-1	Pacific white shrimp	YRGGY	TGPIP	RPPPI	GRPPL	RLVVC	ACYRL	SVSDA	RNCCI	114
		KFGSC	CHLVK							
Penaeidin-3	Pacific white shrimp	EVYKG	GYTRP	IPRPP	PFVRP	LPGGP	IGPYN	GCPVS	CRGIS	114
		FSQAR	SCCSR	LGRCC	HVGKG	YS				
Molluscs										
MGD 1	Mediterranean mussel	GFGCP	NNYQC	HRHCK	SIPGR	CGGYC	GGWHR	LRCTC	YRCG	122
MGD 2	Mediterranean mussel	GFGCP	NNYAC	HQHCK	SIRGY	CGGYC	AGWFR	LRCTC	YRCK	122
Mytilin A	Common mussel	GCASR	CKAKC	AGRRC	KGWAS	ASFRG	RCYCK	CFRC		121
Mytilin D	Mediterranean mussel	GCASR	CKAKC	AGRRC	KGWAS	ASFRR	RCYCK	CFRC		123
Myticin A	Mediterranean mussel	HSHAC	TSYWC	GKFCG	TASCT	HYLCR	VLHPG	KMCAC	VHCSA	123
Myticin B	Mediterranean mussel	HPHVC	TSYYC	SKFCG	TAGCT	RYGCR	NLHRG	KLCFC	LHCSR	123

exhibited a bacterial agglutinating property. Tachycitin and big defensin have been shown to have synergistic antimicrobial action (111). Tachystatins consist of a family of isoforms with 41 to 44 amino acid residues. Like other horseshoe crab antimicrobial peptides, techystatins exhibit antimicrobial activity against Gram-negative and Gram-positive bacteria as well as fungi (112). Tachystatin A isoform have been shown to fold into a triple-stranded β-sheet and two β-turn, which is quite different from the structures of the other three peptides from horseshoe crab (113).

Penaeidins from the Pacific white shrimp (*Panaeus vannamei*) have also been isolated and characterized (114). The penaeidins are composed of 50 to 62 amino acid residues, highly cationic, and secreted into the blood upon immune response stimulation (114) at various development stages of the shrimp *Penaeus vannamei* (115). Furthermore, these peptides are composed of a proline-rich domain at the amino terminal and six cysteine residues forming three intramolecular disulfide bridges at the carboxyl terminal (114). The proline-rich domain shows striking similarities to the 6.5-kDa peptide from *C. maenas* and other proline-rich antimicrobial peptides, including the presence of the Pro-Arg-Pro motif (116). Penaeidins also have demonstrated a wide antibacterial activity against Gram-positive bacteria as well as antifungal properties (117). However, unlike other proline-rich peptides, the penaeidins do not display strong activity against Gram-negative bacteria (116).

Recent advances in genome knowledge have enabled the detection of two 11.5 kDa antimicrobial peptides from penaeid shrimps (*Litopenaeus vannamei* and *Litopenaeus setiferus*) (118). These two peptides have shown great homology to the 11.5 kDa peptide isolated from *C. maenas* but no homology with other known antimicrobial peptides. Hence, a new term, crustin, has been given to this new family of 11.5 kDa antimicrobial peptides (118).

7.2.3.3 Antimicrobial Peptides from Molluscs

Antimicrobial peptides have also been characterized from mussels. These small (4 kDa), cationic, cysteine-rich antimicrobial peptides were classified into three groups, namely, defensins, mytilins, and myticins, according to their primary sequence and the organization of their cysteines (119,120). The first group comprises two *Mytilus galloprovincialis* defensins (MGDs) characterized from common mussels (*Mytilus edulis*) (121) and Mediterranean mussels (*Mytilus galloprovincialis*) (122,123). Containing six cysteine residues, the MGDs have a helical part and two antiparallel β-strands, together giving rise to the common cysteine-stabilized α-β motif stabilized by four disulfide bonds as observed in other defensins (124). The second group of molecules, mytilins, consists of five isoforms (A, B, C, D, and G-1). The isoforms A and B were isolated from *M. edulis* (121), and isoforms B, C, D, and G-1 from *M. galloprovincialis* (123). The third group of peptides includes myticins A and B, which were characterized from hemocytes and plasma of *M. galloprovincialis* (123). In addition, a 6.5 kDa antifungal peptide containing 12 cysteines has been partially characterized from *M. edulis* plasma (121). Antimicrobial activity has also been detected in the plasma of oysters (*Crassostrea virginica*). However, the identity of the agent responsible for the antimicrobial activity has not yet been determined (125).

7.2.4 Antimicrobial Peptides Derived from Enzymatic Hydrolyzates of Food Proteins

7.2.4.1 Dairy Proteins

Milk contains a wide array of antimicrobial peptides that are of value not only for the maintenance of its nutritional integrity but also for the protection of the newborn and the lactating mother (126). Hydrolytic reactions such as those catalyzed by digestive enzymes result in the release of these peptides in the digestive tract from the intact proteins in milk.

Casein comprises approximately 80% of the protein in bovine milk. Although casein does not exhibit any antimicrobial property, chymosin digestion of casein lead to the formation of many polycationic low-molecular mass peptides known as casecidins (see reviews 127 and 128). Casecidins are derived from the proteolysis of casein by chymosin at pH 6 or 7, and exhibit antimicrobial activity against staphylococci, sarcina, *Bacillus subtilis*, *Diplococcus pneumoniae*, and *Streptococcus pyogenes*. It was later confirmed that these peptides originate from both α_{s1}-casein and κ-caseins.

The digestion of α_{s1}-casein with chymosin at pH 6.4 resulted in the production of isracidin from the N-terminus with the sequence R_1PKHP-IKHQG-LPEQV-LNENL-LRF_{23} (129). Isracidin has been found to be effective against a variety of Gram-negative and Gram-positive bacteria and possesses strong antibacterial activity *in vivo* at concentrations similar to that of known antibiotics (129). This suggests that the effect of isracidin may be indirect. For instance, isracidin has been shown to protect mice against *Candida albicans* by stimulation of both phagocytosis and immune responses (127).

Casocidin-I is a 39-amino-acid fragment derived from acid treated α_{s2}-casein (K_{150}TKLT-EEEKN-RLNFL-KKISQ-RYQKF-ALPQY-LKTVY-$QHQK_{188}$) (130). The peptide inhibits the growth of *Escherichia coli* and *Staphylococcus carnosus*. In further analyses, two distinct cationic antibacterial domains (fragments L_{164}KKIS-QRYQK-FALPQ-Y_{179} and V_{183}YQHQ-KAMKP-WIQPK-TKVIP-$YVRYL_{207}$) were isolated after pepsin digestion of α_{s2}-casein. Depending on the target bacterial strain, these two α_{s2}-casein fragments exhibited minimum inhibitory concentrations between 8 and 99 μM and were more potent than casocidin-I. Being more positively charged, fragment 183–207 also exhibited a consistently higher antibacterial activity than fragment 164–179 (131).

Kappacin, commonly known as bovine glycomacropeptide (GMP), is generated from chymosin treated κ-casein (sequence 106–169) during cheese making (132,133). Kappacin is best known for its ability to improve dental health by inhibiting the growth of oral pathogens such as *Streptococcus mutans* and *Porphyromonas gingivalis* (132). Trypsin digestion of bovine β-lactoglobulin yielded four antimicrobial peptide fragments: V_{15}AGTW-Y_{20}, A_{25}ASDIS

LLDAQ-SAPLR$_{40}$, I$_{78}$PAVF-K$_{83}$, and V$_{92}$LVLD-TDYK$_{100}$ (134). All four exerted antimicrobial effects against Gram-positive bacteria only, and none showed any activity against Gram-negative bacteria or fungi. Unlike most other antimicrobial peptides, these four peptides are all negatively charged. It is therefore suggested that the repulsive forces between the negatively charged peptides and the lipopolysaccharide content of the outer membrane of Gram-negative bacteria may account for the lack of antimicrobial effects against these microorganisms. Indeed, when V$_{92}$LVLDTDYK$_{100}$ was modified by replacing the Asp98 by Arg and adding a Lys residue to the C-terminus, these modifications changed the anionic V$_{92}$LVLDTDYK$_{100}$ into a cationic peptide P*$_{92-101}$ (pI 9.7) with enhanced antimicrobial activity against the Gram-negative bacteria *Escherichia coli* and *Bordetella bronchiseptica*. This result confirms that the cationic character is very important for the bactericidal action of the peptides.

Although native bovine α-lactalbumin did not exhibit any antimicrobial activity, proteolytic digestion of α-lactalbumin by trypsin produced two antimicrobial peptides. One was a pentapeptide with the sequence E$_1$QLTK$_5$ and the other G$_{17}$YGGV SLPEW-VCTTF$_{31}$-A$_{109}$LCSE-K$_{114}$ (disulfide linked). Fragmentation of α-lactalbumin by chymotrypsin yielded C$_{61}$KDDQ-NPH$_{68}$-I$_{75}$SCDK-F$_{80}$ (also disulfide linked). In contrast, pepsin digestion of α-lactalbumin did not produce any active peptide (135). The three peptides were mostly active against Gram-positive bacteria, and *B. subtilis* was the most susceptible among the strains tested. The disulfide-linked GYGGV-SLPEW-VCTTF and ALCSE-K were the most bactericidal, while EQLTK was the least bactericidal. Lactoferricin B, which is probably the most thoroughly studied antimicrobial milk peptide (see a review 136), is released by the pepsin digestion of bovine lactoferrin (137). This 25 amino acid peptide (F$_{17}$KCRR-WQWRM-KKLGA-PSITC-VRRAF$_{41}$) forms a cationic distorted β-sheet joined together by a disulfide bridge with a cationic tail near the N-terminal (138). Lactoferricin B exhibited antimicrobial effect on a wide range of Gram-negative (139) and Gram-positive bacteria (140,141), fungi (142,143), and protozoa (144).

7.2.4.2 Egg Proteins

Hen egg white lysozyme is probably the most widely used preservative agent from a food source. This 129-residue-long protein exerts its bactericidal action by enzymatic action on the bacterial cell wall materials of Gram-positive bacteria. Recently, it was reported that heat treatment leading to the inactivation of the enzymatic action site had no effect on the bactericidal activity of lysozymes (145,146). Furthermore, genetically inactivated lysozymes also showed unaffected bactericidal activity (147). Hence, it has been hypothesized that chicken lysozyme contains peptide sequences that can induce noncatalytic bacterial death differing from enzymatic lysis of cell membrane and the antimicrobial action of lysozymes is due to structural factors (147). Secondary structure derivatives of peptide fragments at the catalytic site of chicken and human lysozymes were synthesized and tested for antimicrobial activity against several bacterial strains. It was found that these peptide fragments were potently antimicrobial against both

Gram-positive and Gram-negative bacteria and the fungus *Candida albicans* (148). The digestion of chicken lysozyme by clostripain produced a pentadecapeptide (I_{98}VSDG-NGMNA-WVAWR$_{112}$) with bactericidal activity against Gram-positive and Gram-negative bacteria (149). It was noted that this fragment forms the second helix in a helix-loop-helix motif, which is situated in the active site of lysozymes and has been observed in other bactericidal peptides (128). Further analysis of this fragment showed that shorter peptides (residues 106–112 and residues 107–112) also demonstrated strong antimicrobial activity against several foodborne pathogens (150).

Ovotransferrin, also named conalbumin, is another antimicrobial protein present in hen egg white and exhibits bactericidal properties against *Pseudomonas* sp., *Escherichia coli*, and *Streptococcus mutans* and fungicidal activity against *C.albicans* (151). Using the trypsin-nicking technique, it was found that a 92-residue cationic peptide (OTAP-92) was the bactericidal domain located at the amino-terminal of ovotransferrin. OTAP-92 exhibited bactericidal activity against Gram-positive (*S. aureus*) as well as Gram-negative (*E. coli* K-12) bacteria (152). The ability of OTAP-92 to cause dissipation in the membrane electrochemical potential of liposomes made from the phospholipids of *E. coli* suggested that OTAP-92 could permeate cell membrane and eventually lead to cell death (153).

7.2.5 Bacteriocin: Antimicrobial Peptides Produced by Bacteria

There is an increasing interest in the understanding of bacteriocins because they have been recognized as safe and effective preservative agents for food and as potential supplements or replacements for currently used antibiotics. The production of small (2- to 6-kDa) antimicrobial peptides by Gram-positive bacteria, especially lactic acid bacteria (LAB), their defense mechanisms against other organisms, and their applications in food have been well researched. Comprehensive reviews on bacteriocins can be found in the literature (154–157).

Bacteriocins are peptides produced by bacteria that display antimicrobial properties against other bacteria (158). In general, bacteriocins have been categorized into four classes (159). Class I bacteriocins (lantibiotics) are small ($<$5 kDa) peptides containing the unusual thioether amino acid lanthionine (Lan) and β-methyllanthionine as part of additional intramolecular rings. Lantibiotics often possess other dehydro amino acids such as dehydroalanine and dehydrobutyrine (160). Examples include nisin and lacticin-481 (161). Class II bacteriocins are small ($<$5 kDa) heat-stable, non-Lan-containing, membrane-active peptides that are further differentiated into subclasses (162). Subclass IIa are *Listeria*-active peptides with the N-terminal consensus sequence YGNGV and include pediocin PA-1 (163), sakacin A (164), and enterocin A (165). Subclass IIb are bacteriocins requiring two peptides for activity whereby one or both of the two peptides is inactive alone but having an enhanced antimicrobial effect in combination (155). Examples of subclass IIb bacteriocin include lactococcin G (166) and lacticin F (167). Subclass IIc are the *sec*-dependent secreted bacteriocins such as acidocin B (168).

Members of Class III are large (>30 kDa) heat-labile proteins such as helveticin J (169). Class IV, known as "complex bacteriocins," require lipid or carbohydrate moiety for activity (155).

Nisin is probably the best known bacteriocin. The prototype LAB lantibiotic nisin was first discovered in 1928, when the metabolites of *Streptococcus lactis* (now reclassified as *Lactococcus lactis*) were shown to be inhibitory to other LAB (170) with reduced or no effect against Gram-negative bacteria. Nisin is a 34-residue-long peptide that is posttranslationally modified such that serine and threonine residues are dehydrated to become dehydroalanine and dehydrobutyrine. Subsequently, five of the dehydrated residues form thioether bonds with upstream cysteine residues and produce the characteristic lanthionine rings (Figure 7.1). Nisin A and nisin Z are two naturally occurring nisin variants differing in a single amino acid at position 27; nevertheless, they have similar antimicrobial activities (170). The carboxyl terminus of nisin is responsible for the initial interaction of nisin, that is, binding to the anionic lipid of the target membrane (171,172). Consequently, the hydrophobic amino terminus would be able to insert into the membrane (173). NMR data indicated that upon association with the membrane, nisin adopts a rod-like amphipathic conformation, rendering the cell permeable to small ionic components (174) and causing cell death from loss of cell integrity upon exposure to nisin (175). Nisin was first introduced commercially as a food preservative in processed cheese products, and since then, numerous other applications in foods and beverages have been identified. It was approved as a "generally regarded as safe" (GRAS) agent by the U.S. Food and Drug Agency in 1988. Currently, nisin is recognized as a safe food preservative in approximately 50 countries (170) and is commonly used in the cheese industry to prevent the outgrowth of *Clostridia* spores and subsequent product spoilage (170).

7.3 MODES OF ANTIMICROBIAL ACTIONS

The primary sequences of the different classes and sources of antimicrobial peptides show little homology, differing in peptide length, amino acid composition,

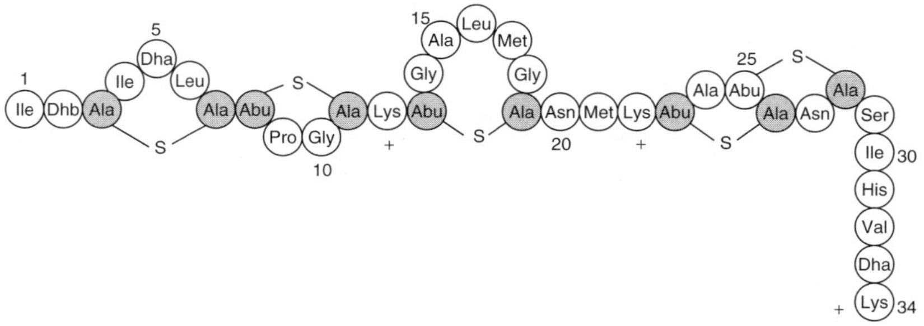

FIGURE 7.1 The primary structure of nisin Z. Abbreviations: Abu, α-butyric acid; Dha, dehydroalanine; Dhb, dehydrobutyrine, Ala–S–Ala, lanthionine; Abu–S–Ala, β-methyllanthionine. (*Source*: From E. Breukink, B. de Kruijff. *Biochim. Biophys. Acta*, 1462, 223–234, 1999. With permission.)

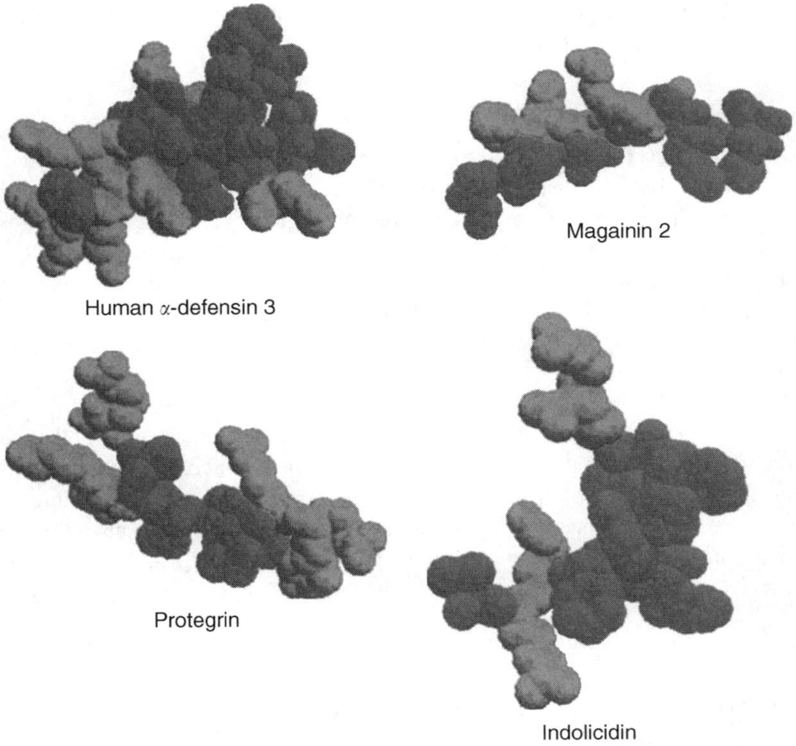

FIGURE 7.2 Amphipathicity of selected antimicrobial peptides. Positively charged amino acids are shown in light grey and hydrophobic amino acids are shown in dark grey. (*Source*: From M. Zasloff. *Nature*, 415, 389–395, 2002. With permission.)

charge, hydrophobicity, and secondary structure. Nevertheless, most antimicrobial peptides are cationic and amphipathic (Figure 7.2). Indeed, these two structural features play key roles in the antimicrobial actions exerted by these peptides.

7.3.1 Membrane Interaction

Before reaching the phospholipid membrane, peptides must traverse the negatively charged outer wall of Gram-negative bacteria, which contains lipopolysaccharides (LPS), or through the outer cell wall of Gram-positive bacteria, which contains acidic polysaccharides (teichoic acids) (176). In this mechanism, the cationic peptides initially interact with the surface LPS, competitively displacing the divalent cations that bridge and partly neutralize the LPS. This causes disruption of the outer membrane that appears as blebs when observed under the microscope (177). Studies with several antimicrobial peptides of different lengths, hydrophobicities, and structures revealed that these blebs are formed only below the minimal inhibitory concentration (MIC). At or above the MIC, however, bacteria are partially lysed and disintegrated (13,178,179).

7.3.2 Membrane Penetration

The initial interaction of cationic antimicrobial peptides with the cytoplasmic membrane involves the insertion of the peptides parallel to the membrane surface into the interface between the phospholipid head groups and fatty acid chains of the outer monolayer of this membrane. Consequently, the membrane can be rendered permeable through the formation of transmembrane pores, causing cell lysis and leading cell death (1,180). Two major models, the "barrel-stave" model and the "carpet-like" model, have been proposed to describe how the peptides interact with the membrane (Figure 7.3) (181,182).

In the "barrel-stave" model, as few as three membrane-bound peptides (amphipathic α-helix, hydrophobic α-helix, β-sheet, or both α-helix and β-sheet structures) recognize each other on the membrane surface, oligomerize, insert themselves into the hydrophilic membrane bilayer, and form transmembrane pores (see Figure 7.3A′). Hence, amphipathic peptides align perpendicular to the membrane to form the staves of a transient barrel of various sizes, forming a hydrophilic pore in its center traversing the cytoplasmic membrane (see Figure 7.3B′). This

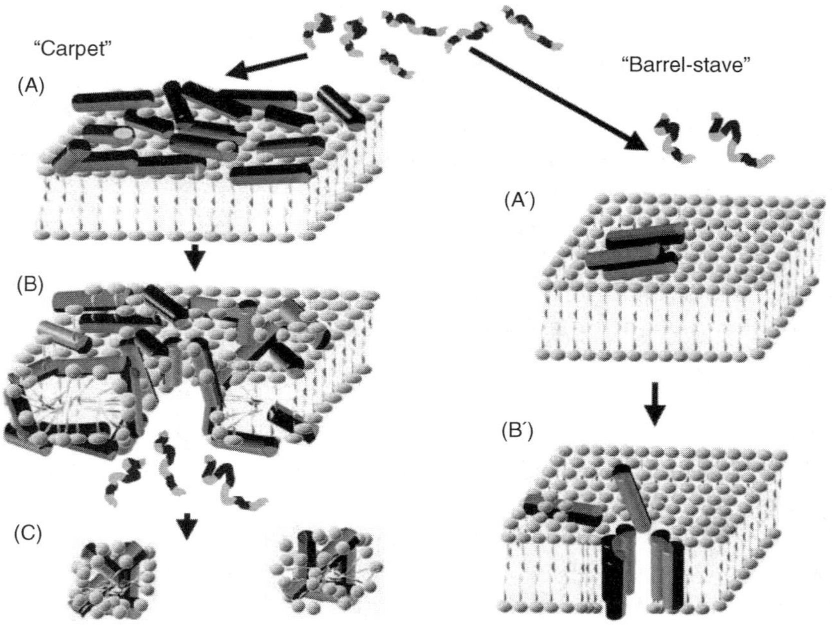

FIGURE 7.3 Membrane penetration models of antimicrobial peptides: the "barrel-stave" model on the right and the "carpet" on the left. (*Source:* From Y. Shai, Z. Oren. *Peptides*, 22, 1629–1641, 2001. With permission.)

would then lead to leakage of the cytoplasmic contents and, subsequently, cell death (181).

In the "carpet-like" model, antimicrobial peptides penetrate into the membrane using the following sequence: (1) binding of peptide monomers to the phospholipid head groups and alignment of the positively charged amino acids of the peptide monomers on the surface of the membrane so that their hydrophilic surface is facing the phospholipid headgroups or water molecules (see Figure 7.3A; (2) rotation of the molecule leading to reorientation of the hydrophobic residues toward the hydrophobic core of the membrane (see Figure 7.3B): and (3) once a threshold concentration is reached, the bilayer curvature is disrupted, leading to permeability and disintegration of the membrane (see Figure 7.3C) (182). Unlike the "barrel-stave" model, no specific peptide structure is needed in the "carpet-like" model.

7.3.3 Interactions with Other Cellular Components

In addition to acting at the cell membrane, antimicrobial peptides may also exhibit their activity against multiple potential targets such as cell division, DNA, ribonucleic acid (RNA) for protein synthesis, and autolysin activation (183). For example, PR-39 kills bacteria by inhibiting protein synthesis and inducing degradation of the proteins required for replication of DNA, rather than by a pore-forming mechanism (184). CP-10A, a derivative of indolicidin induces mesosome-like structures in the cytoplasm of Gram-positive bacteria (185). Studies have shown that DNA could be the intracellular target for tryptophan-containing antibacterial peptides. After indolicidin binds to LPS and renders the outer membrane of Gram-negative bacteria permeable, indolicidin exhibits partial inhibitory actions on DNA synthesis and protein synthesis (186–188).

Defensins have also been shown to potently inhibit phospholipid/Ca^{2+} protein kinase (PKC) and phosphorylation of endogenous proteins (189,190). Furthermore, it is believed that indolicin induces cell death by degeneration of intracellular organization, extensive cytoplasmic vacuolization, and membrane blebbing via a non-apoptotic process as shown by a lack of nuclear fragmentation or DNA laddering or dependence on caspase-like activity. Collectively, these results indicate that in addition to their effects on the membrane, indolicin induces autophagic cell death in its target microbes (180). Finally, antimicrobial bovine neutrophilic cathelicidin-like peptides, BMAP-27 and BMAP-28, have been proposed to induce apoptosis at similar concentrations as those for the antimicrobial effect (191). Whether this applies to other antimicrobial peptides is not known, but these observations could indicate an active role for the peptides in the elimination of tumor cells, virus infected cells, and cells infected with intracellular bacteria (192).

7.4 ROLES OF ANTIMICROBIAL PEPTIDES IN THE IMMUNE SYSTEM

The action of cationic antimicrobial peptides is not limited to their effect on microorganisms. Evidence has shown that antimicrobial peptides may act as chemokines for immature dendritic cells and memory T cells. Hence, antimicrobial peptides may serve as a bridge between the innate and adaptive immune systems (192–194). Figure 7.4 illustrates some of the proposed roles of antimicrobial peptides in inflammation.

After the initial lysis or penetration through the bacterial cells, antimicrobial peptides continue their actions to protect the host organisms (1,195). Antimicrobial peptides could act as signaling molecules to increase the chemotaxis of neutrophils and T helper cells resulting in leukocyte recruitment to the infection site, thus promoting nonopsonic phagocytosis at those infection sites (196). Other examples of intercellular signaling by antibacterial peptides include the stimulation of histamine release from mast cells by defensins (196,197) and cathelicidins (198,199). The effect of antimicrobial peptides on mast cells may be viewed as a coordination of innate immunity and further confirm the roles of antimicrobial peptides in immunity (192). Furthermore, antimicrobial peptides also help in minimizing tissue injury by inhibiting certain proteases. For example, a hCAP18/LL-37-like domain has been shown to exhibit an inhibitory effect on cathepsin (200).

Finally, antimicrobial peptides also play a role in wound healing. Growth-promoting activities that may be involved in wound healing have been demonstrated for neutrophilic antibacterial peptides. It has also been shown that α-defensins have an *in vitro* mitogenic effect on fibroblast and epithelial cells

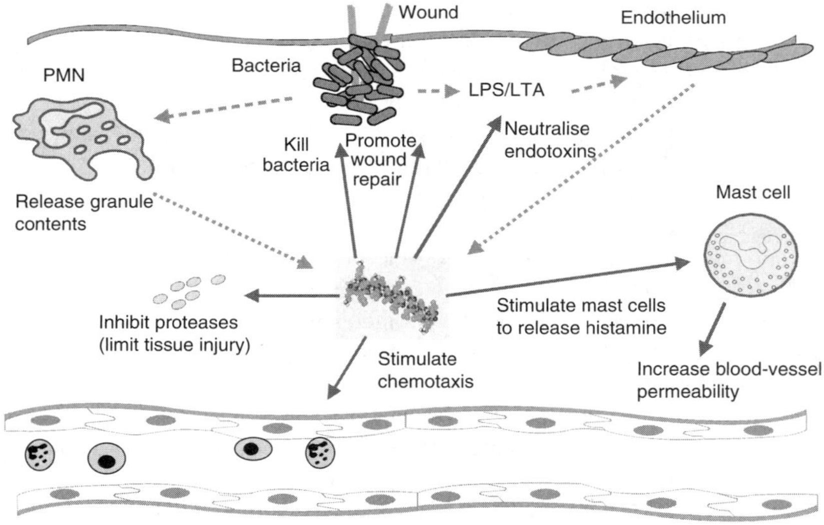

FIGURE 7.4 Proposed roles of antimicrobial peptides in chronic inflammation. Dotted arrows represent events that lead to increased production of extracellular antimicrobial peptides; solid arrows indicate actions of the peptides; and dashed arrows show events due to the bacteria. The overall scheme presented is a mosaic of the separate effects. Abbreviations: LPS, lipopolysaccharide; LTA, lipoteichoic acid; and PMN, polymorphonuclear leucocytes. (*Source*: From R.E. Hancock. *Lancet Infect. Dis.*, 1, 156–164, 2001. With permission.)

(201). More detailed studies have been performed on the porcine cathelicidin PR-39 that has been found to induce the heparan sulfate extracellular matrix proteins syndecan-1 and -4 (202); the PR-39-mediated induction of syndecans mimics their expression in the wound healing process. Neutrophilic peptides such as hCAP-18/LL-37 and human β-defensin-3 are known to be present in human wound fluid (192,203,204), possibly performing a dual role in keeping the wound sterile and promoting the healing process (205).

7.5 CURRENT APPLICATIONS

7.5.1 The Use of Antimicrobial Peptides in Agriculture

Advances in genetic engineering have made possible the development of plants with new phenotypes. Efforts have been made to give antimicrobial properties to plants in order to enhance their resistance against microorganisms and diseases in their natural environment. For instance, genes encoding antimicrobial peptides have been engineered into crops such as potato (206), pear (207), banana (208), and tomato (209) to enhance their resistance against natural fungal and microbial diseases. Among all plants, the tobacco plant (*Nicotiana tabacum*) has been the most intensively studied. Antimicrobial peptides from both nonplants (e.g., tobacco budworm [210], fruit fly [210], and frog [211]) and plants (e.g., barley [212,213], ornamental flowers [214], and pea [215]) have been strategically engineered into tobacco plants, and antimicrobial peptides have been successfully expressed and released by the transgenic plants. Although purified antimicrobial peptides from transgenic plants exhibited similar levels of antimicrobial activity to that of the authentic peptides *in vitro*, transgenic plants often demonstrated unpredictable antimicrobial activity *in vivo*. Only minor or no increase in resistance against pathogens has been observed (210,214).

The poor *in vivo* efficacy has been correlated to peptide degradation caused by plant proteases (216) or inhibition of antimicrobial activity caused by divalent cations (214). The challenge is how to improve the *in planta* efficacy of existing genes to use or to discover molecules with exceptional activity and stability toward plant cell composition. Recent efforts have focused on the use of mutated, hybrid, or synthetic peptides with less sensitivity to plant proteases to enhance the efficacy of transgenic antimicrobial peptides for field applications (217–219).

7.5.2 The Use of Antimicrobial Peptides in Foods

Although many antimicrobial peptides have demonstrated strong *in vitro* antimicrobial activity, the *in vivo* effectiveness of antimicrobial peptides as preservative in food systems is often diminished because of the complexity of the natural environment. For instance, while nisin exhibited antimicrobial activity against *E. coli* and *Salmonella* spp. in buffer (220), no significant effect was observed when trials were conducted on beef (221,222). Similarly, the combined effect of lysozyme and nisin against lactic acid bacteria was enhanced when the growth medium was diluted eightfold, whereas no effect was observed in a pork juice medium under the same conditions (223). The antimicrobial activity of lactoferrin hydrolysate was also inhibited in richer media such as tryptic soy broth, milk-based infant formula, and soy-based infant formula (224) and in the presence of carrot juice (225). The antimicrobial properties could have been inhibited by minerals, proteins, and other substances that are naturally present in foods. New strategies are being used to maximize the potential of antimicrobial peptides in food systems.

The synergistic effects of nisin and other preservative agents such as lysozyme (223,226), lactic acid (227–229), ethylenediaminetetraacetic acid (EDTA) (230), and carbon dioxide (231) have been examined. It was found that a 3:1 mixture of lysozyme and nisin acted as an effective agent against bacteria, even in the presence of salt (223). Subsequently, Cutter and Siragusa (232–235) conducted a series of experiments to assess the use of nisin alone or nisin–lysozyme mixtures to inhibit the growth of pathogenic bacteria on raw meat with different packaging methods and different storage conditions. Data from these studies demonstrated that nisin spray treatments could increase shelf life by suppressing or inhibiting the growth of undesirable bacteria present on fresh beef (232,233). Immobilization of nisin in an edible calcium alginate gel or a meat-binding system could further enhance the antimicrobial effect on fresh beef (233,235). The incorporation of nisin into the edible coating of packaging films was found to be an effective way to inhibit the growth of meat-spoiling organisms and pathogens on meat surfaces (236,237). The addition of nisin into polyethylene and polyethylene oxide films also inhibited the growth of the psychrotrophic, meat-spoiling organism *Brochothrix thermosphacta* on vacuum-packaged beef surfaces under long-term, refrigerated storage (238,239).

A recent study has evaluated the effects of antimicrobial treatments on the sensory properties of fresh pork loins (240). Off-odor intensity and odor acceptability of the loins stored in vacuum were not affected by this treatment. However, the antimicrobial treatment did have an impact on off-odor intensity and odor acceptability after the loin chops were transferred to an aerobic retail case. A more detailed analysis of the population of lactic acid bacteria and *Enterobacteriaceae* during transfer from an anoxic environment to an aerobic one would be needed to fully understand the impact on the sensory properties of products treated with antimicrobial peptides (240).

7.5.3 The Use of Antimicrobial Peptides for Clinical Applications

Antimicrobial peptides are interesting candidates for investigation and development as therapeutic agents for topical or systemic administration. The broad spectrum of activity and low incidence of development of bacterial resistance with repeated exposure are attractive features of antimicrobial peptides as a new type of antibiotic. Reviews on the clinical application of antimicrobial peptides can be found in the literature (241–243). The efficacy of various antimicrobial peptides have been tested in animal models of peritoneal infection (244), fungal infection (245), pneumonia (246), septic shock (247,248), oral mucositis (249), and infection-associated obstructive jaundice (250), in cancer patients undergoing chemotherapy, and in organ transplant patients (251,252). Some of these antimicrobial effects have been investigated in clinical trials (242,243). The use of antimicrobial peptides on the prevention of catheter sepsis caused by microbes and on the "superbug" that now resists most common current antibiotics was investigated by MIGENIX Inc. in phase III clinical trials (http://www.migenix.com/, accessed on May 31, 2005). The effects of these peptides on acne have also been examined. In phase III trials, protegrin-derived peptide IB-367 was used as a topical application to treat oral ulcers (oral mucositis) in cancer patients undergoing chemotherapy or radiotherapy (253). Intrabiotics has completed a phase I clinical trial on the efficacy of the aerosol form of protegrin for its preventative effect against *Pseudomonas* infections in cystic fibrosis patients. An oral solution of protegrin is also being tested in phase II/III trials in pneumonia patients on ventilators (http://www.intrabiotics.com, accessed on May 31, 2005). Other trials have included the use of the lantibiotic peptide nisin to treat *Helicobacter pylori* stomach ulcers and magainin for the treatment of foot-ulcer infections in diabetics (242).

7.6 CONCLUSIONS

While preparing this chapter, we noticed that novel antimicrobial peptides are being reported in the literature almost every week (if not everyday) and that more sophisticated investigations are being conducted to better elucidate the fundamental properties and actions of selective peptides such as nisin, protegrin, and magainin. Unfortunately, although many antimicrobial peptides exhibit effects against microbes under *in vitro* conditions, to date, only a few have demonstrated promising protective effects to the host under *in vivo* environments. Innovative strategies are therefore needed to identify new peptide candidates and design antimicrobial peptides with enhanced stability in the host in order to maximize their potential in agriculture, in food systems, and in medical applications.

ACKNOWLEDGMENTS

The authors gratefully acknowledge funding from the Natural Sciences and Engineering Research Council of Canada and the Dairy Farmers of Canada, as well as a University Graduate Fellowship from the University of British Columbia to J.C.K. Chan.

REFERENCES

1. M. Zasloff. Antimicrobial peptides of multicellular organisms. *Nature*, 415: 389–395, 2002.
2. H.I. Zeya, J.K. Spitznagel. Cationic proteins of polymorphonuclear leukocyte lysosomes. II. Composition, properties, and mechanism of antibacterial action. *J. Bacteriol.*, 91: 755–762, 1966.

3. H.I. Zeya, J.K. Spitznagel. Cationic proteins of polymorphonuclear leukocyte lysosomes. I. Resolution of antibacterial and enzymatic activities. *J. Bacteriol.*, 91: 750–754, 1966.
4. R.I. Lehrer, T. Ganz. Defensins of vertebrate animals. *Curr. Opin. Immunol.*, 14: 96–102, 2002.
5. A.J. Ouellette, C.L. Bevins. Paneth cell defensins and innate immunity of the small bowel. *Inflamm. Bowel Dis.*, 7: 43–50, 2001.
6. M.V. Sawai, H.P. Jia, L. Liu, V. Aseyev, J.M. Wiencek, P.B.J. McCray, T. Ganz, W.R. Kearney, B.F. Tack. The NMR structure of human beta-defensin-2 reveals a novel alpha-helical 14 segment. *Biochemistry*, 40: 3810–3816, 2001.
7. C. Hill, J. Yee, M.E. Selsted, D. Eisenberg. Crystal structure of defensin HNP-3, an amphiphilic dimer: mechanisms of membrane permeabilization. *Science*, 251: 5000, 1991.
8. Y.-Q. Tang, J. Yuan, G. Ösapay, K. Ösapay, D. Tran, C.J. Miller, A.J. Ouellette, M.E. Selsted. A cyclic antimicrobial peptide produced in primate leukocytes by the ligation of two 19 truncated alpha-defensin. *Science*, 286: 498–502, 1999.
9. M. Trabi, H.J. Schirra, D.J. Craik. Three-dimensional structure of RTD-1, a cyclic antimicrobial defensin from Rhesus macaque leukocytes. *Biochemistry*, 40: 4211–4221, 2001.
10. W. Wang, A.M. Cole, T. Hong, A.J. Waring, R.I. Lehrer. Retrocyclin, an antiretroviral theta-defensin, is a lectin. *J. Immunol.*, 170: 4708–4716, 2003.
11. M. Zanetti, R. Gennaro, D. Romeo. Cathelicidins: a novel protein family with a common proregion and a variable C-terminal antimicrobial domain. *FEBS Lett.*, 374: 1–5, 1995.
12. A. Tossi, L. Sandri, A. Giangaspero. Amphipathic, alpha-helical antimicrobial peptides. *Biopolymers*, 55: 4–30, 2000.
13. B. Ramanathan, E.G. Davis, C.R. Ross, F. Blecha. Cathelicidins: microbicidal activity, mechanisms of actions, and roles in innate immunity. *Microb. Infect.*, 4: 361–372, 2002.
14. P. Storici, M. Scocchi, A. Tossi, R. Gennaro, M. Zanetti. Chemical synthesis and biological activity of a novel antibacterial peptide deduced from a pig myeloid cDNA. *FEBS Lett.*, 337: 303–307, 1994.
15. H. Wu, G. Zhang, C.R. Ross, F. Blecha. Cathelicidin gene expression in porcine tissues: roles in ontogeny and tissue specificity. *Infect. Immun.*, 67: 439–442, 1999.
16. G. Zhang, C.R. Ross, F. Blecha. Porcine antimicrobial peptides: new prospects for ancient molecules of host defense. *Vet. Res.*, 31: 277–296, 2000.
17. M.M. Mahoney, A.Y. Lee, D.J. Brezinski-Caliguri, K.M. Huttner. Molecular analysis of the 16 sheep cathelin family reveals a novel antimicrobial peptide. *FEBS Lett.*, 377: 519–522, 1995.
18. B. Skerlavaj, M. Benincasa, A. Risso, M. Zanetti, R. Gennaro. SMAP-29: a potent antibacterial and antifungal peptide from sheep leukocytes. *FEBS Lett.*, 463: 58–62, 1999.
19. J. Larrick, M. Hirata, Y. Shimomoura, M. Yoshida, H. Zheng, J. Zhong, S. Wright. Antimicrobial activity of rabbit CAP18-derived peptides. *Antimicrob. Agents Chemother.*, 37: 2534–2539, 1993.
20. K.A. Zarember, S.S. Katz, B.F. Tack, L. Doukhan, J. Weiss, P.H. Elsbach. Host defense 1 functions of proteolytically processed and parent (unprocessed) cathelicidins of rabbit granulocytes. *Infect. Immun.*, 70: 569–576, 2002.

21. R. Gennaro, M. Scocchi, L. Merluzzi, M. Zanetti. Biological characterization of a novel mammalian antimicrobial peptide. *Biochim. Biophys. Acta*, 1425: 361–368, 1998.
22. J.W. Larrick, M. Hirata, R.F. Balint, J. Lee, J. Zhong, S.C. Wright. Human CAP18: a novel antimicrobial lipopolysaccharide-binding protein. *Infect. Immun.*, 63: 1291–1297, 1995.
23. J.B. Cowland, A.H. Johnsen, N. Borregaard. hCAP-18, a cathelin/pro-bactenecin-like protein of human neutrophil specific granules. *FEBS Lett.*, 368: 173–176, 1995.
24. A.E. Popsueva, M.V. Zinovjeva, J.W. Visser, J.M. Zijlmans, W.E. Fibbe, A.V. Belyavsky. A novel murine cathelin-like protein expressed in bone marrow. *FEBS Lett.*, 391: 5–8, 1996.
25. R.L. Gallo, K.J. Kim, M. Bernfield, C.A. Kozak, M. Zanetti, L. Merluzzi, R. Gennaro. Identification of CRAMP, a cathelin-related antimicrobial peptide expressed in the embryonic and adult mouse. *J. Biol. Chem.*, 272: 13088–13093, 1997.
26. R.W. Frank, R. Gennaro, K. Schneider, M. Przybylski, D. Romeo. Amino acid sequences of two proline-rich bactenecins. Antimicrobial peptides of bovine neutrophils. *J. Biol. Chem.*, 265: 18871–18874, 1990.
27. B. Agerberth, J.Y. Lee, T. Bergman, M. Carlquist, H.G. Boman, V. Mutt, H. Jornvall. Amino acid sequence of PR-39. Isolation from pig intestine of a new member of the family of proline-arginine-rich antibacterial peptides. *Eur. J. Biochem.*, 202: 849–854, 1991.
28. D. Romeo, B. Skerlavaj, M. Bolognesi, R. Gennaro. Structure and bactericidal activity of an antibiotic dodecapeptide purified from bovine neutrophils. *J. Biol. Chem.*, 263: 9573–9575, 1988.
29. V.N. Kokryakov, S.S.L. Harwig, E.A. Panyutich, A.A. Schevchenko, G.M. Aleshina, O.V. Shamova, H.A. Korneva, R.I. Lehrer. Protegrins: leukocyte antimicrobial peptides that combine features of corticostatic defensins and tachyplesins. *FEBS Lett.*, 327: 231–236, 1993.
30. M.E. Selsted, M.J. Novotny, W.L. Morris, Y.-Q. Tang, W. Smith, J.S. Cullor. Indolicidin, a novel bactericidal tredecapeptide amide from neutrophils. *J. Biol. Chem.*, 267, 1992.
31. S.M. Travis, N.N. Anderson, W.R. Forsyth, C. Espiritu, B.-A.D. Conway, E.P. Greenberg, P.B.J. McCray, R.I. Lehrer, M.J. Welsh, B.F. Tack. Bactericidal activity of mammalian cathelicidin-derived peptides. *Infect. Immun.*, 68: 2748–2755, 2000.
32. L. Saiman, S. Tabibi, T.D. Starner, P. San Gabriel, P.L. Winokur, H.P. Jia, P.B. McCray, Jr., B.F. Tack. Cathelicidin peptides inhibit multiply antibiotic-resistant pathogens from patients with cystic fibrosis. *Antimicrob. Agents Chemother.*, 45: 2838–2844, 2001.
33. F. García-Olmedo, A. Molina, J.M. Alamillo, P. Rodríguez-Palenzuéla. Plant defense peptides. *Biopolymers (Pept. Sci.)*, 47: 479–491, 1998.
34. F. García-Olmedo, P. Rodríguez-Palenzuela, A. Molina, J.M. Alamill, E. López-Solanilla, M. Berrocal-Lobo, C. Poza-Carrión. Antibiotic activities of peptide, hydrogen peroxide and peroxynitrite in plant defence. *FEBS Lett.*, 498: 219–222, 2001.
35. W.F. Broekaert, F.R.G. Terras, B.P.A. Cammue, R.W. Osborn. Plant defensins: novel antimicrobial peptides as components of the host defense system. *Plant Physiol.*, 108: 1353–1358, 1995.

36. F.J. Colilla, A. Rocher, E. Mendez. Gamma-purothionins: amino acid sequence of two polypeptides of a new family of thionins from wheat endosperm. *FEBS Lett.*, 270: 191–194, 1990.
37. E. Mendez, A. Moreno, F. Colilla, F. Pelaez, G.G. Limas, R. Mendez, F. Soriano, M. Salinas, C. De Haro. Primary structure and inhibition of protein synthesis in eukaryotic cell-free system of a novel thionin, γ-hordothionin, from barley endosperm. *Eur. J. Biochem.*, 194: 533–539, 1990.
38. Y. Zhang, K. Lewis. Fabatins: new antimicrobial plant peptides. *FEMS Microbiol. Lett.*, 149: 59–64, 1997.
39. F.R.G. Terras, H.M.E. Schoofs, M.F.C. De Bolle, F. Van Leuven, S.B. Rees, J. Vanderleyden, B.P.A. Cammue, W.F. Broekaert. Analysis of two novel classes of plant antifungal proteins from radish (*Raphanus sativus* L.) seeds. *J. Biol. Chem.*, 267: 15301–15309, 1992.
40. A.L. Vilas Alves, G.W. Samblanx, F.R.G. Terras, B.P.A. Cammue, W. Broekaert. Expression of functional *Raphanus sativus* antifungal protein in yeast. *FEBS Lett.*, 348: 228–232, 1994.
41. A. Segura, M. Moreno, A. Molina, F. García-Olmedo. Novel defensin subfamily from spinach (*Spinacia oleracea*). *FEBS Lett.*, 435: 159–162, 1998.
42. M.S. Almeida, K.M.S. Cabral, R.B. Zingali, E. Kurtenbach. Characterization of two novel defense peptides from pea (*Pisum sativum*) seeds. *Arch. Biochem. Biophys.*, 378: 278–286, 2000.
43. M.S. Almeida, K.M.S. Cabral, E. Kurtenbach, F.C.L. Almeida, A.P. Valente. Solution structure of *Pisum sativum* defensin 1 by high resolution NMR: plant defensins, identical backbone with different mechanisms of action. *J. Mol. Biol.*, 315: 749–757, 2002.
44. M. Moreno, A. Segura, F. García-Olmedo. Pseudothionin-St1, a potato peptide active against potato pathogen. *Eur. J. Biochem.*, 223: 135–139, 1994.
45. W.J. Stiekema, F. Heidekamp, W.G. Dirkse, J. van Beckum, P. de Haan, C. ten Bosch, J.D. Louwerse. Molecular cloning and analysis of four potato tuber mRNAs. *Plant Mol. Biol.*, 11: 255–269, 1988.
46. B. Meyer, G. Houlne, J. Pozueta-Romero, M.L. Schantz, R. Schantz. Fruit-specific expression of a defensin-type gene family in bell pepper. *Plant Physiol.*, 112: 615–622, 1996.
47. M. Aluru, J. Curry, M.A. O'Connell. Nucleotide sequence of a defensin of gamma-thionin-like gene from habanera chili. *Plant Physiol.*, 120: 633, 1999.
48. Q. Gu, E.E. Kawata, M.-J. Morse, H.-M. Wu, A.Y. Cheung. A flower-specific cDNA encoding a novel thionin in tobacco. *Mol. Gen. Genet.*, 234: 89–96, 1992.
49. F.T. Lay, F. Brugliera, M.A. Anderson. Isolation and properties of floral defensins from ornamental tobacco and petunia. *Plant Physiol.*, 131: 1283–1293, 2003.
50. B. Karunanandaa, A. Singh, T.-H. Kao. Characterization of a predominantly pistil-expressed gene encoding a gamma-thionin-like protein of *Petunia inflata*. *Plant Mol. Biol.*, 26, 1994.
51. J.-P. Douliez, T. Michon, K. Elmorjani, D. Marion. Structure, biological and technological functions of lipid fransfer proteins and indolines, the major lipid binding proteins from cereal kernels. *J. Cereal Sci.*, 32: 1–20, 2000.
52. K.K. Nielsen, J.E. Nielsen, S.M. Madrid, J.D. Mikkelsen. New antifungal proteins from sugar beet (*Beta vulgaris* L.) showing homology to non-specific lipid transfer proteins. *Plant Mol. Biol.*, 31: 539–552, 1996.

53. J.-C. Kader. Lipid-transfer proteins in plants. *Annu. Rev. Plant Physiol. Plant Mol. Biol.*, 47: 627–654, 1996.
54. M.S. Castro, I.R. Gerhardt, S. Orrù, P. Pucci, C. Bloch Jr. Purification and characterization of a small (7.3 kDa) putative lipid transfer protein from maize seeds. *J. Chromatrogr.*, B 794: 109–114, 2003.
55. A. Molina, A. Segura, F. García-Olmedo. Lipid transfer protein (nsLTPs) from barley and maize leaves are potent inhibitors of bacterial and fungal plant pathogens. *FEBS Lett.*, 316: 119–122, 1993.
56. A. Segura, M. Moreno, F. García-Olmedo. Purification and antipathogenic activity of lipid transfer proteins from the leaves of *Arabidopsis* and spinach. *FEBS Lett.*, 3: 243–246, 1993.
57. J. Pyee, H. Yu, P.E. Kolattukudy. Identification of a lipid transfer protein as the major protein in the surface wax of broccoli (*Brassica oleracea*) leaves. *Arch. Biochem. Biophys.*, 311: 460–468, 1994.
58. B.L. Archer. The proteins of *Hevea brasiliensis* latex. Isolation and characterization of crystalline hevein. *Biochemistry*, J 75: 236–240, 1960.
59. J.C. Koo, S.Y. Lee, H.J. Chun, Y.H. Cheong, J.S. Choi, S. Kawabata, M. Miyagi, S. Tsunasawa, K.S. Ha, D.W. Bae, C.-D. Han, B.L. Lee, M.J. Cho. Two hevein homologs isolated from the seed of *Pharbitis nil L.* exhibit potent antifungal activity. *Biochim. Biophys. Acta*, 1382: 80–90, 1998.
60. X. Huang, W.-J. Xie, Z.-Z. Gong. Characteristics and antifungal activity of a chitin binding protein from *Ginkgo biloba*. *FEBS Lett.*, 478: 123–126, 2000.
61. S.-S. Li, P. Claeson. Cys/Gly-rich proteins with a putative single chitin-binding domain from oat (*Avena sativa*) seeds. *Phytochemistry*, 63: 249–255, 2003.
62. K.P.B. Van den Bergh, P. Proost, J. Van Damme, J. Coosemans, E.J.M.V. Damme, W.J. Peumans. Five disulfide bridges stabilize a hevein-type antimicrobial peptide from the bark of spindle tree (*Euonymus europaeus* L.). *FEBS Lett.*, 530: 181–185, 2002.
63. A. Kiba, H. Saitoh, M. Nishihara, K. Omiya, S. Yamamura. C-Terminal domain of a hevein- like protein from *Wasabia japonica* has potent antimicrobial activity. *Plant Cell. Physiol.*, 44: 296–303, 2003.
64. B.P.A. Cammue, M.F.C. De Bolle, F.R.G. Terras, P. Proost, J. Van Damme, S.B. Rees, J. Vanderleyden, W.F. Broekaert. Isolation and characterization of a novel class of plant antimicrobial peptides from *Mirabilis jalapa* L. seeds. *J. Biol. Chem.*, 267: 2228–2233, 1992.
65. M.F. De Bolle, K. Eggermont, R.E. Duncan, R.W. Osborn, F.R. Terras, W.F. Broekaert. Cloning and characterization of two cDNA clones encoding seed-specific antimicrobial peptides from *Mirabilis jalapa* L. *Plant Mol. Biol.*, 28: 713–721, 1995.
66. F. Shao, Z. Hu, Y.-M. Xiong, Q.-Z. Huang, C.-G. Wang, R.-H. Zhu, D.-C. Wang. A new antifungal peptide from the seeds of *Phytolacca americana*: characterization, amino acid sequence and cDNA cloning. *Biochim. Biophys. Acta*, 1430: 262–268, 1999.
67. G.-H. Gao, W. Liu, J.-X. Dai, J.-F. Wang, Z. Hu, Y. Zhang, D.-C. Wang. Solution structure of PAFP-s: a new knottin-type antifungal peptide from the seeds of *Phytolacca americana*. *Biochemistry*, 40: 10973–10978, 2001.
68. R.H. Tailor, D.P. Acland, S. Attenborough, B.P.A. Cammue, I.J. Evans, R.W. Osborn, J.A. Ray, S.B. Rees, W.F. Broekaert. A novel family of small cysteine-rich antimicrobial peptides from seed of *Impatiens balsamina* is derived from a single precursor protein. *J. Biol. Chem.*, 272: 24480–24487, 1997.

69. S.U. Patel, R.W. Osborn, S. Rees, J.M. Thornton. Structural studies of *Impatiens balsamina* antimicrobial protein (Ib-AMP1). *Biochemistry*, 37: 983–990, 1998.
70. A. Segura, M. Moreno, F. Madueno, A. Molina, F. García-Olmedo. Snakin-1, a peptide from potato that is active against plant pathogens. *Mol. Plant-Microb. Interact.*, 12: 16–23, 1999.
71. M. Berrocal-Lobo, A. Segura, M. Moreno, G. López, F. García-Olmedo, A. Molina. Snakin-2, an antimicrobial peptide from potato whose gene is locally induced by wounding and responds to pathogen infection. *Plant Physiol.*, 128: 951–961, 2002.
72. T. Ando, S. Ishii, M. Yamasaki, K. Iwai, C. Hashimoto, F. Sawada. Protamines. I. Amino acid composition and homogeneity of clupeine, salmine, and iridine. *J. Biochem. (Tokyo)*, 44: 275–288, 1957.
73. K. Iwai, C. Nakahara, T. Ando. Protamines. XV. Complete amino acid sequence of the Z component of clupeine. Application of N→O acyl rearrangement and selective hydrolysis in sequence determination. *J. Biochem. (Tokyo)*, 69: 493–509, 1971.
74. K. Suzuki, T. Ando. Protamines. XVI. Complete amino acid sequence of clupeine YII. *J. Biochem. (Tokyo)*, 72: 1419–1432, 1972.
75. K. Suzuki, T. Ando. Protamines. XVII. Complete amino acid sequence of clupeine YI. *J. Biochem. (Tokyo)*, 72: 1433–1446, 1972.
76. N.M.D. Islam, T. Itakura, T. Motohiro. Antibacterial spectra and minimum inhibition concentration of clupeine and salmine. *Bull. Jpn. Soc. Sci. Fish.*, 50: 1705–1708, 1984.
77. M. Uyttendaele, J. Debevere. Evaluation of the antimicrobial activity of protamine. *Food Microbiol.*, 11: 417–427, 1994.
78. L. Truelstrup Hansen, T.A. Gill. Solubility and antimicrobial efficacy of protamine on *Listeria monocytogenes* and *Escherichia coli* as influenced by pH. *J. Appl. Microbiol.*, 88: 1049–1055, 2000.
79. L. Truelstrup Hansen, J.W. Austin, T.A. Gill. Antibacterial effect of protamine in combination with EDTA and refrigeration. *Int. J. Food Microbiol.*, 66: 149–161, 2001.
80. C. Johansen, T.A. Gill, L. Gram. Changes in cell morphology of *Listeria monocytogenes* and *Shewanella putrefaciens* resulting from the action of protamine. *Appl. Environ. Microbiol.*, 62: 1058–1064, 1996.
81. C. Johansen, A. Verheul, L. Gram, T.A. Gill, T. Abee. Protamine-induced permeabilization of cell envelopes of Gram-positive and Gram-negative bacteria. *Appl. Environ. Microbiol.*, 63: 1155–1159, 1997.
82. S.A. Thompson, K. Tachibana, K. Nakanishi, I. Kubota. Melittin-like peptides from the shark-repelling defense secretion of the sole *Pardachirus pavoninus*. *Science*, 233: 341–343, 1986.
83. R. Pal, Y. Barenholz, R.R. Wagner. Pardaxin, a hydrophobic toxin of the Red Sea flatfish, disassembles the intact membrane of vesicular stomatitis virus. *J. Biol. Chem.*, 256: 10209–10212, 1981.
84. Y. Shai, J. Fox, C. Caratsch, Y.-L. Shih, C. Edwards, P. Lazarovici. Sequencing and synthesis of pardaxin, a polypeptide from the Red Sea Moses sole with ionophore activity. *FEBS Lett.*, 242: 161–166, 1988.
85. P. Lazarovici. The structure and function of pardaxin. *J. Toxicol. Toxin Rev.*, 21: 391–421, 2002.
86. Z. Oren, Y. Shai. A class of potent antibacterial peptides derived from pardaxin, a pore forming peptide isolated from Moses sole, *Paradachirus marmoratus*. *Eur. J. Biochem.*, 237: 303–310, 1996.

87. A.M. Cole, P. Weis, G. Diamond. Isolation and characterization of pleurocidin, an antimicrobial peptide in the skin secretions of winter flounder. *J. Biol. Chem.*, 272: 12008–12103, 1997.
88. S.E. Douglas, J.W. Gallant, Z.-Z. Gong, C. Hew. Cloning and developmental expression of a family of pleurocidin-like antimicrobial peptides from winter flounder, *Pleuronectes americanus* (Walbaum). *Dev. Comp. Immunol.*, 25: 137–147, 2001.
89. A.M. Cole, R.O. Darouiche, D. Legarda, N. Connell, G. Diamond. Characterization of a fish antimicrobial peptide: gene expression, subcellular localization, and spectrum of activity. *Antimicrob. Agents. Chemother.*, 44: 2039–2045, 2000.
90. N. Saint, H. Cadiou, Y. Bessin, G. Molle. Antibacterial peptide pleurocidin forms ion channels in planar lipid bilayers. *Biochim. Biophys. Acta*, 1564: 359–364, 2002.
91. K. Yoshida, Y. Mukai, T. Niidome, T. Hatakeyama, H. Aoyagi. Property and activity of amphiphilic antibacterial peptide pleurocidin. *Pept. Sci.*, 35, 1999.
92. S. Inoue, B.-H. Nam, I. Hirono, T. Aoki. A survey of expressed genes in Japanese flounder (*Paralichthys olivaceus*) liver and spleen. *Mol. Mar. Biol. Biotechnol.*, 6: 376–380, 1997.
93. S.E. Douglas, J.W. Gallant, C.E. Bullerwell, C. Wolff, J. Munholland, M.E. Reith. Winter flounder expressed sequence tags: establishment of an EST database and identification of novel fish genes. *Mar. Biotechnol.*, 1: 458–464, 1999.
94. S.E. Douglas, J.W. Gallant, R.S. Liebscher, A. Dacanay, S.C.M. Tsoi. Identification and expression analysis of hepcidin-like antimicrobial peptides in bony fish. *Dev. Comp. Immunol.*, 27: 589–601, 2003.
95. A.Y. Gracey, J.V. Troll, G.N. Somero. Hypoxia-induced gene expression profiling in the euryoxic fish *Gillichthys mirabilis*. *Proc. Natl. Acad. Sci. USA*, 98: 1993–1998, 2001.
96. H. Shike, X. Lauth, M.E. Westerman, V.E. Ostland, J.M. Carlberg, J.C. Van Olst, C. Shimizu, P. Bulet, J.C. Burns. Bass hepcidin is a novel antimicrobial peptide induced by bacterial challenge. *Eur. J. Biochem.*, 269: 2232–2237, 2002.
97. C.B. Park, J.M. Lee, I.Y. Park, M.S. Kim, S.C. Kim. A novel antimicrobial peptide from the loach, *Misgurnus anguillicaudatus*. *FEBS Lett.*, 411: 173–178, 1997.
98. I.Y. Park, C.B. Park, M.S. Kim, S.C. Kim. Parasin I, an antimicrobial peptide derived from histone H2A in the catfish, *Parasilurus asotus*. *FEBS Lett.*, 437: 258–262, 1998.
99. G. Basañez, A.E. Shinnar, J. Zimmerberg. Interaction of hagfish cathelicidin antimicrobial peptides with model lipid membranes. *FEBS Lett.*, 532: 115–120, 2002.
100. X. Lauth, H. Shike, J.C. Burns, M.E. Westerman, V.E. Ostland, J.M. Carlberg, J.C.V. Olst, V. Nizet, S.W. Taylor, C. Shimizu, P. Bulet. Discovery and characterization of two isoforms of moronecidin, a novel antimicrobial peptide from hybrid striped bass. *J. Biol. Chem.*, 277: 5030–5039, 2002.
101. E.J. Noga, T.A. Arroll, R.A. Bullis, L. Khoo. Antibacterial activity in hemolymph of white shrimp *Penaeus setiferus*. *J. Mar. Biotechnol.*, 4: 181–184, 1996.
102. J. Pan, A. Kurosky, B. Xu, A.K. Chopra, D.H. Coppenhaver, I.P. Singh, S. Baron. Broad antiviral activity in tissue of crustaceans. *Antivir. Res.*, 48: 39–47, 2000.
103. J.R.S. Chisholm, V.J. Smith. Camparison of antibacterial activity in the hemocytes of different crustacean species. *Comp. Biochem. Physiol.*, A110: 39–45, 1995.
104. E.J. Noga, T.A. Arroll, Z. Fan. Specificity and some physiochemical characteristics of the antibacterial activity from the blue crab *Callinectes sapidus*. *Fish Shellfish Immunol.*, 6, 1996.

105. T. Haug, A.K. Kjuul, K. Stensvåga, E. Sandsdalen, O.B. Styrvold. Antibacterial activity in four marine crustacean decapods. *Fish Shellfish Immunol.*, 12: 371–385, 2002.
106. D. Schnapp, G.D. Kemp, V.J. Smith. Purification and characterization of a proline-rich antibacterial peptide with sequence similarity to bactenecin-7, from the haemocytes of the shore crab, *Carcinus maenas*. *Eur. J. Biochem.*, 240: 532–539, 1996.
107. J.M. Relf, J.R.S. Chisholm, G.D. Kemp, V.J. Smith. Purification and characterization of a cysteine-rich 5-kDa antibacterial protein from the granular haemocytes of the shore crab, *Carcinus maenas*. *Eur. J. Biochem.*, 264: 350–357, 1999.
108. L. Khoo, D.W. Robinette, E.J. Noga. Callinectin an antibacterial peptide from the blue crab, *Callinectes sapidus*, hemocytes. *Mar. Biotechnol.*, 1: 44–51, 1999.
109. T. Nakamura, H. Furunaka, T. Miyata, F. Tokunaga, T. Muta, S. Iwanaga, M. Niwa, T. Takao, Y. Shimonishi. Tachyplesin, a class of antimicrobial peptide from the hemocytes of the horseshoe crab (*Tachypleus tridentatus*). Isolation and chemical structure. *J. Biol. Chem.*, 263: 16709–16713, 1988.
110. T. Saito, S. Kawabata, T. Shigenaga, Y. Takayenoki, J. Cho, H. Nakajima, M. Hirata, S. Iwanaga. A novel big defensin identified in horseshoe crab hemocytes: isolation, amino acid sequence, and antibacterial activity. *J. Biochem.*, 117: 1131–1137, 1995.
111. S. Kawabata, R. Nagayama, M. Hirata, T. Shigenaga, K.L. Agarwala, T. Saito, J. Cho, H. Nakajima, T. Takagi, S. Iwanaga. Tachycitin, a small granular component in horseshoe crab hemocytes, is an antimicrobial protein with chitin-binding activity. *J. Biochem.*, 120: 1253–1260, 1996.
112. T. Osaki, M. Omotezako, R. Nagayama, M. Hirata, S. Iwanaga, J. Kasahara, J. Hattori, I. Ito, H. Sugiyama, S. Kawabata. Horseshoe crab hemocyte-derived antimicrobial polypeptides, tachystatins, with sequence similarity to spider neurotoxins. *J. Biol. Chem.*, 274: 26172–26178, 1999.
113. N. Fujitani, S.-I. Kawabata, T. Osaki, Y. Kumaki, M. Demura, K. Nitta, K. Kawano. Structure of the antimicrobial peptide tachystatin A. *J. Biol. Chem.*, 277: 23651–23657, 2002.
114. D. Destoumieux, P. Bulet, D. Loew, A. Van Dorsselaer, J. Rodriguez, E. Bachère. Penaeidins, a new family of antimicrobial peptides isolated from the shrimp *Penaeus vannamei* (Decapoda). *J. Biol. Chem.*, 272: 28398–28406, 1997.
115. M. Muñoz, F. Vandenbulcke, Y. Gueguen, E. Bachère. Expression of penaeidin antimicrobial peptides in early larval stages of the shrimp *Penaeus vannamei*. *Dev. Comp. Immunol.*, 27: 283–289, 2003.
116. E. Bachère, D. Destoumieux, P. Bulet. Penaeidins, antimicrobial peptides of shrimp: a comparison with other effectors of innate immunity. *Aquaculture*, 191: 71–88, 2000.
117. D. Destoumieux, P. Bulet, J.-M. Strub, A. van Dorsselaer, E. Bachèr. Recombinant expression and range of activity of penaeidins, antimicrobial peptides from penaeid shrimp. *Eur. J. Biochem.*, 266: 335–346, 1999.
118. T.C. Bartlett, B.J. Cuthbertson, E.F. Shepard, R.W. Chapman, P.S. Gross, G.W. Warr. Crustins, homologues of an 11.5-kDa antibacterial peptide, from two species of penaeid shrimp, *Litopenaeus vannamei* and *Litopenaeus setiferus*. *Mar. Biotechnol.*, 4: 278–293, 2002.
119. G. Mitta, F. Hubert, E.A. Dyrynda, P. Boudry, P. Roch. Mytilin B and MGD2, two antimicrobial peptides of marine mussels: gene structure and expression analysis. *Dev. Comp. Immunol.*, 24: 381–393, 2000.

120. G. Mitta, F. Vandenbulcke, P. Roch. Original involvement of antimicrobial peptides in mussel innate immunity. *FEBS Lett.*, 486: 185–190, 2000.
121. M. Charlet, S. Chernysh, H. Philippe, C. Hetru, J.A. Hoffmann, P. Bulet. Isolation of several cysteine-rich antimicrobial peptides from the blood of a mollusc, *Myitlus edulis. Innate. Immun.*, 271: 21808–21813, 1996.
122. F. Hubert, T. Noël, P. Roch. A member of the arthropod defensin family from edible Mediterranean mussels (*Mytilus galloprovincialis*). *Eur. J. Biochem.*, 240: 302–306, 1996.
123. G. Mitta, F. Hubert, T. Noël, P. Roch. Myticin, a novel cysteine-rich antimicrobial peptide isolated from haemocytes and plasma of the mussel *Mytilus galloprovincialis. Eur. J. Biochem.*, 265: 71–78, 1999.
124. Y.-S. Yang, G. Mitta, A. Chavanieu, B. Calas, J.F. Sanchez, P. Roch, A. Aumelas. Solution 17 structure and activity of the synthetic four-disulfide bond Mediterranean mussel defensin (MGD-1). *Biochemistry*, 39: 14436–14447, 2000.
125. R.S. Anderson, A.E. Beaven. Antibacterial activities of oyster *(Crassostrea virginica)* and mussel *(Mytilus edulis* and *Geukensia demissa)* plasma. *Aquat. Living Resour.*, 14: 343–349, 2001.
126. H. Meisel. Biochemical properties of bioactive peptides derived from milk proteins: potential nutraceuticals for food and pharmaceutical applications. *Livestock Prod. Sci.*, 50: 125–138, 1997.
127. E. Lahov, W. Regelson. Antibacterial and immunostimulating casein-derived substances from milk: casecidin, isracidin peptides. *Food Chem. Toxicol.*, 34: 131–145, 1996.
128. A. Pellegrini. Antimicrobial peptides from food proteins. *Curr. Pharm. Des.*, 9: 1225–1238, 2003.
129. R.D. Hill, E. Lahav, D. Givol. A rennin-sensitive bond in alpha-s1 beta-casein. *J. Dairy Res.*, 41: 147–153, 1974.
130. H.-D. Zucht, M. Raida, K. Adermann, H.-J. Mägert, W.-G. Forssmann. Casocidin-I: a casein-αs2 derived peptide exhibits antibacterial activity. *FEBS Lett.*, 372: 185–188, 1995.
131. I. Recio, S. Visser. Identification of two distinct antibacterial domains within the sequence of bovine αs2-casein. *Biochim. Biophys. Acta*, 1428: 314–326, 1999.
132. M. Malkoski, S.G. Dashper, N.M. O'Brien-Simpson, G.H. Talbo, M. Macris, K.J. Cross, E.C. Reynolds. Kappacin, a novel antibacterial peptide from bovine milk. *Antimicrob. Agents. Chemother.*, 45: 2309–2315, 2001.
133. G.H. Talbo, D. Suckau, M. Malkoski, E.C. Reynolds. MALDI-PSD-MS analysis of the phosphorylation sites of caseinomacropeptide. *Peptides*, 22: 1093–1098, 2001.
134. A. Pellegrini, C. Dettling, U. Thomas, P. Hunziker. Isolation and characterization of four bactericidal domains in the bovine β-lactoglobulin. *Biochim. Biophys. Acta*, 1526: 131–140, 2001.
135. A. Pellegrini, U. Thomas, N. Bramaz, P. Hunziker, R. von Fellenberg. Isolation and identification of three bactericidal domains in the bovine alpha-lactalbumin molecule. *Biochim. Biophys. Acta*, 1426: 439–448, 1999.
136. H. Wakabayashi, M. Takase, M. Tomita. Lactoferricin derived from milk protein lactoferrin. *Curr. Pharm. Des.*, 9: 1277–1287, 2003.
137. M. Tomita, W. Bellamy, M. Takase, K. Yamauchi, H. Wakabayashi, K. Kawase. Potent antibacterial peptides generated by pepsin digestion of bovine lactoferrin. *J. Dairy Sci.*, 74: 4137–4142, 1991.

138. P.M. Hwang, N. Zhou, X. Shan, C.H. Arrowsmith, H.J. Vogel. Three-dimensional solution structure of lactoferricin B, an antimicrobial peptide derived from bovine lactoferrin. *Biochemistry*, 37: 4288–4298, 1998.
139. R.T. Ellison, T.J. Giehl, F.M. LaForce. Damage of the outer membrane of enteric gram-negative bacteria by lactoferrin and transferrin. *Infect. Immun.*, 56: 2774–2781, 1988.
140. K. Yamauchi, M. Tomita, T.J. Giehl, R.T. Ellison. Antibacterial activity of lactoferrin and a pepsin-derived lactoferrin peptide fragment. *Infect. Immun.*, 61: 719–728, 1993.
141. W. Bellamy, M. Takase, H. Wakabayashi, K. Kawase, M. Tomita. Antibacterial spectrum of lacteferricin B, a potent bactericidal peptide derived from the N-terminal region of bovine lactoferrin. *J. Appl. Bacteriol.*, 73: 472–479, 1992.
142. W. Bellamy, H. Wakabayashi, M. Takase, K. Kawase, S. Shimamura, M. Tomita. Killing of *Candida albicans* by lactoferricin B, a potent antimicrobial peptide derived from the N-terminal region of bovine lactoferrin. *Med. Microbiol. Immunol.*, 182: 97–105, 1993. 21
143. H. Wakabayashi, S. Abe, S. Teraguchi, H. Hayasawa, H. Yamaguchi. Inhibition of hyphal growth of azole-resistant strains of *Candida albicans* by triazole antifungal agents in the presence of lactoferrin-related compounds. *Antimicrob. Agents Chemother.*, 42: 1587–1591, 1998.
144. Y. Omata, M. Satake, R. Maeda, A. Saito, K. Shimazaki, K. Yamaguchi, Y. Uzuka, S. Tanabe, T. Sarashina, T. Mikami. Reduction of the infectivity of *Toxoplasma gondii* and *Eimeria stiedae* sporozoites by treatment with bovine lactoferricin. *J. Vet. Med. Sci.*, 63: 187–190, 2001.
145. H.R. Ibrahim, S. Higashiguchi, L.R. Juneja, M. Kim, T. Yamamoto. A structural phase of heat-denatured lysozyme with novel antimicrobial action. *J. Agric. Food Chem.*, 44: 1416–1423, 1996.
146. K. Düring, P. Porsch, A. Mahn, O. Brinkmann, W. Gieffers. The non-enzymatic microbicidal activity of lysozymes. *FEBS Lett.*, 449: 93–100, 1999.
147. H.R. Ibrahim, T. Matsuzaki, T. Aoki. Genetic evidence that antibacterial activity of lysozyme is independent of its catalytic function. *FEBS Lett.*, 506: 27–32, 2001.
148. H.R. Ibrahim, U. Thomas, A. Pellegrini. A helix-loop-helix peptide at the upper lip of the active site cleft of lysozyme confers potent antimicrobial activity with membrane permeabilization action. *J. Biol. Chem.*, 276: 43767–43774, 2001.
149. A. Pellegrini, U. Thomas, N. Bramaz, P. Hunziker, R. von Fellenberg. Identification and isolation of a bactericidal domain in chicken egg white lysozyme. *J. Appl. Microbiol.*, 82: 372–378, 1997.
150. A. Pellegrini, S. Schumacher, R. Stephan. *In-vitro* activity of various antimicrobial peptides developed from the bactericidal domains of lysozyme and beta-lactoglobulin with respect to *Listeria monocytogenes*, *Escherichia coli* O157, *Salmonella* spp. and *Staphylococcus aureus*. *Arch. Lebensmittelhyg.*, 54: 34–36, 2003.
151. P. Valenti, G. Antonini, C. Von Hunolstein, P. Visca, N. Orsi, E. Antonini. Studies on the antimicrobial activity of ovotransferrin. *Int. J. Tissue Reac.*, 5: 97–105, 1983.
152. H.R. Ibrahim, E. Iwamori, Y. Sugimoto, T. Aoki. Identification of a distinct antibacterial domain within the N-lobe of ovotransferrin. *Biochim. Biophys. Acta*, 1401: 289–303, 1998.
153. H.R. Ibrahim, Y. Sugimoto, T. Aoki. Ovotransferrin antimicrobial peptide (OTAP-92) kills bacteria through a membrane damage mechanism. *Biochim. Biophys. Acta*, 1523: 196–205, 2000.

154. O. McAuliffe, R.P. Ross, C. Hill. Lantibiotics: structure, biosynthesis and mode of action. *FEMS Microbiol. Rev.*, 25: 285–308, 2001.
155. S. Garneau, N.I. Martin, J.C. Vederas. Two-peptide bacteriocins produced by lactic acid bacteria. *Biochimie*, 84: 577–592, 2002.
156. M.A. Riley, J.E. Wertz. Bacteriocin diversity: ecological and evolutionary perspectives. *Biochimie*, 84: 357–364, 2002.
157. J. Cleveland, T.J. Montville, I.F. Nes, M.L. Chikindas. Bacteriocins: safe, natural antimicrobials for food preservation. *Int. J. Food Microbiol.*, 71: 1–20, 2001.
158. J.R. Tagg, A.S. Dajani, L.W. Wannamaker. Bacteriocins of Gram-positive bacteria. *Microbiol. Rev.*, 40: 722–756, 1976.
159. T.R. Klaenhammer. Genetics of bacteriocins produced by lactic acid bacteria. *FEMS Microbiol. Rev.*, 12: 39–86, 1993.
160. C. van Kraaij, W.M. de Vos, R.J. Siezen, O.P. Kuipers. Lantibiotics: biosynthesis, mode of action and applications. *Natl. Prod. Rep.*, 16: 575–587, 1999.
161. J.C. Piard, O.P. Kuipers, H.S. Rollema, M.J. Desmazeuad, W.M. de Vos. Structure, organization and expression of the *lct* gene for lacticin 481, a novel lantibiotic produced by *Lactococcus lactis*. *J. Biol. Chem.*, 268: 16361–16368, 1993.
162. M.J. van Belkum, M.E. Stiles. Nonlantibiotic antibacterial peptides from lactic acid bacteria. *Natl. Prod. Rep.*, 17: 323–335, 2000.
163. J.T. Henderson, A.L. Chopko, P.D. van Wassenaar. Purification and primary structure of pediocin PA-1 produced by *Pediococcus acidilactici* PAC-1.0. *Arch. Biochem. Biophys.*, 295: 5–12, 1992.
164. A.L. Holck, L. Axelsson, S. Birkeland, T. Aukrust, H. Blom. Purification and amino acid sequence of sakacin A, a bacteriocin from *Lactobacillus sake* Lb706. *J. Gen. Microbiol.*, 138: 2715–2720, 1992.
165. T. Aymerich, H. Holo, L.S. Håvarstein, M. Hugas, M. Garriga, I.F. Nes. Biochemical and genetic characterization of enterocin A from *Enterococcus faecium*, a new antilisterial bacteriocin in the pediocin family of bacteriocins. *Appl. Environ. Microbiol.*, 62: 1676–1682, 1996.
166. J. Nissen-Meyer, H. Holo, L.S. Håvarstein, K. Sletten, I.F. Nes. A novel lactococcal bacteriocin whose activity depends on the complementary action of two peptides. *J. Bacteriol.*, 174: 5686–5692, 1992.
167. P.M. Muriana, T.R. Klaenhammer. Purification and partial characterization of lactacin F, a bacteriocin produced by *Lactobacillus acidophilus* 11088. *Appl. Environ. Microbiol.*, 57: 114–121, 1991.
168. R.J. Leer, J.M.B.M. van der Vossen, M. van Giezen, J.M. van Noort, P.H. Pouwels. Genetic analysis of acidocin B, a novel bacteriocin produced by *Lactobacillus acidophilus*. *Microbiology*, 141: 1629–1635, 1995.
169. M.C. Joerger, T.R. Klaenhammer. Characterization and purification of helveticin J and evidence for a chromosomally determined bacteriocin produced by *Lactobacillus heleveticus* 481. *J. Bacteriol.*, 167: 439–446, 1986.
170. J. Delves-Broughton, P. Blackburn, R.J. Evans, J. Hugenholtz. Applications of bacteriocin, nisin. *Int. J. Gen. Mol. Microbiol.*, 69: 193–202, 1996.
171. E. Breukink, C. van Kraaij, R.A. Demel, R.J. Siezen, O.P. Kuipers, B.D. Kruijff. The C-terminal region of nisin is responsible for the initial interaction of nisin with the target membrane. *Biochemistry*, 36: 6968–6976, 1997.
172. H. Brötz, M. Josten, I. Wiedemann, U. Schneider, F. Götz, G. Bierbaum, H.-G. Sahl. Role of lipid-bound peptidoglycan precursors in the formation of pores by nisin, epidermin and other lantibiotics. *Mol. Microbiol.*, 30: 317–327, 1998.

173. L. Lins, P. Ducarme, E. Breukink, R. Brasseur. Computational study of nisin interaction with model membrane. *Biochim. Biophys. Acta*, 1420: 111–120, 1999.
174. M.E.C. Bruno, A. Kaiser, T.J. Montville. Depletion of proton motive force by nisin in Listeria monocytogenes cells. *Appl. Environ. Microbiol.*, 58: 2255–2259, 1992.
175. E. Breukink, B. de Kruijff. The lantibiotic nisin, a special case or not? *Biochim. Biophys. Acta*, 1462: 223–234, 1999.
176. L. Voland, H. Ulvatne, Ø. Rekdal, J. Svendsen. Initial binding sites of antimicrobial peptides in *Staphylococcus aureus* and *Escherichia coli*. *Scand. J. Infect. Dis.*, 31: 467–473, 1999.
177. H. Ulvatne, H. Haukland, Ø. Olsvik, L. Vorland. Lactoferricin B causes depolarization of the cytoplasmic membrane of *Escherichia coli* ATCC 26922 and fusion of negatively charged liposomes. *FEBS Lett.*, 492: 62–65, 2001.
178. Z. Oren, Y. Shai. Selective lysis of bacteria but not mammalian cells by diastereomers of melittin: structure-function study. *Biochemistry*, 36: 1826–1835, 1997.
179. Z. Oren, Y. Shai. Cyclization of a cytolytic amphipathic alpha-helical peptide and its diastereomer: effect on structure, interaction with model membranes, and biological function. *Biochemistry*, 39: 6103–6114, 2000.
180. A. Bera, S. Singh, R. Nagaraj, T. Vaidya. Induction of autophagic cell death in *Leishmania donovani* by antimicrobial peptides. *Mol. Biochem. Parasitol.*, 127: 23–35, 2003.
181. H.W. Huang. Action of antimicrobial peptide: two-state model. *Biochemistry*, 39: 8347–8452, 2000.
182. Y. Shai, Z. Oren. From "carpet" mechanism to de-novo design diastereomeric cell-selective antimicrobial peptides. *Peptides*, 22: 1629–1641, 2001.
183. M. Cudic, L.J. Otvos. Intracellular targets of antibacterial peptides. *Curr. Drug Targets*, 3: 101–106, 2002.
184. H.G. Boman, B. Agerberth, A. Boman. Mechanisms of action on *Escherichia coli* of cecropin P1 and PR-39, two antibacterial peptides from pig intestine. *Infect. Immun.*, 61: 2978–2984, 1993.
185. C.L. Friedrich, A. Rozek, A. Patrzykat, R.E. Hancock. Structure and mechanism of action of an indolicidin peptide derivative with improved activity against gram-positive bacteria. *J. Biol. Chem.*, 276: 24015–24022, 2001.
186. T.J. Falla, D.N. Karunaratne, R.E.W. Hancock. Mode of action of the antimicrobial peptide indolicidin. *J. Biol. Chem.*, 271: 19298–19303, 1996.
187. C. Subbalakshmi, N. Krishnakumari, R. Nagaraj, N. Sitaram. Requirements for antibacterial and hemolytic activities in the bovine neutrophil derived 13-residue peptide indolicidin. *FEBS Lett.*, 395: 48–52, 1996.
188. C. Subbalakshmi, N. Sitaram. Mechanism of antimicrobial action of indolicidin. *FEMS Microbiol. Lett.*, 160: 91–96, 1998.
189. P.A. Charp, W.G. Rice, R.L. Raynor, E. Reimund, J.M. Kinkade, J.T. Ganz, M.E. Selsted, R.I. Lehrer, J.F. Kuo. Inhibition of protein kinase C by defensins, antibiotic peptides form human neutrophils. *Biochem. Pharmacol.*, 37: 951–956, 1988.
190. A. Hobta, I. Lisovskiy, S. Mikhalap, D. Kolybo, S. Romanyuk, M. Soldatkina, N. Markeyeva, L. Garmanchouk, S.P. Sidorenko, P.V. Pogrebnoy. Epidermoid carcinoma-derived antimicrobial peptide (ECAP) inhibits phosphorylation by protein kinases *in vitro*. *Cell Biochem. Funct.*, 19: 291–298, 2001.
191. A. Risso, M. Zanetti, R. Gennaro. Cytotoxicity and apoptosis mediated by two peptides of innate immunity. *Cell Immunol.*, 189: 107–115, 1998.

192. G.H. Gudmundsson, B. Agerberth. Neutrophil antibacterial peptides, multifunctional effector molecules in the mammalian immune system. *J. Immunol. Methods*, 232: 45–54, 1999.
193. R.E.W. Hancock, G. Diamond. The role of cationic antimicrobial peptides in innate host defences. *Trends Microbiol.*, 8: 402–410, 2000.
194. R.E. Hancock. Cationic peptides: effectors in innate immunity and novel antimicrobials. *Lancet Infect. Dis.*, 1: 156–164, 2001.
195. M. Salzet. Antimicrobial peptides signaling molecules. *Trends Immunol.*, 23: 283–284, 2002.
196. D. Yang, O. Chertov, J.J. Oppenheim. Participation of mammalian defensins and cathelicidins in anti-microbial immunity: receptors and activities of human defensins and cathelicidin (LL-37). *J. Leukocyte Biol.*, 69: 691–697, 2001.
197. A.D. Befus, C. Mowat, M. Gilchrist, J. Hu, S. Solomon, A. Bateman. Neutrophil defensins induce histamine secretion from mast cells: mechanisms of action. *J. Immunol.*, 163: 947–953, 1999.
198. A. Di Nardo, A. Vitiello, R.L. Gallo. Cutting edge: mast cell antimicrobial activity is mediated by expression of cathelicin antimicrobial peptide. *J. Immunol.*, 170: 2274–2278, 2003.
199. S. Yomogida, I. Nagaoka, T. Yamashita. Comparative studies on the extracellular release and biological activity of guinea pig neutrophil cationic antibacterial polypeptide of 11 kDa (CAP11) and defensin. *Comp. Biochem. Physiol. B. Biochem. Mol. Biol.*, 116: 99–107, 1997.
200. M. Zaiou, V. Nizet, R.L. Gallo. Antimicrobial and protease inhibitory functions of the human cathelicidin (hCAP18/LL-37) prosequence. *J. Invest. Dermatol.*, 120: 810–816, 2003.
201. C.J. Murphy, B.A. Foster, M.J. Mannis, M.E. Selsted, T.W. Reid. Defensins are mitogenic for epithelial cells and fibroblasts. *J. Cell Physiol.*, 155: 408–413, 1993.
202. R.L. Gallo, M. Ono, T. Povsic, C. Page, E. Eriksson, M. Klagsbrun, M. Bernfield. Syndecans, cell surface heparan sulfate proteoglycans, are induced by a proline-rich antimicrobial peptide from wounds. *Prod. Natl. Acad. Sci. USA*, 91: 11035–11039, 1994.
203. M. Frohm, B. Agerberth, G. Ahangari, M. Stahle-Backdahl, S. Liden, H. Wigzell, G.H. Gudmundsson. The expression of the gene coding for the antibacterial peptide LL-37 is induced in human keratinocytes during inflammatory disorders. *J. Biol. Chem.*, 272: 15258–15263, 1997.
204. M. Frohm, H. Gunne, A.C. Bergman, B. Agerberth, T. Bergman, A. Boman, S. Liden, H. Jornvall, H.G. Boman. Biochemical and antibacterial analysis of human wound and blister fluid. *Eur. J. Biochem.*, 237: 86–92, 1996.
205. O.E. Sørensen, J.B. Cowland, K. Theilgaard-Mönch, L. Liu, T. Ganz, N. Borregaard. Wound healing and expression of antimicrobial peptides/polypeptides in human keratinocytes, a consequence of common growth factors. *J. Immunol.*, 170: 5583–5589, 2003.
206. A.G. Gao, S.M. Hakim, C.A. Mittanck, Y. Wu, B.M. Woerner, D.M. Stark, D.M. Shah, J. Liang, C.M. Rommens. Fungal pathogen protection in potato by expression of a plant defensin peptide. *Nature Biotechnol.*, 18: 1307–1310, 2000.
207. V.G. Lebedev, S.V. Dolgov, N. Lavrova, V.G. Lunin, B.S. Naroditski. Plant-defensin genes introduction for improvement of pear phytopathogen resistance. *Acta Hortic.*, 596: 167–172, 2002.

208. A. Chakrabarti, T.R. Ganapathi, P.K. Mukherjee, V.A. Bapat. Msl-99, a magainin analogue, impact enhanced disease resistance in transgenic tobacco and banana. *Planta*, 216: 587–596, 2003.
209. O.S. Lee, B. Lee, N. Park, J.C. Koo, Y.H. Kim, T. Prasad, C. Karigar, H.J. Chun, B.R. Jeong, D.H. Kim, J. Nam, J.-G. Yun, S.-S. Kwak, M.J. Cho, D.-J. Yun. Pn-AMPs, the hevein-like proteins from *Pharbitis nil* confers disease resistance against phytopathogenic fungi in tomato, *Lycopersicum esculentum*. *Phytochemistry*, 62: 1073–1079, 2003.
210. N. Banzet, M.-P. Latorse, P. Bulet, E. François, C. Derpierre, M. Dubald. Expression of insect cysteine-rich antifungal peptides in transgenic tobacco enhances resistance to a fungal disease. *Plant Sci.*, 162: 995–1006, 2002.
211. D. Ponti, M. Luisa Mangoni, G. Mignogna, M. Simmaco, D. Barra. An amphibian antimicrobial peptide variant expressed in *Nicotiana tabacum* confers resistance to phytopathogens. *Biochem. J.*, 370: 121–127, 2003.
212. S. Holtorf, K. Apel, H. Bohlmann. Specific and different expression patterns of two members of the leaf thionin multigene family of barley in transgenic tobacco. *Plant Sci.*, 111: 27–37, 1995.
213. M.J. Carmona, A. Molina, J.A. Fernandez, J.J. Lopez-Fando, F. García-Olmedo. Expression of the alpha-thionin from barley in tobacco confers enhanced resistance to bacterial pathogens. *Plant J.*, 3: 457–462, 1993.
214. M.F.C. De Bolle, R.W. Osborn, I.J. Goderis, L. Noe, D. Acland, C.A. Hart, S. Torrekens, F. Van Leuven, W.F. Broekaert. Antimicrobial peptides from *Mirabilis jalapa* and *Amaranthus caudatus*: expression, processing, localization and biological activity in transgenic tobacco. *Plant Mol. Biol.*, 31: 993–1008, 1996.
215. F.-M. Lai, C. DeLong, K. Mei, T. Wignes, P.R. Fobert. Analysis of the DRR230 family of pea defensins: gene expression pattern and evidence of broad host-range antifungal activity. *Plant Sci.*, 163: 855–864, 2002.
216. N.P. Everett. Design of antifungal peptides for agricultural applications. *ACS Symp. Ser.*, 551: 278–291, 1994.
217. L.D. Owens, T.M. Heutte. A single amino acid substitution in the antimicrobial defense protein cecropin B is associated with diminished degradation by leaf intercellular fluid. *Mol. Plant-Microbe Interact.*, 10: 525–528, 1997.
218. J.W. Cary, K. Rajasekaran, J.M. Jaynes, T.E. Cleveland. Transgenic expression of a gene encoding a synthetic antimicrobial peptide result in inhibition of fungal growth in vitro and in planta. *Plant Sci.*, 154: 171–181, 2000.
219. M. Osusky, G. Zhou, L. Osuska, R.E. Hancock, W.W. Kay, S. Misra. Trangenic plants expressing cationic peptide chimeras exhibit broad-spectrum resistance to phytopathogens. *Nat. Biotechnol.*, 18: 1162–1166, 2000.
220. C.N. Cutter, G.R. Siragusa. Population reductions of Gram negative pathogens following treatments with nisin and chelators under various conditions. *J. Food Prot.*, 58: 977–983, 1995.
221. C.N. Cutter, G.R. Siragusa. Treatments with nisin and chelators to reduce *Salmonella* and *Escherichia coli* on beef. *J. Food Prot.*, 58: 1028–1030, 1995.
222. G.A. Dykes, S.M. Moorhead. Combined antimicrobial effect of nisin and a listeriophage against *Listeria monocytogenes* in broth but not in buffer or on raw beef. *Int. J. Food Microbiol.*, 73: 71–81, 2002.
223. W. Chung, R.E.W. Hancock. Action of lysozyme and nisin mixtures against lactic acid bacteria. *Int. J. Food Microbiol.*, 60: 25–32, 2000.

224. M.J. Facon, B.J. Skura. Antibacterial activity of lactoferrin, lysozyme and EDTA against *Salmonella enteritidis*. *Int. Dairy J.*, 6: 303–313, 1996.
225. P. Chantaysakorn, R.L. Richter. Antimicrobial properties of pepsin-digested lactoferrin added carrot juice and filtrates of carrot juice. *J. Food Prot.*, 63: 376–380, 2000.
226. F.M. Nattress, C.K. Yost, L.P. Baker. Evaluation of the ability of lysozyme and nisin to control meat spoilage bacteria. *Int. J. Food Microbiol.*, 70: 111–119, 2001.
227. T. Ariyapitipun, A. Mustapha, A.D. Clarke. Microbial shelf life determination of vacuum-packaged fresh beef treated with polylactic acid, lactic acid, and nisin solutions. *J. Food Prot.*, 62: 913–920, 1999.
228. T. Ariyapitipun, A. Mustapha, A.D. Clarke. Survival of *Listeria monocytogenes* Scott A on vacuum-packaged raw beef treated with polylactic acid, lactic acid, and nisin. *J. Food Prot.*, 63: 131–136, 2000.
229. M.L. Cabo, M.A. Murado, M.P. Gonzalez, L. Pastoriza. Dose-response relationships. A model for describing interactions, and its application to the combined effect of nisin and lactic acid on *Leuconostoc mesenteroides*. *J. Appl. Microbiol.*, 88: 756–763, 2000.
230. S. Zhang, A. Mustapha. Reduction of *Listeria monosytogenes* and *Escherichia coli* O157:H7 numbers on vacuum-packaged fresh beef treated with nisin or nisin combined with EDTA. *J. Food Prot.*, 62: 1123–1127, 1999.
231. L. Nilsson, Y. Chen, M.L. Chikindas, H.H. Huss, L. Gram, T.J. Montville. Carbon dioxide and nisin act synergistically on *Listeria monocytogenes*. *Appl. Environ. Microbiol.*, 66: 769–774, 2000.
232. C.N. Cutter, G.R. Siragusa. Reductions of *Listeria innocua* and *Brochothrix thermosphacta* on beef following nisin spray treatments and vacuum packaging. *Food Microbiol.*, 13: 23–33, 1996.
233. C.N. Cutter, G.R. Siragusa. Reduction of *Brochothrix rheumosphacta* on beef surfaces following immobilization of nisin in calcium alginate gels. *Lett. Appl. Microbiol.*, 23: 9–12, 1996.
234. C.N. Cutter, G.R. Siragusa. Growth of *Brochothrix thermosphacta* in ground beef following treatment with nisin in calcium alginate gels. *Food Microbiol.*, 14: 425–430, 1997.
235. C.N. Cutter, G.R. Siragusa. Incorporation of nisin into a meat binding system to inhibit bacteria on beef surfaces. *Lett. Appl. Microbiol.*, 27: 19–23, 1998.
236. P.L. Dawson, I.Y. Han, T.R. Padgett. Effect of lauric acid on nisin activity in edible protein packaging films. *Poult. Sci.*, 76 (S): 74, 1997.
237. T. Padgett, I.Y. Han, P.L. Dawson. Incorporation of food-grade antimicrobial compounds into biodegradable packaging films. *J. Food Prot.*, 61: 1330–1335, 1998.
238. G.R. Siragusa, C.N. Cutter, J.L. Willett. Incorporation of bacteriocin in plastic retains activity and inhibits surface growth of bacteria on meat. *Food Microbiol.*, 16: 229–235, 1999.
239. C.N. Cutter, J.L. Willett, G.R. Siragusa. Improved antimicrobial activity of nisin-incorporated polymer films by formulation change and addition of food grade chelator. *Lett. Appl. Microbiol.*, 33: 325–328, 2001.
240. F.M. Nattress, L.P. Baker. Effects of treatment with lysozyme and nisin on the microflora and sensory properties of commercial pork. *Int. J. Food Microbiol.*, 85: 259–267, 2003.
241. R.E.W. Hancock, R.I. Lehrer. Cationic peptides: a new source of antibiotics. *Trends Biotechnol.*, 16: 82–88, 1998.

242. T. Gura. Innate immnunity: ancient system gets new respect. *Science*, 291: 2068–2070, 2001.
243. A.R. Koczulla, R. Bals. Antimicrobial peptides: current status and therapeutic potential. *Drugs*, 63: 389–406, 2003.
244. M. Gough, R.E. Hancock, N.M. Kelly. Antiendotoxin activity of cationic peptide antimicrobial agents. *Infect. Immun.*, 64: 4922–4927, 1996.
245. I. Ahmad, W.R. Perkins, D.M. Lupan, M.E. Selsted, A.S. Janoff. Liposomal entrapment of the neutrophil-derived peptide indolicidin endows it with *in vivo* antifungal activity. *Biochim. Biophys. Acta*, 1237: 109–114, 1995.
246. D.A. Steinberg, M.A. Hurst, C.A. Fujii, A.H. Kung, J.F. Ho, F.C. Cheng, D.J. Loury, J.C. Fiddes. Protegrin-1: a broad-spectrum, rapidly microbicidal peptide with *in vivo* activity. *Antimicrob. Agents Chemother.*, 41: 1738–1742, 1997.
247. T. Kirikae, M. Hirata, H. Yamasu, F. Kirikae, H. Tamura, F. Kayama, K. Nakatsuka, T. Yokochi, M. Nakano. Protective effects of a human 18-kilodalton cationic antimicrobial protein (CAP18)-derived peptide against murine endotoxemia. *Infect. Immun.*, 66: 1861–1868, 1998.
248. A. Giacometti, O. Cirioni, R. Ghiselli, F. Mocchegiani, M.S. Del Prete, C. Viticchi, W. Kamysz, E. Lempicka, V. Saba, G. Scalise. Potential therapeutic role of cationic peptides in three experimental models of septic shock. *Antimicrob. Agents Chemother.*, 46: 2132–2136, 2002.
249. D. Loury, J.R. Embree, D.A. Steinberg, S.T. Sonis, J.C. Fiddes. Effect of local application of the antimicrobial peptide IB-367 on the incidence and severity of oral mucositis in hamsters. *Oral Surg. Oral Med. Oral Pathol. Oral Radiol. Endod.*, 87: 544–551, 1999.
250. A. Giacometti, O. Cirioni, R. Ghiselli, F. Mocchegiani, G. D'Amato, M.S. Del Prete, F. Orlando, W. Kamysz, J. Lukasiak, V. Saba, G. Scalise. Adminstration of protegrin peptide IB-367 to prevent endotoxin induced mortality in bile duct ligated rats. *Gut*, 52: 874–878, 2003.
251. Y. Ge, D.L. MacDonald, K.J. Holroyd, C. Thornsberry, H. Wexler, M. Zasloff. *In vitro* antibacterial properties of pexiganan, an analog of magainin. *Antimicrob. Agents Chemother.*, 43: 782–788, 1999.
252. M. Cazzola, A. Sanduzzi, M.G. Matera. Novelties in the field of antimicrobial compounds for the treatment of lower respiratory tract infections. *Pulm. Pharmacol. Ther.*, 16: 131–145, 2003.
253. D.A. Mosca, M.A. Hurst, W. So, B.S.C. Viajar, C.A. Fujii, T.J. Falla. IB-367, a protegrin peptide with *in vitro* and *in vivo* activities against the microflora associated with oral mucositis. *Antimicrob. Agents Chemother.*, 44: 1803–1808, 2000.

8 Bovine Milk Antibodies for Protection against Microbial Human Diseases

Hannu Korhonen and Pertti Marnila

CONTENTS

8.1 INTRODUCTION

Gastrointestinal diseases continue to represent a major threat to human health on a global scale. Factors such as malnutrition, human immunodeficiency virus (HIV) immunocompromization, and lack of clean water have exacerbated the incidence of acute and chronic acquired gastrointestinal infections, while the increase in global travel has ensured that new emerging strains of enteric pathogens can rapidly spread and become established in other continents. Furthermore, the increasing use of broad-spectrum antibiotics has created new multiresistant bacteria, which often cause endemic hospital infections. The possibilities — if any — of preventing these problems are currently limited at best. In fact, there is a real need to develop new means of combating gastrointestinal tract infections and systemic infections originating from the intestinal tract. Exploiting the molecules of immune defense from an immunized animal may provide an appropriate strategy to this end.

It has been recognized for more than 100 years that maternal milk can offer passive protection to a newborn infant against enteric pathogens, primarily via the transfer of immunoglobulins (Igs) and associated factors from mother to infant (1). The concept of "immune milk" dates back to the 1950s when Petersen and Campbell first suggested that orally administered bovine colostrum could provide passive immune protection for humans (2). Since the 1980s, an increasing number of studies have demonstrated that immune milk preparations, based on bovine antibodies derived from the milk or colostrum of immunized cows, can be effective in the prevention and treatment of human and animal diseases caused by enteropathogenic microbes (for reviews, see 3–11). The efficacy of bovine immune milk products is mainly based on the antimicrobial activity of specific antibodies. Other antimicrobial substances such as complement components, lactoferrin, lysozyme and lactoperoxidase, which occur naturally in colostrum and milk, may also contribute to the antimicrobial spectrum of the preparation. This chapter reviews the current state of knowledge about the properties of bovine Igs, their utilization in the development of bovine immune milk preparations, and the application of these preparations in the prevention and treatment of microbial infections in humans.

8.2 BIOCHEMICAL CHARACTERISTICS OF BOVINE IMMUNOGLOBULINS

Immunoglobulins, also called antibodies, are present in the colostrum and milk of all lactating species. Igs are divided into different classes on the basis of physicochemical structures and biological activities. The major classes in bovine and human lacteal secretions are IgG, IgM, and IgA. The basic structure of all Igs is similar and is composed of two identical light chains (23-kDa) and two identical heavy chains (53-kDa). These four chains are joined together by disulfide bonds. The complete Ig (or antibody) molecule possesses stereochemically a Y-shaped structure and has a molecular weight of about 160 kDa. Each Ig molecule has two identical antigen-binding sites (termed F[ab]'2), which are formed by the N-terminal part of one heavy chain and one light chain. The C-terminal region of the heavy chain is termed the Fc region. The bovine IgG molecule occurs predominantly in two subclasses: IgG-1 and IgG-2. The basic structure of monomeric IgM and IgA is similar to that of IgG, except for the addition of a C-terminal octapeptide to the heavy chains. IgA occurs as a monomer or dimer, the latter comprising two IgA molecules joined together by a J-chain and a secretory component. This complex, called secretory IgA (SIgA), has a molecular weight of about 380 kDa. Except for the lacteal secretions of ruminants, IgA is the dominating Ig in all external secretions of the body. Similar to dimeric IgA, IgM consists of five subunits that are linked together in a circular mode by disulfide bonds and a J-chain; the molecular weight of pentameric IgM is approximately 900 kDa (for reviews see 12–15).

Cows secrete colostrum during the first five to seven days after delivery of the calf. The concentration of the various Igs in the bovine serum and lacteal secretions varies according to the breed, age, health status, and stage of lactation of the animal (16,17). Igs make up 70 to 80% of the total protein content in colostrum, whereas in mature milk, they account for only 1 to 2% of total protein (12,18). The concentration of Igs in the first colostrum can vary considerably, from 20 to 200 g l^{-1}, being 60 g l^{-1} on average. IgG-1 comprises over 75% of the Igs in colostral whey, followed by IgM, IgA, and IgG-2, respectively. Total Ig levels in milk decline rapidly following parturition to around 0.7 to 1.0 g l^{-1}. IgG-1, however, remains the predominant subclass in these secretions (Table 8.1) (15).

8.3 TECHNOLOGICAL PROPERTIES OF BOVINE IMMUNOGLOBULINS

Since the 1980s, the rapid development of membrane separation and chromatographic techniques has made it possible to isolate or concentrate Igs from bovine colostrum and milk on an industrial scale. Numerous patented methods based on ultrafiltration (UF) or a combination of UF and ion exchange chromatography have been designed for this purpose (for reviews see 14,19). A combination of different membrane technologies has proven to be the most cost-effective approach for the commercial production of crude Ig preparations. Even though

TABLE 8.1
Concentration of Immunoglobulins in Bovine Colostrum and Milk

	Concentration g l^{-1}			
	Colostrum		Milk	
Ig class	Average	Range	Average	Range
IgG_1	46.4	15–180	0.35	0.3–0.6
IgG_2	2.9	1–3	0.05	0.06–0.12
IgM	6.8	3–9	0.09	0.04–1.0
IgA	5.4	1–6	0.08	0.05–0.1
Ig Total	61.5	20–200	0.57	0.4–1.0

Source: Data from J.E. Butler. *Vet. Immun. Immunopathol.*, 4, 43–152, 1983; and P. Marnila, H. Korhonen, in *Encyclopedia of Dairy Sciences.* Academic Press, London, 2002, 1950–1956. With permission.

TABLE 8.2
Basic Chemical Composition of Bovine Immune Colostral Whey Preparations

Component	Percent (w/w)
Total protein	67–81
Immunoglobulin G	38–58
β-lactoglobulin	12–16
α-lactalbumin	2–4
Carbohydrates	0.5–11
Ash	<5
Fat	<1
Water	<5

colostrum is the richest source of Igs per volume unit, cheese whey is often considered a more reliable source of these components, since it is available in large volumes despite its relatively low Ig content. The Ig concentration of such whey-derived preparations has been reported to range mostly between 40 and 70% of dry weight, depending on the raw material and isolation method used. Table 8.2 shows the basic chemical composition of various bovine immune colostral whey concentrates.

Among dairy technological processes, thermal processes have been found to affect the properties of Igs the most. Earlier studies (20) considered Igs rather heat labile as compared with other whey proteins. More recent studies (21–23), however, have established that ordinary milk heat treatments (high-temperature short-time [HTST], 72°C 15 s^{-1}) or batch pasteurization (63°C 30 min^{-1}) destroy only 0.5 to 10% of the antibody activity of Igs. Considerable antibody activity has also been detected after ultra-high temperature (UHT) treatment (138°C 4 s^{-1}) and low-temperature spray-drying of milk, whereas the evaporation process appears to fully destroy the specific immune activity of milk (21). However, bovine IgG added to UHT milk has been shown to retain its specific immune activity almost totally for more than five months when stored at different temperatures up to 35°C (24). Moreover, in a dried form, Ig molecules seem to retain their specific activity quite well, irrespective of the storage temperature, since storage of a freeze-dried Ig concentrate from colostral whey at 4, 20, and 37°C for up to 1 year has been found to reduce the antibody activity only marginally (25).

8.4 DIGESTION OF ANTIBODIES

Antibodies ingested by humans are normally degraded by proteolytic enzymes in the stomach and intestine into small peptides and amino acids, which are subsequently absorbed. Orally administered bovine milk Igs are degraded in the human gastrointestinal tract by the intestinal proteases pepsin, trypsin, chymotrypsin, carboxypeptidase, and elastase into $F(ab')_2$, Fab, and Fc fragments. The $F(ab')_2$ and Fab fragments retain some of the neutralizing and adhesion-inhibiting activities of the intact antibody. The secretory piece component of the IgA molecule makes SIgA more resistant to the activities of proteolytic digestive enzymes than in the case of other Ig classes (26). In humans, this property is manifested by the passage of a considerable proportion (20 to 80%) of undigested active IgA from human colostrum through the gut of the neonate (27). In comparison, it has been shown in many *in vitro* studies that bovine IgG is also relatively resistant to proteolysis by pepsin and is only partially inactivated by gastric acid (28–30). Petschow and Talbot (31) showed that a low pH of human gastric acid, rather than pepsin or pancreatic trypsin, causes a significant reduction in the rotavirus-neutralizing activity of bovine milk Ig concentrates.

Bovine Igs have been detected in the feces of human infants fed bovine immune milk (26). Studies (32) with human volunteers showed that 10 to 30% of orally administered Igs could be recovered from adult feces in the form of $F(ab')_2$, Fab, and Fc fragments. In another study (33) with adult volunteers, some 49% of the ingested 2.1 g of bovine colostral Igs was recovered in the ileal fluid. Pacyna et al. (34) administered 100 ml of whole milk supplemented with hyperimmune colostrum containing different levels of rotavirus antibodies to 105 children three times a day for 6 days. Using the virus reduction ELISA method, rotavirus antibody activity was detected in 86% of the 602 fecal specimens obtained during the study. Antibody activity was detected from 8 hours after ingestion of hyperimmune colostrum up to 72 hours after consumption had ceased. The titers of the administered rotavirus antibody correlated with the antibody titers in feces, showing that anti-rotavirus activity survived the passage through the gut. Kelly et al. (35) measured the survival of orally administered bovine specific Ig concentrate against *Clostridium difficile* toxins in the human gastrointestinal tract. Without encapsulation, the mean fecal bovine Ig content of 3-day stools was 1.6 to 3.8% of the administered Ig (45-g single dose), whereas Ig (8-g single dose) administered in enteric capsules resulted in a 32.7% recovery in feces. An appropriate delivery system and controlled release of Igs in the desired part of the gastrointestinal tract thus appear to have a significant effect on the pharmacological efficacy of the ingested Igs. In successful clinical trials, the effective daily dose of powdered immune milk (mainly in the form of colostral or milk whey concentrate) has varied between 5 and 30 g corresponding to 1 to 5 g of total Igs (8,9).

8.5 FUNCTIONS AND ACTIVITIES OF BOVINE MILK ANTIBODIES

Immunoglobulins function as flexible adaptors linking together various parts of the cellular and humoral immune systems. While one part (F(ab)′2) of an antibody molecule binds to the antigen, other parts (mostly the Fc region) interact with other elements. In addition to antigen binding, all Igs exhibit one or more effector functions, the mediated functions depending on the Ig class. Colostral and milk Igs are able to prevent the adhesion of microbes to surfaces, inhibit bacterial metabolism by blocking enzymes, agglutinate bacteria, and neutralize toxins and viruses. Another vital function of Igs is opsonization — that is, augmenting the recognition and phagocytosis of bacteria by leukocytes. In blood and tissues, their most important function is possibly the activation of complement-mediated bacteriolytic reactions, but the significance of this mechanism in milk or in colostrum remains obscure. The IgG antibodies, which comprise 80 to 90% of total Igs in colostrum, have a multitude of functions, including fixation of complement, opsonization, agglutination of bacteria, and neutralization of toxins and viruses. IgM antibodies, which are produced in smaller amounts in mature milk than are IgG, are considerably more efficient than IgG in most of the above-mentioned activities, especially in complement fixation and complement-mediated opsonization and bacteriolysis. In contrast, the IgA class does not fix complement or opsonize bacteria but agglutinates antigens, neutralizes viruses and bacterial toxins, and prevents the adhesion of enteropathogenic bacteria to mucosal epithelial cells (15).

The efficacy of colostral and milk Ig preparations against microbes such as enteropathogenic *Escherichia coli*, rotavirus, and *Cryptosporidium* has been studied *in vitro* and *in vivo* with variable success in a variety of laboratory and clinical trials. Controlled clinical trials have been carried out also using hyperimmune bovine colostrum containing specific antibodies against *Shigella flexneri, Helicobacter pylori, Vibrio cholerae,* and caries-promoting streptococci. These studies have been extensively reviewed, for example, by Hammarström et al. (4), Davidson (6), Weiner et al. (8), Korhonen et al. (9), and Lilius and Marnila (10). The following section presents an overview of the main functions and mechanisms of specific milk antibodies with representative examples of the results achieved in clinical trials. Table 8.3 lists some of these clinical studies in humans.

8.5.1 Inhibition of Enzyme Activity

Blocking enzymes provide for Igs a means of inhibiting bacterial metabolism and reducing the production of harmful components such as toxins. Using this approach, Loimaranta et al. (36) studied *in vitro* efficacy of a bovine immune colostral preparation containing high titers of IgG antibodies against the human cariogenic bacteria *Streptococcus mutans* and *Streptococcus sobrinus* to inhibit the metabolic activity of these bacteria. The immune preparation significantly inhibited the incorporation of [14C] glucose by both *S. mutans* and *S. sobrinus* in quite low concentrations (0.5% w/v). A higher concentration ($>$1%) was needed to inhibit the activities of glucosyltransferase and fructosyltransferase enzymes, which are responsible for producing sticky capsule polysaccharides by these bacteria. No such inhibitory effects were observed with the control preparation from nonimmunized cows, indicating that bovine immune colostrum has significant inhibitory potential against dental-caries-promoting streptococci. Further studies by these researchers demonstrated that the same immune colostrum preparation does not inhibit the activity of the lactoperoxidase enzyme but, rather, has a synergistic effect, with the lactoperoxidase–thiocyanate–hydrogen peroxide antimicrobial system occurring naturally in saliva (37).

TABLE 8.3
Prophylactic and Therapeutic Efficacy of Bovine Immune Colostral and Milk Preparations against Microbial Infections in Humans

Microorganism Used	Target Disease	Treatment Dose/Period	Clinical Effect	Reference
Prophylactic Studies				
Escherichia coli	Diarrhea	5 g cw d^{-1} for 7 d	Prevented infection in adults after experimental challenge	59
Shigella flexneri	Dysenteria	30 g cw d^{-1} for 7 d	Prevented infection in adults after experimental challenge	60
Streptococcus mutans	Dental caries	3 rinses d^{-1} with 5% solution for 3 d	Reduced acidogenicity and number of mutans streptococci in dental plaque of adults	43
Rotavirus	Diarrhea	20–50 ml c d^{-1} for 3 d	Prevented infection in healthy children	71
Rotavirus	Diarrhea	50 ml c d^{-1} for 1 d	Prevented infection in healthy children	72
Cryptosporidium parvum	Diarrhea	30 g cw d^{-1} for 5 d	Reduced diarrhoea and oocyst exretion in adults after experimental challenge	48
Therapeutic Studies				
Escherichia coli	Diarrhea	20 g mw d^{-1} for 4 d	No alleviation of symptoms and duration of diarrhoea in infected children	66
Helicobacter pylori	Gastritis	12 g cw d^{-1} for 21 d	Reduced chronic inflammation and number of *H. pylori* in gastric antrum of infected children, no eradication of *H. pylori*	57
Helicobacter pylori	Gastritis	1 g cw d^{-1} for 30 d	No eradication of *H.pylori* in infected children	58
Cryptosporidium parvum	Diarrhea	200–500 ml c d^{-1} for 21 d	Reduced or ceased diarrhoea in infected adults	45
Rotavirus	Diarrhea	10 g cw/d^{-1} for 4 d	Shortened duration and decreased severity of diarrhea in infected children	78

Abbreviations used: c = colostrum, cw = colostral whey, mw = milk whey

8.5.2 Prevention of Adhesion

The attachment of a pathogenic organism to the epithelial lining or to tooth enamel is the most critical step in the establishment of infection. Thus, the prevention of the adhesion of microbes to surfaces may be the most important mechanism of colostral and milk antibodies in preventing an infection. Again, the ability of antibodies to agglutinate bacteria *in vivo* is likely to reduce the capability of bacteria to adhere to epithelial surfaces. Normal colostrum and milk are known to contain natural agglutinating antibodies to a large number of pathogenic and nonpathogenic microorganisms, including lipopolysaccharide (LPS) from *E. coli*, *Salmonella enteriditis*, *S. typhimurium*, and *Sh. flexneri* (38). Below is a review of studies that have investigated the efficacy of colostral and milk antibody preparations in preventing adhesion and subsequent infection induced by certain pathogenic organisms under either *in vitro* or *in vivo* conditions.

8.5.2.1 Caries Streptococci

Both adherence and aggregation are important phenomena in the bacterial colonization of the human oral cavity. Due to the risk of potentially harmful effects related to active immunization, the use of specific antibodies produced in milk or eggs has been suggested as an alternative means of controlling the colonization of oral microflora and preventing dental caries (for a review, see 39). Loimaranta et al. (40) showed that a bovine immune colostral preparation produced against caries-promoting *S. mutans* and *S. sobrinus* strongly inhibited the adherence of these bacteria to parotid saliva–coated hydroxyapatite (SHA) *in vitro*. The adherence of 35S-labeled *S. mutans* cells was dose-dependently inhibited by both the immune and control preparations when SHA was coated with either product before exposure to bacteria, but markedly lower concentrations of the immune preparation than the control preparation were needed for inhibition. When the bacterial cells were pretreated with the immune or control preparations, only the immune preparation strongly and dose-dependently inhibited the adherence of caries streptococci to SHA. Another recent study (41) showed that the above immune preparation added to milk and subsequently fermented with a probiotic *Lactobacillus rhamnosus* strain (LGG) had a slightly synergistic inhibitory effect on the adherence of caries streptococci to SHA. The activity of the specific antibodies remained unchanged for more than 2 weeks when stored at + 4°C, indicating that the lactic fermentation did not interfere with antibody activity. These results suggest that it might be beneficial to add milk-derived specific antibodies to fermented milk when developing liquid immune milk with an extended shelf life, specifically for the prevention of dental caries.

Antibodies against caries streptococci have proven effective also *in vivo*. Filler et al. (42) examined the use of bovine milk containing specific antibodies to *S. mutans* as a mouth rinse. Rinsing resulted in an initial reduction in the numbers of recoverable *S. mutans* in the entire group of nine individuals, whereas the 10 subjects consuming control milk had variable levels of plaque *S. mutans*. In another short-term study (43), an immune colostral preparation, a control preparation from nonimmunized cows, and water were used as mouth rinse for 3 days after initial antimicrobial cleaning of the dentition of the test subjects. The use of the immune preparation resulted in a higher resting pH in dental plaque as compared with the control preparations. The resting pH values of plaques with low "innate" pH increased after a period of rinsing with the immune product. The absolute number of cultivable facultative flora and total streptococci were unaffected by different rinsings, but the relative number of mutans streptococci significantly decreased after the rinsing period with the immune product when compared with the period of rinsing with the control product. These results are supported by another recent study (44), which examined the efficacy of an immune milk product containing specific antibodies against a fusion protein of the functional domains of two major *S.mutans* colonization factors, namely, a cell-surface protein antigen and the water-insoluble glucan-synthesizing enzyme I. Mouth rinsing with immune milk significantly inhibited recolonization of *S. mutans* in saliva and dental plaque after initial cleaning of the dentition of the test persons with cetylpyridinium. In contrast, the number of *S. mutans* cells in saliva and dental plaque in the control group increased immediately after the cetylpyridinium treatment and surpassed the baseline level at 42 and 28 days, respectively, after treatment. Also, the ratios of *S. mutans* to total streptococci in saliva and dental plaque of the group receiving immune milk were lower than those in the control group. Thus, rinsing with immune milk preparations containing specific anticaries antibodies seem to provide favorable effects on dental plaque by controlling *S. mutans* in the human oral cavity.

8.5.2.2 *Cryptosporidium parvum*

Immune milk preparations have proven effective *in vivo* in the prevention of infection induced by the enteric protozoan parasite *Cryptosporidium parvum* in both immunocompromised (45–47) and healthy humans (48). Immune milk was found to reduce the duration and severity of diarrhea and the number of oocysts in feces. Permanent eradication of *C. parvum* in human patients has also been reported (for reviews see 8,9). These results can be considered highly important, since there is no effective therapy at present for cryptosporidiosis, one of the contributors to mortality in patients with acquired immune deficiency syndrome (AIDS) . In contrast, nonimmune colostrum and nonspecific bovine Ig concentrate have been shown to afford only marginal protection against *Cryptosporidium* infection, again emphasizing the importance of specific antibodies (49–51). The preventive and therapeutic efficacy of specific antibodies is obviously due to their ability to attach onto the sporozoites released from the oocysts in the intestinal lumen, thus preventing them from attaching to and penetrating epithelial cells (52).

8.5.2.3 *Helicobacter*

Helicobacter pylori has been identified as the major etiological agent of active chronic gastritis and peptic ulcer disease (53). Many animal model and human clinical studies have been undertaken to establish the potential prophylactic and therapeutic efficacy of specific antibodies raised in bovine colostrum or milk against *H. pylori* or *H. felis*. A recent mouse-model study (54) demonstrated that an immune colostral preparation containing specific *H. felis* antibodies protected mice against experimental *H. felis* infection. The protection was dependent on the presence of specific antibodies in the preparation, the control preparation providing no protective effect. These results are supported by an earlier human study (55) that reported a strong negative correlation between the occurrence of *H. pylori* antibodies in the milk of Gambian mothers and the incidence of *H. pylori* infection in their small children. Casswall et al. (56) recently showed *in vitro* that a specific bovine immune preparation blocked almost 90% of *H. pylori* adherence to the human gastric mucosal tissue section, inhibited 95% of the binding of *H. pylori* to Lewis (b) glycoconjugate, and inhibited hemagglutination of *H. pylori* and human red blood cells. The promising results about the protective effect of bovine antibodies suggest that the incidence of *Helicobacter* infection could perhaps be lowered in countries having a high prevalence by adding specific bovine antibodies, for example, to infant formulas. However, disappointing results have been reported on the efficacy of specific immune colostrum preparations in the eradication of *Helicobacter* infection in infants and adult humans (57,58) (see section 8.5.4). It appears that the complete eradication of *Helicobacter* cannot be achieved with oral immunotherapy using bovine anti-*H. pylori* antibodies but that a significant reduction in the rate of *Helicobacter* colonization and the mitigation of clinical symptoms can be achieved.

8.5.2.4 *Escherichia coli*

Oral administration of 5 g d^{-1} of a bovine immune colostral whey concentrate containing specific antibodies that block the H antigen on the fimbrae of enteropathogenic *E. coli* (EPEC) strains has proven beneficial in preventing diarrhea caused by *E. coli* in humans (59) (see section 8.5.3). Further, natural antibodies to a colonization factor antigen (CFA-1) of human enterotoxigenic *E. coli* (ETEC) (3) as well as natural agglutinating and neutralizing antibodies to LPS from *E. coli* (38) have been found in normal bovine colostrum and milk.

8.5.2.5 *Shigella*

Promising results were obtained in a well-controlled study by Tacket et al. (60), who administered 30 g d^{-1} of bovine immune colostral whey concentrate containing specific antibodies against LPS from *Sh. flexneri* 2a to human volunteers for 3 days before and 4 days after experimental challenge with a virulent strain of the same pathogen. The immune milk preparation prevented the outcome of the illness in all 10 subjects, whereas 5 out of 11 subjects fell ill in the control group that had been similarly treated with an Ig preparation from nonimmunized cows. However, another double-blind, randomized, controlled trial (61) with children suffering from an already established *Sh. dysenteriae* infection showed that the administration of hyperimmune bovine colostrum against *Sh. dysenteriae* antigen-I in combination with an antibiotic agent failed to produce beneficial results when compared with results from patients treated with the control preparation and an antibiotic. These studies suggest that specific colostral antibodies can prevent shigellosis but are not effective in treating an already established infection.

8.5.2.6 *Candida albicans*

It has been reported that up to 76% of immunocompromised bone marrow transplantation patients develop invasive *Candida* infections. *Candida* colonization is endogenous, and infections occur during the neutropenic phase after transplantation. In a human clinical study by Tollemar et al. (62), patients with high buccal colonization (>100 colony-forming units (CFU) ml^{-1} saliva) were given 3.3 g of immune milk with specific antibodies against several *Candida* species three times daily for various periods before and after transplantation. Fifty-four percent of the patients were colonized with *C. albicans* before transplantation, and 72% were colonized at some point. In 7 out of 10 patients, buccal colonization was reduced, correlating with the length of the treatment period, and one patient became negative, whereas all patients in the control group remained colonized.

8.5.3 Neutralization of Toxins

Orally administered antibodies, also partially degraded ones, can benefit humans by attaching to toxins produced by harmful bacteria. Milk or colostral antibodies that neutralize or inactivate toxins have been shown to be effective in the treatment of diseases caused by toxin-producing *E. coli, C. difficile* strains, and *Vibrio cholerae*. Mietens et al. (63) treated 60 infants having diarrhea induced by EPEC for 10 days with 1 g kg^{-1} body weight of a bovine milk Ig concentrate containing specific EPEC antibodies. The treatment eliminated the pathogen in 43 out of 51 children infected with the strains present in the vaccine. The control group was treated similarly but was naturally infected with the strain not included in the vaccine. In the control group, the treatment was effective in 1 patient out of 9. In a study by Tacket et al. (59), a lyophilized and buffered colostral Ig concentrate (prepared from the colostrum of cows immunized with several ETEC strains and enterotoxins) was administered (5 g d^{-1} for 7 days) to volunteers, followed by a dose of 10^9 CFU of the ETEC/H10407 strain. This strain produces a colonization factor antigen-I and heat-labile and heat-stable enterotoxins. None of the 10 volunteers receiving the anti-*E. coli* Ig concentrate developed diarrhea, but 9 out of the 10 control subjects had symptoms of diarrhea. All volunteers, however, excreted *E. coli* H10407. It was concluded that both the antibodies against the colonization factor and the toxin-neutralizing antibodies had a decisive role in preventing the development of diarrhea. Recently, Huppertz et al. (64) successfully treated children infected by Shiga-toxin-producing *E. coli*, an intimin-expressing or an enterohemorrhagic *E. coli* (EHEC) infection, using bovine immune colostrum rich in specific antibodies against Shiga toxin and EHEC-hemolysin. This treatment resulted in reduced stool frequencies as compared with the control group treated with nonimmune colostrum. The above studies suggest that immune colostral or milk preparations could be useful, for example, in the prevention of traveler's diarrhea and infantile gastroenteritis caused by pathogenic *E. coli*.

However, field trials carried out so far show that a sufficient concentration of the specific antibodies is crucial for successful treatment. In a small-scale field trial (65) carried out in Chile, no protective effect was observed when an infant formula containing 1% (w/w) bovine milk Ig concentrate from cows hyperimmunized with human rotaviruses and the major EPEC serotypes was administered to infants (3 to 6 months old) for 6 months. The researchers suggested that the result might be attributed to the low level of antibody in the formula used. In a recent well-controlled clinical study (66) conducted in Bangladesh, no significant therapeutic benefit was established in infants (4 to 24 months old) suffering from acute diarrhea caused by EPEC/ETEC strains when the children were administered 20 g of bovine immune colostral whey concentrate containing anti-EPEC/ETEC antibodies for 4 consecutive days.

In animal model studies, bovine antibody preparations have been shown to neutralize bacterial toxins in the gastrointestinal tract. McClead and Gregory (30) reported that a specific bovine colostral Ig preparation against cholera enterotoxin decreased mortality and intestinal fluid responses in rabbits exposed to the cholera enterotoxin. Also, a colostral Ig preparation produced by hyperimmunizing cows with toxoids of *C. difficile* protected hamsters against *C. difficile*-induced diarrhea (67). Kink and Williams (68) showed that an Ig preparation from bovine immune colostrum was effective in the treatment of diarrhea caused by *C. difficile* in hamsters. In another study (69), a basically similar preparation neutralized the cytotoxic effects of *C. difficile* toxins on rat ileum both *in vitro* and *in vivo*. Thus, passive oral immunotherapy with colostrum-derived products containing anti-*C. difficile* antibodies may be a useful nonantibiotic approach to the prevention and treatment of *C. difficile* diarrhea and colitis.

8.5.4 Neutralization of Viruses

The virus-neutralizing property of colostral Igs has, thus far, been utilized almost solely against human rotavirus. Colostrum and milk are known to contain natural neutralizing antibodies to human rotavirus (70). Many clinical studies have demonstrated that bovine immune colostral preparations can protect infants from rotavirus and other viral infections (for reviews see 6,8,9). Such antibody preparations appear to be useful also in the treatment of rotavirus-infected children. Ebina et al. (71) demonstrated that a dosage of 20 to 50 ml d^{-1} of an immune colostrum containing antirotavirus antibodies for 3 days protected against infection, while infants fed commercial milk or purified bovine Igs (IgG, IgM, and IgA) were not protected against rotavirus. Davidson et al. (72) examined the efficacy of a 10-day course of administration (50 ml d^{-1}) of bovine colostrum having high antibody titers against the four known human rotavirus serotypes in protecting children (3 to 15 months old) against rotavirus infection. Nine out of 65 control children but none of the 55 treated children acquired rotavirus infection during the treatment period. A similar study was carried out in Hong Kong and India, confirming the above findings (73). It was concluded that the antibody titer is important for protection and that the immune colostrum could protect against more than one rotavirus serotype. Turner and Kelsey (74) showed that passive immunization of healthy infants with immune colostrum had no effect on the actual incidence of rotavirus infection but was effective in reducing symptoms in infected children. Hilpert et al. (75) tested the efficacy of a bovine immune colostral whey concentrate in 75 children hospitalized for acute rotavirus gastroenteritis. The children received the concentrate in a daily dose of 2 g kg^{-1} body weight for 5 days. A decrease in the duration of rotavirus excretion was noted, but there was no associated effect on clinical symptoms.

Other encouraging results were reported in a double-blind, placebo-controlled clinical study by Mitra et al. (76), carried out in Bangladesh with a group of rotavirus-infected children aged 6 to 24 months. They were administered 100 ml of immune bovine colostrum three times daily for 3 days. As compared with the control group children, who received nonimmune colostrum, the children treated with immune colostrum showed a significant reduction in the duration and severity of diarrhea. Ylitalo et al. (77) used a similar mode of administration (100 ml of hyperimmune colostrum four times per day for 4 days) for treating rotavirus-infected children and observed indicative but statistically nonsignificant improvement in all evaluated variables (weight gain, duration of diarrhea, and number of stools). In another successful study carried out in Bangladesh, Sarker et al. (78) treated children aged 4 to 24 months with 10 g of an Ig concentrate (made from the colostrum of rotavirus-immunized cows) in 20 ml of water four times per day for 4 days and achieved a significant reduction in the daily and total stool output as well as in the duration of diarrhea. It can be concluded from the above studies that the specific antibodies of immune colostral preparations can neutralize rotavirus in the human gastrointestinal tract and, as a result, provide effective protection against infection. Also, these preparations can be beneficial in the treatment of rotavirus infections, but an adequate intake of specific antibodies is needed in order to achieve positive results.

8.5.5 Opsonization

Opsonization — that is, augmenting the phagocytizing leukocytes to recognize, bind, activate, ingest, and to kill the pathogen — is the most important task of Igs, in addition to complement activation. Since the leukocyte receptor and antibody structures vary between species, it is not self-evident that Igs of bovine origin can augment the recognition and phagocytosis of human leukocytes. However, Loimaranta et al. (79) showed that an immune colostral whey protein preparation from cows immunized against mutans streptococci stimulated *in vitro* the phagocytosis and killing of mutans streptococci by human leucocytes when the same bacteria were opsonized with the immune preparation. Neutrophils, eosinophils, and monocytes weakly phagocytozed the nonopsonized bacteria. Also, bacteria opsonized with a control product from nonimmunized cows were poorly phagocytozed. Binding, onset of oxidative burst, and ingestion of *S. mutans*, in contrast, were more efficient when the immune preparation was used for opsonization in the test system. Thus, specific antimutans antibodies may be clinically useful, especially in preventing the colonization of newly erupted teeth in children by mutans streptococci. Due to the potential negative side effects of active immunization against cariogenic mutans streptococci, oral administration of passively derived antibodies could be a more acceptable way of reducing the colonization and virulence of these bacteria in human dentition.

In a recent preliminary experiment (Marnila et al., unpublished results) a whey protein concentrate made from bovine immune colostrum containing specific antibodies against *H. pylori* increased, in a dose-dependent manner, the activation of murine granulocytes against *H. pylori* cells measured as oxidative burst activity. Nonopsonized bacteria and bacteria opsonized with a control product from nonimmunized cows, however, activated the murine granulocytes only weakly. The apparent ability of bovine antibodies to support phagocytosis may augment the host immune system to resist *Helicobacter* infection. This hypothesis is supported by an earlier clinical trial on children and recent mouse-model studies. A clinical trial carried out in Estonia by Oona et al. (57) on a group of 16 children (aged 10 to 17 years) having an established *H. pylori* infection and suffering from gastritis symptoms showed that a treatment for 3 weeks with a daily dose of 12 g of a bovine immune colostral whey preparation having anti-*H. pylori* antibodies decreased in most subjects the severity of experienced symptoms and the degree of gastric inflammation as determined from gastric biopsies. Although the degree of *Helicobacter* colonization in the stomach mucosa was clearly reduced after the treatment, total eradication of *Helicobacter* infection was not observed in any of the treated children. In an earlier study by Korhonen et al. (80), the same immune preparation was found to be highly bactericidal against the *H. pylori* strain used in the immunization. The bactericidal activity was associated with the presence of specific antibodies and the active complement system. The alleviation of gastric inflammation suggests that the specific *H. pylori* antibodies help neutralize the proinflammatory components secreted by *Helicobacter* and reduce bacterial colonization by opsonization in the gastric mucosa and direct killing of bacteria. These results are consistent with those of Casswall et al. (58), who showed that *H. pylori* was not eradicated from any of a group of 11 infants (4 to 29 months old) treated for 30 days with a daily dose of 1 g of a bovine immune colostral whey concentrate containing specific *H. pylori* antibodies.

In a *H. pylori* mouse model, Casswall et al. (56) treated infected mice with various dosages of a bovine immune colostral preparation having specific anti-*H. pylori* antibodies. In the group of mice treated with the highest concentration of the immune preparation, only 7 out of 40 were infected after the treatment, whereas 25 out of 48 were infected in the control group. Thus, the efficacy of the therapeutic treatment seemed to be related to the concentration of the specific antibodies administered. Marnila et al. (54) studied the therapeutic efficacy of a bovine immune colostral preparation having specific anti-*H. felis* antibodies in a *H. felis* mouse model. In the first trial, a significant decrease in *H. felis* colonization in the gastric antra of infected mice was observed, and in the second trial, a respective decrease in the grade of gastric inflammation was recorded. However, a statistically significant eradication of *H . felis* was observed (*H. felis* found in none of the nine mice) only when the immune preparation was used in combination with antibiotic (amoxycillin) treatment. This antibiotic alone had a statistically indicative effect (*H. felis* found in two out of nine mice) only. In infected control groups receiving water, no significant eradication was seen (*H. felis* found in 33 out of 39 mice). Although different animal-pathogen models and treatment regimens yield somewhat variable results, it is obvious that the specific antibodies produced, and possibly also complement factors, in bovine colostrum and milk can support host immune defense against *Helicobacter* infection but that effective curing may require combining the specific antibodies with antibiotics.

8.6 SYNERGISTIC EFFECTS OF MILK ANTIBODIES WITH OTHER ANTIMICROBIAL FACTORS

Bovine colostrum and milk are known to contain a large number of naturally occurring antimicrobial substances which, unlike antibodies, are nonspecific by nature (17,81–84). The best-documented noncellular components are lactoferrin, the lactoperoxidase–thiocyanate–hydrogen peroxide system, and lysozyme. The specific antibodies in the colostrum of cows immunized with human pathogens have been found to exert synergistic effects *in vitro* with all these components (37,85). To what extent such interactions take place *in vivo* both in the mammary gland of the cow and in the human gastrointestinal tract upon ingestion of immune colostrum remains to be elucidated. All the above nonspecific components are known to retain at least some of their activity during the passage through the gastrointestinal tract and are known to modulate its resident microflora (86,87).

The most important task of Igs — in addition to opsonization — is the activation of the complement system, which consists of more than 20 different proteins acting in concert and leading eventually to a bacteriolytic effect in the case of sensitive bacteria (14,15). Bovine colostrum regularly (and milk irregularly) contains individual complement components or the complete complement system; both, in association with antibodies, form the major agent for the bactericidal activity found especially in colostrum against a great number of Gram-negative bacteria (for a review see 83). The natural antibody-mediated bacteriolytic activity of complement can be retained in freeze-dried colostral whey concentrate for a long period (more than 2 years) (54). It is, however, not known whether the complement system is effective *in vivo* against bacterial infections in the gut of bovine or human newborns, since the bactericidal activity of colostrum can be readily destroyed by low pH, trypsin, or pancreatic juice. However, it has been demonstrated experimentally that the loss of complement bioactivity due to trypsin can be prevented by the addition of excess bovine colostral trypsin inhibitor. This enzyme is known to occur naturally in colostrum (for a review see 14).

In a study by Korhonen et al. (80), all colostrum samples derived from normal healthy cows were naturally bactericidal against H. pylori, while none of the milk samples from the same animals showed bactericidal activity. In contrast, bactericidal activity was detected in 43% of the milk samples obtained from cows immunized prepartum with a *H. pylori* vaccine. The latter finding clearly demonstrates that the antibacterial activity of milk can be increased by systemic immunization of cows against a defined pathogen. Consistently, the bactericidal property of immune colostrum against complement-sensitive pathogens seems to be associated with a rise in the titer of pathogen-specific Igs (88,89). In view of the potential exploitation of the active complement system in formulating commercial bovine colostrum-based immune preparations, it should be observed that heating at 56°C for 30 minutes inactivates the complement system and abolishes the bactericidal antibody activity. Therefore, nonthermal treatments such as membrane filtration techniques should be applied as an alternative.

8.7 HOW CAN BOVINE MILK ANTIBODIES BENEFIT HUMAN HEALTH IN THE FUTURE

As described above, orally administered bovine milk and colostral Igs appear to have significant potential in the prophylaxis of many orally mediated infections. Furthermore, in case of already established infections, limited beneficial therapeutic effects have been reported in certain microbial diseases. The targeted disease conditions for potential applications of immune milk products may be grouped as follows:

1. The infection is maintained through a repeated reattachment and recolonization, for example, inside the oral cavity or the gastrointestinal lumen (examples are cariogenic streptococci, rotavirus, *E. coli, C. albicans, C. parvum, H. pylori*)
2. Toxins or inflammatory compounds that can be neutralized by the specific antibodies (e.g., *C. difficile, E. coli*) are involved.
3. More effective opsonization is needed due to a low local Ig level, for example, in the oral cavity or esophagus (e.g., *H. pylori,* cariogenic streptococci).

An emerging group of microbes that present an increasing problem worldwide in today's healthcare are antibiotic-resistant strains causing endemic hospital infections. The development of appropriate immune milk products to combat these infections appears as a highly interesting challenge for future research. Thus far, no well-controlled study on the efficacy of such intervention has been reported. Table 8.4 lists examples of potential areas of application for immune milk preparations.

In spite of many encouraging studies carried out so far, the future development of immune milk products is still challenged by a number of biological, technological, and even regulatory constraints. A few of them will be addressed briefly below. Orally ingested immune milk products are considered a pleasant way to decrease the risk of infectious diseases. In trials published hitherto, the preparations have been well tolerated, and no side effects have been observed. Milk allergy may, in some cases, be a limiting factor for the use of these products, since bovine IgG belongs to the whey proteins having allergenic potential (90). The potential allergenicity of Igs can, however, be reduced by a treatment with pepsin, which cleaves Ig in two F(ab)′2 fragments that are less immunogenic than the intact molecule but retain almost fully their antibody properties (91).

TABLE 8.4
Potential Applications of Immune Milk Preparations

Disease	Target Microorganism
Dental caries	*Streptococcus mutans, S. sobrinus*
Diarrhoea	Rotavirus
Diarrhoea	*E.coli* (EHEC, EPEC, ETEC)
Diarrhoea	*Campylobacter jejuni*
Diarrhoea	*Vibrio cholerae*
Parasite diarrhoea	*Cryptosporidium, Amoeba, Giardia*
Dysenteria	*Shigella flexneri, S. sonnei*
Antibiotic-associated diarrhoea	*Clostridium difficile*
Gastritis	*Helicobacter pylori*
Infantile otitis	*Haemophilus influenzae, Streptococcus pneumoniae*
Candidiosis	*Candida albicans*
Crohn's disease	Anaerobic intestinal bacteria
Rheumatoid arthritis	Anaerobic intestinal bacteria
Hospital infections	Antibiotic resistant bacteria, e.g., enterococci and staphylococci

However, the degradation of Igs in the stomach and intestines by gastric acid, bile juice, and proteolytic enzymes needs to be considered in the development of immune milk preparations. To this end, supplementary buffering and encapsulation techniques as well as specific vehicle agents should be developed and applied to maximize the access of antibodies in an active form to the desired locus in the gastrointestinal tract. For industrial large-scale production of Igs, milk and colostrum are ideal sources because of their ready availability and safety as compared, for example, with blood-derived analogues. Producing specific antibodies in cell cultures or in plants is currently too expensive, and there are no technologies available for the scale of production required for orally used commercial products.

The main limitation of the clinical use of bovine milk antibodies in humans is that they originate from a foreign species and can, thus, be used only against oral and gastrointestinal pathogens or for topical applications. In order to overcome this limitation, it may be necessary in the future to develop a chimeric cow that is able to produce human or humanized IgA or IgG antibodies. New immunotherapies based on chimeric monoclonal antibodies against proinflammatory cytokines and adhesion molecules have proven overwhelmingly effective in the treatment of chronic inflammatory diseases such as rheumatic arthritis and inflammatory bowel disease. These new-generation antibody-based pharmaceuticals provide huge possibilities in the future for the treatment of inflammatory and autoimmune diseases such as asthma, allergies, and many other diseases. At present, chimeric monoclonal antibodies are generated in cell cultures, which makes their testing and clinical use very expensive. Cow's milk may provide one solution to this problem because of its superior antibody production capacity.

The food and pharmaceutical industries have an increasing interest in developing products for the manipulation of oral and intestinal microflora. The possible benefits obtained from the dietary application of probiotic bacteria in combination with specific antibodies may prove useful and warrants thorough investigation. In conclusion, it is anticipated that immune colostral or milk-based antibody preparations, targeted at specific consumer or patient groups, may in the future have remarkable potential to contribute to human health, both as part of a health-promoting diet and as an alternative or a supplement to the medical cure of specified human diseases.

REFERENCES

1. P. Ehrlich. Über Immunität durch Vererbung und Säugung. *Z. Hyg Infektionskrankh.*, 12: 183, 1892.
2. B. Campbell, W.E. Petersen. Immune milk-a historical survey. *Dairy Sci. Abstr.*, 25: 345–358, 1963.
3. M. Facon, B.J. Skura, S. Nakai. Potential immunological supplementation of foods. *Food Agric. Immunol.*, 5: 85–91, 1993.
4. L. Hammarström, A. Gardulf, V. Hammarström, A. Janson, K. Lindberg, C.I. Smith. Systemic and topical immunoglobulin treatment in immunocompromised patients. *Immunol. Rev.*, 139: 43–70, 1994.
5. L.P. Ruiz. Antibodies from milk for the prevention and treatment of diarrheal disease. *Indigen. Antimicrob. Agents Milk-Recent Dev.*, (IDF Special Issue) No. 9404: 108–121, 1994.
6. G.P. Davidson. Passive protection against diarrhoeal disease. *J. Ped. Gastroenterol. Nutr.*, 23: 207–212, 1996.
7. A.K. Bogstedt, K. Johansen, H. Hatta, M. Kim, T. Casswall, L. Svensson, L. Hammarström. Passive immunity against diarrhoea. *Acta Paediatr.*, 85: 125–128, 1996.
8. C. Weiner, Q. Pan, M. Hurtig, T. Boren, E. Bostwick, L. Hammarström. Passive immunity against human pathogens using bovine antibodies. *Clin. Exp. Immunol.*, 116: 193–205, 1999.
9. H. Korhonen, P. Marnila, H. Gill. Bovine milk antibodies for health: a review. *Br. J. Nutr.*, 84 (Suppl 1): 135–146, 2000.
10. E-M. Lilius, P. Marnila. The role of colostral antibodies in prevention of microbial infections. *Curr. Opin. Infect. Dis.*, 14: 295–300, 2001.
11. R.A. Hoerr, E.F. Bostwick. Commercializing colostrum-based products: a case study of *Gala. Gen. Inc. IDF Bull.*, 375: 33–46, 2002.
12. B.L. Larson. Immunoglobulins of the mammary secretions. In: P.F. Fox (Ed.) *Advanced Dairy Chemistry*, 1-Proteins. London: Elsevier Science Publishers, 1992, 231–254.
13. J.E. Butler. Immunoglobulin diversity, B-cell and antibody repertoire development in large animals. *Vet. Immun. Immunopathol.*, 4, 43–152, 1983.
14. H. Korhonen, P. Marnila, H.S. Gill. Milk immunoglobulins and complement factors. *Br. J. Nutr.*, 84(Suppl 1): S75–S80, 2000.
15. P. Marnila, H. Korhonen. Immunoglobulins. In: H. Roginski, J.W. Fuquay, P.F. Fox (Eds.) *Encyclopedia of Dairy Sciences*. London: Academic Press, 2002, 1950–1956.

16. J.D. Quigley, J.J. Drewry. Nutrient and immunity transfer from cow to calf pre- and postcalving. *J. Dairy Sci.*, 81: 2779–2790, 1998.
17. P. Marnila, H. Korhonen. Colostrum. In: H. Roginski, J.W. Fuquay, P.F. Fox (Eds.) *Encyclopedia of Dairy Sciences*. London: Academic Press, 2002, 473–478.
18. D. Levieux, A. Ollier. Bovine immunoglobulin G, β-lactoglobulin, α-lactalbumin and serum albumin in colostrum and milk during the early *post partum* period. *J. Dairy Sci.*, 66: 421–430, 1999.
19. P.M. Kelly, D. McDonagh. Innovative dairy ingredients. *World Food Ingred.*, Oct/Nov: 24–32, 2000.
20. P. Lindström, M. Paulsson, T. Nylander, U. Elofsson, H. Lindmark-Månsson. The effect of heat treatment on bovine immunoglobulins. *Milchwiss*, 49: 67–70, 1994.
21. E. Li-Chan, A. Kummer, J.N. Losso, D.D. Kitts, S. Nakai. Stability of bovine immunoglobulins to thermal treatment and processing. *Food Res. Int.*, 28: 9–16, 1995.
22. E. Dominguez, M.D. Perez, M. Calvo. Effect of heat treatment on the antigen-binding activity of anti-peroxidase immunoglobulins in bovine colostrum. *J. Dairy Sci.*, 80: 3182–3187, 1998.
23. G. Mainer, E. Dominguez, M. Randrup, L. Sanchez, M. Calvo. Effect of heat treatment on anti-rotavirus activity of bovine milk. *J. Dairy Res.*, 66: 131–137, 1999.
24. L.R. Fukumoto, B.J. Skura, S. Nakai. Stability of membrane-sterilized bovine immunoglobulins aseptically added to UHT milk. *J. Food Sci.*, 59: 757–759, 1994.
25. J. Husu, E-L. Syväoja, H. Ahola-Luttila, H. Kalsta, S. Sivelä, T.U. Kosunen. Production of hyperimmune bovine colostrum against *Campylobacter jejuni*. *J. Appl. Bacteriol.*, 74: 564–569, 1993.
26. R.M. Reilly, R. Domingo, J. Sandhu. Oral delivery of antibodies. *Clin. Pharmacokinet.*, 32: 313–323, 1997.
27. A.S. Goldman. The immune system of human milk: antimicrobial, antiinflammatory and immunomodulating properties. *Pediatr. Infect. Dis. J.*, 12: 664–671, 1993.
28. O. de Rham, H. Isliker. Proteolysis of bovine immunoglobulins. *Int. Arch. Allergy Appl. Immunol.*, 55: 61–69, 1977.
29. J.H. Brock, A. Pineiro, F. Lampreave. The effect of trypsin and chymotrypsin on the antibacterial activity of complement, antibodies, and lactoferrin and transferrin in bovine colostrum. *Ann. Rech. Vèt.*, 9: 287–294, 1978.
30. R.E. McClead, S.A. Gregory. Resistance of bovine colostral anti-cholera toxin antibody to *in vitro* and *in vivo* proteolysis. *Infect. Immun.*, 44: 474–478, 1984.
31. B.W. Petschow, R.D. Talbott. Reduction in virus-neutralizing activity of a bovine colostrum immunoglobulin concentrate by gastric acid and digestive enzymes. *J. Pediatr. Gastroenterol. Nutr.*, 19: 228–235, 1994.
32. N. Roos, S. Mahé, R. Benamouzig, H. Sick, J. Rautureau, D. Tomé. 15-N- labeled immunoglobulins from bovine colostrum are partially resistant to digestion in human intestine. *J. Nutr.*, 125: 1238–1244, 1995.
33. M. Warny, A. Fatimi, E.F. Bostwick, D.C. Laine, F. Lebel, J.T. LaMont, C. Pothoulakis, C.P. Kelly. Bovine immunoglobulin concentrate — *Clostridium difficile* retains *C difficile* toxin neutralising activity after passage through the human stomach and small intestine. *Gut*, 44: 212–217, 1999.
34. J. Pacyna, K. Siwek, S.J. Terry, E.S. Robertson, R.B. Johnson, G.P. Davidson. Survival of rotavirus antibody activity derived from bovine colostrum after passage through the human gastrointestinal tract. *J. Pediatr. Gastroenterol. Nutr.*, 32: 162–167, 2001.

35. C.P. Kelly, S. Chetman, S. Keates, E. Bostwick, A.M. Roush, I. Castagliolo, J.T. Lamont, C. Pothoulakis. Survival of anti-*Clostridium difficile* bovine immunoglobulin concentrate in the human gastrointestinal tract. *Antimicrob. Agents Chemother.*, 41: 236–241, 1997.
36. V. Loimaranta, J. Tenovuo, S. Virtanen, P. Marnila, E. Syväoja, T. Tupasela, H. Korhonen. Generation of bovine immune colostrum against *Streptococcus mutans* and *Streptococcus sobrinus* and its effect on glucose uptake and extracellullar polysaccharide formation by mutans streptococci. *Vaccine*, 15: 1261–1268, 1997.
37. V. Loimaranta, J. Tenovuo, H. Korhonen. Combined inhibitory effect of bovine immune whey and peroxidase-generated hypothiocyanate against glucose uptake by *Streptococcus mutans.* Short communication. *Oral Microbiol. Immunol.*, 13: 378–381, 1998.
38. J.N. Losso, J. Dhar, A. Kummer, E. Li-Chan, S. Nakai. Detection of antibody specificity of raw bovine and human milk to bacterial lipopolysaccharides using PCFIA. *Food Agric. Immunol.*, 5: 231–239, 1993.
39. T. Koga, T. Oho, Y. Shimazaki, Y. Nakano. Immunization against dental caries. *Vaccine*, 20: 2027–2044, 2002.
40. V. Loimaranta, A. Carlen, J. Olsson, J. Tenovuo, E-L. Syväoja, H. Korhonen. Concentrated bovine colostral whey proteins from *Streptococcus mutans/Strep. sobrinus* immunized cows inhibit the adherence of *Strep. mutans* and promote the aggregation of mutans streptococci. *J. Dairy Res.*, 65: 599–607, 1998.
41. H. Wei, V. Loimaranta, J. Tenovuo, S. Rokka, E.-L. Syväoja, H. Korhonen, V. Joutsjoki, P. Marnila. Stability and activity of specific antibodies against *Streptococcus mutans* and *Streptococcus sobrinus* in bovine milk fermented with *Lactobacillus rhamnosus* strain GG or treated at ultra-high temperature. *Oral Microbiol. Immunol.*, 17: 9–15, 2002.
42. S.J. Filler, R.L. Gregory, S.M. Michalek, J. Katz, J.R. McGhee. Effect of immune bovine milk on *Streptococcus mutans* in human dental plaque. *Arch. Oral Biol.*, 36: 41–47, 1991.
43. V. Loimaranta, M. Laine, E. Söderling, E. Vasara, S. Rokka, P. Marnila, H. Korhonen, O. Tossavainen, J. Tenovuo. Effects of bovine immune- and non-immune whey preparations on the composition and pH response of human dental plaque. *Eur. J. Oral Sci.*, 107: 244–250, 1999.
44. Y. Shimazaki, M. Mitoma, T. Oho, Y. Nakano, Y. Yamashita, K. Okano, Y. Nakano, M. Fukuyama, N. Fujihara, Y. Nada, T. Koga. Passive immunization with milk produced from an immunized cow prevents oral recolonization by *Streptococcus mutans*. *Clin. Diagn. Lab. Immunol.*, 8: 1136–1139, 2001.
45. S. Tzipori, D. Robertson, D.A. Cooper, L. White. Chronic cryptosporidial diarrhoea and hyperimmune cow colostrum. *Lancet*, 2 (8554): 344–345, 1987.
46. J. Nord, P. Ma, D. DiJohn, S. Tzipori, C.O. Tacket. Treatment with bovine hyperimmune colostrum of cryptosporidial diarrhea in AIDS patients. *AIDS*, 4: 581–584, 1990.
47. P.D. Greenberg, J.P. Cello. Treatment of severe diarrhea caused by *Cryptosporidium parvum* with oral bovine immunoglobulin concentrate in patients with AIDS. *J. Acquir. Immun. Def. Syndr. Hum. Retrovirol.*, 13: 348–354, 1996.
48. P.C. Okhuysen, C.L. Chappell, J. Crabb, L.M. Valdez, E.T. Douglass, H.L. DuPont. Prophylactic effect of bovine anti-*Cryptosporidium* hyperimmune colostrum immunoglobulin in healthy volunteers challenged with *Cryptosporidium parvum.* *Clin. Infect. Dis.*, 26: 1324–1329, 1998.

49. A. Saxon, W. Weinstein. Oral administration of bovine colostrum anti-*Cryptosporidium* antibody fails to alter the course of human cryptosporidiosis. *J. Parasitol.*, 73: 413–415, 1987.
50. J.A. Rump, R. Arndt, A. Arnold, C. Bendick, H. Dichtelmuller, M. Franke, E.B. Helm, H. Jager, B. Kampmann, P. Kolb. Treatment of diarrhoea in human immunodeficiency virus-infected patients with immunoglobulins from bovine colostrum. *Clin. Invest.*, 70: 588–594, 1992.
51. A. Plettenberg, A. Stoehr, H.J. Stellbrink, H. Albrecht, W. Meigel. A preparation from bovine colostrum in the treatment of HIV-positive patients with chronic diarrhea. *Clin. Invest.*, 71: 42–45, 1993.
52. L.E. Perryman, M.W. Riggs, P.H. Mason, R. Fayer. Kinetics of *Cryptosporidium parvum* sporozoite neutralization by monoclonal antibodies, immune bovine serum, and immune bovine colostrum. *Infect. Immun.*, 58: 257–259, 1990.
53. R.M. Peek, M.J. Blaser. Pathophysiology of *Helicobacter pylori*–induced gastritis and peptic ulcer disease. *Am. J. Med.*, 102: 200–207, 1997.
54. P. Marnila, S. Rokka, L. Rehnberg-Laiho, P. Kärkkäinen, T.U. Kosunen, H. Rautelin, M.L. Hänninen, E.L. Syväoja, H. Korhonen. Prevention and suppression of *Helicobacter felis* infection in mice using colostral preparation with specific antibodies. *Helicobacter.*, 8: 192–201, 2003.
55. J.E. Thomas, S. Austin, A. Dale, P. McClean, M. Harding, W.A. Coward, L.T. Weaver. Protection by human milk IgA against *Helicobacter pylori* infection in infancy. *Lancet*, 342 (8863): 121, 1993.
56. T.H. Casswall, H.O. Nilsson, L. Björck, S. Sjöstedt, L. Xu, C.K. Nord, T. Boren, T. Wadström, L. Hammarström. Bovine anti-*Helicobacter pylori* antibodies for oral immunotherapy. *Scand. J. Gastroenterol.*, 37: 1380–1385, 2002.
57. M. Oona, T. Rägo, H-I. Maaroos, M. Mickelsaar, K. Lõivukene, S. Salminen, H. Korhonen. *Helicobacter pylori* in children with abdominal complaints: has immune bovine colostrum some influence on gastritis? *Alpe. Adria. Microbiol. J.*, 6: 49–57, 1997.
58. T.H. Casswall, S.A. Sarker, M.J. Albert, G.J. Fuchs, M. Bergström, L. Björck, L. Hammarström. Treatment of *Helicobacter pylori* infection in infants in rural Bangladesh with oral immunoglobulins from hyperimmune bovine colostrum. *Aliment. Pharmacol. Ther.*, 12: 563–568, 1998.
59. C.O. Tacket, G. Losonsky, H. Link, Y. Hoang, P. Guesry, H. Hilpert, M.M. Levine. Protection by milk immunoglobulin concentrate against oral challenge with enterotoxigenic *Escherichia coli*. *N. Engl. J. Med.*, 318: 1240–1243, 1988.
60. C.O. Tacket, S.B. Binion, E. Bostwick, G. Losonsky, M.J. Roy, R. Edelman. Efficacy of bovine milk immunoglobulin concentrate in preventing illness after *Shigella flexneri* challenge. *Am. J. Trop. Med. Hyg.*, 47: 276–283, 1992.
61. H. Ashraf, D. Mahalanabis, A.K. Mitra, S. Tzipori, G.J. Fuchs. Hyperimmune bovine colostrum in the treatment of shigellosis in children: a double-blind, randomized, controlled trial. *Acta Paediatr.*, 90: 1373–1378, 2001.
62. J. Tollemar, N. Gross, N. Dolgiras, C. Jarstrand, O. Ringdén, L. Hammarström. Fungal prophylaxis by reduction of fungal colonization by oral administration of bovine anti-*Candida* antibodies in bone marrow transplant recipients. *Bone Marrow Transplant.*, 23: 283–290, 1999.
63. C. Mietens, H. Kleinhorst, H. Hilpert, H. Gerber, H. Amster, J.J. Pahud. Treatment of infantile *E.coli* gastroenteritis with specific bovine anti-*E. coli* milk immunoglobulins. *Eur. J. Pediatr.*, 132: 239–252, 1979.

64. H.I. Huppertz, S. Rutkowski, D.H. Busch, R. Eisebit, R. Lissner, H. Karch. Bovine colostrum ameliorates diarrhea in infection with diarrheagenic *Escherichia coli*, shiga toxin-producing *E. coli*, and *E. coli* expressing intimin and hemolysin. *J. Pediatr. Gastroenterol. Nutr.*, 29: 452–456, 1999.
65. O. Brunser, J. Espinoza, G. Figueroa, M. Araya, E. Spencer, H. Hilpert, H. Link-Amster, H. Brüssow. Field trial of an infant formula containing anti-rotavirus and anti-*Escherichia coli* milk antibodies from hyperimmunized cows. *J. Pediatr. Gastroenterol. Nutr.*, 15: 63–72, 1992.
66. T.H. Casswall, S.A. Sarker, S.M. Faruque, A. Weintraub, M.J. Albert, G.J. Fuchs, N.H. Alam, A.K. Dahlström, H. Link, H. Brüssow, L. Hammarström. Treatment of enterotoxigenic and enteropathogenic *Escherichia coli*–induced diarrhoea in children with bovine immunoglobulin milk concentrate from hyperimmunized cows: a double-blind, placebo-controlled, clinical trial. *Scand. J. Gastroenterol.*, 35: 711–718, 2000.
67. D.M. Lyerly, E.F. Bostwick, S.B. Binion, T.D. Wilkins. Passive immunization of hamsters against disease caused by *Clostridium difficile* by use of bovine immunoglobulin G concentrate. *Infect. Immun.*, 59: 2215–2218, 1991.
68. J.A. Kink, J.A. Williams. Antibodies to recombinant *Clostridium difficile* toxins A and B are an effective treatment and prevent relapse of *C. difficile*-associated disease in a hamster model of infection. *Infect. Immun.*, 66: 2018–2025, 1998.
69. C.P. Kelly, C. Pothoulakis, F. Vavva, I. Castagliolo, E. Bostwick, C. O'Keane, S. Keates, T. LaMont. Anti-*Clostridium difficile* bovine immunoglobulin concentrate inhibits cytotoxicity and enterotoxicity of *C. difficile* toxins. *Antimicrob. Agents Chemother.*, 40: 373–379, 1996.
70. R.H. Yolken, G. Losonsky, S. Vonderfecht, F. Leister, S.B. Wee. Antibody to human rotavirus in cow's milk. *N. Engl. J. Med.*, 312: 605–610, 1985.
71. T. Ebina, A. Sato, K. Umezu, N. Ishida, S. Ohyama, A. Oizumi, K. Aikawa, S. Katagiri, N. Katsushima, A. Imai, S. Kitaoka, H. Suzuki, T. Konno. Prevention of rotavirus infection by oral administration of colostrum containing anti-human rotavirus antibody. *Med. Microbiol. Immunol. (Berl.)*, 174: 177–185, 1985.
72. G.P. Davidson, E. Daniels, H. Nunan, A.G. Moore, P.B.D. Whyte, K. Franklin, P.I. McCloud, D.J. Moore. Passive immunisation of children with bovine colostrum containing antibodies to human rotavirus. *Lancet*, 23 (8665): 709–712, 1989.
73. G.P. Davidson, J. Tam, C. Kirubakaran. Passive protection against symptomatic hospital acquired rotavirus infection in India and Hong Kong. *J. Ped. Gastroenterol.*, 19: 351, 1994.
74. R.B. Turner, D.K. Kelsey. Passive immunization for prevention of rotavirus illness in healthy infants. *Pediatr. Infect. Dis. J.*, 12: 718–722, 1993.
75. H. Hilpert, H. Brüssow, C. Mietens, J. Sidoti, L. Lerner, H. Werchau. Use of bovine milk concentrate containing antibody to rotavirus to treat rotavirus gastroenteritis in infants. *J. Infect. Dis.*, 156: 158–166, 1987.
76. A.K. Mitra, D. Mahalanibus, H. Ashraf, L. Unicomb, R. Eeckels, S. Tzipori. Hyperimmune cow colostrum reduces diarrhoea due to rotavirus : a double-blind, controlled clinical trial. *Acta Pediatr.*, 84: 996–1001, 1995.
77. S. Ylitalo, M. Uhari, S. Rasi, J. Pudas, J. Leppäluoto. Rotaviral antibodies in the treatment of acute rotaviral gastroenteritis. *Acta Paediatr.*, 87: 264–267, 1998.
78. S.A. Sarker, T.H. Casswall, D. Mahalanabis, N.H. Alam, M.J. Albert, H. Brüssow, G.J. Fuchs, L. Hammarström. Successful treatment of rotavirus diarrhea in

children with immunoglobulin from immunized bovine colostrum. *Pediatr. Infect. Dis. J.*, 17: 1149–1154, 1998.
79. V. Loimaranta, J. Nuutila, P. Marnila, J. Tenovuo, H. Korhonen, E-M. Lilius. Colostral proteins from cows immunised with *Streptococcus mutans/Strep. sobrinus* support the phagocytosis and killing of mutans streptococci by human leukocytes. *J. Med. Microbiol.*, 48: 917–926, 1999.
80. H. Korhonen, E-L. Syväoja, H. Ahola-Luttila, S. Sivelä, S. Kopola, J. Husu, T.U. Kosunen. Bactericidal effect of bovine normal and immune serum, colostrum and milk against *Helicobacter pylori. J. Appl. Bacteriol.*, 78: 655–662, 1995.
81. R. Pakkanen, J. Aalto. Growth factors and antimicrobial factors of bovine colostrum. *Int. Dairy J.*, 7: 285–297, 1997.
82. G.O. Regester, G.W. Smithers, I.R. Mitchell, G.H. McIntosh, D.A. Dionysis. Bioactive factors in milk: natural and induced. In: R.A.S. Welch, D.J.W. Burns, S.R. Davis, A.I. Popay, C.G. Prosser (Eds.) *Milk Composition, Production and Biotechnology*. Wallingford UK: Cab International, 1997, 119–132.
83. H. Korhonen. Antibacterial and antiviral activities of whey proteins. In: *The Importance of Whey and Whey Components in Food and Nutrition.* Proceedings of the.Third International Whey Conference, Munich. Hamburg: B. Behr's Verlag GmbH&Co, 2002, 303–321.
84. D.A. Clare, G.L. Catignani, H.E. Swaisgood. Biodefense properties of milk: the role of antimicrobial proteins and peptides. *Curr. Pharm. Des.*, 9: 1239–1255, 2003.
85. N. Takahashi, G. Eisenhuth, I. Lee, N. Laible, S. Binion, C. Schachtele. Immunoglobulins in milk from cows immunized with oral strains of *Actinomyces*, *Prevotella*, *Porphyromonas, and Fusobacterium. J. Dental Res.*, 71: 1509–1515, 1992.
86. N.P. Shah. Effects of milk-derived bioactives: an overview. *Br. J. Nutr.*, 84 (Suppl 1): S3–S10, 2000.
87. A.C.M. van Hooijdonk, K.D. Kussendrager, J.M. Steijns. *In vivo* antimicrobial and antiviral activity of components in bovine milk and colostrum involved in non-specific defence. *Br. J. Nutr.*, 84 (Suppl 1): S127–S134, 2000.
88. H. Korhonen, E-L. Syväoja, H. Ahola-Luttila, S. Sivelä, S. Kopola, J. Husu, T.U. Kosunen. *Helicobacter pylori*-specific antibodies and bactericidal activity in serum, colostrum and milk of immunized and non-immunized cows. In: *Indigenous Antimicrobial Agents of Milk — Recent Developments.* (IDF Special Issue). 9404: 151–163, 1994.
89. E.M. Early, H. Hardy, T. Forde, M. Kane. Bactericidal effect of a whey protein concentrate with anti-*Helicobacter pylori* activity. *J. Appl. Microbiol.*, 90: 741–748, 2001.
90. J.M. Bernhisel-Broadbent, R.H.M. Yolken, H.A.M. Sampson. Allergenicity of orally administered immunoglobulin preparations in food-allergic children. *Pediatrics*, 87: 208–214, 1991.
91. C. Lefranc-Millot, D. Vercaigne-Marko, J.M. Wal, A. Leprete, G. Peltre, P. Dhulster, D. Guillochon. Comparison of the IgE titers to bovine colostral G immunoglobulins and their F(ab′)2 fragments in sera of patients allergic to milk. *Int. Arch. Allergy Immunol.*, 110: 156–162, 1996.
92. D.J. Freedman, C.O. Tacket, A. Delehanty, D.R. Maneval, J. Nataro, J.H. Crabb. Milk immunoglobulin with specific activity against purified colonization factor antigens can protect against oral challenge with enterotoxigenic *Escherichia coli. J. Infect. Dis.*, 177: 662–667, 1998.

9 Avian Immunoglobulin Y and Its Application in Human Health and Disease

Jennifer Kovacs-Nolan, Yoshinori Mine, and Hajime Hatta

CONTENTS

9.1 INTRODUCTION

The existence of an immunoglobulin G (IgG)-like molecule in chickens was reported by Klemperer (1) over 100 years ago. This molecule was determined to be the major serum immunoglobulin in hens and was named IgY due to its deposition in the egg yolk to provide passive protection for the developing chick. The immune system of the chicken, and in particular IgY, has since been studied extensively, revealing the many important characteristics and applications of IgY.

Due to its favorable immunochemical characteristics, IgY has found many applications in the medicine and research, including diagnostics, and as an immunoaffinity ligand. It has received the most attention for its potential in immunotherapy applications. Passive immunotherapy — the administration of preformed antibodies to prevent infectious diseases or treat existing conditions — may be one of the most valuable applications of antibodies. It is currently driven by the need to find alternatives to traditional antibiotic therapy because of an increasing number of antibiotic-resistant organisms and pathogens (2,3). It is often limited, however, due to the large amounts of antibody often required for treatment (4). Eggs present an ideal source of specific antibodies for immunotherapeutic applications, as large quantities of IgY are transferred to the egg yolk and can easily be extracted and purified. Also, since eggs are common dietary components, they should be an acceptable and safe alternative source of antibodies for passive immunotherapy techniques (5). The structural, physical, and immunochemical properties of IgY as well as its applications in immunotherapy, in diagnosis, and as an immunoaffinity ligand will be discussed in this chapter.

9.2 EGG YOLK AS A SOURCE OF ANTIBODIES

9.2.1 Antibodies and Avian Immunity

Hens confer passive immunity on their offspring by transferring serum immunoglobulins to their eggs (6,7). IgY is the functional equivalent of IgG, the major serum antibody found in mammals, and it is selectively transferred to the egg yolk via a receptor on the surface of the yolk membrane specific for IgY translocation (8–10) (Figure 9.1). IgY makes up approximately 75% of the total antibody population (3). IgA and IgM have also been found to exist in the chicken and are believed to be synthesized in the oviduct and secreted directly into the egg white (11). The serum concentrations of IgY, IgA, and IgM have been reported to be 5.0, 1.25, and 0.61 mg ml^{-1}, respectively (12). The egg white contains IgA and IgM at concentrations of around 0.15 and 0.7 mg ml^{-1}, respectively, whereas the yolk may contain from 5 to 25 mg ml^{-1} of IgY (11,13,14).

9.2.2 Advantages of Eggs as an Antibody Source

There are several advantages to the production of polyclonal antibodies in chickens. The evolutionary distance between chickens and mammals makes the chicken an excellent host for the production of antibodies against highly conserved mammalian proteins, which would otherwise not be possible in mammals, and much less antigen is required to produce an efficient immune response (15). These benefits were demonstrated by Knecht et al. (16), who created and compared IgY and IgG raised against dihydroorotate dehydrogenase.

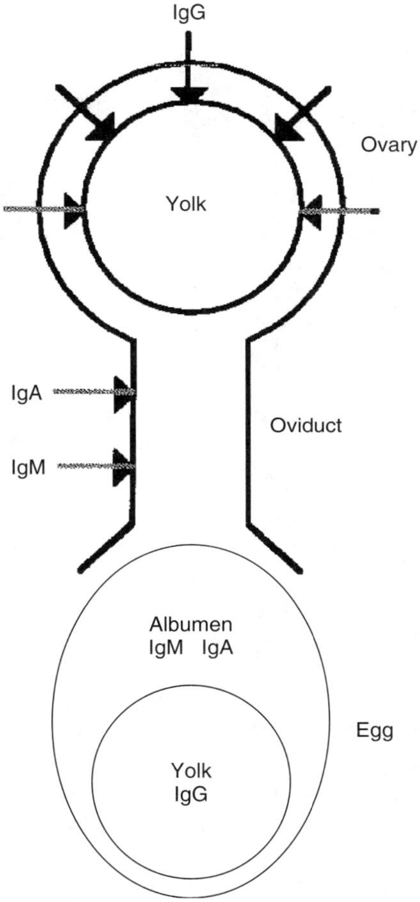

FIGURE 9.1 Distribution of immunoglobulins in hen egg. (Adapted from H. Hatta, M. Ozeki, K. Tsuda,in *Hen Eggs: Their Basic and Applied Science*. New York: CRC Press, 151–178, 1997. With permission.).

The method of producing antibodies in hens is less invasive, requiring only the collection of eggs, rather than the collection of blood or sacrificing of animals, as is the case with antibody production in mammals, and is therefore less stressful on the animal (13). Also, sustained high titers in chickens reduce the need for frequent injections (17). The eggs can be easily harvested, and the IgY may be purified from the yolk by simple precipitation techniques (17). The difference in antibody preparation procedures from rabbits and hens is illustrated in Figure 9.2. Large quantities of IgY are transferred to the egg yolk; it has been estimated that the productivity of antibodies in hens may be as much as 30 times greater than that in rabbits, based on the amount of antibody produced per animal per year (4).

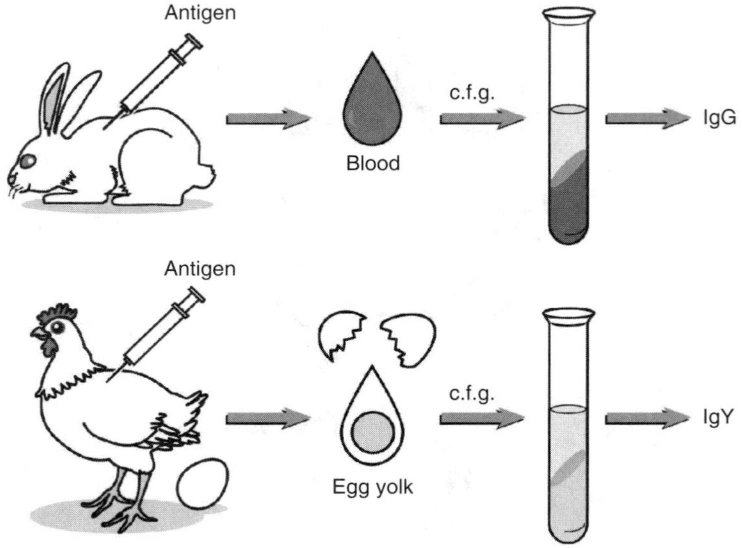

FIGURE 9.2 Preparation of specific antibody from rabbit and hen. (Adapted from H. Hatta, M. Ozeki, K. Tsuda,in *Hen Eggs: Their Basic and Applied Science*. New York: CRC Press, 151–178, 1997. With permission.)

The systematic vaccination of large numbers of chickens, long used to control avian infections in flocks (18), and the industrial-scale automation of egg collection and separation (19) render possible the large-scale production of IgY. The animal-care costs are also lower for chickens when compared with those for mammals (3). In the egg, IgY is stable for months, and once purified, it may be kept for years in cold storage (20).

9.2.3 Production and Purification of IgY

Egg yolk contains a considerable amount of IgY, around 100 to 150 mg per egg (11). A laying hen produces an average of 240 eggs per year (21). Therefore, an immunized hen may produce over 30 g of IgY a year.

Egg yolk is a fluid emulsion of a continuous phase of water-soluble protein and a dispersed phase of lipoprotein particles (4). The purification of IgY has been extensively studied; the major problem in isolating the IgY is the separation of the lipoproteins from the egg yolk prior to purification.

Several purification methods using water dilution, followed by centrifugation or ultrafiltration, to isolate the water soluble fraction (WSF) have been reported (22–24). These methods rely on the aggregation of yolk lipoproteins at low ionic strengths (25). Freezing and thawing of diluted yolk has also been employed, resulting in the formation of lipid aggregates that are large enough to be removed by conventional low-speed centrifugation (26). These methods are often performed as preliminary purification steps and are highly dependent on pH and extent of dilution. Nakai et al. (27) found that the best results were obtained using a six-fold water dilution, at pH 5. Ammonium sulfate precipitation has also been reported for the purification of IgY from WSF, following lipoprotein precipitation (22,28). Other methods of IgY separation include: delipidation by organic solvents (28–35) and lipoprotein precipitation by polyethylene glycol (28,36,37), sodium dextran sulfate (25,37), and dextran blue (38). Natural gums such as xanthan gum (37,39), sodium alginate (39), and carrageenan (39) have also been used for IgY purification. Chang et al. (40) reported the precipitation of over 90% of lipoproteins from yolk using λ-carrageenan, sodium alginate, carboxymethyl cellulose, and pectin. Ion exchange chromatography has been reported as a final step in IgY purification (22,41–43), as well as hydrophobic interaction chromatography (44) and preparative electrophoresis (45).

Because of the failure of IgY to bind proteins A and G and its sensitivity to traditional affinity purification conditions, alternative methods for the affinity purification of IgY have been examined, including immobilized metal ion affinity chromatography (46,47), thiophilic interaction chromatography (48), affinity chromatography using alkaline conditions (49), and synthetic peptide ligands, designed specifically for immobilizing antibodies (50,51).

9.3 STRUCTURE AND IMMUNOCHEMICAL PROPERTIES OF IgY

Although IgY and IgG share a similar function, they differ markedly in structure and immunochemical properties. IgY is composed of two heavy (H) chains and two light (L) chains and has a molecular mass of 180 kDa, larger than that of mammalian IgG (150 kDa) (Figure 9.3). This is due, in part, to the fact that the H-chain of IgY has a larger molecular mass (68 kDa) than that of IgG (50 kDa). The H-chain of IgG consists of four domains: the variable domain (V_H) and three constant domains (Cγ1, Cγ2, and Cγ3). The Cγ1 domain of the IgG H-chain is separated from the Cγ2 domain by a hinge region, which confers flexibility to the Fab fragments. The H-chain of IgY, in contrast, does not have a hinge region and possesses four constant domains (Cυ1 to Cυ4) (52).

Sequence comparisons between IgG and IgY have indicated that the Cγ2 and Cγ3 domains of IgG are closely related to the Cυ3 and Cυ4 domains of IgY, while the equivalent of the Cυ2 domain is absent in the IgG chain, having been replaced by the hinge region. IgY also contains two additional cysteine residues, Cys331 and Cys 338, in the Cυ2–Cυ3 junction, which likely participate in inter–υ chain disulfide linkages (52). IgY and IgG both contain Asn-linked oligosaccharides; however, the structures of oligosaccharides in IgY differ from those of any mammalian IgG, containing unusual monoglucosylated oligomannose type oligosaccharides with Glc_1–Man_{7-9}–$GlcNAc_2$ structure (53,54).

The isoelectric point of IgY is lower than that of IgG (55), and it has been suggested that IgY may be a more hydrophobic molecule than IgG (56). Unlike IgG, IgY does not have the ability to precipitate multivalent antigens, possibly due to the steric hindrance caused by the closely aligned Fab arms (57).

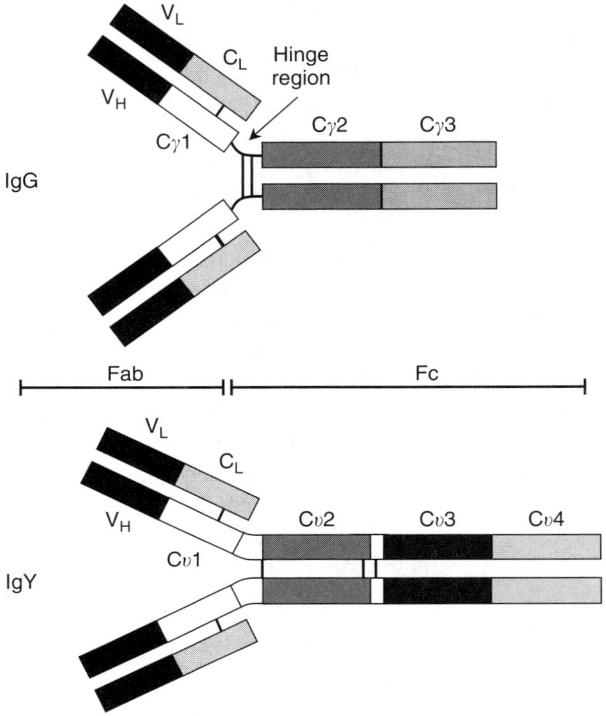

FIGURE 9.3 Structure of IgG and IgY. Disulfide bonds are indicated by lines connecting the two chains. (Adapted from G.W. Warr, K.E. Magor, D.A. Higgins. *Immunol. Today,* 16, 392–398, 1995. With permission.).

IgY does not associate with mammalian complement (58) or rheumatoid factors (RF) (59), and the binding of IgY to human and bacterial Fc receptors on cell surfaces is less than that of IgG (60). As well, IgY does not bind to *Staphylococcus* protein A or *Streptococcus* protein G (7,61) as does IgG. The main differences between IgG and IgY are summarized in Table 9.1.

9.4 MOLECULAR PROPERTIES OF IgY

9.4.1 Heat and pH Stability

IgY and IgG differ in their stability to extremes in pH and heat. IgY has been shown to be more sensitive to heating than rabbit IgG. Shimizu et al. (62) found that the activity of IgY was decreased by heating for 15 minutes at 70°C or higher, whereas that of the IgG did not decrease until 75 to 80°C or higher. Using differential scanning calorimetry (DSC), it was determined that the temperature corresponding to

TABLE 9.1
Comparison of the Structural and Functional Properties of IgG and IgY

	Mammals (IgG)	Chicken (IgY)
Source of antibody	Blood plasma	Egg yolk
Antibody sampling	Noninvasive	
Antibody amount	200 mg bleed^{-1} (40 ml blood)	50–100 mg egg^{-1} (5–7 eggs week^{-1})
Specific antibody	~5%	2–10%
Molecular weight	150 kDa	180 kDa
Protein A/G binding	Yes	No
Interference with rheumatoid factor	Yes	No
Activation of mammalian complement	Yes	No
Binding to Fc Receptors	Yes	No

(Adapted from R. Schade, C. Pfister, R. Halatsch, P. Henklein. *ATLA*, 19, 403–419, 1991. With permission.)

the maximum of the denaturation endotherm (T_{max}) of IgY was 73.9°C, in contrast to that of rabbit IgG, which was 77.0°C (63). It has been found, however, that the addition of stabilizers such as sucrose (64), maltose, and glycerol (65) reduced the extent of heat denaturation of IgY.

Although the stability of both immunoglobulins was similar when subjected to alkaline conditions, IgY showed much less stability than did rabbit IgG to acid denaturation. Shimizu et al. (62,66) found that the activity of IgY was decreased by incubating at pH 3.5 or lower and was completely lost at pH 3. IgG, on the other hand, did not demonstrate a loss of activity when the pH was lowered to 2, at which point some activity still remained. Similar results were observed using IgY produced against human rotavirus (63). The acid denaturation of IgY was found to be reduced with the addition of sucrose (64) and sorbitol (67).

The differences in stability between IgY and IgG have been attributed to structural differences between the two molecules, and it has been suggested that the lower β-structure content of IgY may indicate that its conformation is more disordered and therefore less stable than that of IgG (62). As well, the lack of a hinge region in IgY may be another factor affecting molecular stability. The lower flexibility of the Cυ1 and Cυ2 domains of IgY, as compared with the hinge region of IgG, may cause the inactivation of the antibody by the various treatments because the flexibility of the hinge region is considered to influence the overall properties of immunoglobulin molecules (68).

9.4.2 Stability against Proteolytic Enzymes

IgY, like IgG, has been found to be sensitive to pepsin digestion but relatively resistant to trypsin and chymotrypsin (69). IgY is, however, more susceptible to digestion with trypsin, chymotrypsin, and pepsin than is IgG (70). Using IgY produced against human rotavirus, it was found that almost all of the IgY activity was lost following digestion with pepsin; however, activity remained even after 8 hours of incubation with trypsin or chymotrypsin (71). It was observed that the digestion of IgY with pepsin at pH 2 resulted in complete degradation of IgY, leaving only small peptides; however, IgY digested with pepsin at pH 4 retained some of its activity, with the H- and L-chains still intact as was the case with smaller peptides (71). Shimizu et al. (69) observed similar results following the digestion of IgY with gastrointestinal enzymes. *In vivo* proteolytic digestion has also produced similar results. The feeding of IgY to calves (72) as well as to pigs (73) has been described, and in both cases, antibody activity had decreased significantly following passage through the gastrointestinal tract.

The degradation of IgY by gastrointestinal enzymes presents a significant challenge for the oral passive immunotherapy applications of IgY. Chang et al. (65) suggested that the addition of gum arabic to an egg yolk solution prior to lyophilization may increase the stability of IgY against proteolytic enzymes. Methods of encapsulating IgY, including using lecithin/cholesterol liposomes (74) and hydroxypropyl methylcellulose phthalate (72), a coating substance used for drugs, have been examined and found to increase the stability of IgY to gastric enzymes.

9.5 APPLICATIONS OF IgY

9.5.1 Immunotherapeutic Applications of IgY

IgY immunotherapy presents an attractive alternative to traditional treatments. Several of the immunochemical properties of IgY make it especially suited to immunotherapy applications. As mentioned previously, it does not activate the human complement system and does not react with rheumatoid factors, human Fc receptors, or human antimouse IgG antibodies (HAMA), which are all well-known cell activators and mediators of inflammation (5,20). As well, eggs are normal dietary components, and so there is little risk of toxic side effects of IgY, or the production of anti-IgY antibody production, when administered orally (5). The immunotherapeutic applications of IgY are summarized in Table 9.2.

9.5.1.1 Oral Administration of IgY

The oral administration of specific antibodies as an alternative to antibiotics and antimicrobial chemotherapy, for the localized treatment of infections, has drawn increasing interest. This is due to an increase in the number of antibiotic-resistant bacteria and viral pathogens and an escalating number of immunocompromised individuals who do not respond to traditional vaccination (2,5,75,76).

Although passive immunization is a relatively recent concept in human health, it is well established in animals. As early as 1963, the use of IgY for passive immunization for the protection of chickens against Newcastle disease virus and Marek's disease virus was described (77–79). The oral administration of IgY remains an attractive approach for the establishment of passive immunity (5), and potential applications of IgY for the prevention and treatment of infections caused by pathogenic bacteria and viruses in both humans and animals have been studied extensively and are discussed below.

9.5.1.1.1 *Escherichia coli*

Enterotoxigenic *Escherichia coli* (ETEC) is a major health concern in humans and animals. It is the most common cause of enteric colibacillosis encountered in children in the developing countries, accounting for one billion diarrheal episodes annually (80) and is also a problem in neonatal calves (81) and piglets (82).

The production of IgY against several fimbrial antigens of porcine ETEC, associated with intestinal colonization, have been described. Anti-K88, K99, and 987P antibodies were found to inhibit the binding of *E. coli* K88+, K99+, and 987P+ strains to porcine epithelial cells (83) and to porcine intestinal mucus (84) *in vitro*. These antifimbrial antibodies, when administered orally to piglets, were found to protect against infection with each of the three strains of *E. coli* in passive immunization trials, in a dose-dependent manner (83). Similarly, Marquardt et al. (85) found that the oral administration of anti-K88 IgY resulted in 100% survival of ETEC-infected piglets, compared with a control group, and a reduction in the incidence and severity of diarrhea. Imberechts et al. (86) isolated IgY against *E. coli* F18ab fimbriae and found that the F18ab antibodies inhibited the attachment of F18ab+ bacteria to porcine intestinal mucosa *in vitro* and, when administered orally, reduced the number of cases of diarrhea and death in infected piglets.

The passive protective effect of anti-ETEC IgY, in neonatal calves, against fatal enteric colibacillosis, has also been studied (87). Calves fed milk containing IgY experienced only transient diarrhea, 100% survival, and good body weight gain during the course of the study. O'Farrelly et al. (88) also reported the prevention of ETEC in rabbits through the oral administration of anti-ETEC IgY.

The production of IgY against enteropathogenic *E. coli* (EPEC), another major cause of childhood diarrhea, and its antigens has also been examined (89). de Almeida et al. (90) produced IgY against a recombinant EPEC protein (BfpA), one of the virulent factors required for EPEC pathogenesis, and found that the anti-BfpA antibodies were capable of inhibiting the colonization of cells by EPECand also *E. coli* growth *in vitro*.

Because anti-*E. coli* IgY has been shown to be successful both *in vitro* against *E. coli* and in the treatment of *E. coli* infections of animals, the clinical application of passive immunization of IgY against diarrhea to prevent and treat *E. coli* infection in infants is now being examined.

9.5.1.1.2 *Salmonella spp.*

Approximately two to four million cases of salmonellosis occur in the United States annually (91). Symptoms may include fever, abdominal pain, headache, malaise,

TABLE 9.2
Action of Specific IgY against Various Target Organisms and Proteins

Target	Effect of IgY	Reference
Escherichia coli	Prevention of K88+, K99+, 987P+ ETEC infection in neonatal piglets	83
	Inhibition of adhesion of ETEC K88 to piglet intestinal mucosa, *in vitro*	84
	Prevention of ETEC K88+ infection in neonatal and early weaned piglets	85
	Inhibition of attachment of F18+ *E. coli* to porcine intestinal mucosa *in vitro*, and reduction in diarrheal cases in infected piglets	86
	Protection of neonatal calves from fatal enteric colibacillosis	87
	Prevention of diarrhea in rabbits challenged with ETEC	88
	Inhibition of growth of *E. coli* and inhibition of colonization of cells by EPEC, using anti-BfpA IgY	90
Salmonella spp.	Inhibition of SE and ST growth *in vitro*	92
	Inhibition of adhesion of SE to human intestinal cells *in vitro*	93, 94
	Protection of mice challenged with SE or ST from experimental salmonellosis, using IgY against OMP, LPS, and Fla	101
	Protection of mice challenged with with SE from experimental salmonellosis, using IgY against SEF-14	102
	Prevention of fatal salmonellosis in neonatal calves exposed to ST or *S. dublin*	103
Rotavirus	Prevention of murine rotavirus in mice	104
	Prevention of HRV-induced gastroenteritis in mice	109
	Prevention and treatment of HRV-induced gastroenteritis using murine model	71
	Prevention of HRV infection *in vitro*, using IgY against recombinant protein VP8	110

	Prevention of BRV infection in mouse model	113
	Protection of calves from BRV-induced diarrhea	114
Streptococcus mutans	Reduction of caries development in animal model	116
	Prevention of *S. mutans* adhesion *in vitro*, and *in vivo* in humans	118
	Prevention of *S. mutans* accumulation and reduction of caries in rats using IgY against *S. mutans* glucan-binding protein (GBP)	119
Helicobacter pylori	Inhibition of growth of *H. pylori* and reduction in binding to human gastric cells, *in vitro*	121, 122
	Decrease in gastric mucosal injury and inflammation in a gerbil model of *H. pylori*-induced gastritis	121, 122
Pseudomonas	Inhibition of *P. aeruginosa* adhesion to epithelial cells *in vitro*	126
Aeruginosa	Prolongation of *P. aeruginosa* colonization, and prevention of chronic infection in CF patients	125, 127
	Anti-*P. aeruginosa* IgY stable in saliva for 8 h	128
Crohn's disease	Neutralization of TNF *in vitro*, and treatment of colitis in rats	130
*Yersinia rucker*i	Protection of rainbow trout against *Y. ruckeri* infection	133
Edwardsiella tarda	Prevention of edwardsiellosis of Japanese eels infected with *E. tarda*	134
IBDV	Protection of chicks from infectious bursal disease virus	136
PEDV	Protection of piglets against porcine epidemic diarrhea virus	137
Bovine coronavirus	Protection of neonatal calves from bovine coronavirus (BCV)-induced diarrhea	138
Cryptosporidium	Prevention of cryptosporidiosis in mice	139
Staphylococcus aureus	Inhibition of production of *Staphylococcus aureus* enterotoxin-A	93
	Protection of monkeys from enterotoxin-B (SEB)-induced toxic shock syndrome	146
PERV	Reduction of infectivity of porcine endogenous retrovirus, using anti-aGal IgY	150

lethargy, skin rash, constipation, and changes in mental state. The elderly, infants, and the immunocompromised may develop more severe symptoms, in which case the infection may spread from the intestines to the blood stream and then to other sites in the body, leading to death. *Salmonella enteritidis* (SE) and *Salmonella typhimurium* (ST), in particular, are the major agents of salmonellosis (91).

In vitro, anti-SE Ig and anti-ST IgY have been found to inhibit bacterial growth when added to culture medium (92) and to inhibit the adhesion of SE to human intestinal cells (93,94).

Salmonella possesses several surface components that are virulence related, including outer membrane protein (OMP) (95,96), lipopolysaccharides (LPS) (97,98), flagella (Fla), and in some strains, fimbrial antigens (99,100). The ability of IgY specific for OMP, LPS, or flagella (Fla) to passively protect against experimental salmonellosis in mice has been demonstrated. In mice challenged with SE or ST antibody treatment with OMP-, LPS-, or Fla-specific IgY resulted in a significantly higher survival rate, as compared with that in control mice (101). A novel fimbrial antigen, SEF-14, produced mainly by SE and *S. dublin* strains, was described by Peralta et al. (102). Mice challenged with SE and treated with anti-SEF-14 IgY had a survival rate of 77.8%, compared with a 32% survival rate in the control mice, fed nonimmunized egg yolk.

Salmonella infection in calves is also a significant problem, with ST and *S. dublin* accounting for most cases of salmonellosis. Passive protection against ST and *S. dublin* was investigated by challenging calves with ST or *S. dublin*, and then orally administering IgY against ST or *S. dublin*. All control calves died within 7 to 10 days, whereas only fever and diarrhea were observed in calves given the high-titre IgY (103).

These results demonstrate that IgY specific for *Salmonella* spp. is protective against fatal salmonellosis and may be clinically useful during a salmonellosis outbreak.

9.5.1.1.3 Rotavirus

Human rotavirus (HRV) is the major etiologic agent of acute infantile gastroenteritis (104), infecting up to 90% of children under the age of 3 years (105) and resulting in around 600,000 infant deaths annually (106). The infection is localized to the epithelial cells of the gastrointestinal tract (107) and causes a shortening and atrophy of the villi of the small intestine; this results in decreased water absorption, leading to severe diarrhea, vomiting, and eventually death due to dehydration if left untreated (105,107,108).

The oral administration of anti-rotavirus IgY has been found to passively protect against rotavirus infection in neonatal mouse models. IgY produced against three different serotypes of rotavirus (mouse, human, and monkey) was found to be capable of preventing rotavirus-induced diarrhea in mice infected with murine rotavirus, whereas IgY isolated from the eggs of nonimmunized chickens had no effect (104). Protection from HRV infection in mice has also been demonstrated using IgY produced against HRV (109). It has been reported that anti-HRV IgY decreased the incidence of rotavirus-induced diarrhea in mice, using a neonatal mouse model of HRV infection, when administered both before and after HRV challenge, suggesting its use in the prevention and treatment of rotavirus gastroenteritis (71).

The recombinant HRV coat protein VP8, a cleavage product of the rotavirus spike protein VP4, which is involved in viral infectivity and neutralization of the virus, has been reported for the induction of antibodies against HRV (110). The resulting anti-VP8 IgY exhibited significant neutralizing activity, *in vitro*, against the Wa strain of HRV, suggesting its use for the prevention and treatment of HRV infection. Oral administration of anti-HRV egg yolk to children afflicted with HRV resulted in only a modest improvement in HRV-related symptoms; however, further studies are required (111).

Neonatal calf diarrhea, caused by bovine rotavirus (BRV), is a significant cause of mortality in cattle (112). Using a mouse model of BRV infection, Kuroki et al. (113) observed protection against two strains of BRV using orally administered anti-BRV IgY. Passive protection of calves against BRV infection, using anti-BRV IgY, has also been demonstrated (114).

9.5.1.1.4 Streptococcus mutans

Streptococcus mutans is believed to be the principal causative bacterium of dental caries in humans. The pathogenesis of *S. mutans*-associated caries involves a series of binding steps, leading to the accumulation of sufficient numbers of these cariogenic bacteria to cause disease (115).

IgY against *S. mutans* MT-8148 serotype c or cell-associated glucosyltransferase has been examined for the passive prevention of dental caries (65,116,117). It was found that the consumption of a cariogenic diet containing more than 2% IgY-containing yolk powder resulted in significantly lower caries scores (116) and prevented the colonization of *S. mutans* in the oral cavity. Mouth rinse containing IgY specific to *S. mutans* was found to be effective in preventing dental plaque *in vitro*, using saliva-coated hydroxyapatite disks, and *in vivo* in humans (118).

IgY produced against the *S. mutans* glucan-binding protein B (GBP-B), which is believed to be involved in *S. mutans* biofilm development, has also been examined. Using a rat model of dental caries, Smith et al. (119) observed a decrease in *S. mutans* accumulation in rats treated with anti-GBP-B IgY, as well as a decrease in the overall amount of dental caries, as compared with control rats.

These studies suggest that IgY against *S. mutans*, or its components, may inhibit *S. mutans* accumulation and control plaque and the subsequent oral health problems associated with plaque accumulation.

9.5.1.1.5. Helicobacter pylori

Helicobacter pylori is the most common cause of gastritis and gastric ulcers and has been implicated in the development of gastric carcinomas. Around 25 to 50% of the population carry *H. pylori;* however, this number is higher in the developing countries (120). While antibiotic therapy is often employed to treat *H. pylori* infection, it has been found to fail in some cases due to the development of antibiotic resistance (67).

Anti-*H. pylori* IgY was shown to significantly inhibit the growth and urease activity of *H. pylori, in vitro*, as well as reduce its binding to human gastric cells. *In vivo*, oral administration of the specific IgY decreased gastric mucosal injury and inflammation in a gerbil model of *H. pylori*-induced gastritis (121,122), suggesting the usefulness of anti-*H. pylori* antibodies in the treatment of *H. pylori*-associated gastric diseases.

It has been suggested that antibodies produced against whole-cell *H. pylori* might cross-react with other bacteria, including normal human flora. To this end, Shin et al. (123) have described the identification of immunodominant *H. pylori* proteins, which would result in the production of increasingly specific IgY against *H. pylori*.

9.5.1.1.6 Pseudomonas aeruginosa

Cystic fibrosis (CF) is the most common fatal genetic disease in Caucasians. Colonization of the airways of CF patients by *Pseudomonas aeruginosa* is the principal cause of morbidity and mortality (124), and once a chronic infection has been established, it is very difficult to eliminate it, even with the use of antibiotics (125).

Carlander et al. (126) found that IgY raised against *P. aeruginosa* was capable of inhibiting the adhesion of *P. aeruginosa* to epithelial cells *in vitro* but not bacterial growth, suggesting that specific IgY might be capable of interfering with the bacterial infection process and preventing colonization in CF patients.

The results of *in vivo* studies with CF patients suggested that anti-*P. aeruginosa* IgY, when administered orally to CF patients on a continuous basis, is capable of prolonging the time before *P. aeruginosa* recolonization, delaying or preventing chronic *P. aeruginosa* infection and reducing the intermittent colonization rate and thus the need for antibiotic treatment (125,127).

The stability, over time, of anti-*P. aeruginosa* IgY in the saliva of healthy individuals following a mouth rinse with an aqueous IgY solution was also examined (128). Antibody activity in the saliva remained after 8 hours. After 24 hours, antibody activity had decreased significantly but was still detectable in some subjects, indicating that oral treatment with specific IgY for various local infections such as common cold, tonsillitis, and so on might be possible.

9.5.1.1.7 Inflammatory bowel disease

Crohn's disease and ulcerative colitis are chronic inflammatory bowel diseases and are an increasing burden to hospitals and society in terms of the cost and lost time due to illness (129). Standard medical care for these diseases includes anti-inflammatory drugs, immunosuppressants, and antibiotics, but their use is limited by side effects and incomplete efficacy. Tumor necrosis factor (TNF) has been implicated in the pathogenesis of inflammatory bowel disease (130). Immunotherapy using monoclonal mouse antibodies directed against TNF has been approved for use; however, it can be costly, and adverse side effects have been reported in patients receiving systemic anti-TNF therapy (131). Worledge et al. (130) reported that anti-TNF antibodies produced in chickens were capable of effectively treating acute and chronic phases of colitis in rats and were also found to neutralize human TNF *in vitro*, indicating its possible use for the treatment of inflammatory bowel disease in humans, in place of the currently used parenterally administered anti-TNF antibodies.

9.5.1.1.8 Fish diseases

Yersinia ruckeri is the causative agent of enteric redmouth disease, a systemic bacterial septicemia of salmonid fish (132), and the persistence of *Y. ruckeri* in carrier fish and shedding of bacteria in feces can present a continuing source of infection. The passive immunization of rainbow trout against infection with *Y. ruckeri* using IgY has been studied (133). The fish fed anti-*Y. ruckeri* IgY prior to challenge with *Y. ruckeri* showed a lower mortality rate and, based on organ and intestine cultures, demonstrated a lower *Y. ruckeri* infection rate. This infection rate appeared lower, regardless of whether IgY was administered prior to or after the challenge.

Edwardsiella tarda is a fish pathogen spread by infection through the intestinal mucosa. Edwardsiellosis in Japanese eels is a serious problem for the eel-farming industry. Specific IgY has been investigated for its prevention, since treatment with antibiotics has been found to promote the growth of bacterial-resistant strains (134). Eels challenged with *E. tarda* and then administered anti-*E. tarda* IgY survived without any symptoms of *E. tarda* infection, unlike the control eels, which died within 15 days (134,135).

The oral administration of specific IgY against fish pathogens would provide an alternative to antibiotic and chemotherapy treatments for the prevention of fish diseases in fish farms and would provide a cost-effective alternative to slaughtering stocks of fish the pose a health risk.

9.5.1.1.9 Others

In addition, specific IgY has been shown to be effective in preventing and treating several other pathogens, for example, in the passive protection of chicks against infectious bursal disease virus (IBDV) (136), of piglets against porcine epidemic virus (PEDV) (137), and of calves against Bovine coronavirus (BCV) (138) and cryptosporidiosis due to Cryptosporidium infection (139). Finally, IgY has also been produced against *Bordetella bronchiseptica*, *Pasteurella multocida*, and *Actinobacillus pleuropneumoniae*, which are causative agents of swine respiratory diseases, and has been proposed as an alternative method to control infectious respiratory diseases in swine (140,141). Yang et al. (142) described the production of IgY against P110, a protein purified from human stomach cancer cells, and suggested its use as a carrier for antitumorigenic drugs to target gastrointestinal system cancers.

9.5.1.2 Systemic Administration of IgY

9.5.1.2.1 Neutralization of venom

An estimated 1.7 million people a year are bitten or stung by venomous snakes, scorpions, jelly fish, or spiders, and of these, approximately 40,000 to 50,000 cases result in fatalities. The most widely used treatment is the administration of specific antivenoms to neutralize the toxic and potentially lethal effects of the venom. Chicken antivenom IgY has been found to have a higher bioactivity than antivenoms traditionally raised in horses (143,144). IgY also has a lower likelihood of producing side effects such as serum sickness and anaphylactic shock, which can occur upon administration of mammalian serum proteins (20,143).

9.5.1.2.2 Staphylococcus aureus

Staphylococcal enterotoxins are a family of bacterial superantigens produced by *Staphylococcus aureus* and are associated with a number of serious diseases, including food poisoning, bacterial arthritis, and toxic shock syndrome (145). Sugita-Konishi et al. (93) found that a specific IgY was capable of inhibiting the production of *S. aureus* enterotoxin-A *in vitro*. Le Claire et al. (146) reported the production of IgY against *S. aureus* enterotoxin-B (SEB) and found that systemically administered anti-SEB IgY provided both pre- and postexposure protection to SEB challenge and protected monkeys from toxic shock syndrome in a rhesus monkey model of SEB-induced lethal shock. These results suggest that antienterotoxin IgY may provide protection as a prophylactic agent as well as a therapeutic agent against lethal doses of *S. aureus* enterotoxins and could be used to reduce or eliminate enterotoxin-mediated disorders.

9.5.1.2.3 Xenotransplantation

Transplantation of pig organs into humans (xenotransplantation) is a serious consideration because of the shortage of human donors for organ transplants. However, the problem with such xenografts is hyperacute rejection, mediated by natural antibodies in humans against pig antigens, complement fixation, and the rapid onset of intravascular coagulation (147). The major target of these natural antibodies is the carbohydrate epitope, Galα1-3Gal, which is expressed by all mammals, except humans, apes, and some monkeys. Besides humans and monkeys, birds, especially chickens, also lack Galα1-3Gal expression (148). Since IgY does not bind human complement or Fc receptors, anti-αGal IgY is a good candidate for use as blocking antibody to inhibit the interactions that may contribute to xenograft rejection. Fryer et al. (149) demonstrated, *in vitro*, that anti-αGal IgY blocked human xenoreactive natural antibody binding to both porcine and rat tissues and also inhibited cell lysis of porcine cells by human serum, suggesting that IgY could be of potential use in inhibiting pig-to-human xenograft rejection. Furthermore, anti-αGal IgY was also found to significantly reduce the infectivity of porcine endogenous retrovirus (PERV), an α-Gal-bearing virus that has emerged as a potential zoonotic agent, with possible pig-to-human transmission (150).

9.5.2 Application of IgY as an Immunological Tool

9.5.2.1 Diagnostic Applications

Besides their therapeutic value, antibodies are important in biological and medical research, where they serve as essential components in a variety of diagnostic systems used for the qualitative and quantitative determination of a wide range of substances (151).

Polyclonal IgG has traditionally been employed for such diagnostic applications (20). The antigen-binding specificity of IgY is comparable with that of IgG, and both IgG and IgY can detect antigens with high specificity (4). As mentioned previously, however, the immunochemical properties of IgY present several advantages over IgG. Evolutionary differences allow the production of IgY against conserved mammalian proteins, resulting in an enhanced immune response not possible in mammals, and can minimize cross-reactivity normally observed between mammalian IgG (7).

The use of IgY in immunological assays for clinical testing can also eliminate interference and false positives normally experienced when using IgG. Freshly obtained human serum samples often contain an active complement system, which would be activated by mammalian antibodies. IgY, which does not activate the human complement system, eliminates the interference that would otherwise be caused by IgG (152). Serum samples may also contain rheumatoid factor (RF) and human antimouse IgG antibodies (HAMA), which are well-known causes of false-positive reactions in immunological assays (7). RF reacts with the Fc portion of IgG, and HAMA, occasionally found occurring naturally in human serum, will bind to any mouse antibodies being used in an immunoassay, producing false-positive results (7). IgY does not react with RF (153) or HAMA (154), and its use has been suggested in place of IgG for immunological assays dealing with human serum. The Fc portion of mammalian antibodies may also interact with the Fc receptor found on many types of blood cells and bacteria. IgY does not interact with Fc receptors and can therefore be used to avoid interference due to Fc binding (7).

IgY has been used in several diagnostic and medical applications, including the diagnosis of gastric cancer (155), the detection of breast and ovarian cancer markers (156–158), the detection of African horsesickness virus (159), *Campylobacter* fetus diagnosis (160), and human hemoclassification of blood group antigens (161). IgY has been produced against the human thymidine kinase-1 (TK-1) enzyme and suggested for the early diagnosis of cancer and for monitoring patients undergoing cancer treatment (162). Other applications of IgY include screening for human papillomavirus for the early detection of cervical cancer (163), and the detection of the protein YRL-40 as a marker for disease models of arthritis, cancer, atherosclerosis, and liver fibrosis (164).

9.5.2.2 Immunoaffinity Ligands

Immunoaffinity chromatography is based on the specific affinity between an antigen and its corresponding antibody. Due to the highly specific nature of the antibody–antigen interaction, immunoaffinity chromatography allows for the purification of the molecule of interest from complex starting materials. The use of this process on a large scale is often limited by the high cost of the technique and problems relating to the production of antibodies and the efficiency of immobilization (165). Specific IgY, which can easily be produced on a large scale and is suitable for industrial applications, would provide an ideal replacement for other polyclonal or monoclonal antibodies currently used in immunoaffinity chromatography.

Standard affinity practices require a low pH to effect dissociation of the molecule from the affinity ligand. Hatta et al. (4) compared the effect of pH on the dissociation efficiency of IgY and rabbit IgG, both directed against mouse IgG, and found that when using IgG as an immunoaffinity ligand, only half of the mouse IgG dissociated at pH 4 while the remaining IgG eluting at pH 2. In contrast, however, 97% of mouse IgG dissociated at pH 4 when using IgY as ligand. These results indicate that using IgY as an immunoaffinity ligand may permit the use of less harsh elution conditions.

Although IgY is more sensitive to low pH than is IgG, it was found that an IgY immunoaffinity column was capable of retaining stability when subjected to standard affinity chromatography conditions and could be reused over 50 times without any significant decrease in binding capacity (166). To extend the use of IgY immunoaffinity columns, Kim et al. (167) examined the reusability of avidin-biotinylated IgY columns, in which biotinylated IgY was held by strong noncovalent interaction on columns containing immobilized avidin. It was found that when the antibody binding activity had been reduced due to prolonged use, the column could be regenerated by dissociating the avidin-biotinylated IgY complex, and applying new biotinylated IgY, thereby restoring the binding activity of the column.

Immobilized yolk antibodies have been used for the isolation of value-added proteins from dairy products, including the purification of lactoferrin (168), for the isolation and separation of IgG subclasses from colostrum, milk, and cheese whey (166), and for the purification of biological molecules from human serum.

9.6 FUTURE APPLICATIONS OF IGY

Despite the evidence supporting the many potential immunotherapeutic applications of IgY, and the advantages it provides when compared with IgG for immunodiagnostics and immunoaffinity applications, mammalian antibodies continue to predominate, perhaps because of a lack of knowledge of the many benefits of IgY technology, unfamiliarity with chicken husbandry, and the techniques involved in IgY production or simply due to convention (151,169). It has been suggested, however, that given the many benefits of IgY technology and its universal application in both research and medicine, it will play an increasing role in research, diagnostics, and immunotherapy in the future (170).

New methods of production and potential applications of egg yolk antibodies continue to be explored. Romito et al. (171) have studied the immunization of chickens with naked DNA, thereby eliminating the protein expression and purification steps and allowing the production of IgY against pathogenic or toxic antigens that would otherwise be harmful to hens. The production of monoclonal IgY, generated by fusing spleen cells from immunized chickens with chicken B cells, has also been examined (172–176). The resultant hybridoma is capable of secreting a consistent supply of monoclonal IgY of a single and known specificity and homogeneous structure and would be and an ideal replacement for mouse monoclonal antibodies in diagnostic applications. Using transgenic chickens, Mohammed et al. (177) demonstrated that recombinant human antibodies could also be deposited into the egg yolk, technology that could further advance the application of egg yolk immunotherapy.

REFERENCES

1. F. Klemperer. Ueber naturliche Immunitat und ihre Verwertung fur die Immunisierungstherapie. *Archiv fur Experimentelle Pathologie und Pharmakologie* 31: 356–382, 1893.
2. R.M. Reilly, R. Domingo, J. Sandhu. Oral delivery of antibodies: future pharmacokinetic trends. *Clin. Pharmacokinet.*, 4: 313–323, 1997.
3. D. Carlander, H. Kollberg, P.-E. Wejaker, A. Larsson. Peroral immunotherapy with yolk antibodies for the prevention and treatment of enteric infections. *Immunol. Res.*, 21: 1–6, 2000.
4. H. Hatta, M. Ozeki, K. Tsuda. Egg yolk antibody IgG and its application. In: T. Yamamoto, L.R. Juneja, H. Hatta, M. Kim (Eds.) *Hen Eggs: Their Basic and Applied Science*. New York: CRC Press, 151–178, 1997.
5. A. Larsson, D. Carlander. Oral immunotherapy with yolk antibodies to prevent infections in humans and animals. *Ups. J. Med. Sci.*, 108: 129–140, 2003.
6. J.A. Wooley, J. Landon. Comparison of antibody production to human interleukin-6 (IL-6) by sheep and chickens. *J. Immunol. Meth.*, 178: 253–265, 1995.
7. D. Carlander, J. Stalberg, A. Larsson. Chicken antibodies: a clinical chemistry perspective. *Ups. J. Med. Sci.*, 104: 179–190, 1999.
8. M.R. Loeken, T.F. Roth. Analysis of maternal IgG subpopulations which are transported into the chicken oocyte. *Immunology*, 49: 21–28, 1983.
9. R.L. Tressler, T.F. Roth. IgG receptors on the embryonic chick yolk sac. *J. Biol. Chem.*, 262: 15406–15412, 1987.
10. S.L. Morrison, M.S. Mohammed, L.A. Wims, R. Trinh, R. Etches. Sequences in antibody molecules important for receptor-mediated transport into the chicken egg yolk. *Mol. Immunol.*, 38: 619–625, 2002.
11. M.E. Rose, E. Orlans, N. Buttress. Immunoglobulin classes in the hen's eggs: their segregation in yolk and white. *Eur. J. Immunol.*, 4: 521–523, 1974.
12. G.A. Leslie, L.N. Martin. Studies on the secretory immunologic system of fowl. Serum and secretory IgA of the chicken. *J. Immunol.*, 110: 1–9, 1973.
13. R. Schade, C. Pfister, R. Halatsch, P. Henklein. Polyclonal IgY antibodies from chicken egg yolk-an alternative to the production of mammalian IgG type antibodies in rabbits. *ATLA*, 19: 403–419, 1991.
14. X. Li, T. Nakano, H.H. Sunwoo, B.H. Paek, H.S. Chae, J.S. Sim. Effects of egg and yolk weights on yolk antibody (IgY) production in laying chickens. *Poult. Sci.*, 77: 266–270, 1997.
15. A. Larsson, D. Carlander, M. Wilhelmsson. Antibody response in laying hens with small amounts of antigen. *Food Agric. Immunol.*, 10: 29–36, 1998.
16. W. Knecht, R. Kohler, M. Minet, M. Loffler. Anti-peptide immunoglobulins from rabbit and chicken eggs recognize recombinant human dihydroorotate dehydrogenase and a 44-kDa protein from rat liver mitochondria. *Eur. J. Biochem.*, 236: 6009–6013, 1996.

17. M. Gassmann, P. Thommes, T. Weiser, U. Hubscher. Efficient production of chicken egg yolk antibodies against a conserved mammalian protein. *FASEB J.*, 4: 2528–2532, 1990.
18. J.M. Sharma. Introduction to poultry vaccines and immunity. *Adv. Vet. Med.*, 41: 481–494, 1999.
19. O.J. Cotterill, L.E. McBee. Egg breaking. In: W.J. Stadelman, O.J. Cotterill (Eds.) *Egg Science and Technology*, 4th Edition. New York: The Haworth Press, Inc., 1995, 231–263.
20. A. Larsson, R. Balow, T.L. Lindahl, P. Forsberg. Chicken antibodies: taking advantage of evolution; a review. *Poult. Sci.*, 72: 1807–1812, 1993.
21. J.S. Sim, H.H. Sunwoo, E.N. Lee. Ovoglobulin IgY. In: A.S. Naidu (Ed.) *Natural Food Antimicrobial Systems*. New York: CRC Press, 2000, 227–252.
22. E.M. Akita, S. Nakai. Immunoglobulins from egg yolk: isolation and purification. *J. Food Sci.*, 57: 629–634, 1992.
23. H. Kim, S. Nakai. Immunoglobulin separation from egg yolk: a serial filtration system. *J. Food Sci.*, 61: 510–523, 1996.
24. H. Kim, S. Nakai. Simple separation of immunoglobulin from egg yolk by ultrafiltration. *J. Food Sci.*, 63: 485–490, 1998.
25. J.C. Jensenius, I. Anderson, J. Hau, M. Crone, C. Koch. Eggs: conveniently packaged antibodies. Method for purification of yolk IgG. *J. Immunol. Meth.*, 46: 63–68, 1981.
26. J.C. Jensenius, C. Koch. On the purification of IgG from egg yolk. *J. Immunol. Meth.*, 164: 141–142, 1993.
27. S. Nakai, E. Li-Chan, K.V. Lo. Separation of immunoglobulin from egg yolk, In: J.S. Sim, S. Nakai (Eds.) *Egg Uses and Processing Technologies. New Developments*. Wallingford, UK: CAB International, 1994, 94–105.
28. L. Svendsen, A. Crowley, L.H. Ostergaard, G. Stodulski, J. Haul, Development and comparison of purification strategies for chicken antibodies from egg yolk. *Lab. Anim. Sci.*, 45: 89–93, 1995.
29. H. Bade, H. Stegemann. Rapid method of extraction of antibodies from hen egg yolk. *J. Immunol. Meth.*, 71: 421–426, 1984.
30. A. Polson, T. Coetzer, J. Kruger, E. von Maltzahn, K.J. van der Merwe. Improvements in the isolation of IgY from the yolks of eggs laid by immunized hens. *Immunol. Invest.*, 14: 323–327, 1985.
31. H. Hatta, J.S. Sim, S. Nakai. Separation of phospholipids from egg yolk and recovery of water-soluble proteins. *J. Food Sci.*, 53: 425–431, 1988.
32. A. Polson. Isolation of IgY from the yolks of eggs by a chloroform polyethylene glycol procedure. *Immunol. Invest.*, 19: 253–258, 1990.
33. L. Kwan, E. Li-Chan, N. Helbig, S. Nakai. Fractionation of water-soluble and -insoluble components from egg yolk with minimum use of organic solvents. *J. Food Sci.*, 56: 1537–1541, 1991.
34. T. Horikoshi, J. Hiraoka, M. Saito, S. Hamada. IgG antibody from hen egg yolk: purification by ethanol fractionation. *J. Food Sci.*, 58: 739–742, 993.
35. R.D. McLaren, C.G. Prosser, R.C.J. Grieve, M. Borissenko. The use of caprylic acid for the extraction of the immunoglobulin fraction from egg yolk of chickens immunised with bovine α-lactalbumin. *J. Immunol. Meth.*, 177: 175–184, 1994.
36. A. Polson, M.B. von Wechmar, M.H. van Regenmortel. Isolation of viral IgY antibodies from yolks of immunized hens. *Immunol. Commn.*, 9: 475–493, 1980.

37. E.M. Akita, S. Nakai. Comparison of four purification methods for the production of immunoglobulins from eggs laid by hens immunized with an enterotoxigenic *E. coli* strain. *J. Immunol. Meth.*, 160: 207–214, 1993.
38. G. Bizhanov, G. Vyshniauskis. A comparison of three methods for extracting IgY from the egg yolk of hens immunized with Sendai virus. *Vet. Res. Commn.*, 24: 103–113, 2000.
39. H. Hatta, M. Kim, T. Yamamoto. A novel isolation method for hen egg yolk antibody "IgY". *Agric. Biol. Chem.*, 54: 2531–2535, 1990.
40. H.M. Chang, T.C. Lu, C.C. Chen, Y.Y. Tu, J.Y. Hwang. Isolation of immunoglobulin from egg yolk by anionic polysaccharides. *J. Agric. Food Chem.*, 48: 995–999, 2000.
41. A.A. McCannel, S. Nakai. Separation of egg yolk immunoglobulins into subpopulations using DEAE-ion exchange chromatography. *Can. Inst. Food Sci. Technol. J.*, 23: 42–46, 1990.
42. J. Fichtali, E.A. Charter, K.V. Lo, S. Nakai. Separation of egg yolk immunoglobulins using an automated liquid chromatography system. *Biotech. Bioeng.*, 40: 1388–1394, 1992.
43. J. Fichtali, E.A. Charter, K.V. Lo, S. Nakai. Purification of antibodies from industrially separated egg yolk. *J. Food Sci.*, 58: 1282–1290, 1993.
44. A. Hassl, H. Aspock. Purification of egg yolk immunoglobulins. A two-step procedure using hydrophobic interaction chromatography and gel filtration. *J. Immunol. Meth.*, 110: 225–228, 1988.
45. S.C. Gee, I.M. Bate, T.M. Thomas, D.B. Rylatt. The purification of IgY from chicken egg yolk by preparative electrophoresis. *Protein Expr. Purif.*, 30: 151–155, 2003.
46. A.A. McCannel, S. Nakai. Isolation of egg yolk immunoglobulin-rich fractions using copper-loaded metal chelate interaction chromatography. *Can. Inst. Food Sci. Technol. J.*, 22: 487–490, 1989.
47. C.R. Greene, P.S. Holt. An improved chromatographic method for the separation of egg yolk IgG into subpopulations utilizing immobilized metal ion (Fe^{3+}) affinity chromatography. *J. Immunol. Meth.*, 209: 155–164, 1997.
48. P. Hansen, J.A. Scoble, B. Hanson, N.J. Hoogenraad. Isolation and purification of immunoglobulins from chicken eggs using thiophilic interaction chromatography. *J. Immunol. Meth.*, 215: 1–7, 1998.
49. I. Kuronen, H. Kokko, I. Mononen, M. Parviainen. Hen egg yolk antibodies purified by antigen affinity under highly alkaline conditions provide new tools for diagnostics. Human intact parathyrin as a model antigen. *Eur. J. Clin. Chem. Clin. Biochem.*, 35: 435–440, 1997.
50. G. Fassina, A. Verdoliva, G. Palombo, M. Ruvo, G. Cassani, Immunoglobulin specificity of TG19318: a novel synthetic ligand for antibody affinity purification. *J. Mol. Recognit.*, 11: 128–133, 1998.
51. A. Verdoliva, G. Basile, G. Fassina. Affinity purification of immunoglobulins from chicken egg yolk using a new synthetic ligand. *J. Chromatogr. B. Biomed. Sci. Appl.*, 749: 233–242, 2000.
52. G.W. Warr, K.E. Magor, D.A. Higgins. IgY: clues to the origins of modern antibodies. *Immunol. Today*, 16: 392–398, 1995.
53. M. Ohta, J. Hamako, S. Yamamoto, H. Hatta, M. Kim, T. Yamamoto, S. Oka, T. Mizuochi, F. Matsuura. Structure of asparagine-linked oligosaccharides from hen egg yolk antibody (IgY). Occurrence of unusual glucosylated oligo-mannose type oligosaccharides in a mature glycoprotein. *Glycoconj. J.*, 8: 400–413, 1991.

54. F. Matsuura, M. Ohta, K. Murakami, Y. Matsuki. Structures of asparagines linked oligosaccharides of immunoglobulins (IgY) isolated from egg-yolk of Japanese quail. *Glycoconj. J.*, 10: 202–213, 1993.
55. A. Polson, M.B. von Wechmar, G. Fazakerley. Antibodies to proteins from yolk of immunized hens. *Immunol. Commn.*, 9: 495–514, 1980.
56. L. Davalos-Pantoja, J.L. Ortega-Vinuesa, D. Bastos-Gonzalez, R. Hidalgo-Alvarez. A comparative study between the adsorption of IgY and IgG on latex particles. *J. Biomater. Sci. Polym. Ed.*, 11: 657–673, 2000.
57. R.T. Hersh, A.A. Benedict. Aggregation of chicken gamma-G immunoglobulin in 1.5 M sodium chloride solution. *Biochim. Biophys. Acta*, 115: 242–244, 1966.
58. D. Carlander, A. Larsson. Avian antibodies can eliminate interference due to complement activation in ELISA. *Ups. J. Med. Sci.*, 106: 189–195, 2001.
59. A Larsson, J Sjöquist. Chicken antibodies: A tool to avoid false positive results by rheumatoid factor in latex fixation tests. *J. Immunol. Meth.*, 108: 205–208, 1988.
60. P.S. Gardner, S. Kaya. Egg globulin in rapid virus diagnosis. *J. Virol. Meth.*, 4: 257–262, 1982.
61. G. Kronvall, U.S. Seal, S. Svensson, R.C. Williams Jr. Phylogenetic aspect of staphylococcal protein A-reactive serum globulins in birds and mammals. *Acta Pathol. Microbiol. Scand.*, Section B 82: 12–18, 1974.
62. M. Shimizu, H. Nagashima, K. Sano, K. Hashimoto, M. Ozeki, K. Tsuda, H. Hatta. Molecular stability of chicken and rabbit immunoglobulin G. *Biosci. Biotechnol. Biochem.*, 56: 270–274, 1992.
63. H. Hatta, K. Tsuda, S. Akachi, M. Kim, T. Yamamoto. Productivity and some properties of egg yolk antibody (IgY) against human rotavirus compared with rabbit IgG. *Biosci. Biotechnol. Biochem.*, 57: 450–454, 1993.
64. M. Shimizu, H. Nagashima, K. Hashimoto, T. Suzuki. Egg yolk antibody (IgY) stability in aqueous solution with high sugar concentrations. *J. Food Sci.*, 59: 763–766, 1994.
65. H.M. Chang, R.F. Ou-Yang, Y.T. Chen, C.C. Chen. Productivity and some properties of immunoglobulin specific against Streptococcus mutans serotype c in chicken egg yolk (IgY). *J. Agric. Food Chem.*, 47: 61–66, 1999.
66. M. Shimizu, H. Nagashima, K. Hashimoto. Comparative studies on molecular stability of immunoglobulin G from different species. *Comp. Biochem. Physiol.*, 106B: 255–261, 1993.
67. K.A. Lee, S.K. Chang, Y.J. Lee, J.H. Lee, N.S. Koo. Acid stability of anti-*Helicobacter pylori* IgY in aqueous polyol solution. *J. Biochem. Mol. Biol.*, 35: 488–493, 2002.
68. I. Pilz, E. Schwarz, W. Palm. Small-angle X-ray studies of the human immunoglobulin molecule. *Eur. J. Biochem.*, 75: 195–199, 1977.
69. M. Shimizu, R.C. Fitzsimmons, S. Nakai. Anti-*E. coli* immunoglobulin Y isolated from egg yolk of immunized chickens as a potential food ingredient. *J. Food Sci.*, 53: 1360–1366, 1988.
70. H. Otani, K. Matsumoto, A. Saeki, A. Hosono. Comparative studies on properties of hen egg yolk IgY and rabbit serum IgG antibodies. *Lebensm. Wiss. Technol.*, 24: 152–158, 1991.
71. H. Hatta, K. Tsuda, S. Akachi, M. Kim, T. Yamamoto, T. Ebina. Oral passive immunization effect of anti-human rotavirus IgY and its behavior against proteolytic enzymes. *Biosci. Biotechnol. Biochem.*, 57: 1077–1081, 1993.
72. Y. Ikemori, M. Ohta, K. Umeda, R.C. Peralta, M. Kuroki, H. Yokoyama, Y. Kodama. Passage of chicken egg yolk antibodies treated with hydroxypropyl

methylcellulose phthalate in the gastrointestinal tract of calves. *J. Vet. Med. Sci.*, 58: 365–367, 1996.

73. H. Yokoyama, R.C. Peralta, S. Sendo, Y. Ikemori, Y. Kodama. Detection of passage and absorption of chicken egg yolk immunoglobulins in the gastrointestinal tract of pigs by use of enzyme-linked immunosorbent assay and fluorescent antibody testing. *Am. J. Vet. Res.*, 54: 867–872, 1993.
74. M. Shimizu, Y. Miwa, K. Hashimoto, A. Goto. Encapsulation of chicken egg yolk immunoglobulin G (IgY) by liposomes. *Biosci. Biotechnol. Biochem.*, 57: 1445–1449, 1993.
75. A. Casadevall, M.D. Scharff. Return to the past: the case for antibody-based therapies in infectious diseases. *Clin. Infect. Dis.*, 21: 150–161, 1995.
76. M. Wierup. The control of microbial diseases in animals: alternatives to the use of antibodies. *Int. J. Antimicrob. Agents*, 14: 315–319, 2000.
77. F.K. Wills, R.E. Luginbuhl. The use of egg yolk for passive immunization of chickens against Newcastle disease. *Avian Dis.*, 7: 5–12, 1963.
78. P.G. Box, R.A. Stedman, L. Singleton. Newcastle disease. I. The use of egg yolk derived antibody for passive immunization of chickens. *J. Comp. Pathol.*, 79: 495–506, 1969.
79. V. Kermani-Arab, T. Moll, W.C. Davis, B.R. Cho, Y.S. Lu, G.A. Leslie. Immunoglobulins and anti-Marek's disease virus antibody synthesis in chickens after passive immunization with immunoglobulin Y anti-Marek's disease virus antibody. *Am. J. Vet. Res.*, 36: 1655–1661, 1975.
80. R.B. Sack. Antimicrobial prophylaxis of travellers' diarrhea: a selected summary. *Rev. Infect. Dis.*, 8: S160–S166, 1986.
81. H.W. Moon, S.C. Whipp, S.M. Skartvedt. Etiologic diagnosis of diarrheal diseases of calves: frequency and methods for detecting enterotoxic and K99 antigen produced by *Escherichia coli*. *Am. J. Vet. Res.*, 37: 1025–1029, 1976.
82. J.A. Morris, W.J. Sojka. *Escherichia coli* as a pathogen in animals. In: M. Sussman (Ed.) *The Virulence of Escherichia coli.* London: Academic Press, Inc., 1985, 47–77.
83. H. Yokoyama, R.C. Peralta, R. Diaz, S. Sendo, Y. Ikemori, Y. Kodama. Passive protective effect of chicken egg yolk immunoglobulins against experimental enterotoxigenic *Escherichia coli* infection in neonatal piglets. *Infect. Immun.*, 60: 998–1007, 1992.
84. L.Z. Jin, K. Samuel, K. Baidoo, R.R. Marquardt, A.A. Frohlich. *In vitro* inhibition of adhesion of enterotoxigenic *Escherichia coli* K88 to piglet intestinal mucus by egg yolk antibodies. *FEMS Immunol. Med. Microbiol.*, 21: 313–321, 1998.
85. R.R. Marquardt, L.Z. Jin, J.W. Kim, L. Fang, A.A. Frohlich, S.K. Baidoo. Passive protective effect of egg yolk antibodies against enterotoxigenic *Escherichia coli* K88+ infection in neonatal and early weaned piglets. *FEMS Immunol. Med. Microbiol.*, 23: 283–288, 1999.
86. H. Imberechts, P. Deprez, E. Van Driessche, P. Pohl. Chicken egg yolk antibodies against F18ab fimbriae of *Escherichia coli* inhibit shedding of F18 positive *E. coli* by experimentally infected pigs. *Vet. Microbiol.*, 54: 329–341, 1997.
87. Y. Ikemori, M. Kuroki, R.C. Peralta, H. Yokoyama, Y. Kodama. Protection of neonatal calves against fatal enteric colibacillosis by administration of egg yolk powder from hens immunized with K99-piliated enterotoxigenic *Escherichia coli*. *Am. Vet. Res.*, 53: 2005–2008, 1992.
88. C. O'Farrelly, D. Branton, C.A. Wanke. Oral ingestion of egg yolk immunoglobulin from hens immunized with an enterotoxigenic *Escherichia coli* strain prevents

diarrhea in rabbits challenged with the same strain. *Infect. Immun.*, 60: 2593–2597, 1992.
89. J.A. Amaral, M. Tino De Franco, M.M. Carneiro-Sampaio, S.B. Carbonare. Anti-enteropathogenic *Escherichia coli* immunoglobulin Y isolated from eggs laid by immunised Leghorn chickens. *Res. Vet. Sci.*, 72: 229–234, 2002.
90. C.M. de Almeida, V.M. Quintana-Flores, E. Medina-Acosta, A. Schriefer, M. Barral-Netto, W. Dias da Silva. Egg yolk anti-BfpA antibodies as a tool for recognizing and identifying enteropathogenic *Escherichia coli. Scand. J. Immunol.*, 57: 573–582, 2003.
91. C. Bell, A. Kyriakides. *Salmonella: A Practical Approach to the Organism and Its Control in Foods*. New York: Blackie, Academic & Professional, 1998.
92. E.N. Lee, H.H. Sunwoo, K. Menninen, J.S. Sim. *In vitro* studies of chicken egg yolk antibodies (IgY) against *Salmonella enteritidis* and *Salmonella typhimurium. Poult. Sci.*, 81: 632–641, 2002.
93. Y. Sugita-Konishi, K. Shibata, S.S. Yun, H.K. Yukiko, K. Yamaguchi, S. Kumagai. Immune functions of immunoglobulin Y isolated from egg yolk of hens immunized with various infectious bacteria. *Biosci. Biotechnol. Biochem.*, 60: 886–888, 1996.
94. Y. Sugita-Konishi, M. Ogawa, S. Arai, S. Kumagai, S. Igimi, M. Shimizu, Blockade of *Salmonella enteritidis* passage across the basolateral barriers of human intestinal epithelial cells by specific antibody. *Microbiol. Immunol.*, 44: 473–479, 2000.
95. A. Isibasi, V. Ortiz, M. Vargas, J. Paniagua, C. Gonzales, J. Moreno, J. Kumate. Protection against *Salmonella typhi* infection in mice after immunization with outer membrane proteins isolated from *Salmonella typhi* 9, 12, d, VI. *Infect. Immun.*, 56: 2953–2959, 1988.
96. V. Udhayakumar, V.R. Muthukkaruppan. Protective immunity induced by outer membrane proteins of *Salmonella typhimurium* in mice. *Infect. Immun.*, 55: 816–821, 1987.
97. H.H. Sunwoo, T. Nakano, T. Dixon, J.S. Sim. Immune responses in chicken against lipopolysaccharides of *Escherichia coli* and *Salmonella typhimurium. Poult. Sci.*, 75: 342–345, 1996.
98. Y. Mine. Separation of *Salmonella enteritidis* from experimentally contaminated liquid eggs using a hen IgY immobilized immunomagnetic separation system. *J. Agric. Food Chem.*, 45: 3723–3727, 1997.
99. C.J. Thorns, M.G. Sojka, D. Chasey. Detection of a novel fimbrial structure on the surface of *Salmonella enteritidis* by using a monoclonal antibody. *J. Clin. Microbiol.*, 28: 2409–2414, 1990.
100. C.J. Thorns, M.G. Sojka, M. McLaren, M. Dibb-Fuller. Characterization of monoclonal antibodies against a fimbrial structure of Salmonella enteritidis and certain other serogroup *D salmonellae* and their application as serotyping reagents. *Res. Vet. Sci.*, 53: 300–308, 1992.
101. H. Yokoyama, K. Umeda, R.C. Peralta, T. Hashi, F. Icatlo, M. Kuroki, Y. Ikemori, Y. Kodama. Oral passive immunization against experimental salmonellosis in mice using chicken egg yolk antibodies specific for *Salmonella enteritidis* and *S. typhimurium. Vaccine*, 16: 388–393, 1998.
102. R.C. Peralta, H. Yokoyama, Y. Ikemori, M. Kuroki, Y. Kodama. Passive immunization against experimental salmonellosis in mice by orally administered hen egg yolk antibodies specific for 14-kDa fimbriae of *Salmonella enteritidis. J. Med. Microbiol.*, 41: 29–35, 1994.
103. H. Yokoyama, R.C. Peralta, K. Umeda, T. Hashi, F.C. Icatlo, M. Kuroki, Y. Ikemori, Y. Kodama. Prevention of fatal salmonellosis in neonatal calves, using

orally administered chicken egg yolk *Salmonella*-specific antibodies. *Am. J. Vet. Res.*, 59: 416–420, 1998.

104. R.H. Yolken, F. Leister, S.B. Wee, R. Miskuff, S. Vonderfecht. Antibodies to rotavirus in chickens' eggs: a potential source of antiviral immunoglobulins suitable for human consumption. *Pediatrics*, 81: 291–295, 1988.
105. J.E. Ludert, A.A. Krishnaney, J.W. Burns, P.T. Vo, H.B. Greenberg. Cleavage of rotavirus VP4 *in vivo*. *J. Gen. Virol.*, 77: 391–395, 1996.
106. H.F. Clark, R.I. Glass, P.A. Offit. Rotavirus vaccines. In: S.A. Plotkin, W.A. Orentstein (Eds.) *Vaccines*, 3rd Edition. Philadelphia: W.B. Saunders Company, 1999, 987–1005.
107. A.Z. Kapikian, R.M. Chanock. Rotaviruses. In: B.N. Fields, D.N. Knipe, P.M. Howley, R.M. Chanock, J.L. Melnick, T.P. Monath, B. Rolzman, S.E. Strauss (Eds.) *Fields Virology*, 3rd Edition. New York: Raven Press, 1996, 1657–1708.
108. C. Hochwald, L. Kivela. Rotavirus vaccine, live, oral, tetravalent (RotaShield). *Pediatr. Nurs.*, 25: 203–204, 1999.
109. T. Ebina. Prophylaxis of rotavirus gastroenteritis using immunoglobulin. *Arch. Virol.* (Suppl.) 12: 217–223, 1996.
110. J. Kovacs-Nolan, E. Sasaki, D. Yoo, Y. Mine. Cloning and expression of human rotavirus spike protein, VP8*, in *Escherichia coli*. *Biochem. Biophys. Res. Commn.*, 282: 1183–1188, 2001.
111. S.A. Sarker, T.H. Casswall, L.R. Juneja, E. Hoq, E. Hossain, G.J. Fuchs, L. Hammarstrom. Randomized, placebo-controlled, clinical trial of hyperimmunized chicken egg yolk immunoglobulin in children with rotavirus diarrhea. *J. Pediatr. Gastroenterol. Nutr.*, 32: 19–25, 2001.
112. J. Lee, L.A. Babiuk, R. Harland, E. Gibbons, Y. Elazhary, D. Yoo. Immunological response to recombinant VP8* subunit protein of bovine rotavirus in pregnant cattle. *J. Gen. Virol.*, 76: 2477–2483, 1995.
113. M. Kuroki, Y. Ikemori, H. Yokoyama, R.C. Peralta, F.C. Icatlo Jr., Y. Kodama. Passive protection against bovine rotavirus-induced diarrhea in murine model by specific immunoglobulins from chicken egg yolk. *Vet. Microbiol.*, 37: 135–146, 1993.
114. M. Kuroki, Y. Ohta, Y. Ikemori, R.C. Peralta, H. Yokoyama, Y. Kodama. Passive protection against bovine rotavirus in calves by specific immunoglobulins from chicken egg yolk. *Arch. Virol.*, 138: 143–148, 1994.
115. S. Hamada, H.D. Slade. Biology, immunology, and cariogenicity of *Streptococcus mutans*. *Microbiol. Rev.*, 44: 331–384, 1980.
116. S. Otake, Y. Nishihara, M. Makimura, H. Hatta, M. Kim, T. Yamamoto, M. Hirasawa. Protection of rats against dental caries by passive immunization with hen egg yolk antibody (IgY). *J. Dent. Res.*, 70: 162–166, 1991.
117. S. Hamada, T. Horikoshi, T. Minami, S. Kawabata, J. Hiraoka, T. Fujiwara, T. Ooshima. Oral passive immunization against dental caries in rats by use of hen egg yolk antibodies specific for cell-associated glucosyltransferase of *Streptococcus mutans*. *Infect. Immun.*, 59: 4161–4167, 1991.
118. H. Hatta, K. Tsuda, M. Ozeki, M. Kim, T. Yamamoto, S. Otake, M. Hirosawa, J. Katz, N.K. Childers, S.M. Michalek. Passive immunization against dental plaque formation in humans: effect of a mouth rinse containing egg yolk antibodies (IgY) specific to *Streptococcus mutans*. *Caries Res.*, 31: 268–274, 1997.
119. D.J. Smith, W.F. King, R. Godiska. Passive transfer of immunoglobulin Y to *Streptococcus mutans* glucan binding protein B can confer protection against experimental dental caries. *Infect. Immun.*, 69: 3135–3142, 2001.

120. B.E. Dunn, H. Cohen, M.J. Blaser. *Helicobacter pylori. Clin. Microbiol. Rev.*, 10: 720–741, 1997.
121. J.-H. Shin, M. Yang, S.W. Nam, J.T. Kim, N.H. Myung, W.-G. Bang, I.H. Roe. Use of egg yolk-derived immunoglobulin as an alternative to antibiotic treatment for control of *Helicobacter pylori* infection. *Clin. Diagn. Lab. Immunol.*, 9: 1061–1066, 2002.
122. I.H. Roe, S.W. Nam, M.R. Yang, N.H. Myung, J.T. Kim, J.H. Shin. The promising effect of egg yolk antibody (immunoglobulin yolk) on the treatment of *Helicobacter pylori*-associated gastric diseases. *Korean J. Gastroenterol.*, 39: 260–268, 2002.
123. J.-H. Shin, S.-W. Nam, J.-T. Kim, J.-B. Yoon, W.-G. Bang, I.-H. Roe. Identification of immunodominant *Helicobacter pylori* proteins with reactivity to *H. pylori*-specific egg-yolk immunoglobulin. *J. Med. Microbiol.*, 52: 217–222, 2003.
124. D.J. Shale, J.S. Elborn. Lung injury. In: D.J. Shale (Ed.) *Cystic Fibrosis*. London: BMJ Publishing Group, 1996, 62–78.
125. H. Kollberg, D. Carlander, H. Olesen, P.-E. Wejaker, M. Johannesson, A. Larsson. Oral administration of specific yolk antibodies (IgY) may prevent *Pseudomonas aeruginosa* infections in patients with cystic fibrosis: a phase I feasibility study. *Pediatr. Pulmonol.*, 35: 433–440, 2003.
126. D.O. Carlander, J. Sundstrom, A. Berglund, A. Larsson, B. Wretlind, H.O. Kollberg. Immunoglobulin Y (IgY) — a new tool for the prophylaxis against *Pseudomonas aeruginosa* in cystic fibrosis patients. *Pediatr. Pulmonol.*, (Suppl) 19: 241 (Abstract), 1999.
127. D. Carlander, H. Kollberg, P.E. Wejaker, A. Larsson. Prevention of chronic *Pseudomonas aeruginosa* colonization by gargling with specific antibodies: a preliminary report. In: J.S. Sim, S. Nakai, W. Guenter (Eds.) *Egg Nutrition and Biotechnology*. New York: CABI Publishing, 2000, 371–374.
128. D. Carlander, H. Kollberg, A. Larsson. Retention of specific yolk IgY in the human oral cavity. BioDrugs 16: 433–437, 2002.
129. JW Hay, AR Hay. Inflammatory bowel disease: costs-of-illness. *J. Clin. Gastroenterol.*, 14: 309–317, 1992.
130. K.L. Worledge, R. Godiska, T.A. Barrett, J.A. Kink. Oral administration of avian tumor necrosis factor antibodies effectively treats experimental colitis in rats. *Dig. Dis. Sci.*, 45: 2298–2305, 2000.
131. W.J. Sandborn, S.B. Hanauer. Antitumor necrosis factor therapy for inflammatory bowel disease: a review of agents, pharmacology, clinical results, and safety. *Inflamm. Bowel Dis.*, 5: 119–133, 1999.
132. R.M.W. Stevenson, D. Flett, B.T. Raymond. Enteric redmouth (ERM) and other enterobacterial infections of fish. In: V Inglis, RJ Roberts, NR Bromage (Eds.) *Bacterial Diseases of Fish*. Oxford: Blackwell Scientific Publications, Oxford, 1993, 80–105.
133. S.B. Lee, Y. Mine, R.M.W. Stevenson. Effects of hen egg yolk immunoglobulin in passive protection of rainbow trout against *Yersinia ruckeri*. *J. Agric. Food Chem.*, 48: 110–115, 2000.
134. H. Hatta, K. Mabe, M. Kim, T. Yamamoto, M.A. Gutierrez, T. Miyazaki. Prevention of fish disease using egg yolk antibody In: J.S. Sim, S. Nakai (Eds.) *Egg Uses and Processing Technologies, New Developments*. Oxon, UK: CAB International, 1994, 241–249.

135. M.A. Gutierrez, T. Miyazaki, H. Hatta, M. Kim. Protective properties of egg yolk IgY containing anti-*Edwardsiella tarda* antibody against paracolo disease in the Japanese eel, *Anguilla japanica J. Fish Dis.*, 16: 113–122, 1993.
136. N. Eterradossi, D. Toquin, H. Abbassi, G. Rivallan, J.P. Cotte, M. Guittet. Passive protection of specific pathogen free chicks against infectious bursal disease by in-ovo injection of semi-purified egg-yolk antiviral immunoglobulins. *J. Vet. Med.*, B44: 371–383, 1997.
137. C.-H. Kweon, B.-J. Kwon, S.-R. Woo, J.-M. Kim, G.-H. Woo, D.-H. Son, W. Hur, Y.-S. Lee. Immunoprophylactic effect of chicken egg yolk immunoglobulin (IgY) against porcine epidemic diarrhea virus (PEDV) in piglets. *J. Vet. Med. Sci.*, 62: 961–964, 2000.
138. Y. Ikemori, M. Ohta, K. Umeda, F.C. Icatlo Jr, M. Kuroki, H. Yokoyama, Y. Kodama. Passive protection of neonatal calves against bovine coronavirus-induced diarrhea by administration of egg yolk or colostrum antibody powder. *Vet. Microbiol.*, 58: 105–111, 1997.
139. V.A. Cama, C.R. Sterling. Hyperimmune hens as a novel source of anti-*Cryptosporidium* antibodies suitable for passive immune transfer. *J. Protozool.*, 38: 42S–43S, 1991.
140. Y.S. Ling, Y.J. Guo, J.D. Li, L.K. Yang, Y.X. Luo, S.X. Yu, L.Q. Zhen, S.B. Qiu, G.F. Zhu. Serum and egg yolk IgG antibody titres from laying chickens vaccinated with *Pasteurella multocida. Avian Dis.*, 42: 186–189, 1998.
141. N.R. Shin, I.S. Choi, J.M. Kim, W. Hur, H.S. Yoo. Effective methods for the production of immunoglobulin Y using immunogens of *Bordatella bronchiseptica*, *Pasteurella multocida* and *Actinobacillus pleuropneumoniae. J. Vet. Sci.*, 3: 47–57, 2002.
142. J. Yang, Z. Jin, Q. Yu, T. Yang, H. Wang, L. Liu. The selective recognition of antibody IgY for digestive system cancers. *Chin. J. Biotechnol.*, 13: 85–90, 1997.
143. B.S. Thalley, S.B. Carroll. Rattlesnakes and scorpion antivenoms from the egg yolk of immunized hens. *Biotechnology*, 8: 934–937, 1990.
144. C.M. Almeida, M.M. Kanashiro, F.B. Rangel Filho, M.F. Mata, T.L. Kipnis. Development of snake antivenom antibodies in chickens and the purification from yolk. *Vet. Rec.*, 143: 579–584, 1998.
145. J. Fraser, V. Arcus, P. Kong, E. Baker, T. Proft. Superantigens-powerful modifiers of the immune system. *Mol. Med. Today*, 6: 125–13, 2000.
146. R.D. LeClaire, R.E. Hunt, S. Bavari. Protection against bacterial superantigen staphylococcal enterotoxin B by passive immunization. *Infect. Immun.*, 70: 2278–2281, 2002.
147. M.S. Sandrin, I.F. McKenzie. Gal alpha(1,3)Gal, the major xenoantigen(s) recognised in pigs by human natural antibodies. *Immunol. Rev.*, 141: 169–190, 1994.
148. J.-F. Bouhours, C. Richard, N. Ruvoen, N. Barreau, J. Naulet, D. Bouhours. Characterization of a polyclonal anti-Galα1-3Gal antibody from chicken. *Glycoconj. J.*, 15: 93–99, 1998.
149. J.P. Fryer, J. Firca, J.R. Leventhal, B. Blondin, A. Malcolm, D. Ivancic, R. Gandhi, A. Shah, W. Pao, M. Abecassis, D.B. Kaufman, F. Stuart, B. Anderson. IgY antiporcine endothelial cell antibodies effectively block human antiporcine xenoantibody binding. *Xenotransplantation*, 56: 98–109, 1999.
150. J.R. Leventhal, A. Su, D.B. Kaufman, M.I. Abecassis, F.P. Stuart, B. Anderson, J.P. Fryer. Altered infectivity of porcine endogenous retrovirus by "protective"

avian antibodies: implications for pig-to-human xenotransplantation. *Transplant. Proc.*, 33: 690, 2001.
151. R. Schade, C. Staak, C. Hendrikson, M. Erhard, H. Hugl, G. Koch, A. Larsson, W. Pollmann, M. van Regenmortel, E. Rijke, H. Spielmann, H. Steinbush, D. Straughan. The production of avian (egg yolk) antibodies: IgY. *ATLA*, 24: 925–934, 1996.
152. A. Larsson, A. Karlsson-Parra, J. Sjöquist. Use of chicken antibodies in enzyme immunoassays to avoid interference by rheumatoid factors. *Clin. Chem.*, 37: 411–414, 1991.
153. A. Larsson, H. Mellstedt. Chicken antibodies: a tool to avoid interference by human anti-mouse antibodies in ELISA after *in vivo* treatment with murine monoclonal antibodies. *Hybridoma*, 11: 33–39, 1992.
154. F. Noack, D. Helmecke, R. Rosenberg, S. Thorban, H. Nekarda, U. Fink, J. Lewald, M. Stich, K. Schutze, N. Harbeck, V. Magdolen, H. Graeff, M. Schmitt. CD87-positive tumor cells in bone marrow aspirates identified by confocal laser scanning fluorescence microscopy. *Int. J. Oncol.*, 15: 617–623, 1999.
155. N. Grebenschikov, A. Geurts-Moespot, H. De Witte, J. Heuvel, R. Leake, F. Sweep, T. Benraad. A sensitive and robust assay for urokinase and tissue-type plasminogen activators (uPA and tPA) and their inhibitor type I (PAI-1) in breast tumor cytosols. *Int. J. Biol. Markers.*, 12: 6–14, 1997.
156. G.J. Lemamy, P. Roger, J.C. Mani, M. Robert, H. Rochefort, J.P. Brouillet. High-affinity antibodies from hen's egg-yolks against human mannose-6-phosphate/insulin-like growth-factor-II receptor (M6P/IGFII-R): characterization and potential use in clinical cancer studies. *Int. J. Cancer*, 80: 896–902, 1999.
157. S. Al-Haddad, Z. Zhang, E. Leygue, L. Snell, A. Huang, Y. Niu, T. Hiller-Hitchcock, K. Hole, L.C. Murphy, Psoriasin (S100A7) expression and invasive breast cancer. *Am. J. Pathol.*, 155: 2057–2066, 1999.
158. D.H. Du Plessis, W. Van Wyngaardt, M. Romito, M. Du Plessis, S. Maree. The use of chicken IgY in a double sandwich ELISA for detecting African horsesickness virus. Onderstepoort. *J. Vet. Res.*, 66: 25–28, 1999.
159. A. Cipolla, J. Cordeviola, H. Terzolo, G. Combessies, J. Bardon, N. Ramon, A. Martinez, D. Medina, C. Morsella, R. Malena. *Campylobacter* fetus diagnosis: direct immunofluorescence comparing chicken IgY and rabbit IgG conjugates. *ALTEX*, 18: 165–170, 2001.
160. E. Gutierrez Calzado, R.M. Garcia Garrido, R. Schade. Human haemoclassification by use of specific yolk antibodies obtained after immunisation of chickens against human blood group antigens. *Alter. Lab. Anim.*, 29: 717–726, 2001.
161. C. Wu, R. Yang, J. Zhou, S. Bao, L. Zou, P. Zhang, Y. Mao, J. Wu, Q. He. Production and characterisation of a novel chicken IgY antibody raised against C-terminal peptide from human thymidine kinase 1. *J. Immunol. Meth.*, 277: 157–169, 2003.
162. A. Di Lonardo, M. Luisa Marcante, F. Poggiali, E. Hamsøikovà, A. Venuti. Egg yolk antibodies against the E7 oncogenic protein of human papillomavirus type 16. *Arch. Virol.*, 146: 117–125, 2001.
163. F. De Ceuninck, P. Pastoureau, S. Agnellet, J. Bonnet, P.M. Vanhoutte. Development of an enzyme-linked immunoassay for the quantification of YKL-40 (cartilage gp-39) in guinea pig serum using hen egg yolk antibodies. *J. Immunol. Meth.*, 252: 153–161, 2001.
164. E.C.Y. Li-Chan. Applications of egg immunoglobulins in immunoaffinity chromatography. In: J.S. Sim, S. Nakai, W. Guenter (Eds.) *Egg Nutrition and Biotechnology*. New York: CAB International, 2000, 323–339.

165. E.M. Akita, E.C.Y. Li-Chan. Isolation of bovine immunoglobulin G subclasses from milk, colostrums, and whey using immobilized egg yolk antibodies. *J. Dairy Sci.*, 81: 54–63, 1998.
166. H. Kim, T.D. Durance, E.C.Y. Li-Chan. Reusability of avidin-biotinylated immunoglobulin Y columns in immunoaffinity chromatography. *Anal. Biochem.*, 268: 383–397, 1999.
167. E.C.Y. Li-Chan, S.S. Ler, A. Kummer, E.M. Akita. Isolation of lactoferrin by immunoaffinity chromatography using yolk antibodies. *J. Food Biochem.*, 22: 179–195, 1998.
168. J. Kovacs-Nolan, Y. Mine. Avian egg antibodies: basic and potential applications. *Avian Poult. Biol. Rev.*, 15: 25–46, 2004.
169. M. Tini, U.R. Jewell, G. Camenisch, D. Chilov, M. Gassmann. Generation and application of chicken egg-yolk antibodies. *Comp. Biochem. Physiol. A. Mol. Integr .Physiol.*, 131: 569–574, 2002.
170. M. Romito, G.J. Viljoen, D.H. Du Plessis. Eliciting antigen-specific egg-yolk IgY with naked DNA. *Biotechniques*, 31: 670–675, 2001.
171. S. Nishinaka, T. Suzuki, H. Matsuda, M. Murata. A new cell line for the production of chicken monoclonal antibody by hybridoma technology. *J. Immunol. Meth.*, 139: 217–222, 1991.
172. H. Asaoka, S. Nishinaka, N. Wakamiya, H. Matsuda, M. Murata. Two chicken monoclonal antibodies specific for heterophil Hanganutziu-Deicher antigens. *Immunol. Lett.*, 32: 91–96, 1992.
173. H.S. Lillehoj, K. Sasai. Development and characterization of chicken-chicken B cell hybridomas secreting monoclonal antibodies that detect sporozoite and merozoite antigens of Eimeria. *Poult. Sci.*, 73: 1685–1693, 1994.
174. S. Nishinaka, H. Akiba, M. Nakamura, K. Suzuki, T. Suzuki, K. Tsubokura, H. Horiuchi, S. Furusawa, H. Matsuda. Two chicken B cell lines resistant to ouabain for the production of chicken monoclonal antibodies. *J. Vet. Med. Sci.*, 58: 1053–1056, 1996.
175. K. Matsushita, H. Horiuchi, S. Furusawa, M. Horiuchi, M. Shinagawa, H. Matsuda. Chicken monoclonal antibodies against synthetic bovine prion protein peptide. *J. Vet. Med. Sci.*, 60: 777–779, 1998.
176. H. Matsuda, H. Mitsuda, N. Nakamura, S. Furusawa, S. Mohri, T. Kitamoto. A chicken monoclonal antibody with specificity for the N-terminal of human prion protein. *FEMS Immunol. Med. Microbiol.*, 23: 189–194, 1999.
177. S.M. Mohammed, S. Morrison, L. Wims, K. Ryan Trinh, A.G. Wildeman, J. Bonselaar, R.J. Etches. Deposition of genetically engineered human antibodies into the egg yolk of hens. *Immunotechnology*, 4: 115–125, 1998.

10 Antiangiogenic Proteins, Peptides, and Amino Acids

Jack N. Losso and Hiba A. Bawadi

CONTENTS

10.1 INTRODUCTION

Angiogenesis, or neovascularization, is the formation of new blood vessels from previous vascular ones as conduits to supply blood and nutrients to and remove metabolic wastes from living tissues. Vessel growth can occur by vasculogenesis, angiogenesis, and arteriogenesis. Vasculogenesis refers to the formation of blood vessels by endothelial progenitors arising from various embryonic regions or the adult bone marrow; and arteriogenesis refers to the stabilization of blood vessels formed by angiogenesis by mural cells (1). Vasculogenesis is particularly critical during development of the embryo. Angiogenesis is an important physiological process that encompasses the body embryonic development, exercise, and repair such as wound healing and the female ovulatory cycle. However, excessive or insufficient angiogenesis is pathological and is a common feature of many neoplastic and nonneoplastic diseases such as solid and hematological tumor growth and metastasis, diabetic retinopathy and wound healing, and chronic inflammatory disorders (Table 10.1). A number of angiogenic regulators, which include hypoxia, growth factors, enzymes, hormones, metals, adhesion molecules and other proteins, and oncogenes, have been identified (Table 10.2). Building on the fundamentals of normal and aberrant cell proliferation, scientists have begun to translate the fundamentals of angiogenesis and targets into prevention and therapeutic initiatives (2–4). Whereas antiangiogenic drugs such angiostatin, thrombospondin, endostatin, combrestatin, and VEGF inhibitors have worked well by reducing or eliminating tumors in animal models, the promises have been slow to realize in humans. It has been pointed out that the progress of antiangiogenic compounds may have been slow for many reasons, which include among others (a) the age of the animals involved in the studies, (b) the toxicity of the drugs which may lead to fatality, and (c) the accumulation of additional malignancy-driving mutations with advanced disease (5–9). Early intervention with antiangiogenic compounds when there are few stimulators of angiogenesis may be more effective than current medical practices that involve curing a disease that may have developed resistance or metastasis (10). The disappointment with certain dietary supplements such as shark cartilage may be explained by factors such as (a) the toxicity of the dietary supplement which may lead to fatality, (b) the accumulation of additional malignancy-driving mutations with advanced disease that cause resistance to the dietary supplements, or (c) individuals with a smoking history or other unhealthy lifestyles.

Functional foods that are mostly for disease prevention are being touted as potentially effective against the early stages of the angiogenic process (9–11). This chapter serves to document the emerging concept of bioactive amino acids, peptides, and polypeptides as antiangiogenic functional foods with clear nutritional and clinical potentials.

TABLE 10.1
Pathological Angiogenesis-Associated Human Diseases

Pathological Angiogenesis	
Insufficient	**Excessive**
Delayed wound healing	Rheumatoid arthritis
Ischemia (myocardial, peripheral, cerebral)	Multiple sclerosis
Stroke	Age-related macular degeneration
Heart disease	Diabetes retinopathy
Scleroderma	AIDS complications
Infertility	Kaposi's sarcoma
Diabetic neuropathy	Tumor growth and metastasis
Systemic sclerosis	Osteoporosis
Coronary heart disease	Alzheimer's disease
Placental insufficiency	Parkinson's disease
Impaired healing of fracture	Obesity
Pulmonary and systemic hypertension	Psoriasis
Vascular dementia	Hepatitis
Lymphedema	Asthma
Liver regeneration	Thyroid enlargement
Impaired collateral vessel formation	Ocular neovascularization
Chronic nonhealing ulcer	Retinopathy of prematurity
Churg-Strauss syndrome	Synovitis
	Osteomyelitis
	Nasal polyps
	Liver regeneration
	Panus growth
	Bone/cartilage destruction
	Endometriosis
	Hematological malignancies
	Hemangioma
	Adipose tissue
	Chronic inflammation
	Pulmonary hypertension
	Pulmonary vasoconstriction
	Pox virus
	Herpes virus
	Atherosclerosis

10.2 CLINICAL SIGNIFICANCE AND CLINICAL TRIALS

The process of angiogenesis is depicted in Figure 10.1 and involves (1) stimulation of angiogenic factors and protease-catalyzed dissolution of basement membrane, (2) endothelial cell migration and proliferation, (3) cell proliferation, and (4) recruitment of pericytes and smooth muscle cells, which stabilize the newly formed vessel. Figure 10.2 represents a real-life situation where a tumor has created a

TABLE 10.2
Stimulators of Pathological Angiogenesis

Mitochondrial Defects

Hypoxia

Hypoxia-inducible factor-1 (HIF-1)

Proteins

Growth Factors :VEGF, aFGF, bFGF, EGF, HGF, Granulocyte colony-stimulating growth factor, HGF, PDGF-BB, PDEGF, PGF, TGFα, TNFα, angiogenin, angiotropin
Enzymes and Signal Transduction Enzymes: aspartic, cysteine, matrix metallo- and serine proteinases, thymidine phosphorylase, ornithine decarboxylase, cyclooxygenase, nitric oxide synthase, farnesyl transferase, geranylgeranyl transferase
Cytokines: IL-1, IL-6, IL-8
Integrins and Adhesion Molecules: $\alpha_v\beta_3$ integrin, fibronectin
Other proteins: angiopoietin-1, ceruloplasmin, endothelin; osteopontin (ETB receptor)
Oncogenes: *c-myc*, *ras*, *c-src*, *v-raf*, *c-jun*

Small molecules

Adenosine
1-Butyrylglycerol
Prostaglandin E1 and E2
Metals: Copper, Cadmium, Zinc

network of new blood vessels for its growth and has spread throughout the body. Angiogenesis is important for reproduction, development, repair, and in the context of angiogenic-dependent diseases. Physiological angiogenesis such as a developing child in a mother's womb is tightly controlled by a network of stimulators and inhibitors that work in balance with each other (Table 10.2). Pathological angiogenesis may be excessive or insufficient. Excessive angiogenesis occurs in cancer, inflammatory bowel diseases, diabetic blindness, age-related macular degeneration, rheumatoid arthritis, psoriasis, and many others (see Table 10.2). Insufficient angiogenesis occurs in coronary artery disease, peripheral artery disease, stroke, infertility, and incomplete wound healing when there is insufficient blood supply and oxygen to the tissues (see Table 10.2). The onset and progression of excessive pathological angiogenesis is catalyzed by increased activities of enzymes (serine, cysteine, proteasome, aspartic, and matrix metalloproteinases), growth factors (vascular endothelial growth factor [VEGF], fibroblast growth factor [FGF], epidermal growth factor, hepatocyte growth factor, platelet-derived growth factor, and many others), adhesion molecules (fibronectin, vitronectin, and many others), and oncogenes (*ras*, *c*-myc, *fas*, and others).

Folkman (2) suggested that the inhibition of angiogenesis was the most appropriate therapeutic means to control the growth of tumors. Folkman and Ingber (12) described three strategies for angiogenesis inhibition: (a) inhibition of release of

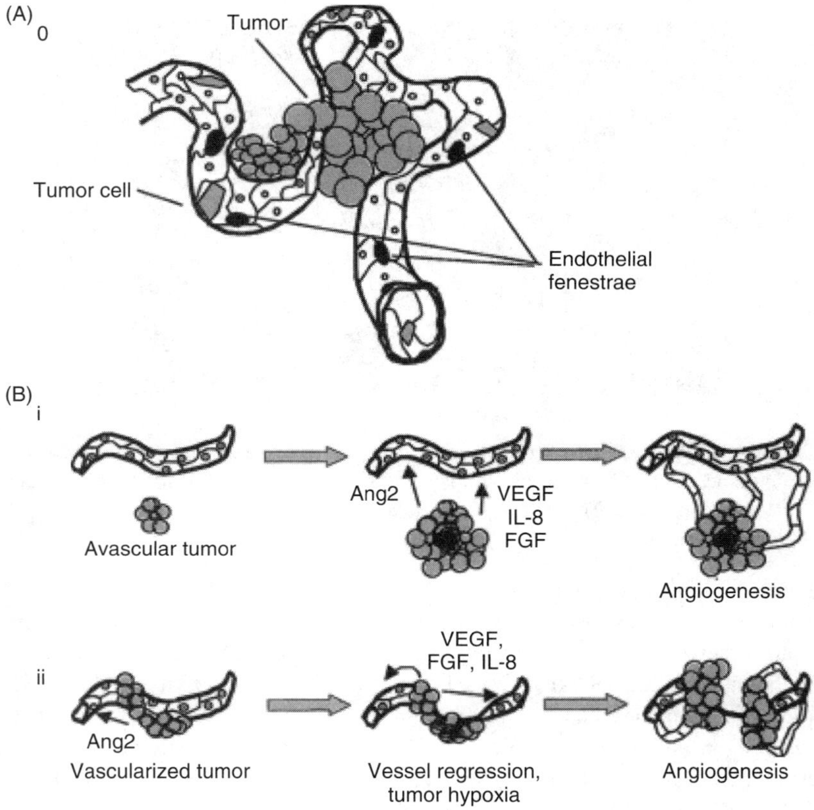

FIGURE 10.1 The process of tumor angiogenesis. (A) Schematic of tumor blood vessel (gray, normal tumor cells; black, necrotic tumor cells). (B) Model of tumor-induced neovascularization. In i, an initially avascular tumor grows until inner regions become hypoxic and upregulate production of angiogenic factors such as VEGF, FGF, and interleukin (IL)-8. In ii, a tumor grows on an existing blood vessel. Soon the tumor induces Ang-2 expression in the preexisting vessel, and it regresses due to endothelial cell apoptosis. The tumor is now avascular, and by upregulating angiogenic factors, as in i, it induces the production of a new blood supply. (From Papetti and Herman. *Am. J. Physiol. Cell Physiol.*, 282(5): C947–C970, 2002. With permission.).

angiogenic molecules from tumor cells; (b) neutralization of angiogenic molecules that have already been released; and (c) inhibition of vascular endothelial cells from responding to angiogenic stimulation. Clinically, compounds that halt new blood vessel growth are candidates for prevention and therapy of excessive angiogenesis and compounds that stimulate new blood vessel growth are good candidates for the treatment of insufficient angiogenesis. Whereas the inhibition of angiogenesis is expected to provide therapy for excessive angiogenic diseases, stimulation of angiogenesis may initiate and exacerbate pathological conditions such as cancer. VEGFs,

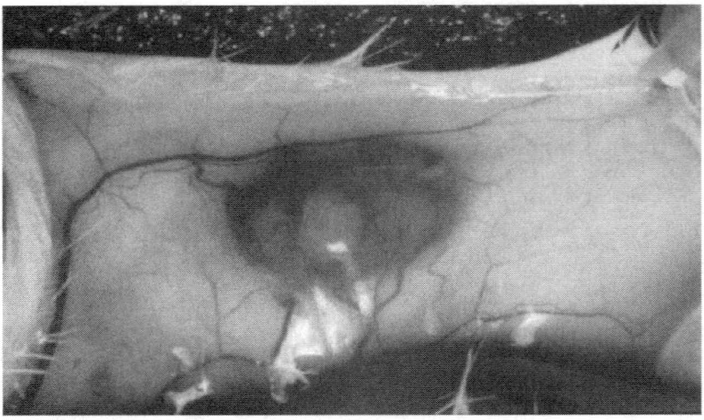
Blood vessels feed a tumor in the skin

FIGURE 10.2 *In vivo* angiogenesis showing blood vessels feeding a tumor in a mouse skin. (With permission from Drs. M. Achen and S. Stacker, Ludwig Institute, Angiogenesis Laboratory, Melbourne, Australia.)

the major stimulators of angiogenesis are endothelial cell mitogens *in vitro* and also stimulate angiogenesis *in vivo* (13). VEGFR-1 (Flt-1) and VEGFR-2 (kinase domain receptors) are two tyrosine kinase receptors expressed in most vascular endothelial cells through which VEGF exerts its biological activity. VEGF is a biomarker of angiogenesis and expressed in a large number of human tumors such as breast, colon, and, pancreatic, non–small-cell lung cancer, small-cell lung cancer, mesothelioma, hematological, prostate, ovarian, and many other cancer cells (14–17). In growing tumors, VEGF expression is upregulated by hypoxia, growth factors, proteases, metals, proteins, and oncogenes (1).

10.3 SCREENING ANTIANGIOGENIC COMPOUNDS

The mechanism of action of antiangiogenic substances include (a) interference with angiogenic stimulators; (b) interference with angiogenic receptors; (c) interference with the extracellular matrix; (d) interference with the control of angiogenesis by hypoxic signaling; (e) interference with proteolysis; (f) vascular targeting (18). Angiogenesis screening assays include *in vitro* and *in vivo* methods. Among the *in vitro* assays, the endothelial cell assays that look at proliferation, migration, and tube formation as indicators of the ability of a compound to inhibit angiogenesis, are the simplest models.

1. *Cell proliferation assay*: *In vitro* models of angiogenesis use endothelial cells, isolated from capillaries or large vessels to provide information on the molecular and cellular biology of angiogenesis. Endothelial cell proliferation, migration, and differentiation studies are based on

thymidine incorporation, cell labeling reagent 5-bromo-2′-deoxyuridine (BrdU), or cell death using Tunel assay, or adenosine triphosphate (ATP) as an indication of actively growing cells [19–21].

2. *Endothelial cell migration*: The Boyden Chamber, which consists of upper and lower wells separated by a membrane filter, is used to evaluate endothelial cell proliferation. A chemoattractant factor such as VEGF or FGF is added to the lower chamber, and the endothelial cells are introduced in the upper cell. The cells will cross the membrane filter and move down toward the chemotactic stimulus in the lower chamber. After an incubation period, cells in the lower chamber are counted using a microscope. An inhibitor of angiogenesis will block endothelial cell chemotaxis. Several references are available from the literature regarding the use of the Boyden Chamber. A modified Boyden Chamber is available from Chemicon International (Temecula, CA) and many other manufacturers.
3. *Tube formation*: Tube formation is a multistep process involving cell adhesion, migration, differentiation, and growth. Experimentally, the assay is carried as follows. The wells of a 96-well microtiter plate or any other chamber slide are coated with 60 to 125 μl of ice-cold Matrigel and allowed to polymerize at 37°C for at least 30 minutes. The following day, endothelial or cancer cells (10^4 to 10^5 cells ml^{-1}) are layered on top of the gel in EGM-2 (Clonetics), 5% serum, 10 μM $ZnCl_2$ supplemented with 10 ng ml^{-1} growth factor such as FGF-2. Control medium or the compound to be tested is added to individual wells. Each treatment is carried out in triplicate. Pictures are taken 48 hours later (22).
4. *Microarray techniques*: Northern and dot blot analysis, reverse-phase transcriptase-coupled–polymerase chain reaction (RT-PCR), differential display, and serial analysis of gene expression (SAGE) have been used to study gene expression at the mRNA levels. These techniques are time consuming and have other shortcomings such as being labor-intensive and requiring complex sample preparation. Microarray technology or DNA Chips technology permits rapid, simultaneous parallel screening of expression analysis of thousands of genes on a global level in biological processes and facilitates the comparative analysis of a large number of samples. The availability of the entire human genome and the power of microarray technology are promising to identify genes with specific or enhanced expression in cancer and other angiogenic diseased cells (23). The proteins produced by these genes that are specifically present on cancer and other chronic angiogenic disease cells are being used as biomarkers for disease therapy or prevention. The advantages of microarray over conventional techniques include its high sensitivity, use of nontoxic chemicals, ability to simultaneously analyze thousands of genes with microliters of samples, and miniaturized size of the equipment. However, the need for careful interpretation of the data minefield is highly critical for the technology to be productive.

10.3.1 Pitfalls of *In Vitro* Angiogenesis Assays

Understanding the process of angiogenesis and its regulation requires identification and use of models for the study of vascular development and regulation that are truly representative of the *in vivo* complexity of the vasculature. Whereas *in vitro* models facilitate ascertaining specific information regarding endothelial cell biology, they do not recreate the microenvironment of an intact organism (24,25).

10.3.1.1 *In vivo* Angiogenesis Assays

The chick chorioallantoic membrane, the cornea assay, sponge implant models, the matrigel plug assay, the orthotopic mouse model, and subcutaneous tumor models are classic *in vivo* angiogenesis assays.

10.3.1.1.1 The Chick Chorioallantoic Membrane (CAM) Assay

The CAM assay represents an *in vivo* vascular network. The chorioallantoic membrane surrounds the developing chicken embryo, and the developing embryo becomes vascularized as the embryo develops. New vessels form from the existing ones. The inhibitor of angiogenesis will cause a regression of the vessels. The major advantage of the CAM assay is that it is simple to perform and has therefore become the preferred screening method (24,25).

The major drawback of the CAM assay is that the fertilized egg is an already well-developed vascular network by the time of the assay. The presence of such a well-developed network of blood vessels makes it difficult to distinguish between newly formed capillaries and previously existing ones. However, the disappearance of capillaries can easily be seen (Figure 10.3). Another major drawback is the need to analyze several fertilized eggs before drawing a conclusion.

10.3.1.1.2 The Cornea Assay

Unlike the CAM assay, which uses a well-developed vascular system, the cornea assay is an *in vivo* avascular tissue. Angiogenic or antiangiogenic activity of a bioactive compound is determined as follows. The bioactive compound is encapsulated along with an angiogenic substance (i.e., FGF-2 or VEGF) in polymer pellets (26). The pellets are implanted in pockets in the corneal stroma of a rabbit, rat, or mouse. A corneal micropocket is created in the eye with a modified von Graefe cataract knife. A micropellet (0.35×0.35 mm) of sucrose aluminum sulfate coated with hydron polymer containing ~80 ng of FGF-2 or VEGF is implanted into each pocket. The pellet is positioned 0.8 mm from the corneal limbus. After implantation, erythromycin or ophthalmic ointment is applied to each eye (27). Angiogenesis or antiangiogenesis is determined using a slit-lamp biomicroscope after 4 to 6 days, by examining new vessels penetrating from the limbus into the corneal stroma (Figure 10.4). The cornea can also be perfused with India ink and analyzed by computer image analysis. The method is highly reliable, but not practical for screening because it is technically demanding and more expensive than the CAM assay.

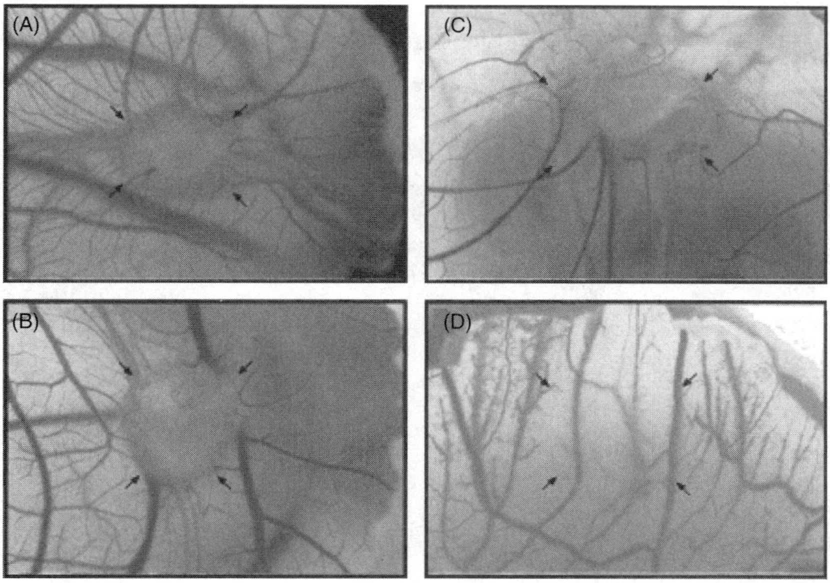

FIGURE 10.3 Motuporamine C inhibits angiogenesis *in vivo*. Photographs of developing CAMs incubated for 2 days with VEGF (A) or VEGF and motuporamine C at 2.5 μM (B), 5 μM (C), or 10 μM (D). The *arrows* indicate the corners of the gelatin sponges containing VEGF and the compounds. (From Roskelley et al. *Cancer Res.*, 61, 6788–6794, 2001. With permission.)

10.3.1.2 *In vivo* Matrigel Plug Assay

The Matrigel is a complex mixture of proteins and growth factors such as laminin, collagen type IV, heparan sulfate, fibrin, EGF, TGF-b, PDGF, and IGF-1, and proteolytic enzymes (plasminogen, tPA, MMPs) that occur normally in Engelbreth Holm-Swarm (EHS) mouse tumor (28). The mixture is a liquid at 4°C and forms a gel at 37°C (Figure 10.5). A growth factor is entrapped inside the gel and is slowly released by the gel. In this assay, the matrigel (500 μL) on ice is mixed with either a growth factor alone or cancer cells at 106 cells, or with a growth factor (or cancer cells) and the potential anti- or pro-angiogenic compound to be tested. The mixture is injected s.c. into a 4 to 8-week-old mouse at sites near the abdominal midline or into the flank. The number of injections allows each animal to simultaneously receive a positive control plug (growth factor and heparin), a negative control plug (heparin plus buffer), and a plug containing the treatment to be tested (growth factor such as FGF-2, heparin, and anti- or pro-angiogenic compound to be tested). All treatments are tested in triplicate. Animals are sacrificed 5 to 7 days after injection. The mouse skin is detached along the abdominal midline, and the Matrigel plugs are recovered and scanned immediately at high resolution. Plugs are then dispersed in water and hemoglobin levels are determined

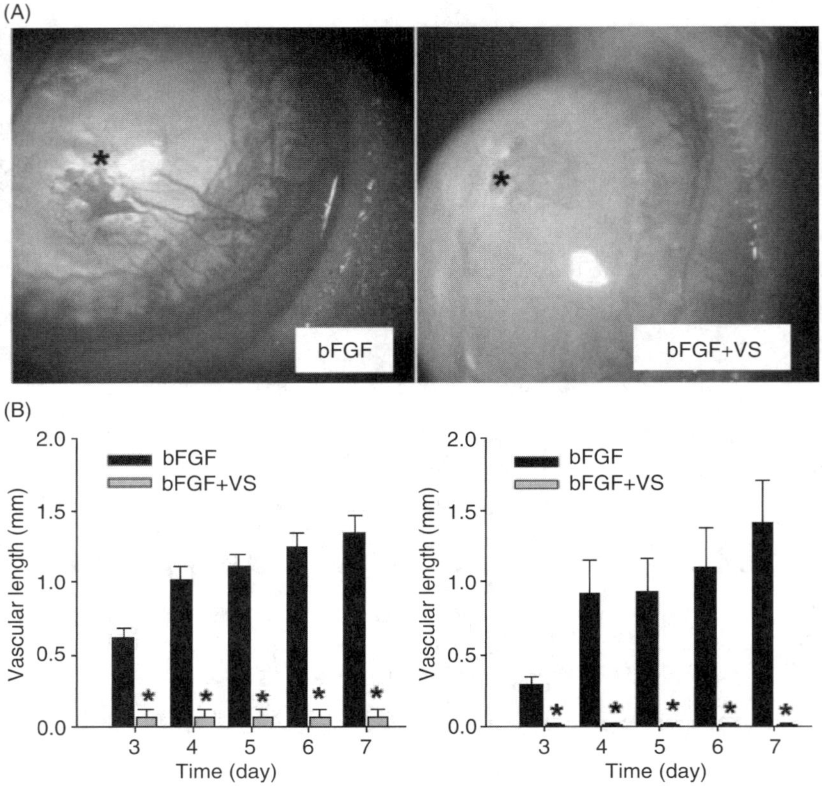

FIGURE 10.4 Inhibition of mouse corneal neovascularization. Part A. A pellet containing 90 ng of bFGF (left panel) or 90 ng of bFGF + 180 ng of vasostatin (right panel) was implanted into the corneal of a rat for 7 days. The asterisk represents the place of hydron pellet implantation. Part B. Neovascularization in rat corneas after implantation bFGF hydron pellets or pellets containing bFGF and vasostatin (VS). The length (left panel) and area (right panel) of neovascularized vessels in rat corneas were measured after implantation for 3–7 days. Bars represent the means and the error bars represent the SEM. The asterisk indicates statistical significance ($p<0.001$). From Wu et al. (2005). *Mol. Vis.*, 11, 28–35. With permission.

using Drabkin's solution (Sigma) according to the manufacturer's instructions for the quantitation of blood vessel formation. The concentration of hemoglobin is calculated from a known amount of hemoglobin assayed in parallel (29). The advantage of the Matrigel is that it provides a natural environment that mimics angiogenesis *in vivo*. However, it is expensive.

10.3.1.3 The Zebrafish

The zebrafish, even though a new model with many unknowns, has recently emerged as an experimentally and genetically accessible alternative model for

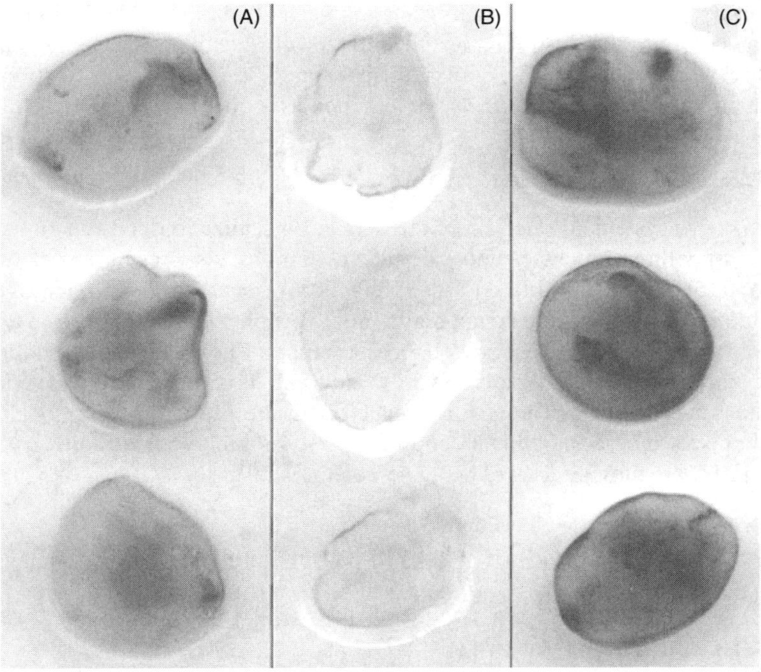

FIGURE 10.5 The histidine-proline-rich glycoprotein (HPRG) plasma protein domain inhibits hemoglobin content and tumor cell growth of 3LL cells. Two million cells of the mouse lung carcinoma cell line 3LL were mixed with Matrigel plus 1.8 μM H/P domain and then injected into the flank of a mouse. Seven days later, the animals were sacrificed and the plugs were removed, weighed, and scanned at high resolution. Panel A: 1.8 μM H/P domain with 2×10^6 3LL cells; Panel B: saline buffer; Panel C: 2×10^6 3LL cells. A robust angiogenic response can be seen in the Matrigel plugs containing 3LL cells *versus* control. From Juarez et al. 2002. *Cancer Research* 62, 5344–5350. With permission.

studying vertebrate vascular formation both developmentally and in adult tissues. However, the advantages of using zebrafish in angiogenesis research are that (a) zebrafish has a short breeding cycle (every 1 to 2 weeks), short generation time (3 months), gives a high number of progenies (100s per clutch); (b) its embryos are transparent vertebrates that develop outside the mother's body. The fish changes from an egg to a well-developed transparent embryo within 24 hours, and the entire process as well as the blood vessels can be watched under a stereomicroscope. The fish embryos develop organs that are similar to those in humans, such as a central nervous system, pancreas, and thymus, and quickly form blood vessels and beating hearts; (c) the 2- to 3-day assay time with zebrafish is similar to the time required for a cell-based assay, that is, low cost, whereas a mouse tumor assay requires at least 4 weeks; (d) a single dosing regimen is used compared with a daily dosing one for a mouse; and (e) it serves as an *in vivo* toxicological model and allows the use of enough vertebrates to obtain good replicates and easily test statistically significant numbers of animals. The inhibition of angiogenesis is zebrafish mimics the effects observed in mice and humans because genes such as VEGF, Flk-1, Tie-1, and Tie-2 have been identified in zebrafish and shown to have the same function as in mammals (25,30,31).

10.3.2 Pitfalls of *in vivo* Models

Most *in vivo* systems do not allow the detailed visualization and functional dissection of cellular processes under defined conditions. As a result, these systems do not allow reproducibility and interpretability of results. Most animal models used are at a young age and do not really mimic the physiology of the aged human body in which most degenerative diseases erupt. Whereas *in vivo* models are tested under well-defined experimental conditions, it is difficult to carry similar tests in living humans. The cost associated with the use of *in vivo* models that could closely mimic the human conditions is so prohibitive that more and more researchers are moving toward using cell cultures instead.

10.4 ANTIANGIOGENIC POLYPEPTIDES, PEPTIDES, AND AMINO ACIDS

10.4.1 Cartilage

Bioavailable shark and bovine cartilage blocks endothelial and tumor cell invasion and proliferation by inhibiting the activity of several matrix metalloproteinases (MMPs) (32,33). Troponin I, is the molecule responsible for the inhibition of the actomyosin ATPase during muscle contraction. Peptide Glu94–Leu123 from troponin I has been reported as the fragment responsible for the antiangiogenic activity of troponin (34). The presence of troponin I in human cartilage has been reported (35). The compound AE-941 (Neovostat trade mark; AEterna) is an orally bioavailable shark-cartilage-derived peptide that molecularly inhibits angiogenesis, possibly by four different potential mechanisms of action: (a) modulation of matrix metalloproteinase activity; (b) inhibition of vascular endothelial growth factor binding to its receptor; (c) induction of endothelial cell apoptosis; and (d) stimulation of angiostatin production (36). AE-941 has been tested in a variety of *in vitro* and *in vivo* models of angiogenesis and is undergoing phase III clinical trials in patients with metastatic cell carcinoma of the kidney (36). Liang and Wong (37) showed that the antiangiogenic compound in shark cartilage was a 10-kilodaltons (kDa) proteoglycan-containing keratan sulfate units. Side effects associated with the use of shark cartilage at clinical levels include hypercalcemia, nausea, diarrhea, vomiting, flatulence, constipation, acne, and rash (38,39).

10.4.2 Collagen

Oral administration of glycosylated undenatured type II (UC-II) chicken collagen (10 mg day^{-1}) for 42 days to five female subjects (58 to 78 years old) suffering from significant joint pain markedly reduced pain associated with joint inflammatory conditions and symptoms of osteoarthritis and rheumatoid arthritis (40). The glycosylated UC-II presented active epitopes, with the correct three-dimensional structures, to Peyer's patches, which influenced the signaling required for the development of immune tolerance in osteoarthritis and rheumatoid arthritis. The addition of type-I collagen to fibrin matrices dose-dependently inhibited tube-formation by human microvascular endothelial cells (41).

10.4.3 α2-Macroglobulin

The alpha 2-macroglobulin (α_2M) is a water-soluble homotetrameric protein of 718,000 Da occurring in the plasma of mammals and aquatic species. α_2M is a general protease inhibitor that inactivates proteinases from all four mechanistic classes by a mechanism of action termed as "trap hypothesis" (42). The inhibition involves a conformational change and compaction of the native α_2M structure and, as a result, the exposure of receptor-binding domains that mediate the interactions of α_2M and its receptors. α_2M can be activated by small molecules such as ammonia or methylamine and by proteinases. Activated α_2M, termed α_2M*, is functionally equivalent to α_2M. α_2M has been shown to regulate cellular growth by binding numerous cytokines and growth factors such as TGF-β and PDGF *in vitro* and *in vivo,* and in circulation, α_2M is often associated with 80 to 90% of these growth factors (43). However, α_2M or α_2M* has the ability to bind to certain members of the FGF family, such as FGF-1, -2, -4, and -6, but does not significantly bind to FGF-5, -7, -9, and -10 (43). These researchers also reported that FGF factors bind to α_2M* with higher affinity than to α_2M. *In vivo*, α_2M* dose-dependently inhibited FGF-2 stimulated endothelial cell proliferation. However, in the presence of basement membrane substances such as collagen and ECM substances such as heparin sulfate proteoglycan that allow presentation of FGF-2 to FGF receptor tyrosine kinase on the surface of the cells, FGF-2 evades α_2M*-mediated inhibition.

10.4.4 Bowman-Birk Inhibitor

The Bowman–Birk inhibitor (BBI) is a water-soluble protein occurring in soybean, legumes, and many monocotyledonous seeds. Kennedy et al.(44), Armstrong et al. (45), Lippman et al. (46), Larionova et al. (47–49), and many others have reported on the anticarcinogenic effect of BBI in animal models and clinical trials. Heparin-enhanced gelatin zymography, quenched fluorescence substrate hydrolysis analysis, and the Biotrak assay of the interaction of soybean BBI with the matrix metalloproteinase-1 (MMP-1) demonstrated that demineralized BBI at 30 nM inhibited MMP-1 activity, whereas BBI was inhibitory at 115 nM (50). Soybean BBI at 0.02 to 0.4 mg ml^{-1}, concentration-dependently inhibited trypsin activation of pro-MMP-9, and rice BBI at concentrations of 0.08 to 0.352 mg ml^{-1} dose-dependently inhibited the *in vitro* activation of pro-MMP-1 by trypsin (51).

10.4.5 Protamine

Protamine, a naturally occurring cationic polypeptide of about 30 to 65 amino acid residues and very rich in arginine residues inhibits Gram-positive and Gram-negative bacteria (52). Human and experimental gliomas spread and grow in response to both paracrine and autocrine release of endothelial, fibroblast, and platelet growth factors. Protamine dose-dependently reduced tumor volume, mitotic index, vascular density, and cell viability of highly malignant C-6 glioblastoma in Wistar rats at a dose lower than toxic dose of suramin (53). Many types of carcinoma accumulate large numbers of degranulating mast cells that will eventually release heparin. Protamine binds to heparin and neutralizes heparin's anticoagulant effect and may therefore induce selective tumor cell thrombosis. Intravenously injected protamine induced selective thrombosis in tumors, and the effect lasted for several hours (54). Antithrombic compounds can also prevent lipid-rich plaque rupture in diabetic and dyslipidemic patients. Ornithine decarboxylase (ODC), which is associated with the onset and progression of a variety of cancers, including colon, prostate, and breast cancers, was inhibited by protamine (55). The inhibition of ODC leads to polyamine depletion in cells, a cytostatic effect on proliferating endothelial cells, and the inhibition of angiogenesis (56). Nitric oxide (NO) is a pivotal factor for gastric ulcer healing, and arginine (Arg) is a precursor of NO. The abundance of Arg residues in the protamine molecule potentiates protamine's healing effect.

Thrombin can activate many of the cellular events such as receptors of VEGF, MMP-2, and MMP-9 involved in the angiogenic cascade. Thrombin inhibitors such as phosvitin and protamine may be used as functional foods. Protamine sulfate, obtained when sulfuric acid is used during protamine purification, dose-dependently and reversibly inhibits thrombin and prevents the conversion of fibrinogen to fibrin (57).

10.4.6 Lactoferrin

Yoo et al. (58) transplanted highly metastatic murine melanoma or lymphoma cells into mice, subsequently administered bovine lactoferrin (bLf) or bovine lactoferricin (bLfcin) subcutaneously, and reported that lung metastasis and angiogenesis were inhibited. bLfcin has also shown metastasis inhibition of colon carcinoma 26 cell line transplanted into mice (59). Bovine Lf exhibited dose-dependent inhibition of angiogenesis on 4- to 6-day-old chick embryo chorioallantoic membranes (CAMs), but the inhibition was reversed when bLf was simultaneously treated with bFGF (60). Orally (not intraperitoneally) administered bLf at doses of 100 or 300 mg kg^{-1} day^{-1} inhibited angiogenesis on experimental metastasis by augmenting CD4+, CD8+, and asialoGM1+ in the spleen and peripheral blood and by stimulating the production of IL-18 in intestinal mucosa (61).

Oral administration of iron-unsaturated bLf inhibited $VEGF_{165}$- and IL-1α induced angiogenesis in rats (62). Lf was also shown to exert an antiproliferative effect on endothelial cells *in vitro*. Since Lf is an endogenous dietary protein available from bovine milk and because of its ability to inhibit angiogenesis following oral administration, Lf draws interest for further study to elucidate its mechanism.

10.4.7 Lectins

Lectins extracted from mistletoe have both cytotoxic and immunomodulatory activities and are associated with anticarcinogenicity. Park et al. (63) used the chicken chorioallantoic membrane (CAM) assay to demonstrate that the antimetastatic activity of Korean mistletoe lectin (*Viscum album* L.) against B16-BL6 melanoma cells in C57-BL6 mice was a result of angiogenesis inhibition and increased apoptosis. Mistletoe lectins therefore call for further study in the development of cancer therapy. Van Huyen et al. (64) investigated the effect of various mistletoe preparations on both human venous endothelial cell (HUVEC) and immortalized human venous endothelial cell (IVEC) lines using morphological assessment of endothelial cells, FACScan analysis after propidium iodine and annexin V labeling, and detection of cleavage of poly(A)DP-ribose polymerase (PARP). All tested mistletoe preparations were cytotoxic in IVEC and induced endothelial apoptosis. It was suggested that further *in vivo* investigations are needed for these antiangiogenic extracts.

Morphological studies of the small bowel of tumor-bearing mice and orally fed mistletoe lectin beginning 3 days after inoculation indicated that (a) the lectin induced hyperplasia, (b) the lectin bound avidly to lymphoid tissue of Peyer's patches, and (c) histological evidence of viable tumor disappeared in 25% of mice fed the ML-1 diet for 11 days (65). A preparation of mistletoe lectin (ML-3) added to the diet of mice 3 days after inoculation of tumor cells slowed down further growth of an established tumor and indicated that orally administered mistletoe lectins can induce powerful anticancer effects. Park et al. (63) investigated the mechanisms of the anticancer and antimetastatic actions of the purified Korean mistletoe lectin (*Viscum album* L. coloratum agglutinin, VCA) by inoculating C57-BL6 mice with B16-BL6 melanoma cells, followed by treatment with VCA to assess for survival and metastasis. Mice treated with VCA had a prolonged survival, and *in vivo* angiogenesis using the CAM assay showed a decrease in vessel growth. The melanoma cells died by apoptosis.

10.4.8 Biliproteins

C-phycocyanin (C-PC), a major biliprotein from the blue algae *Spirulina platensis* is a 37,468-Da water-soluble protein pigment containing α- and β-chains of 18,186, and 19,283 Da, respectively. C-PC is a highly nutritious algal protein with significant health-enhancing activities (66,67). Antioxidative, hepatoprotective, antiarthritic, antiinflammatory, radical scavenging, and selective cyclooxygenase-2-inhibiting activities have all been ascribed to C-PC (68–73). The apoptotic activity of C-PC was associated with the inhibition of prostaglandin biosynthesis, cytochrome *c* release from the mitochondrion, and PARP cleavage (74).

10.5 OTHER PROTEINS, PEPTIDES, AND AMINO ACIDS

A lipoprotein fraction extracted from rice bran (RBF) was reported to inhibit human endometrial adenocarcinoma cell proliferation at a concentration as low as 100 μg ml^{-1} (75). Fragment 142–148 from beta-lactoglobulin has angiotensin converting enzyme inhibitory activity (76).

10.5.1 Opioid Growth Factors

Many peptides encrypted in and inactive within the sequence of protein molecules have been associated with various activities *in vivo*, with the cardiovascular, endocrine, immune, and nervous systems, and with nutrient utilization. The amino acid sequence of the peptides often dictates the ensuing biological activities. Many of these properties are attributed to physiologically active peptides encrypted in protein molecules. Particularly rich sources of such peptides are milk (α-, β-, or γ-casein; α-lactalbumin, β-lactoglobulin, or lactotransferrin) and animal blood (bovine albumin), but they are also found in meats of various kinds as well as many in plants (HMW wheat glutenin, oat hordenin, maize zein, soy α-protein, rice albumin, and many others). These peptides are inactive within the sequence of the parent protein and can be released during processing (77). While there is increasing commercial interest in the production of bioactive peptides from various sources, large-scale production of such peptides has been hampered by the lack of suitable technologies, and very few of these products are commercially available. There is a need to investigate new technologies for isolation such as protein fermentation and expanded bed adsorption (EBA) chromatography to increase the yield and potential utilization in human foods.

10.5.2 L-Arginine

Bioavailable arginine is deposited in endothelial cells, where it is converted to nitric oxide (NO). Ingested arginine has the ability to reduce endothelial cell dysfunction in hypercholesterolemic patients, smokers, hypertensive individuals, diabetics, obese people, old people, and in those with coronary artery disease, ischemia/reperfusion, or congestive heart failure (78). Arginine-rich hexapeptide RRKRRR was found to block the growth and metastasis of VEGF-secreting HM-7 human colon carcinoma cells in nude mice and may be effective for the treatment of various human tumors and other angiogenesis-dependent diseases that are related to the action of VEGF (79).

10.5.3 Glycine

Glycine at 5% inhibited the growth of new vascular cells in C57-BL/6 mice subcutaneously implanted with B-16 tumor cells (80). Glycine at 0.01 to 10 mM inhibited endothelial cell proliferation, angiogenesis, and metastasis of B-16 tumor cells and proliferation of endothelial cells.

10.5.4 Phenylalanine Metabolite

Sodium phenylacetate (NaPa), a nontoxic physiological product of phenylalanine metabolism, present in micromolar concentrations in human plasma, has been shown to restore E-cadherin (a powerful invasion suppressor) function in various human mammary carcinoma cells, indicating that NaPa could be a novel therapeutic agent in breast cancer treatment (81,82).

10.5.5 Selenocysteine

Selenocysteine, an amino acid that contains selenium atom in place of the usual sulfur atom, is a good antiangiogenic compound effective against colon, prostate, and lung cancers.

The effects of the chemopreventive levels of Se on the intratumoral microvessel density and the expression of vascular endothelial growth factor in 1-methyl-1-nitrosourea-induced rat mammary carcinomas and on the proliferation and survival and matrix metalloproteinase activity of human umbilical vein endothelial cells *in vitro* were examined. Increased Se intake as Se-enriched garlic and sodium selenite or Se-methylselenocysteine fed for 7 weeks to rats bearing mammary carcinoma led to lower levels of vascular endothelial growth factor expression compared with those in healthy control animals (83).

10.6 CHALLENGES TO ANTIANGIOGENIC PROTEIN, PEPTIDE, AND AMINO ACID FUNCTIONAL FOODS

Food must be ingested orally, except for medical foods that may be administered intravenously or by other means. Low-molecular-weight antiangiogenic compounds such as phenolics, acids, vitamins, terpenes, small carbohydrates, peptides, and amino acids may be more bioavailable than are large molecules such as proteins, polypeptides, and large carbohydrate molecules that are marginally absorbed in the human gastrointestinal tract. The gastrointestinal tract constitutes a major challenge to dietary antiangiogenic peptides, polypeptides, and proteins: (a) the acidity of the stomach represents the first barrier for suppressing the activity of potential antiangiogenic polypeptides; (b) the digestive enzymes in the gastrointestinal tract are the next challenge to antiangiogenic polypeptides; and (c) the bioavailability and metabolism of peptides, polypeptides, and proteins are also challenges to the effectiveness of these compounds as inhibitors of angiogenesis. However, large molecules such as proteins and carbohydrates can be effective against angiogenesis-dependent diseases of the gastrointestinal tract, while low-molecular-weight compounds such as phenolics may be useful in other organs such as the liver. A combination of more angiogenesis inhibitors may provide more antiangiogenic inhibitory activity than an inhibitor of one angiogenic stimulator. Concerns about herb–nutrient–drug interactions, product quality, and standardization emphasize the need for rigorous research (84).

10.7 BIOAVAILABILITY AND SYNERGY OF ANTIANGIOGENIC POLYPEPTIDE AND AMINO-ACID FUNCTIONAL FOODS

Although oral administration is the major route for functional foods, since food can also be ingested through buccal absorption, this route has never been exploited by the food industry. Usually, the intravenous and intraperitoneal routes are used for medical food delivery. Oral bioavailability, with potency equal to intraperitoneal injection, as well as low or no *in vivo* toxicity for proliferating normal and nonvascular cells is the desirable attribute of antiangiogenic functional foods. Orally ingested BBI was absorbed by human subjects but did not develop an increase in the serum level of anti-BBI antibodies (85). Shark cartilage is one of the most studied antiangiogenic proteoglycans and has been shown to effectively inhibit angiogenesis and tumor growth in a variety of *in vivo* models. Besides these molecules that appear to be bioavailable and despite the great promise of the *in vitro* antiangiogenic activities of peptides, the fact remains that peptides can have bioavailability problems and may result in secondary immune responses. Additionally, peptides have difficulty crossing the blood–brain barrier to prevent inflammation. Accordingly, it is believed that some classes of peptides such as protein transduction domains (PTD) — which are short peptides 10 to 16 residues in length and containing numerous positively charged lysine (KKKKKKKK) and arginine (RRRRRRRR) polypeptides groups — are promising (86). PTD are part of cell penetrating peptides (CPP) that include synthetic cell-permeable peptides, PTD, and membrane-translocating sequences (MTS) that have the ability to translocate the membrane and gain access to the cellular interior. These peptides can enter cells or even cross the blood–brain barrier by a mechanism other than endocytosis (87). Numerous examples of PTD-mediated liposome or large protein transport across the cell membrane exist (88–90). Overcoming these bioavailability and cell permeability problems associated with antiangiogenic polypeptide functional foods would enhance the effectiveness of the existing antiangiogenic polypeptides and broaden the scope of viable antiangiogenetic therapeutic strategies. By linking PTD to bioactive and nonimmunogenic proteins such as BBI, a potent and cell permeable antiangiogenic molecule may be generated.

Antiangiogenic compounds, in general, should be selective and nontoxic. They should be effective against newly forming vessels while sparing existing ones. Mice bearing B-16 melanoma grafts were treated with protamine or heparinase and showed evidence of selective thrombosis of the blood vessels within the protamine- and heparinase-treated melanoma grafts (91). Tumors grown in the protamine-treated animals were significantly smaller than the tumors in control (untreated) mice. The action of antiangiogenic compounds must be reversible in the event angiogenesis is needed, for example, to support tissue repair after injury and after a woman's menstrual cycle. Most bioactive compounds of food origin such as protamine display reversible activity.

10.8 FUTURE DIRECTIONS

The identification of antiangiogenic amino acids, peptides, and proteins should also be associated with the identification of peptides or proteins that are stimulators of angiogenesis. Such polypeptides do exist in certain foods and need to be studied for their potential disease-causing effects (92). The postgenomic promises of functional foods will need to cover the whole spectrum from basic research to clinical trials in order to ascertain the efficacy or disease-causing/toxic effects of functional foods. However, understanding the stimulator modulation of angiogenesis will be key to unlocking the body's full potential to fight off these diseases. Most angiogenic diseases are proteomic diseases. Since genomics, transcriptomics, and proteomics identify drug targets, the discovery of antiangiogenic polypeptides and amino acids is suited to exploiting the output of these technologies. The identification of amino acid(s), peptides, or polypeptides that neutralize the activity of a target angiogenic stimulator may be a strategy for disease prevention. Herceptin as inhibitor of Her2/neu, avastin as inhibitor of VEGF, and Enbrel as inhibitor of TNF, are typical examples of antibody and protein inhibitors of angiogenic stimulators. Another advantage of a potential antiangiogenic polypeptide is that these molecules, like their counterparts (protein drugs such as monoclonal antibodies (mAbs)), are more selective and can be promoted with shorter and less expensive clinical trials (93). However, antiangiogenic polypeptides will need to be optimized as for their bioavailability, organ tropism, and activity *in vivo* by reducing or selecting for compounds without potential immunogenicity. The challenge to food scientists includes the development of food products with optimal antiangiogenic compounds, the delivery of these bioactive compounds to different organs and tissues, and the monitoring of the effects of these compounds so that appropriate dosing and delivery mechanism are provided. The training of future food scientists interested in functional food research will require a solid understanding of protein biochemistry, cell growth in health and disease, fundamentals of pharmacokinetics, signal transduction, metabolomics, proteomics and other "omics" technologies, microarrays, and any newly designed technology for drug design and delivery optimization.

10.9 CONCLUSIONS

The wish to address the failures of modern dietary patterns, which have been associated with an increase in the "diseases of civilization" (obesity, type 2 diabetes, high cholesterol, physical inactivity, osteoporosis, arthritis, cancer, AD, PD, and many others) in the Western world, has given rise to a greater awareness of the role of diet and nutrition in health. Since food has been part of the cause of these diseases, it may also be part of the answer.

Functional foods are being touted as a potential alternative to curve the progression of these diseases. Amino acids, peptides, and proteins with demonstrated activities against *in vitro* and *in vivo* angiogenesis (mostly in animal models) have been identified. It is suggested that these bioactive compounds be screened further in preclinical trials and in humans to ascertain their usefulness against angiogenic diseases. People are dying and standard drugs for most chronic angiogenic diseases are yet to be available.

REFERENCES

1. P. Carmeliet. Manipulating angiogenesis in medicine. *J. Intern. Med.*, 255: 538–561, 2004.
2. J. Folkman. Tumor angiogenesis: therapeutic implications. *N. Engl. J. Med.*, 285: 1182–1186, 1971.
3. J. Folkman. The role of angiogenesis in tumor growth. *Semin. Cancer Biol.*, 3: 65–71, 1992.
4. P. Carmeliet, R.K. Jain. Angiogenesis in cancer and other diseases. *Nature*, 407: 249–257, 2000.
5. R. Longo, R. Sarmiento, M. Fanelli, B. Capaccetti, D. Gattuso, G. Gasparini. Antiangiogenic therapy: rationale, challenges and clinical studies. *Angiogenesis*, 5: 237–256, 2002.
6. P. Workman, S.B. Kaye. Translating basic cancer research into new cancer therapeutics. *Trends Mol. Med.*, 8 (Suppl.4): S1–S9, 2002.
7. C.J. Sweeney, K.D. Miller, G.W. Sledge Jr. Resistance in the anti-angiogenic era: nay-saying or a word of caution? *Trends Mol. Med.*, 9: 24–29, 2003.
8. F.A. Scappaticci. Mechanisms and future directions for angiogenesis-based cancer therapies. *J. Clin. Oncol.*, 20: 3906–3927, 2002.
9. J.N. Losso. Preventing degenerative diseases by anti-angiogenic functional food. *Food Technol.*, 56: 78–88, 2002.
10. J.N. Losso. Targeting excessive angiogenesis with functional foods and nutraceuticals. *Trends Food Sci. Technol.*, 14: 455–468, 2003.
11. M. Atalay, G. Gordillo, S. Roy, B. Rovin, D. Bagchi, M. Bagchi, C.K. Sen. Antiangiogenic property of edible berry in a model of hemangioma. *FEBS Lett.*, 544: 252–257, 2003.
12. J. Folkman, D. Ingber. Inhibition of angiogenesis. *Semin. Cancer Biol.*, 3: 89–96, 1992.
13. M. Papetti, I.M. Herman. Mechanisms of normal and tumor-derived angiogenesis. *Am. J. Physiol. Cell Physiol.*, 282: C947–C970, 2002.
14. J. Dai, Y. Kitagawa, J. Zhang, Z. Yao, A. Mizokami, S. Cheng, J. Nor, L. McCauley, R.S. Kaichman, E.T. Keller. Vascular endothelial growth factor contributes to the prostate cancer-induced osteoblast differentiation mediated by bone morphogenetic protein. *Urol. Oncol.,* 22: 437–439, 2004.
15. M. Elkin, A. Orgel, H.K. Kleinman. An angiogenic switch in breast cancer involves estrogen and soluble vascular endothelial growth factor receptor 1. *J. Natl. Cancer Inst.*, 96: 875–878, 2004.
16. R.S. Kerbel. Antiangiogenic drugs and current strategies for the treatment of lung cancer. *Semin. Oncol.,* 31(1 Suppl 1): 54–60, 2004.
17. S.K. Kassim, E.M. El-Salahy, S.T. Fayed, S.A. Helal, T. Helal, El-D. Azzam, A. Khalifa. Vascular endothelial growth factor and interleukin-8 are associated with poor prognosis in epithelial ovarian cancer patients. *Clin. Biochem.,* 37: 363–369, 2004.

18. D. Albo, T.N. Wang, G.P. Tuszynski. Antiangiogenic therapy. *Curr. Pharm. Des.*, 10: 27–37, 2004.
19. R.A. White, N.H. Terry, K.A. Baggerly, M.L. Meistrich. Measuring cell proliferation by relative movement. I. Introduction and *in vitro* studies. *Cell Prolif.*, 24: 257–270, 1991.
20. G.W. Cockerill, J.R. Gamble, M.A. Vadas. Angiogenesis: models and modulators. *Int. Rev. Cytol.*, 159: 113–60, 1995.
21. D.A. Bradbury, T.D. Simmons, K.J. Slater, S.P. Crouch. Measurement of the ADP:ATP ratio in human leukaemic cell lines can be used as an indicator of cell viability, necrosis and apoptosis. *J. Immunol. Meth.*, 240: 79–92, 2000.
22. M.V. Volin, J.M. Woods, M.A. Amin, M.A. Connors, L.A. Harlow, A.E. Koch. Fractalkine: a novel angiogenic chemokine in rheumatoid arthritis. *Am. J. Pathol.*, 159: 1521–1530, 2001.
23. P. Liang, A.B. Pardee. Analysing differential gene expression in cancer. *Natl. Rev. Cance*r, 3: 869–876, 2003.
24. R. Auerbach, R. Lewis, B. Shinners, L. Kubai, N. Akhtar. Angiogenesis assays: a critical overview. *Clin. Chem.,* 49: 32–40, 2003.
25. C.A. Staton, S.M. Stribbling, S. Tazzyman, R. Hughes, N.J. Brown, C.E. Lewis. Current methods for assaying angiogenesis *in vitro* and *in vivo. Int. J. Exp. Pathol.*, 85: 233–248, 2004.
26. M. Presta, M. Rusnati, M. Belleri, L. Morbidelli, M. Ziche, D. Ribatti. Purine analogue 6-methylmercaptopurine riboside inhibits early and late phases of the angiogenesis process. *Cancer Res.*, 59: 2417–2424, 1999.
27. P.C. Wu, L.C. Yang, K.H. Kuo, C.C. Huang, P.R. Lin, P.C. Wu, S.J. Shin, M.H. Tai. Inhibition of corneal angiogenesis by local application of vasostatin. *Mol. Vis.*, 11(1): 28–35, 2005.
28. T.J. Lawley, Y. Kubota. Induction of morphologic differentiation of endothelial cells in culture. *J. Invest. Dermatol.*, 93(2 Suppl): 59S–61S, 1989.
29. A. Passaniti, R.M. Taylor, R. Pili, Y. Guo, P.V. Long, J.A. Haney, R.R. Pauly, D.S. Grant, G.R. Martin. A simple, quantitative method for assessing angiogenesis and antiangiogenic agents using reconstituted basement membrane, heparin, and fibroblast growth factor. *Lab. Invest.*, 67: 519–528, 1992.
30. G. Taraboletti, R. Giavazzi. Modelling approaches for angiogenesis. *Eur. J. Cance*r, 40: 881–889, 2004.
31. L.M. Cross, M.A. Cook, S. Lin, J.N. Chen, A.L. Rubinstein. Rapid analysis of angiogenesis drugs in a live fluorescent zebrafish assay. *Arterioscler. Thromb. Vasc. Biol.*, 23: 911–912, 2003.
32. D. Gingras, D. Boivin, C. Deckers, S. Gendron, C. Barthomeuf, R. Beliveau. Neovastat — a novel antiangiogenic drug for cancer therapy. *Anticancer Drugs*, 14: 91–96, 2003.
33 J.R. Sheu, C.C. Fu, M.L. Tsai, W.J. Chung. Effect of U-995, a potent shark cartilage-derived angiogenesis inhibitor, on anti-angiogenesis and anti-tumor activities. *Anticancer Res.*, 18 (6A): 4435–4441, 1998.
34. B.E. Kern, J.H. Balcom, B.A. Antoniu, A.L. Warshaw, C. Fernandez-del Castillo. Troponin I peptide (Glu94-Leu123), a cartilage-derived angiogenesis inhibitor: *in vitro* and *in vivo* effects on human endothelial cells and on pancreatic cancer. *J. Gastrointest. Surg.*, 7: 961–968, 2003; discussion 969.
35. M.A. Moses, D. Wiederschain, I. Wu, C.A. Fernandez, V. Ghazizadeh, W.S. Lane, E. Flynn, A. Sytkowski, T. Tao, R. Langer. Troponin I is present in human

cartilage and inhibits angiogenesis. *Proc. Natl. Acad. Sci. USA*, 96: 2645–2650, 1999.
36. R.M. Bukowski. AE-941, a multifunctional antiangiogenic compound: trials in renal cell carcinoma. *Expert Opin. Invest. Drugs*, 12: 1403–1411, 2003.
37. J.H. Liang, K.P. Wong. The characterization of angiogenesis inhibitor from shark cartilage. *Adv. Exp. Med. Biol.*, 476: 209–223, 2000.
38. R. Lagman, D. Walsh. Dangerous nutrition? Calcium, vitamin D, and shark cartilage nutritional supplements and cancer-related hypercalcemia. *Support Care Cancer*, 11: 232–235, 2003.
39. D.N. Sauder, J. Dekoven, P. Champagne, D. Croteau, E. Dupont. Neovastat (AE-941), an inhibitor of angiogenesis: randomized phase I/II clinical trial results in patients with plaque psoriasis. *J. Am. Acad. Dermatol.*, 47: 535–541, 2002.
40. D. Bagchi, B. Misner, M. Bagchi, S.C. Kothari, B.W. Downs, R.D. Fafard, H.G. Preuss. Effects of orally administered undenatured type II collagen against arthritic inflammatory diseases: a mechanistic exploration. *Int. J. Clin. Pharmacol. Res.*, 22: 101–110, 2002.
41. M.E. Kroon, M.L. van Schie, B. van der Vecht, V.W. van Hinsbergh, P. Koolwijk. Collagen type 1 retards tube formation by human microvascular endothelial cells in a fibrin matrix. *Angiogenesis*, 5: 257–265, 2002.
42. P.B. Armstrong, J.P. Quigley. A role for protease inhibitors in immunity of long-lived animals. *Adv. Exp. Med. Biol.*, 484: 141–160, 2001.
43. I.R. Asplin, S.M. Wu, S. Mathew, G. Bhattacharjee, S.V. Pizzo. Differential regulation of the fibroblast growth factor (FGF) family by alpha(2)-macroglobulin: evidence for selective modulation of FGF-2-induced angiogenesis. *Blood*, 97: 3450–3457, 2001.
44. A.R. Kennedy, D. Kritchevsky, W.C. Shen. Effects of spermine-conjugated Bowman-Birk inhibitor (spermine-BBI) on carcinogenesis and cholesterol biosynthesis in mice. *Pharm. Res.*, 20: 1908–1910, 2003.
45. W.B. Armstrong, X.S. Wan, A.R. Kennedy, T.H. Taylor, F.L. Meyskens Jr. Development of the Bowman-Birk inhibitor for oral cancer chemoprevention and analysis of Neu immunohistochemical staining intensity with Bowman–Birk inhibitor concentrate treatment. *Laryngoscope*, 113: 1687–1702, 2003.
46. S.M. Lippman, L.M. Matrisian. Protease inhibitors in oral carcinogenesis and chemoprevention. *Clin. Cancer Res.*, 6: 4599–4603, 2000.
47. N.I. Larionova, I.P. Gladysheva, D.P. Gladyshev. Human leukocyte elastase inhibition by Bowman-Birk soybean inhibitor. Discrimination of the inhibition mechanisms. *FEBS Lett.*, 404: 245–248, 1997.
48.. N.I. Larionova, S.S. Vartanov, N.V. Sorokina, I.P. Gladysheva, S.D. Varfolomeyev. Conjugation of the Bowman–Birk soybean proteinase inhibitor with hydroxyethylstarch. *Appl. Biochem. Biotechnol.*, 62: 175–182, 1997.
49. N.I. Larionova, N.G. Balabushevitch, P.A. Zhatikov. Interaction of human leukocyte elastase with plasma fibronectin and its inhibition by soybean Bowman–Birk protease inhibitor. *Biochemistry (Mosc.)*, 63: 1078–1082, 1998.
50. J.N. Losso, C.N. Munene, R.R. Bansode, H.A. Bawadi. Inhibition of matrix metalloproteinase-1 by the soybean Bowman–Birk inhibitor. *Biotechnol. Lett.*, 26: 901–905, 2004.
51. H.A. Bawadi, T.M.S. Antunes, F. Shih, J.N. Losso. *In vitro* inhibition of the activation of pro-matrix metalloproteinase-1 (pro-MMP-1) and pro-matrix metalloproteinase-9 (pro-MMP-9) by rice and soybean Bowman–Birk inhibitors. *J. Agric. Food Chem.*, 52: 4730–4736, 2004.

52. C. Johansen, A. Verheul, L. Gram, T. Gill, T. Abee. Protamine-induced permeabilization of cell envelopes of gram-positive and gram-negative bacteria. *Appl. Environ. Microbiol.*, 63: 1155–1159, 1997.
53. O. Arrieta, P. Guevara, S. Reyes, A. Ortiz, D. Rembao, J. Sotelo. Protamine inhibits angiogenesis and growth of C6 rat glioma; a synergistic effect when combined with carmustine. *Eur. J. Cancer*, 34: 2101–2106, 1998.
54. M.Y. Su, M.K. Samoszuk, J. Wang, O. Nalcioglu. Assessment of protamine-induced thrombosis of tumor vessels for cancer therapy using dynamic contrast-enhanced MRI. *NMR Biomed.*, 15: 106–113, 2002.
55. F. Flamigni, C. Guarnieri, S. Marmiroli, C.M. Caldarera. Inhibition of rat heart ornithine decarboxylase by basic polypeptides. *Biochem. J.*, 229: 807–810, 1985.
56. K. Kashiwagi, K. Igarashi. Nonspecific inhibition of *Escherichia coli* ornithine decarboxylase by various ribosomal proteins: detection of a new ribosomal protein possessing strong antizyme activity. *Biochim. Biophys. Acta.*, 911: 180–190, 1987.
57. R.J. Cobel-Geard, H.I. Hassouna. Interaction of protamine sulfate with thrombin. *Am J. Hematol.*, 14: 227–233, 1983.
58. Y.C. Yoo, S. Watanabe, R. Watanabe, K. Hata, K. Shimazaki, I. Azuma. Bovine lactoferrin and lactoferricin, a peptide derived from bovine lactoferrin, inhibit tumor metastasis in mice. *Jpn. J. Cancer Res.*, 88: 184–190, 1997.
59. M. Iigo, T. Kuhara, Y. Ushida, K. Sekine, M.A. Moore, H. Tsuda,. Inhibitory effects of bovine lactoferrin on colon carcinoma 26 lung metastasis in mice. *Clin. Exp. Metastasis*, 17: 35–40, 1999.
60. M. Shimamura, Y. Yamamoto, H. Ashino, T. Oikawa, T. Hazato, H. Tsuda, M. Iigo. Bovine lactoferrin inhibits tumor-induced angiogenesis. *Int. J. Cancer*, 111: 111–116, 2004.
61. T. Kuhara, M. Iigo, T. Itoh, Y. Ushida, K. Sekine, N. Terada, H. Okamura, H. Tsuda. Orally administered lactoferrin exerts an antimetastatic effect and enhances production of IL-18 in the intestinal epithelium. *Nutr. Cancer*, 38: 192–199, 2000.
62. K. Norrby, I. Mattsby-Baltzer, M. Innocenti, S. Tuneberg. Orally administered bovine lactoferrin systemically inhibits VEGF(165)-mediated angiogenesis in the rat. *Int. J. Cancer*, 91: 236–240, 2001.
63. W.B. Park, S.Y. Lyu, J.H. Kim, S.H. Choi, H.K. Chung, S.H. Ahn, S.Y. Hong, T.J. Yoon, M.J. Choi. Inhibition of tumor growth and metastasis by Korean mistletoe lectin is associated with apoptosis and antiangiogenesis. *Cancer Biother. Radiopharm.*, 16: 439–447, 2001.
64. J.P. Van Huyen, J. Bayry, S. Delignat, A.T. Gaston, O. Michel, P. Bruneval, M.D. Kazatchkine, A. Nicoletti, S.V. Kaveri. Induction of apoptosis of endothelial cells by *Viscum album*: a role for anti-tumoral properties of mistletoe lectins. *Mol. Med.*, 8: 600–606, 2002.
65. I.F. Pryme, S. Bardocz, A. Pusztai, S.W. Ewen. Dietary mistletoe lectin supplementation and reduced growth of a murine non-Hodgkin lymphoma. *Histol. Histopathol.*, 17: 261–271, 2002.
66. B.I. Bockow. United States Patent No. 05709855.
67. R.A. Kay. Microalgae as food and supplement. *Crit. Rev. Food Sci. Nutr.*, 30: 555–573, 1991.
68. V.B. Bhat, K.M. Madyastha. C-Phycocyanin: a potent peroxyl radical scavenger *in vivo* and *in vitro*. *Biochem. Biophys. Res. Commn.*, 275: 20–25, 2000.
69. R. Gonzalez, S. Rodriguez, C. Romay, O. Ancheta, A. Gonzalez, J. Armesto, D. Remirez, N. Merino. Anti-inflammatory activity of phycocyanin extract in acetic acid-induced colitis in rats. *Pharmacol. Res.*, 39: 55–59, 1999.

70. Ch. Romay, R. Gonzalez, N. Ledon, D. Remirez, V. Rimbau. C-phycocyanin: a biliprotein with antioxidant, anti-inflammatory and neuroprotective effects. *Curr. Protein Pept. Sci.*, 4: 207–216, 2003.
71. C. Romay, N. Ledon, R. Gonzalez. Effects of phycocyanin extract on prostaglandin E2 levels in mouse ear inflammation test. *Arzneimittelforschung*, 50: 1106–1109, 2000.
72. B.B. Vadiraja, N.W. Giakwad, K.M. Madyastha. Hepatoprotective effect of c-pycocyanin: protection for carbon tetrachloride and R-(+)-pulegone-mediated hepatotoxicity in rats. *Biochem. Biophys. Res. Commn.*, 249: 428–431, 1998.
73. C.M. Reddy, V.B. Bhat, G. Kiranmai, M.N. Reddy, P. Reddanna, K.M. Madyastha. Selective inhibition of cyclooxygenase-2 by C-phycocyanin, a biliprotein from *Spirulina platensis*. *Biochem. Biophys. Res. Commn.*, 277: 599–603, 2000.
74. M.C. Reddy, J. Subhashini, S.V. Mahipal, V.B. Bhat, P. Srinivas Reddy, G. Kiranmai, K.M. Madyastha, P. Reddanna. C-Phycocyanin, a selective cyclooxygenase-2 inhibitor, induces apoptosis in lipopolysaccharide-stimulated RAW 264.7 macrophages. *Biochem. Biophys. Res. Commn.*, 304: 385–392, 2003.
75. H. Fan, T. Morioka, E. Ito. Induction of apoptosis and growth inhibition of cultured human endometrial adenocarcinoma cells (Sawano) by an antitumor lipoprotein fraction of rice bran. *Gynecol. Oncol.*, 76: 170–175, 2000.
76. R.J. FitzGerald, H. Meisel. Lactokinins: whey protein-derived ACE inhibitory peptides. *Nahrung*, 43: 165–167, 1999.
77. H. Teschemacher. Opioid receptor ligands derived from food proteins. *Curr. Pharm. Des.*, 9: 1331–1344, 2003.
78. G. Wu, C.J. Meininger. Arginine nutrition and cardiovascular function. *J. Nutr.*, 130: 2626–2629, 2000.
79. D.G. Bae, Y.S. Gho, W.H. Yoon, C.B. Chae. Arginine-rich anti-vascular endothelial growth factor peptides inhibit tumor growth and metastasis by blocking angiogenesis. *J. Biol. Chem.*, 275: 13588–13596, 2000.
80. M.L. Rose, J. Madren, H. Bunzendahl, R.G. Thurman. Dietary glycine inhibits the growth of B16 melanoma tumors in mice. *Carcinogenesis*, 20: 793–798, 1999.
81. D. Thibout, M. Kraemer, M. Di Benedetto, L. Saffar, L. Gattegno, C. Derbin, M. Crepin. Sodium phenylacetate (NaPa) induces modifications of the proliferation, the adhesion and the cell cycle of tumoral epithelial breast cells. *Anticancer Res.*, 19(3A): 2121–2126, 1999.
82. M. Vasse, D. Thibout, J. Paysant, E. Legrand, C. Soria, M. Crepin. Decrease of breast cancer cell invasiveness by sodium phenylacetate (NaPa) is associated with an increased expression of adhesive molecules. *Br. J. Cancer*, 84: 802–807, 2001.
83. C. Jiang, W. Jiang, C. Ip, H. Ganther, J. Lu. Selenium-induced inhibition of angiogenesis in mammary cancer at chemopreventive levels of intake. *Mol. Carcinogr.*, 26: 213–225, 1999.
84. M.A. Richardson. Biopharmacologic and herbal therapies for cancer: research update from *NCCAM. J. Nutr.*, 131(11 Suppl): 3037S–3040S, 2001.
85. X.S. Wan, D.G. Serota, J.H. Ware, J.A. Crowell, A.R. Kennedy. Detection of Bowman-Birk inhibitor and anti-Bowman-Birk inhibitor antibodies in sera of humans and animals treated with Bowman-Birk inhibitor concentrate. *Nutr. Cancer*, 43: 167–173, 2002.
86. S.R. Schwarze, S.F. Dowdy. *In vivo* protein transduction: intracellular delivery of biologically active proteins, compounds and DNA. *Trends Pharmacol. Sci.*, 21: 45–48, 2000.

87. L. Bonetta. Getting protein into cells. *Scientist*, 16: 38–40, 2002.
88. A. Ho, S.R. Schwarze, S.J. Mermelstein, G. Waksman, S.F. Dowdy. Synthetic protein transduction domains: enhanced transduction potential *in vitro* and *in vivo*. *Cancer Res.*, 61: 474–477, 2001.
89. S.R. Schwarze, K.A. Hruska, S.F. Dowdy. Protein transduction: unrestricted delivery into all cells? *Trends Cell Biol.*, 10: 290–295, 2000.
90. V.P. Torchilin, R. Rammohan, V. Weissig, T.S. Levchenko. TAT peptide on the surface of liposomes affords their efficient intracellular delivery even at low temperature and in the presence of metabolic inhibitors. *Proc. Natl. Acad. Sci. USA*, 98: 8786–8791, 2001.
91. M. Samoszuk, M. Corwin, H. Yu, J. Wang, O. Nalcioglu, M.Y. Su. Inhibition of thrombosis in melanoma allografts in mice by endogenous mast cell heparin. *Thromb. Haemost.*, 90: 351–360, 2003.
92. O.N. Shcheglovitova, E.V. Maksyanina, I.I. Ionova, Y.L. Rustam'yan, G.S. Komolova. Cow milk angiogenin induces cytokine production in human blood leukocytes. *Bull. Exp. Biol. Med.*, 135: 158–160, 2003.
93. J.M. Reichert. Monoclonal antibodies in the clinic. *Natl. Biotechnol.*, 19: 819–822, 2001.

11 Relevance of Growth Factors for the Gastrointestinal Tract and Other Organs

Subrata Ghosh and Raymond J. Playford

CONTENTS

11.1 INTRODUCTION

The gastrointestinal (GI) tract possesses the remarkable ability to remain intact despite being constantly bathed in acid and proteolytic enzymes that can digest virtually any form of food, including the GI tract of other animals. When a superficial mucosal injury occurs, as in direct physical trauma or ingestion of noxious agents such as aspirin or alcohol, it is rapidly healed. The gut therefore possesses powerful mucosal defense and repair mechanisms.

The healing process following injury can be considered in three phases (Figure 11.1). There is an initial rapid response involving migration of surviving cells from the wound edge to cover the denuded area. This begins to occur within the first hour following injury and the process is usually referred to as *restitution* (1). This is followed by a much slower increase in proliferation and differentiation, which only begins 1 to 2 days after the injury has occurred. The final stage consists of epithelial remodeling and may take from months to years. Peptides produced within the mucosa can influence all or some of these effects. Importantly, some of these peptides (such as the trefoil peptides) appear to stimulate the repair process without influencing proliferation. These are therefore not *growth factors* as such but may, nevertheless, play an important physiological role and offer potential as a therapeutic strategy.

Mucosal integrity is maintained when the damaging effects of aggressive factors such as hydrochloric acid and proteolytic enzymes are counterbalanced by mucosal defense mechanisms. These include a high rate of cellular turnover (second only to the hemopoietic 'bone marrow' system), an efficient mucosal blood flow, a continuous adherent mucus layer, and the presence of regulatory peptides that can directly stimulate repair and also influence all of the other protective factors.

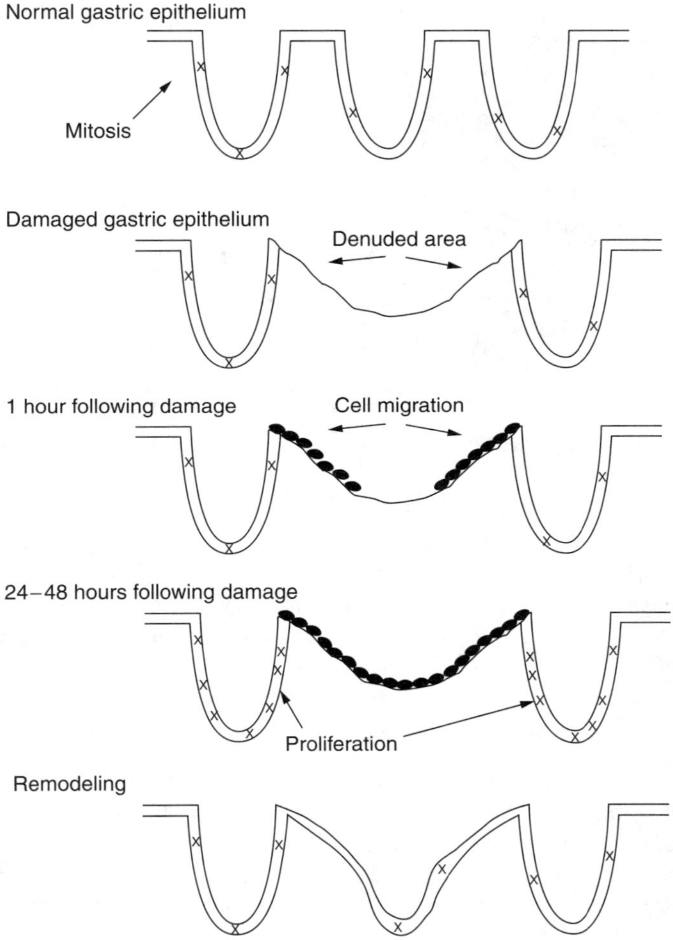

FIGURE 11.1 Schematic diagram of the three phases of the healing process following an injury.

When the system breaks down, possibly due to a combination of increased aggressive factors in subjects with a predisposition to illness due to innate weakened defenses, excessive inflammation or ulceration occurs. Therapeutic intervention, with medical therapy or other approaches, therefore aims to reduce the actions of the aggressive factors or strengthen the defense mechanisms (Figure 11.2).

11.2 WHAT ARE GROWTH FACTORS?

Several molecules can be loosely classified as "growth factors" as described below.

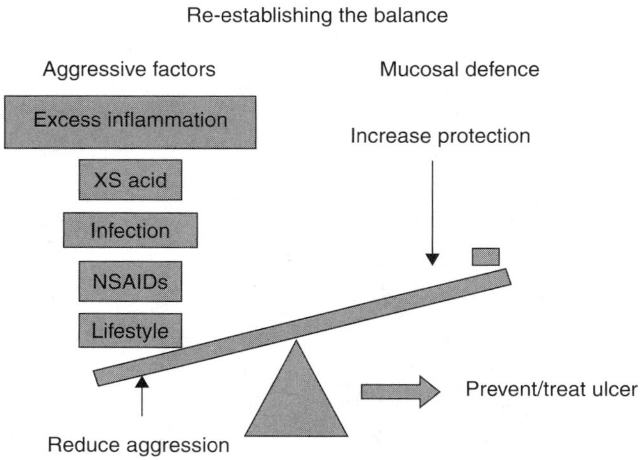

FIGURE 11.2 Gut integrity usually relies on a dynamic balance between aggressive factors and innate defense mechanisms. Excessive aggression (some examples shown) in subjects with weakened defenses results in injury/ulceration. NSAID = nonsteroidal antiinflammatory drugs (strong pain killers). Medical and alternative approaches act by either reducing these aggressive factors or improving defense processes. (Reproduced from R.J. Playford. Using the full spectrum of protective factors on the GI tract. *Functional Foods and Nutraceuticals*. May 2003, 52–58. With permission.)

11.2.1 Nonpeptide "Trophic Factors"

When several nonpeptide constituents of food, are added to cells *in vitro* or infused into animal models, they undergo increased proliferation. These include glutamine, polyamines, and nucleotides. It is debatable whether they should be considered growth factors, as the increased proliferation is not mediated by the "classic" receptor-ligand, secondary-messenger system. Factors such as glutamine are therefore often termed "preferred substrates." Nevertheless, they play an important role in maintaining GI mucosal mass and modulating the immune system via multiple mechanisms, including alteration of intestinal flora, and influencing the actions of growth factors. For example, the trophic response of epidermal growth factor (EGF) on the rat small intestinal cell line, IE-6 requires the presence of glutamine within the medium (2). These areas are reviewed further by Levy (3) and Carver and Barness (4).

11.2.2 Hormones

Hormones are peptides that are produced at a distant site, circulate via the blood stream, and act via the classic receptors. Although there is no doubt that several peptides are important for GI function (e.g., cholecystokinin for gall bladder and pancreatic function), the importance of these peptides in controlling gut growth is less clear. One peptide hormone that is of particular interest is glucagon-like peptide-2 (GLP-2), which is discussed later in the text.

11.2.3 Cytokines

The protein molecules known as cytokines have a broad range of cellular functions and are active in the picomolar to nanomolar concentration range. In general, cytokines do not regulate normal cellular homeostasis but act to alter cellular metabolism during times of perturbation, for example, in response to inflammation (5). Cytokines trigger acute cellular responses such as chemotaxis, protein synthesis, and cellular differentiation.

Although often considered separately, it is important to appreciate that the distinction between cytokines and growth factors is sometimes blurred. For example, interleukin-8 (IL-8) has been shown to stimulate migration of the human colonic epithelial cell line LIM-1215 (6), an effect which is usually attributed to growth factors such as EGF and transforming growth factor-β (TGF-β). In addition, recent studies have shown "cross-talk" between cytokines and growth factors. For example, Yasunaga and co-workers have examined the molecular mechanisms underlying *Helicobacter pylori (H. pylori*)-induced gastric hyperproliferation in patients with large-fold gastritis. The presence of *H.pylori* caused the gastric mucosa to release the cytokine Interleukin-1β (IL-1β), which, in turn, resulted in the local production of hepatocyte growth factor (7). Further information regarding the functions of cytokines within the GI tract is provided in the literature (8).

11.2.4 Growth Factors

Growth factors are so called because historically they were identified by their ability to stimulate the growth of various cell lines *in vitro,* but, in reality, the functions of these peptide-based molecules are considerably more diverse (Table 11.1). Various nomenclatures have been ascribed to molecular species as they have been identified. As characterization has become more sophisticated, however, it is apparent that some of these differently named species are structurally and functionally similar and may, in fact, be identical.

With regard to intrinsic growth factors, at least 30 different peptides have been identified, and the list continues to grow. It has been difficult to understand how they can act in an integrated fashion because their sites of production, mechanisms of action, and changes in concentration at sites of injury appear to vary between one peptide and another. In an attempt to put these factors into a physiological context, we have previously suggested that they may be usefully considered as belonging to one of three main groups based on their role *in vivo*: mucosal integrity peptides, luminal surveilance peptides, and rapid response peptides (9).

11.2.4.1 Mucosal Integrity Peptides

These peptides are expressed in the normal mucosa and are involved in maintaining normal mucosal integrity, including the control of proliferation and other functions of the nondamaged bowel. Examples of peptides that belong to this group include transforming growth factor-α (TGF-α), which acts directly on the enterocytes to stimulate proliferation and migration (10), and pancreatic secretory

TABLE 11.1
Actions of the Growth Factor EGF

Action	Effect	Possible Secondary Message
Proliferation	↑	
Acid secretion	↑	Protein kinase C cAMP
Bicarbonate	↑	Prostaglandins
NaCl and glucose uptake	↑	Brush border area, Na +− glucose co-transporter, Lipids
Chloride secretion	↑/↓	Phosphatidylinositol 3-kinase
Amylase secretion-pancreas	↑&↓	cAMP phospholipase C
Mucus sercretion	↑	Prostaglandins
GI blood flow	↑	β-ardrenergic NO prostaglandins
Smooth muscle contraction (longitudinal)	↑	Prostaglandins
Smooth muscle contraction (circular)	↑	(Densentities) not prostaglandins
Gastric emptying	↓	
Restitution	↑	Cell migration prostaglandins
Permeability	↑	
Muscosal protection	↑	Profileration, ployamines, mucus, trefoil peptides

Source: Adapted from J.M. Uribe, K.E. Barrett. Nonmitogenic actions of growth factors: an integrated view of their role in intestinal physiology and pathophysiology. *Gastroenterology*, 112: 255–268, 1997.

trypsin inhibitor (PSTI), which protects the overlying mucus protective layer from excessive digestion by luminal proteases (11).

TGF-α is produced in the mucosa throughout the GI tract. It is synthesized as a 160-amino acid precursor molecule that spans the cell membrane; subsequent exposure of the external domains to specific proteases releases a soluble 50-amino acid form. The biological function of the membrane-bound form is unclear but may be associated with stimulation of EGF/TGF-α receptors via a paracrine mechanism. TGF-α is trophic to a variety of cell lines and stimulates gut growth when infused systemically (10). Most studies suggest that its role in the mucosa is to maintain normal epithelial integrity. In support of this idea, mice that have had the TGF-α gene "knocked out" have an increased sensitivity to colonic injury (12).

Pancreatic secretory trypsin inhibitor (also known as SPINK-1; serine protease inhibitor, Kazal type-1) is a potent protease inhibitor that was originally identified in the pancreas. It has subsequently been shown to be present in mucus-secreting cells throughout the GI tract and also in the kidney, lung, and breast. Its major roles likely are preventing premature activation of pancreatic proteases and decreasing the rate of mucus digestion by luminal proteases within the stomach and colon. In addition, PSTI increases the proliferation of a variety of cell lines and stimulates cell migration, suggesting that PSTI may also be involved in both the early and late phases of the healing response following injury. Reduced levels of PSTI have been demonstrated in the colon of patients with ulcerative colitis during both acute attacks and when the mucosa appears to have healed (13).

Although it is unlikely that abnormalities of PSTI are the primary cause of this disease, this long-term deficiency of a mucosal defence peptide may explain some of the abnormalities of mucus found and the well-known clinical phenomenon of extension of disease more proximally over time.

11.2.4.2 Luminal Surveillance Peptides

These peptides are continuously secreted into the lumen, but because their receptors are not present on the apical (luminal) aspects of the enterocytes, they can only stimulate growth/repair at sites of injury, where they can gain access to their receptor. Their predominant function is therefore to be readily available to stimulate repair at sites of GI damage. The prime example of this is EGF (9).

EGF is a 53-amino acid peptide (molecular weight 6400 Da) that is continuously secreted into the gut lumen by the salivary glands and Brunner's glands of the duodenum. EGF is present in high concentrations in human gastric juice, at about 500 ng l^{-1} (14), which is sufficient to stimulate growth of various cells lines *in vitro*. EGF is a potent stimulant of proliferation for a variety of GI cell lines *in vitro* and is trophic to the bowel when infused intravenously (15). It is also a potent cytoprotective agent, able to stabilize cells against noxious agents such as ethanol or nonsteroidal antiinflammatory drugs, via acid-independent means (14).

The mechanisms underlying cytoprotection are poorly understood but probably include stimulation of goblet cells to release mucus (16) and increasing gastric blood flow (17). The fact that EGF is present in the gut lumen while its receptor is only present on the basolateral receptors of the bowel (18), strongly supports the idea that it belongs to and functions as a luminal surveillance peptide (Figure 11.3). Further details of some of the actions of the action of EGF can be found in Table 11.1.

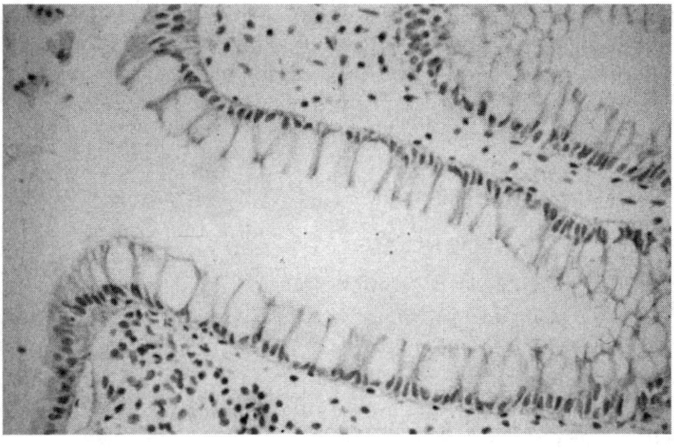

FIGURE 11.3 Expression of EGF receptor on the basolateral but not apical surface of the intestinal epithelium.

11.2.4.3 Rapid Response Peptides

These peptides are present in relatively low concentrations under normal circumstances but their production is rapidly upregulated at sites of injury. A prime example of this group is a member of the trefoil factor family (TFF), TFF-2, which is upregulated within 1 hour of injury and contributes to the early phase of the repair process, particularly restitution (19). The trefoil peptide family in mammals consists of a group of small proteins, each containing one or two copies of the trefoil motif. This motif comprises a three-loop (trefoil) configuration with six highly conserved cysteine residues allowing three intrachain disulfide bridges. In humans, the trefoil peptides are found in the mucus-producing epithelium in the stomach and small and large intestines. Trefoil peptides are thought to have two distinct functions in the GI tract: (a) under basal circumstances, they play a role in mucus stabilization; and (b) when an acute injury occurs, their rapid upregulation is important in stimulating the repair process, particularly that of restitution. Human spasmolytic polypeptide (hSP) may therefore play a key role in early stages of mucosal repair by stimulating the initial reepithelialization by cell migration (19).

11.3 OVERVIEW OF OTHER PEPTIDES

Space prohibits a detailed discussion of every growth factor but some of the key peptides of particular interest to gastroenterologists are discussed below.

11.3.1 Transforming Growth Factor-βs (TGF-βs)

TGF-βs represent a ubiquitous group of cytokines structurally distinct from TGF-α, which exert a multiplicity of influences in a wide array of tissue and cell types. In the GI tract, TGF-βs inhibit proliferation in normal epithelial cell populations but promote it in many mesenchymal tissues (20,21). Because of its distribution in the nonproliferative zones of the intestinal villi, it has been suggested that it may function as a "mitotic brake" to prevent continued proliferation of the enterocytes as they leave the crypt zone. TGF-βs are also thought to play a key role as an intermediate signal peptide in the promigratory effects of several other peptides, such as EGF, involved in healing (22).

11.3.2 Insulin-Like Growth Factors (IGF, Somatomedins) and Their Binding Proteins

IGF-I and IGF-II promote cell proliferation and differentiation (23). They are similar in structure to proinsulin, and it is possible that they also exert insulin-like effects at high concentrations. The liver is a major site of IGF synthesis (24); IGF-I and II are both also expressed at particularly high levels in the developing human fetal stomach and small intestine, with expression reaching a maximum soon after birth (25).

IGF-I and -II are both present also in milk and colostrum, where they are thought to play a role in the suckling neonate's growth and development (for a detailed specific review of growth factor and colostrum, see reference 26). Bovine colostrum contains much higher concentrations of IGF-I than does human colostrum (500 *versus* 18 ng ml^{-1}) (27, 28), with lower levels being present in mature bovine milk (10 ng ml^{-1}) (29). These growth factors are relatively stable to both heat and acid conditions. They therefore survive the harsh conditions of both commercial milk processing and gastric acid to maintain their biological activity (30). IGF-I is known to promote protein accretion, that is, it is an anabolic agent (31) and is at least partly responsible for mediating the growth-promoting activity of the growth hormone. IGF-II is present in bovine milk and colostrum at much lower concentrations than IGF-I but, like IGF-I, has anabolic activity and has been shown to reduce the catabolic state in starved animals (32).

The IGFs in bovine and human colostrum and milk are present in both free and bound forms. The amount of free IGF varies during the perinatal period with most of the IGF-I in bovine colostrum being present in the free form (i.e., not associated with its binding protein), whereas the reverse is true in the antepartum period and in mature milk (33). Six IGF-binding proteins (IGFBPs) have been identified and cloned. Their main function was initially thought to be as carrier proteins, reducing the proteolytic digestion of IGF and limiting its biological activity, as only the free forms of IGF are thought to have any major proliferative activity. Additional roles for IGFBPs have been suggested as a result of the findings that the different IGFBPs show distinct patterns of distribution in different tissues and their levels are altered in response to hormone or nutrient status. Examples include the observation that administration of dexamethasone to rats increases hepatic production of IGFBP-1 (34) and malnutrition of neonatal rats resulted in reduced serum IGF-I and IGF-II but caused an elevation in serum IGFBP-2 (35). The detailed functions of IGFBPs are unclear, although it is probable that one of the roles of secreted or soluble IGFBPs is to inhibit IGF-mediated proliferation or amino acid uptake by limiting the availability of free IGF to bind to its receptors. Conversely, cell surface/matrix-associated IGFBPs may potentiate the actions of IGF by increasing local concentrations of the IGF-I and IGF-II next to their receptors. (For a detailed review of IGFBPs, see reference 36, and for a general review of the role of IGFs and IGFBPs see reference 37.) The changes in secretion and mammary uptake of IGF-related peptides in the peripartum period of dairy cows have also been described (38).

11.3.3 Platelet-Derived Growth Factor (PDGF)

PDGF is an acid-stable molecule, which was originally identified from platelets but is also synthesized and secreted by macrophages. It consists of two disulfide-linked polypeptides: chain A (14 kDa) and chain B (17 kDa). The dimer therefore exists in three isoforms (AA, AB, and BB), which bind to tyrosine-kinase-type receptors. PDGF is a potent mitogen for fibroblasts and arterial smooth muscle cells, and exogenous PDGF has been shown to facilitate ulcer healing when administered orally to animals. Although PDGF is present in human and bovine milk and colostrum, the majority of the PDGF-like mitogenic activity in bovine milk is actually derived from bovine colostral growth factor, which shares sequence homology with PDGF (39,40). (For a general review of the effects of PDGF, see reference 41.)

11.3.4 Vascular Endothelial Growth Factor (VEGF, Vasculotropin)

VEGF is a homodimeric 34-kDa to 42-kDa heparin-binding glycoprotein with potent angiogenic-, mitogenic-, and vascular-permeability-enhancing factors and is related to PDGF (42). It is present in human breast milk at a concentration of about 75 ng ml^{-1} during the first week of lactation, and levels fall to about 25 ng ml^{-1} during the second postnatal week (43). Specific receptors for VEGF have been identified on the apical membranes of the human colonic cell line Caco-2 (43) and also on the human H-4 cell line. Although VEGF bound to these cell lines; it does not induce a proliferative response (43). The pathophysiological role of VEGF is therefore unclear, although its angiogenic activity may play an important role in the healing of conditions such as peptic ulceration.

11.3.5 Keratinocyte Growth Factor (KGF)

KGF, also known as FGF-7, is a member of the fibroblast growth factor family. It is produced in the mesenchymal tissue of the GI tract and binds to a splice variant of the FGF 2 receptor located on the gut epithelium (44). This has led to the theory that it normally functions in a paracrine manner. It is a potent stimulant of proliferation of a variety of epithelial cell lines *in vitro* and stimulates gut proliferation in a dose-responsive manner in rats when administered via continuous intravenous infusion (45). Studies utilizing animal models of injury suggest that it may also be of value in the treatment of chemotherapy- and radiation-induced GI injury (46), inflammatory bowel disease (47), and short-bowel syndrome (48). Animal studies with recombinant human KGF-1 (rHuKGF) demonstrated a beneficial effect in the dextran sodium sulfate (DSS) mice colitis model and in the CD45RB Hi mice colitis model. Animal studies with a homologue of KGF-1, KGF-2 (Repifermin) showed a beneficial effect in DSS colitis (49) and in the indomethacin small intestinal injury rat model (50); a phase II trial of Repifermin in patients with active ulcerative colitis has been completed, but the results have not been reported yet. Preliminary reports, however, are not very encouraging.

11.3.6 Basic Fibroblast Growth Factor (bFGF)

bFGF is a heparin-binding factor with potent angiogenic properties. In a clinical trial, administration of recombinantly produced, acid-stable bFGF decreased the amount of duodenal, but not gastric, injury (erosions) induced by coadministration of aspirin. In the second component of the study, indomethacin was administered to the same subjects after cessation of treatment with bFGF. Subjects who had received bFGF in the first part of the study were found to have a reduced susceptibility to indomethacin-induced relapse of artificially induced mini-ulcers. This suggests that the mucosa was somehow "stronger" in subjects who had been treated with bFGF when compared with those who had received a placebo (51).

11.4 SOURCES OF GROWTH FACTORS

Although much time is dedicated to the study of recombinant peptides, there is currently a resurgence in interest in the use of "natural" bioactive products possessing growth factor activity for the prevention and treatment of GI problems.

The initial question is what constitutes natural therapy? This is far from clear, and we would consider it more of a continuum from natural products to entirely artificially derived compounds (Table 11.2). Further discussion of natural sources of products with potential GI applications can also be found in reviews by Playford (52) and Ghosh and Playford (53). Space limitation prohibits a detailed discussion of all factors, and therefore examples are taken from each category.

11.4.1 NATURAL PRODUCTS

Aloe vera has been used as a medicine since before Roman times. It contains several potentially bioactive compounds, including salicylates, magnesium lactate, acemannan, lupeol, campestrol, β-sitosterol, γ-linolenic acid, aloctin A, and anthraquinones. It is currently available as an ingredient in a number of healthcare products, as both leaf exudate and gel. Aloe vera possesses multiple activities that may have beneficial effects, including anti-inflammatory, analgesic, and prohealing effects, in treating GI diseases. For example, inflammation produced by application of croton oil to mouse ear or rat hind paw is decreased after topical application of aloe gel (54). In addition, a component of aloe vera, acemannan, has been reported to accelerate healing and reduce pain in aphthous stomatitis (55) and prevent stress-induced gastric ulceration in rats (56). The wound-healing properties of aloe vera is probably not solely a consequence of its antiinflammatory action, as *in vivo* and *in vitro* studies suggest that aloe vera or mannin derivatives can stimulate collagen synthesis and

TABLE 11.2
Sources of Growth Factors

Essentially unmodified products: e.g., aloe vera
Partly "purified" natural products: e.g., colostrum
Isolated individual/combination factors from natural products: e.g., plant/herbal extracts
Synthetically produced compound based on natural product: e.g., zinc carnosine, epidermal growth factor (artificially produced)
Artificially modified product based around natural material: e.g., growth-factor-enhanced transgenic sheep
Completely artificial compound with growth-factor-like actions: e.g., naftazone

fibroblast activity (57,58). A glycoprotein fraction G1G1M1DI2 isolated from aloe vera accelerated wound healing on a monolayer of human keratinocytes (59) and enhanced wound healing in hairless mice by day 8 after injury, with significant cell proliferation (59). Cinical trials of aloe vera in inflammatory bowel disease are ongoing.

11.4.2 Partly Purified Natural Products

Colostrum is the milk produced during the first few days following birth. Bovine colostrum is a commercial by-product of the milk industry and is already available in some health-food shops. It is currently marketed as a general health-food supplement, and several formulations are currently available. The claims made for some colostrum preparations are wide and include antiaging, antiinflammatory, and antimicrobial effects and athletic enhancement. However, some of the claims made for these preparations are based on limited evidence, with major extrapolation from cell culture systems to whole body physiology. Nevertheless, studies have confirmed the maintenance of biological growth factor activity in several of these products when studied both *in vitro* (cell culture) and *in vivo* (including some limited clinical trials). For example, one form of colostrum has been shown to prevent the increase in small intestinal permeability caused by ingestion of the pain-killing drug indomethacin (60) (Figure 11.4). In addition, our group recently showed, in a randomized clinical trial, that colostrum enema therapy was efficacious in reducing inflammation and improving symptoms in patients suffering from ulcerative colitis (61) (Figure 11.5).

The identity of the factors responsible for such changes are currently unclear but may include EGF, TGF-α, and TGF-β. Cell culture studies suggest that more than one factor is probably involved, as size exclusion studies show both promigratory (restitution) and proproliferative effects in more than one fraction (62).

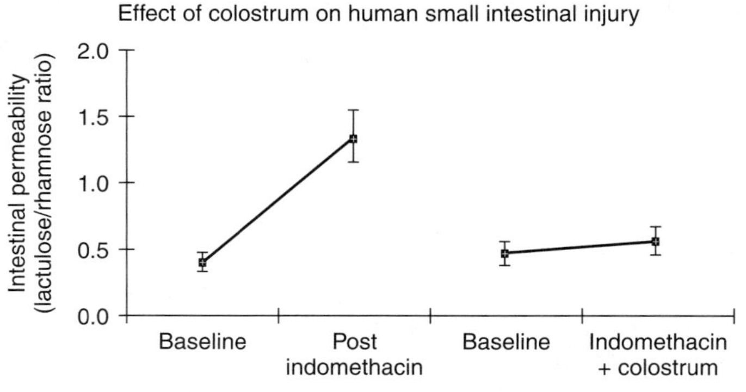

FIGURE 11.4 Coadministration of colostrum prevents indomethacin-induced increase in small intestinal permeability determined by differential sugar absorption test. (From R.J. Playford, et al. *Clin. Sci.*, 6, 627–633, 2001. With permission.)

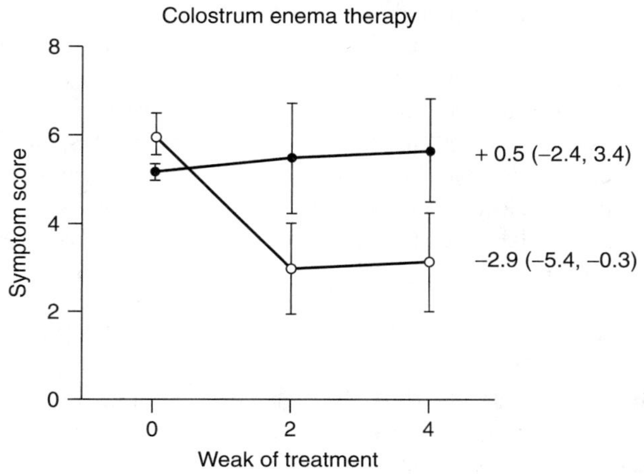

FIGURE 11.5 Administration of colostrum enema (*open circles*) significantly improves symptom score in left-sided colitis compared with placebo enema (*closed circles*). Numbers represent median change (interquartile range). (From Z. Khan, C. Macdonald, A.C. Wicks, M.P. Holt, D. Floyd, S. Ghosh, N.A. Wright, R.J. Playford. *Aliment Pharmacol. Ther.*, 16, 1917–1922, 2002. With permission.)

11.4.3 Isolated Individual Factors from Natural Products

As stated earlier, EGF is a 53-amino acid peptide present in colostrum and also produced by the salivary glands and duodenal glands of humans and other animals. It is a potent stimulant of growth and has been shown to be beneficial in several models of upper and lower GI injury. There have been relatively few clinical trials of EGF, although these are currently ongoing. One notable example is of an individual case report of pediatric necrotizing enterocolitis, which is a severe ulcerative condition of the bowel with a high death rate. A 7-day infusion of EGF caused the histology of the small intestine to revert to an essentially normal appearance, and the child subsequently went on to make a good recovery (63). Although EGF is a constituent of colostrum, the costs and difficulty of isolating pure factors from these natural sources means that most clinical trials will be using artificially produced (usually recombinant) forms of the peptide. Although there appears to be little biological difference between the natural peptide and artificially produced peptides, there is the concern that artificially produced peptides cannot be considered as a "natural" approach.

11.4.4 Synthetically Produced Compounds Based on Natural Products

Carnosine is a dipeptide comprising beta-alanine and L-histidine. It is naturally present in long-living cells such as muscle and nerves, where, among other actions, it also plays a role as an antioxidant.

Zinc carnosine is an artificially produced derivative of carnosine, where it is linked in a one-to-one ratio to provide a polymeric structure. Zinc carnosine has been extensively studied in cell culture and animal models and also as a potential antiulcer drug in clinical trials. Zinc carnosine does appear to possess antioxidant effects (64) and to stimulate proliferation in a variety of cell lines, including human umbilical vein endothelial cells, human foreskin fibroblast cells, and human MCT3-E1 cells (osteoblasts). However, no increase in proliferation was seen in guinea pig gastric mucosal epithelial cells (65). Zinc carnosine has been shown to have beneficial effects in various animal models of gut injury, including water immersion stress and hydrochloric acid–ethanol gastric damage models (66), stress-induced gastric ulceration (67), and the trinitrobenzene sulfonic acid (TNBS) model of colitis (68). Clinical trials for the treatment of gastric ulcer have also shown a beneficial effect above that of placebo (69), although, perhaps not surprisingly, it does not appear as efficacious in healing ulceration as the powerful proton pump inhibitor therapy used by most clinicians to treat severe gastric ulceration.

11.4.5 Artificially Modified Products Based on Natural Materials

Recent advances in molecular biology have also allowed the modification of animal milk to produce increased quantities of growth factors and also to qualitatively change the composition, including making the human homologues (equivalent) of various peptides. This usually requires raising transgenic sheep, in which the relevant peptide is linked to a genetic *promoter* causing expression within the milk (70).

11.4.6 Completely Artificial Compound with Growth Factor Actions

The development of substances that enhance endothelial cell growth and proliferation is of interest, as restoration of an intact surface after endothelial damage by angioplasty or endarterectomy is necessary for the ultimate success of these procedures. Naftazone, an original synthetic molecule distinct from growth factor peptides, has been shown to accelerate human saphenous vein endothelial cell proliferation *in vitro* at concentrations that did not alter the hemostatic balance. This is therefore a promising candidate for accelerating vascular repair (71).

11.5 GROWTH FACTOR THERAPY FOR GUT DISEASE

11.5.1 Short Bowel Syndrome

Some patients have an insufficient length of bowel to digest and absorb food adequately, usually as a result of massive intestinal resection for vascular insufficiency or following repeated operations for inflammatory bowel disease. Present therapeutic options, for example, to instigate long-term parenteral (intravenous) feeding or, in a few selected cases, to perform small bowel transplantation, are unpleasant and associated with a high risk of morbidity or mortality. After massive small bowel loss, the remnant intestine adapts to compensate for the reduced mucosal absorptive and digestive surface area. In patients with 100 cm of intestinal length, 45% will require lifelong parenteral nutrition, with a 5-year mortality rate of approximately 25%. Strategies to optimize the function of residual bowel and to ultimately wean patients off total parenteral nutrition would therefore be of great benefit. There is evidence that growth factors could be instrumental in achieving this goal; for example, systemic administration of individual growth factors such as EGF have been shown to stimulate bowel growth in rats receiving total parenteral nutrition (72). In addition, administration of oral EGF helped restore glucose transport and phlorizin binding in rabbit intestines following jejunal resection (73), and colostrum supplementation of piglet feeding regimens resulted in a significant increase in intestinal proliferation (74).

Another peptide that is of particular interest is glucagon-like peptide-2 (GLP-2), which is a potent trophic peptide that enhances recovery following intestinal recovery. GLP-2 enhances intestinal epithelial restitution in rat jejunum following acid induced injury (75) and stimulates proliferation of the human cell lines Caco-2, T84, and IEC-6 (76). GLP-2 has also been shown to improve intestinal adaptation and nutrient absorption after major small bowel resection in rats (77). A recent study has demonstrated that GLP-2 also acts as a tropic factor in humans with short bowel syndrome, causing a significant trophic effect on the intestine by stimulating enterocyte proliferation, at the same time reducing rates of enterocyte apoptosis (78). After an initial, extensive balance study, GLP-2 was administered for 35 days by a twice-daily subcutaneous injection. Balance studies on these patients were then repeated, and GLP-2 was found to have resulted in significantly greater intestinal absorption of energy, water, and nitrogen. Patients also demonstrated increases in lean body mass, body weight, and reduced gastric emptying. Further, larger, trials are ongoing.

11.5.2 Nonsteroidal Antiinflammatory Drug (NSAID)-Induced Gut Injury

NSAIDs are widely prescribed and effective for the treatment of musculoskeletal injury and chronic arthritic conditions. Nevertheless, approximately 2% of subjects taking NSAIDs for 1 year will suffer from GI side effects, which include bleeding, perforation, and stricture formation of the stomach and intestine (79). Acid suppressants and prostaglandin analogues have been shown to be effective in reducing gastric injury induced by NSAIDs but are less efficacious in preventing small intestinal injury. Novel therapeutic approaches to deal with these problems, such as using recombinant peptides, are therefore still required. A recent series of *in vivo* and *in vitro* studies support this idea; EGF (14), TGF-α, and TGF-β (62) have all been shown to reduce NSAID-induced gastric injury. The beneficial effects of recombinant growth factors on NSAID-induced small and large intestinal injury is, however, less well documented. It has recently been shown that a defatted colostrum preparation, which is rich in the growth factors discussed earlier, reduced small intestinal permeability used, as a marker of intestinal damage, in human volunteers taking clinically relevant doses of the drug indomethacin (80) (Figure 11.4). Clinical trials involving patients taking long-term NSAIDs are presently under way.

11.5.3 Chemotherapy-Induced Mucositis

Current regimens for the treatment of cancers require patients to take much higher doses of chemotherapeutic agents than were used previously. As a result of this escalation in dosing, toxic side effects on the bone marrow or GI tract can be the factor limiting the dose or duration of treatment. Strategies to protect these tissues and encourage their recovery may facilitate the use of higher dosage with greater potential for cure. Examples include the findings that EGF enhances repair of rat intestinal mucosa damaged by methotrexate (81), TGF-β ameliorates chemotherapy-induced mucositis (82) and administration of a cheese whey-derived preparation reduced methotrexate-induced gut injury in mice (83). Not all studies have shown favorable results, however, as EGF had only a minor beneficial effect in reducing mouth ulceration in a phase I clinical study of patients undergoing chemotherapy (84).

If peptides with growth stimulatory or inhibitory effects are to be used, timing of administration is likely to be critical; growth-arresting factors might protect bone marrow or gut from the damaging effects of chemotherapy (which tends to affect areas with the highest cell turnover) if given prior to chemotherapy. In contrast, growth-stimulating factors might "rescue" recovery of injured areas if administered following chemotherapy. This latter approach is already being used clinically as *colony-stimulating growth factor* is used to stimulate bone marrow recovery following chemotherapy. GLP-2 may also be effective in chemotherapy-induced mucositis.

11.5.4 Inflammatory Bowel Disease

The etiology of ulcerative colitis and Crohn's disease is unknown, and current treatment of these severe, incapacitating conditions therefore has to be on an empiric basis. Studies examining the effect of administration of EGF, PDGF, TGF-β, or IGF-I in animal models of colitis have given encouraging results (85), and a cheese whey growth factor extract containing several of these growth factors has also shown positive results in a similar model (86). The recently published trial of colostrum enema gave positive results (61), and other peptides that are undergoing study as potential therapeutic agents for these conditions include keratinocyte growth factor (47) and trefoil peptides (87). However, most of these are still at a very early (animal model) stage and are unlikely to be in standard clinical usage for many years.

11.5.5 Necrotizing Enterocolitis

Necrotizing enterocolitis (NEC) is a severe life-threatening illness of young children, which causes severe ulceration of the small and large bowel. Its etiology is unclear, although multiple risk factors, including prematurity, enteric infections, intestinal ischemia, and abnormal immune responses, may all be relevant. Although many proinflammatory molecules are likely to be involved in the etiology of NEC, there is currently interest in the role of the phospholipid mediator platelet activating factor (PAF), produced by intestinal flora and inflammatory cells, in the development of NEC. The finding that human colostrum contains the enzyme PAF acetylhydrolase (PAF-AH) (88), which degrades PAF, might therefore be relevant in explaining why human milk feeds protect against the development of NEC. (For a further discussion, see references 88–90.) Although the molecular mechanisms underlying the development of NEC are unclear, there is no doubt that once NEC is established, it is associated with a very high mortality rate. At the present time, treatment consists of general supportive measures consisting of fluid replacement and antibiotic therapy, although intestinal resection is often required. There is therefore a need for novel therapeutic approaches, for example, using peptides to stimulate the repair process. Support for this idea comes from a recent case study that reported that a continuous infusion of EGF resulted in a remarkable restorative effect on gut histology in a child with NEC (63).

11.6 GROWTH FACTOR PEPTIDES IN THERAPY OF NONGASTROINTESTINAL DISEASES

11.6.1 Blocking of Angiogenesis as Therapy of Cancer

Angiogenesis is a process of development and growth of new capillary blood vessels from preexisting vessels. When pathological, it contributes to the development of numerous types of tumors and the formation of metastases. In order to grow, areas of carcinoma need new blood vessels to form so that they can feed themselves. Therefore, it is now generally accepted that in most cancers, development of new vessel formation is an important component of cancer pathophysiology. This concept makes the control of tumoral angiogenesis an exciting novel therapeutic strategy. Vascular endothelial growth factor (VEGF) is found in several types of tumors and posseses tumor angiogenic activity both *in vitro* and *in vivo* (91). Therefore, intereference with VEGF-mediated angiogenesis is an area of active research.

11.6.2 Promoting Angiogenesis in Ischemic Disease

Recombinant human vascular endothelial growth factor protein (rhVEGF) stimulates angiogenesis in animal models. It has therefore been suggested that factors such as this may be useful to revascularize ischemic areas such as those occurring in cardiac tissue due to coronary vessel disease. In a phase I clinical trial VIVA (vascular endothelial growth factor in ischemia for vascular angiogenesis), rhVEGF was investigated using a double-blind, placebo-controlled protocol trial designed to evaluate the safety and efficacy of intracoronary and intravenous infusions of rhVEGF. By day 120, high-dose rhVEGF resulted in significant improvement in angina and favorable trends in exercise treadmill test (ETT) time and angina frequency (92).

11.6.3 Repair of Articular Cartilage

Loss of cartilage is a major component of joint degeneration and methods to stimulate regeneration therefore have major therapeutic potential for conditions such as osteoarthritis. One example of this approach is an innovative study where stimulation of perichondrium-derived mesenchymal cells was attempted using transfer of growth factor cDNA (adenoviral vectors carrying bone morphogenetic protein-2 or insulin-like growth factor–1) in a partial-thickness defect model. Extremely encouraging results were seen (93).

11.6.4 Use of Growth Factors to Stimulate Repair Following Eye Injury

In alkali-injured rabbit cornea, a low concentration of human EGF (hRGF) (5 μg ml^{-1}) selectively enhanced epithelial healing without affecting endothelial healing (94). Similarly, in a rat corneal epithelial injury model, nerve growth factor (NGF) applied exogenously accelerated healing of corneal epithelium (95).

11.7 ROUTES OF ADMINISTRATION

One of the major areas of concern of using potent growth factor peptides in the clinical setting is that systemic administration of growth factors may act to promote tumor growth elsewhere in the body. It is partly for this reason that there is increasing interest in the potential value of trefoil peptides to treat human GI disease, as they appear to stimulate repair without increasing proliferation (19). Oral, as opposed to systemic, administration of growth factors provides one possible approach to this problem. However, for EGF and trefoil peptides, the oral doses required to treat GI damage may be up to a thousand times greater than when the peptide is given systemically, making oral therapy economically unrealistic. This may be due to the therapeutic peptides (like any other ingested food product) coming into contact with luminal proteases and being digested into inactive forms (96).

Various strategies are available to attempt to stabilize ingested peptides from proteolysis. In peptides with a single pepsin cleavage site, altering the recombinant sequence may decrease the susceptibility to acid and pepsin degradation (but may also result in loss of biological activity). More simply, use of proton pump inhibitor drugs to raise the gastric pH above 4 markedly reduces peptic activity, stabilizing peptides within the gastric lumen (14). Stabilizing molecules against digestion in the small intestine is likely to be more difficult due to the many different proteolytic enzymes produced by the pancreas. The peptide may have to be coadministered with nonspecific serine protease inhibitors or in a site-specific release formulation to overcome these problems.

11.8 SUMMARY

The diverse repertoir of actions of growth factors on physiology and pathology is both exciting and confusing. The source of such factors is diverse and the division between foods and drugs sometimes artificial (hence nutriceuticals). Over the next 10 years, the clinical importance of several of these factors will be much clearer and at least some of the front runners are likely to find a place in standard clinical practice. This area is likely to remain a growth industry for the foreseeable future.

ACKNOWLEDGMENTS

Professor Playford has a research group examining the actions of a variety of neutraceuticals. Funding was received (by RJP) for this research into the value of colostrum for GI disease from Scientific Hospital Supplies, research grants from the Medical Research Council for studies of epidermal growth factor, and lecture fees from various companies that market proton pump inhibitors, including Wyeth, Astra, and Byk (UK), as well as from Lonza (US), which markets zinc carnosine. Research funding was received from Scientific Hospital Supplies (by SG). SG served on the advisory committee for Schering-Plough. He has received honorariums for lecturing in educational meetings sponsored by Wyeth, Astra-Zeneca, and Schering-Plough.

REFERENCES

1. R.J. Playford. Leading article: Peptides and gastrointestinal mucosal integrity. *Gut*, 37: 595–597, 1995a.
2. T.C. Ko, R.D. Beauchamp, C.M. Townsend Jr, J.C. Thompson. Glutamine is essential for epidermal growth factor-stimulated intestinal cell proliferation. *Surgery*, 114: 147–153, 1993.
3. J. Levy. Immunonutrition: the paediatric experience. *Nutrition*, 14: 641–647, 1998.
4. J.D. Carver, L.A. Barness. Trophic factors for the gastrointestinal tract. *Neonat. Gastroenterol.*, 23: 265–285, 1996.
5. C.A. Dinarello. The interleukin-1 family: 10 years of discovery. *FASEB J.*, 8: 1314–1325, 1994.
6. A.J. Wilson, K. Byron, P.R. Gibson. Interleukin 8 stimulates the migration of human colonic epithelial cells *in vitro*. *Clin. Sci.*, 97: 385–390, 1999.
7. Y. Yasunaga, Y. Shinomura, S. Kanayama, Y. Higashimoto, M. Yabu, Y. Miyazaki, S. Kondo, Y. Murayama, H. Nishibayashi, S. Kitamura, Y. Matsuzawa. Increased production of interleukin 1β and hepatocyte growth factor may contribute to foveolar hyperplasia in enlarged fold gastritis. *Gut*, 39: 787–794, 1996.
8. R.T. Przemioslo, P.J. Ciclitira. Cytokines and gastrointestinal disease mechanisms. In: R.A. Goodlad and N.A. Wright, (Ed.). *Bailliere's Clinical Gastroenterology* Volume 10/Number 1: Cytokines and growth factors in gastroenterology. London: Bailliere Tindall Press, 1996, 17–32.
9. R.J. Playford. Leading article: Peptides and gastrointestinal mucosal integrity. *Gut*, 37: 595–597, 1995.
10. J.A. Barnard, R.D. Beauchamp, W.E. Russell, R.N. Dubois, R.J. Coffey. Epidermal growth factor-related peptides and their relevance to gastrointestinal pathophysiology. *Gastroenterology*, 108: 564–580, 1995.

11. R.J. Playford, J.J. Batton, T.C. Freeman, K. Beardshall, D.A. Vesey, G.C. Fenn, J.H. Baron, J. Calam. Gastric output of pancreatic secretory trypsin inhibitor is increased by misoprostol. *Gut*, 32: 1396–1400, 1991.
12. B. Egger, F. Procaccino, J. Lakshmanan, M. Reinshagen, P. Hoffmann, A. Patel, W. Reuben, S. Gnanakkan, L. Liu, L. Barajas, V.E. Eysselein. Mice lacking transforming growth factor β have an increased susceptibility to dextran sulphate-induced colitis. *Gastroenterology*, 113: 825–832, 1997.
13. R.J. Playford, A.M.,Hanby, K. Patel, J. Calam. Influence of inflammatory bowel disease on the distribution and concentration of pancreatic secretary trypsin inhibitor within the colon. *Am. J. Pathol.*, 146: 310–316;1995.
14. R.J. Playford, T. Marchbank, D.P. Calnan, J. Calam, P. Royston, J.J. Batten, H.F. Hansen. Epidermal growth factor is digested to smaller, less active forms in acidic gastric juice. *Gastroenterology*, 108: 92–101, 1995.
15. H.S. Park, R.A. Goodlad, D.J. Ahnen, A. Winnett, P. Sasieni, C.Y. Lee, N.A. Wright. Effects of epidermal growth factor and dimethylhydrazine on crypt size, cell proliferation and crypt fission in the rat colon. Cell proliferation and crypt fission are controlled independently. *Am. J. Pathol.*, 151: 843–852, 1997.
16. S. Ishikawa, G. Cepinskas, R.D. Specian, M. Itoh, P.R. Kvietys. Epidermal growth factor attenuates jejunal mucosal injury induced by oleic acid: role of mucus. *Am. J. Physiol.*, 267: G1067–G1077, 1994.
17. W.M. Hui, B.W. Chen, A.W. Kung, C.H. Cho, C.T. Luk, S.K. Lam. Effect of epidermal growth factor on gastric blood flow in rats:possible role in mucosal protection. *Gastroenterology*, 104: 1605–1610, 1993.
18. R.J. Playford, A.M. Hanby, S. Gschmeissner, L.P. Peiffer, N.A. Wright, T. McGarrity. The epidermal growth factor receptor (EGF-R) is present on the basolateral, but not the apical, surface of enterocytes in the human gastrointestinal tract. *Gut*, 39: 262–266, 1996.
19. R.J. Playford, T. Marchbank, R. Chinery, R. Evison, M. Pignatelli, R.A. Boulton, L. Thim, A.M. Hanby. Human spasmolytic polypeptide is a cytoprotective agent that stimulates cell migration. *Gastroenterology*, 108: 108–116, 1995.
20. S.A. Lamprecht, B. Schwartz, A. Glicksman. Transforming growth factor-beta in intestinal epithelial differentiation and neoplasia. *Anticancer Res.*, 9: 1877–1881, 1989.
21. G.S. Baldwin, R.H. Whitehead. Gut hormones, growth and malignancy. *Baillieres Clin. Endocrinol. Metabol.*, 8: 185–214, 1994.
22. A. Dignass, D.K. Podolsky. Cytokine modulation of intestinal epithelial cell restitution:central role of transforming growth factor beta. *Gastroenterology*, 105: 1323–1332, 1993.
23. W.H. Daughaday, P. Rotwein. Insulin-like growth factors I & II. Peptide messenger RNA-like structures, serum and tissue concentrations. *Endocrin. Rev.*, 10: 68–91, 1989.
24. P.K. Lund, E.M. Zimmerman. Insulin-like growth factors and inflammatory bowel disease. *Baillieres Clin. Gastroenterol.*, 10: 83–96, 1996.
25. V.K.M. Han, A.J. D'Ercole, P.K. Lund. Cellular localisation of somatomedin (insulin-like growth factor) messenger RNA in the human fetus. *Science*, 236: 193, 1987.
26. R.J. Playford, C.E. Macdonald, W.S. Johnson. Colostrum and milk-derived growth factors for the treatment of gastrointestinal disorders. *Am. J. Clin. Nutr.*, 72: 5–14, 2000.

27. R.C. Baxter, Z. Zaltsman, J.R. Turtle. Immunoreactive somatomedin-C/insulin-like growth factor I and its binding protein in human milk. *J. Clin. Endo. Metab.*, 58: 955–959, 1984.
28. P.Y. Vacher, J.W. Blum. Age dependency of insulin like growth factor 1, insulin protein and immunoglobulin concentrations and gamma glutamyl transferase activity in first colostrum of dairy cows. *Milchwissenshaft*, 48: 423–425, 1993.
29. R.J. Collier, M.A. Mille, J.R. Hildebrant, A.R. Torkelson, T.C. White, K.S. Madsen, J.L. Vicini, P.J. Eppard, G.M. Lanza. Factors affecting insulin-like growth factor I concentration in Bovine colostrum. *J. Dairy Sci.*, 74: 2905–2911, 1991.
30. W.L. Lowe. Biological actions of the insulin-like growth factors. In: D LeRoith (Ed.) Insulin-like growth factors: molecular and cellular aspects. Boca Raton: CRC Press, 1991, 49–85.
31. H-C. Lo, P.S. Hinton, H. Yang, T.G. Unterman, D.M. Ney. Insulin-like growth factor-I but not growth hormone attenuates dexamethasone-induced catabolism in parenterally fed rats. *J. Parenter. Enter. Nutr.*, 20: 171–177, 1996.
32. P.D. Gluckman, D.J. Mellor. Use of growth factor IGF-II. International patent application 93/25227, 1993.
33. D. Schams, R. Einspanier. Growth hormone, IGF-I and insulin in mammary gland secretion before and after parturition and possibility of their transfer into the calf. *Endocrin. Reg.*, 25: 139–43, 1991.
34. D.S. Suh, M.M. Rechler. Hepatocyte nuclear factor 1 and the glucocorticoid receptor synergistical activate transcription of the rat insulin-like growth factor binding protein-1 gene. *Mol. Endocrinol.*, 11: 1822–1831, 1997.
35. S.M. Donovan, L.C. Atilano, R.L. Hintz, D.M. Wilson, R.G. Rosenfeld. Differential regulation of the insulin-like growth factors (IGF-I and –II) and IGF binding proteins during malnutrition in the neonatal rat. *Endocrinology*, 129: 149–157, 1991.
36. M.M. Rechler. Insulin-like growth factor binding proteins. *Vitam. Horm.*, 47: 1–114, 1993.
37. P.K. Lund, E.M. Zimmerman. Insulin-like growth factors and inflammatory bowel disease. *Baillieres Clin. Gastroenterol.*, 10: 83–96, 1996.
38. P.V. Malven, H.H. Head, R.J. Collier, F.C. Buonomo. Periparturient changes in secretion and mammary uptake of insulin and in concentrations of insulin and insulin-like growth factors in milk of dairy cows. *J. Dairy Sci.*, 70: 2254–2265, 1987.
39. Y. Shing, M. Klagsbrun. Purification and characterization of a bovine colostrum-derived growth factor. *Mol. Endocrinol.*, 1: 335–338, 1987.
40. Y.W. Shing, M. Klagsbrun. Human and bovine milk contain different sets of growth factors. *Endocrinology*, 115: 273–282, 1984.
41. S. Szabo, Z. Sandor. Basic fibroblast growth factor and PDGF in GI diseases. In: R.A. Goodlad, N.A. Wright (Eds.) *Bailliere's Clinical Gastroenterology*, Volume 10/Number 1 Cytokines and growth factors in gastroenterology. London: Bailliere Tindall Press, 1996, 97–112.
42. P.J. Keck, S.D. Hauser, G. Krivi, K. Sanzo, T. Warren, J. Feder, D.T. Connolly. Vascular permeability factor, an endothelial cell mitogen related to PDGF. *Science*, 246: 1309–1312, 1989.
43. C.G. Siafakas, F. Anatolitou, R.D. Fusunyan, W.A. Walker, I.R. Sanderson. Vascular endothelial growth factor (VEGF) is present in human breast milk and its receptor is present on intestinal epithelial cells. *Pediatr. Res.*, 45: 652–657, 1999.

44. R.M. Housley, C.F. Morris, W. Boyle, B. Ring, R. Biltz, J.E. Tarpley, S.L. Aukerman, P.L. Devine, R.H. Whitehead, G.F. Pierce. Keratinocyte growth factor induces proliferation of hepatocytes and epithelial cells throughout the rat gastrointestinal tract. *J. Clin. Invest.*, 94: 1764–1777, 1994.
45. R.J. Playford, T. Marchbank, N. Mandir, A. Higham, K. Meeran, M.A. Ghatei, S.R. Bloom, R.A. Goodlad. Effects of keratinocyte growth factor (KGF) on gut growth and repair. *J. Pathol.*, 184: 316–322, 1998.
46. C.L. Farrell, J.V. Bready, K.L. Rex, J.N. Chen, C.R. DiPalma, K.L. Whitcomb, S. Yin, D.C. Hill, B. Wiemann, C.O. Starnes, A.M. Havill, Z.N. Lu, S.L. Aukerman, G.F. Pierce, A. Thomason, C.S. Potten, T.R. Ulich, D.L. Lacey. Keratinocyte growth factor protects mice from chemotherapy and radiation-induced gastrointestinal injury and mortality. *Cancer Res.*, 58: 933–939, 1998.
47. J.M. Zeeh, F. Procaccino, P. Hoffmann, S.L., Aukerman, J.A. McRoberts, S. Soltani, G.F. Pierce, J. Lakshmanan, D. Lacey, V.E. Eysselein. Keratinocyte growth factor ameliorates mucosal injury in an experimental model of colitis in rats. *Gastroenterology*, 110: 1077–1083, 1996.
48. W.F. Johnson, C.R. Dipalma, T.R. Ziegler, C.L. Farrell. Keratinocyte growth-factor (KGF) enhances early regrowth in a rat model of short-bowel syndrome. *Gastroenterology*, 114**:** G3631–G3631, 1998.
49. B. Egger, F. Procaccino, I. Sarosi, J. Tolmos, M.W. Buchler, V.E. Eysselein. Keratinocyte growth factor ameliorates dextran sodium suphate colitis in mice. *Dig. Dis. Sci.*, 44: 836–844, 1999.
50. D.S. Han, F. Li, K. Connolly, M. Hubert, R. Miceli, Z. Okoye, G. Santiago, K. Windle, E. Wong, R.B. Sartor. Keratinocyte growth factor-2 (FGF-10) promotes healing of experimental small intestinal ulceration in rats. *Am. J. Physiol. Gastrointest. Liver Physiol.*, 279: G1011–G1022, 2000.
51. M.A. Hull, A. Knifton, B. Filipowicz, J.L. Brough, G. Vautier, C.J. Hawkey. Healing with basic fibroblast growth factor is associated with reduced indomethacin induced relapse in a human model of gastric ulceration. *Gut*, 40: 204–210, 1997.
52. R.J. Playford. Using the full spectrum of protective factors on the GI tract. *Functional Foods and Nutraceuticals*. May 2003, 52–58.
53. S. Ghosh, R.J. Playford. Bioactive natural compounds for the treatment of gastrointestinal disorders. *Clin. Sci.*, 104: 547–556, 2003.
54. R.H. Davis, M.G. Leitner, J.M. Russo, M.E. Byrne. Anti-inflammatory activity of aloe vera against a spectrum of irritants. *J. Am. Podiatr. Med. Assoc.*, 79: 263–276, 1989.
55. T. Reynolds, A.C. Dweck. Aloe vera leaf gel: a review update. *J. Ethnopharm.*, 68: 3–37, 1999.
56. E.E. Galal, A. Kandil, R. Hegazy. Aloe vera and gastrogenic ulceration. *J. Drug Res.*, 7: 73–76, 1975.
57. P. Chithra, G.B. Sajithlal, G. Chandrakasan. Influence of aloe vera on the glycosaminoglycans in the matrix of healing dermal wounds in rats. *J. Ethnopharmacol.*, 59: 179–186, 1998.
58. B.H. McAnalley. Process for preparation of aloe products, produced thereby and composition thereof. *US4*, 735, 935, 1988.
59. S.W. Choi, B.W. Son, Y.S. Son, Y.I. Park, S.K. Lee, M.H. Chung. The wound healing effect of a glycoprotein fraction isolated from aloe vera. *Br. J. Dermatol.*, 145: 535–545, 2001.

60. R.J. Playford, C.E. MacDonald, D.P. Calnan, D.N. Floyd, T. Podas, W. Johnson, A.C. Wicks, O. Bashir, T. Marchbank. Co-administration of the health food supplement, bovine colostrum, reduces the acute NSAID-induced increase in intestinal permeability. *Clin. Sci.*, 6: 627–633, 2001.
61. Z. Khan, C. Macdonald, A.C. Wicks, M.P. Holt, D. Floyd, S. Ghosh, N.A. Wright, R.J. Playford. Use of the 'nutraceutical', bovine colostrum, for the treatment of distal colitis. *Aliment Pharmacol. Ther.*, 16: 1917–1922, 2002.
62. R.J. Playford, D.N. Floyd, C.E. Macdonald, D.P. Calnan, R.O. Adenekan, W. Johnson, R.A. Goodlad, T. Marchbank. Bovine colostrum is a health food supplement which prevents NSAID-induced gut damage. *Gut*, 44: 653–658, 1999.
63. P.B. Sullivan, M.J. Brueton, Z. Tabara, R.A. Goodlad, C.Y. Lee, N.A. Wright. Epidermal growth factor in necrotizing enterocolitis. *Lancet*, 338: 53–54, 1991.
64. H. Hirashi, T. Sasai, T. Oinuma, T. Shimada, H. Sugaya, A. Terano. Polaprezinc protects gastric mucosal cells from noxious agents through antioxidant properties *in vitro*. *Aliment. Pharmacol. Ther.*, 13: 261–269, 1999.
65. K. Seto, T. Yoneta, H. Suda, H. Tamaki. Effect of polaprezinc (N-3-aminopropionyl-L-histidinato zinc), a novel antiulcer agent containing zinc, on cellular proliferation: role of insulin-like growth factor I. *Biochem. Pharmacol.*, 58: 245–250, 1999.
66. T. Matsukura, H. Tanaka. Applicability of zinc complex of L-carnosine for medical use. *Biochemistry (Mosc.)*, 65: 817–823, 2000.
67. C.H. Cho, C.T. Luk, C.W. Ogle. The membrane-stabilizing action of zinc carnosine (Z-103) in stress-induced gastric ulceration in rats. *Life Sci.*, 49: 189–194, 1991.
68. T. Yoshikawa, T. Yamaguchi, N. Yoshida, H. Yamamoto, S. Kitazumi, S. Takahashi, Y. Naito, M. Kondo. Effect of Z-103 on TNB-induced colitis in rats. *Digestion*, 58: 464–468, 1997.
69. A. Miyoshi, H. Matsuo, T. Miwa, M. Namiki, A. Yanai, K. Masamune, S. Asagi, H. Mori, A. Iwasaki, S. Nakazawa, J. Kobayashi, S. Ooshiba, T. Takemoto, S. Kishi, K. Hayakawa, M. Nakajima. Clinical trial of Z-103 on gastric ulcer. *Jpn. Pharmacol. Ther.*, 20:1–14, 1992.
70. M.A. Dalrymple, I. Garner. Genetically modified livestock for the production of human proteins in milk. *Biotechnol. Genet. Eng. Rev.*, 15: 33–49, 1998.
71. C. Klein-Soyer, C. Bloy, G. Archipoff, A. Beretz, J.P. Cazenave. Naftazone accelerates human saphenous vein endothelial cell proliferation *in vitro*. *Nouv. Rev. Fr. Hematol.*, 37: 187–192, 1995.
72. R.J. Playford, R. Boulton, M.A. Ghatei, S.R. Bloom, N.A. Wright, R.A. Goodlad. Comparison of the effects of TGFα and EGF on gastrointestinal proliferation and hormone release. *Digestion*, 57: 362–367, 1996.
73. W. O'Loughlin, M. Winter, A. Shun, J.A. Hardin, D.G. Gall. Structural and functional adaptation following jejunal resection in rabbits: effect of epidermal growth factor. *Gastroenterology*, 107: 87–93, 1994.
74. D. Kelly, T.P. King, M. McFadyen, A.G.P. Coutts. Effect of preclosure colostrum intake on the development of the intestinal epithelium of artificially reared piglets. *Biol. Neonate*, 64**:** 235–244, 1993.
75. A.P. Ramsanahie, A. Perez, A.U. Duensing, M.J. Zinner, S.W. Ashley, E.E. Whang. Glucagon-like peptide 2 enhances intestinal epithelial restitution. *J. Surg. Res.*, 107: 44–49, 2002.
76. J. Jasleen, S.W. Ashley, N. Shimoda, M.J. Zinner, E.E. Whang. Glucagon-like peptide 2 stimulates intestinal proliferation *in vitro*. *Dig. Dis. Sci.*, 47: 1135–1140, 2002.

77. D.J. Drucker. Biological actions and therapeutic potential of the glucagon-like peptides. *Gastroenterology*, 122: 531–544, 2002.
78. P.B. Jeppesen, B. Hartman, J. Thulesen, B.S. Hansen, J.J. Holst, S.S. Poulsen, P.B. Mortensen. Elevated plasma glucagons-like peptide 1 and 2 concentrations in ileum resected short bowel patients with a preserved colon. *Gut*, 47: 370–376, 2002.
79. T.M. MacDonald, S.V. Morant, G.C. Robinson, M.J. Shield, M.M. McGilchrist, F.E. Murray, D.G. McDevitt. Association of upper gastrointestinal toxicity of non-steroidal anti-inflammatory drugs with continued exposure: cohort study. *Br. Med. J.*, 315: 1333–1337, 1997.
80. C.E. Macdonald, D.P. Calnan, T. Podas, W. Johnson, R.J Playford. Clinical trial of colostrum for protection against NSAID induced enteropathy. *Gastroenterology*, 114: G0854, 1998.
81. M. Hirano, R. Iweakiri, K. Fujimoto, H. Sakata, T. Ohyama, T. Sakai, T. Joh, M. Itoh. Epidermal growth factor enhances repair of rat intestinal mucosa damaged after oral administration of methotrexate. *J. Gastroenterol.*, 30: 169–176, 1995.
82. S.T. Sonis, L. Lindquist, A. Van Vugt, A.A. Stewart, K. Stam, G.Y. Qu, K.K. Iwata, J.D. Haley. Prevention of chemotherapy-induced ulcerative mucositis by transforming growth factor beta 3. *Cancer Res.*, 54: 1135–1138, 1994.
83. G.S. Howarth, G.L. Francis, J.C. Cool, R.W. Ballard, L.C. Read. Milk growth factors enriched from cheese whey ameliorate intestinal damage by methotrexate when administered orally to rats. *J. Nutr.*, 126: 2519–2530, 1996.
84. N.M. Gordler, M. McGurk, S. Aqual, M. Prince. The effect of EGF mouthwash on cytotoxic-induced oral ulceration. *Am. J. Clin. Oncol.*, 18: 403–406, 1995.
85. F. Procaccino, M. Reinshagen, P. Hoffman, J.M. Zeeh, J. Lakshmanan, J.A. McRoberts, A. Patel, S. French, V.E. Eysselein. Protective effect of epidermal growth factor in an experimental model of colitis in rats. *Gastroenterology*, 107: 12–17, 1994.
86. S.N. Porter, G.S. Howarth, R.N. Butler. An orally administered growth factor extract derived from bovine whey suppresses breath ethane in colitic rats. *Scand. J. Gastroenterol.*, 33: 967–974, 1998.
87. H. Mashimo, C. Wu, M.C. Fishman, D.K. Podolsky. Protection and healing of intestinal mucosa: gene-targeted disruption of intestinal trefoil factor impairs defense of mucosal integrity. *Gastroenterology*, 110: A959, 1996.
88. F.R. Moya, H. Eguchi, B. Zhao, M. Furukawa, J. Sfeir, M ,Osorio et al. Platelet-activating factor acetylhydrolase in term and preterm human milk: a preliminary report. *J. Pediatr. Gastroenterol. Nutr.*, 19: 236–239;1994.
89. R.M. Kliegman, W.A. Walker, R.H. Yolken. Necrotizing enterocolitis: research agenda for a disease of unknown etiology and pathogenesis. *Pediatr. Res.*, 34: 701–708, 1993.
90. M.S. Caplan, M. Lickerman, L. Adler, G.N. Dietsch, A. Yu. The role of recombinant platelet-activating factor acetylhydrolase in a neonatal rat model of necrotizing enterocolitis. *Pediatr. Res.*, 42: 779–783, 1997.
91. S.V. Lutsenko, S.M. Kiselev, S.E. Severin. Molecular mechanisms of tumor angiogenesis. *Biochemistry*, 68: 286–300, 2003.
92. T.D. Henry, B.H. Annex, G.R. McKendall, M.A. Azrin, J.J. Lopez, F.J. Giordano, P.K. Shah, J.T. Willerson, R.L. Benza, D.S. Berman, C.M. Gibson, A. Bajamonde, A.C. Rundle, J. Fine, E.R. McCluskey. The VIVA trial: vascular endothelial growth factor in ischemia for vascular angiogenesis. *Circulation*, 107: 1359–1365, 2003.

93. K. Gelse, K. von der Mark, T. Aigner, J. Park, H. Schneider. Articular cartilage repair by gene therapy using growth factor-producing mesenchymal cells. *Arthr. Rheum.*, 48: 430–441, 2003.
94. M.J. Kim, R.M. Jun, W.K. Kim, H.J. Hann, Y.H. Chong, H.Y. Park, J.H. Chung. Optimal concentration of human epidermal growth factor (hEGF) for epithelial healing in experimental corneal alkali wounds. *Curr. Eye Res.*, 22: 272–279, 2001.
95. A. Lambiase, L. Manni, S. Bonini, P. Rama, A. Micera, L. Aloe. Nerve growth factor promotes corneal healing: structural. Biochemical, and molecular analyses of rat and human corneas. *Invest. Ophthalmol. Vis. Sci.*, 41: 1063–1069, 2000.
96. R.J. Playford, A.C. Woodman, P. Clark, P. Watanapa, D. Vesey, P.H. Deprez, R.C. Williamson, J. Calam. Effect of luminal growth factor preservation on intestinal growth. *Lancet*, 341: 843–848, 1993.

12 Cystatin: A Novel Bioactive Protein

Soichiro Nakamura

CONTENTS

12.1 INTRODUCTION

Proteolytic enzymes are potentially hazardous to their protein environment, and thus their activity must be carefully controlled. Proteolytic activity can be regulated by modulating the synthesis or degradation of the enzyme or necessary cofactors or via interactions with activators or inhibitors (1). Living organisms use inhibitors as a major tool in regulating the proteolytic activity of proteinases. Indeed, proteinases and their respective inhibitors are found in almost every system where proteolysis occurs. Among the four major classes of proteolytic enzymes — serine, cysteine, aspartic, and metalloproteinases — special attention must be given to the class of cysteine proteinases, especially to the papain-like cysteine proteinase that forms the largest subfamily among them. Papain-like proteinases are widely expressed throughout the animal and plant kingdoms and have also been identified in viruses and bacteria (2). Mammalian papain-like cysteine proteinases are known as thiol-dependent cathepsins (3). According to the human genome database (http://www.genome. ad.jp/dbget/dbget), 11 cathepsins are expressed in the human genome as papain-like enzymes. Currently, evidence suggests that papain-like cathepsins carry out specific functions in extracellular matrix turnover, antigen presentation, and processing events in the human body (2). Consequently, investigations of papain-like proteinase inhibitors have been designed to reduce pathogen infectivity, prevent muscular dystrophies and joint destruction, and inhibit tumorigenesis and metastasis. In addition to the commercial applications of the inhibitors for controlling the deterioration of protein gels during food processing (4,5), understanding their properties for regulation of the proteinases would be helpful in therapeutic applications, in which they can represent viable drug candidate for major diseases such as osteoporosis, arthritis, immune-related diseases, tumors, cancers, and a wide variety of parasitic infections.

In 1968, an inhibitor of ficin and papain was found in chicken egg-white (6) and further characterized by the inhibitory activities toward cathepsins B and C (7,8). This protein was later named "cystatin" because of its distinctive inhibitory activity on cysteine proteinases (9). A number of investigations on the biochemical properties and chemical structures of cystatins have been conducted mostly in the fields of medicine and physiology. Ohkubo et al. (10) found that multiple cystatin-like sequences were present in the kininogens and that the stefins were related to both the cystatins and the repeats in the kininogens, which resulted in the emergence of the concept of a "cystatin superfamily" in 1986 (11). The cystatin superfamily encompasses a large array of proteins with various sizes, often containing multiple cystatin-like sequences; consequently, it has been dramatically expanded. Recently, based on the partial amino acid homology, several new members of the superfamily have been identified, including proteins from vertebrates, insects, and plants (12). It appears that the superfamily comprises at least six evolutionarily linked families of proteins: (a) the cytoplasmic stefins, (b) the cell-secreted cystatins, (c) the plasma kininogens, (d) the cell-secreted cystatin-related proteins, (e) the plasma fetuins, and (f) the histidine-rich plasma glycoproteins, in which (a), (b), and (c) are active (which gave rise to the name of the whole superfamily), while others are not significant (12). According to their biological and structural properties, the cystatin superfamily could be subdivided, for the time being, into four groups, that is, three families — the stefins, the cystatins, the kininogens — and others. For example, the cystatins are a family of tight and reversible binding inhibitors comprising approximately 120 amino acids and containing four conserved cysteine residues known to form two disulfide bonds (13,14).

In the last decade, tremendous progress in the field of *proteomics* has been achieved, which offers a new insight into the process of the proteinaceous inhibitor-related area. As of today, more than 200 kinds of complementary and genomic DNA sequences of the proteins belonging to the cystatin superfamily have been reported in the gene-bank database. This chapter will provide reliable and advanced information on the biochemistry of the proteinous inhibitors and their recombinants as nutraceutical aids.

12.2 FORMATION OF THE CYSTATIN SUPERFAMILY

Many new cystatin-like proteinase inhibitors such as fetuins, histidine-rich glycoproteins, cystatin-related proteins, divergent cystatins, and plant seed cystatins have been discovered in addition to the original cystatins, stefins, and kininogens; however, some of them may have lost the inhibitory activities upon evolution or perhaps never acquired it (12). Thus, the stefins (family I), the cystatins (family II) and the kininogens (family III) are recognized as the main family members of the cystatin superfamily (Table 12.1).

Family I includes single-chain proteins that lack intramolecular disulfide bonds and carbohydrates and have approximately 100 amino acids and an MW of about 11 kDa (15). This family can be further divided into two subfamilies, stefin A (cystatin α) and B (cystatin β), depending on the presence or absence of cytsteine residues. The variants in both human and rat have been studied in detail. High concentrations of human stefin A and rat cystatin-α have been found in various types of epithelial cells and leukocytes. The most abundant source of stefin A is from polymorphonuclear leukocytes in the liver. It is also found in extracts of squamous epithelia from the mouth and esophagus and has been localized to the cytoplasm of the strata corneum and granulosum of the epidermis (16–21). On the other hand, stefin Bs are found to be broadly distributed in different cells and tissues as an intracellular cysteine proteinase inhibitor (18,19,22,23). They were originally discovered as an inhibitor of cathepsins B in a variety of tissues (22). It is noteworthy that stefin B forms disulfide-linked dimers with MW of about 25 kDa through cysteine residues at position 3. Both stefins A and B are potent reversible and competitive inhibitors of papain-like proteinases (Table 12.2). Human stefin B binds tightly and rapidly to papain with an inhibition constant of 1.9×10^{-11} M (24). The selective distribution of the inhibitor correlates with tissues that constitute a "first line of defense" against pathogenic organisms (20). Stefins may thus provide a protective function by acting as an inhibitor of cysteine proteinases, which are used as invasive tools by many infectious agents.

Family II comprises single-chain proteins with approximately 120 amino acids and 13 to 15 kDa MW, each of which has two intramolecular disulfide bonds located toward the carboxyl terminus, which constitute the primary characteristic of this family (11). This family includes proteins that have been called chicken cystatin (9,25), cystatin C (26–29), cystatin S/SA/SN (30–33), cystatin E/M (34–36), cystatin D (37,38), cystatin F (39), colostrum cystatin (40), and snake venom cystatin (41). Cystatin F is termed leukocystatin (39). Recently, a new cysteine proteinase inhibitor that lacks some of the conserved motifs, believed to be important for inhibition of cysteine proteinase activity, has been cloned from mouse testis and named cystatin T (42). The same characteristics are seen in two other recently cloned genes, CRES and testatin (43,44), which, including cystatin T, appear to fall within a subgroup of the family II cystatins, as shown in Table 12.1. Chicken cystatin is the representative of the cystatin family (6,11) as the three-dimensional structure of the protein has been well-defined (25). Alternatively, the most investigated inhibitor of human origin is cystatin C. Production of cathepsins B and L is augmented in many transformed cell lines and in some cases correlates with tumor cell invasion (45). Cystatin C could also be useful in the therapy of certain metastatic tumors because it is effective not only against cathepsins L but also against cathepsins B belonging to exopeptidase, as shown in Table 12.2. A potent modulator of

TABLE 12.1
Representative Members of the Cystatin Superfamily

Name	Source	References
Family I (Stefin family)		
Stefin A (cystatin α)	Human polymorphonuclear granulocytes	16, 17
	Human liver	18
	Human epithelium	19
	Rat epidermis	20
Stefin B (cystatin β)	Mammalian tissues	21
	Human liver	18
	Human spleen	19
	Rat liver	22
Family II (Cystatin family)		
Chicken cystatin	Hen egg white	6–9, 25
Cystatin C	Human cerebrospinal fluid	26
	Rat urine vesicles	27
	Carp ovarian fluid	28
	Chum salmon liver	29
Cystatin S/SA/SN	Human saliva	30–33
Cystatin E/M	Human amniotic and epithelial cells	34–36
Cystatin D	Human saliva	37, 38
Cystatin F (Leukocystatin)	Human and rat hematopoietic cells	39
Colostrums cystatin	Bovine colostrum	40
Venom cystatin	African puff adder	41
Subgroup		
CRES	Mouse epididymis	42
Testatin	Mouse testis	43
Cystatin T	Mouse testis	44
Family III (Kininogen family)		
HMW-kininogen	Human and rat plasma	50
LMW-kininogen	Human and rat plasma	50–52
T-kininogen (Thiostatin)	Rat plasma	53
Others		
Fetuins	Human plasma	55
	Bovine plasma	56, 57
Histidine-rich glycoprotein (HRG)	Human plasma	58
Cystatin-related proteins (CRPs)	Rat ventral prostate	59, 60
Sarcocystatin	Flesh fly larvae	61
Drosophila cystatin	Fruit fly head	62
Phytocystatin[a]	Rice seed	64, 65
	Corn seed	66
	Cowpea seed	67
	Soy been	68
	Carrot seed	69

(Continued)

TABLE 12.1 (Continued)

Name	Source	References
	Sunflower seed	70
	Apple fruit	71
	Potato	72
	Avocado fruit	73
	Herbs	74

[a] Phytocystatins from rice seed, soy been, and herbs are called orizacystatin, soyacystatin, and chelidocystatin, respectively.

TABLE 12.2
Apparent Inhibition Constants[a] of Members of the Cystatin Superfamily to Papain-Like Proteinases

Name	Papain	Cathepsin B	Cathpsin H	Cathpsin L	Cathpsin C	Ficin	Reference
Family I							
Stefin A	<0.005	1.7	0.064	0.019	0.35	NA[b]	24
Stefin B	0.0019	8.2	0.31	1.3	3	NA	24
Family II							
Chicken cystatin	0.12	73	0.58	0.23	0.23	ND[c]	24
Cystatin C	<0.005	0.25	0.28	<0.005	3.5	ND	24
Cystatin S	108.2	ND	ND	ND	>630.2	4.2	30
Cystatin SA	0.12	ND	ND	ND	80.6	0.074	31
Cystatin SN	21.4	ND	ND	ND	26.3	4.4	32
Cystatin E/M	0.39	32	ND	NA	NA	NA	39
Cystatin D	0.9	>1000	7.5	18	0.27	NA	38
Cystatin F	1.1	>1000	NA	0.31	NA	NA	39
Family III							
HWW kininogen	0.07	12	15	0.7	NA	NA	45
LMW kininogen	0.05	23	44	1.8	NA	NA	45
T-kininogen	0.19	170	470	2.8	NA	NA	45
Others							
Orizacystatin I	32	ND	790	ND	NA	NA	65
Orizacystatin II	830	ND	10	ND	NA	NA	65

[a] Data are expressed in nM units. [b] NA and [c] ND represent "not assessed" and "not determined," respectively.

neutrophil migration, cystatin C prevents destruction of connective tissue by intracellular enzymes leaking from dying cells or being misrouted from secretion from neoplastic tissue (1). In addition, the antiviral effects of chicken cystatin, cystatin C, and cystatin D have been reported in cases of polio, herps simplex, and coronal virus-infected cultures (46–48). Synthesized cystatin C peptide-derivatives also exhibit potent bacteriostatic effects against a large number of bacterial strains belonging to 13 different species (49).

Family III comprises plasma kininogens with higher molecular weight, which play key roles in acting as the precursors of vasoactive kinins and participate in the blood coagulation cascade. High-molecular-weight (HMW)-kininogen, low-molecular-weight (LMW)-kininogen, and T-kininogen are the representative members of this family (50–53). T-kininogen is known as thiostatin or as a major acute-phase alpha-protein, which is only detected in rat plasma (53). Both HMW- and LMW-kininogens are single-chain glycoproteins carrying a bradykinin moiety in the interior of the polypeptide chains that are bridged by a disulfide linkage and thus consist of three parts: an amino-terminal heavy chain, the bradykinin moiety, and a carboxyl-terminal light chain. They differ in the carboxyl-terminal chain due to differential processing of the primary transcript (50). The NH_2-terminal chain is glycosylated with additional disulfide bonds. Cysteine proteinase inhibitory activity rises in this heavy chain, whose structure is made up of tandemly repeated cystatin-like domains. LMW-kininogen can be subjected to limited proteolysis with trypsin and divided into three fragments, D-I, D-II and D-III, as shown in Figure 12.1. Among them, only D-I has no inhibitory activity on cysteine proteinases; D-II appears highly inhibitory against *m*-calpain, papain, and cathepsin L, whereas D-III cannot inhibit *m*-calpain but inhibits papain and cathepsin L strongly (52). It is assumed that these activities may control inflammation by inhibiting cysteine proteinases released from damaged tissues.

In addition to the above three families, it has been reported that many proteins have sequence homology with the established members of the cystatin superfamily (12,54). Fetuins have been found in both human and bovine plasmas (55–57). Human fetuin, first characterized as α_2-HS glycoprotein, has a close link with the kininogens (55). Histidine-rich glycoprotein (HRG), which is structurally related to HMW-kininogen, generally occurs in mammalian plasma (58). Analysis of gene structure showed that HRG has cystatin domains that are each encoded by three exons as they are in many cystatins. The genomic structures of cystatin-related proteins (CRPs) found in rat ventral prostate (59) are homologous to the typical cystatins (60). Sarcocystatin (61) and *Drosophila* cystatin (62), isolated from the hemolymph of *Sacrophaga peregrina* larvae and fruit flies and *Drosophila melanogaster*, respectively, are characterized as inhibitors of the cysteine proteinases. A series of cysteine proteinase inhibitors has been studied from plant sources and named as "phytocystatin" (63). Phytocystatins have been found in seeds of rice (64,65), corn (66), cowpea (67), soy been (68), carrot (69), sunflower (70), apple (71), potato (72), avocado (73), herbs (74), and so on. Among them, phytocystatins from rice seed, soy bean, and herbs have been called orizacystatin, soyacystatin, and chelidocystatin, respectively. In general, the affinities of phytocystatins are much weaker than those of the original members of the cystatin superfamily (see Table 12.2). They may, however, be modified to have strong inhibiting activities through the information from the recently determined three-dimensional NMR structure of oryzacystatin

```
Stefin A (human)    --------------MIPGGLSEAKPATPEIQEIVDKVKPQLEEK-TNETY—GKLEAVQYKTQVVAG    50
Stefin B (human)    --------------MMCGAPSATQPATAETQHIADQVRSQLEEK-YNKKF--PVFKAVSFKSQVVAG    50
Cystatin C (human)  -------SSPGKPPRLVGGPMDASVEEEGVRRALDFAVGEYNKA-SNDMYHSRALQVVRARKQIVAG    59
Cystatin D (human)  ------GSASAQSRTLAGGIHATDLNDKSVQRALDFAISEYNKVINKDEYYSRPLQVMAAYQQIVGG    61
Cystatin E (human)  ----------RPQERMVGELRDLSPDDPQVQKAAQAAVASYNMG-SNSIYYFRDTHIIKAQSQLVAG    56
Cystatin F (human)  GPSPDTCSQDLNSRVKPGFPKTIKTNDPGVLQAARYSVEKFNN*C-TNDMFLFKESRITRALVQIVKG    66
Cystatin S (human)  ------SSSKEENRIIPGGIYDADLNDEWVQRALHFAISEYNKA-TEDEYYRRPLQVLRAREQTFGG    60
Cystatin SA (human) -----WSPQEEDRIIEGGIYDADLNDERVQRALHFVISEYNKA-TEDEYYRRLLRVLRAREQIVGG    60
Cystatin SN (human) -----WSPKEEDRIIPGGIYNADLNDEWVQRALHFAISEYNKA-TKDDYYRRPLRVLRARQQTVGG    60
Chicken cystatin    -----------SEDRSRLLGAPVPVDENDEGLQRALQFAMAEYNRA-SNDKYSSRVVRVISAKRQLVSG    57
Mouse cystatin C    -----------------MLGAPEEADANEEGVRRALDFAVSEYNKGS-NDAYHSRAIQVVRARKQLVAG    51
Rat cystatin C      -----------------LLGAPQEADESEEGVQRALDFAVSEYNKGS-NDAYHSRAIQVVRARKQLVAG    51
Kininogen DI       -----------------QESQSEEIDCNDKDLFKAVDAALKKYNSQNQSNNQFVLYRITEATKTVGSD    51
Kininogen DII       ---AEGPVVTAQYDCLGCVHPISTQSPDLEPILRHGIQYFNN*N-TQHSSLFMLNEVKRAQRQVVAG    62
Kininogen DIII      ---GKDFVQPPTKICVGCPRDIPTNSPELEETLTHTITKLNAE-NN*ATFYFKIDNVKKARVQVVAG    62

51 TNYYIKVRAGDNKYMHLKVFKSL---------------------------PGQNEDLVLTGYQVDKNKDDELTGF    98
51 TNYFIKVHVGDEDFVHLRVFQSL---------------------------PHENKPLTLSNYQTNKAKHDELTYF    98
60 VNYFLDVELGRTTCTKT-----QPNLDNCPFHDQPHLKRKAFCSFQIYAVPWQGTMTLSKSTCQDA    120
62 VNYYFNVKFGRTTCTKS-----QPNLDNCPFNDQPKLKEEEFCSFQINEVPWEDKISILNYKCRKV    122
57 IKYFLTMEMGSTDCRKTRVTGDHVDLTTCPLAAGAQQEK-LRCDFEVLVVPWQN*SSQLLKHNCVQM    121
67 LKYMLEVEIGRTTCKKN--- QHLRLDDCDFQTN*HTLKQTLSCYSEVWVVPWLQHFEVPVLRCH    126
61 VNYFFDVEVGRTICTKS-----QPNLDTCAFHEQPELQKKQLCSFEIYEVPWEDRMSLVNSRCQEA    121
61 VNYFFDIEVGRTICTKS-----QPNLDTCAFHEQPELQKKQLCSFQIYEVPWEDRMSLVNSRCQEA    121
61 VNYFFDVEVGRTICTKS-----QPNLDTCAFHEQPELQKKQLCSFEIYEVPWENRRSLVKSRCQES    121
58 IKYILQVEIGRTTCPKS-----SGDLQSCEFHDEPEMAKYTTCTFVVYSIPWLNQIKLLESKCQ    116
48 VNYFFDVEMGRTTCTKSQ---- TN*LTDCPFHDQPHLMRKALCSFQIYSVPWKGTHSLTKFSCKNA    112
47 INYYLDVEMGRTTCTKSQ---- TN*LTNCPFHDQPHLMRKALCSFQIYSVPWKGTHTLTKSSCKNA    112
52 TFYSFKYEIKEGDCPVQSGKT----WQDCEYKDAAKAAT-GECTATVGKRSST-KFSVATQTCQITP    112
63 LNFRITYSIVQTN*CSKENFLF---LTPDCK-SLWNGDT--GECTDNAYIDIQLRIASF-SQNCDIYP    122
63 KKYFIDFVARETTCSKESNEE---LTESCE-TKKLGQS--LDCNAEVYVVPWEKKIYP-TVNCQPLGM    123
```

FIGURE 12.1 Alignment of amino acid sequences of the members belonging to the cystatin superfamily. Numbering is according to the chicken cystatin sequence using the single letter code for amino acids. The most highly conserved sequences (9, 53–57, and 103–104) are double underlined. The four cysteine residues that presumably form two disulfide bridges are underlined. The Asn residues of a theoretical N-glycosylation site (Asn-X-Ser/Thr) are marked with an asterisk.

I (75). The resulting recombinant inhibitors will be used for constructing transgenic plants with improved bioactivities and expression levels capable of controlling insect pests.

12.3 PRODUCTION AND USE OF RECOMBINANT CYSTATINS

12.3.1 Application of Genetically Glycosylated Cystatin to Surimi Manufacturing

Tenderness, one of the most important quality attributes of meat and seafood, is mainly caused by enzymatic breakdown of the contractile proteins that have profound effects on the myofibrillar toughness of muscles. Cysteine proteinases

— cathepsins B, L, and L-like cathepsins — have proteolytic activity for a variety of protein substrates and are considered to participate in postmortem muscle tenderization. In addition, they play important roles in the disintegration of "*surimi*" gels — for example, gel softening — by the hydrolysis of myosin heavy chain, light chains, actin, and troponins (5). Surimi, a water-leached, cryostabilized fish mince muscle, is a raw material for making gelled seafood products, such as "*kamaboko*" and "*imitation crab meat.*" With increasing demand for surimi products, attention is now being focused on the utilization of underutilized fish, such as roe-herring and late-run salmon. As a result, proteinase inhibitors are increasingly sought out to make good surimi-based products from the underutilized fish because most of them contain large amounts of endogenous muscle proteinases, which weaken gel strength by degrading myosin heavy chain during the cooking of surimi (2–5). Since major lysosomal proteinases in animal cells are papain-like enzymes, members belonging to the cystatin superfamily would be useful ingredients in making good surimi-based products. Cathepsins D, however, is capable of degrading these protein inhibitors. The proteins are easily inactivated by peptide bond cleavages at hydrophobic amino acid residues due to the action of the aspartic proteinase cathepsin D (76).

Genetic techniques can be employed for improving the molecular stability of the inhibitors. Since it has been reported that carbohydrate moieties contributed to the protection of polypeptide chains of glycoproteins from proteolysis (77,78); site-specific glycosylation occurring in yeast could be employed for improving the molecular stability of the inhibitors. The technique is based on the finding that the secretion of recombinant proteins in yeasts induces glycosylation at asparagine sites with the recognition sequences of Asn-X-Ser/Thr of polypeptides. It has been proposed that this technique be a new approach in enhancing the molecular stability of inhibitors against heating and proteolysis (79). Cystatin C has well-known molecular structure of a relatively small size, having a great potential for food applications and as pharmaceutical aids. Therefore, we employed the protein as a template for the biosynthesis of neoglycoprotein by using a yeast expression system and applied it to make good surimi products with high gel strength from roe-herring (80,81). A possible site for N-linked glycosylation has been found in mouse cystatin C, cloned from mouse embryo, beginning from the position 79 (82), as shown in Figure 12.1. Thus, by selecting a mouse cystatin C gene, which possesses a potential site for N-linked glycosylation at Asn-79, we successfully produced natural glycoproteins of cystatin C (polymannosyl cystatin) without specific selection of the glycosylation site.

Although cystatin C is resistant to heating due to its tightly packed conformation (83), further enhancement of the thermostability was achieved in the polymannosyl cystatin (80). Approximately 60% of the residual papain-inhibiting activity was retained in the glycosylated protein after heating at 95°C for 30 minutes, whereas that of the nonglycosylated wild-type cystatin C was completely inactivated during heating. In addition, the papain-inhibiting activity of the polymannosylated cystatin C was 50% of the initial activity after incubation with cathepsin D for 40 minutes, while that of the nonglycosylated one was rapidly inactivated with increasing incubation time. The polysaccharide chain could disturb the physical accessibility of cathepsin D to the proteolytic cleavage sites in the polymannosylated cystatin (81). Since it is known that cysteine proteinase inhibitors are the most active agents compared with other seryl- and aspartyl-proteinases inhibitors (4), improvement in the conformational stability against cathepsin D should play a key role in the maintenance of the gel strength of surimi gel. Thus, enhancement of the molecular stability of cystatin C is required for protection against cathepsin D, such as that by genetic glycosylation.

The recombinant cystatin with a polymannosyl chain was added to roe-herring surimi for preventing gel weakening due to autolysis during cooking. As a result, proteolysis of myosin heavy chain in the surimi was effectively suppressed, while cooking at 90°C for 20 minutes after preincubation at 40°C for 30 minutes. The gel strength of roe-herring surimi was drastically improved by adding 10 μg inhibitors per gram surimi, while the herring surimi did not form a gel strong enough to be measured by the punch test without these inhibitors (81). It was assumed that the gel formation was interfered by proteolysis of myosin heavy chain, due to sufficient proteinases remaining in the herring surimi. As shown in Figure 12.2, the herring surimi gel with the polymannosyl cystatin has 2.5 times higher gel strength than that of the nonglycosylated cystatin (81). In conclusion, the glycoprotein was effective in maintaining good gelling properties in roe-herring surimi because of its inhibitory activity for sulfhydryl proteinases that enhanced molecular stability. The recombinant cystatin may also be useful for making kamaboko from surimi derived from other under-utilized fish.

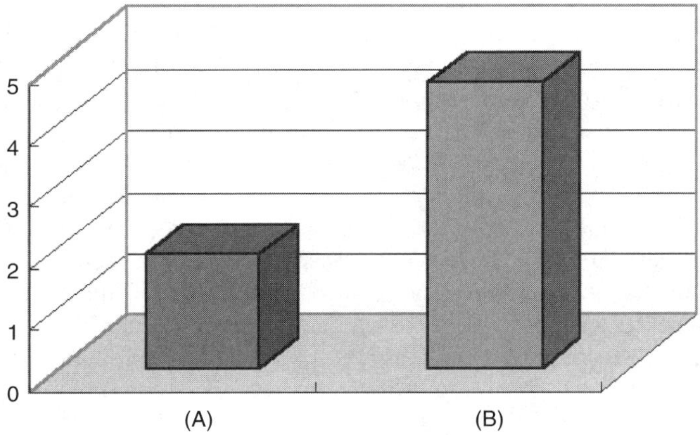

FIGURE 12.2 Gel strength of heat-induced herring surimi gels. Breaking force of the gels was measured by the punch test using TA-XT2 texture analyzer (Stable Micro Systems, Surrey, England) equipped with a 5-mm cylinder plunger. The breaking strength represented as the load value (N) at the breaking point when the surimi gel was compressed at a crosshead speed of 30 mm min^{-1}. Gel strength was determined by multiplying the breaking strength (N) and deformation (mm). A and B represent gel strengths of the surimi products containing nonglycosylated and polymannosylated cystatin Cs , respectively.

12.3.2 Anti-*Salmonella* and Anti-Rotavirus Effects of Genetically Glycosylated Cystatin

Cysteine proteinases play important roles in the pathology derived from a wide variety of organisms such as staphylococci, streptococci, fungi, arthropods, nematodes, protozoa, intestinal flagellates, trypanosomes, and also certain viruses, including polio, herpes, corona, and HIV (human immunodeficiency virus); subsequently their counterparts, cystatins, are regarded as significant for the treatment and prevention of diseases caused by these microbes (49,84–94). Human cystatin C must have anti-*Salmonella* and anti-rotavirus effects. Thus, an investigation has been designed to reduce the risk of food and waterborne illness from *Salmonella enteritidis* (95), which emerged as an egg-associated pathogen in the late 20th century, is associated with a significant number of human illnesses, and continues to be a public health concern. Rotavirus infection has been associated with a high rate of infant mortality totaling over one million in the developing countries and are now the leading cause of dehydrating diarrhea throughout the world (96–98).

A synthetic gene encoding human cystatin C (HCC) was constructed on the basis of its amino acid sequence and *Pichia pastoris* preferred codons (99). Four oligonucleotides were synthesized, hybridized, and ligated, including a *Xho*-I site at 5′-end and a *Xba*-I site at the 3′-end (95). The resulting synthetic gene was cut with *Xho*-I and *Xba*-I, and ligated into a *Pichia pastoris* expression vector, pPICZα-A (Invitrogen, Carlsbad, CA). Site-directed mutagenesis for the creation of the Asn_{35}–Lys_{36}–Ser_{37} sequence (A37S), to introduce the N-linked glycosylation site at position 35, was performed according to the QuikChange Site-Directed Mutagenesis Kit (Stratagene, La Jolla, CA). Oligonucleotide primers used for the mutation were 5′-GGTGAGTACAACAAGTCCTCTAACGACATG-3′ and its antisense. The resulting pPICZα-A vector carrying the A37S gene was linearized with *Dra*-I and transformed *P. pastoris* X-33. A single colony of transformants was inoculated into 2 ml of YPD medium and incubated at 30°C and with shaking at 250 rpm. Cells were grown in 60 ml of YPD. Following the collection of cells by centrifugation at 1500 $\times$ g, the resultant cell pellet yeast was suspended in 24 ml of BMMY (1% yeast extract, 2% peptone, 100 mM potassium phosphate buffer at pH 7, 1.34 % yeast nitrogen base, 4 $\times$ 10^{-5} % biotin, and 0.5% methanol as the sole carbon source). Methanol was added to a final concentration of 0.5% every 24 hours to maintain induction. After 2-days' induction with methanol, recombinant proteins were purified by hydrophobic interaction chromatography using Hiprep 16/10 Phenyl low-subcolumn (Amersham Pharmacia Biotech, Japan, Tokyo) followed by size-exclusion chromatography using Superdex 200 HR 10/30 column (Amersham Pharmacia Biotech). Glycosylated protein was finally purified using affinity chromatography using concanavalin A-Sepharose (Amersham Pharmacia Biotech). As shown in Figure 12.3, A37S mutant HCC was oligoglycosylated with an N-linked polymannosyl chain. Oligomannosylated HCC with a carbohydrate chain of $Man_{10}GlcNAc_2$ produced by the *Pichia* transformant was stable against proteolysis with an aspartic protease, cathepsin D, and at the lower ionic strength condition of 50 mM

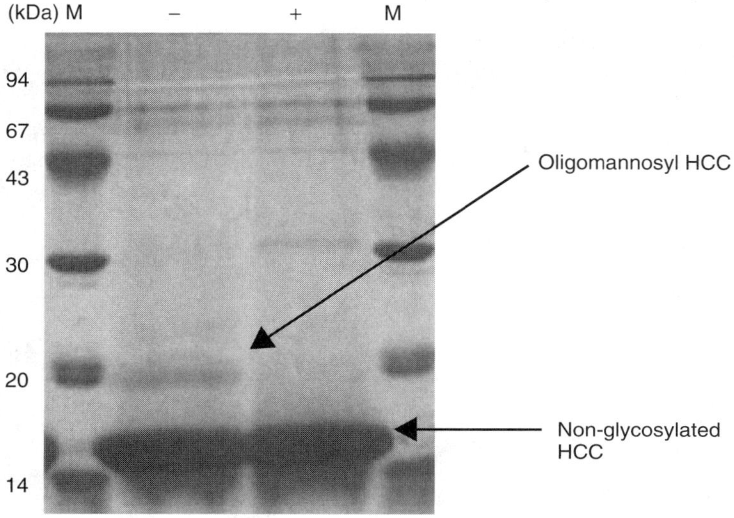

FIGURE 12.3 SDS-poly acrylamide gel electrophoresis patterns of A37S mutant human cystatin C. A37 human cystatin C was treated with (+) and without (−) endo-α-N-acetyl-glucosaminidase (endo-H), and then subjected to SDS-PAGE. The gel sheet was stained with Coomassie brilliant blue. *M* represents molecular weight marker.

potassium phosphate buffer, pH 7, without any additional NaCl, compared with the wild-type HCC.

Anti-*Salmonella* and anti-rotavirus activities of the oligomannosyl HCC were investigated using *Salmonella enteritidis (inv* A) and human rotavirus type-2 Wa strain (95). One hundred microgram of the protein was dissolved in 1 ml of minimum essential medium (MEM). Samples were further diluted to contain 75 μg ml^{-1}, 50 μg $ml^{-1}$and 25 μg ml^{-1}. CV-1 and MA-104 cells were used as host cells for *S. enteritidis* and rotavirus Wa, respectively, which are both kidney cell lines from African green monkeys. The mammalian cells were regularly subcultured on MEM containing 10% fetal calf serum and incubated under 5% carbon dioxide (CO_2). They did not show any significant cytotoxicity at a concentration of 100 μg ml^{-1}. The antiinvasive activity was assessed by an invasion system with CV-1 cells between passages 1 to 20 using mid-log-phase bacteria as a challenge. CV-1 cells were disrupted with Triton-X after invasion, and then the internalized *Salmonella* that was released from the host cells was serially diluted and cultured on DHL agar plate. After incubation at 37°C for 24 hours, the colonies that appeared on the plate were counted to compare the negative control. Results were recorded as a percentage of the invasion. Antiinvasion effects of *Salmonella* showed a first-order dependence on the concentrations of both wild-type and oligomannosylated HCCs (Table 12.3). The effect of the oligomannosyl HCC was significantly ($p > .05$) stronger than that of wild-type HCC. In contrast, MTT assay was used to determine antirotavirus effects of the recombinant HCCs. As

TABLE 12.3
Anti-*Salmonella* Effects* of Wild-Type and Oligomannosyl Human Cystatin Cs

	Protein concentration (μg ml^{-1})			
	25	50	75	100
Control (without inhibitor)	100	100	100	100
Wild-type HCC	80.1 ± 1.8	62.9 ± 1.6	41.1 ± 1.2	26.5 ± 1.4
Oligomannosyl HCC	63.2 ± 2.5	50.3 ± 1.5	21.5 ± 1.5	11.7 ± 1.2

*Data indicate percentages of bacterial numbers detected from the host CV-1 cells under the coexisting systems with wild-type or oligomannosyl human cystatin Cs to that from the control system without these inhibitors.

TABLE 12.4
Anti-Rotavirus Effects* of Wild-Type and Oligomannosyl Human Cystatin Cs

	Protein concentration (μg ml^{-1})			
	25	50	75	100
Control (without rotavirus)	100	100 ± 0.2	99.7 ± 0.3	99.7 ± 0.5
Wild-type HCC	39.5±1.5	60.3 ± 1.2	68.1 ± 0.8	76.7 ± 1.4
Oligomannosyl HCC	65.2±1.7	77.5 ± 1.3	90.8 ± 1.6	93.1 ± 1.1

*Data indicate percentages of viabilities of host MA-104 cells to that of in the control system without rotavirus and with 25 μg ml^{-1} wild-type HCC.

the result, more than 80% viability of the host cell infected with 1.0×10^5 ml^{-1} of rotavirus was conserved under the condition coexisting with 75 μg $ml^{-1}$of the oligomannosyl HCC and was 15.2% higher than that of wild-type HCC, as shown in Table 12.4. Thus, the *in vitro,* anti-*Salmonella,* and anti-rotavirus assays indicated that the supplement of a proper amount of the oligomannosyl HCC can be used as anti-*Salmonella* and anti-rotavirus agents.

12.3.3 Anticancer Effects of Double-Mutated Cystatin

Recently, a new human cystatin named cystatin E/M, with 40% homology to human cystatin C, was discovered (36). Since no cystatin E/M was detected in metastatic mammary epithelial tumor cells, the loss of expression of the inhibitor is likely associated with the progression of a primary tumor to metastatic phenotype (36). In human colorectal cancer, it has been reported that differences in the suppression ratio of cathepsins and cystatin C between noncancer and cancer tissues occurs especially in the early stages of cancer progression (100,101). Studies have suggested that latent cathepsins also participate in the invasion and metastasis of cancer cells through the degradation of extracellular matrix (101,102). Cathepsins can play an important role in modulating cancer cell growth, invasion, and metastasis (103). Recent studies have also suggested that cystatins induce synthesis of cytokines such as TNF-α, interleukins-6 and -10 (104,105). These findings led us to examine the effect of proteinase inhibitors on human colon carcinoma cells.

Meanwhile, a novel and unique computer program, random-centroid optimization for genetics (RCG), has been designed to select mutation which is optimal for improving the stability of biologically active protein (106). RCG seeks both mutation sites and amino acids for substitution to improve a given protein without any preliminary knowledge about its conformation. RCG was thus applied to two-site-directed mutagenesis of human cystatin C to increase its stability and activity. A favorable variant, the double-mutated human cystatin C, G12W/H86V that has five times the activity of wild-type HCC, was successfully created after 23 trials (107). Growth-inhibition and anti-invasion activities toward human colon carcinoma cell lines of G12W/H86V were investigated using Caco-2 and HCT-116 cells. The mutated inhibitor caused more than 10% growth inhibition of Caco-2 cells at 5.6 nM. *In vitro* anti-invasion test using HCT-116 cells showed that the recombinant inhibitor significantly suppressed the cell invasion by 15% (Table 12.5), while its wild–type cystatin, aspartic proteinase inhibitor pepstatin A, and matrix metalloproteinase inhibitor MMP-I did not. G12W/H86V showed a significantly strong activity of inhibiting cathepsin L, which may cause the suppression of growth and invasion of the human colon carcinoma cells. It is well known that a balance of proteinase and its inhibitor is of great importance for changes in malignant cancer cell viability. High concentrations of proteinase inhibitors may induce adverse side effects. Synthetic anticancer agents have been approved for treating patients; however, they frequently cause troublesome side effects on patients. Alternatively, natural proteinase inhibitors such as recombinant cystatin Cs could have acceptable use for therapeutic or preventive purposes because of their biodegradability.

TABLE 12.5
Anticancer Effects Toward Caco-2 Cells and HCT-116 cells

Control	Wild-type	G12W/H86V	
Invasion index[*1]	1.1 ± 0.02^{a}	0.99 ± 0.03	0.93 ± 0.02^{b}
Cathepsin B activity[*2]	65 ± 0.5^{a}	59 ± 0.4^{b}	60 ± 0.5^{b}
Cathepsin L activity[*2]	43 ± 0.3^{a}	44 ± 0.3	35 ± 0.3^{b}

[*1] Effects of protease inhibitors on matrigel invasion by HCT-116 cells. The invasion of HCT-116 cells was set at 1 (control), and the ratios of the invasion in the presence of various inhibitors (5.6 nM) to the control are represented as mean ± SD (n = 3).

[*2] Cellular cathepsin activity in Caco-2 cells treated with cystatins. Cathepsin B and cathepsin L activities were determined by substrates Z-Arg-Arg-Nmec and Z-Phe-Arg-NMec, respectively. Data represent as mean ± SD (n = 3).

[a-c] Means within the same column with different superscript are significantly different ($p < .05$).

12.4 HIGH-LEVEL PRODUCTION FOR FURTHER APPLICATIONS

Since there are many useful applications for the members of the cystatin superfamily, development of low-cost expression systems of these proteins will be required for medical treatments. The concept of chemoprevention is based on diet, which makes low cost an essential property of these agents. We chose two expression systems for large productions of cystatins, that of the *Pichia* expression system and the *Bacillus* expression system, because production using higher eukaryotes such as silk worm and mammalian cell may be excessively expensive.

The high-level expression of human cystatin C in *P. pastoris* has been attempted using an Inceltech LH bioreactor with a two-liter working volume and controlled modules of pH, temperature, and dissolved oxygen, as shown in Figure 12.4. *Pichia* transformant was precultured for 24 hours at 30°C in shake flasks (250 rpm) in YPD medium (10 g l^{-1} yeast extract, 20 g l^{-1} peptone, and 20 g l^{-1} glycerol), and then directly added to one liter of Basal Salts Medium (26.7 ml l^{-1} 85% o-phosphoric acid, 0.93 g l^{-1} calcium sulfate·$2H_2O$, 18.2 g l^{-1} potassium sulfate, 14.9 g l^{-1} magnesium sulfate·$7H_2O$, 4.13 g l^{-1} potassium hydroxide, 40 g l^{-1} glycerol) containing 4.35 ml l^{-1} of trace-salt solution (6 g l^{-1} cupric sulfate·$5H_2O$, 0.08 g l^{-1} sodium iodide, 3 g l^{-1} manganese sulfate·H_2O, 0.2 g l^{-1} sodium molybdate·$2H_2O$, 0.02 g l^{-1} boric acid, 0.5 g l^{-1} cobalt chloride, 20 g l^{-1} zinc chloride, 65 g l^{-1} ferrous sulfate·$7H_2O$, 5 ml l^{-1} sulfuric acid, and 0.2 g l^{-1} D-biotin), where the pH was adjusted to 5 by the addition of 28% (w/w) ammonium hydroxide. The dissolved oxygen concentration was above 30% of air saturation. A methanol feed was also maintained for as long as 96 hours at a constant rate of 2.25 ml l^{-1} h^{-1}. As a result, an average yield of recombinant HCCs was found to be 200 mg l^{-1}, although maximization of the expressed proteins still has some room for improvement by changing conditions, for example, addition of more glycerol while setting pH, oxygen amount, and temperature constant (107).

The methylotrophic yeast, *P. pastoris* is known for its high-level expression of heterologous proteins (108,109). pPCIZ vectors (Invitrogen, Carlsbad, CA) are tightly regulated by alcohol oxidase-1 (AOX-1) gene promoter so that heterogeneous proteins can be induced by the presence of alcohol (110). It is well known that there are two phenotypes, Mut^+ and Mut^s (methanol utilization slow), for *Pichia* transformant with the vectors. We used a Mut^s transformant that exhibits slower growth on methanol, compared with the wild strain (Mut^+), because it will not require higher dissolved oxygen. The protein expression with Mut^s strains requires long induction times (approximately 100 hours) for maximal protein expression (111), which may cause some negative results for the amyloidogenic protein, human cystatin C (46). Thus, an investigation was conducted to compare with two expression systems for *Pichia pastoris*, by using a pPCIZα vector with an AOX promoter and pGAPZα vector with a glyceraldehydes-3-phosphate dehydrogenase (GAP) promoter, in order to obtain large amounts of recombinant human stefin A belonging to family I of the cystatin superfamily. The protein is a monomeric protein with 98 amino acid residues and no disulfide bond and is also

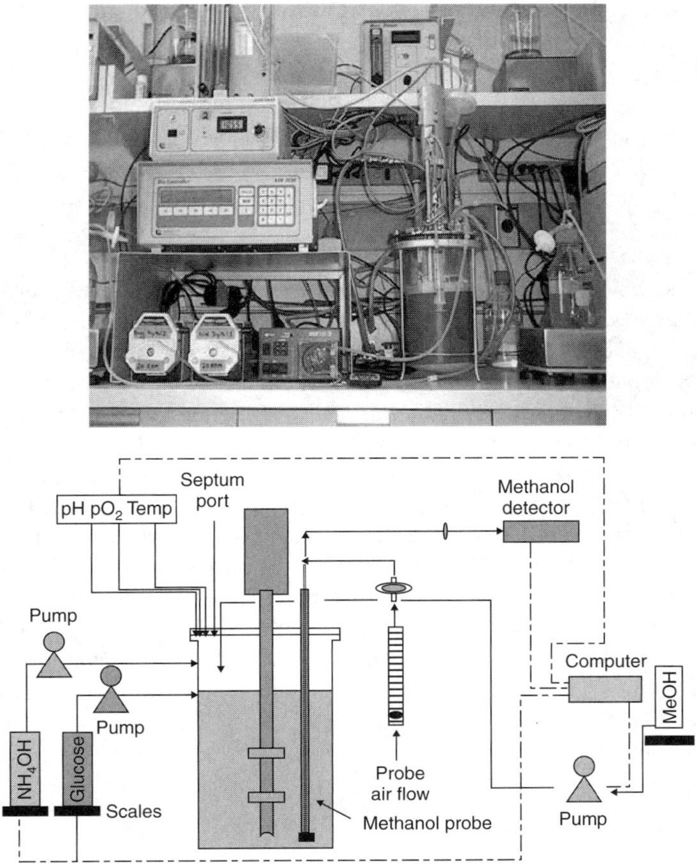

FIGURE 12.4 Picture (*left*) and a schematic diagram (*right*) of computed fermentation system with methanol sensor.

well recognized as amyloidgenic (112). As the result, *P. pastoris* X-33 transformed with pGAPZα-A vector produced 120 mg l^{-1} pGAPZα-A as an active form by the general shaking flask procedure for 3 days, while the yield from the *Pichia* transformant with pPICZα-A vector was less than 20 mg l^{-1}. Unique characteristics of the expression system using the pPICZα-A vector requiring methanol in the culture media may play an important role in the suppressed production of the amyloidgenic protein in *Pichia* transformants. A pilot-plant application using a fermentor to obtain high cell densities and product concentration has been anticipated for the production of the recombinant human stefin A (112).

Developing *Bacillus subtilis* strains that export heterologous proteins into the extracellular culture medium has been of considerable interest in protein engineering. Human interleukin-2 (113), human granzyme K (114), interferon (115), and human growth hormone (116) have been secreted into the medium by fusing the structural genes of these proteins and signal sequence coding region of the genes from the *Bacillus* species. Recently, we have also been successful in expressing cystatin in *B. subtilis.* Unlike the expression in *Escherichia coli,* which forms inclusion body, *Bacillus* has only cytoplasmic membrane, which means that the expressed proteins can be directly released into the culture medium (117). The method using nonpathogenic bacteria has great potential in the large-scale production of premium proteins. Thus, the establishment of a *Bacillus* expression system was tried using the human cystatin C homologue gene (107). For the construction of the *Bacillus* expression vector, a 437 bp *Nco*-1/*Xba*-1 fragment of the human cystatin C structural region was ligated into the multicloning site of pUC-19 in order to create *Alw-44*1 and *Mul*-1 site using polymerase chain reaction (PCR) mutagenesis. The resulting 0.5 kbp *Alw-44*1/*Mul*-1 fragment was inserted into a 5.8-kbp template of pDMP vector containing the entire promoter and pre-pro-region of the *B. stearothermophilus* neutral proteinase gene (118). The fusion vector between the sequences encoding the human cystatin C and the signal peptide and pro-region from the neutral proteinase was transformed with *B. subtilis* ANA-1 as described by Harday (119). Kanamycin-resistant transformants were selected and cultivated in SD medium containing 50 mM of MEM (117) and extracellular protein was collected from a CM-papain Sepharose affinity column. Our preliminary experiments showed that recombinant cystatin was successfully overexpressed in the *Bacillus* culture media with a 10 mg ml^{-1} yield. Unfortunately, the activity of the *Bacillus* cystatin C was, however, considerably lower than that of the *Pichia* cystatin C. The approach (106) used for optimizing the thermal stability of *B. stearothermophilus* neutral proteinase could be useful for improving the properties of the recombinant cystatin.

12.5 CONCLUSIONS

The cystatin superfamily includes a number of chicken cystatin-like cysteine proteinase inhibitors that are widely distributed in mammalian tissues and body fluids (11). Family I is involved in the control of the intracellular protein breakdown system, family II is suggested to conduct defense against the invasion of pathogens and parasites, and family III is kininogen itself. The cystatin superfamily is assumed to have great potential applications in foods and pharmaceuticals. Prospective nutraceutical usages of the cysteine proteinase inhibitors produced by genetic techniques were presented in this chapter.

Many studies suggest that considerably high-level activities of cathepsins remain in surimi sol after the grinding process, are stable, and are very difficult to be simply inactivated or removed during surimi processing (4,5). Accordingly, an application was conducted to produce good surimi products from roe-herring by using a genetically glycosylated cystatin C with molecular stability against heating and proteolysis. The recombinant cystatin showed potent inhibitory effect on the digestion of myosin heavy chains in roe-herring surimi, resulting in the improvement of gel formation. The protein inhibitor could be used as a gel-strengthening agent for surimi products from underutilized fish such as roe-herring and late-run salmon.

The cysteine proteinase inhibitors have been anticipated to be important in protecting the host from invading organisms and tissue destruction caused by endogenous cysteine proteinases during maladies such as inflammation, tumor growth, and metastasis (84–94,100–103). Therefore, the biological activities of recombinant cystatins were investigated on anti-rotavirus, anti-*Salmonella*, and anticancer effects. As a result, it has been revealed that oligomannosyl human cystatin C show significantly strong anti-invasive actions against both rotavirus and *S. enteritidis* (*inv* A). A double-mutated human cystatin C also showed growth inhibition effects toward Caco-2 cells and *in vitro* anti-invasion effects toward HCT-116 cells.

Since human cystatin C showed some capabilities as a therapeutic protein for treating viral and bacterial infections, cancers, as well as a quality controlling agent for fish surimi product, large-scale productions of recombinant cystatins have been designed. A computer-controlled 2-liter-scale fermentor produced 200 mg l^{-1} of recombinant human cystatin C, while 120 mg l^{-1} recombinant human stefin A was produced from a *Pichia pstoris* transformed with pGAPZ vector by the general shake flasks. The establishment of the *Bacillus* expression system will afford us a bright and successful future for high-level expression of the protein inhibitor. In conclusion, production using genetic fermentation of nutraceutical proteins may be the most efficient and cost-effective method. Once the production costs have been successfully reduced, the use of these materials as quality control agents as well as chemopreventives in the diet becomes feasible.

ACKNOWLEDGMENT

Most results with respect to the bioactivities of cystatins are from the collaborative works with Drs. Shuryo Nakai and Masahiro Ogawa (University of British Columbia), supported by both NSERC (Natural Sciences and Engineering Research Council) of Canada and JSPS (Japan Society of the Promotion of Science). The author is indebted to them.

REFERENCES

1. R. Colella, A.F. Chambers, D.T. Denhardt. Anticarcinogenic activates of naturally occurring cysteine proteinase inhibitors. In: W. Toroll, A.R. Kennedy (Ed.) *Protease Inhibitors as Cancer Chemopreventive Agents*. New York and London: Plenum Press, 1993, 199–214.
2. F. Lecaille, J. Kaleta, D. Bromme. Human and parasitic papain-like cysteine proteases: their role in physiology and pathology and recent developments in inhibitor design. *Chem. Rev.*, 102: 4459–88, 2002.
3. A.J. Barrett, H. Kirschke. Cathepsin B, cathepsin H, and cathepsin L. *Meth. Enzymol.*, 80: 535–561, 1981.
4. S.T. Jiang, J.J. Lee, H.C. Chen. Purification and characterization of a novel cysteine proteinase from mackerel (*Scomber australasicus*). *J. Agric. Food Chem.*, 42: 1639–1646, 1994.

5. J.J. Lee, H.C. Chen, S.T. Jiang. Comparison of the kinetics of cathepsins B, L, L-like and X from the dorsal muscle of mackerel on the hydrolysis of methylcoumarylamind substrates. *J. Agric. Food Chem.*, 44: 774–778, 1996.
6. K. Fossum, J.R. Whitaker. Ficin and papain inhibitor from chicken egg white. *Arch. Biochem. Biophys.*, 125: 367–375, 1968.
7. H. Keilova, V. Tomasek. Effect of papain inhibitor from chicken egg white on cathepsin B1. *Biochem. Biophys. Acta*, 334: 179–186, 1974.
8. H. Keilova, V. Tomasek. Inhibition of cathepsin C by papain inhibitor from chicken egg white and by complex of this inhibitor with cathepsin B1. *Coll. Czech. Chem. Commn.*, 40: 218–224, 1975.
9. A.J. Barrett. Cystatin, the egg white inhibitor of cysteine proteinases. *Meth. Enzymol.*, 80: 771–778, 1981.
10. I. Ohkubo, K. Kurachi, T. Takasawa, H. Shiokawa, M. Sasaki. Isolation of a human cDNA for a thiol proteinase inhibitor and its identity with low molecular weight kininogen. *Biochemistry*, 23: 5691–5697, 1984.
11. A.J. Barrett, H. Fritz, A. Grubb, S. Isemura, M. Jarvinen, N. Katunuma, W. Machleidt, W. Muller-Esterl, M. Sasaki, V. Turk. Nomenclature and classification of the proteins homologous with the cysteine-proteinase inhibitor chicken cystatin. *Biochem. J.*, 236: 312, 1986.
12. W.M. Brown, K.M. Dziegielewska. Friends and relation of the cystatin superfamily — new members and their evolution. *Prot. Sci.*, 6: 5–12, 1997.
13. A. Grubb, H. Lofberg, A.J. Barrett. The disulphide bridges of human cystatin C (g-trace) and chicken cystatin. *FEBS Lett.*, 170: 370–374, 1984.
14. V. Turk, W. Bode. The cystatins: protein inhibitors of cysteine proteinases. *FEBS Lett.*, 285: 213–219, 1991.
15. A.J. Barrett, N.D. Rawlings, M.E. Davies, W. Machleidt, G. Salvesen, V. Turk. Cysteine protease inhibitors of the cystatin superfamily. In: J.J. Barrett, G. Salvesen, (Ed.) *Proteinase Inhibitors.* Amsterdam: Elsevier, 1986, 515–569.
16. J. Brzin, M. Kopitar, V. Turk, W. Machleidt. Isolation and characterization of stefin, a cytosolic inhibitor of cysteine proteinase from human polymorphonuclear granulocytes. *Hoppe-Seylers Z. Physiol. Chem.*, 363: 1475–1480, 1983.
17. W. Machleidt, U. Borchart, H. Fritz, J. Brzin, A. Ritonja, V. Turk. Protein inhibitors of cysteine proteinases. II. Primary structure of stefin, a cytosolic protein inhibitor of cysteine proteinases from human polymorphonuclear granulocytes. *Hoppe-Seylers Z. Physiol. Chem.*, 364: 1481–1486, 1983.
18. G.D.J. Green, A.A. Kembhavi, M.E. Davies, A.J. Barrett. Cysteine-like cysteine proteinase inhibitors from human liver. *Biochemistry*, 218: 939–946, 1984.
19. M. Jarvinen, A. Rinne. Human spleen cysteine proteinase inhibitor. Purification, fractionation into isoelectric variants and some properties of the variants. *Biochim. Biophys. Acta*, 708: 210–217, 1982.
20. K. Takio, K. Kiminami, Y. Bando, N. Katunuma, K. Titani. Amino acid sequence of rat epidermal thiol proteinase inhibitor. *Biochem. Biophys. Res. Commn.*, 121: 149–154, 1984.
21. J.F. Lenney, J.R. Tolan, W.J. Sugai, A.G. Lee. Thermostable endogenous inhibitors of cathepsins B and H. *Eur. J. Biochem.*, 101: 153–161, 1979.
22. K. Takio, E. Kominami, N. Wakamatsu, N. Katunuma, K. Titani. Amino acid sequence of rat liver thiol proteinase inhibitor. *Biochem. Biophys. Res. Commn.*, 115: 902–908, 1983.

23. M. Jarvinen, A. Rinne, V.K. Hopsu-Havu. Human cystatins in normal and diseased tissues — a review. *Acta, Histochem.*, 82: 5–18, 1987.
24. A.J. Barrett, M.E. Davies, A. Grubb. The place of human gamma-trace (cystatin C) amongst the cysteine proteinase inhibitors. *Biochem. Biophys. Res. Commn.*, 120: 631–636, 1984.
25. W. Bode, R. Engh, D. Musil, U. Thiele, R. Huber, A. Karshikov, J. Brzin, J. Kos, V. Turk. The 2.0 A X-ray crystal structure of chicken egg white cystatin and its possible mode of interaction with cysteine proteinases. *EMBO J.*, 7: 2593–2599, 1988.
26. A. Grubb, H. Lofberg. Human gamma-trace, a basic microprotein: amino acid sequence and presence in the adenohypophysis. *Proc. Natl. Acad. Sci. USA*, 79: 3024–3027, 1982.
27. A. Esnard, F. Esnard, D. Faucher, F. Gauthier. Two rat homologues of human cystatin C. *FEBS Lett.*, 236: 475–478, 1988.
28. Y-J. Tsai, G-D. Chang, C-J. Hwang, Y-S. Chang, F-L. Hwang. Purification and molecular cloning of carp ovarian cystatin. *Comp. Biochem. Physiol.*, 113B: 573–580, 1996.
29. M. Yamashita, S. Konagaya. Molecular cloning and gene expression of chum salmon cystatin. *J. Biochem.*, 120: 483–487, 1996.
30. S. Isemura, E. Saitoh, K. Sanada. Isolation and amino acid sequence of SAP-1, an acidic protein of human whole saliva, and sequence homology with human gamma-trace. *J. Biochem. (Tokyo)*, 96: 489–498, 1984.
31. S. Isemura, E. Saitoh, K. Sanada. Characterization of a new cysteine proteinase inhibitor of human saliva, cystatin SN, which is immunologically related to cystatin S. *FEBS Lett.*, 198: 145–149, 1986.
32. S. Isemura, E. Saitoh, K. Sanada. Characterization and amino acid sequence of a new acidic cysteine proteinase inhibitor (cystatin SA) structurally closely related to cystatin S, from human whole saliva. *J. Biochem. (Tokyo)*, 102: 693–704, 1987.
33. L.A. Bobek, A. Aguirre, M.J. Levine. Human salivary cystatin S. Cloning, sequence analysis, hybridization in situ and immunocytochemistry. *Biochem. J.*, 278: 627–635, 1991.
34. J. Ni, M. Abrahamson, M. Zhang, M.A. Fernandez, A. Grubb, J. Su, G.L. Yu, Y. Li, D. Parmelee, L. Xing, T.A. Coleman, S. Gentz, R. Thotakura, N. Nguyen, M. Hesselberg, R. Gentz. Cystatin E is a novel human cysteine proteinase inhibitor with structural resemblance to family 2 cystatins. *J. Biol. Chem.*, 272: 10853–10858, 1997.
35. G. Sotiropoulou, A. Anisowicz, R. Sager. Identification, cloning, and characterization of cystatin M, a novel cysteine proteinase inhibitor, down-regulated in breast cancer. *J. Biol. Chem.*, 272: 903–910, 1997.
36. P.L. Zeeuwen, I.M. van Vlijmen-Willems, B.J. Jansen, G. Sotiropoulou, J.H. Curfs, J.F. Meis. Cystatin M/E expression is restricted to differentiated epidermal keratinocytes and sweat glands: a new skin-specific proteinase inhibitor that is a target for cross-linking by transglutaminase. *J. Invest. Dermatol.*, 116: 693–701, 2001.
37. J.P. Freije, M. Balbin, M. Abrahamson, G. Velasco, H. Dalboge, A. Grubb, C. Lopez-Otin. Human cystatin D. cDNA cloning, characterization of the *Escherichia coli* expressed inhibitor, and identification of the native protein in saliva. *J. Biol. Chem.*, 268: 15737–15744, 1993.
38. A. Hall, I. Ekiel, R.W. Mason, F. Kasprzykowski, A. Grubb, M. Abrahamson. Structural basis for different inhibitory specificities of human cystatins C and D. *Biochemistry*, 37: 4071–4079, 1998.

39. J. Ni, M.A. Fernandez, L. Danielsson, R.A. Chillakuru, J. Zhang, A. Grubb, J. Su, R. Gentz, M. Abrahamson. Cystatin F is a glycosylated human low molecular weight cysteine proteinase inhibitor. *J. Biol. Chem.*, 273: 24797–24804, 1998.
40. M. Hirado, S. Tsunasawa, F. Sakiyama, M. Niinobe, S. Fujii. Complete amino acid sequence of bovine colostrum low-Mr cysteine proteinase inhibitor. *FEBS Lett.*, 186: 41–45, 1985.
41. A. Ritonja, H.J. Evans, W. Machleidt, A.J. Barrett. Amino acid sequence of a cystatin from venom of the African puff adder (*Bitis arietans*). *Biochem. J.*, 246: 799–802, 1987.
42. K. Shoemaker, J.L. Holloway, T.E. Whitmore, M. Maurer, A.L. Feldhaus. Molecular cloning, chromosome mapping and characterization of a testis-specific cystatin-like cDNA, cystatin T. *Gene*, 245: 103–108, 2000.
43. G.A. Cornwall, M.C. Orgebib-Crist, S.R. Hann, The Cres gene: a unique testis-regulated gene related to the cystatin family is highly restricted in its expression to the proximal region of the mouse epididymis. *Mol. Endocrin.*, 6: 1653–1664, 1992.
44. V. Tohonen, C. Osterlund, K. Nordqvist, Testatin: a cystatin-related gene expressed during early testis development. *Proc. Natl. Acad. Sci. USA*, 95: 14208–14213, 1998.
45. T. Sueyoshi, K. Enjyoji, T. Shimada, H. Kato, S. Iwanaga, Y. Bando, E. Kominami, N. Katunuma. A new function of kininogens as thiol-proteinase inhibitors: inhibition of papain and cathepsins B, H and L by bovine, rat and human plasma kininogens. *FEBS Lett.*, 182: 193–195, 1985.
46. M. Abrahamson, A.J. Barrett, G. Salvesen, A. Grubb. Isolation of six cysteine proteinase inhibitors from human urine. Their physicochemical and enzyme kinetic properties and concentrations in biological fluids. *J. Biol. Chem.*, 261: 11282–11289, 1986.
47. P.A. Shaw, T. Barka. Beta-adrenergic induction of a cysteine-proteinase-inhibitor mRNA in rat salivary glands. *Biochem. J.*, 257: 685–689, 1989.
48. P.A. Shaw, T. Barka, A. Woodin, B.S. Schacter, J.L. Cox. Expression and induction by beta-adrenergic agonists of the cystatin S gene in submandibular glands of developing rats. *Biochem. J.*, 265: 115–120, 1990.
49. L. Bjorck, P. Akesson, M. Bohus, T. J. Rojnar, M. Abrahamson, I. Olafsson, A. Grubb. Bacterial growth blocked by a synthetic peptide based on the structure of a human proteinase inhibitor. *Nature*, 337: 385–386, 1989.
50. Y. Takagaki, N. Kitamura, S. Nakanishi. Cloning and sequence analysis of cDNAs for human high molecular weight and low molecular weight prekininogens. Primary structures of two human prekininogens. *J. Biol. Chem.*, 260: 8601–8609, 1985.
51. I. Ohkubo, K. Kurachi, T. Takasawa, H. Shiokawa, M. Sasaki. Isolation of a human cDNA for alpha 2-thiol proteinase inhibitor and its identity with low molecular weight kininogen. *Biochemistry*, 23: 5691–5697, 1984.
52. G. Salvesen, C. Parkes, M. Abrahamson, A. Grubb, A.J. Barrett. Human low-Mr kininogen contains three copies of a cystatin sequence that are divergent in structure and in inhibitory activity for cysteine proteinases. *Biochem. J.*, 234: 429–434, 1986.
53. T. Cole, A. Inglis, M. Nagashima, G. Schreiber. Major acute-phase alpha (1) protein in the rat: structure, molecular cloning, and regulation of mRNA levels. *Biochem. Biophys. Res. Commn.*, 126: 719–724, 1985.
54. S. Arai, I. Matsumoto, Y. Emori, K. Abe. Plant seed cystatins and their target enzymes of endogenous and exogenous origin. *J. Agric. Food Chem.*, 50: 6612–6617, 2002.

55. K.M. Dziegielewska, K. Mollgard, M.L. Reynolds, N.R. Saunders. A fetuin-related glycoprotein (α_2-HS) in human embryonic and fetal development. *Cell Tissue Res.*, 248: 33–41, 1987.
56. F.A. Elzanowski, W.C. Barker, L.T. Hunt, E. Seibel-Ross. Cystatin domains in alpha-2-HS-glycoprotein and fetuin. *FEBS Lett.*, 227: 167–170, 1988.
57. K.M. Dziegielewska, W.M. Brown, S.-J. Casey, D.L. Christie, R.C. Foreman, R.M. Hill, N.R. Saunders. The complete amino acid and nucleotide sequence of bovine fetuin: its sequence homology with α_2-HS glycoprotein and its relationship to other members of the cystatin superfamily. *J. Biol. Chem.*, 265: 4354–4357, 1990.
58. T. Koide, D. Foster, S. Yoshitake, E.W. Davie. Amino acid sequence of human histidine-rich glycoprotein derived from the nucleotide sequence of its cDNA. *Biochemistry*, 25: 2220–2225, 1986.
59. M.G. Parker, G.T. Scrace, W.I.A. Mainwaring. Testosterone regulates the synthesis of major proteins in rat ventral prostate. *Biochemistry*, 170: 115–121, 1978.
60. A. Devos, N. de Clercq, I. Vercaeren, W. Heyns, W. Rombauts, B. Peeters. Structure of rat genes encoding androgen-regulated cystatin-related proteins (CRPs): a new member of the cystatin superfamily. *Gene*, 125: 159–167, 1993.
61. T. Suzuki, S. Natori. Purification and characterization of an inhibitor of the cysteine protease from the hemolymph of *Sacrophaga peregrina* larvae. *J. Biol. Chem.*, 260: 5115–5120, 1985.
62. M.L. Delbridge, L.E. Kelly. Sequence analysis, and chromosomal localization of a gene encoding a cystatin-like protein from *Drosophila melanogaster. FEBS Lett.*, 274: 141–145, 1990.
63. K. Abe, H. Kondo, H. Watanabe, Y. Emori, S. Arai. Oryzacystatins as the first well-defined cystatins of plant origin and their target proteinases in rice seeds. *Biomed. Biochim. Acta*, 50: 637–641, 1991.
64. K. Abe, Y. Emori, H. Kondo, K. Suzuki, S. Arai. Molecular cloning of a cysteine proteinase inhibitor of rice (oryzacystatin). *J. Biol. Chem.*, 262: 16793–16797, 1987.
65. H. Kondo, K. Abe, I. Nishimura, H. Watanabe, Y. Emori, S. Arai. Two distinct cystatin species in rice seeds with different specificities against cysteine proteinases. Molecular cloning, expression, and biochemical studies on oryzacystatin-II. *J. Biol. Chem.*, 265: 15832–15837, 1990.
66. M. Abe, K. Abe, M. Kuroda, S. Arai. Corn kernel cysteine proteinase inhibitor as a novel cystatin superfamily member of plant origin. *Eur. J. Biochem.*, 209: 933–937, 1992.
67. K.V.S. Fernandes, P.A. Sabelli, D.H.P. Barratt, M. Richardson, J. Xavier-Filho, P.R. Shewry. The resistance of cowpea seeds to bruchid beetles is not related to level of cysteine proteinase inhibitors. *Plant Mol. Biol.*, 23: 215–219, 1993.
68. T. Misaka, M. Kuroda, K. Iwabuchi, K. Abe, S. Arai. Soyacystatin, a novel cysteine proteinase inhibitor in soybean, is distinct in protein structure and gene organization from other cystatins of animal and plant origin. *Eur. J. Biochem.*, 240: 609–614, 1996.
69. A. Ojima, H. Shiota, K. Higashi, H. Kamada, Y. Shimma, M. Wada, S. Satoh. An extracellular insoluble inhibitor of cysteine proteinases in cell cultures and seeds of carrot. *Plant Mol. Biol.*, 34: 99–109, 1997.
70. K. Doi-Kawano, Y. Kouzuma, N. Yamasaki, M. Kimura. Molecular cloning, functional expression, and mutagenesis of cDNA encoding a cysteine proteinase inhibitor from sunflower seeds. *J. Biochem.*, 124: 911–916, 1998.

71. S.N. Ryan, W.A. Laing, M.T. McManus. A cysteine proteinase inhibitor purified from apple fruit. *Phytochemistry*, 49: 957–963, 1998.
72. C. Waldron, L.M. Wegrich, P.A.O. Merlo, T.A. Walsh. Characterization of a genomic sequence coding for potato multicystatin, an eight-domain cysteine proteinase inhibitor. *Plant Mol. Biol.*, 23: 801–812, 1993.
73. M. Kimura, T. Ikeda, D. Fukumoto, N. Yamasaki, M. Yonekura. Primary structure of a cysteine proteinase inhibitor from the fruit of avocado (*Persea americana* Mill). *Biosci. Biotechnol. Biochem.*, 59: 2328–2329, 1995.
74. B. Rogelj, T. Popovic, A. Ritonja, B. Strukelj, J. Brzin. Chelidocystatin, a novel phytocystatin from *Chelidonium majus*. *Phytochemistry*, 49: 1645–1649, 1998.
75. K. Nagata, N. Kudo, K. Abe, S. Arai, M. Tanokura. Three-dimensional solution structure of oryzacystatin-I, a cysteine proteinase inhibitor of the rice, *Oryza sative* L. *japonica*. *Biochemistry*, 39: 14753–14760, 2000.
76. B. Lenarcic, M. Kraöovec, A. Ritonja, I. Olafsson, V. Turk. Inactivation of human cystatin C and kininogen by human cathepsins D. *FEBS Lett.*, 280: 211–215, 1991.
77. K. Olden, B.A. Bernet, M.J. Humphries, T.-K. Yeo, K.-T. Yeo, S.L. White, S.A. Newton, H.C. Bauer, J.B. Parent. Function of glycoprotein glycans. *Trends Biochem. Sci.*, 10: 78–82, 1985.
78. J. Gu, T. Matsuda, R. Nakamura, H. Ishiguro, I. Ohkubo, M. Sasaki, N. Takahashi. Chemical deglycosylation of hen ovomucoid: protective effect of carbohydrate moiety on tryptic hydrolysis and heat denaturation. *J. Biochem.*, 106: 66–70, 1989.
79. S. Nakamura, H. Takasaki, K. Kobayashi, A. Kato. Hyperglycosylation of hen egg white lysozyme in yeast. *J. Biol. Chem.*, 268: 12706–12712, 1993.
80. S. Nakamura, M. Ogawa, S. Nakai. Effects of polymannosylation of recombinant cystatin C in yeast on its stability and activity. *J. Agric. Food Chem.*, 46: 2882–2887, 1998.
81. S. Nakamura, M. Ogawa, S. Saito, S. Nakai. Application of polymannosylated cystatin to surimi from roe-herring to prevent gel weakening. *FEBS Lett.*, 427: 252–254, 1998.
82. M. Solem, C. Rawson, K. Lindburg, D. Barnes. Transforming growth factor beta regulates cystatin C in serum-free mouse embryo (SFME) cells. *Biochem. Biophys. Res. Commn.*, 172: 945–951, 1990.
83. A. Hall, K. Håkansson, R.W. Mason, A. Grubb, M. Abrahamson. Structural basis for the biological specificity of cystatin C. Identification of leucine 9 in the N-terminal binding region as a selectivity-conferring residue in the inhibition of mammalian cysteine peptidases. *J. Biol. Chem.*, 270: 5115–5121, 1995.
84. B.D. Korant, J. Brzin, V. Turk. Cystatin, a protein inhibitor of cysteine proteases alters viral protein cleavages in infected human cells. *Biochem. Biophys. Res. Commn.*, 127: 1072–1076, 1985.
85. L. Bjorck, A. Grubb, L. Kjellen. Cystatin C, a human proteinase inhibitor, blocks replication of herpes simplex virus. *J. Virol.*, 64: 941–943, 1990.
86. A.R. Collins, A. Grubb. Inhibitory effects of recombinant human cystatin C on human coronaviruses. *Antimicrob. Agents Chemother.*, 35: 2444–2446, 1991.
87. M. Takahashi, T. Tezuka, H. Kakegawa, N. Katunuma. Linkage between phosphorylated cystatin α and filaggrin by epidermal transglutaminase as a model of cornified envelop and inhibition of cathepsins L activity by cornified envelop and conjugated cystatin α. *FEBS Lett.*, 340: 173–176, 1994.

88. M. Takahashi, T. Tezuka, N. Katunuma. Inhibition of growth and cysteine proteinase activity of *Staphylococcus aureus* V8 by phosphorylated cystatin α in skin cornified envelop. *FEBS Lett.*, 355: 275–278, 1994.
89. B.M. Franke de Cazzulo, J. Martinez, M.J. North, G.H. Coombs, J.J. Cazzulo. Effects of proteinase inhibitors on the growth and differentiation of *Trypanosoma cruzi*. *FEMS Microbiol. Lett.*, 124: 81–86, 1994.
90. M.M. Wasilewski, K.C. Lim, J. Phillips, J.H. McKerrow. Cysteine protease inhibitors block schistosome hemoglobin degradation in vitro and decrease worm burden and egg production in vivo. *Mol. Biochem. Parasitol.*, 81: 179–189, 1996.
91. M. Pernas, R. Sanchez-Monge, L. Gomez, G. Salcedo. A chestnut seed cystatin differentially effective against cysteine proteinases from closely related pests. *Plant Mol. Biol.*, 38: 1235–1242, 1998.
92. S. Visal, M.A. Taylor, D. Michaud. The proregion of papaya proteinase IV inhibits Colorado potato beetle digestive cysteine proteinases. *FEBS Lett.*, 434: 401–405, 1998.
93. A.R. Collins, A. Grubb. Cystatin D, a natural salivary cysteine protease inhibitor, inhibits coronavirus replication at its physiologic concentration. *Oral Microbiol. Immunol.*, 13: 59–61, 1998.
94. M. Takahashi, T. Tezuka, B. Korant, N. Katunuma. Inhibition of cysteine protease and growth of *Staphylococcus aureus* V8 and poliovirus by phosphorylated cystatin α conjugate of skin. *BioFactors*, 10: 339–345, 1999.
95. S. Nakamura, J. Hata, M. Kawamukai, H. Matsuda, K. Nakamura, M. Ogawa, S. Nakai. Molecular stability and bioactivity of genetically glycosylated human cystatin C in *Pichia pastoris*. Proceedings of 2nd Food Protein Symposium, Vancouver, BC, 2003, 25.
96. K. Tatti, J. Gentsch, W. Shieh, T. Ferebee-Harris, M. Lynch, J. Bresee, B. Jiang, S. Zaki, R. Glass. Molecular and immunological methods to detect rotavirus in formalin-fixed tissue. *J. Virol. Meth.*, 105: 305–319, 2002.
97. T.K. Fischer, P. Valentiner-Branth, H. Steinsland, M. Perch, G. Santos, P. Aaby, K. Molbak, H. Sommerfelt. Protective immunity after natural rotavirus infection: a community cohort study of newborn children in Guinea-Bissau, west Africa. *J. Infect. Dis.*, 186: 593–597, 2002.
98. E. Rosa, M.L. Silva, I. Pires de Carvalho, V. Gouvea. 1998–1999 rotavirus seasons in Juiz de Fora, *Minas Gerais*, Brazil: detection of an unusual G3P epidemic strain. *J. Clin. Microbiol.*, 40: 2837–2842, 2002.
99. K. Sreekrishna, K.E. Kropp. *Pichia pastoris*. In: K. Wolf (Ed.) *Nonconventional Yeasts in Biotechnology*. Berlin: Springer-Verlag, 1996, 203–253.
100. K. Hirai, M. Yokoyama, G. Asano, S. Tanaka. Expression of cathepsin B and cystatin C in human colorectal cancer. *Hum. Pathol.*, 30: 680–686, 1999.
101. O. Corticchiato, J.-F. Cajot, M. Abrahamson, S.J. Chan, D. Keppler, B. Sordat. Cystatin C and cathepsin B in human colon carcinoma: expression by cell lines and matrix degradation. *Int. J. Cancer*, 52: 645–652, 1992.
102. S. Coulibaly, H. Schwihla, M. Abrahamson, A. Albini, C. Cerni, J.L. Clark, K.M. Ng, N. Katunuma, O. Schlappack, J. Glössl, L. Mach. Modulation of invasive properties of murine squamous carcinoma cells by heterologous expression of cathepsin B and cystatin C. *Int. J. Cancer*, 83: 526–531, 1999.
103. I.M. Berquin, B.F. Sloane. Cysteine proteases and tumor progression. *Perspect. Drug Disc. Des.*, 2: 371–388, 1994.

104. L. Verdot, G. Lalmanach, V. Vercruysse, J. Hoebeke, F. Gauthier, B. Vray. Chicken cystatin stimulates nitric oxide release from interferon-gamma-activated mouse peritoneal macrophages via cytokine synthesis. *Eur. J. Biochem.*, 266: 1111–1117, 1999.
105. T. Kato, T. Imatani, T. Miura, K. Minaguchi, E. Saitoh, K. Okuda. Cytokine-inducing activity of family 2 cystatins. *Biol. Chem.*, 381: 1143–1147, 2000.
106. S. Nakai, S. Nakamura, C.H. Scaman. Optimization of site-directed mutagenesis. 2. Application of random-centroid optimization to one-site mutation of *Bacillus stearothermophilus* neutral protease to improve thermostability. *J. Agric. Food Chem.*, 46: 1655–1661, 1998.
107. M. Ogawa, S. Nakamura, C.H. Scaman, H. Jing, D.D. Kitts, J. Dou, S. Nakai. Enhancement of proteinase inhibitory activity of recombinant human cystatin C using random-centroid optimization. *Biochim. Biophys. Acta*, 1599: 115–124, 2002.
108. J. Tschopp, G. Sverlow, R. Kosson, W. Craig, L. Grinna. High-level secretion of glycosylated invertase in the methylotrophic yeast, *Pichia pastoris. Biotechnology*, 5: 1305–1308, 1987.
109. J.J. Clare, F.B. Rayment, S.P. Ballantine, K. Sreekrishna, M.A. Romanos. High-level expression of tetanus toxin fragment C in *Pichia pastoris* strains containing multiple tandem integrations of the gene. *Biotechnology*, 9: 455–460, 1991.
110. J.F. Tschopp, P.F. Brust, J.M. Cregg, C.A. Stillman, T. Gingeras. Expression of the lacZ gene from two methanol-regulated promoters in *Pichia pastoris. Nucleic Acids Res.*, 15: 3859–3876, 1987.
111. Y. Chen, J. Cino, G. Hart, D. Freedman, C. White, E.A. Komives. High protein expression in fermentation of recombinant *Pichia pastoris* by a fed-batch process. *Process Biochem.*, 32: 107–111, 1997.
112. S. Nakamura, K. Nakamura, M. Ogawa, S. Nakai. Comparison of *Pichia* expression vectors, pPICZα and pGAPZα, on the yield of recombinant human stefin A. *Proceedings of 2nd Food protein symposium*, Vancouver, BC, 2003, 25.
113. Y. Takimura, M. Kato, T. Ohta, H. Yamagata, S. Udaka. Secretion of human interleukin-2 in biologically active form by *Bacillus brevis* directly into culture medium. *Biosci. Biotechnol. Biochem.*, 61: 1858–1861, 1997.
114. M.L. Babe, S. Yosat, M. Dreyer, B.F. Schmidt. Heterologous expression of human granzyme K in *Bacillus subtilis* and characterization of its hydrolytic activity *in vitro. Biotechnol. Appl. Biochem.*, 27: 117–124, 1998.
115. C.H. Schein, K. Kashiwagi, A. Fujisawa, C. Weisswann. Secretion of mature IFN-α_2 and accumulation of uncleared precursor by *Bacillus subtilis* transformed with a hybrid α-amylase signal sequence-IFN-α_2 gene. *Biotechnology*, 4: 719–725, 1986.
116. Y. Sagiya, H. Yamagata, S. Udata. Direct high-level secretion into the culture medium of tuna growth hormone in biologically active form by *Bacillus brevis. Appl. Microbiol. Biotechnol.*, 42: 358–363, 1994.
117. S. Nakamura, T. Tanaka, R. Yada, S. Nakai. Improving the thermostability of *Bacillus stearothermophilus* neutral protease by introducing proline into the active site helix. *Protein Eng.*, 10: 1263–1269, 1997.
118. T. Imanaka, M. Shibazaki, M. Takagi. A new way of enhancing the thermostability of proteases. *Nature*, 324: 695–697, 1986.
119. K.G. Hardy. *Bacillus* cloning methods. In: DM Glover (Ed.) *DNA Cloning, a Practical Approach.* Oxford: IRL Press, 1985, 1–16.

Section III

The Body's Regulating System

13 ACE Inhibitory Peptides

Hans Meisel, Daniel J. Walsh, Brian Murray, and Richard J. FitzGerald

CONTENTS

13.1 INTRODUCTION

Angiotensin-I-converting enzyme (ACE) is a key enzyme in the regulation of peripheral blood pressure and electrolyte homeostasis. ACE, a membrane-anchored dipeptide-liberating carboxypeptidase (peptidyldipeptide hydrolase, EC 3.4.15.1) classically associated with the renin-angiotensin system, converts angiotensin I to the highly potent vasoconstrictor octapeptide angiotensin II (1). This enzyme also plays a key physiological role in the regulation of local levels of several endogenous bioactive peptides such as encephalins, substance P, and bradykinin, which are inhibitors and competitive substrates for ACE (2,3). In the kinin–kallikrein system, for example, ACE inactivates the vasodilatory peptide bradykinin (4). Thus, ACE inhibition would be expected to prevent the formation of the vasoconstrictory (hypertensive) agent angiotensin II and to potentiate the vasodilatory (hypotensive) properties of bradykinin, leading to a concerted lowering of the blood pressure. Inhibitors of ACE are therefore widely used in therapy for hypertension, heart failure, myocardial infarction, and diabetic nephropathy.

Exogenous ACE inhibitors having an antihypertensive effect *in vivo* were first discovered in snake venom (5). ACE-inhibitory peptides are also present in the amino acid sequences of several food proteins (6–10). The intrinsic bioactivities of the peptides encrypted in food proteins are latent until they are released and activated by enzymatic hydrolysis, for example, during gastrointestinal digestion and food processing. Activated peptides are potential modulators of various regulatory processes in the living system. Therefore, food protein-derived inhibitors of ACE represent natural, physiologically active, food-grade components, which may provide health benefits beyond basic nutrition. In particular, food protein-derived peptides may contribute to reducing the risk of developing cardiovascular disease through the consumption of ACE inhibitors as functional food ingredients.

13.2 FOOD-PROTEIN-DERIVED ACE-INHIBITORY PEPTIDES

Three strategies have generally been used in the identification and characterization of ACE-inhibitory peptides: (a) isolation from *in vitro* enzymatic or (b) from *in vivo* gastrointestinal digests of precursor proteins, and (c) chemical synthesis of peptides having identical or similar structures to those known to be bioactive. In some cases ACE-inhibitory peptides may be isolated from a food source without prior enzymatic processing, for example, from garlic (11).

A wide variety of ACE inhibitory peptides have been identified and characterized from milk, animal (nonmilk), plant, and miscellaneous protein sources (Table 13.1) (2,6,11–49). The potency of an ACE-inhibitory peptide is usually expressed as an IC_{50} value, which is equivalent to the concentration of peptide mediating a 50% inhibition of activity. In the majority of cases, the most frequently used analytical method to determine IC_{50} is based on the hydrolysis of hippuryl-histidine-leucine (50). With the creation of new artificial substrates for ACE, alternative methods have been developed to quantify the IC_{50} of ACE inhibitory peptides (51–54). Unfortunately, the use of various modifications of the method of Cushman and Cheung (50) has made the comparison of IC_{50} values from different studies difficult because some reports do not detail the number of enzyme units used in the inhibition analyses or include an IC_{50} value for an ACE inhibitory standard such as Captopril®.

As can be seen in Table 13.1, the majority of peptides are of low molecular mass corresponding to relatively short-chain peptides. This is in agreement with the results of Natesh et al. (55) which demonstrated from crystallography studies that the active site of ACE cannot accommodate large peptide molecules. The peptides listed differ significantly in their ACE *in vitro* inhibitory potencies. Several structural features that appear to influence the ultimate potency of an ACE inhibitory peptide have been identified (see 13.3).

TABLE 13.1
Characteristics of Some Angiotensin-I-Converting Enzyme (ACE) Inhibitory Peptides Mainly Derived from Food

Peptide Sequence[a]	Fragment	Source Protein	IC_{50} (μmol l^{-1})[b]	Preparation	Reference
Bovine Milk					
VAP	(*f*25–27)	α_{S1}-casein	2.0	Synthetic	2
FFVAP	(*f*23–27)	α_{S1}-casein	6.0	Enzymatic	12
FVAP	(*f*24–27)	α_{S1}-casein	10.0	Synthetic	2
TTMPLW	(*f*194–199)	α_{S1}-casein	16.0	Enzymatic	13
YKVPQL	(*f*104–109)	α_{S1}-casein	22.0	Synthetic	14
PLW	(*f*197–199)	α_{S1}-casein	36.0	Synthetic	13
LW	(*f*198–199)	α_{S1}-casein	50.0	Synthetic	13
LAYFYP	(*f*142–147)	α_{S1}-casein	65.0	Enzymatic	15
FFVAPFPEVFGK	(*f*23–34)	α_{S1}-casein	77.0	Enzymatic	16
DAYPSGAW	(*f*157–164)	α_{S1}-casein	98.0	Enzymatic	15
FPEVFGK	(*f*28–34)	α_{S1}-casein	140.0	Synthetic	2
FGK	(*f*32–34)	α_{S1}-casein	160.0	Synthetic	2
PFPE	(*f*27–30)	α_{S1}-casein	>1000.0	Synthetic	2
AYFYP	(*f*143–147)	α_{S1}-casein	>1000.0	Synthetic	14
LFRQ	(*f*136–139)	α_{S1}-casein	17.0[c]	Enzymatic	18
AYFYPE	(*f*143–148)	α_{S1}-casein	106.0[c]	Enzymatic	17
FALPQY	(*f*174–179)	α_{S2}-casein	4.3	Enzymatic	19
FALPQYLK	(*f*174–181)	α_{S2}-casein	4.3	Enzymatic	19
FPQYLQY	(*f*92–98)	α_{S2}-casein	14.0	Enzymatic	19
TVY	(*f*182–184)	α_{S2}-casein	15.0	Enzymatic	19
NMAINPSK	(*f*25–32)	α_{S2}-casein	60.0	Enzymatic	19
ALNEINQFY	(*f*81–89)	α_{S2}-casein	219.0	Enzymatic	19
ALNEINQFYQK	(*f*81–91)	α_{S2}-casein	264.0	Enzymatic	19
MKPWIQPK	(*f*190–197)	α_{S2}-casein	300.0	Synthetic	14
TKVIP	(*f*198–202)	α_{S2}-casein	400.0	Synthetic	14
AMKPW	(*f*189–192)	α_{S2}-casein	580.0	Synthetic	14
AMKPWIQPK	(f189–197)	α_{S2}-casein	600.0	Synthetic	14
IPP	(*f*74–76)	*β*-casein	5.0	Enzymatic	20
KVLPVP	(*f*169–174)	*β*-casein	5.0	Synthetic	14
VPP[d]	(*f*84–86)	*β*-casein	9.0	Enzymatic	20
AVPYPQR	(*f*177–183)	*β*-casein	15.0	Enzymatic	2
IHPFAQTQSLVYP	(*f*49–61)	*β*-casein	19.0	Synthetic	21
FAQTQSLVYP	(*f*52–61)	*β*-casein	25.0	Synthetic	21
HPFAQTQSLVYP	(*f*50–61)	*β*-casein	26.0	Synthetic	21
KIHPFAQTQSLVYP	(*f*48–61)	*β*-casein	39.0	Synthetic	21
SLVYP	(*f*57–61)	*β*-casein	40.0	Synthetic	21
QSLVYP	(*f*56–61)	*β*-casein	41.0	Synthetic	21
VYP	(*f*59–61)	*β*-casein	44.0	Synthetic	21
TQSLVYP	(*f*55–61)	*β*-casein	64.0	Synthetic	21

(*Continued*)

TABLE 13.1 (Continued)

Peptide Sequence[a]	Fragment	Source Protein	IC_{50} ($\mu mol\ l^{-1}$)[b]	Preparation	Reference
QTQSLVYP	(*f*54–61)	*β*-casein	73.0	Synthetic	21
AQTQSLVYP	(*f*53–61)	*β*-casein	76.0	Synthetic	21
AVPYP	(*f*177–181)	*β*-casein	80.0	Synthetic	2
LVYP	(*f*58–61)	*β*-casein	170.0	Synthetic	21
PYP[e]	(*f*179–181)	*β*-casein	220.0	Synthetic	2
VYPFPG	(*f*59–64)	*β*-casein	221.0	Synthetic	22
AVPYPQR	(*f*177–183)	*β*-casein	274.0	Enzymatic	15
YQEPVL	(*f*193–198)	*β*-casein	280.0	Enzymatic	15
VYP	(*f*59–61)	*β*-casein	288.0	Synthetic	22
YQEPVLQPVR	(*f*193–202)	*β*-casein	300.0	Synthetic	23
AVP	(*f*177–179)	*β*-casein	340.0	Synthetic	2
EMPFPK	(*f*108–113)	*β*-casein	423.0	Enzymatic	15
LQSW	(*f*140–143)	*β*-casein	500.0	Synthetic	14
YPFPGPI	(*f*60–66)	*β*-casein	500.0	Synthetic	23
TPVVVPPFLQP	(*f*80–90)	*β*-casein	749.0	Synthetic	22
LLYQQPV	(*f*191–197)	*β*-casein	>1000	Synthetic	14
LHLPLP	(*f*133–138)	*β*-casein	7.0[c]	Enzymatic	18
LVYPFPGPIP-NSLPQNIPP	(*f*58–76)	*β*-casein	19.0[c]	Enzymatic	18
VTSTAV	(*f*185–190)[g]	*κ*-casein	30.0[c]	Enzymatic	18
YP[f]	(*f*58–59)	*κ*-casein	720.0	Synthetic	24
WLAHK	(*f*104–108)	*α*-lactalbumin	77.0	Enzymatic	25
VGINYWLAHK	(*f*99–108)	*α*-lactalbumin	327.0	Enzymatic	25
YGL	(*f*50–52)	*α*-lactalbumin	409.0	Enzymatic	25
LAHKAL	(*f*105–110)	*α*-lactalbumin	621.0	Enzymatic	15
YGLF	(*f*50–53)	*α*-lactalbumin	733.3	Synthetic	26
YG	(*f*50–51)	*α*-lactalbumin	1522.6	Synthetic	26
LF	(*f*52–53)	*α*-lactalbumin	349.0	Synthetic	26
ALPMHIR	(*f*142–148)	*β*-lactoglobulin	42.6	Enzymatic	27
YL	(*f*102–103)	*β*-lactoglobulin	122.1	Synthetic	26
IPA	(*f*78–80)	*β*-lactoglobulin	141.0	Synthetic	22
YLLF	(*f*102–105)	*β*-lactoglobulin	171.8	Synthetic	26
LF	(*f*104–105)	*β*-lactoglobulin	349.1	Synthetic	26
ALPMH	(*f*142–146)	*β*-lactoglobulin	521.0	Enzymatic	25
VAGTW	(*f*15–19)	*β*-lactoglobulin	534.0	Enzymatic	25
GLDIQK	(*f*9–14)	*β*-lactoglobulin	580.0	Enzymatic	15
LDAQSAPLR	(*f*34–40)	*β*-lactoglobulin	635.0	Enzymatic	25
RL	(*f*148–149)	*β*-lactoglobulin	2438.9	Synthetic	26
CMENSA	(*f*106–111)	*β*-lactoglobulin	788.0	Enzymatic	25
HIR	(*f*146–148)	*β*-lactoglobulin	953.0	Synthetic	28
VLDTDYK	(*f*94–100)	*β*-lactoglobulin	946.0	Enzymatic	25
VFK	(*f*81–83)	*β*-lactoglobulin	1029.0	Enzymatic	25
LAMA	(*f*22–25)	*β*-lactoglobulin	1062.0	Enzymatic	25

(Continued)

TABLE 13.1 (Continued)

Peptide Sequence[a]	Fragment	Source Protein	IC_{50} (μmol l^{-1})[b]	Preparation	Reference
HIRL	(*f*146–149)	*β*-lactoglobulin	1153.2	Synthetic	26
IR	(*f*147–148)	*β*-lactoglobulin	695.5	Synthetic	26
VAGTWY	(*f*15–20)	*β*-lactoglobulin	1682.0	Enzymatic	15
LDIQK	(*f*10–14)	*β*-lactoglobulin	17.0[c]	Enzymatic	18
LIVTQ	(*f*1–5)	*β*-lactoglobulin	17.0[c]	Enzymatic	18
VF	(*f*81–82)	*β*-lactoglobulin	19.0[c]	Enzymatic	18
MKG	(*f*7–9)	*β*-lactoglobulin	24.0[c]	Enzymatic	18
ALKAWSVAR	(*f*208–216)	Serum albumin	3.0	Synthetic	29
FP	(*f*221–222)	Serum albumin	315.0	Synthetic	22
GKP	(*f*18–20)	β_2-microglobulin	352.0	Enzymatic	22
GPAGAHyP	—	Gelatin	8.3	Collagenase	41
GPPGAHyP	—	Gelatin	8.6	Collagenase	41
GPHyPGTDGAHyP	—	Gelatin	10.5	Collagenase	41
GPHyPGAPHyP	—	Gelatin	11.3	Collagenase	41
GPIGSVGAHyP	—	Gelatin	31.8	Collagenase	41
GPAGAPGAA	—	Gelatin	37.0	Collagenase	41
GPIVGPHyPA	—	Gelatin	123.4	Collagenase	41
GPHyPGAIGP	—	Gelatin	146.5	Collagenase	41
GPHyP	—	Gelatin	250.0	Synthetic	41
GPA	—	Gelatin	405.0	Synthetic	41
Human Milk					
YYPQIMQY	(*f*136–143)	Human α_{S1}-casein	24.8	Trypsin	30
NNVMLQW	(*f*164–170)	Human α_{S1}-casein	41.0	Trypsin	30
YIPIQYVLSR	(*f*8–11)	Human α_{S1}-casein	132.5	Trypsin	30
SFQPQPLIYP	(*f*43–52)	Human *β*-casein	1.4	Synthetic	21
KIYPSFQPQPLIYP	(*f*39–52)	Human *β*-casein	8.6	Synthetic	21
VRP	(*f*63–65)	Human *κ*-casein	2.2	Synthetic	31
TAP	(*f*18–20)	Human *κ*-casein	3.5	Synthetic	31
IPP	(*f*96–98)	Human *κ*-casein	5.0	Synthetic	31
YANPNVVRP	(*f*57–65)	Human *κ*-casein	7.8	Synthetic	31
MYY	(*f*24–26)	Human *κ*-casein	9.6	Synthetic	31
PAVVRP	(*f*60–65)	Human *κ*-casein	18.0	Synthetic	31
NPAVVRP	(*f*59–65)	Human *κ*-casein	19.0	Synthetic	31
ANPNVVRP	(*f*58–65)	Human *κ*-casein	25.0	Synthetic	31
QKTAP	(*f*16–20)	Human *κ*-casein	30.0	Synthetic	31
PTPAP	(*f*118–122)	Human *κ*-casein	33.0	Synthetic	31
KTAP	(*f*17–20)	Human *κ*-casein	37.0	Synthetic	31
AVVRP	(*f*61–65)	Human *κ*-casein	74.0	Synthetic	31
VVRP	(*f*62–65)	Human *κ*-casein	81.0	Synthetic	31
PAP	(*f*120–122)	Human *κ*-casein	87.0	Synthetic	31
YVP	(*f*21–23)	Human *κ*-casein	200.0	Synthetic	31
VAV	(*f*151–153)	Human *κ*-casein	260.0	Synthetic	31

(*Continued*)

TABLE 13.1 (Continued)

Peptide Sequence[a]	Fragment	Source Protein	IC_{50} (μmol l^{-1})[b]	Preparation	Reference
SHP	(*f*78–80)	Human κ-casein	280.0	Synthetic	31
AIP	(*f*95–97)	Human κ-casein	350.0	Synthetic	31
IAIP	(*f*94–97)	Human κ-casein	470.0	Synthetic	31
AIPP	(*f*95–98)	Human κ-casein	900.0	Synthetic	31
IAIPP	(*f*94–98)	Human κ-casein	900.0	Synthetic	31
PNSHP	(f76–80)	Human k-casein	>1000	Synthetic	31
Murine Milk					
PQAFP	(*f*55–59)	Mouse *β*-casein	300.0	Synthetic	21
SIQSQPQAFP	(*f*50–59)	Mouse *β*-casein	560.0	Synthetic	21
AFP	(*f*57–59)	Mouse *β*-casein	610.0	Synthetic	21
QPQAFP	(*f*54–59)	Mouse *β*-casein	610.0	Synthetic	21
IQSQPQAFP	(*f*51–59)	Mouse *β*-casein	630.0	Synthetic	21
QSQPQAFP	(*f*52–59)	Mouse *β*-casein	700.0	Synthetic	21
SQPQAFP	(*f*53–59)	Mouse *β*-casein	710.0	Synthetic	21
QAFP	(*f*56–59)	Mouse *β*-casein	>1000	Synthetic	21
SSIQSQPQAFP	(*f*49–59)	Mouse *β*-casein	>1000		21
HSSIQSQPQAFP	(*f*48–59)	Mouse *β*-casein	>1000		21
VHSSIQSQPQAFP	(*f*47–59)	Mouse *β*-casein	>1000		21
KVHSSIQSQPQAFP	(*f*46–59)	Mouse *β*-casein	>1000		21
Rat Milk					
PKAIP	(*f*55–59)	Rat *β*-casein	185.0	Synthetic	21
KAIP	(*f*956–59)	Rat *β*-casein	330.0	Synthetic	21
EPKAIP	(*f*54–59)	Rat *β*-casein	380.0	Synthetic	21
SEPKAIP	(*f*53–59)	Rat *β*-casein	430.0	Synthetic	21
GIQSEPKAIP	(*f*50–59)	Rat *β*-casein	430.0	Synthetic	21
FHSGIQSEPKAIP	(*f*47–59)	Rat *β*-casein	520.0	Synthetic	21
IQSEPKAIP	(*f*51–59)	Rat *β*-casein	570.0	Synthetic	21
SGIQSEPKAIP	(*f*49–59)	Rat *β*-casein	650.0	Synthetic	21
AIP	(*f*57–59)	Rat *β*-casein	670.0	Synthetic	21
QSEPKAIP	(*f*52–59)	Rat *β*-casein	750.0	Synthetic	21
HSGIQSEPKAIP	(*f*48–59)	Rat *β*-casein	>1000	Synthetic	21
KFHSGIQSEPKAIP	(*f*46–59)	Rat *β*-casein	>1000	Synthetic	21
Animal (Nonmilk)					
Chicken					
LKP	—	Chicken muscle	0.3	Thermolysin	32
FKGRYYP	—	Chicken muscle	0.6	Thermolysin	32
IVGRPRHQG	—	Chicken muscle	2.4	Thermolysin	32
LAP	—	Chicken muscle	3.5	Thermolysin	32
LKA	—	Chicken muscle	8.5	Thermolysin	32
FQKPKR	—	Chicken muscle	14.0	Thermolysin	32

(*Continued*)

TABLE 13.1 (Continued)

Peptide Sequence[a]	Fragment	Source Protein	IC$_{50}$ (μmol l^{-1})[b]	Preparation	Reference
Fish					
LKP	—	Dried bonito	0.32	Thermolysin	33
IKP	—	Dried bonito	1.6	Thermolysin	32
IY	—	Dried bonito	2.1	Thermolysin	32
LKPMN	—	Dried bonito	2.4	Thermolysin	33
IWH	—	Dried bonito	3.5	ACE	32
IWHHT	—	Dried bonito	5.8	Thermolysin	32
IKPLNY	—	Dried bonito	43.0	Thermolysin	6
LRP	—	Bonito bowels	1.0	Autolysate	34
IRP	—	Bonito bowels	1.8	Autolysate	34
VRP	—	Bonito bowels	2.2	Autolysate	34
YRPY	—	Bonito bowels	320.0	Autolysate	34
GHF	—	Bonito bowels	>1000	Autolysate	34
LKVGVKQY	—	Sardine	11.0	—	6
KVLAGM	—	Sardine	30.0	—	6
HQAAGW	—	Sardine	60.0	—	6
VKAGF	—	Sardine	83.0	—	6
LKL	—	Sardine	188.0	—	6
YALPHA	—	Squid	9.8	Autolysate	35
GYALPHA	—	Squid	27.3	Autolysate	35
IF	—	Tuna muscle	70.0	—	36
VWIG	—	Tuna muscle	110.0	—	36
LTF	—	Tuna muscle	330.0	—	36
IFG	—	Tuna muscle	>1000	—	36
Porcine					
PANIKWGD	—	Porcine GADPH	21.0	Autolysate	6
PSKIKWGD	—	Porcine GADPH	68.0	Synthetic	6
GKKVLQ	—	Porcine hemoglobin	1.90	Alcalase 0.6L	37
FQKVVAK	—	Porcine hemoglobin	2.10	Alcalase 0.6L	37
FQKVVA	—	Porcine hemoglobin	5.8	Alcalase 0.6L	37
FQKVVAG	—	Porcine hemoglobin	7.4	Alcalase 0.6L	37
MNP	(*f*79–81)	Porcine myosin	66.6	Synthetic	38
TNP	(*f*308–310)	Porcine myosin	207.4	Synthetic	38
NPP	(*f*80–82)	Porcine myosin	290.5	Synthetic	38
ITTNP	(*f*306–310)	Porcine myosin	549.0	Thermolysin	38
TTN	(*f*307–309)	Porcine myosin	672.7	Synthetic	38
ITT	(*f*306–308)	Porcine myosin	678.5	Synthetic	38
MNPPK	(*f*79–83)	Porcine myosin	945.5	Thermolysin	38
PPK	(*f*981–83)	Porcine myosin	>1000	Synthetic	38

(*Continued*)

TABLE 13.1 (Continued)

Peptide Sequence[a]	Fragment	Source Protein	IC_{50} (μmol l^{-1})[b]	Preparation	Reference
Plant					
LGP	—	Pea albumin	0.7	Synthetic	39
YW	—	Pea albumin	10.0	Synthetic	39
VY	—	Pea albumin	10.0	Synthetic	39
DG	—	Pea albumin	12.0	Synthetic	39
LY	—	Pea albumin	18.0	Synthetic	39
MF	—	Pea albumin	45.0	Synthetic	39
GP	—	Pea albumin	450.0	Synthetic	39
GS	—	Pea albumin	>1000	Synthetic	39
GK	—	Pea albumin	>1000	Synthetic	39
LKP	—	Pea vicilin	0.3	Synthetic	39
IY	—	Pea vicilin	2.1	Synthetic	39
VK	—	Pea vicilin	13.0	Synthetic	39
AF	—	Pea vicilin	15.2	Synthetic	39
GYK	—	Pea vicilin	160.0	Synthetic	39
IR	—	Pea vicilin	696.0	Synthetic	39
QK	—	Pea vicilin	885.0	Synthetic	39
FG	—	Pea vicilin	>1000	Synthetic	39
SG	—	Pea vicilin	>1000	Synthetic	39
GK	—	Pea vicilin	>1000	Synthetic	39
LRP	—	α-Zein, Maize endosperm	0.3	Thermolysin	40
LSP	—	α-Zein, Maize endosperm	1.7	Thermolysin	40
LQP	—	α-Zein, Maize endosperm	1.9	Thermolysin	40
LAY	—	α-Zein, Maize endosperm	3.9	Thermolysin	40
IRA	—	α-Zein, Maize endosperm	6.4	Thermolysin	40
VSP	—	α-Zein, Maize endosperm	10.0	Thermolysin	40
LAA	—	α-Zein, Maize endosperm	13.0	Thermolysin	40
VAA	—	α-Zein, Maize endosperm	13.0	Thermolysin	40
VAY	—	α-Zein, Maize endosperm	16.0	Thermolysin	40
FY	—	α-Zein, Maize endosperm	25.0	Thermolysin	40
LNP	—	α-Zein, Maize endosperm	43.0	Thermolysin	40

(Continued)

TABLE 13.1 (Continued)

Peptide Sequence[a]	Fragment	Source Protein	IC_{50} (μmol l^{-1})[b]	Preparation	Reference
LLP	—	α-Zein, Maize endosperm	57.0	Thermolysin	40
LQQ	—	α-Zein, Maize endosperm	100.0	Thermolysin	40
IRAQQ	—	α-Zein, Maize endosperm	160.0	Thermolysin	40
DLP	—	Soy protein	4.8	Hydrolysis	42
HHL	—	Soy protein	5.0	Fermentation	43
DG	—	Soy protein	12.3	Hydrolysis	42
FY	—	Garlic	3.7	Synthetic	11
NY	—	Garlic	32.6	Synthetic	11
NF	—	Garlic	46.3	Synthetic	11
SY	—	Garlic	66.3	Synthetic	11
GY	—	Garlic	72.1	Synthetic	11
SF	—	Garlic	130.2	Synthetic	11
GF	—	Garlic	277.9	Synthetic	11
IVY	—	Wheat germ	0.48	Alcalase 2	45
VY	—	Wheat germ	5.2	Alcalase 2	45
LYPVK	—	Fig tree latex	4.5	Autolysate	6
AVNPIR	—	Fig tree latex	13.0	Autolysate	6
LVR	—	Fig tree latex	14.0	Autolysate	6
GGY	—	Sake	1.3	Synthetic	45
VY	—	Sake	7.1	Fermentation	45
HY	—	Sake	26.1	Fermentation	45
GY	—	Sake	259.0	Synthetic	45
SS	—	Sake	>1000	Synthetic	45
YGG	—	Sake	>1000	Synthetic	45
VW	—	Sake lees	1.4	Fermentation	45
IY	—	Sake lees	2.4	Synthetic	45
PRY	—	Sake lees	2.5	Synthetic	45
PR	—	Sake lees	4.1	Synthetic	45
VWY	—	Sake lees	9.4	Fermentation	45
IYPR	—	Sake lees	10.0	Synthetic	45
RY	—	Sake lees	10.5	Synthetic	45
YW	—	Sake lees	10.5	Fermentation	45
YPR	—	Sake lees	16.5	Synthetic	45
YPRY	—	Sake lees	17.4	Synthetic	45
FWN	—	Sake lees	18.3	Fermentation	45
RF	—	Sake lees	93.0	Fermentation	45
Miscellaneous					
FFGRCVSP	—	Ovalbumin	0.4	Pepsin	32
ERKIKVYL	—	Ovalbumin	1.2	Pepsin	32

(*Continued*)

TABLE 13.1 (Continued)

Peptide Sequence[a]	Fragment	Source Protein	IC_{50} (μmol l^{-1})[b]	Preparation	Reference
FGRCVSP	—	Ovalbumin	6.2	Pepsin	32
LW	—	Ovalbumin	6.8	Pepsin	32
FCF	—	Ovalbumin	11.0	Pepsin	32
NIFYCP	—	Ovalbumin	15.0	Pepsin	32
AKYSY	—	Red algae	1.5	Hydrolysis	46
IY	—	Red algae	2.7	Hydrolysis	46
LRY	—	Red algae	5.1	Hydrolysis	46
MKY	—	Red algae	7.3	Hydrolysis	46
IY	—	Wakame	2.7	Synthetic	47
FY	—	Wakame	3.7	Synthetic	47
KY	—	Wakame	13.0	Synthetic	47
YNKL	—	Wakame	21.0	Pepsin	47
KF	—	Wakame	28.3	Synthetic	47
YKY	—	Wakame	43.5	Synthetic	47
KFY	—	Wakame	45.0	Synthetic	47
KL	—	Wakame	50.2	Synthetic	47
YKYY	—	Wakame	64.2	Pepsin	47
KYY	—	Wakame	79.0	Synthetic	47
NKL	—	Wakame	88.0	Synthetic	47
KFYG	—	Wakame	90.5	Pepsin	47
YNK	—	Wakame	125.0	Synthetic	47
IYK	—	Wakame	177.0	Synthetic	47
AIYK	—	Wakame	213.0	Pepsin	47
YK	—	Wakame	610.0	Synthetic	47
NK	—	Wakame	810.0	Synthetic	47
GHKIATFQER	—	Yeast GADPH	0.4	Autolysate	6
KKIATYQER	—	Yeast GADPH	2.0	Autolysate	6
PANLPWGSSNV	—	Yeast GADPH	18.0	Autolysate	6
LIY	(*f*518–520)	Human plasma	0.8	Trypsin	48
YLYEIAR	(*f*138–144)	Human plasma	16.0	Trypsin	49
YLYEIARR	(*f*138–145)	Human plasma	86.0	Trypsin	49
YLYEIA	(f138–143)	Human plasma	500.0	Trypsin	49

[a] One letter amino acid code; HyP, hydroxyproline

[b] IC_{50}, concentration of peptide material mediating a 50% inhibition of ACE activity

[c] IC_{50} values quoted are expressed as mg l^{-1}

[d] This sequence also occurs in κ-casein (*f*108–110)

[e] This sequence also occurs in κ-casein (*f*57–59)

[f] This sequence also occurs in α_{S1}-casein (*f*146–147) and (*f*159–160) and in β-casein (*f*114–115)

[g] This sequence corresponds to glycomacropeptide (*f*59–64) of bovine κ-casein; ACE, angiotensin-I-converting enzyme

13.3 STRUCTURAL FEATURES OF ACE INHIBITION

Inhibitors of ACE were developed for therapy of human hypertension without knowledge of the structure of human ACE, but were designed on the basis of an assumed mechanistic homology with carboxypeptidase A. Recently, analysis of the three-dimensional structure of ACE has shown that it resembles zinc metallopeptidases such as neurolysin; however, there is no detectable sequence similarity with ACE (55). Three different forms of ACE have been identified. The somatic form of ACE consists of two homologous domains (N- and C-domain; Figure 13.1) (56), each of which contains an active site which catalyzes the hydrolysis of angiotensin I (57). Testicular ACE contains the C-terminal active site of somatic ACE (see Figure 13.1). Recently, human genome studies have isolated a third form of ACE, an ACE homologue (ACEH) or ACE2 (58,59) which also contains one active site corresponding to the N-terminal domain of somatic ACE. ACEH is unaffected by ACE inhibitors such as Captopril® or Enalaprilat®, and its specificity is distinct from that of ACE. Furthermore, ACEH does not hydrolyse bradykinin (BK). ACE is a transmembrane peptidase, which can be bound to the external surface of the plasma membrane of cells by a hydrophobic anchor (60,61). An enzyme known as ACE secretase cleaves the membrane-bound form of ACE to release ACE into plasma. ACE contains an arginine residue that confers chloride sensitivity to the enzyme.

ACE inhibitors may preferentially act on either ACE domain. However, the C-domain seems to be necessary for controlling blood pressure, suggesting that this domain is the dominant angiotensin-converting site (55). Although there is no known specific physiological substrate of the C-domain, the C-domain activity can be assessed specifically *in vitro* by use of the synthetic substrate hippury-histidyl-leucine (HHL) (57,62).

Although the structure–activity relationship of food-derived ACE-inbibitory peptides has not yet been established, these peptides show some common features. Structure-activity correlations among different peptide inhibitors of ACE indicate that binding to the enzyme is strongly influenced by the C-terminal

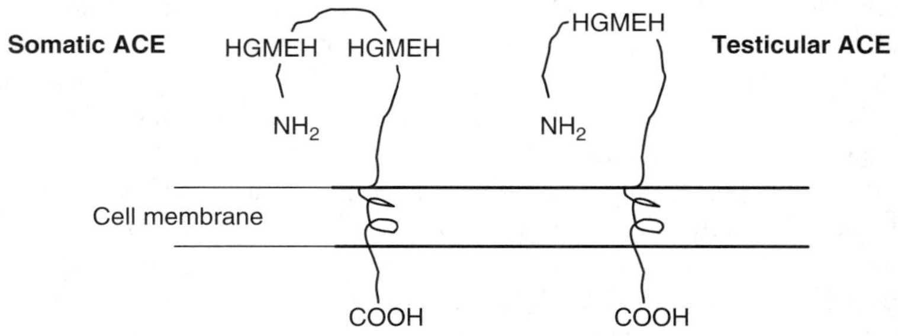

FIGURE 13.1 Schematic representation of angiotensin-I-converting enzyme (ACE) structure in somatic and testicular cells. (Adapted from T. Inagami. *Essays Biochem.*, 28, 147–164, 1992.)

tripeptide sequence of the substrate. ACE appears to prefer substrates or competitive inhibitors containing hydrophobic (aromatic or branched side chains) amino acid residues at each of the three C-terminal positions and many naturally occurring peptidic inhibitors contain proline at the C-terminus (see Table 13.1). In principle, this applies also for the highly active short-chain peptides: the majority of di- and tripeptide inhibitors have a Tyr, Phe, Trp, or Pro residue at the C-terminal end; Trp seems to be most effective in increasing the ACE-inhibitory potential. Among the N-terminal amino acids of di- and tripeptides, the branched-chain aliphatic amino acids Ile and Val are predominant in highly active inhibitors (see Table 13.1). Nevertheless, for long-chain peptides, it is expected that peptide conformation, that is, the structure adopted in a specific environment of the binding site, should contribute to ACE inhibitor potency.

Several ACE-inhibitory peptides have lysine or arginine as the C-terminal residue (see Table 13.1). Structure-activity data suggest that the positive charge on the guanidino or the ε-amino group of the C-terminal arginine and lysine side chain, respectively, contribute substantially to the inhibitory potency (5–7,63). The replacement of the arginine residue at the C-terminal end can result in essentially inactive analogues (Figure 13.2). This is substantiated by the results of theoretical studies of peptide structures using methods for computer-assisted molecular modeling. It has been found that the molecular electrostatic potentials (MEP) of ACE-inhibitory peptides possess a characteristic pattern different from that of inactive molecules, in which a similar positive potential is located at the C-terminal end. Thus, it has been postulated that the mechanism of ACE inhibition could also involve inhibitor interaction with an anionic binding site that is distinct from the catalytic site (64). A few inhibitory peptides have a C-terminal Glu residue, a structural feature that is not in accordance with the structure-activity data given above. Possibly, the Glu residue has a chelating effect on the zinc ion which is bound at the active site as an important catalytic component of ACE (57).

Structure-activity studies with artificial neural networks (ANN) have been established recently as a mathematical model that can help to find out the structure–activity relationships of ACE-inhibitory peptides and thus can be used to identify inhibitory sequences (H. Meisel, unpublished work, 2003). Because dipeptides represent the bioactive minimal structure, the final ANN input frame can be restricted to dipeptide sequences (Figure 13.3). Amino acids that carry insignificant information as input variables were excluded during ANN training. Thus, the resulting ANN recognized those amino acids at the respective sequence position that turned out to be important for the ACE-inhibitory activity and enabled the error-free identification of potent bioactive dipeptides having IC_{50} values below 100 $\mu mol\ l^{-1}$. Finally, only 22 amino acids were needed as an input frame, 12 to encode the N-terminal and 10 to encode the C-terminal amino acid (see Figure 13.3); the most important N- and C-terminal residues proved to be Val and Trp, respectively.

Remarkably, several peptides are effective ACE-inhibitors *in vitro* but noneffective hypotensive peptides *in vivo*. Some peptides may be susceptible to degradation by gastrointestinal enzymes and by blood serum and intracellular peptidases, respectively, or to modification in the liver (*see also* 13.5 and 13.7).

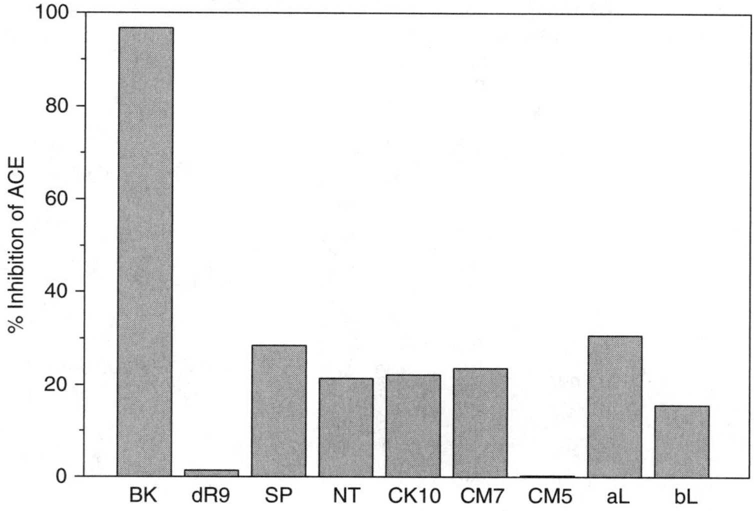

FIGURE 13.2 Inhibition of ACE activity by endogenous and milk-protein-derived peptides, respectively, at a 100 μmol l^{-1} concentration. The data represent the percentage of inhibition of activity compared with control (no peptide inhibitor present) (23). ACE-inhibitory activity was measured spectrophotometrically using hippuryl-L-histidyl-L-leucine as substrate according to Cushman and Cheung (50). BK = bradykinin (RPPGFSPFR); dR9 = desArg9-Bradykinin; SP = substance P (RPKPQQFFGLM); NT = neurotensin (pyroELYENKPRPYIL); CK10 = *β*-casokinin-10 (sequence 18); *β*CM7 = *β*-casomorphin-7 (YPFPGPI); CM5 = *β*-casomorphin-5 (YPFPG); aL = *α*-lactorphin (YGLF); bL = *β*-lactorphin (YLLF).

On the other hand, the hypotensive activity of a long-chain ACE-inhibitory peptide may be caused by peptide fragments generated by gastrointestinal enzymes. In particular, the bioactive potential of di- and tripeptides is high because small peptides can be absorbed in the intestine without being decomposed by digestive enzymes and then reach target sites in the body.

Several peptides have been reported to have hypotensive activities *in vivo* without sufficient ACE-inhibitory activities *in vitro*, that is, the IC_{50} value is not always directly related to the *in vivo* hypotensive effect. The hypotensive dipeptide Tyr-Pro did not show significant ACE-inhibitory activity (IC_{50} = 720 μmol l^{-1}). In the case of oligopeptides, active fragments could be generated from a long-chain precursor peptide by gastrointestinal digestion (*see also* 13.5). Another possibilty is that a blood-pressure-regulating mechanism other than the system with ACE is related to the hypotensive effect, for example, through receptors expressed in the gastrointestinal tract (66) or opioid receptors which are located in the nervous, endocrine and immune systems as well as in the intestinal

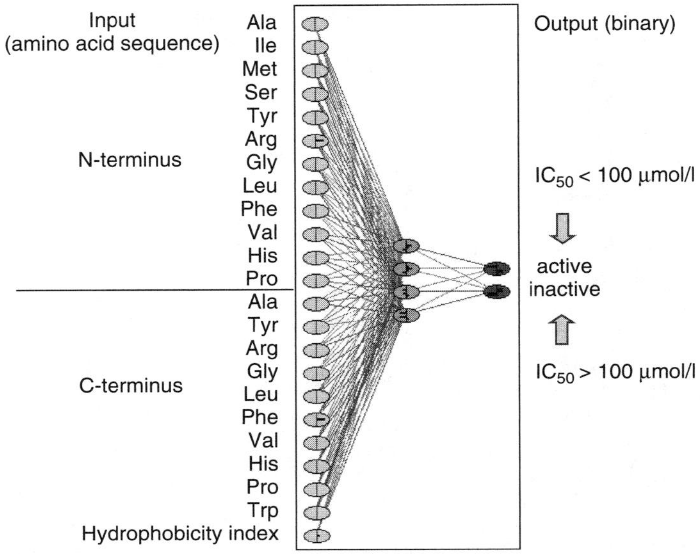

FIGURE 13.3 Topology (23 × 4 × 2) of an Artificial Neural Network (ANN) for the idendification of ACE-inhibitory dipeptides having an IC_{50} <100 μmol l^{-1}. ANN was calculated with the software EasyNN-plus 2.0c (© 2002-2003 by Stephen Wolstenholme, UK) using the back-propagation algorithm (65). The network was reduced to the minimum size by using pruning while learning, that is, amino acids at nonrelevant sequence positions in regard to ACE-inhibitoty activity were removed from the network. The data set of dipeptides used for ANN training comprised 12 active (IC_{50} 1.7 to 77.4 μmol l^{-1}) and 23 inactive peptides (IC_{50} 122.1 to 21000 μmol l^{-1}).

tract of the mammalian organism (67). Whether the presence of two active catalytic sites in ACE (domains N and C) may account for the discrepancy observed for the *in vitro* and *in vivo* effects of the same peptide is not known. A detailed knowledge of the mechanism of interaction with ACE should lead to a better understanding of the hypotensive potential of food-protein-derived peptides.

13.4 PHENOMENON OF MULTIFUNCTIONALITY OF MILK-PROTEIN-DERIVED ACE INHIBITORS

Many milk protein-derived ACE inhibitors reveal multifunctional properties, that is, specific peptide sequences having two or more different biological activities *in vitro* have been reported (68). In particular, regions in the primary structure of caseins contain overlapping peptide sequences that exert different biological effects. These regions have been considered "strategic zones" that are partially protected from proteolytic breakdown (68–70). The "phenomenon of multifunctionality" makes it sometimes difficult to strictly differentiate the different groups of food-derived bioactive peptides. For example, it is known that opioid peptides reveal a multifunctional activity as both opioid agonists and ACE inhibitors. The opioid fragment β-casomorphin-7, a heptapeptide derived from bovine β-casein, has a low ACE-inhibitory activity ($IC_{50} = 500$ μmol l^{-1}) (23). Whey protein-derived opioid tetrapeptides, α-lactorphin (Tyr-Gly-Leu-Phe) and β-lactorphin (Tyr-Leu-Leu-Phe), have been identified as ACE-inhibitory peptides with low ($IC_{50} = 733$ μmol l^{-1}) and moderate ($IC_{50} = 172$ μmol l^{-1}) activities, respectively (26). Notably, 230-fold lower doses of the weak ACE-inhibitor α-lactorphin were needed to produce a hypotensive effect that was comparable with the synthetic drug Captopril®, which is a very potent ACE-inhibitor having a 120,000-fold higher ACE-inhibitory activity than α-lactorphin (71). Since the hypotensive response of the tetrapeptide was abolished by pretreatment with the specific opioid receptor antagonist naloxone, the mechanism of the blood pressure-lowering effect seems to be due to an interaction with opioid receptors. α-Lactorphin and endogenous opioid peptides share an important structural motif that fits into the binding site of the opioid receptors, that is, the presence of a tyrosine residue at the amino-terminal end and the presence of another aromatic residue, phenylalanine or tyrosine, in the third or fourth position of the opioid sequence (67). Interestingly, it has been shown that the active sites on the somatic ACE also interact with endogenous opioid encephalins (C domain > N domain; see 13.3) (72), which may be an explanation for the ACE-inhibitory activity of opioid peptides and their possible function as competitive inhibitors of ACE.

Besides their function as ACE-inhibitors, some casokinins as well as β-casomorphin-7 have been found to be also cytomodulatory peptides with immunomodulatory, cell growth-stimulating and apoptosis-inducing activities, respectively. The molecular mechanism by which these milk protein-derived peptides exert cytomodulatory effects is not yet defined.

The β-casokinin-10 Tyr-Gln-Gln-Pro-Val-Leu-Gly-Pro-Val-Arg, β-CN (*f*193–202), and the immunocasokinin Thr-Thr-Met-Pro-Leu-Trp, α_{S1}-CN (*f*194–199), function as cytomodulatory peptides by stimulating phagocytosis of sheep red blood cells by murine peritoneal macrophages and by exerting a protective effect against *Klebsiella pneumoniae* infection in mice after intravenous administration of peptides (73). The C-terminal β-casein sequence 193–209 containing β-casokinin decapeptide induced a significant proliferative response in rat lymphocytes (74). For β-casokinin-10 (*f*193–202) and β-casomorphin-7, Kayser and Meisel (75) reported that proliferation of human peripheral blood lymphocytes was suppressed at lower concentrations ($<10^{-7}$ mol l^{-1}) but stimulated at higher concentrations *in vitro*. β-Casomorphin-7 inhibits the proliferation of human colonic lamina propria lymphocytes, where the antiproliferative effect of micromolar concentrations was reversed by the opiate receptor antagonist naloxone (76), that is, opioid peptides may affect the immunoreactivity via the opiate receptors located on lymphocytes. Immunomodulatory peptides may contribute to an overall immunostimulatory response and thus may be favorable to improve immunological function. The oral ingestion of immunomodulatory peptides is claimed to enhance mucosal immunity (77), and the immunostimulating activities may have a direct effect on the resistance to bacterial and viral infections of newborn and adult humans.

The casokinin β-CN(*f*177–183) has been shown to exert a cytomodulatory, DNA synthesis-stimulating activity in mouse fibroblast cells BALB/c3T3 (78). It is possible that cell growth-promoting peptides derived from casein as well as other growth factors in milk cooperatively stimulate the development of the digestive tract of neonatal infants.

There is increasing evidence of the possible involvement of milk protein-derived peptides as specific signals that can trigger viability of cancer cells, that is, proliferation and apoptosis, in different human cell culture models (79,80). Accordingly, apoptosis of human leukemia cells (HL-60) was induced by β-casomorphin-7, which can therefore be classsified as an opioid and an ACE-inhibitory/-cytomodulatory casein peptide. Milk protein-derived inducers of apoptosis may be of significance in the control of malignant cell proliferation. A vast majority of tumor promoters are potent inhibitors of apoptosis (81), and therefore apoptosis-inducing peptides can be classified as probable human anticarcinogens. The primary target site for possible apoptotic stimuli against malignant cells could be the gastrointestinal tract. The antiproliferative effects of cytomodulatory peptides, including multifunctional ACE inhibitors, in colon cancer cell lines suggests that they could have a role in the prevention of colon cancer by blocking hyperproliferation of the epithelium and by promoting apoptosis (79,82,83).

13.5 PROTEOLYTIC ACTIVATION OF ENCRYPTED ACE-INHIBITORY PEPTIDES

The release of bioactive peptides from their immediate precursor sequence is a prerequisite for any functional role in the living system. Several reports definitively show that bioactive peptides are produced *in vivo* following intake of milk proteins. β-Casomorphins (84–86) and caseinophosphopeptides (87,88) were found in the intestinal contents of mammals fed milk or casein-containing diets. After ingestion of milk or yogurt by adult humans, Chabance et al. (89) detected many casein-derived peptides in gastrointestinal samples, including a fragment of α_{S1}-casokinin (residues 25–32), several β-caseinophosphopeptides, two lactoferrin peptides, and the minimal antithrombotic sequence κ-CN (*f*112–116). Most recently, the presence of caseinophosphopetides in the human distal small intestine (ileum) has been demonstrated for the first time after ingestion of milk or CPP preparations (90).

In addition to their possible liberation during intestinal digestion, ACE-inhibitory peptides may already be generated during manufacture of several milk products and thus may be ingested as food components (91). A number of bacterial species used in food production may produce various bioactive peptides (92,93). The proteolytic system of lactic acid bacteria can contribute to the liberation of bioactive peptides. Peptide inhibitors of ACE have been reported in milk fermented with cheese (94–100) and yogurt (15,25,93) starter cultures. *In vitro*, the purified cell wall proteinase of *Lactococcus lactis* was shown to liberate oligopeptides from β- and α-caseins, which contain amino acid sequences present in casokinins as well as casomorphins and immunopeptides (101). The further degradation of these peptides by endopeptidases and exopeptidases of lactic acid bacteria could lead to the liberation of bioactive peptides in fermented milk products. Recently, Leclerc et al. (102) have shown that caseins appear to be a better substrate than whey proteins for extracellular proteinases, and proteolysis is enhanced in a higher protein content medium of reconstituted milk.

The proteolytic activity of commercial lactic acid starters followed by further proteolysis with pepsin and trypsin was needed to liberate ACE-inhibitory peptides from casein or whey *in vitro* (15). However, bacterial proteolysis appears to be essential to produce ACE inhibitors in fermented milk (102), and some strains of *Lactobacillus helveticus* are known to produce potent ACE-inhibitors during milk fermentation (*see* "Origin of Peptide" in Table 13.2).

Cheese undergoes a series of complex changes during ripening, which are caused by indigenous milk proteinases, milk-clotting enzymes, microbial proteinases from lactic starter cultures, and other microorganisms that are adventitious or are added (103). Protein conversion to peptides and amino acids is the most characteristic process in cheese manufacture. During ripening of cheese, the casein proteins are continuously broken down to yield a variety of peptides, and the total amount of peptides progressively increases during ripening. The peptide patterns, especially in the low-molecular-mass range, are clearly dependent on ripening time and temperature, manufacturing history, and water and salt content.

It has been shown that secondary proteolysis during cheese ripening leads to the formation of ACE-inhibitory peptides (94). The ACE-inhibitory activity, which is mainly the result of a complex mixture of low-molecular-weight water-soluble peptides, increases during cheese maturation but decreases when the proteolysis exceeds a certain level. Saito et al. (97) evaluated the ACE-inhibtory and hypotensive activities of water-soluble peptides from several types of ripened cheeses (Gouda, Emmental, Blue, Camembert, Edam, Havarti). The strongest inhibitory effect *in vitro* as well as *in vivo* was detected in 8-month-aged Gouda cheese. The moderately active casokinins β-CN (*f*58–72) and β-CN (*f*193/194–209) were identified in the water-soluble extract from Norvegia (Gouda-type) cheese (104). The nonapeptide β-CN (*f*60–68), possessing both opioid and ACE-inhibitory activities, has been found in Cheddar cheese after a 6-month ripening period (105), and several peptides containing β-casomorphin sequences have been identified in Parmesan cheese (106). Given the above, ACE-inhibitory peptides exist naturally in fermented milk products and in ripened cheese types. Hence, ripened-type cheeses may be considered as "naturally occurring functional foods" (*see* 13.8).

In vitro digestion of milk proteins with various endoproteinase activities such as trypsin, chymotrypsin, subtilisin (Alcalase™), and proteinase K results in the release of ACE-inhibitory peptides (15,22,28). The most frequently used process for enrichment of the ACE-inhibitory peptides in food-protein hydrolyzates is membrane processing. Processing through systems fitted with membranes having defined molecular mass cut-off values is an efficient approach to enriching ACE-

TABLE 13.2
Hypotensive Effects of Some Food Protein Derived Peptides and Food Protein Hydrolyzates or Fractions in Spontaneously Hypertensive Rats (SHR)

Origin of Peptide	Main Bioactive Peptides[a]	Dose[g]	Study Design[h]	Results[i]	Ref.
Lactobacillus helveticus LBK 16H fermented skim milk	Mix containing VPP and IPP	*Ad libitum*	SHR, 6 wk old, Finland, Oral, 12 wk administration and 4 wk follow-up.	SBP - 17mmHg ($p<0.001$) after 12 wks	135
	VPP (IC_{50}: 9 µmol l^{-1}) + IPP (IC_{50}: 5 µmol l^{-1})	c. 2.5–3.5 mg kg^{-1} BW day^{-1}	SHR, 6 wk old, Finland, Oral, Intakes were adjusted to approximate that consumed in fermented milk.	SBP - 12 mmHg ($p<0.001$) after 12 wks	
Lactobacillus helveticus and *Saccharomyces cerevisiae* fermented skim milk	Fermented sour milk containing VPP and IPP	5 ml kg^{-1} BW	SHR, 22–26 wk old, Japan, SOD, Placebo controlled	SBP - 20.0 ± 5.2, -21.8 ± 4.2, -17.7 ± 3.5 mmHg ($p<0.05$) after 4, 6, and 8 h, respectively. SBP returned to normal after 24 h. No significant effect on WKY SBP	20
	IPP (IC_{50}: 5 µmol l^{-1})	0.3 mg kg^{-1} BW	SHR, 22–26 wk old, Japan, SOD, Placebo controlled	SBP - 16.2 ± 5.8, -21.7 ± 4.1, -24.1 ± 2.2, -28.3 ± 4.8 mmHg ($p< 0.05$) after 2, 4, 6, and 8 h, respectively. SBP returned to normal after 24 h. Dose dependent response	
	VPP (IC_{50}: 9 µmol l^{-1})	0.6 mg kg^{-1} BW	SHR, 22–26 wk old, Japan, SOD, Placebo controlled	SBP - 24.6 ± 3.5, -32.1 ± 5.3, -28.8 ± 3.6, -26.7 ± 5.3 mmHg ($p< 0.01$) after 2, 4, 6, and 8 h, respectively. SBP returned to normal after 24 h. Dose dependent response	

Caseinate-enriched milk	n.d.	1.0–2.5 g/kg BW	SHR, 18 wk old, Canada, SOD	MAP - 13.4 to -19.0 mmHg ($p < 0.05$) after 5–7 h	102
L. helveticus R211 fermented caseinate-enriched milk	n.d.	0.5 g kg^{-1} BW	SHR, 18 wk old, Canada, SOD	MAP - 13.0 mmHg ($p < 0.05$) after 5–7 h	
		1.0 g kg^{-1} BW	SHR, 18 wk old, Canada, SOD	MAP - 12.5 mmHg ($p < 0.05$) after 5–7 h	
		2.5 g kg^{-1} BW	SHR, 18 wk old, Canada, SOD	MAP - 24.9 mmHg ($p < 0.05$) after 5–7 h	
Skim milk fermented with *L. helveticus* LBK 16H	IPP, VPP + other peptides	0.4 mg day^{-1} IPP 0.6 mg day^{-1} VPP	SHR, 6 wk old, Finland, Placebo controlled 14 wk feeding trial	SBP - 21 mmHg ($p < 0.001$) after 14 wks. Plasma rennin activity increased. Significant differences in mineral levels existed between test samples and controls	136
Casein hydrolyzed by extracellular proteinase of *L. helveticus* CP790	KVLPVPQ, β-CN (*f*169–175), IC_{50}: 1000 μmol l^{-1}	1.0 mg kg^{-1} BW	SHR, 18–25 wk old, Japan, SOD	SBP - 24.1 ± 7.8 mmHg after 6 h. SOD of 2 mg kg^{-1} BW of this peptide resulted in a SBP reduction of -31.5 ± 5.6 mmHg	14
	LQSW, β-CN (*f*140–143), IC_{50}: 500 μmol l^{-1}	1.0 mg kg^{-1} BW		SBP - 1.6 ± 3.4 mmHg after 6 h	
	RELEEL, β-CN (*f*1–6), IC_{50}: >1000 μmol l^{-1}		SHR, 18–25 wk old, Japan, SOD	SBP + 1.3 ± 3.0 mmHg after 6 h	
	TKVIP, α_{S2}-CN (*f*198–202), IC_{50}: 400 μmol l^{-1}		SHR, 18–25 wk old, Japan, SOD	SBP - 9.2 ± 7.0 mmHg after 6 h	
	AMKPW, α_{S2}-CN (*f*189–192), IC_{50}: 580 μmol l^{-1}		SHR, 18–25 wk old, Japan, SOD	SBP - 4.6 ± 5.8 mmHg after 6 h	
	YKVPQL, α_{S1}-CN (*f*104–109), IC_{50}: 22 μmol l^{-1}		SHR, 18–25 wk old, Japan, SOD	SBP - 12.5 ± 5.2 mmHg after 6 h	
	MKPWIQPK, α_{S2}-CN (*f*190–197), IC_{50}: 300 μmol l^{-1}		SHR, 18–25 wk old, Japan, SOD	SBP - 2.9 ± 5.9 mmHg after 6 h	
	LLYQQPV, β-CN (*f*191–197), IC_{50}: >1000 μmol l^{-1}		SHR, 18–25 wk old, Japan, SOD	SBP + 1.2 ± 4.4 mmHg after 6 h	

(*Continued*)

TABLE 13.2 (Continued)

Origin of Peptide	Main Bioactive Peptides[a]	Dose[g]	Study Design[h]	Results[i]	Ref.
	AYFYP, α_{S1}-CN (*f*143–147), IC_{50}: >1000 μmol l^{-1}		SHR, 18–25 wk old, Japan, SOD	SBP - 4.3 ± 3.3 mmHg after 6 h	
	AMKPWIQPK, α_{S2}-CN (*f*189–197), IC_{50}: 600 μmol l^{-1}		SHR, 18–25 wk old, Japan, SOD	SBP - 0.6 ± 4.2 mmHg after 6 h	
Whey fraction of skim milk fermented by *L. helveticus* CPN4	YP α_{S1}-CN (*f*146–147), β-CN (*f*114–115), κ-Cn (*f*58–59)IC_{50}: 720 μmol l^{-1}	0.1 mg kg^{-1} BW	SHR, 18–21 wk old, Japan, SOD	SBP - 10.2 ± 4.0 mmHg ($p<0.05$) after 6 h	24
		0.3 mg kg^{-1} BW	SHR, 18–21 wk old, Japan, SOD	SBP - 20.8 ± 4.1 mmHg ($p<0.001$) after 6 h	
		1.0 mg kg^{-1} BW	SHR, 18–21 wk old, Japan, SOD	SBP - 27.4 ± 7.1 mmHg ($p<0.001$) after 6 h	
		3.0 mg kg^{-1} BW	SHR, 18–21 wk old, Japan, SOD	SBP - 30.5 ± 6.7 mmHg ($p<0.001$) after 6 h	
		10.0 mg kg^{-1} BW	SHR, 18–21 wk old, Japan, SOD	SBP - 32.1 ± 7.4 mmHg ($p<0.001$) after 6 h, antihypertensive response was time dependent	
α-Lactalbumin	α-Lactorphin (α-LA) YGLF IC_{50}: 733.3 μmol l^{-1}	0.1 mg kg^{-1} BW	SHR, 18–26 wk, Finland, Subcutaneous administration	SBP - 23 ± 4 mmHg ($p<0.05$) - and DBP 17 ± 4 mmHg ($p<0.05$) after 50–100 min. No further BP reductions observed at higher doses. No significant differences in BP response to α-LA by SHR or WKY. Nalaoxone inhibited α-LA induced BP reduction	71
Water-soluble Gouda cheese peptides	RPKHPIKHQ α_{S1}-CN (*f*1–9), IC_{50}: 13.4 μmol l^{-1}	6.1–7.5 mg kg^{-1} BW	SHR, 12–16 wks, Japan, SOD	SBP - 9.3 ± 4.8 mmHg	97
	YPFPGPIPN β-CN (*f*60–68), IC_{50}: 14.8 μmol l^{-1}	6.1–7.5 mg kg^{-1} BW	SHR, 12–16 wk, Japan, SOD	SBP - 7.0 ± 3.8 mmHg	

Tryptic hydrolysate of casein	Whole hydrolysate (including peptides below)	610 mg kg^{-1} BW day^{-1}	SHR, 9 wk old, Japan, 4 wk trial	SBP - 14.0 mmHg ($p < 0.05$) after 4 wks	137
	FFVAPFPEVFGK, α_{S1}-CN (*f*23–34), IC$_{50}$: 18 μmol l^{-1}	100 mg kg^{-1}	SHR, 18–22 wk old, Japan, SOD	SBP - 34.0 ± 13.0 mmHg after 3–5 h	
	TTMPLW, α_{S1}-CN (*f*194–199), IC$_{50}$: 12 μmol l^{-1}	100 mg kg^{-1}	SHR, 18–22 wk old, Japan, SOD	SBP - 13.6 ± 3.6 mmHg after 3 h	
	AVPYPQR, *β*-CN, IC$_{50}$: 15 μmol l^{-1}	100 mg kg^{-1}	SHR, 18–22 wk old, Japan, SOD	SBP - 10.0 ± 0.7 mmHg after 3 h	
Whey protein hydrolyzates and derived peptides	Unhydrolyzed whey protein	8 mg kg^{-1} BW	SHR, 12 wk old, Japan, SOD	SBP - 38.0 ± 1.7 mmHg after 6 h	22
	Pepsin hydrolysate of whey	8 mg kg^{-1} BW	SHR, 12 wk old, Japan, SOD	SBP - 47.0 ± 2.6 mmHg after 6 h	
	Trypsin hydrolysate of whey	8 mg kg^{-1} BW	SHR, 12w k old, Japan, SOD	SBP - 51.0 ± 3.5 mmHg ($p \leq 0.05$) after 6 h	
	Chymotrypsin hydrolysate of whey	8 mg kg^{-1} BW	SHR, 12 wk old, Japan, SOD	SBP - 40.0 ± 4.4 mmHg after 6 h	
	Proteinase K hydrolysate of whey	8 mg kg^{-1} BW	SHR, 12 wk old, Japan, SOD	SBP - 55.0 ± 2.6 mmHg ($p \leq 0.01$) after 6 h	
	Actinase E hydrolysate of whey	8 mg kg^{-1} BW	SHR, 12 wk old, Japan, SOD	SBP - 55.0 ± 4.4 mmHg ($p \leq 0.01$) after 6 h	
	Thermolysin hydrolysate of whey	8 mg kg^{-1} BW	SHR, 12 wk old, Japan, SOD	SBP - 42.0 ± 3.5 mmHg after 6 h	
	Papain hydrolysate of whey	8 mg kg^{-1} BW	SHR, 12 wk old, Japan, SOD	SBP - 47.0 ± 3.6 mmHg after 6 h	
	Proteinase K–liberated whey peptides	8 mg kg^{-1} BW	SHR, 12 wk old, Japan, SOD	SBP - 22.0 ± 4.6 mmHg ($p \leq 0.05$) after 6 h	
	VYPFPG, *β*-CN[b] (*f*59–64); IC$_{50}$: 221.0 μmol l^{-1}	8 mg kg^{-1} BW	SHR, 12 wk old, Japan, SOD	SBP - 26.0 ± 4.4 mmHg ($p \leq 0.05$) after 6 h	

(Continued)

TABLE 13.2 (Continued)

Origin of Peptide	Main Bioactive Peptides[a]	Dose[g]	Study Design[h]	Results[i]	Ref.
	GKP, β_2-m[c] (*f*18–20); IC_{50}: 352.0 µmol l^{-1}	8 mg kg^{-1} BW	SHR, 12 wk old, Japan, SOD	SBP - 31.0 ± 6.1 mmHg ($p \leq 0.01$) after 6 h	
	IPA, *β*-lactoglobulin (*f*78–80), IC_{50}: 141.0 µmol l^{-1}	8 mg kg^{-1} BW	SHR, 12 wk old, Japan, SOD	SBP - 27.0 ± 4.4 mmHg ($p \leq 0.05$) after 6 h	
	FP [BSA (*f*221–222) and [b], IC_{50}: 315.0 µmol l^{-1}	8 mg kg^{-1} BW	SHR, 12 wk old, Japan, SOD	SBP - 21.0 ± 4.6 mmHg ($p \leq 0.05$) after 6 h	
	VYP, *β*-CN[b] (*f*59–61), IC_{50}: 288.0 µmol l^{-1}	8 mg kg^{-1} BW	SHR, 12 wk old, Japan, SOD	SBP - 8.0 ± 2.6 mmHg after 6 h	
	TPVVVPPFLQP, *β*-CN[b] (*f*80–90), IC_{50}: 749.0 µmol l^{-1}	8 mg kg^{-1} BW	SHR, 12 wk old, Japan, SOD		
Casein and soy protein	Standard rat chow 1:4 casein : standard rat chow 1:4 soy protein : standard rat chow	*Ad libitum* *Ad libitum* *Ad libitum*	SHR, 7 wk old, Male and Female, Finland, 5 wk study	No BP values detailed in report. However, no significant (p = 0.19) reduction in SBP of female SH after 5 weeks consuming either casein or soy blend diet. SBP was significantly (p<0.05) reduced only in male SHR that were fed soy-rat chow blend	138
Katsuo-bushi (dried bonito) hydrolyzed by thermolysin (FOSHU)	LKPNM, IC_{50}: 2.4 µmol l^{-1}	8 mg kg^{-1} BW	SHR, 16–25 wk old, Japan, SOD	SBP - 23 mmHg at 4 h	139
	LKP, IC_{50}: 0.32 µmol l^{-1}	2.25 mg kg^{-1} BW	SHR, 16–25 wk old, Japan, SOD	SBP - 18 mmHg at 2 h	
	LKPNM	100 mg kg^{-1} BW	SHR, 16–25 wk old, Japan, i.v.	SBP - 30 mmHg	
	LKP	30 mg kg^{-1} BW	SHR, 16–25 wk old, Japan, i.v.	SBP - 50 mmHg	
Sardine muscle hydrolyzed with *Bacillus licheniformis* alkaline protease	VY, IC_{50}: 7.1 µmol l^{-1}	20 mg kg^{-1} BW	SHR, 12 wk old, Japan, i.v.	SBP - 7.2 mmHg & DBP - 28.8 mmHg (p<0.01)	140
		50 mg kg^{-1} BW	SHR, 12 wk old, Japan, i.v.	SBP - 18 mmHg (p<0.05) and DBP - 35 mmHg (p<0.01)	

Bonito muscle hydrolyzed by thermolysin	IY, IC_{50}: 2.1 μmol l^{-1}	10 mg kg^{-1} BW by i.v. and 60 mg kg^{-1} BW for SOD	SHR, Japan, i.v. and SOD	SBP - 45.0 mmHg (i.v.), SBP - 19.0 mmHg (SOD + 2 h)	32
	IW, IC_{50}: 5.1 μmol l^{-1}			SBP - 55.0 mmHg (i.v.), SBP - 22.0 mmHg (SOD + 2 h)	
	IKP, IC_{50}: 1.6 μmol l^{-1}			SBP - 70.0 mmHg (i.v.), SBP - 20.0 mmHg (SOD + 6 h)	
	IWH, IC_{50}: 3.5 μmol l^{-1}			SBP - 70.0 mmHg (i.v.), SBP - 30.0 mmHg (SOD + 4 h)	
	IVGRPR, IC_{50}: 300.0 μmol l^{-1}			SBP - 25.0 mmHg (i.v.), SBP - 17.0 mmHg (SOD + 6 h)	
	LKPNM[d], IC_{50}: 2.4 μmol l^{-1}			SBP - 80.0 mmHg (i.v.), SBP - 23.0 mmHg (SOD + 6 h)	
	IWHHT[e], IC_{50}: 5.8 μmol l^{-1}			SBP - 60.0 mmHg (i.v.), SBP - 26.0 mmHg (SOD + 6 h)	
	IVGRPRHQG[f], IC_{50}: 2.4 μmol l^{-1}			SBP 0.00 mmHg (i.v.), SBP - 14.0 mmHg (SOD + 8 h)	
Porcine hemoglobin hydrolyzed with Alcalase®	FQKVVA, IC_{50}: 5.8 μmol l^{-1}	50 mg kg^{-1} BW	SHR, 12–13 wk old, Japan, SOD	SBP - 30 mmHg ($p<0.01$) after 180 min	37
	GKKVLQ, IC_{50}: 1.9 μmol l^{-1}	50 mg kg^{-1} BW	SHR, 12–13 wk old, Japan, SOD	SBP - 30 mmHg ($p<0.01$) after 180 min In this study 10 mg kg^{-1} Captopril SBP of SHR by -22 mmHg in 30 min	37
Chicken muscle hydrolyzed by thermolysin	LAP, IC_{50}: 3.5 μmol l^{-1}	10 mg kg^{-1} BW by i.v. and 60 mg kg^{-1} BW for SOD	SHR, Japan, i.v. and SOD	SBP - 40.0 mmHg (i.v.)	32
	IKW, IC_{50}: 0.21 μmol l^{-1}			SBP - 50.0 mmHg (i.v.), SBP - 17.0 mmHg (SOD + 4 h)	
	LKP, IC_{50}: 0.32 μmol l^{-1}			SBP - 75.0 mmHg (i.v.), SBP - 18.0 mmHg (SOD + 4 h)	
	FKGRYYP, IC_{50}: 0.55 μmol l^{-1}			No change of SBP after i.v. or SOD	
Royal jelly hydrolyzed	Peptide mixture, IC_{50}: 353 mg l^{-1}	1 g kg^{-1} BW	SHR, 10 wk old, Japan, SOD	SBP -22.7 mmHg ($p<0.05$) after 2 h and maintained for up to 6 h,	141

(Continued)

TABLE 13.2 (Continued)

Origin of Peptide	Main Bioactive Peptides[a]	Dose[g]	Study Design[h]	Results[i]	Ref.
by pepsin, trypsin, and chymotrypsin				heart rate did not change	
Extract of *Lactobacillus casei* YIT9018 cell lysate	Polysaccharide-glycopeptide complex	10 mg kg^{-1} BW	SHR, 16–18 wk old, Japan, SOD	SBP - 7.2 ± 3.3 mmHg ($p<0.05$) at 3 h SBP - 14.4 ± 2.2 mmHg ($p<0.01$) at 6 h SBP - 15.2 ± 3.3 mmHg ($p<0.05$) at 12 h SBP - 10.0 ± 2.3 mmHg ($p<0.05$) at 24 h	142
Wheat germ alkaline protease hydrolysate	Ion-exchange fraction containing the peptide IVY	25 mg kg^{-1} BW	SHR, 12 wk old, Japan, i.v.	MAP - 8.5 mmHg after 5 min	44
	IVY, IC_{50}: 0.48 µmol l^{-1}	50 mg kg^{-1} BW	SHR, 12 wk old, Japan, i.v.	MAP - 10.3 mmHg ($p<0.01$) after 13 min	
		2 mg kg^{-1} BW	SHR, 12 wk old, Japan, i.v.	MAP - 5.7 mmHg after 13 min	
		5 mg kg^{-1} BW	SHR, 12 wk old, Japan, i.v.	MAP - 19.2 mmHg after 8 min	
α-Zein hydrolyzed by thermolysin	LRP, IC_{50}: 0.27 µmol l^{-1}	30 mg kg^{-1} BW	SHR, 11 wk old, Japan, i.v.	Max BP reduction was –15.0 mmHg after 2 min; BP returned to baseline levels after 5 min	40
Sake and Sake lees	YGGY, IC_{50}: 16.2 µmol l^{-1}	100 mg kg^{-1} BW	SHR, 9 wk old, Japan, SOD	No change to SBP	45
	IYPRY, IC_{50}: 4.1 µmol l^{-1}		SHR, 9 wk old, Japan, SOD	max SBP - 21.0 mmHg ($p<0.01$) after 6 h	
	IY, IC_{50}: 2.4 µmol l^{-1}		SHR, 9 wk old, Japan, SOD	max SBP - 18.0 mmHg ($p<0.01$) after 6 h	
	YP, IC_{50}: ?		SHR, 9 wk old, Japan, SOD	max SBP - 13.0 mmHg ($p<0.05$) after 6 h	
	RY, IC_{50}: 10.5 µmol l^{-1}		SHR, 9 wk old, Japan, SOD	max SBP - 17.0 mmHg ($p<0.01$) after 6 and 24 h	
Soy protein meal hydrolyzed with Alcalase 2.4L®	Peptide mixture	100 mg kg^{-1} BW day^{-1} 500 mg kg^{-1} BW day^{-1}	SHR, 7 wk old, China, 30 day trial	SBP - 37.8 mmHg ($p<0.05$) after 30 days SBP - 31.5 mmHg ($p<0.05$) after 24 days	143

		1000 mg kg^{-1} BW day^{-1}		SBP - 39.0 mmHg ($p<0.05$) after 30 days	
		50 mg kg^{-1} day^{-1} Captopril		SBP - 46.0 mmHg ($p<0.05$) after 30 days	
Soybean paste	HHL	3 × (5 mg kg^{-1} BW)	SHR, Korea, i.v.	SBP - 61.0 mmHg ($p<0.01$) after third injection	43
Garlic	SY, IC_{50}: 66.3 GY, IC_{50}: 72.1 FY, IC_{50}: 3.74 NY, IC_{50}: 32.6 SF , IC_{50}: 130.2 GF, IC_{50}: 277.9 NF, IC_{50}: 46.3	200 mg kg^{-1} BW	SHR, 12 wk old, Japan, SOD	No specific change of SBP values detailed in report. Maximal decreases in SBP were observed within 1 to 4 h after administration of each dipeptide Blood pressure lowering abilities of each dipeptide in SHRs were reported to be comparable to that which could be attained following administration of 10 mg kg^{-1} BW of Captopril	46
Ovalbumin hydrolyzed with bovine chymotrypsin	Ovokinin (*f*2–7) RADHPF IC_{50}: 1000 μmol l^{-1}	20 mg kg^{-1} BW	SHR, 15–30 wk old, Japan, SOD	SBP - 12.5 mmHg ($p<0.001$) after 6 h and was dose-dependent. Vasorelaxive effects attributed to nitric oxide release from endothelial cells. More potent effects seen with SHR than WKYs.	144
Ovalbumin hydrolyzed by pepsin	LW, IC_{50}: 6.8 μmol l^{-1}	10 mg kg^{-1} BW by i.v. and 60 mg kg^{-1} BW for SOD	SHR, Japan, i.v. and SOD	SBP -40.0 mmHg (i.v.)	32
	FFGRCVSP, IC_{50}: 0.4 μmol l^{-1}			SBP -50.0 mmHg (i.v.), ΔSBP -17.0 mmHg (SOD + 4 h)	

(*Continued*)

TABLE 13.2 (Continued)

Origin of Peptide	Main Bioactive Peptides[a]	Dose[g]	Study Design[h]	Results[i]	Ref.
	ERKIKVYL, IC_{50}: 1.2 μmol l^{-1}			SBP -75.0 mmHg (i.v.), ΔSBP -18.0 mmHg (SOD + 4 h) No change of SBP after i.v. or SOD	
Chicken egg yolk hydrolysate	Newlase F hydrolysate of yolk	20, 100, 500 mg kg^{-1} BW day^{-1}	SHR, 5 wk old, Japan, SOD, 12 wk trial	No SBP values provided. However, SBP, mean BP and DBP were lower than the control from week 3 up to week 12	145
	Newlase F + pepsin hydrolysate of yolk	20, 100, 500 mg kg^{-1} BW day^{-1}			
	Newlase F + trypsin hydrolysate of yolk	20, 100, 500 mg kg^{-1} BW day^{-1}			
	Newlase F + chymotrypsin hydrolysate of yolk	20, 100, 500 mg kg^{-1} BW day^{-1}			
	Newlase F + pepsin, trypsin, chymotrypsin hydrolysate of yolk				

[a] One letter amino acid code; IC_{50}, concentration of peptide or peptide material mediating a 50% inhibition of ACE activity
[b] β-casein fragment hydrolyzed by plasmin existing as a proteose-peptone in cheese whey
[c] β_2-microglobulin; BSA, bovine serum albumin
[d] Prodrug activated by ACE
[e] Prodrug activated by chymotrypsin
[f] Prodrug activated by trypsin
[g] BW, body weight; SOD, single oral dose
[h] SHR, spontaneously hypertensive rats; SOD, single oral dose; i.v., intravenous administration
[i] SBP, systolic blood pressure; DBP, diastolic blood pressure; MAP, mean arterial blood pressure calculated using the equation MAP = DBP + [(SBP – DBP)/3]; Values expressed ± standard error; mm Hg, millimeters of mercury; WKY, Wistar-Kyoto rats

inhibitory peptides in permeate streams. This has, for example, been demonstrated in the case of bonito-bowel autolysates (34), milk-protein hydrolyzates (25,28), cod frame-protein hydrolyzates (107), plasma protein hydrolyzates (108), and soy protein hydrolyzates (42). It generally appears that the most potent ACE-inhibitory peptides are associated with permeates collected using low-molecular-mass cut-off membranes.

The bioactive potential of ACE-inhibitors that have been produced by limited proteolysis during processing and intestinal digestion of milk proteins is dependent on the ability of these peptides to reach their target sites without being inactivated. These peptides therefore need to survive degradation by gastrointestinal proteinases and peptidases; they need to pass from the intestine to the serum where they may be susceptible to brush-border and intracellular peptidase activities; and they need to be resistant to degradation by serum peptidases (109). In any case, ACE inhibitors could be further digested by intestinal proteinases or brush-border peptidases which may lead to the release of further bioactive peptide fragments. It is also likely that nonactive precursors containing the bioactive sequence could enter the blood stream and reach potential sites of action in the body to elicit effects after proteolytic release, for example, in plasma, of the protected active sequence from the precursor molecule.

ACE-inhibitory peptides may be absorbed via carrier-mediated transport or via paracellular transport. The paracellular permeability is regulated by intestinal tight junctions, and modulation of tight-junction structure by other food substances facilitates paracellular transport (110). Small ACE-inhibitory di- and tripeptides can directly pass across the intestine to reach peripheral target sites without being decomposed by digestive enzymes (111). Once they are activated and liberated in the body, bioactive peptides may function as exogenous regulatory substances that may act on different intestinal and peripheral target sites of the mammalian organism as outlined in section 13.6.

13.6 HYPOTENSIVE MECHANISMS

Blood pressure (BP) is controlled by a number of different biochemical pathways within the body and can be increased or decreased depending on which pathway predominates. The main biochemical pathways involved in BP responses are the renin–angiotensin system (RAS), the renin–chymase system (RCS), the kinin–nitric oxide system (KNOS), and the neutral endopeptidase system (NEPS) (61,112–114). Figure 13.4 summarizes the different pathways associated with these systems and clearly illustrates the central role of ACE in BP regulation. Together, these systems generate a multitude of endogenous bioactive peptides that regulate BP as well as fluid and electrolyte balance via membrane-bound receptors located on different tissues throughout the body.

13.6.1 The Renin–Angiotensin System (RAS)

The RAS regulates BP, electrolyte balance, and renal, neuronal, and endocrine functions associated with cardiovascular control in the body. RASs specific to the brain

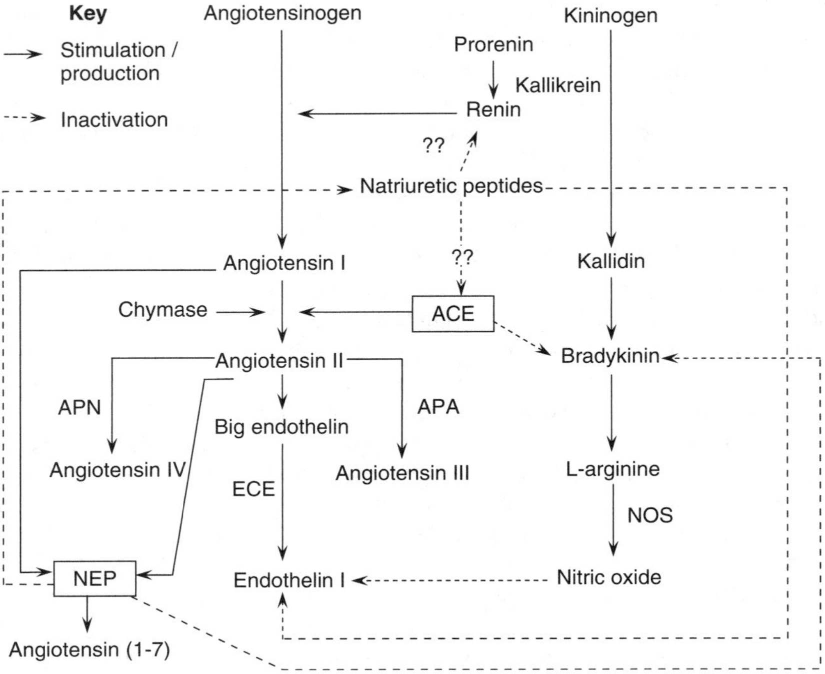

FIGURE 13.4 Schematic of the biochemical pathways involved in the regulation of blood pressure (Angiotensin-converting enzyme [ACE], aminopeptidase N [APN], aminopeptidase A [APA], endothelin converting enzyme [ECE], nitric oxide synthase [NOS], neutral endopeptidase [NEP].

(115,116), placenta (117), bone marrow (118), and pancreas (119) have been identified. In the RAS, renin (EC 3.4.22.15) is released from the precursor compound prorenin through the action of kallikrein (EC 3.4.21.34) (61,120). Renin cleaves angiotensinogen to release angiotensin I (Ang-I, see Figure 13.4). Ang-I (a decapeptide, Asp-Arg-Val-Tyr-Ile-His-Pro-Phe-His-Leu) is hydrolyzed by ACE removing the C-terminal dipeptide HL, yielding the potent vasoconstrictor angiotensin II (Ang-II). ACE also hydrolyses bradykinin (BK) to inactive fragments. In addition to its hydrolytic activity, ACE has been hypothesized to act as a mechanosensor that senses shear stress or changes in blood velocity in regions of turbulent blood flow (121). Ang-II controls blood pressure, body fluid volume, the activity of gonadotrophic-hormone-releasing hormone (GHRH) and pituitary hormones during the reproductive cycle and pregnancy, in addition to controlling neurotransmitter interactions (115). The main receptors of Ang-II are $AT_{1,2,3}$. The AT_1 receptors are responsible for vasoconstriction, proliferation, and hypertrophy, while AT_2 receptors have vasodilatory, antiproliferative, and apototic functions (59). The role of the AT_3 receptor is not currently well defined. Ang-II also activates endothelin-converting enzyme (ECE, EC 3.4.24.71), which produces endothelin I (End-I), the most potent naturally occurring vasoconstrictor, within the endothelial cells of blood vessels (122) (see Figure 13.4). End-I enhances the formation of Ang-II from Ang-I, indicating that End-I in some way modifies ACE activity (123). End-I also plays a role in lipid metabolism (124). Aminopeptidase A (APA), a glutamyl aminopeptidase (EC 3.4.11.7), hydrolyses Ang-II to produce des-Asp-Ang-II or Ang-III, Arg-Val-Tyr-Ile-His-Pro-Phe (125) (see Figure 13.4). Ang-III acts as a vasoconstrictory substance via binding to the AT_1 and AT_2 receptors. Angiotensin (Ang-IV), Val-Tyr-Ile-His-Pro-Phe, is formed by the action of aminopeptidase N (APN, EC 3.4.11.2) on Ang-II (see Figure 13.4). This hexapeptide can increase renal blood flow and natriuresis and may also be involved in memory retention/retrieval in the brain (59). Ang-IV acts through the AT_4 receptor. ACEH has been shown to hydrolyse Ang-I and Ang-II to Ang (1–9). ACE, in turn, hydrolyses Ang (1–9) to Ang (1–7). Ang (1–7) appears to have vasodilatory and antiproliferative effects mediated through binding to a novel, not yet defined, Ang (1–7) receptor (126). This peptide specifically binds to the N-terminal domain of the ACE (55).

13.6.2 The Renin–Chymase System (RCS)

The RCS contains all the components associated with the RAS (113). The main exceptions are that the RCS is mainly located in human heart tissue and ACE activity is replaced by chymase (EC 3.4.21.39), which forms Ang-II from Ang-I within this BP-controlling pathway (see Figure 13.4). Chymase does not hydrolyse BK (127).

13.6.3 The Kinin–Nitric Oxide System (KNOS)

α_2-Globulins or kininogens are cleaved by numerous enzymes such as kallikrein to give rise to the kinins, which, in turn, give rise to kallidin, Lys-Arg-Pro-Pro-Gly-Phe-Ser-Pro-Phe-Arg (128) (see Figure 13.4). Kinins are located in vascular and other tissues within the body. Kallidin is further degraded to BK, Arg-Pro-Pro-Gly-Phe-Ser-Pro-Phe-Arg, $[\text{des-Arg}]^9$-BK and $[\text{des-Arg}]^{10}$-kallidin, the latter two peptides being hydrolyzed by ACEH (126). BK is a potent vasodilator, whose effect is mediated via the release of nitric oxide (NO) from L-arginine by the action of NO synthase (NOS) (112,129,130). This vasodilatory pathway is inhibited by the action of ACE, which cleaves Phe-Arg from the C-terminal of BK and thus increases vasoconstriction (131). Ang-II and NO signaling pathways have been shown to interact with each other in a number of different ways. Ang-II acts at both AT_1 and AT_2 receptors to stimulate production of NO. NO generation, in turn, opposes the action of Ang-II and also downregulates the expression of the AT_1 receptor (132). The release of NO from the surface of the vascular endothelium within the blood vessel walls can inhibit the release of End-I (129,133) (see Figure 13.4).

13.6.4 The Neutral Endopeptidase System (NEPS)

Neutral endopeptidase (NEP, EC 3.4.24.11) causes vasoconstriction through a number of actions. These include the hydrolysis of BK to inactive fragments (see

Figure 13.4). NEP hydrolyses natriuretic peptides, which are involved in vasodilatation, and this results in vasoconstriction. NEP can hydrolyse both Ang-I and Ang-II to Ang-1–7, a vasodilatory peptide (114,134) (see Figure 13.4).

Natriuretic peptides are a group of peptides present in the body that can exhibit vasodilatory and tissue protective properties. These responses suggest that natriuretic peptides may be ACE- or renin-inhibitory peptides produced naturally within the body (114). Several types of natriuretic peptides have been identified, that is, atrial-type, brain B-type, and brain C-type, which are found in various locations around the body. Atrial natriuretic peptide has been reported to inhibit production of End-I. It is therefore evident that control of BP is a very complex process involving a number of interacting metabolic pathways.

13.7 HYPOTENSIVE EFFECTS *IN VIVO*

13.7.1 Animal Studies

The ability to inhibit ACE *in vitro* is only indicative of the potential of a given peptide to act as a hypotensive agent *in vivo*. Usually, the first step to investigate the hypotensive potential of ACE-inhibitory peptides *in vivo* is via animal studies using spontaneously hypertensive rats (SHR). Numerous rat studies have been performed to determine the hypotensive effects of food protein-derived ACE inhibitors. The blood pressure responses in SHRs following oral ingestion or intraveneous administration of ACE-inhibitory food protein hydrolyzates and derived peptides are summarized in Table 13.2. As can be seen from this table, systolic BP (SBP) responses ranging from +1.3 to –80 mm Hg have been reported. The wide variation in BP responses may be reflective of variations in sample type, the dosage and duration of administration, the mechanism of delivery, that is, oral or intravenous administration, and the method by which BP was measured.

Animal studies may be used as a tool to verify the action of hypotensive peptides in complex mixtures of hydrolyzates. In a study related to the hypotensive effects of *L. helveticus* LBK 16H fermented skim milk, two tripeptides, that is, Ile-Pro-Pro and Val-Pro-Pro, were administered to SHRs at a dose similiar to that ingested from consumption of the fermented milk drink and were identified as being the main peptides contributing to the hypotensive response (135). A similar approach was employed by Maeno et al. (14) to identify peptides in an *L. helveticus* CP790 extracellular proteinase hydrolysate of casein that mediated hypotensive effects.

The method of peptide delivery appears to have a significant effect on the observed BP response. Notable differences are evident between BP responses in SHRs following oral versus intravenous peptide administration (see Table 13.2). In general, it appears that intraveneously administered samples exert a greater hypotensive response. Although these differences may be, in part, due to dosage differences in some cases, they may also result from gastrointestinal and serum proteinase and peptidase modification of the ingested peptides (*see also* 13.5).

There appears to be no direct relationship between BP reduction and the IC_{50} value for the different peptides examined to date (see Table 13.2). For example, the tetrapeptide Tyr-Gly-Gly-Tyr having an IC_{50} of 16.2 $\mu mol\ l^{-1}$ did not yield any change in systolic BP (SBP) following oral administration at 100 mg kg^{-1} (see Table 13.2). Conversely, Lys-Val-Leu-Pro-Val-Pro-Gln (*β*-casein [*f*169–175]) having an IC_{50} of 1000 $\mu mol\ l^{-1}$ on oral administration at 1 mg kg^{-1} gave a maximal decrease in SPB of –31.5 mm Hg. However, the corresponding hexapeptide Lys-Val-Leu-Pro-Val-Pro, which was obtained after liberation of the C-terminal Gln residue by pancreatic digestion *in vitro*, had a strong ACE-inhibitory activity as well as a remarkable hypotensive effect *in vivo* (14).

As already outlined in Section 13.3, not all hypotensive effects may be attributed to ACE inhibition *per se*. The whey protein-derived peptide *α*-lactorphin (IC_{50}: 733 $\mu mol\ l^{-1}$) demonstrated the ability to reduce SBP and diastolic BP (DBP) by –23 and –17 mm Hg, respectively, following subcutaneous administration of 0.1 mg kg^{-1} body weight in SHRs. However, this BP-reducing effect was inhibited by naloxone, an opioid antagonist (71). Valuable information can be obtained from *in vitro* model systems with respect to the proteolytic/peptideolytic stability and susceptibility to intracellular passage (28,146), however, it is only through *in vivo* studies that the hypotensive effects of a given peptide or peptide preparation can be reliably assessed.

13.7.2 Human Studies

Several human studies with mildly hypertensive subjects definitively demonstrated a significant reduction in BP following daily ingestion of food-derived ACE inhibitors (for details see Table 13.3) (147–155). The origin of hypotensive peptides used for human trials were milk-protein hydrolyzates (casein, whey protein), fermented milk and fish hydrolyzates (sardine, dried bluefish/bonito). Except for the Val-Tyr, isolated from sardine muscle, mixtures of bioactive peptides were administered as ingredients of sour milk or hydrolyzates. It is worth noting that the ingested dose of the main ACE-inhibitory peptides, which were not detectable in the placebo drink and thus are considered to be the active substances, was in the range of 2 to 6 mg. The oral ingestion of such relatively low amounts of peptides may lower BP to an extent that is comparable with a pharmacological treatment. However, no side effects and no significant changes were observed in the BP of normotensive subjects.

13.8 CONSUMER ASPECTS

Various definitions have been used to describe functional foods. The main concept of functional foods is that they are food products or ingredients that are taken as part of the normal diet in order to have beneficial effects beyond those of basic nutrition. Furthermore, functional foods are foods, not drugs, and their role in relation to disease is in "reducing the risk" rather than in "prevention" (156,157). Numerous surveys have demonstrated that consumers are becoming more aware

TABLE 13.3
Human Studies with ACE-Inhibitory Peptide Preparations of Food Origin

Origin of Peptide	Main Bioactive Peptides[a]	Dose	Study Design[b]	Results[b]	References
Milk casein Hydrolyzed by trypsin	n. d.	2 × 10 g hydrolysate day^{-1}	Japan, 4 weeks of treatment, 28 mildly hypertensive subjects, without any drugs and placebo	BP reduction after 4 weeks of treatment: SBP - 4.6 mmHg ($p<0.01$), DBP - 6.6 mmHg ($p<0.01$), pulse rate did not change,	147
Extract of autologous *Lactobacillus casei* YIT9018 cell lysate (lyophilized: LEx)	Polysaccharide-glycopeptide complex (from the cell wall)	2 × 400 mg LEx in white capsules day^{-1}	Japan, randomized double-blind placebo-controlled study (8 weeks), 8 weeks of treatment, 28 mildly hypertensive subjects, SBP/DBP: 169/100 mmHg, antihypertensive drugs were allowed placebo: dextrin	BP reduction after 8 weeks of treatment: SBP - 9 mmHg ($p<0.01$), DBP - 6 mmHg ($p<0.05$), heart rate reduction: - 7 beats min^{-1} ($p < 0.05$)	148
Sour milk (Calpis)[c] fermented with *Lactobacillus helveticus* strains and *Saccharomyces cerevisiae* strains	Ile-Pro-Pro (15 mg l^{-1}) IC_{50}: 5 µmol l^{-1} Val-Pro-Pro (11 mg l^{-1}) IC_{50}: 9 µmol l^{-1}	100 ml sour milk day^{-1} containing 2.6 mg tripeptides	Japan, randomized double-blind placebo controlled study (16 weeks), 8 weeks of treatment, 30 mildly hypertensive subjects: SBP/DBP: 151/87 mmHg, antihypertensive drugs were allowed Placebo: acidified skim milk	BP reduction after 4&8 weeks of treatment: SBP: - 9.4&- 14.1 mmHg ($p<0.05/0.01$), DBP: - 6.6 mmHg ($p<0.05$) after 8 weeks, pulse rate did not change, 4 weeks follow-up period: BP reduction maintained	149
Sardine muscle Hydrolyzed by alkaline protease	Val-Tyr (isolated peptid) IC_{50}: 12.2 µmol l^{-1}	2 × 100 ml drink day^{-1} containing	Japan, randomized double-blind placebo controlled study (11 weeks),	BP reduction after 1&4 weeks of treatment: SBP: - 9.7&- 9.3 mmHg ($p<0.001$),	150

(*Bacillus lichenformis*)		2 × 3 mg dipeptide	4 weeks of treatment, 29 mildly hypertensive subjects: SBP/DBP: 145/92 mmHg, antihypertensive drugs were allowed, Placebo: dink without Val-Tyr	DBP: - 5.3&- 5.2 mmHg (p<0.001), heart rate did not change, blood samples: Val-Tyr ↑ (p<0.05), 4 weeks follow-up period: BP returnted to the pretrial level	
Sequence present in sardine muscle	Val-Tyr (isolated peptid) IC_{50}: 12.2 μmol l^{-1}	100 ml drink containining 6 or 12 mg dipeptide	Japan, randomized double-blind placebo-controlled study (24 h), single oral administration, 12 mildly hypertensive subjects: SBP/DBP: 145/93 mmHg Placebo: drink without Val-Tyr	BP reduction after 1, 2, 4, 8 and 24 h of treatment: no acute BP lowering effect, blood samples: Val-Tyr ↑ Val-Tyr $absorption_{max}$: after 2 h, accumulation: C_{max} after 8 h	151
Dried bluefish/bonito (katsuobushi) Hydrolyzed by thermolysin	Leu-Lys-Pro-Asn-Met IC_{50}: 2.4 μmol l^{-1}	10 tablets day^{-1} containing 1.5 g Katsuobushi Oligopeptid (5 mg Leu-Lys-Pro-Asn-Met)	Japan, randomized double-blind placebo-controlled cross over study (12 weeks), 5 weeks of treatment, 61 borderline and mildly hypertensive subjects: 149/93 mmHg, antihypertensive drugs were not allowed, Placebo: tablets with katsuobushi powder	BP reduction after 5 weeks of treatment: SBP: - 11.7 mmHg (p<0.01), DBP: - 6.9 mmHg (p<0.01), SBP/DBP: 149/93 mmHg, heart rate did not change, 5 weeks follow-up period (placebo administration): BP returnted to the pretrial level	152
Whey protein hydolysate (BioZate® 1)[d]	n. d.	20 g day^{-1} evening solubilized in water	USA, randomized double-blind placebo controlled study (6 weeks), 6 weeks of treatment, 30 borderline hypertensive subjects, antihypertensive drugs were not allowed, Placebo: whey protein isolate (BiPRO®)[d]	BP reduction after 1 week of treatment: SBP: - 11 mmHg, DBP: - 7 mmHg, significant BP reduction	153

(Continued)

TABLE 13.3 (Continued)

Origin of Peptide	Main Bioactive Peptides[a]	Doses	Study Design[b]	Results[b]	References
Sour milk[e] Fermented with *Lactobacillus helveticus* LBK-16 H strains	Ile-Pro-Pro (15 mg l^{-1}) IC_{50}: 5 µmol l^{-1} Val-Pro-Pro (20–25 mg l^{-1}) IC_{50}: 9 µmol l^{-1}	150 ml sour milk day^{-1} containing 5.25 - 6 mg tripeptide	Finland, randomized double-blind placebo controlled study (16 weeks), 8 weeks of treatment, 17 mildly hypertensive subjects: SBP/DBP: 148/93 mmHg, antihypertensive drugs were not allowed, Placebo: sour milk without tripeptides	BP reduction after 8 weeks of treatment: SBP: - 10.1 mmHg ($p<0.05$), DBP: - 9.4 mmHg ($p<0.05$), heart rate did not change, 4 weeks follow-up period: BP returned to the pretrial level	154
Sour milk (Evolus®)[e] Fermented with *Lactobacillus helveticus* LBK-16 H strains	Ile-Pro-Pro (15 mg l^{-1}) IC_{50}: 5 µmol l^{-1} Val-Pro-Pro (20 mg l^{-1}) IC_{50}: 9 µmol l^{-1}	150 ml sour milk day^{-1} morning containing 5.25 mg tripeptide	Finland, randomized double-blind placebo-controlled study (23 weeks: long-term evaluation), 21 weeks of treatment, 39 hypertensive subjects: SBP/DBP: 152/96 mmHg, antihypertensive drugs were allowed, Placebo: sour milk without tripeptides	BP reduction after 21 weeks of treatment: SBP: - 6.7 mmHg ($p=0.03$), DBP: - 3.6 mmHg ($p<0.06$), the effect persist for 21 weeks	155

[a] IC_{50}, concentration of peptide mediating a 50% inhibition of ACE activity using hippuryl-histidyl-leucine as substrate

[b] BP, blood pressure; SBP, systolic blood pressure; DBP, diastolic blood pressure; mm Hg, millimeters of mercury

[c] Calpis produced by Calpis Food Industry co, Ltd, Tokyo, Japan

[d] BioZate® 1 and BiPRO® produced by DAVISCO foods international, INC, Eden Prairie, USA

[e] Sour milk and Evolus® (with blueberry concentrate) produced by Valio Ltd, Valio, Finland.

of the direct link between diet and health (158). Consumers are beginning to look for foods that are capable of improving their health as well as reducing their risk of developing chronic diseases. Therefore, the market for functional foods already exists. For example, the following health conditions listed in decreasing order of importance, that is, coronary vascular disease (49%), cancer (37%), obesity (37%), osteoporosis (27%), gut health (21%), and immunity (17%), were ranked by European food manufacturers as having a very great influence on the market for functional foods (159). However, the widespread uptake of functional foods by consumers is dependent on their efficacy, safety, and sensory properties.

The efficacy of a functional food is of paramount importance. It is essential that human trials demonstrate the beneficial effects of consuming specific functional food products. In the case of fermented milk drinks containing specific tripeptides, information is already in the scientific literature demonstrating antihypertensive effects during the course of double-blind placebo-controlled human studies (8, *see also* Table 13.3). It is also essential that information in relation to these beneficial effects are efficiently communicated to the consumer, as it is well documented that in certain instances, consumers are confused about the plethora of "information" they are receiving on the proposed beneficial health effects of consuming specific foods/food components. The long-term future of functional foods is contingent on the availability of stringent scientific evidence demonstrating efficacy. Furthermore, the future of functional foods is dependent on regulatory authorities developing specific legal frameworks in relation to allowable health claims. Japan was the first country to realize the benefits of functional foods. It introduced the so-called FOSHU (Food for Specified Health Use) licencing system in 1991, where health claims must be verified before FOSHU approval is granted (160). The European Commission has recently adopted a proposed regulation on nutrition and health claims on foods, including food supplements (161). This regulation is designed to provide consumers with legal security in relation to nutrition and health claims. Scientifically substantiated claims will only be permitted following stringent evaluation by the European Food Safety Authority. Health claims in European Union (EU) countries are currently governed by different national guidelines. The aim of the new regulation, if finalized, is to harmonize legislation giving both consumers and manufacturers benefit from the correct use of claims on a EU-wide basis. In the United States, health claims are regulated via the 1990 Nutrition Labeling and Education Act and the 1994 Dietary Supplement Health and Education Act (162). These Acts allow health claims to be made for specific foods/food components which the U.S. Food and Drug Administration (FDA) has scientifically confirmed to be correct. The FDA has recently launched a Task Force on Consumer Health Information for Better Nutrition aimed at fast-tracking the health claims system. This task force has recommended the adoption of unqualified health claims, that is, where the scientific evidence shows significant scientific agreement, and qualified health claims, where there is a lower level of scientific agreement/evidence in support of a food substance/disease relationship (http://www.cfsan.fda.gov/~dms/nuttftoc.html, accessed on July 10th, 2003). All these developments are beneficial to both the consumer and to industry. New functional foods claimed to have hypotensive properties will therefore need to satisfy the legislative requirements in the different geographic regions before they can be marketed with specific health claims.

The sensory attributes of functional foods or ingredients for functional foods can also dictate consumer acceptance. One of the challenges in using hydrolyzed food proteins for the development of hypotensive peptide ingredients is reducing/masking the bitterness associated with food protein hydrolyzates. Bitterness in protein hydrolyzates is associated with the presence of peptides containing hydrophobic amino acid residues (163,164). Many strategies have been suggested for the removal of bitterness including the selective removal of hydrophobic peptides (165) and the use of masking agents such as cyclodextrin (166). However, these strategies have disadvantages such as the potential reduction in essential amino acid content during selective removal of hydrophobic peptides. Clegg (167) was the first to demonstrate the use of exopeptidase activities in casein hydrolysate debittering. Since then, various studies have shown the potential role of starter culture derived exopeptidases in food protein hydrolysate bitterness reduction (168,169). This approach shows much promise in naturally reducing hydrolysate bitterness. The effects of hydrolysis with starter culture exopeptidases on the generation and destruction of hypotensive peptide structures warrants further research.

13.9 POTENTIAL APPLICATIONS

Hypertension is a controllable risk factor in the development of cardiovascular diseases (CVDs) such as stroke, left ventricular hypertrophy, smooth muscle cell hypertrophy, and coronary infarction (170,171). It is estimated that hypertension (SBP >140 and DBP >90 mm Hg) affects approximately 25% of the population. More recently, the Seventh Report of the Joint National Committee (JNC-7) on Prevention, Detection, Evaluation, and Treatment of High Blood Pressure pointed out that patients with "prehypertension" (120/80 to 139/89 mm Hg) are at risk for progression to hypertension (172). In the US alone, hypertension and its associated complications were reported to have led to 35 million medical consultations in 2002 (173). Furthermore, the annual drug costs associated with treatments for hypertension and related diseases was estimated to be US $15 billion per annum (174). Studies have shown that a 5-mm Hg reduction in DBP can lead to a 15% reduction in the risk of developing CVD (175,176). The first recommendation for control of BP is to adopt appropriate dietary and excercise patterns. However, this is not always feasible, and drug-based approaches such as the use of ACE inhibitors, calcium channel blockers, diuretics and Angiotensin II receptor blocker are employed to control BP (177). However, adverse side effects such as hypotension, increased potassium levels, reduced renal function, cough, angiodema, fetal abnormalities, and skin rashes are associated with the use of ACE-inhibitory drugs (178–181). The advantage of using food protein-derived hypotensive peptides is that these side effects can be avoided. Furthermore, it has been documented that these peptides have no effect on normotensive individuals. As a consequence, a number of food protein-derived hypotensive products/ingredients are introduced to the market (see Table 13.3). The widespead use of these and related products/ingredients is dependent on the availability of rigorous scientific data demonstrating their efficacy and safety for human use. As already mentioned, regulatory authorities in Japan, Europe, and the US recognize the importance of having a legistative framework to ensure that consumers are being provided with functional foods that have validated health claims.

ACKNOWLEDGMENT

Financial support from the Commission of the European Communities, EU Fifth Framework Shared-Cost Project, QLK1-2000-00043: "Hypotensive peptides from milk proteins": (http://www.ul.ie/acepeptides/) is gratefully acknowledged. This publication does not necessarily reflect its views and in no way anticipates the Commission's future policy in this area. Anne-Katrin Pentzien is acknowledged for valuable contributions to the review of the human studies.

REFERENCES

1. L.T. Skeegs, J.E. Kahn, N.P. Shumway. The preparation and function of the angiotensin-converting enzyme. *J. Exper. Med.*, 103: 295–299, 1956.
2. S. Maruyama, H. Mitachi, H. Tanaka, N. Tomizuka, H. Suzuki. Studies on the active site and antihypertensive activity of angiotensin I-converting enzyme inhibitors derived from casein. *Agric. Biol. Chem.*, 51: 1581–1586, 1987.
3. B.R. Steve, A. Fernandez, C. Kneer, J.J. Cerda, M.I. Phillips, E.R. Woodward. Human intestinal brush border angiotensin-converting enzyme activity and its inhibition by antihypertensive ramipril. *Gastroenterology*, 94: 942–947, 1988.
4. E.G. Erdös, et al. Potentiation of bradykinin actions by ACE inhibitors. *Trends Endocrinol. Metab.*, 10: 223–229, 1999.
5. M.A. Ondetti, B. Rubin, D.W. Cushman. Design of specific inhibitors of angiotensin-converting enzyme: new class of orally active antihypertensive agents. *Science*, 196: 441–444, 1977.
6. Y. Ariyoshi. Angiotensin-converting enzyme inhibitors derived from food proteins, *Trends Food Sci. Technol.*, 4: 139–144, 1993.
7. H. Meisel. Casokinins as inhibitors of angiotensin-converting-enzyme. In: G. Sawatzki, B. Renner (Eds.) *New Perspectives in Infant Nutrition*. Stuttgart, New York: Thieme, 1993, 153–159.
8. T. Takano. Milk derived peptides and hypertension reduction. *Int. Dairy J.*, 8: 375–381, 1998.
9. R.J. FitzGerald, H. Meisel. Lactokinins: whey protein-derived ACE inhibitory peptides, *Nahrung/Food*, 3: 165–167, 1999.
10. D.A. Clare, H.E. Swaisgood. Bioactive milk peptides: a prospectus. *J. Dairy Sci.*, 83: 1187–1195, 2000.
11. K. Suetsuna. Purification and identification of angiotensin-I-converting enzyme inhibitors from red alga *Porphyra yezoensis*. *J. Mar. Biotechnol.*, 6: 163–167, 1998a.
12. S. Maruyama, K. Nakagomi, N. Tomizuka. Angiotensin-I-converting enzyme inhibitor derived from an enzymatic hydrolysate of casein. II. Isolation and

bradykinin-potentiating activity on the uterus and ileum of rats. *Agric. Biol. Chem.*, 49: 1405–1409, 1985.
13. S. Maruyama, H. Mitachi, J. Awaya, M. Kurono, N. Tomizuka, H. Suzuki. Angiotensin-I-converting enzyme activity of the C-terminal hexapeptide of α_{s1}-casein. *Agric. Biol. Chem.*, 51: 2557–2561, 1987.
14. M. Maeno, Y. Yamamoto, T. Takano. Identification of an antihypertensive peptide from casein hydrolysate produced by a proteinase from *Lactobacillus helveticus* CP790. *J Dairy Sci.*, 79: 1316–1321, 1996.
15. A. Pihlanto-Leppälä, T. Rokka, H. Korhonen. Angiotensin I converting enzyme inhibitory peptides from bovine milk proteins. *Int. Dairy J.*, 8: 325–331, 1998.
16. S. Maruyama, H. Suzuki. A peptide inhibitor of angiotensin I converting enzyme in the tryptic hydrolysate of casein. *Agric. Biol. Chem.*, 46: 1393–1394, 1982.
17. N. Yamamoto, A. Akino, T. Takano. Antihypertensive effects of peptides derived from casein by an extracellular proteinase from *Lactobacillus helveticus* CP790. *J. Dairy Sci.*, 77: 917–922, 1994.
18. R.-C. Schlothauer, L.M. Schollum, A.M. Singh, J.R. Reid. Bioactive whey protein hydrolysate. World Patent, WO 99/65326, 1999.
19. J. Tauzin, L. Miclo, J.L. Gaillard. Angiotensin-I-converting enzyme inhibitory peptides from a tryptic hydrolysate of bovine α_{s2}-casein. *FEBS Lett.*, 531: 369–374, 2002.
20. Y. Nakamura, N. Yamamoto, K. Sakai, A. Okubo, S. Yamazaki, T. Takano. Purification and characterisation of angiotensin-I-converting enzyme inhibitors from sour milk. *J. Dairy Sci.*, 78: 777–783, 1995.
21. M. Kohmura, N. Nio, Y. Ariyoshi. Inhibition of angiotensin-converting enzyme by synthetic peptide fragments of various β-caseins. *Agric. Biol. Chem.*, 54: 1101–1102, 1990.
22. A. Abukabar, T. Saito, H. Kitazawa, Y. Kawai, T. Itoh. Structural analysis of new antihypertensive peptides deived from cheese whey protein by proteinase K digestion. *J. Dairy Sci.*, 81: 3131–3138, 1998.
23. H. Meisel, E. Schlimme. Inhibitors of angiotensin-converting-enzyme derived from bovine casein (casokinins). In: V. Brantl, H. Teschemacher (Eds.) *β-Casomorphins and Related Peptides: Recent Developments.* Weinheim: VCH, 1994, 27–33.
24. N. Yamamoto, T. Takano. Antihypertensive peptides derived from milk proteins. *Nahrung/Food*, 3: 159–164, 1999.
25. A. Pihlanto-Leppälä, P. Koskinen, K. Piilola, T. Tupasela, H. Korhonen. Angiotensin-1-converting enzyme inhibitory peptides of whey protein digests: concentration and characterisation of active peptides. *J. Dairy Res.*, 67: 53, 2000.
26. M.M. Mullally, H. Meisel, R.J. FitzGerald. Synthetic peptides corresponding to α-lactalbumin and β-lactoglobulin sequences with angiotensin-I-converting enzyme inhibitory activity. *Biol. Chem. Hoppe Seyler*, 377: 259–260, 1996.
27. M.M. Mullally, H. Meisel, R.J. FitzGerald. Identification of a novel angiotensin-I-converting enzyme inhibitory peptide corresponding to a tryptic fragment of bovine β-lactoglobulin. *FEBS Lett.*, 402: 99–101, 1997.
28. M.M. Mullally, H. Meisel, R.J. FitzGerald. Angiotensin-I-converting enzyme inhibitory activities of gastric and pancreatic proteinase digests of whey proteins. *Int. Dairy J.*, 7: 299–303, 1997.
29. H. Chiba, M. Yoshikawa. Bioactive peptides derived from food proteins. *Kagaku To Seibutsu*, 29: 454–458, 1991.
30. Y.K. Kim, S. Yoon, D.Y. Yu, B. Lonnerdal, B.H. Chung. Novel angiotensin-I-converting enzyme inhibitory peptides derived from recombinant human α_{s1}-casein expressed in *Escherichia coli*. *J. Dairy Res.*, 66: 431–439, 1999.

31. M. Kohmura, N. Nio, Y. Ariyoshi. Inhibition of angiotensin-converting enzyme by synthetic peptide fragments of human κ-casein. *Agric. Biol. Chem.*, 54: 835–836, 1990.
32. H. Fujita, K. Yokoyama, M. Yoshikawa. Classification and antihypertensive activity of angiotensin-I-converting enzyme inhibitory activity derived from food proteins. *J. Food Sci.*, 65: 564–569, 2000.
33. M. Yoshikawa, H. Fujita, N. Matoba, Y. Takenaka, T. Yamamoto, R. Yamauchi, H. Tsuruki, K. Tahahata. Bioactive peptides derived from food proteins preventing lifestyle related diseases. *Biofactors*, 12: 143–146, 2000.
34. N. Matsumura, M. Fujii, Y. Takeda, K. Sugita, T. Shimizu. Angiotensin-I-converting enzyme inhibitory peptides derived from bonito bowels autolysate. *Biosci. Biotech. Biochem.*, 57: 695–697, 1993.
35. Y. Wako, S. Ishikawa, K. Muramoto. Angiotensin-I-converting enzyme inhibitors in autolysates of squid liver and mantle muscle. *Biosci. Biotechnol. Biochem.*, 60: 1353–1355, 1996.
36. Y. Kohama, S. Matsumoto, H. Oka, T. Teramoto, M. Okabe, T. Mimura. Isolation of angiotensin-converting enzyme inhibitor from tuna muscle. *Biochem. Biophys. Res. Commn.*, 155: 332–337, 1988.
37. K. Mito, M. Fujii, M. Kuwahara, N. Matsumura, T. Shimizu, S. Sugano, H. Karaki. Antihypertensive effect of angiotensin-I-converting enzyme inhibitory peptides derived from hemoglobin. *Eur. J. Pharmacol.*, 304: 93–98, 1996.
38. K. Arihara, Y. Nakashima, T. Mukai, S. Ishikawa, M. Itoh. Peptide inhibitors for angiotensin I-converting enzyme from enzymatic hydrolysis of porcine skeletal muscle proteins. *Meat Sci.*, 57: 319–324, 2001.
39. V. Vermeirssen. Release and activity of ACE inhibitory peptides from pea and whey protein: fermentation *in vitro* digestion and transport. PhD dissertation, Ghent University, Belgium, 2003.
40. S. Miyoshi. Structures and activity of angiotensin-converting enzyme inhibitors in an α-zein hydrolysate. *Agric. Biol. Chem.*, 55: 1313–1318, 1991.
41. G. Oshima, H. Shimabukuro, K. Nagasawa. Peptide inhibitors of angiotensin I-converting enzyme in digests of gelatin by bacterial collegenase. *Biochim. Biophys. Acta*, 566: 128–137, 1979.
42. J. Wu, X. Ding. Characterisation of inhibition and stability of soy-protein-derived angiotensin I-converting enzyme inhibitory peptides. *Food Res. Int.*, 35: 367–375, 2002.
43. Z.I. Shin, R. Yu, S.A. Park, D.K. Chung, C.W. Ahn, H.S. Nam, K.S. Kim, H.J. Lee. His-His-Leu, an angiotensin I converting enzyme inhibitory peptide derived from Korean soybean paste, exerts antihypertensive activity *in vivo*. *J. Agric. Food Chem.*, 49: 3004–3009, 2001.
44. T. Matsui, C.H. Li, T. Tanaka, T. Maki, Y. Osajima, K. Matsumoto. Depressor effect of wheat germ hydrolysate and its novel angiotensin I-converting enzyme inhibitory peptide, Ile-Val-Tyr, and the metabolism in rat and human plasma. *Biol. Pharm. Bull.*, 23: 427–431, 2000.
45. Y. Saito, K. Wanezaki (Nakamura), A. Kawato, S. Imayasu. Structure and activity of angiotensin I converting enzyme inhibitory peptides from sake and sake lees. *Biosci. Biotech. Biochem.*, 58: 1767–1771, 1994.
46. K. Suetsuna. Isolation and characterisation of angiotensin I-converting enzyme inhibitor dipeptides derived from *Allium sativum* L (garlic). *J. Nutr. Biochem.*, 9: 415–419, 1998.

47. K. Suetsuna, T. Nakano. Identification of an antihypertensive peptide from peptic digest of wakame (*Undaria pinnatifida*). *J. Nutr. Biochem.*, 11: 450–454, 2000.
48. K. Nakagomi, R. Yamada, H. Ebisu, Y. Sadakane, T. Akizawa, T. Tanimura. Isolation of Acein-2, a novel angiotensin-I-converting enzyme inhibitory peptide derived from a tryptic hydrolysate of human plasma. *FEBS Lett.*, 467: 235–238, 2000.
49. K. Nakagomi, A. Fujimura, H. Ebisu, T. Sakai, Y. Sadakane, N. Fujii, T. Tanimura. Acein-1, a novel angiotensin-I-converting enzyme inhibitory peptide isolated from tryptic hydrolysate of human plasma. *FEBS Lett.*, 438: 255–257, 1998.
50. W. Cushman, H.S. Cheung. Spectrophotometric assay and properties of the angiotensin-converting enzyme of rabbit lung. *Biochem. Pharmacol.*, 20: 1637–1648, 1971.
51. B. Holmquist, P. Bünning, J.F. Riordan. A continuous spectrophotometric assay for the angiotensin converting enzyme. *Anal. Biochem.*, 95: 540–548, 1979.
52. G. Elbl, H. Wagner. A new method for the *in vitro* screening of inhibitors of angiotensin converting enzyme (ACE), using the chromophore- and fluorophore-labelled substrate, dansyltriglycine. *Planta. Med.*, 57: 137–141, 1994.
53. A.S. Mehanna, M. Dowling. Liquid chromatographic determination of hippuric acid for the evaluation of ethacrynic acid as angiotensin converting enzyme inhibitor. *J. Pharm. Biomed. Anal.*, 19: 967–973, 1999.
54. V. Vermeirssen, J. Van Camp, W. Verstrate. Optimization and validation of an angiotensin-converting enzyme inhibition assay for the screening of bioactive peptides. *J. Biochem. Biophys. Meth.*, 51: 75–87, 2002.
55. R. Natesh, S.L.U. Schwager, E.D. Sturrock, R. Acharya. Crystal structure of the human angiotensin-converting enzyme-lisinopril complex. *Nature*, 421: 551–554, 2003.
56. T. Inagami. The renin-angiotensin system. *Essays Biochem.*, 28: 147–164, 1992.
57. L. Wei, E. Clauser, F. Alhenc-Gelas, P. Corvol. The two homologous domains of human angiotensin I-converting enzyme interact differently with competitive inihibitors. *J. Biol. Chem.*, 267: 13398–13405, 1992.
58. A.J. Turner, N.M. Hooper. The angiotensin-converting enzyme gene family, genomics and pharmacology. *Trends Pharmcol. Sci.*, 23: 177–183, 2002.
59. G.Y. Oudit, M.A. Crackower, P.H. Backx, J.M. Penninger. The role of ACE2 in cardiovascular physiology. *Trends Cardiovasc. Med.*, 13: 93–101, 2003.
60. R.W. Ehlers, J.F. Riordan. Angiotensin-converting enzyme: new concepts concerning its biological role. *Biochemistry*, 28: 5311–5318, 1989.
61. V. Beldent, A. Michaud, L. Wei, M.T. Chauvet, P. Corvol. Proteolytic release of human angiotensin converting enzyme: localization of the cleavage site. *J. Biol. Chem.*, 268: 26428–26434, 1993.
62. M. Azizi, C. Massien, A. Michaud, P. Corvol. *In vitro* and *in vivo* inhibition of the 2 active sites of ACE by omapatrilat, a vasopeptidase inhibitor. *Hypertension*, 35: 1226–1231, 2000.
63. H.S. Cheung, W. Feng-Lai, M.A. Ondetti, E.F. Sabo, D.W. Cushman. Binding of peptide substrates and inhibitors of angiotensin-converting-enyzme. *J. Biol. Chem.*, 255: 401–407, 1980.
64. H. Meisel. Casokinins as bioactive peptides in the primary structure of casein. In: K.D. Schwenke, R. Mothes (Eds.) *Food Proteins — Structure Functionality.* Weinheim, New York: VCH, 1993, 67–75.
65. J. Zupan, J. Gasteiger. *Neural Networks for Chemists.* Weinheim, New York, Basel, Cambridge, Tokyo: VCH, 1993, 119–148.

66. Y. Yamada, N. Matoba, H. Usui, K. Onishi, M. Yoshikawa. Design of a highly potent ant-hypertensive peptide based on ovokinin(2-7). *Biosci. Biotechnol. Biochem.*, 66: 1213–1217, 2002.
67. H. Teschemacher, G. Koch, V. Brantl. Milk protein-derived oipiod receptor ligands. *Biopolymers*, 43: 99–117, 1997.
68. H. Meisel. Overview on milk protein-derived peptides. *Int. Dairy J.*, 8: 363–373, 1998.
69. A.M. Fiat, P. Jollès. Caseins of various origins and biologically active casein peptides and oligosaccharides: structural and physiological aspects. *Mol. Cell Biochem.*, 87: 5–30, 1989.
70. F.L. Schanbacher, R.S. Talhouk, F.A. Murray. *Liv. Prod. Sci.*, 50: 105–123, 1997.
71. M.L. Nurminen, M. Sipola, H. Kaarto, A. Pihlanto-Leppälä, K. Piilola, R. Korpela, O. Tossavainen, H. Korhonen, H. Vapaatalo. α-Lactorphin lowers blood pressure measured by radiotelemetry in normotensive and spontaneously hypertensive rats. *Life Sci.*, 66: 1535–1543, 2000.
72. P.A. Deddish, H.L. Jackman, R.A. Skidgel, E.G. Erdös. *Biochem. Pharmacol.*, 53: 1459–1463, 1997.
73. D. Migliore-Samour, F. Floc'h, P.J. Jollès. *Dairy Res.*, 56: 357–362, 1989.
74. M. Coste, V. Rochet, J. Léonil, D. Mollé, S. Bouhallab, D. Tomé. *Immunol. Lett.*, 33: 41–46, 1992.
75. H. Kayser, H. Meisel. Chemical characterization and opioid activity of an exorphin isolated from *in vivo* digests of casein. *FEBS Lett.*, 196: 223–227, 1986.
76. Y. Elitsur, G.D. Luk. β-Casomorphin (BCM) and human colonic lamina propria lymphocyte proliferation. *Clin. Exp. Immunol.*, 85: 493–497, 1991. *Clin. Exp. Immunol.*, 85:493-497, 1991.
77. H. Otani, Y. Kihara, M. Park. The immunoenhancing property of a dietary casin phosphopeptide preparation in mice. *Food Agric. Immunol.*, 12: 165–173, 2000.
78. S. Nagaune, N. Azuma, Y. Ishino, H. Mori, S. Kaminogawa, K. Yamauchi. DNA-synthesis stimulating peptide from bovine β-casein. *Agric. Biol. Chem.*, 53: 3275–3278, 1989.
79. H. Meisel, S. Günther. Food proteins as precursors of peptides modulating human cell activity. *Nahrung/Food*, 42: 175–176, 1998.
80. R. Hartmann, S. Günther, D. Martin, H. Meisel, A.K. Pentzien, E. Schlimme, N. Scholz. Cytochemical model systems for th detection and characterization of potentially bioactive milk components. *Kieler Milchwirtschaftl Forschungsber*, 52: 61–85, 2000.
81. S.C. Wright, J. Zhong, J.W. Larrick. Inhibition of apoptosis as a mechanism of tumor promotion. *FASEB J.*, 8: 654–660, 1994.
82. R.S. MacDonald, W.H. Thornton, R.T. Marshall. A cell culture model to identify biologically active peptides generated by bacterial hydrolysis of casein. *J. Dairy Sci.*, 77: 1167–1175, 1994.
83. L.S. Ganjam, W.H. Thornton, R.T. Marshall, R.S. MacDonald. Antiproliferative effects of yoghurt fractions obtained by membrane dialysis on cultured mammalian intestinal cells. *J. Dairy Sci.*, 80: 2325–2329, 1997.
84. H. Meisel, R.J. FitzGerald. Opioid peptides encrypted in intact milk protein sequences. *Br. J. Nutr.*, 58 (Suppl. 1): 27–31, 2000.
85. J. Svedberg, J. de Haas, G. Leimenstoll, F. Paul, H. Teschemacher. Demonstration of β-casomorphin immunoreactive materials in *in vitro* digests of bovine milk and in small intestine contents after bovine milk ingestion in adult humans. *Peptides*, 6: 825–830, 1985.

86. H. Meisel, H. Frister. Chemical characterization of bioactive peptides from *in vivo* digests of casein. *J. Dairy Res.*, 56: 343–349, 1989.
87. H. Meisel, H. Frister. Chemical characterization of a caseinophosphopeptide isolated from *in vivo* digests of a casein diet. *Biol. Chem. Hoppe Seyler*, 369: 1275–1279, 1988.
88. T. Kasai, T. Honda, S. Kiriyama. Caseinophosphopeptides (CPP) in feces of rats fed casein diet. *Biosci. Biotechnol. Biochem.*, 56: 1150–1151, 1992.
89. B. Chabance, P. Marteau, J.C. Rambaud, D. Migliore-Samour, M. Boynard, P. Perrotin, R. Guillet, P. Jollès, A.M. Fiat. Casein peptide release and passage to the blood in humans during digestion of milk or yogurt. *Biochimie*, 80: 155–165, 1998.
90. H. Meisel, H. Bernard, S. Fairweather-Tait , R.J. FitzGerald, R. Hartmann, C.N. Lane, D. McDonagh, B. Teucher, J.M. Wal. The occurrence of caseinophosphopeptides (CPPs) in the distal ileum of humans. *Br. J. Nutr.*, 89: 351–358, 2003.
91. H. Meisel. Biochemical properties of regulatory peptides derived from milk proteins. *Biopolymers*, 43: 119–128, 1997.
92. H. Meisel, W. Bockelmann. Bioactive peptides encrypted in milk proteins: proteolytic activation and tropho-finctional properties, *Antonie Van Leeuwenhoek*, 76: 207–215, 1999.
93. M. Gobetti. Functional potentiation by fermentation. In: *Proceedings of IDF World Dairy Summit*, Dresden, Nutrition and Health, 2000, 1–20.
94. H. Meisel, A. Goepfert, S. Günther. Occurrence of ACE inhibitory peptides in milk products. *Milchwissenschaft*, 52: 307–311, 1997.
95. E. Smaachi, M. Gobbetti. Peptides from several Italian cheeses inhibitory to proteolytic enzymes of lactic acid bacteria, *Pseudomonas fluorescens* ATCC 948 and to the angiotensin-1-converting enzyme. *Enzyme Microb. Technol.*, 22: 687–694, 1998.
96. S.S. Haileselassie, B.H. Lee, B.F. Gibbs. Purification and identification of potentially bioactive peptides from enzyme-modified cheese. *J. Dairy Sci.*, 82: 1612–1617, 1999.
97. Saito, T. Nakamura, H. Kitazawa, Y. Kawai, T. Itoh. Isolation and structural analysis of antihypertensive peptides that exist naturally in Gouda cheese. *J. Dairy Sci.*, 83: 1434–1440, 2000.
98. V. Gagnaire, D. Molle, M. Herrouin, J. Leonil. Peptides identified during emmental cheese ripening: origin and proteolytic systems involved. *J. Agric. Food Chem.*, 49: 4402–4423, 2001.
99. E.L. Ryhäne, A. Pihlanto-Leppälä, E. Pahkala. A new type of low fat ripened cheese with bioactive peptides. *Int. Dairy J.*, 11: 441–447, 2001.
100. J.A. Gomez-Ruiz, M. Ramos, I. Recio. Angiotensin-converting enzyme-inhibitory peptides in Manchego cheese manufactured with different starter cultures. *Int. Dairy J.*, 12: 697–706, 2002.
101. V. Juillard, H. Laan, E.R.S. Kunji, C.M. Jeronimus-Stratingh, A.P. Bruins, W.N. Konings. The extracellular PI-type proteinase of *Lactococcus lactis* hydrolyzes β-casein into more than one hundred different oligopeptides. *J. Bacteriol.*, 177: 3472–3478, 1995.
102. P.L. Leclerc, S.F. Gauthier, H. Bachelard, M. Santure, D. Roy. Antihypertensive activity of casein-enriched milk fermented by *Lactobacillus helveticus*. *Int. Dairy J.*, 12: 995–1004, 2002.

103. P.F. Fox, L. Stepaniak. Enzymes in cheese technology. *Int. Dairy J.*, 3: 509–530, 1993.
104. L. Stepaniak, L. Jedrychowski, B. Wróblewska, T. Sørhaug. Immunoreactivity and inhibition of angiotensin converting enzyme and latococcal oligopeptidase by peptides from cheese. *Ital. J. Food Sci.*, 13: 373–381, 2001.
105. K.P. Kaiser, H.D. Belitz, R.J. Fritsch. Monitoring Cheddar cheese ripening by chemical indices of proteolysis. *Z. Lebensm. Unters. Forsch.*, 195: 8–14, 1992.
106. F. Addeo, L. Chianese, A. Salzano, R. Sacchi, U. Cappuccio, P. Ferranti, A. Malorni. Characterization of the 12% trichloric acid soluble oligopeptides of Parmigiano-Reggiano cheese. *J. Dairy Res.*, 59: 401–411, 1992.
107. Y.-J. Jeon, H.-G. Byun, S.-K. Kim. Improvement of functional properties of cod frame protein hydrolyzates using ultrafiltration membranes. *Process Biochem.*, 35: 471–478, 1999.
108. C.-K. Hyun, H.-K. Shin. Utilisation of bovine plasma proteins for the production of angiotensin I converting enzyme inhibitory peptides. *Process Biochem.*, 36: 65–71, 2000.
109. R.J. FitzGerald, H. Meisel. Milk protein hydrolyzates and bioactive peptides. In: P.F. Fox, P.L.H. McSweeney (Eds.) *Advanced Dairy Chemistry*, Volume 1, 3rd Edition, Part B. New York, Boston, Dordrecht, London, Moscow: Kluwer Academic/Plenum Publishers, 2003, 675–698.
110. M. Shimizu. Modulation of intestinal functions by food substances. *Nahrung/Food*, 3: 154–158, 1999.
111. N. Yamamoto. Antihypertensive peptides derived from food proteins. *Biopolymers*, 43: 129–134, 1997.
112. K. Schrör. Role of prostoglandins in the cardiovascular effects of bradykinin and angiotensin converting enzyme inhibitors. *J. Cardiovasc. Pharmacol.*, 20 (Suppl 9): 568–573, 1992.
113. A. Husain. The chymase-angiotensin system in humans. *J. Hypertens.*, 11: 155–1159, 1993.
114. M.A. Weber. Vasopeptidase inhibitors. *Lancet*, 358: 1525–1532, 2001.
115. M.I. Philips. Functions of angiotensin in the central nervous system. *Annu. Rev. Physiol.*, 49: 413–435, 1987.
116. M.J. McKinley, A.L. Albiston, A.M. Allen, M.L. Mathai, C.N. May, R.M. McAllen, B.J. Oldfeild, F.A.O. Mendelson, S.Y. Chai. The brain renin-angiotensin system: location and physiological roles. *Int. J. Biochem. Cell Biol.*, 1418: 1–18, 2003.
117. A.M. Poisner. The human placental renin-angiotensin system. *Frontiers Neuroendocrinol.*, 19: 232–252, 1998.
118. I.C. Haznedaroglu, M.A. Öztürk. Towards the understanding of the local hematopoietic bone marrow renin-angiotensin system. *Int. J. Biochem. Cell. Biol.*, 1418: 1–14, 2003.
119. P.S. Leung. Pancreatic renin-angiotensin system: a novel target for the potential treatment of pancreatic diseases? *JOP*, 4: 89–91, 2003.
120. M.A. Ondetti, D.W. Cushman. Enzymes of the renin-angiotensin system and their inhibitors. *Annu. Rev. Biochem.*, 51: 283–308, 1982.
121. D.W. Moskowitz. Is "somatic" angiotensin I-converting enzyme a mechanosensor. *Diabetes Technol. Ther.*, 6: 841–858, 2002.
122. A. Hemsen. Biochemical and functional characterisation of endothelin peptides with special reference to vascular effects. *Acta, Physiol. Scand.*, (Suppl.) 602: 1–61, 1991.

123. T. Disashi, H. Nonoguchi, T. Iwaoka, S. Naomi, Y. Nakayama, K. Shimada, K. Tanzawa, K. Tomita. Endothelin converting enzyme — 1 gene expression in the kidney of spontaneously hypertensive rats. *Hypertension*, 30: 1591–1597, 1997.
124. M. Barton, R. Carmona, H. Morawietz, L.V. D'Uscio, W. Goettsch, H. Hillen, C.C. Haudenschild, J.E. Krieger, K. Munter, T. Lattmann, T.F. Luscher, S. Shaw. Obesity is associated with tissue specific activation of renal angiotensin converting enzyme *in vivo*: evidence for a regulatory role of endothelin. *Hypertension*, 35: 329–336, 2000.
125. L. Song, D.P. Healy. Kidney aminopeptidase A and hypertension, Part II: Effects of angiotensin II. *Hypertension*, 33: 746–752, 1999.
126. U. Eriksson, U. Danilczyk, J.M. Penninger. Just the beginning: novel functions for the angiotensin converting enzymes. *Curr. Biol.*, 12: R745–R752, 2002.
127. H. Urata, H. Nishimura, D. Ganten. Chymase dependent angiotensin II forming system in humans. *Am. J. Hypertens.*, 9: 277–284, 1996.
128. P. Wohlfart, J. Dedio, K. Wirth, B.A. Schölkens, G. Wiemer. Different B_1 kinin receptor expression and pharmacology in endothelial cells of different origins and species. *J. Pharmacol. Exp. Ther.*, 280: 1109–1116, 1997.
129. F. Cosentino, T.F. Luscher. Maintenance of vascular integrity: role of nitric oxide and other bradykinin mediators. *Eur. Heart J.*, 16 (Suppl K): 4–12, 1995.
130. B. Tom, A. Dendorfer, A.H. Jan Danser. Bradykinin, angiotensin (1-7), and ACE inhibitors: how do they interact? *Int. J. Biochem. Cell Biol.*, 35: 792–801, 2003.
131. H.Y.T. Yang, E.G. Erdös, Y. Levin. A dipeptidyl carboxypeptidase that converts angiotensin I and inactivates bradykinin. *Biochem. Biophys. Acta*, 214: 374–376, 1970.
132. L.J. Millatt, E.M. Abdel-Rahman, H.M. Siragy. Angiotensin II and nitric oxide: a question of balance. *Regul. Pept.*, 81: 1–10, 1999.
133. E. Thorin, S.M. Shreeve, N. Thorin-Trescases, J.A. Bevan. Reversal of endothelin-I release by stimulation of endothelial α_2-adrenoreceptor contributes to cerebral vasorelaxation. *Hypertension*, 30: 830–836, 1997.
134. C. Oliveri, M.P. Ocaranza, X. Campos, S. Lavandero, J.E. Jalil. Angiotensin-I-converting enzyme modulates neutral endopeptidase activity in the rat. *Hypertension*, 38: 650–654, 2001.
135. M. Sipola, P. Finckenberg, J. Santisteban, R. Korpela, H. Vapaatalo, M.L. Nurminen. Long term intake of milk peptides attenuates development of hypertension in spontaneously hypertensive rats. *J. Physiol. Pharmacol.*, 52: 745–754, 2001.
136. M. Sipola, P. Finckenberg, R. Korpela, H. Vapaatalo, M.L. Nurminen. Effect of long-term intake of milk products on blood pressure in hypertensive rats. *J. Dairy Res.*, 69: 103–111, 2002.
137. H. Karaki, K. Doi, S. Sugano, H. Uchiwa, R. Sugai, U. Murakami, S. Takemoto. Antihypertensive effect of tryptic hydrolysate of milk casein in spontaneously hypertensive rats. *Comp. Biochem. Physiol.*, 96: 367–371, 1990.
138. R. Nevala, T. Vaskonen, J. Vehniäinen, R. Korpela, H. Vapaatalo. Soy based diet attenuates the development of hypertension when compared to casein based diet in spontaneously hypertensive rat. *Life Sci.*, 66: 115–124, 2000.
139. H. Fujita, M. Yoshikawa. LKPNM: a prodrug-type ACE inhibitory peptide derived from fish protein. *Immunopharmacology*, 44: 123–127, 1999.
140. H. Matsufuji, T. Matsui, S. Ohshige, T. Kawasaki, K. Osajima, Y. Osajima. Antihypertensive effects of angiotensin fragments in SHR. *Biosci. Biotech. Biochem.*, 59: 1398–1401, 1995.

141. T. Matsui, A. Yukiyoshi, S. Doi, H. Sugimoto, H. Yamada, K. Matsumoto. Gastrointestinal enzyme production of bioactive peptides from royal jelly protein and their antihypertensive ability in SHR. *J. Nutr. Biochem.*, 13: 80–86, 2002.
142. M. Furushiro, H. Sawada, K. Hirai, M. Motoike, H. Sansawa, S. Kobayashi, M. Watanuki, T. Yokokura. Blood pressure-lowering effect of extract from *Lactobacillus casei* in spontaneously hypertensive rats (SHR). *Agric. Biol. Chem.*, 54: 2193–2198, 1990.
143. J. Wu, X. Ding. Hypotensive and physiological effect of angiotensin converting enzyme inhibitory peptides derived from soy proteins on spontaneously hypertensive rats. *J. Agric. Food Chem.*, 49: 501–506, 2001.
144. N. Matoba, H. Usui, H. Fujita, M. Yoshikawa. A novel anti-hypertensive peptide derived from ovalbumin induces nitric oxide-mediated vasorelaxation in an isolated SHR mesenteric artery. *FEBS Lett.*, 452: 181–184, 1999.
145. H. Yoshi, N. Tachi, R. Ohba, O. Sakamura, H. Takeyama, T. Itani. Antihypertensive effect of ACE inhibitory oligopeptides from chicken egg yolks. *Comp. Biochem. Physiol.*, 128: 27–33, 2001.
146. V. Vermeirssen, B. Deplancke, K.A. Tappenden, J. Van Camp, H.R. Gaskins, W. Verstrete. Intestinal transport of the lactokinin Ala-Leu-Pro-Met-His-Ile-Arg through a Caco-2-Bbe monolayer. *J. Peptide Sci.*, 8: 95–100, 2002.
147. S. Sekiya, Y. Kobayashi, E. Kita, Y. Imamura, S. Toyama. Antihypertensive effects of tryptic hydrolysate of casein on normotensive and hypertensive volunteers. *J. Jpn. Soc. Nutr. Food Sci.*, 45: 513–517, 1992.
148. K. Nakajima, Y. Hata, Y. Osono, M. Hamura, S. Kobayashi, M. Watanuki. Antihypertensive effect of extracts of lactobacillus casei in patients with hypertension. *J. Clin. Biochem. Nutr.*, 18: 181–187, 1995.
149. Y. Hata, M. Yamamoto, M. Ohni, K. Nakajima, Y. Nakamura, T. Takano. A placebo-controlled study of the effect of sour milk on blood pressure in hypertensive subjects. *Am. J. Clin. Nutr.*, 64: 767–771, 1996.
150. T. Kawasaki, E. Seki, K Osajima, M. Yoshida, K. Asada, T. Matsui, Y. Osajima. Antihypertensive effect of Valyl-Tyrosine, a short chain peptide derived from sardine muscle hydrolyzate, on mild hypertensive subjects. *J. Hypertens.*, 14: 519–523, 2000.
151. T. Matsui, K. Tamaya, E. Seki, K. Osajima, K. Matsumoto, T. Kawasaki. Absorption of Val-Tyr with *in vitro* angiotensin I-converting enzyme inhibitory activity into the circulating blood system of mild hypertensive subjects. *Biol. Pharm. Bull.*, 25: 1228–1230, 2002.
152. H. Fujita, T. Yamagami, K. Ohshima. Effects of an ace-inhibitory agent, katsuobushi oligopeptide, in the spontaneously hypertensive rat and in borderline and mildly hypertensive subjects. *Nutr. Res.*, 21: 1149–1158, 2001.
153. J.J. Pins, J.M. Keenan. The antihypertensive effects of a hydrolyzed whey protein supplement. *Cardiovasc. Drugs Ther.*, 16: 68, 2002.
154. L. Seppo, O. Kerojoki, T. Suomalainen, R. Korpela. The effect of *Lactobacillus helveticus* LBK-16 H fermented milk on hypertension — a pilot study on humans. *Milchwissenschaft*, 57: 124–127, 2002.
155. L. Seppo, T. Jauhiainen, T. Poussa, R. Korpela. A fermented milk high in bioactive peptides has a blood pressure-lowering effect in hypertensive subjects. *Am. J. Clin. Nutr.*, 77: 326–330, 2003.
156. A.T. Diplock, P.J. Aggett, M. Ashwell, F. Bornet, E.B. Fern, M.B. Roberfroid. Scientific concepts of functional foods in Europe: consensus document. *Br. J. Nutr.*, 81(S1): S1–27, 1999.

157. M.B. Roberfroid. Global view on functional foods: European perspectives. *Br. J. Nutr.*, 88(S2): S133–138, 2002.
158. P.R. Conlin, D. Chow, E.R. Miller, L.P. Svetkey, P.H. Lin, D.W. Harsha, T.J. Moore, F.M. Sacks, L.J. Appel. The effect of dietary patterns on blood pressure control in hypertensive patients: results from the Dietary Approaches to Stop Hypertension (DASH) trial. *Am. J. Hypertens.*, 13: 949–955, 2000.
159. A.E. Sloan. The top ten functional food trends. *Food Technol.*, 54: 33–62, 2000.
160. S. Arai. Global view on functional foods: Asian perspectives. *Br. J. Nutr.*, 88(S2): S139–143, 2002.
161. D. Byrne. Commission proposal on nutrition and health claims to better inform consumers and harmonise the market. European Commission Press Release, DN: IP/03/1022, 2003.
162. J.A. Milner. Functional foods and health: a US perspective. *Br. J. Nutr.*, 88(S2): S151–158, 2002.
163. T. Matoba, T. Hata. Relationship between bitterness of peptides and their chemical structures. *J. Agric. Biol. Chem.*, 36: 1423–1431, 1972.
164. J. Adler-Nissen. Control of the proteolytic reaction and the level of bitterness in protein hydrolysis processes. *J. Chem. Technol. Biotechnol.*, 34: 215–222, 1984.
165. G. Lalasidis, L.B. Sjoberg. Two new methods of debittering protein hydrolyzates and a fraction of hydrolyzates with a high content of essential amino acids. *J. Food Chem.*, 26: 742–749, 1978.
166. M. Tamura, N. Mori, T. Miyoshi, S. Koyama, H. Kohri, H. Okai. Practical debittering using model peptides and related compounds. *J. Agric. Biol. Chem.*, 54: 41–51, 1990.
167. K.M. Clegg. Improvements in or relating to the production of pre-digested forms of protein. British Patent: 1338936, 1973.
168. P. Bouchier, R.J. FitzGerald, G. O'Cuinn. Hydrolysis of α_{s1}- and β-casein-derived peptides with a broad specificity aminopeptidase and proline specific aminopeptidases from Lactococcus lactis supsp. cremoris AM2. *FEBS Lett.*, 445: 321–324, 1999.
169. P.J. Bouchier, G.O'Cuinn, D. Harrington, R.J. FitzGerald. Debittering and hydrolysis of a tryptic hydrolysate of β-casein with purified general and proline specific aminopeptidases from *Lactococcus lactis subsp cremoris* AM2. *J. Food Sci.*, 66: 816–820, 2001.
170. J.M. Neutel, D.H.G. Smith, M.A. Weber. Is high blood pressure a late manifestation of the hypertension syndrome? *Am. J. Hypertens.*, 12: S215–S223, 1999.
171. T. Unger. The role of the renin-angiotensin system in the development of cardiovascular disease. *Am. J. Cardiol.*, 89(2A): 3–9, 2002.
172. A.V. Chobanian, G.L. Bakris, H.R. Black, W.C. Cushman, L.A. Green, J.L. Izzo, D.W. Jones, B.J. Materson, S. Oparil, J.T. Wright, E.J. Roccella. The Seventh Report of the Joint National Committee on Prevention, Detection, Evaluation and Treatment of High Blood Pressure: The JNC 7 Report. *J. Am. Med. Assoc.*, 289: 2560–2572, 2003.
173. D.K. Cherry, D.A. Woodwell. National ambulatory medical care survey: 2000 summary. Advanced data from vital and health statistics. No. 328. Hyattsville, Maryland: National Center for Health Statistics, 2002.
174. S. Frantz. Antihypertensive treatments — ALLHATs off to the golden oldie. *Nat. Rev. Drug Discov.*, 2: 91, 2003.

175. R. Collins, R. Peto, S. MacMahon, P. Herbert, N.H. Fiebach, K.A. Eberlein, J. Godwin, N. Qizilbash, J.O. Taylor, C.H. Hennekens. Blood pressure, stroke and coronary heart disease. Part 2, Short-term reductions in blood pressure: overview of randomised drug trials in their epidemiological context. *Lancet*, 335: 827–838, 1990.
176. S. MacMahon, R. Peto, J. Cutler, R. Collins, P. Sorlie, J. Neaton, R. Abbott, J. Godwin, A. Dyer, J. Stamler. Blood pressure, stroke and coronary heart disease. Part 1, Prolonged differences in blood pressure: prospective observational studies corrected for the regression dilution bias. *Lancet*, 335: 765–774, 1990.
177. P.R. Conlin. Efficacy and safety of angiotensin receptor blockers: a review of Losartan in essential hypertension. *Curr. Ther. Res. Clin. Exp.*, 62: 79–91, 2001.
178. R.P. Ames. Negative effects of diuretic drugs on metabolic risk factors for coronary heart disease. *Am. J. Cardiol.*, 51: 632–638, 1983.
179. S. Seseko, Y. Kaneko. Cough associated with the use of captopril. *Arch. Intern. Med.*, 145:1524, 1985.
180. H. Nakamura. Effects of antihypertensive drugs on plasma lipid. *Am. J. Cardiol.*, 60: 24E–28E, 1987.
181. A. Agostoni, M. Cicardi. Drug-induced angioedema without urticaria. *Drug Saf.*, 24: 599–606, 2001.

14 Regulation of Bone Metabolism: Milk Basic Protein

Yukihiro Takada and Seiichiro Aoe

CONTENTS

14.1 INTRODUCTION

Osteoporosis is an important public health problem in the world. With an increasing population of the aged, the incidence of osteoporosis and bone fracture has been increasing. In bone tissue, bone formation and bone resorption are always occurring, and the integrity of the bone tissue is maintained. For bone health, it is important that the proper balance of bone formation and resorption be kept. While this balance is maintained properly in the young, bone resorption in older individuals exceeds bone formation for various reasons such as menopause, among others. There is thus a need to develop food components that increase bone formation or suppress bone resorption for the improvement of the unbalanced bone metabolism that occurs later in life.

Milk is well known as a safe food that can be taken over a long time and is beneficial for human health. It is also a good source for calcium, considering the absorption of calcium and its bioavailability. It contains several beneficial components for calcium absorption such as lactose and phosphopeptides, the latter being formed in the intestines by the proteolytic digestion of milk casein (1–5). Milk has also a functional role in the growth of newborn animals. Thus, milk protein may have components that affect bone metabolism (6). Milk whey protein (WP) stimulates the proliferation and differentiation of osteoblastic MC3T3-E1 cells (7). WP also suppresses osteoclast cell formation and bone resorption (8). Administration of WP was effective in increasing bone strength and the contents of collagen-typical amino acids, such as a hydroxyproline, in young ovariectomized (OVX) rats (9,10). Therefore, the active components from WP were concentrated and their biological activity estimated. A milk protein fraction with a basic isoelectric point, designated as milk basic protein (MBP), strongly stimulated both bone formation and bone resorption. The effects of MBP on bone metabolism *in vitro*, *in vivo* (rats), and human studies are also presented.

14.2 BONE METABOLISM

Bone is a highly specialized form of connective tissue composed of an organic matrix that is strengthened by deposits of calcium salts. Type I collagen constitutes approximately 95% of the organic matrix, the remaining 5% being composed of proteoglycans and numerous noncollagenous proteins. The crystalline salts deposited in the organic matrix of bone under cellular control are primarily calcium and phosphate in the form of hydroxyapatite. Morphologically, there are two types of bone: cortical and trabecular bone. In cortical bone, densely packed collagen fibrils form concentric lamellae, and the fibrils in adjacent lamellae run in perpendicular planes. Cancellous bone has a loosely organized, porous matrix. In both types of bone, continuous cellular remodeling occurs, which results in the breakdown and reformation of discrete packets of bone. This continual bone remodeling, or bone turnover, is presumably important for maintaining the structural integrity and strength of the bone. Bone tissue consists of a wide variety of cells of bone-forming and bone-resorbing cell lineages. Osteoblasts, osteocytes, and bone-lining cells are present on the bone surfaces, whereas osteocytes permeate the mineralized interior. Especially, both osteoblasts and osteoclasts are important for bone remodeling (Figure 14.1). Osteoblasts originate from local osteoprogenitor cells, whereas osteoclasts arise from the fusion of mononuclear precursors that originate from various hemotopoietic cells. The cellular events

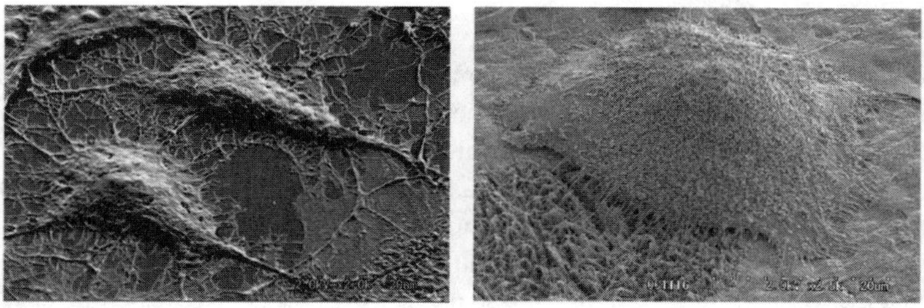

FIGURE 14.1 Bone-forming cells (osteoblast; left) and bone-resorbing cell (osteoclast; right).

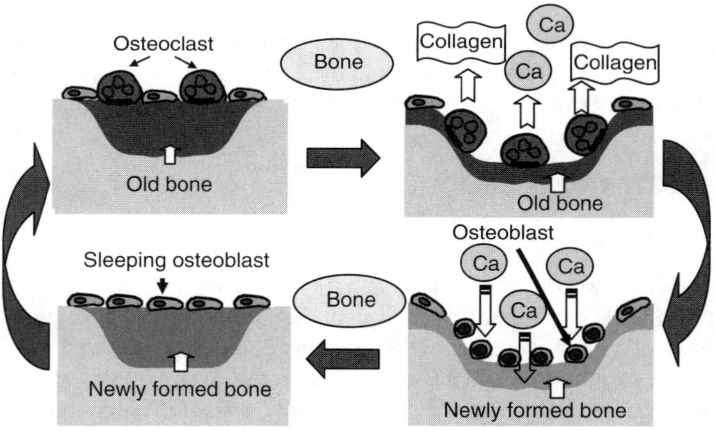

FIGURE 14.2 Bone remodeling by osteoclasts and osteoblasts.

involved in remodeling are the same in both cortical and trabecular bone (Figure 14.2). They begin with a phase of relatively short-lived bone resorption by osteoclasts. These cells are formed from precursors in the hematopoietic bone marrow, which progenitors are common not only to osteoclasts but also to the formed elements of blood. Under the direction of local signals, osteoclasts form, attach to bone, and start to resorb it. The phase of osteoclastic bone resorption lasts about 10 days and is followed by a reversal phase in which the osteoclasts are replaced by mononuclear cells. These mononuclear cells are soon replaced by osteoblast precursors, which are attracted to the site of the defect made by the osteoclasts and stimulated to replicate and differentiate into mature osteoblasts. These cells are capable of forming new bone, which replaces the defect made by the osteoclasts over a period of several months. Many hormones and factors regulate the activities of bone cells. Imbalances in the rates of bone resorption and bone formation lead to metabolic bone diseases such as osteoporosis.

14.3 MBP IN MILK

Milk is a liquid food secreted by the mammary gland for the nourishment of newborns and contains water, fat, proteins, lactose, and minerals. Milk proteins constitute approximately 3.5% of milk. They are composed of a number of fractions, of which casein constitutes approximately 2.9% and whey proteins 0.6%. The main whey protein fractions are acidic and contain β-lactoglobulin, α-lactoglobulin, and serum albumin. MBP is found in a small fraction of whey. MBP is actually composed of several proteins having basic isoelectric points ranging from approximately 7.0 to 10.5, as determined by isoelectric focusing disc gel electrophoresis. From the analysis of the amino acid composition of MBP, the amounts of lysine and arginine (basic amino acids) of MBP were found to be higher than those of casein. Almost all of the growth factors in milk possess basic isoelectric points (11).

MBP is obtained from defatted milk. The defatted milk is loaded onto a column that has been packed with cation exchange resin. The column is then sufficiently washed with deionized water, and the bound proteins are eluted with 1 M NaCl. After dialysis of the eluted fraction, MBP is obtained by freeze drying.

14.4 STUDY ON OSTEOBLASTS

14.4.1 MBP

The cell proliferation of MC3T3-E1 was examined using [^{3}H]-thymidine incorporation methods. After preculture for 18 hours, they were then labeled for 2 hours with 1 μCi of [^{3}H]-thymidine. MBP promoted dose-dependent [^{3}H]-thymidine incorporation into the cells (Table 14.1). This activity was heat stable for 10 minutes at temperatures from 75 to 95°C and was not neutralized by antibodies against epidermal growth factor (EGF), insulin-like growth factor (IGF)-I, IGF-II, or fibroblast growth factor (FGF). The presence of cell-growth-promoting activity in whole bovine milk has been established by using Balb/c 3T3 cells; and such activity toward CHO-K1, L6 myoblasts, and human skin fibroblasts and hybridomas

TABLE 14.1
Effect of MBP on [^{3}H]-Thymidine Incorporation and PICP Contents in Osteo-blastic Cells

		MBP (μg ml^{-1})		
	Control	1	10	100
[^{3}H]-thymidine incorporation (CPM)	1205 ± 316	2467 ± 421*	5231 ± 572*	9463 ± 894*
PICP contents (ng ml^{-1})	465 ± 34	734 ± 45*	853 ± 79*	1324 ± 132*

Values are mean ± SD.
*Significantly different from the control group ($p < 0.05$).

was demonstrated in the whey fraction (12–15). In this context, several mitogenic factors have been identified in bovine colostrum, such as insulin, IGF, transforming growth factor (TGF), FGF, and angiogenin (16–19). MBP contains some active components for the growth promotion of osteoblasts. The effect of MBP on the collagen synthesis of MG 63 cells was also examined by measuring PICP, which is a C-terminal fragment of a precursor protein of type I collagen. After the cells had been precultured for 2 days in the presence of MBP, MBP dose-dependently increased the PICP content in these MG-63 cells (see Table 14.1). Thus, MBP contains some active components for collagen synthesis by osteoblasts.

14.4.2 Kininogen Fragment 1.2

The activity-stimulating osteoblastic proliferation was concentrated in a basic protein fraction by Mono S chromatography, and the molecular weights of the active components were comparatively low as estimated from the results of Superose 12 chromatography (7). The active components of MBP that promoted the proliferation of osteoblastic cells were purified and identified (20). The osteoblast-growth-promoting fractions from cation-exchange chromatography (Mono S) were eluted with 0.4 to 0.6 M NaCl, and the fractions with the highest activity were pooled and loaded onto a MonoQ anion-exchange column. One of the growth-promoting fractions was an internal sequence of bovine high-molecular-weight (HMW) kininogen, kininogen fragment 1.2 (Figure 14.3). The molecular weight of this growth-promoting protein was estimated to be 17 kDa by electrophoresis. The activity of this purified fragment from milk was as high as that of EGF for osteoblastic MC3T3-E1 cells. Although EGF also promoted the proliferation of fibroblast BALB/3T3 cells, the purified protein did not promote the growth of these cells under the same conditions (20), suggesting that the kininogen fragment 1.2 was specific for osteoblasts.

HMW kininogen is well known as a plasma protein that acts to prevent blood coagulation. For this role, HMW kininogen is considered to be cleaved by a specific enzyme, kallikrein. HMW kininogen then releases two peptides, a biologically active peptide, bradykinin, and fragment 1.2. The kininogen fragment 1.2 is rich in basic amino acid residues, containing 22 histidines, 13 lysines, and 3

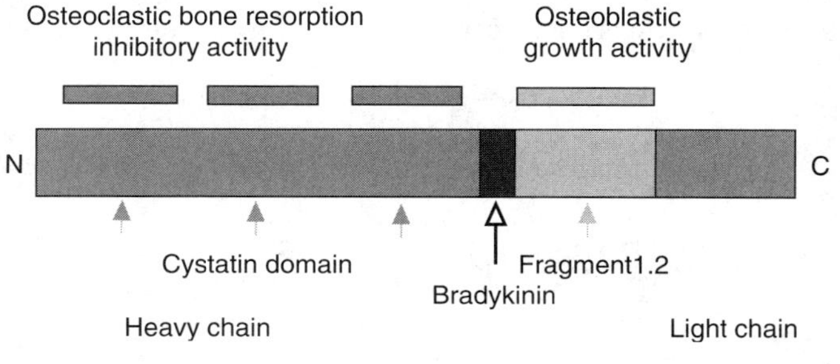

FIGURE 14.3 Active component for osteoblast.

arginines out of a total of 110 residues; and furthermore, it is a potent inhibitor of the contact activation of factor XII (21). The heavy chain of kininogen has cystatin domains that inhibit cysteine proteases. As cathepsin is the major cysteine protease in bone, which is secreted by osteoclasts and acts in bone resorption (22), it might be expected that the heavy chain of kininogen would play a significant role in modulating bone resorption by inhibiting protease activity. Because kininogen fragment 1.2 stimulates the proliferation of osteoblasts, kininogen might be a multifunctional molecule in bone metabolism. Thus, kininogen fragment 1.2 in milk is proposed to play an important role in bone formation by activating osteoblasts.

14.4.3 HMG-like Protein

Another active component of MBP that promoted the proliferation of osteoblastic cells was purified and identified (23). The active component had a molecular weight of 10 kDa, and its sequence was homologous to the amino-terminal sequence of bovine high-mobility-group (HMG) protein. In view of its amino-terminal sequence and molecular weight, the purified protein was named "HMG-like protein" (Figure 14.4).

HMG protein is a well-known nuclear nonhistone chromosomal protein that has been implicated in DNA replication and cellular differentiation (24). There are both acidic and basic domains within the HMG protein molecule. One striking feature of HMG is the presence of a 30-residue polyacidic domain near its carboxyl terminus, which is composed entirely of glutamic and aspartic acid residues. In contrast, the amino-terminal part of the HMG protein contains a basic domain. Therefore, HMG protein is an extremely polar molecule. The amino-terminal amino acid sequence of our purified protein represents this basic domain. HMG protein consists of two HMG boxes that are highly homologous, folded, basic DNA-binding domains, each comprising approximately 80 amino acid residues (25). In view of the molecular weight of the HMG-like protein, it would contain one HMG box. The HMG box is a sequence motif recognized in certain sequence-specific DNA-binding proteins, such as HMG protein and other HMG-like DNA-binding proteins, including eukaryotic upstream binding factor (UBF), and a gene cloned from the sex-determining resin of the human Y chromosome (SRY) (26). The HMG box is defined by a set of highly conserved residues and, in turn, appears to define a novel DNA-binding structural motif. The primary structure of HMG protein is highly conserved among mammals, and this HMG-like

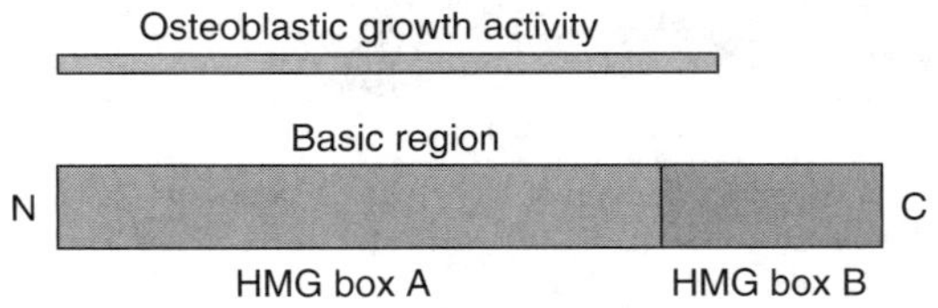

FIGURE 14.4 Active component for osteoblast; High Mobility Group-protein.

protein in milk is considered to be a member of the HMG family. HMG-like protein activity is specific for osteoblasts, not for fibroblast Balb/3T3 or Chinese hamster V79 cells (23). Therefore, it is expected that the HMG-like protein also plays an important role in bone formation by activating osteoblasts.

14.5 STUDY ON OSTEOCLASTS

14.5.1 MBP

The effect of MBP on bone resorption by preexisting and newly formed osteoclasts was examined by using unfractionated bone cells. A biological system for evaluating bone resorption, osteoclast recruitment, and activation by cultures of unfractionated bone cells on dentine slices was established (27). Moreover, since unfractionated bone cells cultured for 6 days were found to contain only a few osteoclasts, only bone resorption by newly formed osteoclasts was evaluated in cultures examined after this time (27). First, the effect of MBP on preexisting osteoclasts among unfractionated bone cells was examined. MBP suppressed the area of pits formed by osteoclasts (Table 14.2). The bone resorption-inhibiting activity of MBP was not affected by heating at 75 to 95°C for 10 minutes. Next, the effect of MBP on bone resorption by newly formed osteoclasts was examined. When MBP was added to the culture at day 7, at which time the cultures contained few TRAP-positive cells, MBP inhibited the area of pits formed by osteoclasts dose-dependently (see Table 14.2). Thus, in this study, MBP suppressed bone resorption by both preexisting and newly formed osteoclasts.

The effect of MBP on isolated osteoclasts was also examined (28). Using isolated osteoclasts, the direct effect of MBP on bone resorption was assessed. The unfractionated cells were plated on collagen gels and incubated for 4 hours. The

TABLE 14.2
Effect of MBP on Bone Resorbing Activity of Osteoclasts (Preexisting and Newly Formed) in Unfractionated Bone Cells and Isolated Osteoclasts

		MBP (μg ml^{-1})		
	Control	1	10	100
Preexisting osteoclasts (Pit area: mm^2)	1.32 ± 0.16	1.23 ± 0.12	0.93 ± 0.09*	0.43 ± 0.07*
Newly formed osteoclasts (Pit area: mm^2)	0.85 ± 0.12	0.84 ± 0.11	0.73 ± 0.09*	0.69 ± 0.09*
Isolated osteoclasts (Pit number)	210 ± 31	167 ± 24*	87 ± 18*	35 ± 16*

Values are mean ± SD.

*Significantly different from the control group ($p < 0.05$).

cultures were next treated with pronase E to remove the nonadherent hematopoietic cells and loosely attached stromal cells from the gel. By this treatment, most of the stromal cells were released, whereas osteoclasts still remained on the gel. After several washes with phosphate buffer saline (PBS), the cultures were digested with collagenase, and the released cells were then collected. The isolated osteoclasts were transferred onto dentine slices and cultured. MBP also dose-dependently suppressed the area of pits formed by isolated osteoclasts (see Table 14.2). This result showed that MBP had direct effects on osteoclasts.

14.5.2 Milk Cystatin

Osteoclast-mediated bone resorption activity was found to be concentrated in the basic protein fraction by Mono S chromatography. The molecular weights of the active components were comparatively low, as estimated from the results of Superose 12 chromatography (9). The active component in MBP was studied and found that one of the active components was milk cystatin (Figure 14.5). Cystatin inhibits cathepsin, which has an important role in bone resorption by osteoclasts (29,30). MBP contained a large amount of milk cystatin as determined by testing for inhibition of cysteine protease activity (31).

Osteoclastic resorption of bone involves both the dissolution of mineral crystals and the enzymatic degradation of bone proteins. With regard to the degradation of bone matrix proteins, lysosomal proteolytic enzymes are important. There are many proteolytic enzymes such as matrix metalloproteinase (MMP), collagenase, gelatinase, stromelysin, and tissue inhibitor of metalloproteinases (TIMP); cathepsin is the main proteinase responsible for osteoclastic bone matrix degradation (29). Milk cystatin has been shown to inhibit bone resorption by inhibiting cathepsin in a dose-dependent manner (31). Some cystatins such as cystatin C are synthesized by bone cells and primarily inhibit the activity of osteoclastic proteolytic enzymes released into the resorption lacunae (29). One of the milk cystatins seems to be cystatin C (31). Therefore, milk cystatin appears to be one of the important components involved in bone resorption by suppressing osteoclastic action.

14.6 TRANSPORT ABILITY OF THE ACTIVE COMPONENTS BY THE SMALL INTESTINE

Oral administration of WP was effective in increasing bone strength and the content of bone proteins such as collagen in OVX rats (9,10), which suggests that the

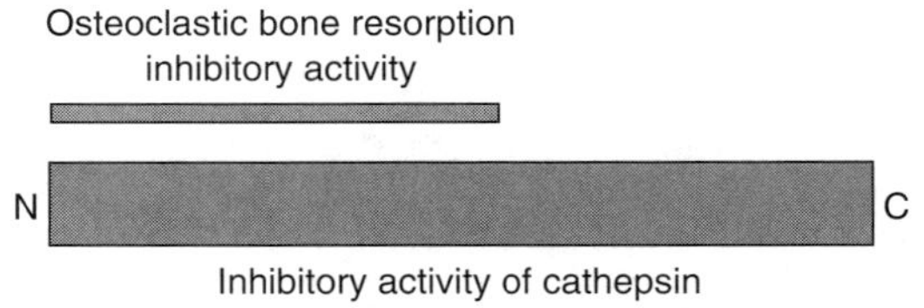

FIGURE 14.5 Active component for osteoclast; milk cystatin.

active components could be absorbed or transported by the small intestine. The active components in WP retained their biological activity when tested *in vitro* with gastrointestinal enzymes (7,8). The transport of the active components was also examined by using everted gut sacs of rat small intestine. In this system, glucose and lactoferrin were detected in the inner solution of the everted gut sac, but lactose and dextran were not (32). After incubating everted gut sacs in solutions of intact basic protein fraction (MBP), pepsin-digested MBP or pepsin-/pancreatin-digested MBP, the inner solutions could still cause an increase in [^{3}H]-thymidine incorporation (7) and also could suppress osteoclast-mediated bone resorption (8). These results suggest that the fragments or active components of MBP could be absorbed or transported by the intestine and remain active.

14.7 *IN VIVO* STUDIES USING RATS

14.7.1 Young OVX Rats

The effects of MBP on bone metabolism in young ovariectomized rats were examined in our study (33). In this model, the bone mineral content and bone mass are rapidly decreased by OVX and a low-calcium diet, and so we could evaluate the rapid recovery of bone formation by the diet. Five-week old female Sprague-Dawley (SD) rats were ovariectomized and fed low-calcium (0.009%) diet *ad libitum* for 5 weeks. Then, the animals were divided into three experimental groups: control group (20% casein), MBP-L group (19.9% casein, 0.1% MBP), and MBP-H group (19% casein, 1% MBP). Groups of rats were fed the control and experimental diets for 3 weeks.

The final body weight, weight gain, food intake, and food efficiency were not significantly different among the three groups. The bone strength (the breaking strength and energy) was measured by the three-point bending method. Bone strength is one of the important determinants for bone fracture. The values of the breaking strength and breaking energy are shown in Table 14.3. Mature bone matrix is a complex of highly mineralized tissue with bone protein as a structural framework. Bone proteins consist of mainly collagen and small amounts of

TABLE 14.3
Effect of MBP on Bone Strength, Hydroxyproline, and Hydroxylysine in Young OVX Rats

	Control	MBP-L	MBP-H
Breaking strength (10^6 dyne)	4.8 ± 0.3	5.1 ± 0.5	5.5 ± 0.3*
Breaking energy (10^5 erg)	4.4 ± 1.0	6.6 ± 1.1*	5.6 ± 0.5*
Hydroxyproline (mg g^{-1} bone)	20.5 ± 1.1	21.6 ± 0.7*	21.9 ± 0.9*
Hydroxylysine (mg g^{-1} bone)	5.9 ± 0.2	6.2 ± 0.1*	6.3 ± 0.2*

Values are mean ± SD.
*Significantly different from the control group ($p < 0.05$).

noncollagenous proteins and glycoproteins. The breaking strength is the maximum power that is required to break this bone structure by the three-point bending method (10,11,40). On the other hand, the breaking energy is an integration value of the power that is required to break the bone. The breaking strength in the MBP-H group was significantly higher than that in the control group. The breaking energy in both the MBP-L and MBP-H groups was significantly higher than that in the control group.

Also, a amounts of femoral proline, hydroxyproline, and hydroxylysine were measured. The amounts of these collagen-typical amino acids in the MBP-H group were higher than those in the control group (see Table 14.3). The amount of femoral hydroxyproline and hydroxylysine in the MBP-L group were also higher than those in the control group. Therefore, an increase in the bone strength by MBP is accompanied by an increase of bone collagen contents. These results show that MBP promoted the recovery of bone, that is, MBP could promote bone formation.

14.7.2 Growing Rats

Bone formation is one of the major aspects of growth. The effect of administration of MBP on bone metabolism, bone mineral content (BMC), bone mineral density (BMD), and cartilage growth in growing rats was examined. Five-week-old female Sprague-Dawley rats were divided into four experimental groups. The growing rats were fed the AIN-76 diet *ad libitum* for 4 weeks, during which time MBP was given by oral administration at 0, 20, 40, or 80 mg kg^{-1} body weight.

After a 4-week feeding period, the femurs were excised from the rats. The BMC of the femurs in the 40- and 80-mg kg^{-1} MBP administration groups was significantly higher than that of the control group. The BMD of the femurs in the 40- and 80-mg kg^{-1} MBP groups was also significantly higher than that of the control group (Table 14.4). These results showed that MBP was effective for increasing BMC and BMD. The breaking strength and breaking energy of the femurs in the 40- and 80-mg kg^{-1} MBP administration groups were likewise significantly higher than those of the control group (see Table 14.4).

TABLE 14.4
Effect of MBP on BMD and Bone Strength in Growing Rats

		MBP		
	Control	**20 mg kg^{-1}**	**40 mg kg^{-1}**	**80 mg kg^{-1}**
BMD (mg cm^{-2})	103 ± 3	104 ± 2	107 ± 2*	109 ± 3*
Breaking strength (10^6 dyne)	12.8 ± 0.5	13.2 ± 0.7	13.9 ± 1.0*	13.9 ± 0.6*
Breaking energy (10^5 erg)	6.0 ± 0.9	6.4 ± 0.6	7.9 ± 1.2*	7.2 ± 0.6*

Values are mean ± SD.
*Significantly different from the control group ($p < 0.05$).

Moreover, the effect of MBP on cartilage growth was examined and the cartilage growth plate of the rat tibia measured. The cartilage growth was activated by the administration of MBP, with the growth in terms of width for the 20-, 40-, and 80-mg kg^{-1} MBP administration groups being significantly greater than that of the control group. In the growth plate, chondrocytes proliferate and synthesize the cartilage matrix and terminally differentiate into the hypertrophic state . The growth of cartilage plays a key role in endochondral bone formation during embryonic development and during the longitudinal growth of bones. MBP could be effective for endochondral bone formation. In addition, MBP the number of chondroclasts and osteoblasts below the cartilage growth plate of the tibia were increased. Chondroclasts and osteoblasts are necessary and important for the embryonic development and during the longitudinal growth of bone. Theses results suggest that MBP could be also effective for cartilage growth.

14.7.3 Aged OVX Rats

The effect of MBP on bone loss in aged OVX rats was examined (35). This is a postmenopause model, in which the bone mass is decreased rapidly by OVX in aged rats. Twenty-one 51-week-old female Sprague-Dawley rats were ovariectomized (OVX), and seven other rats received a sham operation (Sham). After a 4-week recovery period, the OVX rats were separated into three groups, and they were then fed a control diet, a 0.01% MBP diet (0.01% casein of the control diet replaced with MBP), or a 0.1% MBP diet for 17 weeks. The Sham rats were fed the control diet.

The femoral BMD over time was investigated by measuring the BMD before and at weeks 4, 8, 12, and 16 after the start of the experiment. The BMD in the OVX-control group noticeably decreased during the experimental period in comparison with that in the Sham group (Table 14.5). The decline in the OVX-0.01% MBP group was the same as that in the OVX-control group. However, the BMD in the OVX-0.1% MBP group decreased more slowly than that in the OVX-control group and was significantly higher than that in the latter at weeks 12 and 16. After

TABLE 14.5
Effect of MBP on BMD, Bone Strength, and D-Pyr Level in Aged OVX Rats

		OVX		
	Control	Control	0.01% MBP	0.1% MBP
BMD (mg cm^{-2})	133 ± 7	118 ± 7	119 ± 4	124 ± 6*
Breaking strength (× 10^6 dyne)	24.7 ± 4.3	19.7 ± 3.1	19.7 ± 3.0	21.4 ± 4.0*
Breaking energy (× 10^5 erg)	14.8 ± 4.1	9.1 ± 3.0	9.5 ± 2.6*	12.9 ± 3.8*
Urinary D-Pyr (nmol day^{-1})	1797 ± 654	3167 ± 985	2136 ± 507*	1673 ± 358*

Values are mean ± SD.

*Significantly different from the control group ($p < 0.05$).

the 17-week feeding period, the femora were excised from all rats. The BMD of the excised femora in the OVX-control group was significantly lower than that in the Sham group, and the value for the OVX-0.1% MBP group was significantly higher than that for the OVX-control group (see Table 14.5).

The breaking energy and breaking force of the excised femur in the OVX-control group were significantly lower than those in the Sham group. The breaking energy in the OVX-0.1% MBP group was significantly higher than that in the OVX-control group, and the breaking force in this group tended to be higher than that in the OVX-control group, although the difference was not significant (see Table 14.5). These results show that MBP suppressed the weakening of bone strength (the breaking energy) caused by ovariectomy because of its preventive effect on bone loss.

The urinary excretion of deoxypyridinoline (D-Pyr: a collagen-based cross-linker, one of the biochemical markers of bone resorption) was measured at week 16. The D-Pyr of the OVX-control group was higher than that of the Sham group. The D-Pyr excretion in the OVX-0.01% MBP and OVX-0.1% MBP groups was significantly lower than that in the OVX-control group (see Table 14.5). The increased urinary excretion of the D-Pyr caused by ovariectomy was clearly reduced by MBP. These results suggest that MBP suppressed the osteoclast-mediated bone resorption.

The femoral metaphysis contains trabecular bone as well as cortical bone, whereas the femoral diaphysis mainly contains cortical bone. The higher BMD of the metaphysis of the femur of OVX-0.01% MBP and OVX-0.1% MBP groups suggests that MBP especially prevented trabecular bone loss. To confirm that MBP prevented trabecular bone loss, we performed histological examination of the proximal tibia. MBP was found to reduce loss of trabecular bone around the growth plate–metaphyseal junction in the proximal tibia. These results indicate that MBP helps prevent bone loss, that is, bone resorption.

14.8 HUMAN STUDIES

14.8.1 HUMAN STUDY 1

The effect of MBP on the bone metabolism of healthy adult women was examined (36,37). In this 6-month double-blind, placebo-controlled trial, 33 healthy women were randomly assigned to treatment with either a placebo or MBP (40 mg day^{-1}) for 6 months. The study complied with the code of ethics of the World Medical Association (Helsinki Declaration of 1964, revised in 1989). Each subject was measured for urinary and blood parameters every 3 months. Prior to treatment (baseline) and after 6 months of treatment, they were also measured for BMD. The nutrient content of their diet was quantified by using a computer program based on the Standard Tables of Food Composition. During the 6-month study period, none of the 33 women dropped out of the study, and all subjects completed the study according to the protocol. Bone formation markers and bone resorption markers were also measured.

There was no significant difference between the MBP and placebo groups in any of the parameters, that is, age, weight, height, body mass index, and BMD. During the 6-month study period, no bloating, diarrhea, or allergy was observed in either group.

The mean BMD value at 6 months for the MBP group was also significantly increased at the calcaneus and at both the one sixth and one tenth portions from the distal end of the radius, whereas that for the control group did not change (36,37). The BMD gain in the MBP group was also significantly higher than that in the placebo group (Figure 14.6). Thus, a daily MBP supplementation of 40 mg in healthy adult women can significantly increase their BMD.

Biochemical parameters in serum and urine are being used clinically to assess the rate of bone formation and resorption. The mean urinary cross-linked N-teleopeptides of type I collagen excretion (NTX), a bone resorption marker, was lower in the MBP group than in the placebo group at both 3 and 6 months (36). The mean D-Pyr excretion, another bone resorption marker, was also lower in the MBP group than in the placebo group at 6 months. These data indicate that MBP supplementation led to a reduction in bone resorption. The serum levels of bone-specific alkaline phosphatase (B-ALP) and osteocalcin (BGP), both bone formation markers, in both groups changed during the study period, but no difference was observed between the groups. However, bone formation markers were not decreased by the decrease in the bone resorption. The value of BGP/NTX for the MBP group was significantly higher than that of the placebo group at 6 months. This result suggested that MBP promoted bone formation. Other biochemical results were normal and did not change in both groups throughout the study period. There was no significant correlation between gain of BMD and intake of any dietary minerals or vitamins in the placebo and MBP groups. These findings suggest that a significant increase in BMD in the MBP group is independent of dietary intake of minerals (calcium, phosphorus, and magnesium) and vitamins

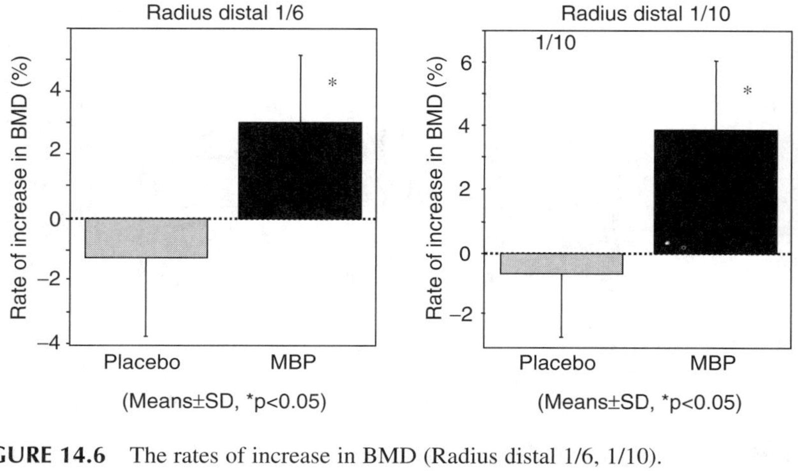

FIGURE 14.6 The rates of increase in BMD (Radius distal 1/6, 1/10).

(vitamins C, D, and K). These findings are consistent with a previous study of MBP. From this study, it appears that 40 mg day^{-1} of MBP supplementation could be beneficial for bone metabolism.

14.8.2 Human Study 2

In a previous controlled trial, MBP supplementation (40 mg of MBP a day) was found to increase BMD and influence bone markers (36,37). Attempts were made to find a clear effect of MBP on bone markers. As biochemical markers of bone turnover that reflect bone changes faster than BMD can are available in serum or urine, the effect of MBP on various biochemical markers of bone metabolism in healthy adult men was examined when slightly more MBP (300 mg of MBP a day) was ingested (38).

Thirty healthy men (mean ± SD age, 36.2 ± 8.5) received daily an experimental beverage containing MBP (300 mg of MBP). The subjects were instructed to drink the beverage daily within any 2-hour period for 16 days and were advised to maintain their usual diet. The study complied with the code of ethics of the World Medical Association. Each participant received a physical checkup every week and had urine and blood measurements taken before and after the 16 days of ingestion. There was no significant difference before and after 16 days of ingestion in either of the parameters of weight and body mass index.

The serum calcium level and urinary calcium excretion remained unchanged after 16 days of ingesting the experimental beverage containing MBP. The BGP had increased significantly after 16 days of ingestion (Figure 14.7), while the PICP level tended to increase after 16 days of the ingestion. The NTX had decreased significantly after 16 days of ingestion (see Figure 14.7). An increase in serum osteocalcin concentration was found in 28 (93%) of the 30 subjects, and a decrease in NTX in 24 (80%) of them. In this study, MBP supplementation was

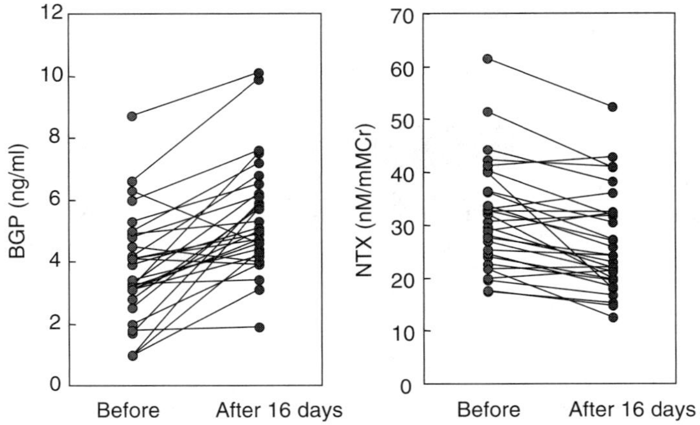

FIGURE 14.7 Change of bone formation marker (BGP) and resorption marker (NTX).

found to increase the BGP and PICP and decrease the NTX excretion in healthy adult men. These results suggest that MBP promotes bone formation and suppresses bone resorption. Results from this human study are consistent with those from *in vitro* and animal studies.

Bones are continuously undergoing a remodeling process through repeated cycles of destruction and rebuilding (39). In healthy young adults, the amount of new bone formation approximately balances the amount of bone resorption. As humans age, however, the balance shifts to favor bone resorption, which can result in debilitating diseases such as osteoporosis. Efforts to treat bone diseases have been primarily concentrated on the development of drugs that block bone resorption, that is, drugs that decrease the formation or activity of osteoclasts (40). However, for the prevention of bone diseases, it might be questionable to block bone resorption aggressively because this will unbalance bone remodeling. Thus, it is important to investigate whether MBP actually causes a loss in the balance of bone remodeling, in light of its suppressive effect on bone resorption. The NTX level was not found to relate to the BGP level before ingestion, but it was related to the serum BGP after 16 days of MBP ingestion (Figure 14.8). These results suggest that MBP promotes bone formation and suppresses bone resorption in healthy adult humans and that it affects bone metabolism while maintaining a balance of bone remodeling.

14.9 CONCLUSIONS

Since the ancient Egyptians first domesticated the dairy cow in 4000 B.C., milk has made a major contribution to human health. The active components in MBP are proposed to play an important role in bone metabolism (Figure 14.9). It is interesting that MBP contain many active components that stimulate both osteoblasts and osteoclasts and they exist in the same fraction. The MBP effect on bone health is likely to be more than can be accounted for by any single

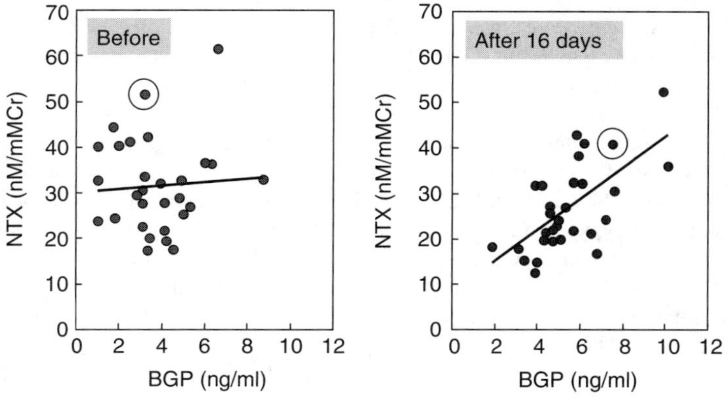

FIGURE 14.8 Relationship between urinary NTX and serum BGP.

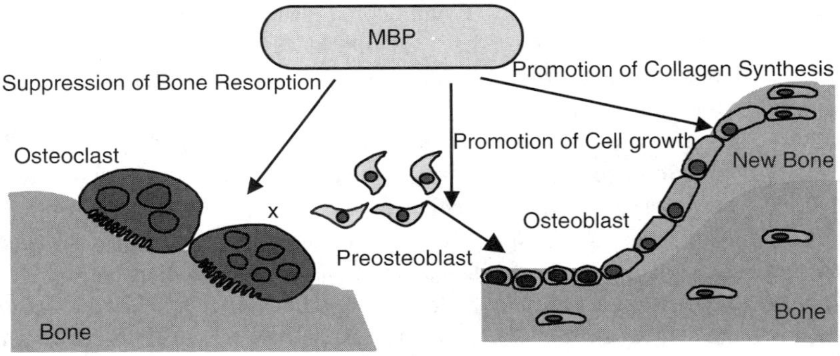

FIGURE 14.9 MBP effects: MBP promotes bone formation and suppresses bone resorption, and increases BMD.

constituent, and the totality of MBP's effect may be more than the sum of the parts. MBP might maintain the balance of bone remodeling because they contain several effective components for both formation and resorption. MBP can be obtained very conveniently because they can be easily separated from milk. MBP extracted from these sources could become a novel, natural, and desirable nutritional supplement, helping improve bone metabolism and thus bone health.

REFERENCES

1. V.K. Kansal, S. Chaudhary. Biological availability of calcium, phosphorus and magnesium from dairy products. *Milchwissenschaft*, 37: 261–263, 1982.
2. Y. Toba, K. Kato, Y. Takada, M. Tanaka, T. Nakano, T. Aoki, S. Aoe. Bioavailability of milk micellar calcium phosphate-phosphopeptide complex in rats. *J. Nutr. Sci. Vitaminol.*, 45: 311–323, 1999.
3. N.P. Wong, D.E. LaCroix. Biological availability of calcium in dairy products. *Nutr. Rep. Intern.*, 21: 673–680, 1980.
4. R.H. Wasserman. Bioavailability of milk micellar calcium phosphate-phosphopeptide complex in rats. *Nature*, 201: 997–999, 1964.
5. Y. Toba, Y. Takada, M. Tanaka, S. Aoe. Comparison of the effects of milk components and calcium source on calcium bioavailability in growing male rats. *Nutr. Res.*, 19: 449–459, 1999.
6. Y. Takada, M. Yahiro, I. Nakajima. Effect of milk components on calcium absorption and bone metabolism. In: K. Yamauchi, T. Imamura, T. Morita (Eds.) Characterization of milk components and health. Tokyo: Kouseikan, 1993, 171–185.
7. Y. Takada, S. Aoe, M. Kumegawa. Whey protein stimulates the proliferation and differentiation of osteoblastic MC3T3-E1 cells. *Biochem. Biophys. Res. Commn.*, 223: 445–449, 1996.
8. Y. Takada, N. Kobayashi, H. Matsuyama, K. Kato, J. Yamamura, M.Yahiro, S. Aoe. Whey protein suppresses the osteoclast-mediated bone resorption and osteoclast cell formation. *Int. Dairy J.*, 7: 821–825, 1997.

9. Y. Takada, N. Kobayashi, K. Kato, H. Matsuyama, M. Yahiro, S. Aoe. Effect of whey protein on calcium and bone metabolism in ovariectomized rats. *J. Nutr. Sci. Vitaminol.*, 43: 199–210, 1997.
10. Y. Takada, H. Matsuyama, K. Kato, N. Kobayashi, J. Yamamura, M. Yahiro, S. Aoe. Milk whey protein enhances the bone breaking force in ovariectomized rats. *Nutr. Res.*, 17: 1709–1720, 1997.
11. G.L. Francis, G.O. Regester, H.A. Webb, F.J. Ballard. Extraction from cheese whey by cation-exchange chromatography of factors that stimulate the growth of mammalian cells. *J. Dairy Sci.*, 23: 1209–1218, 1995.
12. M. Klagsbrun, J. Neumann. The mitogenic activity of human breast milk. *J. Surg. Res.*, 26: 417–422, 1979.
13. R. Pakkanen, A. Kanttinen, L. Satama, J. Aalto. Bovine colostrum fraction as a serum substitute for the cultivation of mouse hybridomas. *Appl. Microbiol. Biotechnol.*, 37: 451–456, 1992.
14. S. Hashizume, K. Kuroda, M. Murakami. Identification of lactoferrin as an essential growth factor for human lymphocyte cell lines in serum-free medium. *Biochim. Biophys. Acta*, 763: 377–382, 1983.
15. F.J. Ballard, M.K. Nield, G.L. Francis, G.W. Dahlenberg, J.C. Wallace. The relationship between insulin content and inhibitory effects in bovine colostrum on protein breakdown in cultured cell. *J. Cell Physiol.*, 110: 249–254, 1982.
16. G.L. Francis, F.M. Upton, F.J. Ballard, K.A. Mcneil, J.C. Wallace. Insulin-like growth factors 1 and 2 in bovine colostrums. Sequences and biological activities compared with those of a potent truncated form. *Biochem. J.*, 251: 95–103, 1988.
17. Y. Jin, D.A. Cox, R. Knecht, F. Raschdorf, N. Cerletti. Separation, purification, and sequence identification of TGF-beta 1 and TGF-beta 2 from bovine milk. *J. Protein Chem.*, 10: 565–575, 1991.
18. M.L. Rogers, C. Goddard, G.O. Regester, F.J. Ballard, D.A. Belford. Transforming growth factor beta in bovine milk: concentration, stability and molecular mass forms. *J. Endocrinol.*, 151: 77–86, 1996.
19. P. Maes, D. Damart, C. Rommens, J. Montreuil, G. Spik, A. Tartar. The complete amino acid sequence of bovine milk angiogenin. *FEBS Lett.*, 241: 41–45, 1988.
20. J. Yamamura, Y. Takada, M. Goto, M. Kumegawa, S. Aoe. Bovine milk kininogen fragment 1.2 promotes the proliferation of osteoblastic MC3T3-E1 cells. *Biochem. Biophys. Res. Commn.*, 269: 628–632, 2000.
21. S. Ohishi, M. Katori, Y. Nam Ham, S. Iwanaga, H. Kato. Possible physiological role of new peptide fragments released from bovine high molecular weight kininogen by plasma kallikrein. *Biochem. Pharmacol.*, 26: 115–120, 1977.
22. S. Asakura, R.W. Hurley, K. Skorstengaard, I. Ohkubo, D.F. Mosher. Inhibition of cell adhesion by high molecular weight kininogen. *J. Cell Biol.*, 116: 465–476, 1992.
23. J. Yamamura, Y. Takada, M. Goto, M. Kumegawa, S. Aoe. High mobility group-like protein in bovine milk stimulates the proliferation of osteoblastic MC3T3-E1 cells. *Biochem. Biophys. Res. Commn.*, 261: 113–117, 1999
24. M. Bustin, D.A. Lehn, D. Landsman. Structural features of the HMG chromosomal proteins and their genes, *Biochim. Biophys. Acta*, 1049: 231–243, 1990.
25. H.M. Wier, P.J. Kraulis, C.S. Hill, A.R. Raine, E.D. Laue, J.O. Thomas. Structure of the HMG box motif in the B-domain of HMG1. *EMBO J.*, 12: 1311–1319, 1993.
26. A.H. Sinclair, P. Berta, M.S. Palmer, J.R. Hawkins, B.L. Griffiths, M.J. Smith, J.W. Foster, A.M. Frischauf,, R. Lovell-Badge, P.N. Goodfellow. A gene from the

human sex-determining region encodes a protein with homology to a conserved DNA-binding motif. *Nature*, 346: 240–244, 1990.

27. Y. Takada, M. Kusuda, K. Hiura, T. Sato, H. Mochizuki, Y. Nagao, M. Tomura, M. Yahiro, Y. Hakeda, H. Kawashima, M. Kumegawa, A simple method to assess osteoclast-mediated bone resorption using unfractionated bone cells. *Bone Miner.*, 17: 347–359, 1992.
28. S. Kakudo, K. Miyazawa, T. Kameda, H. Mano, Y. Mori, T. Yuasa, Y. Nakamaru, M. Shiokawa, K. Nagahira, S. Tokunaga, Y. Hakeda, M. Kumegawa. Isolation of highly enriched rabbit osteoclasts from collagen gels: a new assay system for bone resorbing activity of mature osteoclasts. *J. Bone Miner. Metab.*, 14: 129–136, 1996.
29. U.H. Lerner, L. Johansson, M. Ranjso, J.B. Rosenquist, F.P. Reinholt, A. Grubb. Cystatin C, and inhibitor of bone resorption produced by osteoblasts. *Acta, Physiol. Scand.*, 161: 81–92, 1997.
30. R. Moroi, T. Yamaza, T. Nishiura, Y. Nishimura, Y. Terada, K. Abe, M. Himeno, T. Tanaka. Immunocytochemical study of cathepsin L and rat salivary cystatin-3 in rat osteoclasts treated with E-64 *in vivo*. *Arch. Oral Biol.*, 42: 305–315, 1997.
31. Y. Matsuoka, A. Serizawa, T. Yoshioka, J. Yamamura, Y. Morita, H. Kawakami, Y. Toba, Y. Takada, M. Kumegawa. Cystatin C in milk basic protein (MBP) and its inhibitory effect on bone resorption *in vitro*. *Biosci. Biotechnol. Biochem.*, 66: 2531–2536, 2002.
32. D.L. Martin, H.F. Deluca. Influence of sodium on calcium transport by the rat small intestine. *Am. J. Physiol.*, 216: 1351–1359, 1969.
33. K. Kato, Y. Toba, H. Matsuyama, Y. Takada, J. Yamamura, Y. Matsuoka, H. Kawakami, A. Itabashi, M. Kumegawa S. Aoe, Y. Takada. Milk basic protein enhances the bone strength in ovariectomized rats. *J. Food Biochem.*, 24: 467–476, 2000.
34. I. Ezawa, R. Okada, Y. Nozaki, E. Ogata. Breaking properties and ash contents of femur of growing rats fed a low calcium diets. *J. Jpn. Soc. Nutr. Food Sci.*, 32: 329–335, 1979.
35. Y. Toba, Y. Takada, J. Yamamura, M. Tanaka, Y. Matsuoka, H. Kawakami, A. Itabashi, S. Aoe, M. Kumegawa. Milk basic protein: a novel protective function of milk against osteoporosis. *Bone*, 27: 403–408, 2000.
36. S. Aoe, Y. Toba, J. Yamamura, H. Kawakami, H. Yahiro, M. Kumegawa, A. Itabashi, Y. Takada. A controlled trial of the effect of milk basic protein (MBP) supplementation on bone metabolism in healthy adult women. *Biosci. Biotechnol. Biochem.*, 65: 913–918, 2001.
37. J. Yamamura, S. Aoe, Y. Toba, H. Kawakami, M. Kumegawa, A. Itabashi, Y. Takada. Milk basic protein (MBP) increases radial bone mineral density in healthy adult women. *Biosci. Biotechnol. Biochem.*, 66: 702–714, 2002.
38. Y. Toba,, Y. Takada, Y. Matsuoka, Y. Morita, M. Motouri, T. Hirai, T. Suguri, S. Aoe, H. Kawakami, M. Kumegawa, A. Takeuchi, A. Itabashi. Milk basic protein promotes bone formation and suppresses bone resorption in healthy adult men. *Biosci. Biotechnol. Biochem.*, 65: 1353–1357, 2001
39. S.M. Ott. Theoretical and methodological approach. In: J.P. Bilezikian, L.G. Raisz, G.A. Rodan (Eds.) *A Principles of Bone Biology*. California: Academic Press, 1996, 231–241.
40. G.A. Rodan, T.J. Martin. Therapeutic approaches to bone diseases. *Science*, 289: 1508–1514, 2000.

15 Anticariogenic Peptides

K.J. Cross, N.L. Huq, and
E.C. Reynolds

CONTENTS

15.1 INTRODUCTION – DENTAL CARIES

Dental caries is initiated via the demineralization of dental hard tissue by organic acids from the fermentation of dietary sugar by odontopathogenic bacteria in dental plaque (1). Even though the prevalence of dental caries has decreased through the use of fluorides in most developed countries, the disease remains a major public health problem (2). In the recently published Australian child dental health survey (2), 40.2% of 6-year-olds showed signs of dental caries compared to 48.6% of 12 year olds. Untreated, clinically detectable decay in the combined deciduous and permanent dentition was present in 35.3% of children 5 to 15 years of age, with the greatest severity occurring in the youngest ages (e.g., 9.1% of 5-year-olds had four or more teeth with untreated decay). The level of disease in these high-risk children has decreased only slightly in recent years (e.g., the proportion of 6 year olds with four or more decayed, missing, or filled teeth [DMFT]decreased by only 3.7% between 1989 and 1996). Therefore, although the caries incidence among the general population of children has improved, there still is a significant percentage of high-risk children that require further targeting. Recent dental health surveys in young Australian adults (2) have indicated that the gains in oral health made in childhood are not necessarily carried into later years, as these age cohorts exhibit higher percentages of individuals with high rates of caries incidence. For example, 78% of young adults selected from the electoral roll in Adelaide showed signs of tooth decay with a mean DMFT index of 3.66, with over 10% of these individuals exhibiting a DMFT of 8 or more (2). Demographic changes and changing patterns of oral disease are resulting in larger numbers of older Australians that are increasingly dentate and at high risk of dental caries. A recent survey of Adelaide nursing home residents (2) showed that 34% of the residents were dentate, with 41% of their teeth showing signs of active decay with a mean DMFT of 23.7.

The total cost of providing dental services in Australia in 1998 was estimated at $2.6 billion (2), with over 50% being attributed to treating the consequences of dental caries. This economic burden is higher than for any other diet-related disease in Australia, including coronary heart disease, hypertension, and stroke (2,3). Recent studies have highlighted a number of sociodemographic variables associated with caries risk, a high risk being associated with ethnicity and low socioeconomic status (4). The level of high-risk individuals of all ages has remained relatively constant, even though the overall severity and prevalence of disease in the community has decreased (2,5). Dental caries is therefore still a major public health problem in Australia, particularly in ethnic and lower socioeconomic groups that tend not to use dental services. This highlights the requirement for the development of a nontoxic, anticariogenic agent that could supplement the effects of fluoride in an approach to further lower the rates of caries incidence.

During the caries process, the organic acids produced by plaque bacteria diffuse into the tooth enamel via the water-filled interprismatic spaces and dissolve apatite crystals in a process referred to as *demineralization*. This loss of calcium phosphate from the enamel structure results in the development of an incipient subsurface enamel lesion. At this stage, the caries process is reversible, and it is possible for calcium and phosphate ions to diffuse into the subsurface lesion to restore the lost apatite in a process referred to as *remineralization*. However, use of remineralization solutions containing calcium and phosphate ions has not been successful clinically due to the low solubility of calcium phosphates, particularly in the presence of fluoride ions. Insoluble calcium phosphates are not easily applied, do not localize effectively at the tooth surface, and require acid for solubility. Soluble calcium phosphate ions exist only at low concentrations and also do not substantially incorporate into plaque or localize at the tooth surface. Dairy products contain a highly bioavailable form of calcium phosphate stabilized by the major milk proteins, the caseins; a substantial volume of literature now exists demonstrating the anticariogenic effect of this food group.

15.2 ANTICARIOGENIC FOOD GROUPS

The food group most recognized as exhibiting anticaries activity is dairy products (milk, milk concentrates, powders, and cheeses) (6). Using *in vitro*, animal, and *in situ* caries models, the components largely responsible for this anticariogenic activity have been identified as casein, calcium, and phosphate (6–12). The bovine milk phosphoprotein, casein, which is known to interact with calcium and phosphate (13,14) and is a natural food component, is an obvious candidate as an anticariogenic food and toothpaste additive, however, this is precluded by its organoleptic properties and the very high levels required for activity (6,10,12). Using a human intraoral caries model, Reynolds (11) showed that digestion of caseinate with trypsin did not destroy the protein's ability to prevent enamel subsurface demineralization. Tryptic peptides of casein were found incorporated into the intraoral appliance plaque and were associated with a substantial increase in the plaque's content of calcium and phosphate. It was concluded that the tryptic peptides that were responsible for the anticariogenic activity were the calcium phosphate sequestering phosphopeptides. The casein phosphopeptides (CPP) released by trypsin that sequester calcium phosphate are Bos α_{s1}-casein X-5P (*f*59–79) **[1]**, Bos β-casein X-4P (*f*1–25) **[2]**, Bos α_{s2}-casein X-4P (*f*46–70) **[3]**, and Bos α_{s2}-casein X-4P (*f*1–21) **[4]**. Using Pse to represent a phosphoseryl residue, the sequences of these peptides are as follows:

[1] Gln^{59}-Met-Glu-Ala-Glu-Pse-Ile-Pse-Pse-Pse-Glu-Glu-Ile-Val-Pro-Asn-Pse-Val-Glu-Gln-Lys^{79}. α_{s1}-casein (*f*59–79)

[2] Arg^{1}-Glu-Leu-Glu-Glu-Leu-Asn-Val-Pro-Gly-Glu-Ile-Val-Glu-Pse-Leu-Pse-Pse-Pse-Glu-Glu-Ser-Ile-Thr-Arg^{25}. β-casein (*f*1–25).

[3] Asn^{46}-Ala-Asn-Glu-GluGlu-Tyr-Ser-Ile-Gly-Pse-Pse-Pse-Glu-Glu-Pse-Ala-Glu-Val-Ala-Thr-Glu-Glu-Val-Lys^{70}. α_{s2}-casein (*f*46–70)

[4] Lys^{1}-Asn-Thr-Met-Glu-His-Val-Pse-Pse-Pse-Glu-Glu-Ser-Ile-Ile-Pse-Gln-Glu-Thr-Tyr-Lys^{21}. α_{s2}-casein (*f*1–21)

15.3 ANTICARIOGENIC CASEIN PHOSPHOPEPTIDES

CPP are 10% (w/w) of caseinate and through their multiple phosphoseryl residues, sequester their own weight in amorphous calcium phosphate (ACP) to form colloidal nanocomplexes (13,15,16). As the CPP are not associated with the unpalatability (15) or allergenicity (17) of the caseins and have the potential for a specific anticariogenicity at least 10 times greater on a weight basis, their potential as a food and toothpaste additive is considerably better than that of the intact proteins. CPP–ACP nanocomplexes can be obtained by microfiltration of a solution containing calcium-phosphate-induced complexes of the multiple phosphoseryl-containing peptides in a tryptic digest of casein (18). The major peptides of this preparation are β-casein (*f*1–25) **[2]**, α_{s1}-casein(*f*59–79) **[1],** and its deamidated forms with smaller amounts of α_{s2}-casein(*f*46–70) **[3]** and α_{s2}-casein (*f*1–21) **[4].** All peptides contain the sequence Pse-Pse-Pse-Glu-Glu. The individual peptides of the preparation can be identified by mass spectrometry and amino acid sequence analysis after purification to homogeneity by anion exchange FPLC (fast protein liquid chromatography) and reversed-phase HPLC (high pressure liquid chromatography) (19).

15.3.1 Interaction of CPP with Calcium Phosphate

CPP have a marked ability to stabilize calcium phosphate in solution. Studies with α_{s1}-casein (*f*59–79) **[1]** have shown the peptide to maximally bind 21 Ca and 14 Pi per molecule. The ion activity products for the various calcium phosphate phases (hydroxyapatite [HA]; octacalcium phosphate [OCP]; tricalcium phosphate [TCP]; amorphous calcium phosphate [ACP]; and dicalcium phosphate dihydrate [DCPD]) were determined from the free calcium and phosphate concentrations in these studies using an iterative computational procedure. This procedure calculates the ion activity coefficients using the expanded Debye-Hückel equation and takes into account the ion pairs $CaHPO_4$, $CaH_2PO_4^+$, $CaPO_4^-$, and $CaOH^+$, the dissociation of H_3PO_4 and H_2O, and the ionic strength. The ion activity product that best correlated with calcium phosphate bound to the peptide independently of pH was that corresponding to a basic ACP phase ($Ca_{3.0877}[PO_4]_2[OH]_{0.1754}\cdot xH_2O$) suggesting that this is the phase stabilized by α_{s1}-casein (*f*59–79) (Figure 15.1). These observations are consistent with the suggestion that basic amorphous calcium phosphate is likely to be thermodynamically more stable than acidic amorphous calcium phosphate in casein micelles (20). Because hydroxyl and fluoride ions have similar size and charge, fluoride ions readily substitute for hydroxyl ions in solid-state structures. This is consistent with the stabilization of an amorphous calcium fluoride phosphate phase by CPP (*see below*).

In neutral and alkaline supersaturated calcium phosphate solutions, ACP nuclei form spontaneously. It is proposed that the peptide α_{s1}-casein (*f*59–79) binds to the forming ACP nanoclusters, producing a metastable solution that prevents ACP growth to the critical size required for nucleation and precipitation. From stoichiometric analysis, the stabilized nanocomplexes have the unit formula (α_{s1}-casein [*f*59–79][ACP]$_7$)$_n$ where n is equal to or greater than one. Analysis of α_{s1}-casein (*f*59–79) by PAGE (polyacrylamide liquid chromatography) after cross-linking the nanocomplexes with glutaraldehyde revealed α_{s1}-casein (*f*59–79) multimers with n up to 6. Molecular modeling of the (α_{s1}-casein [*f*59–79][ACP]$_7$)$_6$ nanocomplexes indicated that the diameter of the particles would be ~4 nm.

A 1.0% weight per volume (w/v), CPP solution can stabilize 60 mM $CaCl_2$ and 36 mM sodium phosphate at pH 7 as amorphous calcium phosphate–CPP nanocomplexes (CPP–ACP). At concentrations up to 40% (w/v), CPP–ACP exists as aggregated nanocomplexes in colloidal solution. In more concentrated solutions, a hydrated thixotropic gel network forms.

15.3.2 Anticariogenicity of CPP–ACP in the Rat

The ability of CPP and amorphous calcium phosphate nanocomplexes (CPP–ACP) to reduce caries activity was investigated using specific-pathogen-free rats orally infected with *Streptococcus sobrinus* 6715WT-13 (21). CPP–ACP solutions, applied to the animals' teeth twice daily, significantly reduced caries activity with 1% (w/v) CPP–ACP, producing a 55% reduction relative to the distilled water control. CPP–ACP at 0.5 to 1% (w/v) produced a reduction in caries activity similar to that of the 500 ppm F^- solution. The anticariogenicity of CPP–ACP

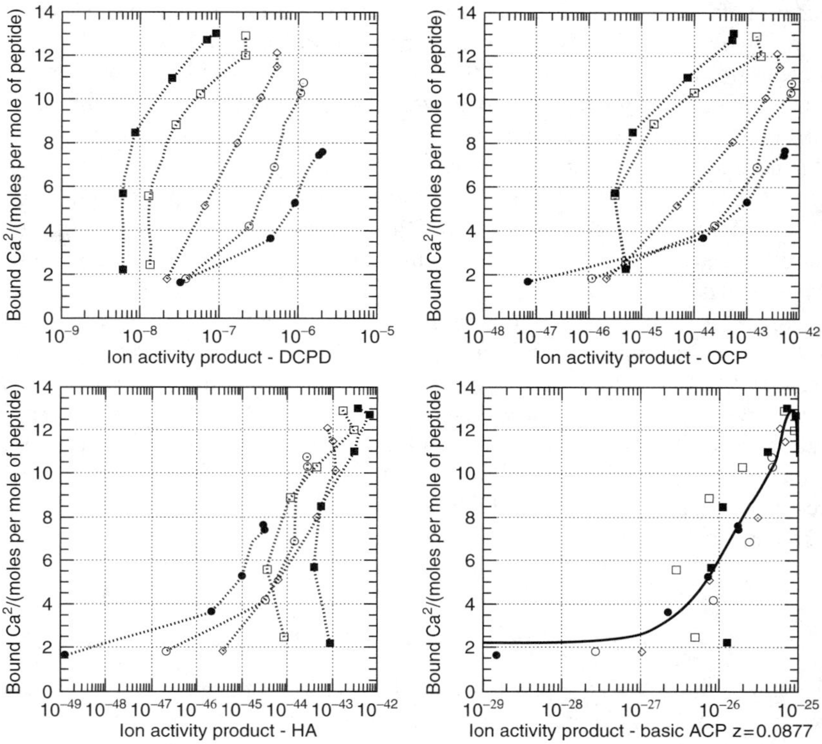

FIGURE 15.1 Bound calcium as a function of the ion activity product for various calcium phosphate phases in equilibrium with α_{S1}-casein (*f*59–79) as a function of pH. The symbols represent data collected at the following pH values 7 (●), 7.5 (○), 8 (◇), 8.5 (□), and 9 (■). The only phase showing a functional relationship between the bound calcium and the ion activity product independent of pH is the basic amorphous calcium phosphate phase, indicating that this is the calcium phosphate phase bound by the peptide.

and fluoride were additive as animals receiving 0.5% CPP–ACP plus 500 ppm F^- had significantly lower caries activity than those animals receiving either CPP–ACP or fluoride alone.

15.3.3 *In vitro* Remineralization of Enamel Lesions by CPP–ACP

Using an *in vitro* remineralization system, the associations between the activities of the various calcium phosphate species in various CPP–ACP solutions and the rate of enamel lesion remineralization have been studied (22). The activity of the neutral ion species $CaHPO_4$ in the various remineralizing solutions was found to be highly correlated with the rate of lesion remineralization (22). The diffusion coefficient for the remineralization process was estimated at 3×10^{-10} m^2 s^{-1}, which is consistent with the diffusion of neutral molecules through a charged matrix. Diffusion of $CaHPO_4$ and associated species into the enamel lesion, followed by the *in situ* formation of Ca^{2+} and PO_4^{3-} ions, would increase the degree of saturation with respect to HA. The formation of HA in the lesion would lead to the generation of acid and phosphate, including H_3PO_4, which would diffuse out of the lesion down a concentration gradient. The results indicate that the CPP-bound ACP acts as a reservoir of the neutral ion species $CaHPO_4$ that is formed in the presence of acid. The acid could be generated by dental plaque bacteria; under these conditions, the CPP-bound ACP would buffer plaque pH and produce calcium and phosphate ions, in particular $CaHPO_4$. The increase in plaque $CaHPO_4$ would offset any fall in pH, thereby preventing enamel demineralization. Acid is also generated in plaque as H_3PO_4 by the formation of HA in the enamel lesion during remineralization. This therefore could explain why the CPP–ACP solutions are such efficient remineralizing solutions, as they would consume the H_3PO_4 produced during enamel lesion remineralization generating more $CaHPO_4$, thus maintaining its concentration gradient into the lesion. These results are therefore consistent with the proposed anticariogenic mechanism of CPP, that is, the inhibition of enamel demineralization and enhancement of remineralization through the localization of ACP at the tooth surface.

15.3.4 Anticariogenicity of CPP–ACP in Human *In Situ* Studies

The ability of a 1% CPP–ACP pH 7 solution to prevent enamel demineralization has been studied in a human *in situ* caries model (11,23). In this model, two exposures of the CPP–ACP solution per day produced a 51 $\pm$ 19% reduction in enamel mineral loss relative to the control enamel and increased plaque calcium and inorganic phosphate contents by 143% and 160%, respectively. CPP were also found in the treated plaque at a level of 2.4 $\pm$ 0.7 mg g^{-1}. The level of the CPP was determined by competitive ELISA (enzyme-linked immunosorbent assay) using an antibody that recognizes both α_{s1}-casein (*f*59–79) and β-casein (*f*1–25).

More recently, an *in situ* study demonstrated the ability of the CPP–ACP complexes in sugar-free chewing gum to remineralize enamel subsurface lesions (24). Thirty subjects in randomized, cross-over, double-blind studies wore removable palatal appliances with six human-enamel half-slab insets containing subsurface demineralized lesions and chewed various sugar-free gums containing CPP–ACP for 20 minutes, four times a day. The studies involved the dose response of the CPP–ACP in sorbitol- and xylitol-based sugar-free gum. Each treatment was of 14 days duration, and at the completion of each treatment, the enamel slabs were removed, paired with their respective demineralized control, embedded, sectioned, and subjected to microradiography and densitometric image analysis to measure the level of remineralization. No significant difference in enamel remineralization was observed between the xylitol- and sorbitol-based gums. Addition of CPP–ACP to either the sorbitol- or xylitol-based gum resulted in a dose-related increase in enamel remineralization with 0.19 mg, 10 mg, 18.8 mg, and 56.4 mg of CPP–ACP producing an increase in enamel remineralization of 9%, 62%,

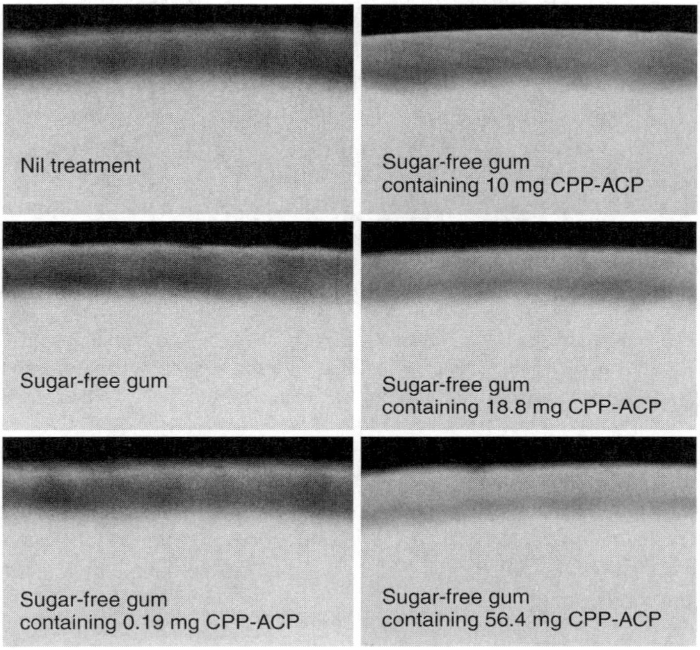

FIGURE 15.2 Representative microradiographs demonstrating a dose response of CPP–ACP with remineralization of enamel subsurface lesions *in situ*. (Adapted from: P. Shen, F. Cai, A. Nowicki, J. Vincent, E.C. Reynolds. *J. Dent. Res.*, 80, 2066–2070, 2001. With permission.)

101%, and 151%, respectively, relative to the control gum, independent of gum weight or type (Figure 15.2) (24).

A similar randomized, double-blind, cross-over study has shown that a mouthrinse containing unstabilized calcium phosphate did not increase plaque Ca and Pi levels significantly, compared with a control mouthrinse consisting of deionized water (16). In contrast, mouthrinses containing 2% and 6% (w/v) CPP–ACP significantly increased the plaque Ca and Pi levels in a dose-dependent fashion (16). Two randomized, double-blind, cross-over remineralization studies were recently conducted with three pellet and three slab sugar-free gums containing different forms of calcium. The gums without CPP–ACP that had $CaHPO_4/CaCO_3$ added to them contained 5 to 13 times the total level of calcium per piece of gum compared with the gums with CPP–ACP. However, the gums with CPP–ACP contained the highest level of water-soluble calcium and inorganic phosphate and produced the highest level of enamel subsurface lesion remineralization (16). CPP were detected using quantitative competitive ELISA in plaque extracts 3 hours after the subjects chewed the CPP–ACP–containing gum. Electron micrographs of immunocytochemically stained sections of plaque revealed localization of the peptide not only

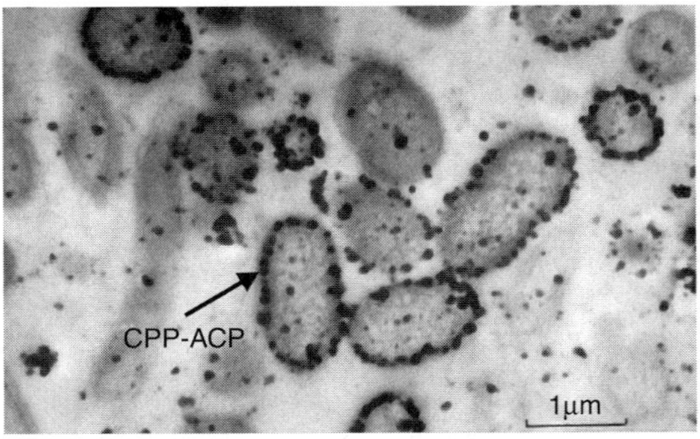

FIGURE 15.3 A representative electron micrograph of supragingival plaque showing CPP–ACP as electron-dense particles associated with the surface of bacteria and the intercellular matrix. (Adapted from E.C. Reynolds, F. Cai, P. Shen, G.D. Walker. *J. Dent. Res.*, 82, 206–211, 2003. With permission.)

on the surface of microorganisms but also in the extracellular matrix (Figure 15.3) (16). These results indicated that CPP were necessary for not only the stabilization of ACP but also for the delivery and localization of the bioavailable calcium phosphate at the tooth surface.

15.3.5 Interaction of CPP–ACP with Fluoride

The additive anticariogenic effect of the 0.5% CPP–ACP and 500 ppm F^- in the rat caries experiments (21) led to the investigation of the potential interaction between CPP–ACP and F^-. Analysis of a solution containing 1% CPP, 60 mM $CaCl_2$, 36 mM sodium phosphate, and 500 ppm F^- (26.3 mM NaF) at pH 7 after ultrafiltration revealed that nearly half of the fluoride ion had incorporated into the ACP phase stabilized by CPP to produce a novel amorphous calcium fluoride phosphate phase. Analysis of the crystallinity of the calcium fluoride phosphate phase stabilized by CPP using powder X-ray diffraction and TEM-EDAX (transmission electron microscopy/electron dispersive X-ray analysis) confirmed that the phase was amorphous. A detailed stoichiometric analysis of the calcium, phosphate and fluoride ions bound by CPP at different pH values suggested that the two phases were stabilized by CPP, ACP ($Ca_3[PO_4]_2xH_2O$), and ACFP ($Ca_2FPO_4xH_2O$). The ratio of the two stabilized phases depended on the mole ratio of the calcium, phosphate and fluoride ions added and the solution pH. A 1:1 ACP:ACFP molar ratio of the two phases (i.e., $Ca_3[PO_4]_2xH_2O$: $Ca_2FPO_4xH_2O$) produces a stoichiometry of $Ca_5(PO_4)_3F$, the same as that of fluorapatite (FA). The identification of the novel amorphous calcium fluoride phosphate (ACFP) phase led us to speculate that the formation of this phase is responsible for the observed additive anticariogenic effect of CPP–ACP and F. The anticariogenic mechanism of fluoride is now proposed to be the localization of the fluoride ion at the tooth surface, particularly in plaque in the presence of Ca and phosphate ions (25). This localization increases the degree of saturation with respect to

FA, thus promoting remineralization of enamel with FA (25). It is clear that for the net formation of FA ($Ca_{10}[PO_4]_6F_2$), calcium and phosphate ions must be co-localized at the tooth surface with the fluoride ion. The additive anticariogenic effect of CPP–ACP and F may therefore be attributable to the localization of ACFP at the tooth surface by CPP, which, in effect, would co-localize Ca, Pi, and F. This hypothesis was tested in a mouthwash study where the ability of a 3% CPP–ACFP mouthwash at pH 7 used three times daily to increase supragingival plaque calcium, inorganic phosphate, and fluoride ions was determined. The 3% CPP–ACFP solution used as a mouthwash contained 192 mM bound calcium ions, 120 mM bound phosphate ions, and 24 mM bound F ions stabilized as $(ACP)_2$:(ACFP). The use of the mouthwash resulted in a 1.9-fold increase in plaque calcium to 336 ± 107 μmol g^{-1}, a 1.5-fold increase in plaque phosphate to 471 ± 113 μmol g^{-1}, and a dramatic 18-fold increase in plaque fluoride ion to 19.9 ± 14.1 μmol g^{-1}. Although these marked increases in plaque calcium, phosphate, and fluoride were found, calculus was not observed in any of the 17 subjects, suggesting that the plaque calcium fluoride phosphate remained stabilized as the amorphous phase by CPP. These increases in the supragingival plaque levels of calcium, phosphate, and fluoride ions are markedly greater than those obtained in a similar study with toothpastes containing 1000 ppm F (MFP and NaF) used twice daily for a similar period (26). These results suggest that CPP are an excellent delivery vehicle to co-localize calcium, fluoride, and phosphate at the tooth surface in a slow-release amorphous form with superior clinical efficacy.

In a recent *in vitro* experiment, the ability of a 2% (w/v) CPP–ACFP solution to remineralize enamel subsurface lesions was compared with that of a 2% CPP–ACP solution at 37°C and different pH values. At pH 5.5, the 2% CPP–ACP solution replaced 41.83 ± 8.47% of the mineral lost, whereas the 2% CPP–ACFP solution replaced 57.67 ± 8.43%, which was significantly greater ($p < 0.001$). At pH 5, the difference in remineralization was even greater with the CPP–ACFP solution replacing 40.03 ± 11.16% of the mineral lost, whereas the CPP–ACP solution only replaced 18.33 ± 8.68%. Analysis of the remineralizing solutions for free and bound calcium, phosphate, and fluoride ions showed that as the pH of the solution was lowered, CPP released more of the bound ions, with over 90% released at pH 4.5. As with our earlier studies, the rate of enamel remineralization was correlated with the activity of the neutral ion pair $CaHPO_4$ in the CPP–ACP solution. For CPP–ACFP, the computational procedure was extended to include the species CaF^+ and HF, and remineralization was still found to be associated with the activity of $CaHPO_4$. The inclusion of F^- resulted in the formation of more $CaHPO_4$ at lower pH values through the following reaction:– $CaH_2PO_4^+ + F^- \rightarrow (CaFH_2PO_4) \rightarrow CaHPO_4 + HF$. The formation of the neutral ions $CaHPO_4$ and HF would facilitate diffusion into the subsurface lesion providing the calcium, phosphate, and fluoride ions for the formation of FA ($Ca_{10}[PO_4]_6F_2$). The higher activity of the neutral ion pair $CaHPO_4$ in the CPP–ACFP solutions at pH 5.5 and 5, together with the presence of significant amounts of HF would explain the greater remineralization obtained with these solutions.

In a recent double-blind, randomized, cross-over, *in situ* study of toothpastes containing CPP–ACP and fluoride, it was shown that the paste containing 2%

CPP–ACP plus 1100 ppm F produced 143% higher levels of remineralization of enamel subsurface lesions than the paste containing 1100 ppm F alone (unpublished). The increased level of remineralization was shown to be associated with the formation of acid-resistant FA throughout the body of the lesion and not predominantly at the surface layer, which occurred with the fluoride-containing paste. The higher levels of remineralization obtained with CPP–ACP plus F suggest that CPP–ACFP may have clinical application for the mineralization of hypomineralized enamel as part of a preventive strategy to target individuals with a high-risk for caries.

15.4 CASEIN MICELLE STRUCTURE

Bovine milk contains ~30 mM Ca and ~22 mM inorganic phosphate in solution with most of the calcium (~68%) and phosphate (~47%) associated with the proteins α_{S1}-, α_{S2}-, and β-caseins in casein micelles (27,28). The α_{S1}-, α_{S2}-, and β-caseins contain the Pse-Pse-Pse-Glu-Glu sequence, which is essential for interaction with calcium phosphate (29).

The casein micelles serve as carriers of calcium phosphate, providing the neonate with a bioavailable source of these ions for bone and teeth formation (29). It has been postulated that the ability of casein to form stable complexes with calcium phosphate is intrinsic to a general mechanism for avoiding pathological calcification and regulating calcium flow in tissues and biological fluids containing high concentrations of calcium (30).

Many techniques have been used to investigate the ultrastructure of the casein micelles. Although the structural details are still being elucidated, the casein micelles are believed to be roughly spherical particles with a radius of about 100 nm (31), dispersed in a continuous phase of water, salt, lactose, and whey proteins. The calcium phosphate isolated after exhaustive hydrazine deproteination of micelles was reported to exhibit, under the electron microscope, a fine and uniform granularity consisting of small subunits with a diameter of approximately 2.5 nm (32,33). Calcium phosphate, present as nanometer-sized clusters, and caseins are not covalently bound, hence the casein micelle is known as an association colloid (34). Nevertheless, the casein micelles are extremely stable and can withstand boiling, freeze-drying, and the addition of salt and ethanol. It is believed that the amphipathic C-terminal end of the κ-casein molecules protrude from the micelle surface forming a so-called "hairy layer" that sterically stabilizes the micelles.

15.5 STRUCTURE OF THE ANTICARIOGENIC CPP

A number of phosphoproteins (e.g., caseins, phosphophoryn, osteopontin, matrix-Gla protein, and statherin) that interact with calcium phosphate in biological fluids and tissues have now been identified (14,35–37). The sequences reveal that these proteins all contain multiple phosphoseryl and acidic residues in clusters. The proposed functions of these proteins are (a) the stabilization of calcium phosphate in solution preventing spontaneous precipitation, and (b) biomineralization, where the protein, bound to a matrix, has been proposed to act as a nucleator/promoter of crystal growth. However, the structures of these proteins in the presence of calcium phosphate have not yet been elucidated. Optical rotatory dispersion (ORD), circular dichroism (CD), and hydrodynamic and ^{31}P nuclear magnetic resonance (NMR) measurements of casein structure all indicate that α_{s1}-casein and β-casein have a rather open structure in solution, with many amino acid side chains exposed to solvent and relatively flexible (15). ^{31}P-NMR relaxation measurements indicate that Pse residues are relatively mobile in β-casein (38). However, none of these measurements was made in the presence of calcium phosphate. Medium- and long-range nuclear Overhauser enhancements (nOes) in 2D ^{1}H NMR spectra of both α_{s1}-casein (*f*59–79) **[1]** and β-casein (*f*1–25) **[2]** in the presence of calcium ions have recently been demonstrated indicating conformational preferences (39,40). Two structured regions were identified in α_{s1}-casein (*f*59–79) **[1]**. Residues Val^{72} to Val^{76} are implicated in a β-turn conformation. Residues Glu^{61} to Pse^{67}, which extend over part of the Pse cluster motif Pse-Pse-Pse-Glu-Glu are involved in a loop-type structure (39). Four structured regions were identified in β-casein (*f*1–25) **[2]**. Residues Arg^1 to Glu^4 form a loop. Residues Val^8 to Glu^{11}, Pse^{18} to Glu^{20}, and Ser^{22} to Thr^{24} are in tight turns (40). The peptide conformation in the region of the Pse-Pse-Pse-Glu-Glu calcium-binding motif is different in these two peptides. Analysis of the ^{1}H NMR spectra of α_{s2}-casein (*f*2–20) **[4]** has shown that several residues, including those around the Pse-Pse-Pse-Glu-Glu motif, adopt a nascent helical structure, a conformation that is distinctly different from that adopted by either α_{s1}-casein (*f*59–79) **[1]** or β-casein (*f*1–25) **[2]** in the presence of calcium (41). Molecular modeling of α_{s1}-casein (*f*59–79), β-casein (*f*1–25), and α_{s2}-casein (*f*2–20) using the constraints derived from the NMR spectroscopy has indicated that the peptides adopt conformations that allow the glutamyl and phosphoseryl side chains to interact collectively with calcium ions.

^{1}H NMR solution spectra of β-casein (*f*1–25) complexed with amorphous calcium phosphate (β-casein [*f*1–25]–ACP) have recently been recorded (42,43). The spectra displayed sharp lines, with $^3J_{NH\alpha}$ coupling constants readily measurable in the amide region (in contrast to the broader resonances observed with the β-casein [*f*1–25]–calcium complex). The signals were intense, with no evidence of a broadened component underlying the spectrum as would be expected for aggregated species. DQ-COSY, TOCSY, and NOESY spectra of the β-casein (*f*1–25)–ACP complex in 90% H_2O and 10% D_2O solution at a temperature of 25°C were recorded. Sequential assignment was completed from the spectra; chemical shifts of resonances showed significant shifts away from random coil values, and differences were compared with the shifts of the β-casein (*f*1–25)–calcium complex.

Using the CLEANEX-PM technique (44) with β-casein (*f*1–25)–ACP, the only solvent-exposed protons detected were the Hα protons of Pro^9, either Ile^{12} or Val^{13}, and a Glu (possibly Glu^{11}). In the case of β-casein (*f*1–25) complexed to calcium, these residues form a β-turn (40). These results suggest that the N- and C-termini are buried, an important observation with implications for the structure of casein micelles.

Using the sLED technique (45), translational diffusion coefficients for the β-casein (*f*1–25)–ACP nanocomplex have been determined. Using the rate of decay of the HDO solvent peak and assuming a hydrodynamic radius of 1.40 Å for the water molecule, the Stokes–Einstein equation was used to estimate the size of the particles giving rise to the observed ^{1}H NMR spectra of β-casein (*f*1–25)–ACP. For solutions containing 1 to 2 mM peptide, hydrodynamic radii of 1.526 ± 0.044 nm at pH 6 increasing to 1.923 ± 0.044 nm at pH 9 for the β-casein (*f*1–25)–ACP complex have been estimated. These dimensions are consistent with the predicted size of the (β-casein [1-25][ACP]$_8$)$_6$ nanocomplexes

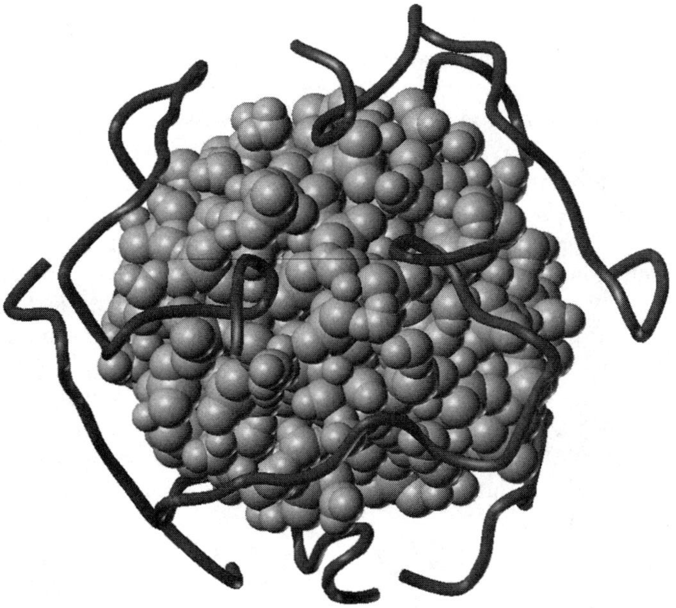

FIGURE 15.4 Model of the β-casein(*f*1–25)–ACP nanocomplex. The model represents the ACP core as an aggregate of CPK atoms representing calcium ions, phosphate ions, and water. The ribbons represent the backbone atoms of the peptides. The model is based on data from 1H NMR experiments constraining the peptide backbone conformation. In addition, the polar side-chains of the peptides were constrained to make close electrostatic interactions with the ACP core. The size of the ACP core is based on studies of calcium and phosphate binding by β-casein (*f*1–25) combined with a knowledge of the number of peptides per nanocomplex from cross-linking peptides in the β-casein (*f*59–79)–ACP nanocomplex using glutaraldehyde. NMR diffusion experiments indicate that the nanocomplexes have hydrodynamic radii of 1.526 ± 0.044 nm at pH 6 increasing to 1.923 ± 0.082 nm at pH 9. The calcium phosphate isolated after exhaustive hydrazine deproteination of casein micelles was reported to exhibit, under the electron microscope, a fine and uniform granularity consisting of small subunits with a diameter of approximately 2.5 nm. The model is consistent with all experimental data currently available (32, 33).

using molecular modeling (*vide infra*). For peptide concentrations of 10 mM or greater, the decay of the echo amplitude due to translational diffusion was multiexponential showing clear evidence of the presence of larger particles that diffused more slowly than the monomeric species. The larger particles were assumed to be aggregates of the β-casein (*f*1–25)–ACP nanocomplexes. Recent studies of bone mineralization centres (46) show that the calcium phosphate phase that is first formed is an amorphous phase similar to that stabilized by CPP. This amorphous calcium phosphate phase of bone mineralization centres exists as ~5 nm diameter subunits that aggregate to form hollow spheres of ~100 nm diameter.

The relationship between CPP structure and the interaction with amorphous calcium phosphate was further investigated using a series of synthetic peptide homologues and analogues (47). These studies showed that the cluster sequence Pse-Pse-Pse-Glu-Glu is effectively responsible for the interaction with ACP and that all three contiguous Pse residues are required for maximal interaction with ACP.

The docking of the peptide Pse-Pse-Pse-Glu-Glu onto three crystallographic planes of HA, {100}, {010}, and {001} — using computer simulation techniques and the unit cell coordinates of synthetic HA — has also been studied (48). These simulation studies revealed that the peptide Pse-Pse-Pse-Glu-Glu is more likely to bind to the {100} surface, followed by the {010} surface. This is in agreement with the finding that bovine phosphophoryn containing multiple Pse-Pse-Pse sequences interacts preferentially with the {100} face of HA. The Pse cluster motif can therefore bind to both {100} and {010} surfaces, thus allowing deposition of calcium, phosphate and hydroxyl ions on the {001} surface and enabling growth of the HA crystal along the *c*-axis only. These results therefore can now explain the *c*-axis growth of HA crystals *in vivo*. Examination of the structure of the most strongly bound conformer reveals that the oxygen atoms of the side-chain carboxyl and phosphoryl groups are orientated to achieve maximal contact with surface calcium atoms. The backbone of the peptide does not adopt a regular structure such as α-helix or β-sheet but rather forms loops consistent with proposed solution three-dimensional structures (48).

Based on the data accumulated on the CPP–ACP complex, a detailed molecular model of the β-casein (*f*1–25)–ACP nanocomplex has been developed (Figure 15.4) (42,43). The model is consistent with the diverse experimental results acquired in our laboratory and elsewhere and these are: (a) the dimensions of the ACP core (32,33), (b) low-angle X-ray scattering data (49), (c) hydrodynamic radii, nOe constraints, and (d) solvent exposure of specific residues (42,43). The model predicts that a strongly lipophilic patch associated with residues Leu^3, Pro^9, Gly^{10}, Ile^{12}, Val^{13}, and Leu^{16} is formed by the β-casein (*f*1–25) peptide when bound to the ACP core particle (Figure 15.5). The formation of a lipophilic patch provides a molecular-level rationalization of properties of the β-casein (*f*1–25)–ACP nanocomplex such as the tendency to form aggregates and to bind to oral surfaces through hydrophobic interactions (16).

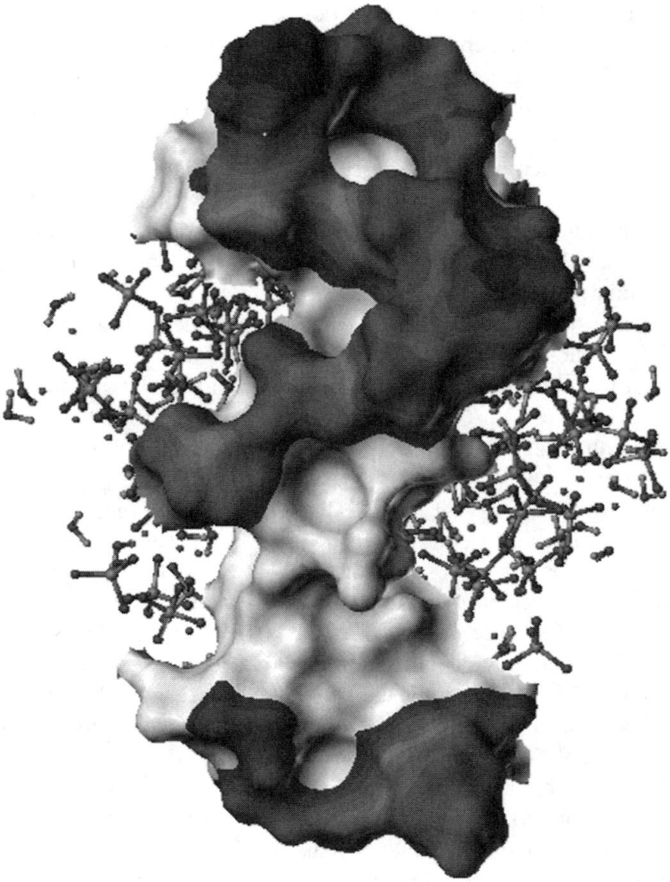

FIGURE 15.5 The solvent accessible surface of one of the β-casein (*f*1–25) peptides as modeled on the ACP core (shown as ball and stick atoms). The surface has been colored by its lipophilicity, the dark area at the top of the diagram represents a strongly lipophilic region associated with Leu^3, Pro^9, Gly^{10}, Ile^{12}, Val^{13}, and Leu^{16}.

15.6 CONCLUSIONS

The CPP–ACP complex has been demonstrated to not only help prevent the development of dental caries but also to repair enamel subsurface demineralized lesions representing early stages of the caries process. Furthermore, CPP–ACP and fluoride have been shown to have an additive effect in reducing caries in animals and in remineralizing enamel subsurface lesions. Upon interaction of CPP–ACP with fluoride ions, a novel amorphous calcium fluoride phosphate (ACFP) phase is formed, which is stabilized by CPP and localized at the tooth surface. At the tooth surface, the CPP–ACFP nanocomplexes act as slow-release reservoirs of soluble calcium, phosphate, and fluoride ions that promote remineralization of subsurface lesions with acid-resistant FA. Physicochemical characterization of the CPP–ACFP nanocomplexes suggests that nanoclusters of $Ca_3(PO_4)_2xH_2O$ and $Ca_2FPO_4xH_2O$ are stabilized by CPP in specific conformations. The CPP–ACP technology has been commercialized under the trademark Recaldent, and functional foods (nutraceuticals) and oral care products containing CPP–ACP and CPP–ACFP, respectively, are now commercially available.

REFERENCES

1. W.J. Loesche. Role of *Streptococcus mutans* in human dental decay. *Microbiol. Rev.*, 50: 353–380, 1986.
2. J. Armfield, K. Roberts-Thomson, A. Spencer. Australia's Health 2000: *The Seventh Biennial report of the Australian Institute of Health & Welfare*. Canberra: Australian Institute of Health & Welfare, 2000.
3. S. Crowley, K. Antioch, R. Crater, L. Conway, C. Matheis. *The Economic Burden of Diet-Related Disease in Australia.* Canberra: National Food and Nutrition Centre for Health Program Evaluation, and Australian Institute of Health, 1992.
4. A.J. Spencer, F.A.C. Wright, L.M. Brown, L.P. Brown. Changing caries experience and risk factors in five- and six-year-old Melbourne children. *Aust. Dent. J.*, 34: 160–165, 1989.
5. F.A.C. Wright. *Ministerial Review of Dental Services in Victoria*, Australia. 1986.
6. E.C. Reynolds. Anticariogenic complexes of amorphous calcium phosphate stabilised by casein phosphopeptides. A review. *J. Spec. Care Dent.*, 18: 8–16, 1998.
7. E.C. Reynolds, I.H. Johnson. Effect of milk on caries incidence and bacterial composition of dental plaque in the rat. *Arch. Oral Biol.*, 26: 445–451, 1981.
8. E.C. Reynolds, A. Del Rio. Effect of casein and whey protein solutions on caries experience of the rat. *Arch. Oral Biol.*, 30: 927–933, 1984.
9. E.C. Reynolds, C.L. Black. Confectionery composition and rat caries. *Caries Res.*, 21: 538–545, 1987.
10. E.C. Reynolds, C.L. Black. The reduction of chocolate's cariogenicity by supplementation with sodium caseinate. *Caries Res.*, 21: 445–451, 1987.
11. E.C. Reynolds. The prevention of sub-surface demineralization of bovine enamel and change in plaque composition by casein in an intra-oral model. *J. Dent. Res.*, 66: 1120–1127, 1987.
12. E.C. Reynolds, C.L. Black. Cariogenicity of a confection supplemented with sodium caseinate at a palatable level. *Caries Res.*, 23: 368–371, 1989.
13. R.E. Reeves, N. Latour. Calcium phosphate sequestering phosphopeptide from casein. *Science*, 128: 472, 1958.
14. C. Holt, M.J.J.M. Van Kemenade. The interaction of phosphoproteins with calcium phosphate. In: D.W.L. Hukins (Ed.). *Calcified Tissue*. Boca Raton, FL: CRC Press, 1989, 175–213.
15. H.E. Swaisgood. Chemistry of milk proteins. In: P.F. Fox (Ed.). *Developments in Dairy Chemistry*. London: Applied Science Publishers, 1982, 1–43.
16. E.C. Reynolds, F. Cai, P. Shen, G.D. Walker. Retention in plaque and remineralization of enamel lesions by various forms of calcium in a mouthrinse or sugar-free chewing gum. *J. Dent. Res.*, 82: 206–211, 2003.

17. A. Ametani, S. Kaminogawa, M. Shimizu, K. Yamauchi. Rapid screening of antigenically reactive fragments of alphaS1-casein using HPLC and ELISA. *J. Biochem.*, 102: 421–425, 1987.
18. E.C. Reynolds. *Production of Phosphopeptides from Casein*. USA Patent 6448374.
19. E.C. Reynolds, P.F. Riley, N.J. Adamson. A selective precipitation procedure for the purification of multiple-phosphoseryl containing peptides and their identification. *Anal. Biochem.*, 217: 277–284, 1994.
20. H.J.M. van Dijk. The properties of casein micelles. 1 The nature of micellar calcium phosphate. *Neth. Milk Dairy J.*, 44: 65–81, 1990.
21. E.C. Reynolds, C.J. Cain, F.L. Webber, C.L. Black, P.F. Riley, I.H. Johnson, J.W. Perich. Anticariogenicity of tryptic casein- and synthetic-phosphopeptides in the rat. *J. Dent. Res.*, 74: 1272–1279, 1995.
22. E.C. Reynolds. Remineralization of enamel subsurface lesions by casein phosphopeptide-stabilized calcium phosphate solutions. *J. Dent. Res.*, 76: 1587–1595, 1997.
23. E.C. Reynolds. *Phosphopeptides*. USA Patent 5015628, 1991.
24. P. Shen, F. Cai, A. Nowicki, J. Vincent, E.C. Reynolds. Remineralization of enamel subsurface lesions by sugar-free chewing gum containing casein phosphopeptide-amorphous calcium phosphate. *J. Dent. Res.*, 80: 2066–2070, 2001.
25. A. Thylstrup, O. Fejerskov. *Textbook of Cariology*. Munksgaard, 1986.
26. A.D. Sidi. Effect of brushing with fluoride toothpastes on the fluoride, calcium and inorganic phosphorus concentrations in approximal plaque of young adults. *Caries Res.*, 23: 268–271, 1989.
27. A.C.M. Van Hooydonk, I.J. Boerrigter, H.G. Hagedoorn. pH-induced physico-chemical changes of casein micelles in milk and their effect on renneting. 2. Effect of pH on renneting of milk. *Neth. Milk Dairy J.*, 40: 297–313, 1986.
28. P. Walstra, R. Jenness. *Dairy Chemistry and Physics*. New York: Wiley, 1984.
29. C. Holt, L. Sawyer. Primary and predicted secondary structures of the caseins in relation to their biological functions. *Protein Eng.*, 2: 251–259, 1988.
30. C. Holt, N.M. Wahlgren, T. Drakenberg. Ability of a beta-casein phosphopeptide to modulate the precipitation of calcium phosphate by forming amorphous dicalcium phosphate nanoclusters. *Biochem. J.*, 314: 1035–1039, 1996.
31. D.G. Schmidt. Association of caseins and casein micelle structure. *Dev. Dairy Chem.*, 1: 61–86, 1982.
32. T.C. McGann, W. Buchheim, R.D. Kearney, T. Richardson. Composition and ultrastructure of calcium phosphate-citrate complexes in bovine milk systems. *Biochim. Biophys. Acta*, 760: 415–420, 1983.
33. T.C. McGann, R.D. Kearney, W. Buchheim, A.S. Posner, F. Betts, N.C. Blumenthal. Amorphous calcium phosphate in casein micelles of bovine milk. *Calcif. Tiss. Int.*, 35: 821–823, 1983.
34. C.G. de Kruif. Casein micelle interactions. *Int. Dairy J.*, 9: 183–188, 1999.
35. E.S. Sorensen, T.S. Petersen. Identification of two phosphorylation motifs in bovine osteopontin. *Biochem. Biophys. Res. Commn.*, 198: 200–205, 1994.
36. P.A. Price, J.S. Rice, M.K. Williamson. Conserved phosphorylation of serines in the Ser-X-Glu/Ser(P) sequences of the vitamin K-dependent matrix Gla protein from shark, lamb, rat, cow and human. *Protein Sci.*, 3: 822–830, 1994.
37. N.L. Huq, K.J. Cross, G.H. Talbo, P.F. Riley, A. Loganathan, M.A. Crossley, J.W. Perich, E.C. Reynolds. N-terminal sequence analysis of bovine dentine phosphophoryn after conversion of phosphoseryl to S-propylcysteinyl residues. *J. Dent. Res.*, 79: 1914–1919, 2000.

38. R.S. Humphrey, K.W. Jolley. ^{31}P-NMR studies of bovine β-casein. *Biochem. Biophys. Acta*, 708: 294–299, 1982.
39. N.L. Huq, K.J. Cross, E.C. Reynolds. A ^{1}H NMR study of the casein phosphopeptide α_{S1}-casein (59-79). *Biochem. Biophys. Acta*, 1247: 201–208, 1995.
40. K.J. Cross, N.L. Huq, W. Bicknell, E.C. Reynolds. Cation-dependent structural features of beta-casein-(1-25). *Biochem. J.*, 356: 277–285, 2001.
41. N.L. Huq, K.J. Cross, E.C. Reynolds. Nascent helix in the multiphosphorylated peptide α_{S2}-CN(2-20). *J. Pept. Sci.*, 9: 386–392, 2003.
42. K.J. Cross, N.L. Huq, D. Eakins, E.C. Reynolds. Structural studies of the β-casein phosphopeptide bound to amorphous calcium phosphate. *J. Dent. Res.*, 80: 588, 2001.
43. K.J. Cross, N.L. Huq, D. Eakins, E.C. Reynolds. Ultrastructural studies of the casein phosphopeptide-amorphous calcium phosphate nanoclusters. *J. Dent. Res.*, 80: 588, 2001.
44. T.L. Hwang, S. Mori, A.J. Shaka, P.C.M. Vanzijl. Application of phase-modulated clean chemical exchange spectroscopy (CLEANEX-PM) to detect water-protein proton exchange and intermolecular NOEs. *J. Am. Chem. Soc.*, 119: 6203–6204, 1997.
45. A.S. Altieri, D.P. Hinton, R.A. Byrd. Association of biomolecular systems via pulsed field gradient NMR self-diffusion measurements. *J. Am. Chem. Soc.*, 117: 7566–7567, 1995.
46. L.N. Wu, B.R. Genge, D.G. Dunkelberger, R.Z. LeGeros, B. Concannon, R.E. Wuthier. Physicochemical characterization of the nucleational core of matrix vesicles. *J. Biol. Chem.*, 272: 4404–4411, 1997.
47. E.C. Reynolds. Anticariogenic casein phosphopeptides. *Prot. Pept. Lett.*, 6: 295–303, 1999.
48. N.L. Huq, K.J. Cross, E.C. Reynolds. Molecular modelling of a multi-phosphorylated sequence motif bound to hydroxyapatite surfaces. *J. Mol. Model*, 6: 35–47, 2000.
49. C. Holt, P.A. Timmins, N. Errington, J. Leaver. A core-shell model of calcium phosphate nanoclusters stabilized by beta-casein phosphopeptides, derived from sedimentation equilibrium and small-angle X-ray and neutron-scattering measurements. *Eur. J. Biochem.*, 252: 73–78, 1998.

16 The Potential Therapeutic Role of Fibrinolytic Enzymes from Food in Cardiovascular Disease

Ada H.-K. Wong and Yoshinori Mine

CONTENTS

16.1 INTRODUCTION – CARDIOVASCULAR DISEASE

Cardiovascular disease, including acute myocardial infarction, ischemic heart disease, valvular heart disease, peripheral vascular disease, arrhythmias, high blood pressure, and stroke, are the leading causes of death throughout the world. In accordance with the data provided by the World Health Organization in 2000 (1), heart disease is responsible for 29% of the total mortality rate in the world (Figure 16.1). Based on the mortality rates of different types of cardiovascular disease, as shown in Figures 16.2a and 16.2b, acute myocardial infarction and ischemic heart disease are the most important heart problems, starting at age 45 for men and 55 for women. However, congestive heart failure and stroke affect older individuals, both men and women, over age 75 (1).

Hemostasis is the tightly regulated process of keeping an optimal balance between coagulation and anticoagulation. Coagulation involves a series of enzymatic reactions in which inactive plasma proteins are converted into active enzymes in each step of the pathway. As shown in Figure 16.3, the cascade is initiated by the release of tissue factor or damaged collagen underneath the vascular

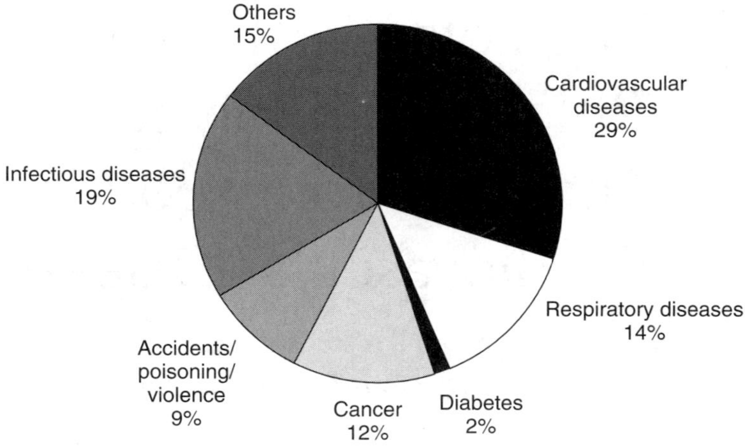

FIGURE 16.1 The proportion of deaths by causes in WHO regions, estimates for 2000. (From World Health Organization. *The World Health Report 2001*. [Accessed on September 2, 2002]. Available from www.who.int. With permission.)

endothelium. The final step involves the formation of a fibrin clot that stabilizes the platelet plug. The fibrin clot is formed from fibrinogen by thrombin, whereas the dissolution of a blood clot is dependent on the action of endogenous plasmin, a serine protease that is activated by tissue plasminogen activator (2).

An imbalance in hemostasis may result in excessive bleeding or in the formation of a thrombus (an inappropriate blood clot) that adheres to the unbroken wall of the blood vessels. Accumulation of fibrin in blood vessels can interfere with blood flow and lead to myocardial infarction and other serious cardiovascular diseases. Unless the blockage is removed promptly, the tissue that is normally supplied with oxygen by the vessel will die or be severely damaged. If the damaged region is large, the normal conduction of electrical signals through the ventricle will be disrupted, leading to irregular heartbeat, cardiac arrest, or death (3). Cardiovascular disease is the leading cause of death in the world, inflicting devastating physical, emotional, and major financial costs on its victims and their families. Thus, significant advances have been made during the past decade in the prevention and treatment of cardiovascular diseases. In general, there are four options for patients as summarized below.

Historically, management of heart disease and stroke has relied on the use of anticoagulant and antiplatelet aggregation drugs. This is because the underlying pathophysiological process in myocardial infarction and stroke is the formation of a thrombus that consists of fibrin and platelets. That is, an optimum antithrombotic prophylactic therapy can, and should be, directed toward both.

Anticoagulants are chemicals that prevent coagulation. Most of them act by blocking one or more steps in the cascade that forms fibrin. Some drugs inhibit the synthesis of clotting factors, while others enhance the anticoagulant activity

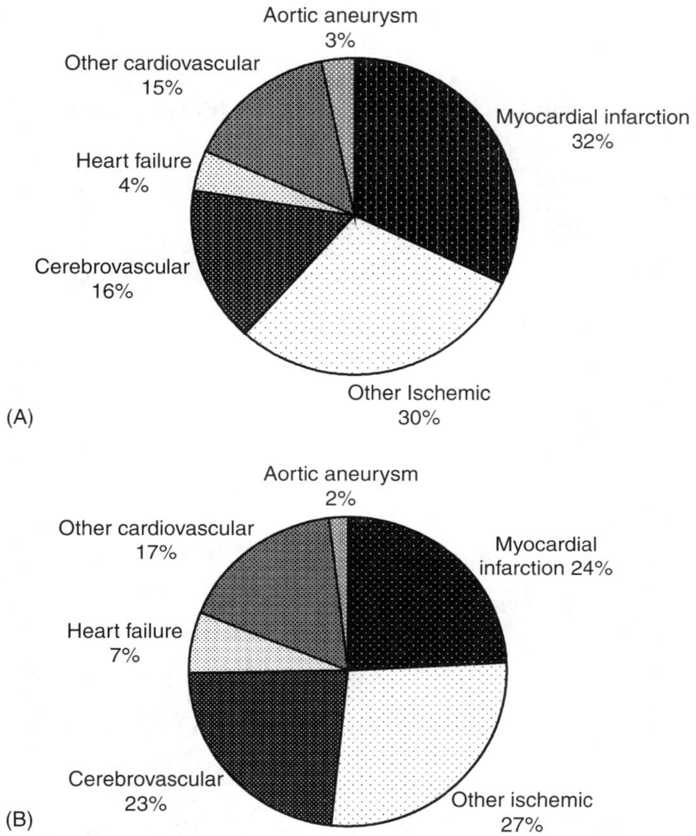

FIGURE 16.2 Prevalence of cardiovascular diseases by types in (A) male and (B) female canadians (Health Canada, 2000).

of the naturally occurring blood factors or prevent platelet plug formation. Indeed, treatment with oral anticoagulants requires a constant balancing between undertreatment and overtreatment. In other words, the intensity of coagulation is monitored within narrow therapeutic margins, and the effect of the daily dose has to be checked regularly because of the influence of disease, food, and other drugs on coagulation (4). Warfarin is the most commonly used anticoagulant in the United Kingdom and the rest of the Western world. It inhibits coagulation by interfering with the incorporation of vitamin K into vitamin K–dependent clotting factors, including factors II, VII, IX, and X. There is a considerable variability in its effect on patients, and its effectiveness can be influenced by age, race, diet, and co-medications such as antibiotics. The drugs can be given orally or intravenously. Adverse effects of warfarin include hemorrhage, hypersensitivity, skin rashes, alopecia, and purpura. Another anticoagulant is heparin, a glycosaminoglycan, whose major anticoagulant effect is accounted for by inhibiting thrombin,

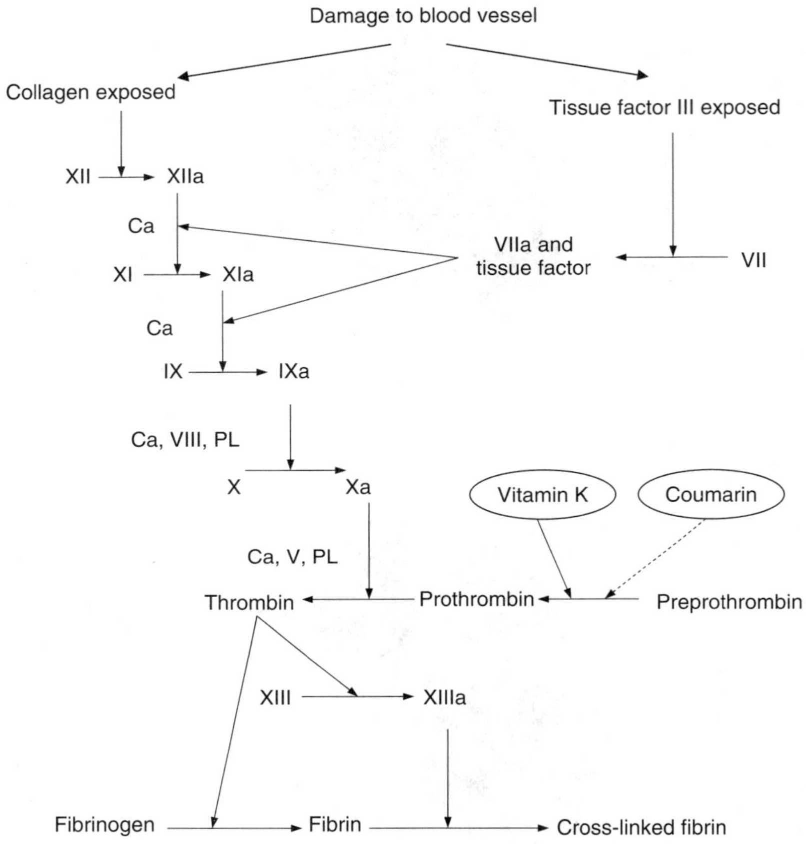

FIGURE 16.3 A diagram of the blood-clotting cascade. The cascade involves a series of enzymatic reactions, in which inactive plasma proteins are converted into active enzymes at each step of the pathway. The final step is the formation of a fibrin mesh that stabilizes the platelet plug. The intrinsic pathway begins with collagen exposure and uses proteins already present in the plasma; while the extrinsic pathway starts when damaged tissues expose tissue factor to the plasma proteins. Solid and dotted arrows represent activation and inhibition of the components, respectively. (Adapted from A.C. Silverthorn, W.C. Ober, C.W. Garrison. *Human Physiology: An Integrated Approach.* USA: Prentice-Hall, Inc., 1998, 465–471. With permission.)

factor IIa, and factor Xa in the coagulation cascade. It has a short half-life, and it must be given parenterally, preferably by continuous intravenous infusion. It is therefore inappropriate for home use. The dosage of this agent is based on the needs of the patients. Commercial heparin products such as unfractionated heparins and low-molecular-weight heparins come from beef lung or pork intestinal mucosa. The use of heparin is associated with hemorrhage, osteoporosis, alopecia, thrombocytopenia, and hypersensitivity (5).

Antiplatelet agents are used to prevent a clot from forming or to prevent a clot from getting larger and occluding the entire vessel. Aspirin is the most widely used antiplatelet drug, which inhibits platelet aggregation for the life of the platelets (7 to 10 days). It is prescribed in the setting of acute myocardial infarction and prophylactically to prevent reinfarction. Antithrombotic doses used in clinical trials vary widely from less than 50 mg day^{-1} to over 1200 mg day^{-1}. Adverse effects of aspirin are similar to those of warfarin. Other antiplatelet drugs such as dipyridamole, clopidogrel and ticlopidine work by inhibiting platelet-activating factor and collagen, and they are often prescribed for patients who have an aspirin allergy. These agents may be used in patients with atherosclerotic disease to prevent heart attacks, strokes, and coronary artery closure in patients undergoing angioplasty. Their usage, however, is associated with bone marrow suppression, in particular leukopenia (6).

Unlike heparin and warfarin, which prevent extension and recurrences of thrombosis, thrombolytic agents (fibrinolytic enzymes) lyze preexisting thrombus. These include urokinase (extracted from kidneys), streptokinase (extracted from bacteria), and genetically engineered tissue plasminogen activator (t-PA). Evidence shows that patients with pulmonary embolism treated with streptokinase and urokinase are three times more likely to show clot resolution than patients taking heparin alone. These enzymes can also prevent some damage if the clot is removed soon after it occurs. Streptokinase, an effective thrombolytic agent for the treatment of acute myocardial infarction and pulmonary thromboembolism, is derived from streptococci. It can potentiate the body's own fibrinolytic pathways by converting plasminogen to plasmin. Being bacteria derived, it is antigenic, and repeated administration may result in neutralizing antibodies and allergic reactions. On the other hand, t-PA is produced by recombinant DNA technology, and it mimics an endogenous molecule that activates the fibrinolytic system. It does not elicit an allergic response and is considered to be more clot-specific. Nevertheless, it has a short half-life and needs to be continuously infused to achieve its greatest efficacy. Because of the lack of site specificity for all of these fibrinolytic enzymes, adverse effects may include gastrointestinal hemorrhage, but severe anaphylaxis is rare (7).

Mechanical and surgical treatments are usually reserved for massive pulmonary embolism, when drug treatments have failed or are contraindicated. Several tests can be done prior to surgery. Heart catheterization and coronary angiography are usually performed to assess the function of the heart muscles and those of the valves within the heart and the small coronary arteries feeding the heart. Patients may also be asked to undergo electrocardiography, echocardiography, exercise test, or holter monitoring before surgery (8). Since the 1970s, coronary bypass surgery has evolved to become one of the most common and most successful operative procedures because of its symptomatic and survival benefits. It is an open-heart operation, in which arteries or veins are taken from another part of the body to channel the needed blood flow directly to the coronary arteries. The surgery is performed not only to improve the patient's symptoms such as chest pain or occasional difficulty in breathing but also to protect the heart against the potential risk of a massive heart attack (8). Another effective modality for acute coronary syndrome is called percutaneous transluminal coronary angioplasty (PTCA). It is a procedure used to dilate the narrowing in the coronary artery. A special catheter with a small balloon is positioned on the end within the narrowed section of the coronary artery. The balloon is then inflated and deflated several times to stretch the artery and to flatten the deposits against the walls of the artery (7). In addition, inferior vena cava filters that are normally inserted via the internal jugular or femoral vein may be used in patients with recurrent symptomatic pulmonary embolism. It is considered a primary prophylaxis of thromboembolism in patients at high risk of bleeding such as patients with extensive trauma or visceral cancer (8).

Some patients with coronary artery disease, however, may have symptoms refractory to percutaneous coronary intervention and coronary artery bypass surgery. Such patients are potential candidates for alternative forms of coronary revascularization such as therapeutic angiogenesis. It is designed to promote the development of supplementary collateral blood vessels that will act as endogenous bypass conduits. Two major avenues for achieving therapeutic angiogenesis are currently under intense investigation, including gene therapy (the introduction of new genetic material into somatic cells to synthesize proteins that are missing, defective, or desired for specific therapeutic purposes) and protein-based therapy (administration of the growth factors instead of the genes encoding for the growth factors responsible for angiogenesis) (8). Although surgical treatment is sometimes physically invasive and traumatic for heart patients, technological advances in endovascular devices are making significant inroads into traditional coronary surgical practice. In fact, surgeons constantly develop new strategies to maximize the effectiveness of coronary surgery and to minimize the injury associated with cardiopulmonary bypass. These new treatment modalities, however, require the use of an anticoagulant as an adjuvant therapy (7).

16.2 FUNCTIONAL FOODS

The research areas of functional foods and nutraceuticals are rapidly expanding throughout the world. Scientists are actively working on the health benefits of foods by identifying the functional constituents, elucidating the biochemical structures, and determining the mechanisms behind the physiological roles. These research findings contribute to a new nutritional paradigm, in which food constituents go beyond their role as dietary essentials for sustaining life and growth, to one of preventing, managing, or delaying the premature onset of chronic disease later in life (9). Research indicates that diets high in saturated fat and sodium may increase the risk of cardiovascular diseases, whereas diets high in soluble fiber and antioxidants may help in preventing the diseases. For instance, the consumption of foods containing soy protein such as soy beverage and tofu in a diet low in saturated fat and cholesterol is associated with a reduced risk of coronary heart disease by lowering blood cholesterol levels. Scientific studies show that consumption of 25 g of soy proteins per day may lower the risk of heart diseases (9–11). As well, plant sterols that are present in small quantities in many fruits, vegetables, nuts, seeds, cereals, and legumes are shown to be beneficial in preventing heart diseases.

It has been shown that 1.3 g of plant sterol esters or 3.4 g of plant stanol esters per day in the diet is needed to show a significant cholesterol-lowering effect in hypercholesterolemic patients (12). In addition, soluble fiber-containing foods such as oats and psyllium are shown to be effective in lowering blood cholesterol levels. Clinical studies prove that dietary fiber reduces blood cholesterol by decreasing the absorption of dietary cholesterol and increasing the excretion of bile acids in the gastrointestinal tract. Soluble fiber may also alter the serum concentration of hormones or short-chain fatty acids that affect lipid metabolism. As part of a low-fat diet, 3 g of soluble fiber daily can help reduce blood cholesterol (13,14). Furthermore, omega-3 fatty acids from cold-water fishes such as salmon, halibut, and tuna have been shown to reduce coronary heart disease mortality rates. This is because long-chain omega-3 fatty acids from fish oil can decrease triglyceride levels, favorably affect platelet function, and decrease blood pressure slightly in hypertensive individuals (15). Ongoing research continues to explore the heart-health benefits of food ingredients such as conjugated linoleic acid and plant phytonutrients (16).

16.3 THE FIBRINOLYSIS SYSTEM

The fibrinolysis system, as shown in Figure 16.4, consists of a proenzyme (plasminogen), enzymes that proteolytically activate plasminogen, and several inhibitors that regulate activation of plasminogen, the activity of plasmin, and the stepwise degradation of fibrin. The structure–function relationships and the mechanisms of activation and inhibition of the main components of the system have been elucidated. Fibrin is a pathological formation and yet a structure that protects from bleeding at the site of vascular injury. Its removal is necessary for the restoration of normal blood flow, but this should occur only after the regeneration of the vessel wall. At the same time, t-PA, which is thought to be the primary initiator of fibrinolysis in the circulation, is the only protease of the hemostatic system that is continuously secreted by the endothelium in active form. Thrombotic occlusion occurs at the site of vessel injury when the growth rate of the thrombus exceeds that of its lysis. Impairments in these mechanisms may predispose patients to bleeding or thrombosis (17).

16.4 FIBRINOLYTIC ENZYMES

Fibrinolytic enzymes are agents that dissolve fibrin clots. The three enzymes that are currently being used for these purposes include urokinase, streptokinase, and genetically engineered t-PA. Fibrinolytic therapy such as intravenous administration of urokinase is expensive, and patients may suffer from undesirable side effects such as resistance to reperfusion, occurrence of acute coronary reocclusion, and bleeding complications (18). Consequently, several lines of investigation are currently being pursued to enhance the efficacy and specificity of fibrinolytic therapy. Recently, fibrinolytic enzymes have been discovered from both food and nonfood sources. These enzymes were proven to be effective, and they have been proposed as one of the potent fibrinolytic regimens.

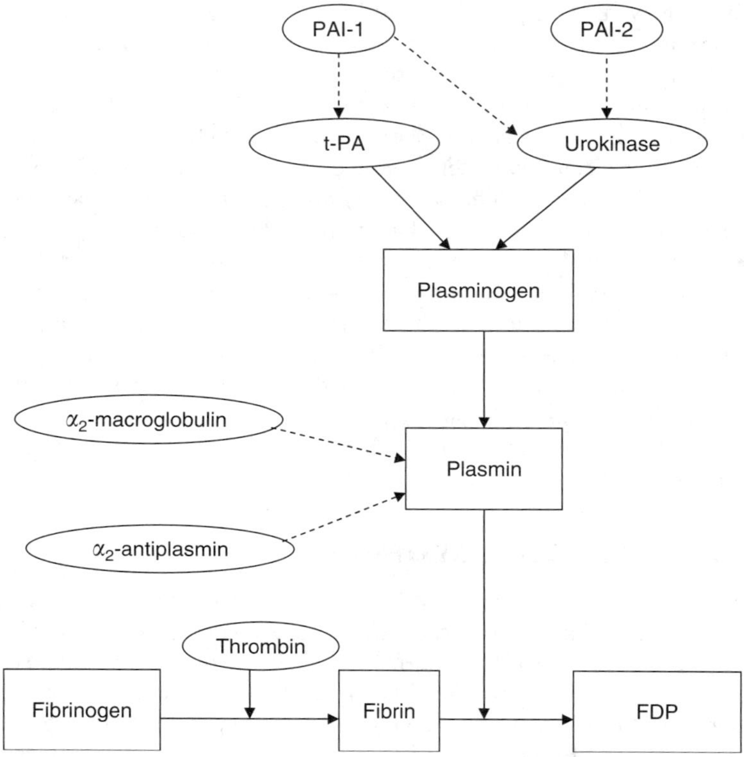

FIGURE 16.4 A simplified diagram of the fibrinolysis system. Fibrin is derived from fibrinogen in response to thrombin. Meanwhile, fibrin is degraded into fibrinogen degradation product (FDP) by endogenous plasmin protease, which is an activated form of plasminogen. t-PA and urokinase are two physiological activators of plasminogen; whereas plasminogen activator inhibitor-1 (PAI-1) negatively regulates t-PA activity and plasminogen activator inhibitor-2 (PAI-2) serves as the primary regulator of urokinase in the extravascular compartment within the body. Alpha-2-antiplasmin and alpha-2-macroglobulin are plasmin inhibitors that terminate fibrinolysis. Solid and dotted arrows represent activation and inhibition of the components, respectively. (Adapted from A.B. Dobrovolsky, E.V. Titaeva. *Biochemistry* (*Moscow*), 67, 99–108, 2002. With permission.)

16.4.1 Non-food Sources

Fibrinolytic enzymes can be found widely in nature. They have been found in hemorrhagic toxin from snake venom (19), earthworm secretions (3), food-grade microorganisms (20,21), marine creatures (22), and herbal medicines (23). In particular, a fibrinolytic protease has been isolated from *Spirodela polyrhiza*, an ingredient of traditional Oriental medicine that has been used for lowering blood pressure and detoxification of snake venom (23). Also, strong fibrinolytic enzymes are produced from *Bacillus* sp. strains that are used in food fermentation, invertebrates such as *Stichopous japonicus*, as well as the seaweed *Codiales codium* (24).

16.4.2 Food Sources

As shown in Table 16.1, fibrinolytic enzymes can be found in a variety of foods such as Japanese natto (25), Korean Chungkook-Jang soy sauce (26), and edible honey mushroom (27). Enzymes have been purified from these foods, and their physiochemical properties have been characterized. Among the food sources, fermented food products have been the focus of research. In particular, oral administration of the fibrinolytic enzyme extracted from Japanese natto can enhance fibrinolysis in dogs with experimentally induced thrombosis. Lysis of thrombus can be observed by angiography (26). More importantly, fibrinolytic activity, the amounts of t-PA, and fibrin degradation by-products in the plasma are all doubled when nattokinase is given to human subjects by oral administration. The underlying mechanism involves the absorption of the administered natto enzyme across the intestinal tract and the release of endogenous plasminogen activator that induces fibrinolysis in the occluded blood vessel (28). Recently, a unique strong fibrinolytic enzyme was discovered in fermented shrimp paste, which is a popular seasoning in Asian countries (29). The enzyme is a monomer with an apparent molecular weight of 18 kDa, and it is composed primarily of β-sheet and random coils. It is a neutral protease with an optimal activity from pH 3 to 7. No inhibition was observed with PMSF, Pepstatin A, E64, and 1,10-phenanthroline, but the enzyme was slightly inhibited by ethylenediaminetetraacetic acid (EDTA) and Cu^{2+}. It was relatively specific to fibrin or fibrinogen as a protein substrate, and yet it hydrolyzed none of the plasma proteins in the studies. *In vitro*, the enzyme was resistant to pepsin and trypsin digestion. It also had anticoagulant

TABLE 16.1
Food Sources of Fibrinolytic Enzymes

Food Source	Origin	Description	Fibrinolytic Enzyme	Reference
Natto	Japan	*Bacillus*-fermented soybean	An extracellular serine protease (nattokinase)	25
Tofuyo	Japan	Fermented bean curd	A soybean milk coagulating enzyme (SMCE)	30
Skipjack Shiokara	Japan	A salt-fermented fish product	An alkaline trypsin-like serine protease (katsuwokinase)	22
Chungkook-Jang	Korea	Fermented soybean sauce	An alkaline serine protease (CK)	26
Kimchi	Korea	Fermented vegetables	A *Bacillus* protease	31
Armillariella mella	Worldwide	An edible honey mushroom	A neutral metalloprotease	27
Shrimp paste	Hong Kong	Fermented shrimp	A neutral protease	29

activity measured with activated partial thrombin time (APTT) and prothrombin time (PT) tests. Kim et al. (26) isolated a *Bacillus* sp. strain that produces a strong fibrinolytic enzyme from Chunkook-Jang, a traditional Korean fermented soybean source. *Bacillus subtilis* for the production of domestic natto in Taiwan is another source of potent fibrinolytic enzyme (20). The potent fibrinolytic enzyme nattokinase (NK) was previously isolated from natto, a traditional fermented food in Japan, by Sumi et al. (25). This enzyme is an extracellular serine protease produced from *Bacillus natto*. Sumi et al. further demonstrated that oral administration of natto or nattokinase (NK) capsules enhance fibrinolysis in canine plasma in an experimental thrombosis model (28). Recently, Jeong et al. also isolated a fibrinolytic enzyme from *Bacillus subtilis* BK-17 strain (21). Thus, the shrimp paste might contain *Bacillus* sp. but produces a novel fibrinolytic enzyme. To confirm the physiological functions of the enzyme, the next step will be to examine *in vitro* the intestinal absorption of the enzyme by using human intestinal epithelial cells and measure the fibrinolytic activity of the enzyme in the blood and organs using animal model systems.

The novel fibrinolytic enzyme derived from a traditional Asian food is useful for thrombolytic therapy like other potent fibrinolytic enzymes such as nattokinase and earthworm enzyme. It will provide an adjunct to the costly fibrinolytic enzymes that are currently used in managing heart disease, since large quantities of enzyme can be conveniently and efficiently produced. In addition, this enzyme has significant potential for food fortification and nutraceutical applications such that their use could effectively prevent cardiovascular diseases.

REFERENCES

1. World Health Organization. *The World Health Report 2001*. [Accessed on September 2, 2002]. Available from www.who.int.
2. A.C. Silverthorn, W.C. Ober, C.W. Garrison. *Human Physiology: An Integrated Approach.* USA: Prentice-Hall, Inc., 1998, 465–471.
3. H. Mihara, H. Sumi, T. Yoneta, H. Mizumoto, R. Ikeda, M. Seiki, M. Maruyama. A novel fibrinolytic enzyme extracted from the earthworm, *Lumbricus rubellus*. *J. Physiol.*, 41: 461–472, 1991.
4. A. Oden, M. Fahlen. Oral anticoagulation and risk of death: a medical record linage study. *Br. Med. J.*, 325: 1073–1075, 2002.
5. M.L. Daviglus, J. Stamler, A.J. Orencia, A.R. Dyer, K. Liu, P. Greenland, M.K. Walsh, D. Morris, R.B. Shekelle. Fish consumption and the 30 year risk of fatal infarction. *N. Engl. J. Med.*, 336: 1046–1053, 1997.
6. A.D. Blann, M.J. Landray, G.Y. Lip. An overview of antithrombotic therapy. *Br. Med. J.*, 325: 762–765, 2002.
7. A.G. Turpie, B.S. Chin, G.Y. Lip. Venous thromboembolism: treatment strategies. *Br. Med. J.*, 325: 948–950, 2002.
8. K.J. Harjai, P. Chowdhury, C.L. Grines. Therapeutic angiogenesis: a fantastic new adventure. *J. Interv. Cardiol.*, 15: 223–329, 2002.
9. M.C. Hasler. Functional foods: their role in disease prevention and health promotion, *Food Technol.*, 52: 63–70, 1998.

10. P. Nestel. Role of soy protein in cholesterol-lowering: how good is it? *Arterioscler. Thromb. Basic Biol.*, 22: 1852–1858, 2002.
11. P. Puska, V. Korpelainen, L.H. Hoie. Soy in hypercholesterolaemia: a double-blind, placebo-controlled trial. *Eur. J. Clin. Nutr.*, 56: 352–357, 2002.
12. L.T. Meade, B.S. Ross, J.M. Blackston. Plant sterol margarines. Nutraceuticals for lowering cholesterol. *Adv. Nurse Pract.*, 9: 55–56, 2001.
13. J.W. Anderson, L.D. Allgood, A. Lawrence, L.A. Altringer, G.R. Jerdack, D.A. Hengegold, J.G. Morel. Cholesterol lowering effects of psyllium intake adjunctive to diet therapy in men and women with hypercholesterolemia: meta-analysis of 8 controlled trials. *Am. J. Clin. Nutr.*, 71: 472–479, 2000.
14. S. Bell, V.M. Goldman, B.R. Bistrian, A.H. Arnold, G. Ostroff, R.A. Forse. Effects of beta-glucan from oats and yeast on serum lipids. *Crit. Rev. Food Sci. Nutr.*, 39: 189–202, 1999.
15. D.A. Fitzmaurice, A.D. Blann, G.Y. Lip. Bleeding risks of antithrombotic therapy. *Br. Med. J.*, 325: 828–831, 2002.
16. L. Chagan, A. Ioselovich, L. Asherova, J.W. Cheng. Use of alternative pharmacotherapy in management of cardiovascular diseases. *Am. J. Manage. Care*, 8: 286–288, 2002.
17. A.B. Dobrovolsky, E.V. Titaeva. The fibrinolysis system: regulation of activity and physiologic functions of its main components. *Biochemistry (Moscow)*, 67: 99–108, 2002.
18. C. Bode, M. Runge, R.W. Smalling. The future of thrombolysis in the treatment of acute myocardial infarction. *Eur. Heart J.*, 17 (Suppl E): 55–60, 1996.
19. T. Nikai, N. Mori, M. Kishida, H. Sugihara, A. Tu. Isolation and biochemical characterization of hemorrhagic toxin f from the venom of Crotalus atrox. *Arch. Biochem. Biophys.*, 231: 309–319, 1984.
20. C.T. Chang, M.H. Fan, F.C. Kuo, H.Y. Sung. Potent fibrinolytic enzyme from a mutant of Bacillus subtilis IMR-NK1. *J. Agric. Food Chem.*, 48: 3210–3216, 2000.
21. Y.K. Jeong, J.U. Park, H. Baek, S.H. Park, I.S. Kong. Purification and biochemical characterization of a fibrinolytic enzyme from Bacillus subtilis BK-17. *World J. Microbiol. Biotechnol.*, 17: 89–92, 2001.
22. H. Sumi, N. Nakajima, H. Mihara. Fibrinolysis relating substances in marine creatures. *Comp. Biochem. Physiol.*, 102B: 163–167, 1992.
23. H.S. Choi, Y.S. Sa. Fibrinolytic and antithrombotic protease from spirodela polyrhiza. *Biosci. Biotechnol. Biochem.*, 65: 781–786, 2001.
24. O.H. Jeon, W.J. Moon, D.S. Kim. An anticoagulant/fibrinolytic protease from *Lumbricus rubellus*. *J. Biochem. Mol. Biol.*, 28: 138–142, 1995.
25. H. Sumi, H. Hamada, H. Tsushima, H. Mihara, H. Muraki. A novel fibrinolytic enzyme (nattokinase) in the vegetable cheese natto; a typical and popular soybean food in the Japanese diet. *Experientia*, 15: 1110–1111, 1987.
26. W. Kim, K. Choi, Y. Kim. Purification and characterization of a fibrinolytic enzyme produced from *Bacillus* sp. Strain CK 11-4 screened from Chungkook-Jang. *Appl. Environ. Microbiol.*, 62: 2482–2488, 1996.
27. J.H. Kim, Y.S. Kim. A fibrinolytic metalloprotease from the fruiting bodies of an edible mushroom, *Armillariella mellea*. *Biosci. Biotechnol. Biochem.*, 63: 2130–2136, 1999.
28. H. Sumi, H. Hamada, K. Nakanishi, H. Hiratani. Enhancement of the fibrinolytic activity in plasma by oral administration of nattokinase. *Acta, Haematol.*, 84: 139–143, 1990.

29. A.H.K. Wong, Y. Mine. Novel fibrinolytic enzyme in fermented shrimp paste, a traditional Asian fermented seasoning. *J. Agric. Food Chem.*, 52: 980–986, 2004.
30. M. Fujita, K. Nomura, K. Hong, Y. Ito, A. Asada, S. Nishimuro. Purification and characterization of a strong fibrinolytic enzyme (nattokinase) in the vegetable cheese NATTO, a popular soybean fermented food in Japan. *Biochem. Biophys. Res. Commn.*, 197: 1340–1347, 1993.
31. K.A. Noh, D.H. Kim, N.S. Choi, S.H. Kim. Isolation of fibrinolytic enzyme producing strains from kimchi. *Korean J. Food Sci. Technol.*, 31: 219–223, 1999.

Section IV

The Body's Nervous System

17 Opioid Peptides

Benjamin Guesdon, Lisa Pichon, and Daniel Tomé

CONTENTS

17.1 INTRODUCTION – OPIOIDS

The existence of opioid receptors, suspected for quite a long time, was simultaneously demonstrated in the mammalian organism in 1973 by three groups (1–3). Shortly thereafter, Kosterlitz's group presented evidence of the presence of endogenous ligands for these receptors in mammals (4). When opioid binding sites were first demonstrated, it was thought that opioid receptors represented a homogeneous group. Since that time, a lot of information has been collected on opioid receptors and their ligands both in mammals and in lower organisms. It was realized that there are, in fact, multiple opioid peptides, each of which could have its own receptor. This chapter is focused on exorphins — exogenous opioid peptides that are derived from food protein and could share several physiological roles in the organism.

17.1.1 The Mammalian Opioidergic System: Opioid Receptors and Endogenous Ligands

The mammalian opioidergic system consists of opioid receptors and their endogenous ligands — endogenous opioid peptides. Depending on their location, their physiological significance appears to be related to a considerable number of neuroendocrine regulatory functions.

Opioids receptors are most abundantly present in the central nervous system (5–7) but have also been localized in many peripheral tissues of the mammalian organism (8). Opioid receptors belong to the big family of G-protein-coupled receptors. They consist of both a recognition site, to which the ligand binds itself, and an effector responsible for translating the binding into intracellular biochemical events that lead to a biological response. Effector systems activated or blocked upon opioid receptor–G-protein interaction are adenyl cyclase, Ca^{2+} channels, K^+ channels, or phosphoinositol turnover. Opioid receptor binding is saturable and of high affinity and displays stereospecificity. The first definitive pharmacological evidence for multiple opioid receptors was published by Martin and colleagues (9,10). There are several receptor types, μ-, δ-, and κ-opioid receptors (μ- for morphine, δ- for deferens, and κ- for ketocyclazocine), which again can be divided into subtypes, that is, μ_1-μ_2 receptors and so on. Further receptors, named the ε-opioid receptor (11) and the ι-opioid receptor (12) (ε- for SFK 100047, N-allyl-normetazocine, and ι for ileum because that is where it was found), have also been reported to occur. Whereas these opioid receptor types and subtypes have been well known for quite some time, yet another member of the opioid receptor family (ORL1, LC132, and so on) has recently been demonstrated by several groups (13,14). This receptor, in contrast to μ-, δ-, and κ-receptors, does not mediate typical opioids but rather has antiopioid effects. Other receptors found in very specific areas — rat brain, murine neuroblastoma cells (λ-receptor and ζ-receptor) — have been demonstrated.

Soon after the first demonstration of opioid receptors in the organism, the first endogenous ligands of these receptors were also identified. These "endogenous opiates," as they were named immediately after their detection, turned out to be peptides. During the following two decades, many of these "opioid peptides" were identified. Originating from only three separate precursor molecules, proencephalin (PENK), prodynorphin (PDYN), and propiomelanocortin (POMC), they are named encephalins, dynorphins, and endorphins, respectively (15). The common structural feature among opioid peptides is the presence of a tyrosine residue at the amino terminal end (except α-casein opioids) and the presence of another aromatic residue, for example, phenylalanine or tyrosine, in the third or fourth position. This is an important structural motif that fits into the binding site of opioid receptors. The negative potential, localized in the vinicity of the phenolic hydroxyl group of tyrosine, seems to be essential for opioid activity. Removal of the tyrosine residue results in total absence of bioactivity (16). Although it is still not clear how these different endogenous peptides are matched with the multiple subtypes of opioid receptors, they seem to be localized within separate neuronal systems : met- and leu- encephalinergic, dynorphinergic, and β-endorphinergic neurons. These opioids are synthesized in many neuronal systems and often have synaptic interaction with other neurotransmitters (acetylcholine, epinephrine, norepinephrin, GABA, substance P, serotonine). Although most of them bind to more than one type of opioid receptor, certain selectivities are obvious, for example, selectivities of dynorphins for κ-receptors, or enkephalins for δ-receptors, or of endorphins not only for μ- or δ- but also for ε-receptors (17).

What can be deduced about the physiological or pathophysiological functions of opioid systems? On the one hand, there are well-defined maps of the distribution of opioid peptidergic systems and multiple receptor types; and on the other hand, there are numerous pharmacological studies investigating the functional consequences of administering diverse opiate agonists or antagonists. The task of integrating these two areas of work is difficult because of striking discrepancies in the anatomical distribution of opioid neuronal systems and receptor types; as well, there are considerable uncertainties in relating the pharmacological effects of agonists or antagonists (or opioid peptides coming from digestion of food protein, see below), acting on one or more particular receptor types to the potential actions of specific opioid neuronal systems that generally have a poorly defined relationship to these receptors. To make matters even more confusing, it appears that opioid precursors are differentially processed in different opioid neuronal systems (18–20). Even for peptides with relatively high selectivity for a particular receptor type, it is likely that "cross-over" to other opioid receptor types occurs if there is a relative paucity of the "preferred" receptor type in a given brain region (21). To add further complexity, there is evidence of differences among species in the processing of opioid peptides and in the relative abundance of different opioid receptors types (22).

There is a plethora of studies on the physiological and behavioral effects of administering different opioid peptides or non-peptidergic opiate agonists or antagonists (23). However, it has still been difficult to draw firm conclusions regarding the functional roles of different opioid receptor types in any neuronal system. Particularly interesting candidates for the functional roles of endogenous opioids would be immunological significance, control of gastrointestinal functions, all kinds of central nervous functions such as stress adaptation, and the control of reproductive mechanisms. Concerning nervous functions, the relatively large concentrations of opioid peptides and their receptors in limbic structures involved in mood regulation has given rise to the notion that opioid peptidergic systems may play a role in the pathophysiology of mental illness and in the activity of psychotropic drugs. Clinically, there is some — although still rather tenuous — evidence that opioid peptides may relieve symptoms in depressed patients and that naloxone may do so in manic patients. The antidepressive action of electroconvulsive treatment may involve the release of endogenous opioid peptides.

17.1.2 Exogenous Opioid Agonists and Antagonists

Opioid receptors can interact with their endogenous ligands as well as with exogenous opioid agonists and antagonists (24). Two types of exogenous agonists and antagonists can be identified in view of their chemical structure : alkaloids and peptides.

Well-known opioid agonist alkaloids are opiates such as morphine, which are widely used for numbing most severe pain syndromes or for anesthetic purposes. Morphine and opiates of similar structure, traditionally extracted from plants or artificially synthesized, can also be biosynthesized in the mammalian organism. Interestingly, a compound whose chemical, pharmacological, and immunological properties were identical to those of morphine has been demonstrated in bovine and human milk (25). Opioid antagonists most frequently used by pharmacologists and clinicians are the well-known synthetic alkaloids naloxone (26) and naltrexone (27). These opioid antagonists have been indispensible tools in opioid research (28). Alkaloid opioids can generally penetrate the blood-brain barrier and therefore can be administered peripherally *in vivo*. They are also less subjected to metabolism than are peptides.

Other kinds of exogenous opioid peptides were discovered, hence Teschemacher proposed to divide the opioid peptides into two groups, the first for the so-called typical opioid peptides, which derivate from the three endogenous precursor molecules, and the second for the atypical opioid peptides comprising all others. Separation of "typical" and "atypical" opioid peptides is based on the following differences: "typical" opioid peptides all have the same N-terminal amino acid sequence Tyr-Gly-Gly-Phe; in contrast, "atypical" opioid peptides are derived from a variety of parent proteins; only a Tyr residue is obligatory for their N-terminal amino acid sequence (29) and, in further contrast to "typical" opioid peptides, N-terminal extensions of the sequence occur beyond the Tyr residue. The main data existing on these exogenous opioid peptides essentially concern "food-derived" opioid peptides, which are "atypical" opioids or antagonist opioids. They are called exorphins. These last peptides can originate from food protein digestion: whereas little is known about the presence and the efficiency of "typical" opioid peptides in this matter, a number of naturally occuring proteins have been shown to contain fragments with opioid activity. Some of these peptide fragments have been demonstrated to be released from their precursors under *in vivo* and *in vitro* conditions; other fragments have only been synthesized. The digestion of these parental proteins can result in the production of opioid peptides in the gastrointestinal tract. The main food-derived opioid peptides studied are the milk protein-derived opioid peptides.

Milk proteins are potential sources of opioid agonistic and antagonistic peptides. They are inactive within the sequence of the precursor milk proteins, but they can be released and thus activated by enzymatic proteolysis, for example, during gastrointestinal digestion or during food processing. In fact, all typical milk proteins, α-casein, β-casein, κ-casein, α-lactalbumin, β-lactoglobulin, and lactoferrin (lactoferrin is not typical in a strict sense), have been shown to contain fragments which — either in natural form or modified to negligible or considerable extent — behave like opioid receptor ligands under *in vitro* or *in vivo* conditions (29–31). They have been named α-casein exorphins or casoxin D (α casein); β-casomorphins or β-casorphin (β-casein); casoxin or casoxin A, B, or C (κ-casein); α-lactorphins (α-lactalbumin); β-lactorphin (β-lactoglobulin); or lactoferroxins (lactoferrin). Most of them are selective for μ-opioid receptors, but some of them also have affinity for δ- or κ-receptors. Most of them are agonists, but antagonists such as casoxins and lactoferroxins also occur. It has been shown that relatively high amounts of bioactive peptides could potentially be produced during the ingestion of 1 g of each of the major casein and whey protein components. Some of these opioids are biogically highly potent. Therefore, even nutritionally insignificant amounts of liberated peptides might be sufficient to exert

TABLE 17.1
Examples of Opioid Peptides Derived from Bovine Milk Proteins

Bioactive Peptide	Sequence[a]	Precursor Protein	Theoretical Yield of Bioactive Peptide Obtainable from 1 g Precursor Protein (mg)
Opioid Agonists			
β-casomorphin-7	YPFPGPI	β-casein (*f*60–66)	33.0
β-casomorphin-5	YPFPG	β-casein (*f*60–64)	24.2
αs1-casein-exorphin	RYLGYLE	αs1-casein (*f*90–96)	38.7
α-lactorphin	YGLF	α-lactalbumin (*f*50–53)	35.2
β-lactorphin	YLLF	β-lactoglobuline (*f*102–105)	30.2
Serorphin	YGFNA	Serum albumin (*f*399–404)	10.5
Opioid Antagonists			
Casoxin A	YPSYGLNY	κ-casein (*f*35–42)	51.4
Casoxin B	YPYY	κ-casein (*f*58–61)	31.8
Casoxin C	YIPIQYVLSR	κ-casein (*f*25–34)	65.8

[a]The one letter amino acid code is used.

Source: H. Meisel, R.J. FitzGerald. *Br. J. Nutr.*, 84 (Suppl 1), S27–S31, 2000. With permission.

physiological effects (Table 17.1). Five opioid peptides showing selectivity for δ-receptor were also isolated from the enzymatic digest of wheat gluten. Their structures were Gly-Tyr-Tyr-Pro-Thr, Gly-Tyr-Tyr-Pro, Tyr-Gly-GlyTrp-Leu, Tyr-Gly-Gly-Trp, and Tyr-Pro-Ile-Ser-Leu, which were named gluten exorphins A5, A4, B5, B4, and C, respectively (32,33). A few studies report that other opioid peptide agonists or antagonists have been found in other food-derived peptides such as peptides derived from blood (34), soy (35), bovine mitochondrial cytochrome and spinach rubisco (36).

In addition to the possible liberation of bioactive peptides during intestinal proteolysis, such peptides may already be generated during the manufacture of several milk products and thus be ingested as food components. For example, partially hydrolyzed milk proteins used for hypoallergenic infant formulas and for clinical applications in enteral nutrition consist exclusively of peptides. Furthermore, it has been demonstrated that a number of caseolytic bacterial species used in the production of some types of cheese and other milk products can produce casomorphins. β-casomorphins have been produced by genetic engineering techniques followed by enzymatic or chemical cleavage of the microbial fusion protein to liberate the required peptide. These recombinant β-casomorphins are intended for oral administration in order to increase animal performance, for example, weight gain or milk yield. As yet, no meaningful application in human nutrition has been described.

17.1.3 Functional Significance of Nutraceutical Opioid Peptides

We have seen that peptides with morphine-like activities, called exorphins, can be isolated from some food proteins. So, neutraceutical opioid peptides could theoretically interact with subepithelial opioid receptors or specific luminal binding sites in the intestinal tract. Furthermore, they may be absorbed and then reach the endogenous opioid receptors. Thus, it might be possible that those atypical opioid peptides that can be found in many foods could also participate in the general opioid regulations. But do these peptides have physiological or even pathophysiological significance? In order for exorphins to function as opioid peptides in the central nervous system *in vivo* they must: (a) survive degradation by intestinal proteases, (b) be absorbed without degradation into the blood stream, (c) cross the blood-brain barrier and thereby reach central opiate receptors, and (d) interact as opiates with their receptors.

The main data existing on this issue concern the milk-derived opioid peptides that appear to very likely have a functional role. Milk is known to contain many bioactive compounds (37), and the well-known nutritive and reproductive significance of milk constitutents appears to match nutritive and reproductive functions, wherein certain endogenous opioid systems appear to be involved. The information collected so far about a functional role, however, varies considerably for the various milk protein-derived opioid peptide groups. There is a growing body of information indicating that milk proteins and their derivatives by no means just serve as food but instead represent messengers with essential functions for their receivers and even their donors. Milk protein derivatives are able to interact with opioid receptors thus fit into this conception, and various physiological or pathophysiological functions have been ascribed to these peptides, in particular to the best-studied group, the β-casomorphins.

Results indicating the liberation of β-casomorphins from β-casein under *in vitro* (38,39) and *in vivo* (40) conditions have been obtained from several studies. Evidence has been obtained for the liberation of β-casomorphins from β-casein into the gastrointestinal lumen of mammals after intake of milk. Peptides were found in the small intestine contents of adult humans following intake of cow's milk and were identified by radioimmunological and chromatographical methods as β-casomorphins (39). Moreover, β-casomorphin-11 has been identified and chemically characterized in the duodenal chyme of Göttingen minipigs after they had been fed with bovine casein (40). It is claimed that β-casomorphins rapidly degrade once they enter the blood stream. However, the presence of β-casomorphin-7 immunoreactive material has been demonstrated in the plasma of newborn calves (41) or puppies (42) following their first milk intake. This material revealed a similar molar mass as β-casomorphin-11 and thus has been considered a β-casomorphin precursor, from which β-casomorphins could be released to elicit effects at any site in the newborn organism. Depending on the situation, the presence of β-casomorphins in the neonate's central nervous system could gain pathophysiological significance: recently, a role for β-casomorphins in sudden infant death syndrome has been proposed (43,44). According to current studies, the absorption of peptides larger than di- or tripeptides is usually highly reduced in adults, but it could be enhanced in some specific situations such as in the neonate or during stress or agression (45,46).

Thus, only the neonatal intestine appears to be permeable to food-derived opioid peptides such as casomorphins; in adult systems, the intestinal brush-border membrane seems to be the main target site for the physiological effects of food-derived opioid peptides. Particularly, as nutrients postulated to modulate gastrointestinal functions, β-casomorphins might fit into the concept of "food hormones." More likely, β-casomorphins interact with endogenous opioid systems in the gastrointestinal wall, in neonates as well as in adult milk consumers. Orally administred milk protein-derived opioid peptides have been demonstrated to influence postprandial metabolism by stimulating secretion of insulin and somatostatin (47,48), modulate intestinal transport of amino acids (49), prolong gastrointestinal transit time, and exert antidiarrheal action (50). Evidence has accumulated that the enhancement of net water and electrolyte absorption by β-casomorphins in the small intestine is a major component of their antidiarrheal action, which could be mediated via subepithelial opioid receptors or through specific luminal binding sites at the brush-border membrane (49,51). β-Casomorphins may also affect the human mucosal immune system, possibly via opiate receptors in lamina propria lymphocytes (52). Modulation of social behaviour (53,54) and analgesic effects (55,56) were observed following intracerebral administration of opioid peptides with agonistic activity, for example, β-casomorphins, to experimental animals, but opioid casein fragments have not been detected in the plasma of adult mammals following oral administration.

REFERENCES

1. E.J. Simon, J.M. Hiller, I. Edelman. Stereospecific binding of the potent narcotic analgesic (3H) Etorphine to rat-brain homogenate. *Proc. Natl. Acad. Sci. USA*, 70: 1947–1949, 1973.
2. C.B. Pert, S.H. Snyder. Opiate receptor: demonstration in nervous tissue. *Science*, 179: 1011–1014, 1973.
3. L. Terenius. Characteristics of the "receptor" for narcotic analgesics in synaptic plasma membrane fraction from rat brain. *Acta, Pharmacol. Toxicol. (Copenh.)*, 177: 377–384, 1973.
4. H.W. Kosterlitz, J. Hughes. Some thoughts on the significance of enkephalin, the endogenous ligand. *Life Sci.*, 17: 91–96, 1975.
5. A. Mansour, R.C. Thompson, H. Akil, S.J. Watson. δ opioid receptor mRNA distribution in the brain: comparison to δ receptor binding and proenkephalin mRNA. *J. Chem. Neuroanat.*, 6: 351–362, 1993.
6. F. Meng, G.X. Xie, R.C. Thompson, A. Mansour, A. Goldstein, S.J. Watson, H. Akil. Cloning and pharmacological characterization of a rat κ opioid receptor. *Proc. Natl. Acad. Sci. USA*, 90: 9954–9958, 1993.
7. R.C. Thompson, A. Mansour, H. Akil, S.J. Watson. Cloning and pharmacological characterization of a rat μ opioid receptor. *Neuron*, 11: 903–913, 1993.

8. G. Wittert, P. Hope, D. Pyle. Tissue distribution of opioid receptor gene expression in the rat. *Biochem. Biophys. Res. Commn.*, 218: 877–881, 1996.
9. P.E. Gilbert, W.R. Martin. The effects of morphine and nalorphine-like drugs in the nondependent, morphine-dependent and cyclazocine-dependent chronic spinal dog. *J. Pharmacol. Exp. Ther.*, 198(1): 66–82, 1976.
10. W.R. Martin, C.G. Eades, J.A. Thompson, R.E. Huppler, P.E. Gilbert. The effects of morphine- and nalorphine-like drugs in the non-dependent and the morphine-dependent chronic spinal dog. *J. Pharmacol. Exp. Ther.*, 197: 517–532, 1976.
11. M. Wüster, R. Schulz, A. Herz. Specificity of opioids towards the μ, δ and ε-opiate receptors. *Neurosci. Lett.*, 15: 193–198, 1978.
12. T. Oka. Enkephalin receptor in the rabbit ileum. *Br. J. Pharmacol.*, 68: 195–198, 1980.
13. C. Mollereau, M. Parmentier, P. Mailleux, J.L. Butour, C. Moisand, P. Chalon, D. Caput, G. Vassart, J.C. Meunier. ORL1, a novel member of the opioid receptor family. Cloning, functional expression and localization. *FEBS Lett.*, 341: 33–38, 1994.
14. M. Satoh, M. Minami. Molecular pharmacology of the opioid receptors. *Pharmacol. Ther.*, 68: 343–364, 1995.
15. V. Höllt. Multiple endogenous opioid peptides. *Trends Neurosci.*, 6: 24–26, 1983.
16. K.J. Chang, A. Lillian, E. Hazum, P. Cuatrecasas, J.K. Chang. Morphiceptin (NH4-tyr-pro-phe-pro-COHN2): a potent and specific agonist for morphine (μ) receptors. *Science*, 212: 75–77, 1981.
17. S.J. Paterson, L.E. Robson, H.W. Kosterlitz. Classification of opioid receptors. *Br. Med. Bull.*, 39: 31–36, 1983.
18. E. Weber, C.J. Evans, J.D. Barchas. Predominance of the amino-terminal β-peptide fragment of dinorphin in rat brain regions. *Nature*, 299: 77–79, 1982.
19. B.R. Seizinger, K. Bovermann, D. Maysinger, V. Hollt, A. Herz. Differential effects of acute and chronic ethanol treatment on particular opioid peptide systems in discrete regions of rat brain and pituitary. *Pharmacol. Biochem. Behav.*, 18(Suppl 1): 361–369, 1983.
20. J.D. White, C.M. Gall, J.F. McKelvy. Pro-enkephalin is processed in a projection-specific manner in the rat central nervous system. *Proc. Natl. Acad. Sci. USA*, 83: 7099–7103, 1986.
21. E.A. Young, J.M. Walker, M.E. Lewis, R.A. Houghten, J.H. Woods, H. Akil. [3H]Dynorphin A binding and selectivity of pro-dynorphin peptides in rat, guinea-pig and monkey brain. *Eur. J. Pharmacol.*, 121: 355–365, 1986.
22. S. Moon Edley, S.L. Hall, M. Herkenham, C.B. Pert. Evolution of striatal opiate receptors. *Brain Res.*, 249: 184–188, 1982.
23. G.A. Olson, R.D. Olson, A.J. Kastin. Endogenous opiates: 1988. Peptides 10: 1253–1280, 1989.
24. H. Teschemacher, G. Koch, V. Brantl. Milk protein derived atypical opioid peptides and related compounds with opioid antagonistic activity. In: V. Brantl, H. Teschemacher (Eds.) *β-Casomorphins and Related Peptides: Recent Developments*. Weinheim: VCH, 1994, 3–17.
25. E. Hazum, J.J. Sabatka, K.J. Chang, D.A. Brent, J.W. Findlay, P. Cuatrecasas. Morphine in cow and human milk: could dietary morphine constitute a ligand for specific morphine (μ) receptors? *Science*, 213: 1010–1012, 1981.
26. J. Sawynok, C. Pinsky, F.S. LaBella. On the specificity of naloxone as an opiate antagonist. *Life Sci.*, 25: 1621–1632, 1979.

27. M.S. Gold, C.A. Dackis, A.L. Pottash, H.H. Sternbach, W.J. Annitto, D. Martin, M.P. Dackis. Naltrexone, opiate addiction, and endorphins. *Med. Res. Rev.*, 2: 211–246, 1982.
28. D.M. Zimmerman, J.D. Leander. Selective opioid receptor agonists and antagonists: research tools and potential therapeutic agents. *J. Med. Chem.*, 33: 895–902, 1990.
29. H. Teschemacher, G. Koch, V. Brantl. Milk protein-derived opioid receptor ligands. *Biopolymers*, 43: 99–117, 1997.
30. H. Meisel. Biochemical properties of regulatory peptides derived from milk proteins. *Biopolymers*, 43: 119–128, 1997.
31. H. Meisel, R.J. FitzGerald. Opioid peptides encrypted in intact milk protein sequences. *Br. J. Nutr.*, 84(Suppl 1): S27–S31, 2000.
32. S. Fukudome, M. Yoshikawa. Opioid peptides derived from wheat gluten: their isolation and characterization. *FEBS Lett.*, 296: 107–111, 1992.
33. S. Fukudome, M. Yoshikawa. Gluten exorphin C. A novel opioid peptide derived from wheat gluten. *FEBS Lett.*, 316: 17–19, 1993.
34. C. Liebmann, U. Schrader, V. Brantl. Opioid receptor affinities of the blood-derived tetrapeptides hemorphin and cytochrophin. *Eur. J. Pharmacol.*, 166: 523–526, 1989.
35. J. Pupovac, G.H. Anderson. Dietary peptides induce satiety via cholecystokinin-A and peripheral opioid receptors in rats. *J. Nutr.*, 132: 2775–2780, 2002.
36. M. Yoshikawa, M. Takahashi, S. Yang. δ opioid peptides derived from plant proteins. *Curr. Pharm. Des.*, 9: 1325–1330, 2003.
37. D. Schams, H. Karg. Hormones in milk. *Ann. NY Acad. Sci.*, 464: 75–86, 1986.
38. P. Petrilli, D. Picone, C. Caporale, F. Addeo, S. Auricchio, G. Marino. Does casomorphin have a functional role? *FEBS Lett.*, 169: 53–56, 1984.
39. J. Svedberg, J. de Haas, G. Leimenstoll, F. Paul, H. Teschemacher. Demonstration of β-casomorphin immunoreactive materials in *in vitro* digests of bovine milk and in small intestine contents after bovine milk ingestion in adult humans. *Peptides*, 6: 825–830, 1985.
40. H. Meisel. Chemical characterization and opioid activity of an exorphin isolated from *in vivo* digests of casein. *FEBS Lett.*, 196: 223–227, 1986.
41. M. Umbach, H. Teschemacher, K. Praetorius, R. Hirschhauser, H. Bostedt. Demonstration of a β-casomorphin immunoreactive material in the plasma of newborn calves after milk intake. *Regul. Pept.*, 12: 223–230, 1985.
42. M. Singh, C.L. Rosen, K.J. Chang, G.G. Haddad. Plasma β-casomorphin-7 immunoreactive peptide increases after milk intake in newborn but not in adult dogs. *Pediatr. Res.*, 26: 34–38, 1989.
43. J. Hedner, T. Hedner. β-Casomorphins induce apnea and irregular breathing in adult rats and newborn rabbits. *Life Sci.*, 41: 2303–2312, 1987.
44. K. Ramabadran, B.E. Moore. Sudden infant death syndrome and opioid peptides from milk. *Am. J. Dis. Child.*, 142: 12–13, 1988.
45. J.B. Meddings, M.G. Swain. Environmental stress-induced gastrointestinal permeability is mediated by endogenous glucocorticoids in the rat. *Gastroenterology*, 119: 1019–1028, 2000.
46. J. Groot, P. Bijlsma, A. Van Kalkeren, A. Kiliaan, P. Saunders, M. Perdue. Stress-induced decrease of the intestinal barrier function. The role of muscarinic receptor activation. *Ann. NY Acad. Sci.*, 915: 237–246, 2000.

47. V. Schusdziarra, R. Schick, A. de la Fuente, A. Holland, V. Brantl, E.F. Pfeiffer. Effect of β-casomorphins on somatostatin release in dogs. *Endocrinology*, 112: 1948–1951, 1983.
48. V. Schusdziarra, A. Schick, A. de la Fuente, J. Specht, M. Klier, V. Brantl, E.F. Pfeiffer. Effect of β-casomorphins and analogs on insulin release in dogs. *Endocrinology*, 112: 885–889, 1983.
49. M. Brandsch, P. Brust, K. Neubert, A. Ermisch. β-casomorphins — chemical signals of intestinal transport systems. In: V. Brantl, H. Teschemacher (Eds.). *β-Casomorphins and Related Peptides: Recent Developments*. Weinheim: VCH, 1994, 207–219.
50. H. Daniel, M. Vohwinkel, G. Rehner. Effect of casein and β-casomorphins on gastrointestinal motility in rats. *J. Nutr.*, 120: 252–257, 1990.
51. D. Tome, A.M. Dumontier, M. Hautefeuille, J.F. Desjeux. Opiate activity and transepithelial passage of intact β-casomorphins in rabbit ileum. *Am. J. Physiol.*, 253(6 Pt 1): G737–G744, 1987.
52. Y. Elitsur, G.D. Luk. β-Casomorphin (BCM) and human colonic lamina propria lymphocyte proliferation. *Clin. Exp. Immunol.*, 85: 493–497, 1991.
53. J. Panksepp, L. Normansell, S. Siviy, J. Rossi III, A.J. Zolovick. Casomorphins reduce separation distress in chicks. *Peptides*, 5: 829–831, 1984.
54. E. Paroli. Opioid peptides from food (the exorphins). *World Rev. Nutr. Diet*, 55: 58–97, 1988.
55. K.J. Chang, P. Cuatrecasas, E.T. Wei, J.K. Chang. Analgesic activity of intracerebroventricular administration of morphiceptin and β-casomorphins: correlation with the morphine (micro) receptor binding affinity. *Life Sci.*, 30: 1547–1551, 1982.
56. H. Matthies, H. Stark, B. Hartrodt, H.L. Ruethrich, H.T. Spieler, A. Barth, K. Neubert. Derivatives of β-casomorphins with high analgesic potency. *Peptides*, 5: 463–470, 1984.

18 Factors that Affect the Body's Nervous System: Relaxation Effects of Tea L-Theanine

M. Ozeki, T.P. Rao, and L.R. Juneja

CONTENTS

18.1 INTRODUCTION

In modern society, people are prone to several kinds of stress and stress-induced diseases. Stress is classified into physical chemical stress, physicochemical stress, and psychological stress. Physicochemical stresses are caused by cold, heat, injury, buzz, exhaust fumes, and so on. Physiological stresses are the result of starvation, infection, overwork, and so on. Psychological stresses arise from anxiety, strain, dissatisfaction, disappointment, anger, and other emotions. Stressful events induce both psychological and physical hypertension, which can lead to diseases such as duodenal ulcer, depression, sleeplessness, and autonomic imbalance. The significance of the correlation between health and stress is increasing, and therefore, relaxation techniques for stress reduction are important. Relaxation methods can calm down the overly tensed organism and prevent stress-induced diseases. Thus, the use of psychosomatic medicines to induce relaxation has become necessary.

Mental and physical health is maintained and promoted by acquiring the habit of relaxation, which induces two essential elements, physical and psychological relaxed states. Mental state of relaxation can be recognized by observing such parameters as oxygen consumption, carbon dioxide elimination, heart rate, respiratory rate, minute ventilation, and arterial blood lactate. In the relaxed state, blood pressure (systolic, diastolic, and mean blood pressure) and renal temperature do not change, while skin resistance markedly increases and skeletal muscle blood flow slightly increases. The electroencephalogram (EEG) demonstrates an increase in the intensity of slow α-waves occasionally some θ-waves activity. These changes are consistent with generalized decreased sympathetic nervous system activity and are distinctly different from the physiological changes noted when a person is sitting quietly or sleeping. Relaxation methods should be effective to induce the above conditions for proper relaxation.

Food factors help relieve stress and induce a feeling of relaxation. For example, drinking tea for relaxation is common in many parts of the world. In Japan, the tradition of drinking green tea is considered beneficial for stress relief and general well being. We investigated the functional substance in green tea that promotes relaxation. L-theanine, an amino acid abundant in green tea leaves (1 to 2% in dry weight), has been recognized as being responsible for creating a feeling of relaxation and for promoting brain health (1,2). L-Theanine exists in free (nonprotein) form, and it is the predominant amino acid (about 50%) of the total free amino acids in green tea leaves (Table 18.1). The presence of L-theanine in tea leaves was discovered in 1949 (3). Its chemical structure was determined as gamma-ethylamino-L-glutamic acid (L-theanine) (Figure 18.1). L-Theanine is the main component responsible for the typical taste of green tea, known as "*umami*." High-quality teas, such as *maccha* and *gyokuro*, contain a higher amount of amino acids as well as L-theanine than do inferior ones (*bancha*). Our research group has established an effective method for the production of L-theanine on an industrial scale. Suntheanine® is the trade name of L-theanine manufactured by Taiyo Kagaku Co., Ltd. We conducted a scientific study of the relaxation effects of Suntheanine® both in animal and human models and proved that Suntheanine® promotes psychological relaxation by inducing brain waves and suppressing the sympathetic nervous system and promotes physiological relaxation by relieving the effects of premenstrual syndrome (PMS). In this chapter, the psychological and physiological relaxation effects of Suntheanine® are discussed.

18.2 PSYCHOLOGICAL RELAXATION EFFECTS

Brain waves, the very weak electrical pulses from the brain surface, are an index for mental condition. Depending on the electrical frequency, the waves are classified as α, β, δ and θ-waves, which represent relaxation, excitation, sound sleep, and doze sleep, respectively (Figure 18.2) (4). The psychological relaxation achieved with the administration of Suntheanine® was determined in humans by determining the production of α-waves in a volunteer test. The test comprised 50

TABLE 18.1
Composition of Amino Acids in Green Tea

Amino Acids	Grade						
	Maccha	Gyokuro		Sencha		Bancha	Houjicha
		High	Normal	High	Normal		
Thenaine	1998.6	2466.1	2007.7	1496.6	652.5	416.7	21.7
Glutamic acid	646.8	449.4	383.5	217.4	214.3	184.5	16.7
Aspartic acid	748.1	432.5	333.0	245.8	172.9	124.9	27.2
Arginine	1046.7	497.5	329.6	198.1	64.2	38.9	14.6
Serine	234.2	352.8	277.9	202.0	114.3	82.4	10.3
Threonine	154.7	128.8	98.3	48.0	29.1	23.0	3.2
Alanine	80.3	54.8	40.8	28.3	24.7	20.2	3.9
Lysine	92.1	39.1	32.7	28.7	14.0	10.1	1.4
Phenylalanine	26.7	37.7	41.2	31.4	24.8	14.2	1.1
Isoleucine	39.7	36.1	28.6	14.4	11.7	7.5	0.9
Leucine	39.4	32.6	26.6	15.5	10.7	3.2	0.6
Tyrosine	31.1	29.6	30.9	25.7	16.3	8.5	1.1
Histidine	59.7	28.2	22.1	10.3	7.6	6.6	0.9
Valine	33.0	25.2	21.5	18.9	14.2	8.2	1.1
Glycine	7.5	4.8	4.3	5.0	3.8	3.0	1.0
Ammonia	—	9.3	9.1	9.2	7.0	7.1	9.4
Total	5238.6	4624.5	3687.8	2595.3	1382.1	959.0	115.1

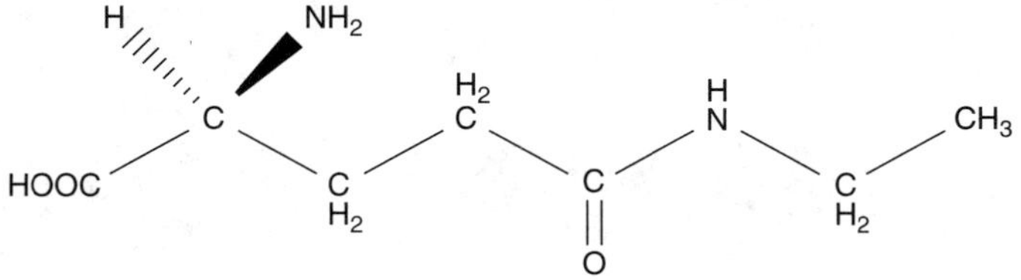

FIGURE 18.1 Chemical structure of L-theanine.

female (18 to 22 years old) volunteers. Since it was expected that the mental responses to Suntheanine® could vary according to anxiety level, the female subjects were divided into two groups, namely, the high-anxiety group and the low-anxiety group, based on the manifest anxiety scale (MAS) (5). Each group was given water, a 50-mg Suntheanine® solution, or a 200-mg Suntheanine® solution once a week, and brain waves were measured for 60 minutes after the administration. All measurements were repeated twice during the two-month test period.

A remarkable production of α-waves was observed from the occipital lobe to the parietal lobe of the brain surface approximately 30 minutes after intake of the Suntheanine® solution. The dose of 200 mg of Suntheanine® resulted in the generation of a large amount of α-waves in the occipital and parietal regions of the

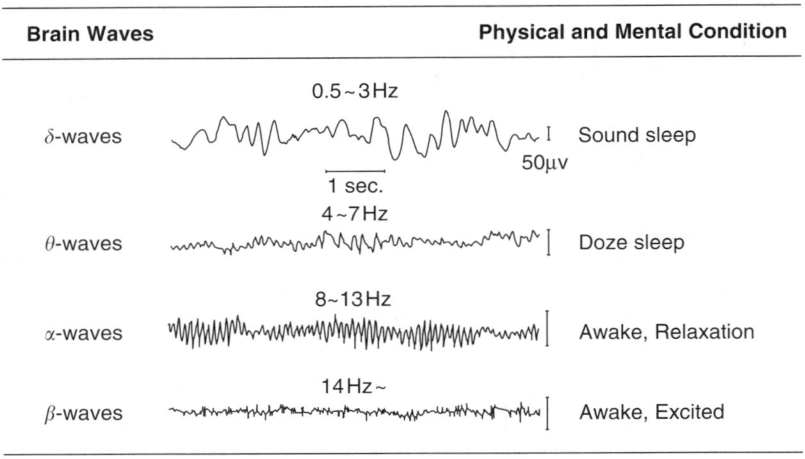

FIGURE 18.2 Classification of brain waves and their relationship with the state of mind in humans.

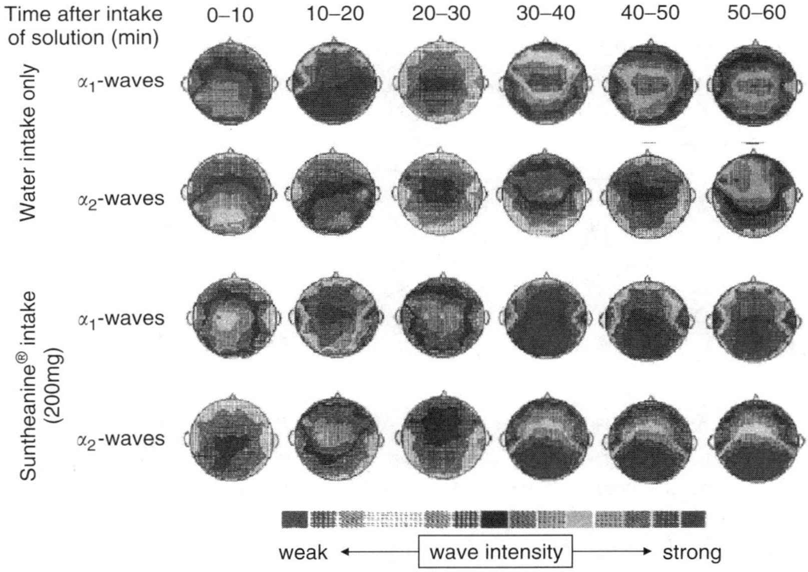

FIGURE 18.3 Topographies converted from data of brain waves on brain surface measured for 60 minutes after intake of water or Suntheanine® solution in human volunteer.

brain, while only a small amount of α-waves was observed with water intake (Figure 18.3). The accumulated intensity of α-waves showed a clear tendency of dose-dependent generation after 30 minutes of Suntheanine® administration (Figure 18.4) (6). It is well known that α-waves are generated during the relaxed

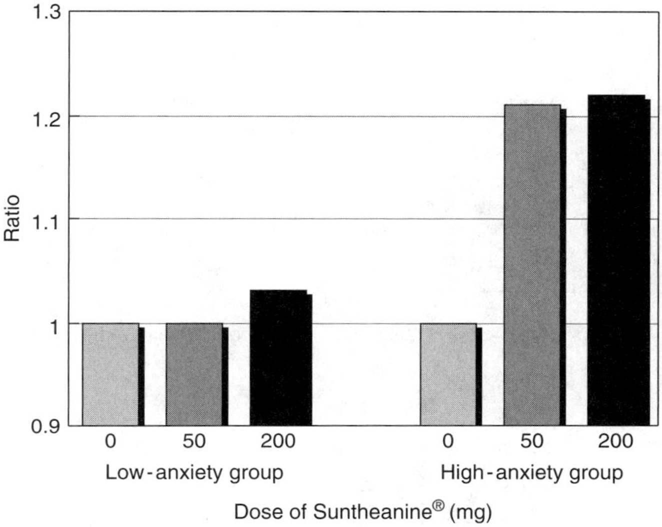

FIGURE 18.4 The ratio of α-wave intensity after intake of Sutheanine® in low-anxiety and high-anxiety groups. Volunteers ingested the solutions containing dissolved Suntheanine® (0, 50, or 200 mg). Number of volunteers of each group.

state and are an index of relaxation (7). Our results show that Suntheanine® can promote the generation of α-waves and thus induce a relaxed state in humans.

Kakuda et al. (8) reported that the relative power of β-wave was increased at 1 to 2 μmol/kg b.w. (0.174 mg/kg or 0.348 mg/kg) of L-theanine administration compared with the control group administrated with the saline. The α1 and α2-waves were also increased at the same dosage. The relative power of β-wave was decreased at more than a dose of 5 μmol/kg b.w. (0.871 mg/kg) of L-theanine, while α-waves kept on high levels (Figure 18.5). The dose of 2 μmol/kg b.w. is equal to 17.4 mg and 5 μmol/kg is equal to 43.5 mg for a 50 kg b.w. of an adult person. These data suggest that it is necessary to use proper dose of L-theanine to achieve effective relaxation effects.

18.3 PHYSIOLOGICAL RELAXATION EFFECTS

18.3.1 Relaxation Effect in PMS

Premenstrual syndrome (PMS) in women appears at the luteal phase, from the ovulation period through the first day of menstruation. In general, the characteristic features of PMS are often reported to be high 1 to 2 days prior to menstruation. The typical symptoms of PMS are generally categorized as mental, physical, and social. In a volunteer study, the effect of Suntheanine® on PMS was examined in 20 subjects (9), who were administered either tablets containing Suntheanine® or placebo. The subjects took two tablets twice a day equal to an amount of 200 mg of Suntheanine®

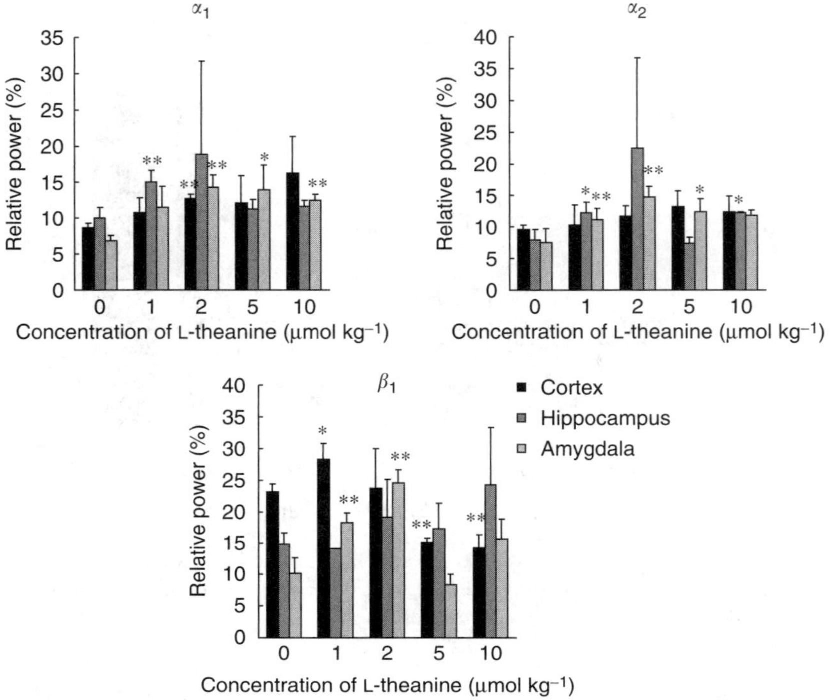

FIGURE 18.5 Changes in relative power of α_1-, α_2-, and β-waves in the cortex, hippocampus, and amygdala at 15 minutes after administration of L-theanine in rats. Values represent mean ± SEM. Number of rats: 6. $*p < 0.01$. $**p < 0.001$.

per day for 2 weeks before the beginning of their menstrual period. The test duration consisted of three menstruation cycles. The first cycle was regarded as the control, during which the subjects were randomly treated with Suntheanine® or placebo. During the second cycle, the subjects were divided into Suntheanine® and placebo groups and treated accordingly. During the third cycle, the subjects in each group were crossed over. Three days prior to the expected date of onset of the menstruation cycle, the test subjects were asked to answer a Menstrual Distress Questionnaire (MDQ) (10). The MDQ survey, developed by Moos et al. (10) contains 47 questions, divided into eight categories, as shown in Table 18.2, of which only six are related to menstrual symptoms. The symptoms were ranked on a scale of 0 to 3, where the high score represents severe symptoms. The survey results indicated that the subjects administered with Suntheanine® reported lower scores in both mental and physical symptoms compared with those administed placebo. These results suggest that Suntheanine® is very effective in alleviating the symptoms of PMS (Figure 18.6).

18.3.2. Antagonistic Effect Against Caffeine

Several beverages, such as coffee, cola, black tea, oolong tea, green tea, and so on, contain caffeine, which is invariably absorbed when these drinks are ingested.

TABLE 18.2
Categories in the Menstrual Distress Questionnaire (MDQ)

Categories	Symptoms
Pain	Muscle stiffness, headache, cramps,backache, etc.
Concentration	Insomnia, forgetfulness, lowered judgement, etc.
Behavioral change	Stay at home, avoid social activities, etc.
Autonomic reactions	Dizziness, cold sweats, vomiting, etc.
Water retention	Weight gain, skin disorders, swelling
Negative affect	Depression, anxiety, lonliness, irribility, etc.
Arousal	Orderliness, excitement, bursts of energy, etc.
Control	Feelings of suffocation, ringing in ears, pounding of heart, etc.

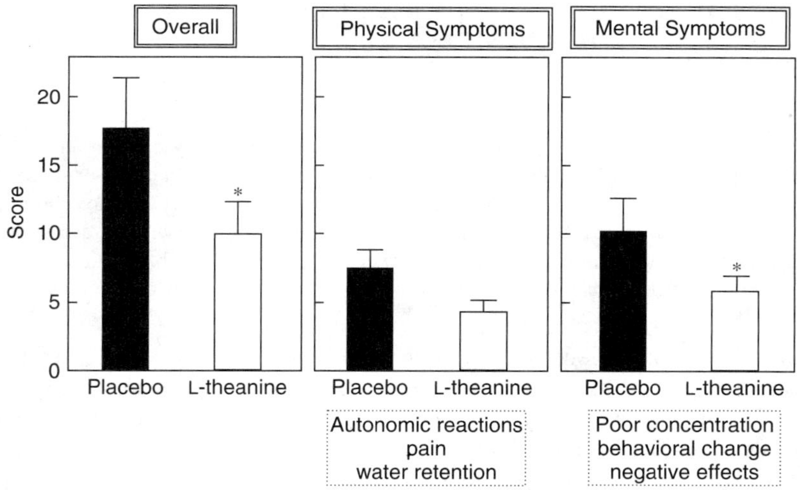

FIGURE 18.6 Effect of L-theanine on the alleviation of PMS symptoms as examined through MDQ score in 20 subjects. $*p < 0.05$ *t*-test.

Caffeine is known to cause excitation of the central nervous system, diuretic effect, acceleration effect, secretion of saliva and gastric fluid, and elevation of blood pressure. L-theanine has been found to act antagonistically against the excitations induced by caffeine. In an *in vivo* study, rats were administrated with L-theanine followed by caffeine at 15-minute intervals, and the incidence of convulsion was determined (11). Saline was administrated to control rats. In rats administered L-theanine, there was a significant drop in the incidence of convulsions as compared with control rats (Figure 18.7A). For the estimation of time interval to onset of tonic convulsion caused by caffeine, the stimulant and the L-theanine was administrated intraperitoneally at the same time. In control rats, saline was administrated along with the stimulant. The number of rats that

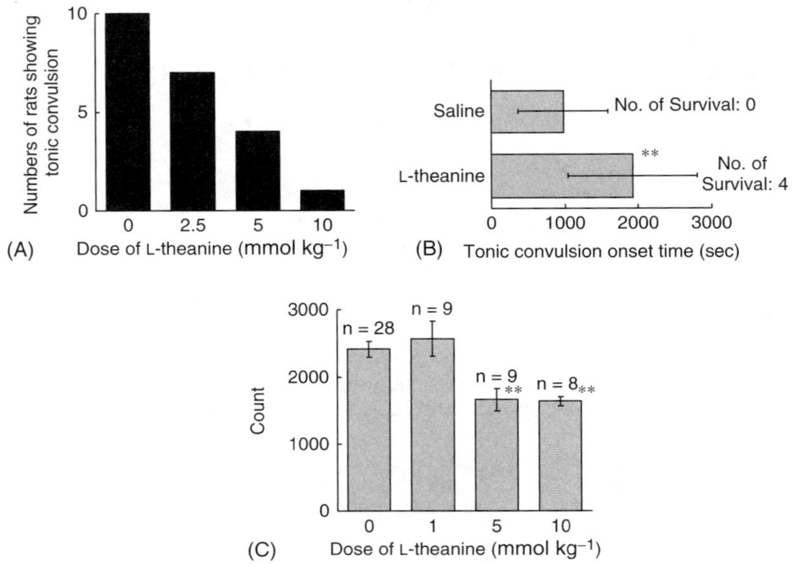

FIGURE 18.7 Effect of L-theanine on the convulsion induced by caffeine in rats. (A) Caffeine was injected 15 minutes after intraperitoneal injection of L-theanine or saline. Ten rats were used for each dose. (B) Number of mice: 14. $^{**}p < 0.01$. Caffeine (300 mg/kg b.w.) was injected intraperitoneally and L-theanine (9 mmol kg^{-1} b.w.) was administered intraperitoneally at the same time. (C) Dose of caffeine: 10 mg kg^{-1} b.w. *S.C.* Total count for 120 minutes (mean ± S.E.). $^{**}p < 0.01$.

showed tonic convulsion within 3 hours after administration of the stimulant is shown in Figure 18.7B. The data suggest that L-theanine significantly delayed the onset of tonic convulsion. Also, a dose-dependent suppression of tonic convulsions was observed with the administration of L-theanine.

In another study, the suppressing effect of L-theanine on caffeine-induced spontaneous activity in rats was examined (12). The caffeine-induced spontaneous activity was measured using the doughnut-type photocell cage. The activity of rats was estimated by counting the number of rotations of the cage made by the rats within 120 minutes. Prior to the running test, caffeine and L-theanine were intraperitoneally administrated to the rats at the same time. L-theanine showed a depressing effect on spontaneous activity at a dose more than 5 mmol kg^{-1} b.w. (Figure 18.7C).

The antagonist effect of L-theanine against caffeine was also measured as brain waves in rats (8). The inhibiting action of L-theanine on the excitation effect of caffeine at a concentration regularly associated with drinking tea was investigated using the electroencephalogram (EEG) in rats. A clear excitation effect of caffeine was observed with the occurrence of β-waves at a dose higher than 5 μmol kg^{-1} b.w. (0.970 mg kg^{-1}), which was determined to be the minimum dose for stimulation. The above dose is almost equal to the amount of caffeine in a cup of tea or coffee, which is equal to 48.7 mg per person (13). In rats, the caffeine-induced excitation was inhibited by intravenous administration of L-theanine at an equivalent molar concentration (5 μmol kg^{-1} b.w.) within 60 minutes. The excitation was dose-dependently reduced with the increase of L-theanine. At a dose of 50 μmol kg^{-1} b.w., the excitation was reduced to less than 50% of that of control rats (Figure 18.8).

18.3.3 Lowering Blood Pressure

It is well known that the regulation of blood pressure is highly dependent on catecholaminergic and serotogenic neurons in both the brain and the peripheral nervous system (14). Thus, neurotransmitter precursors such as tyrosine and tryptophan affect blood pressure. L-Theanine, being a derivative of glutamic acid (a neurotransmitter in the brain) also affects blood pressure. The effect of L-theanine on blood pressure was studied in spontaneously hypertensive rats (SHR) (15). When SHR injected with various amounts of L-theanine (0, 500, 1000, 1500, and 2000 mg kg^{-1} b.w.) intraperitoneally, the changes in blood pressure were dose dependent, and a significant decrease in blood pressure was observed at high doses (1500 and 2000 mg kg^{-1} b.w.) (Figure 18.9A). A dose of 2000 mg kg^{-1} b.w. of L-theanine did not alter the blood pressure of normal Wister Kyoto rats, while the same dose to SHR decreased the blood pressure significantly (Figure 18.9B).

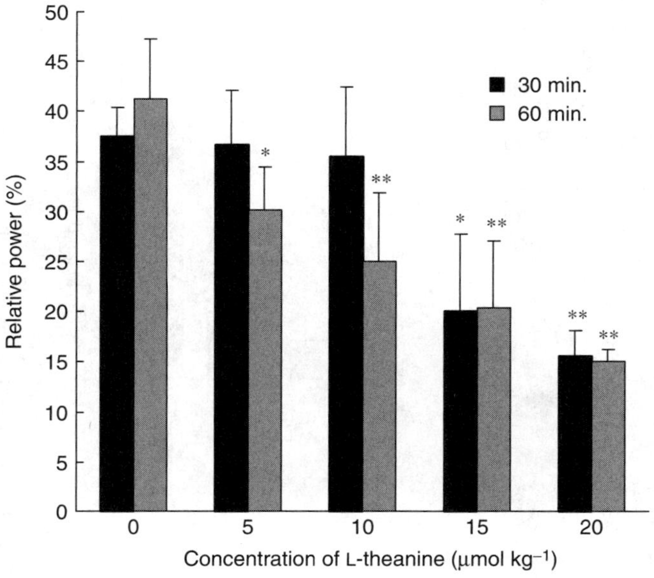

FIGURE 18.8 Changes in relative power of β-waves in the cortex at 30 and 60 minutes after intravenous administration of L-theanine in rats administrated 5 μmol kg^{-1} b.w. of caffeine 10 minutes earlier. Values represent mean ± SEM. Number of rats: 6. $^{*}p < 0.01$. $^{**}p < 0.001$.

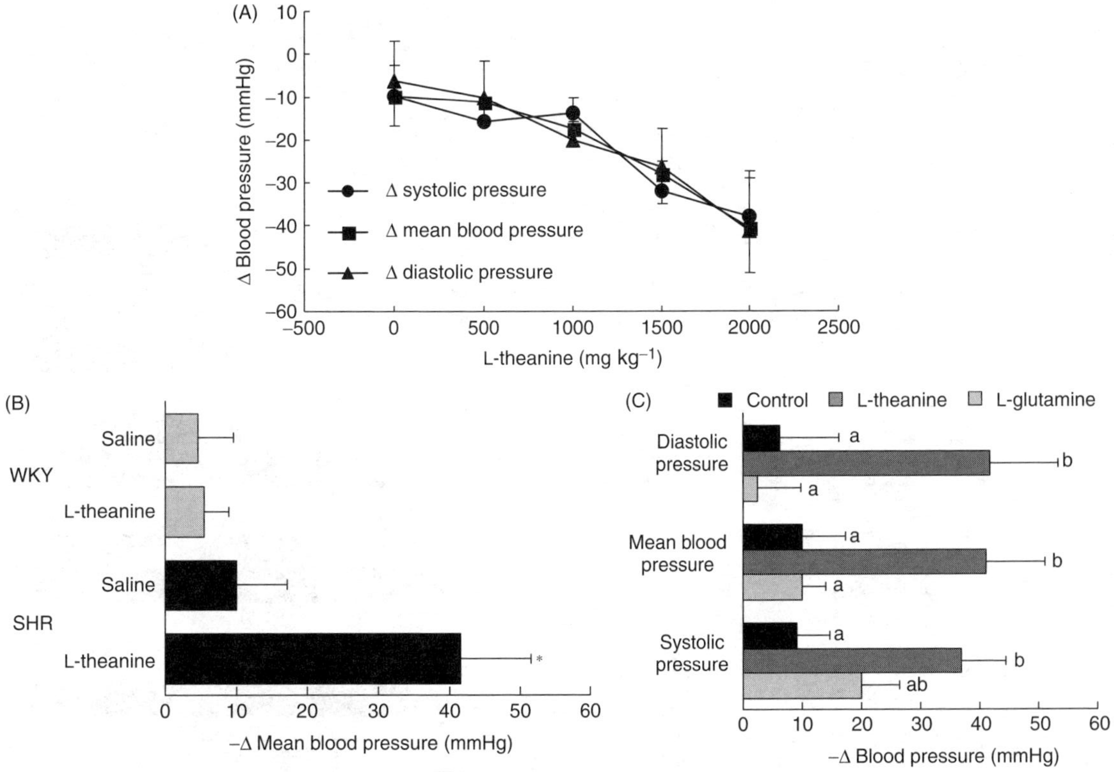

FIGURE 18.9 Effects of L-theanine administration on systolic blood pressure, mean blood pressure, and diastoloc blood pressure of SHR. (A) Number of mice: 5. Significant difference; systolic pressure: 1500 and 2000 mg kg^{-1}; mean blood pressure: 2000 mg kg^{-1}. The blood pressure was measured before and 60 min after the administration of various amounts of L-theanine. Each value represents the mean ± SEM. The baseline for the systolic blood pressure of SHR is 222.7 ± 5.5 mmHg. (B) Number of rats: 5. $*p < 0.05$. Each value represents the mean ± SEM. The baseline for the systolic blood pressure in SHR vs. WKY are 226.9 ± 4.6 mmHg vs. 150.8 ± 7.4 mmHg, respectively. (C) Number of rats: 5. Bars nor followed by the same letter for the systolic blood pressure in SHR is 233.6 ± mmHg.

On the other hand, L-glutamine, which resembles L-theanine in structure, did not exhibit an antihypertensive action on SHR (Figure 18.9C). These results suggest that antihypertensive action was specific to L-theanine.

18.4 ABSORPTION AND METABOLISM OF L-THEANINE

The above psychological and physiological relaxation effects of L-theanine were mainly attributed to its quick absorption, transport in blood, and incorporation into the brain. L-Theanine was found to be incorporated into brain tissue within 30 minutes after intraperitoneal administration, without undergoing any metabolic changes. L-Theanine was accumulated in a dose-dependent manner into the brain of Wistar female rats after intragastric administration of Suntheanine® (Figure 18.10) (16). When L-theanine was orally administrated to rats, it was adsorbed in the intestinal brush-border membrane (17) and incorporated into the brain via the leucine-preferring transport system of the blood–brain barrier. In general, the amino acids circulating in blood are competitively incorporated into the brain via three independent transport systems, namely, L-, A- or ASC-system, depending on their chemical nature, that is, neutral, basic, or acidic. The neutral carrier has a preferential affinity for large neutral essential amino acids in blood and is analogous to the leucine-preferring system, or the L-system, which is sodium independent and insulin insensitive. The alanine-preferring system, or the A-system, is sodium dependent and insulin sensitive and is not present on the lumenal side of the blood–brain barrier. Another sodium-dependent neutral amino acid transport system has a preference for alanine, serine, and cysteine and thus is called the ASC-system; this

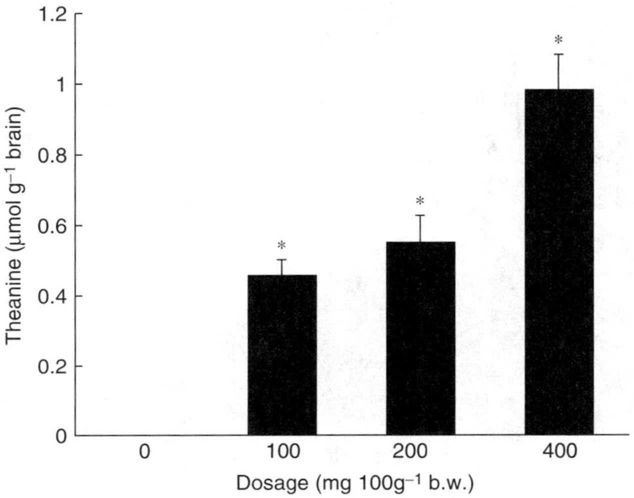

FIGURE 18.10 L-theanine concentration in brain after oral administrartion. Values represent mean ± SEM. Number of rats: 6. $^*p < 0.05$.

system is not present on the blood–barrier. While comparing the concentrations of several amino acids in serum, the concentrations of almost all amino acids transported via the L-system were found to be significantly decreased in the brain with the administration of L-theanine. This shows that L-theanine is incorporated into the brain via the L-system.

The incorporation of L-theanine into various organs was investigated by administrating 4000 mg kg^{-1} b.w. intragastrically to rats (18). The concentrations of L-theanine in the serum, liver, and brain were significantly increased 1 hour after its administration and thereafter decreased gradually. The concentration of L-theanine in brain reached the maximum level in 5 hours (Figure 18.11). This result suggests that the relaxation effects of L-theanine can be maintained for at least 5 hours. Twenty-four hours following administration, L-theanine disappeared completely from all tissues. L-Theanine was detected in the urine 5 hours after administration. This result shows that after administration, L-theanine rapidly incorporates into many tissues and thereafter its levels are reduced within 24 hours, with a concomitant increase in urinary excretion. The quick absorption and sustained release of L-theanine to various organs forms the basis for its strong relaxation effects.

18.5 APPLICATIONS OF SUNTHEANINE®

As mentioned earlier, Suntheanine® is the trade name of L-theanine manufactured by Taiyo Kagaku Co., Ltd. On the basis of its strong relaxation effects, Suntheanine®

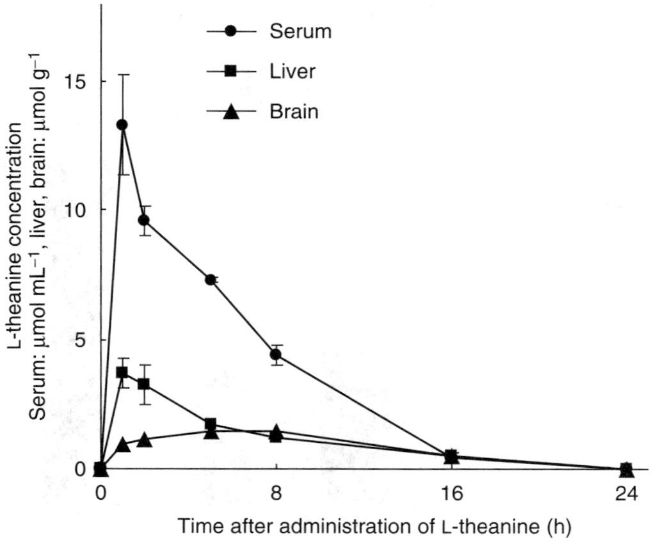

FIGURE 18.11 Time-dependent changes of L-theanine concentrations in the serum, liver, and brain of rats administered L-theanine. Rats were administered L-theanine (4 g kg^{-1} b.w.) Each value is the mean for six rats per group at each experimental time. Vertical bars indicate SEM.

has been recommended for various food applications that aim to achieve relaxation. Suntheanine® in tablet form has been recommended for psychological relaxation for alleviation of PMS and excitation, and for lowering blood pressure. Based on brain-wave studies, a dose ranging from 50 to 200 mg has been recommended to induce the relaxation effect. Suntheanine® also removes the bitter taste in various food materials. It has been used in candies, herbal teas, cocoa drinks, beverages, chocolates, puddings, jellies, chewing gums, and other confectioneries to provide the relaxation effect and for improving taste.

18.6 SAFETY ASPECTS OF SUNTHEANINE®

The results of acute and subacute toxicity tests and a mutagenecity test performed by a public institute have confirmed the safety of Suntheanine®. There are no dietary exposure limits imposed on L-theanine by the Japan Food Additive Association. On the basis of its high LD_{50} (5000 mg kg^{-1} b.w.) and a history of substantial consumption of L-theanine in green tea by a significant number of consumers over long periods of time, L-theanine can be safely recommended.

18.7 CONCLUSIONS

In modern times, stress has become the root cause of many abnormalities in our metal and physical conditions. The state of relaxation is a basic requirement not only to relieve stress but also its side effects. In Japan, drinking green tea for relaxation has been customary from ancient times. Research on green tea has revealed that L-theanine, an abundant amino acid found in green tea leaves, can bring about a state of relaxation. L-Theanine commercially sold as Suntheanine® has been shown to induce both psychological and physiological relaxation. Suntheanine® induces strong α-waves in the brain, which clearly indicates its psychological relaxation effects in humans. Suntheanine® has also been found to alleviate the symptoms of PMS and reduce caffeine-induced excitation as well as blood pressure in hyperactivity by suppressing the central nervous system. Suntheanine® is safe, and no dietary limits have been imposed in Japan. Based on various studies, Suntheanine® has been widely recommended for various food applications, which target relaxation. Suntheanine® tablets (50 mg $tablet^{-1}$) are convenient to use and highly effective.

REFERENCES

1. L.R. Juneja, D.-C. Chu, T. Okubo, Y. Nagato, H. Yokogoshi. L-theanine - a unique amino acid of green tea and its relaxation effect in humans. *Trend. Food Sci. Technol.*, 10: 199–204, 1999.
2. L.R. Juneja, T.P. Rao. Brain health for an ageing society. *Ingr. Health Nutr.*, 6: 10–13, 2003.
3. Y. Sakato. Studies on the chemical constituents of tea. Part III. On a new amide theanine (in Japanese) *J. Agric. Chem. Soc.*, 23: 262–267, 1949.

4. D.R. Morse. Brain wave synchronizers: Part 1–Review of the literature and the first dental anxiety study. *Compend. Contin. Educ. Dent.*, 15: 32–45, 1994.
5. J.A. Taylor. A personality scale of manifest anxiety. *J. Abnorm. Soc. Psychol.*, 48: 285–290, 1953.
6. K. Kobayashi, Y. Nagato, N. Aoi, L.R. Juneja, M. Kim, T. Yamamoto, S. Sugimoto. Effects of theanine on the release of alpha brain waves in human volunteers (in Japanese). *Nippon Nogeikagaku Kaishi*, 72: 153–157, 1998.
7. H. Benson, J.F. Beary, M.P. Carol. The relaxation response. *Psychiatry*, 37: 37–46, 1974.
8. T. Kakuda, A. Nozawa, T. Unno, N. Okumura O. Okai. Inhibiting effects of theanine on caffeine stimulation evaluated by EEG in the rat. *Biosci. Biotech. Biochem.*, 64: 287–293, 2000.
9. T. Ueda, M. Ozeki, T. Okubo, D.-C. Chu, L.R. Juneja, H. Yokogoshi, S. Matsumoto. Improving effect of L-theanine on premenstrual syndrome (in Japanese). *J. JSPOG.*, 6: 234–239, 2001.
10. R.H. Moos. The development of a menstrual distress questionnaire. *Psychosom. Med.*, 30: 853–867, 1968.
11. R. Kimura, T. Murata. Influence of alkylamides of glutamic acid and related compounds on the central nervous system I. Central depressant effect of theanine. *Chem. Pharm. Bull.*, 19: 1257–1261, 1971.
12. R. Kimura, M. Kurita, T. Murata. Influence of alkylamides of glutamic acid and related compounds on the central nervous system III. Effect of theanine on spontaneous activity of mice (in Japanese). *Yakugaku Zasshi.*, 95: 892–895, 1975.
13. T. Shimotoku, et al. Relating between amounts of some ingredients extracted from green tea and brewing conditions (in Japanese). *Tea Res. J.*, 58: 43–50, 1982.
14. D.M. Kuhn, W.A. Wolf, W. Lovenberg. Review of the role of the central serotonergic neuronal system in blood pressure regulation. *Hypertension*, 2: 243–255, 1980.
15. H. Yokogoshi, Y. Kato, M. Sagesaka, T. Takihara-Matsuura, T. Kakuda, N. Takeuchi. Reduction of theanine on blood pressure and brain 5-hydroxyindoles in spontaneously hypertensive rats. *Biosci. Biotech. Biochem.*, 59: 615–618, 1995.
16. H. Yokogoshi, M. Kobayashi, M. Mochizuki, T. Terashima. . Effect of theanine, r-glutamylethylamide, on brain monoamines and striatal dopamine release in conscious rats. *Neurochem. Res.*, 23: 667–673, 1998.
17. S. Kitaoka, H. Hayashi, H. Yokogoshi, Y. Suzuki. Transmural potential changes associated with the *in vitro* absorption of theanine in the guinea pig intestine. *Biosci. Biotech. Biochem.*, 60: 1768–1771, 1996.
18. T. Terashima, J. Takido, H. Yokogoshi. Time-dependent changes of amino acids in the serum, liver, brain, and urine of rats administered with theanine. *Biosci. Biotech. Biochem.*, 63: 615–661, 1999.

Section V

Hypoallergenic Foods

19 Introduction to Food Allergy

Steve L. Taylor and Susan L. Hefle

CONTENTS

19.1 INTRODUCTION

Food allergies are illnesses that occur in some individuals as the result of abnormal immunological responses to a particular food or food component, usually a naturally occurring protein (1–4). Two different types of immunological mechanisms are known to be involved in true food allergies: immediate hypersensitivity reactions and delayed hypersensitivity reactions (3). Immediate hypersensitivity reactions are antibody-mediated; immunoglobulin E (IgE)-mediated food allergies are the only well-recognized form of antibody-mediated, immediate hypersensitivty associated with foods. Delayed hypersensitivity reactions are cell-mediated.

The role of cell-mediated reactions in food allergies is far less well established. The cell-mediated mechanism certainly plays a major role in celiac disease, which will be the only type of delayed hypersensitivity discussed in this chapter.

Other types of individualistic adverse reactions to foods also occur and are often confused with true food allergies. In contrast to true food allergies, these individualistic adverse reactions to foods are not mediated by responses of the immune system, although they do typically involve abnormal physiological responses to a particular food or food component. Like true food allergies, this same food or food component is safe for the vast majority of consumers to ingest. These other types of food sensitivities, sometimes referred to as food intolerances, are beyond the scope of this chapter and are covered in other reviews (3).

In addition, allergy-like intoxications can often be confused with true food allergies because the symptoms are quite similar (3,5). Histamine poisoning is the primary example of an allergy-like intoxication. Unlike true food allergies or food intolerances, allergy-like intoxications can affect anyone in the population who eats a hazardous amount of the causative substance.

19.2 IgE-Mediated Food Allergy

19.2.1 Mechanism

The mechanism involved in IgE-mediated, immediate hypersensitivity reactions to foods is depicted in Figure 19.1. The IgE-mediated reactions are also responsible for allergic reactions to other environmental substances such as pollens, mold spores, animal danders, and bee venoms, except that the source and structure of the allergens are different. In susceptible individuals, B cells produce allergen-specific IgE antibodies in response to an immunological stimulus created by exposure of

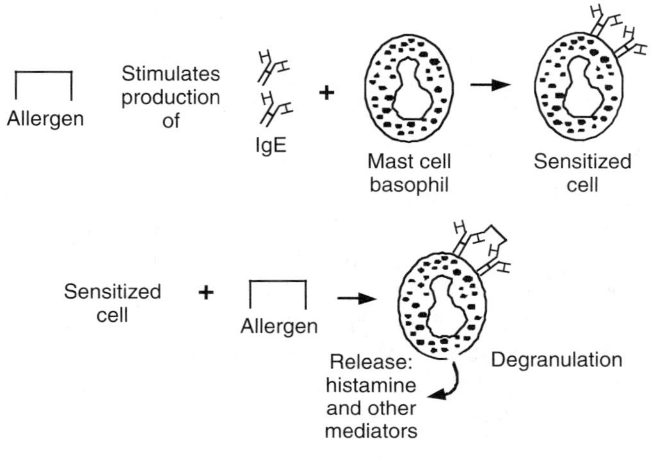

Figure 19.1 Mechanism of IgE-mediated food allergy.

the immune system to the allergen (1). Most food allergens are naturally occurring proteins present in food (4). However, only a few of the many thousands of proteins existing in the food supply are known to be allergens (6). During the sensitization phase of the allergic response, the allergen-specific IgE antibodies bind to the surfaces of the mast cells in tissues and the basophils in blood. No symptoms develop during this sensitization phase. Then, upon subsequent exposure to the specific allergen, the allergen cross-links IgE-antibodies affixed to the sensitized mast cells or basophils. The allergen–antibody interaction induces the sensitized cells to degranulate and release a variety of potent, physiologically active mediators into the blood stream and tissues. Many important mediators of allergic disease are found within the granules of mast cells and basophils, while other non-granule-associated mediators are also released from these cells simultaneously. The mediators are actually responsible for the symptoms of immediate hypersensitivity reactions. Histamine is probably the most important of the mediators released from mast cells and basophils during an allergic reaction. Histamine triggers some of the immediate effects encountered in IgE-mediated food allergies eliciting inflammation, pruritis, and contraction of the smooth muscles in the blood vessels, gastrointestinal tract, and respiratory tract (7). Other important mediators include various leukotrienes and prostaglandins (8). The leukotrienes are associated with some of the symptoms that develop more slowly in IgE-mediated food allergies, such as late-phase asthmatic reactions.

In IgE-mediated food allergies, a susceptible individual forms allergen-specific IgE antibodies after exposure, usually via ingestion, to a specific food protein. However, even among individuals predisposed to allergies, exposure to food proteins does not typically result in the formation of IgE antibodies. More typically, exposure to a food protein in the gastrointestinal tract results in oral tolerance. In oral tolerance, either protein-specific IgG, IgM, or IgA antibodies are formed, or no immunological response whatsoever (clonal anergy) occurs (9,10). Oral tolerance occurs to all dietary proteins in normal individuals and to the vast majority of dietary proteins even among individuals predisposed to development of food allergy. Genetic inheritance is an important factor in predisposing individuals to the development of IgE-mediated allergies, including food allergies (11). Estimates indicate that approximately 65% of patients with clinically documented allergy have first-degree relatives with allergic disease (11). Recent studies with monozygotic and dizygotic twins confirm that genetics is an extremely important parameter and that identical twins may even inherit the likelihood of responding to the same allergenic food, for example, peanuts (12,13). Other coexisting conditions affecting the permeability of the small intestinal mucosa to proteins such as viral gastroenteritis, premature birth, and cystic fibrosis also seem to increase the risk of development of food allergy.

Sensitization to foods usually occurs following exposure from ingestion of the offending food. However, sensitization to foods can occur via respiratory or cutaneous exposure, too. This type of sensitization most often occurs in occupational settings from frequent exposure to the offending food and can result in occupational asthma or contact dermatitis or urticaria (14). Baker's asthma is perhaps the premier example (15). Most individuals with occupational food allergies can safely ingest the offending food, although some exceptions do exist.

Food allergies can also occasionally occur as the result of sensitization to other environmental allergens. For example, sensitization to certain pollens can lead to a typically mild form of food allergy known as oral allergy syndrome (OAS) (16). Sensitization to natural rubber latex usually occurring through occupational exposure to latex gloves can sometimes result in allergies to certain foods (17). These food allergies occur because of structural similarities between the allergens in the food and allergens present in either the specific pollens or latex.

19.2.2 Symptoms

With IgE-mediated food allergies, symptoms typically develop within a few minutes to several hours after the consumption of the offending food. As shown in Table 19.1, the symptoms of IgE-mediated food allergies range from mild and annoying to severe and life threatening. Considerable variability exists in the nature of the symptoms suffered by food-allergic individuals upon exposure to the offending food. Most will suffer from only a few of the symptoms noted in Table 19.1. The type of symptoms and their severity depend upon several factors, including the individual, the amount of the offending food ingested, the tissue receptors that are affected, and the length of time since the last exposure (3).

TABLE 19.1
Symptoms Associated with IgE-mediated Food Allergy

Gastrointestinal:
Diarrhea
Vomiting
Abdominal cramps
Nausea
Gastroesophogeal reflux
Cutaneous:
Urticaria
Dermatitis or eczema
Angioedema
Pruritus
Respiratory:
Rhinitis
Asthma
Laryngeal edema
Generalized:
Anaphylactic shock

Cutaneous and gastrointestinal symptoms (see Table 19.1) are commonly associated with IgE-mediated food allergies. Respiratory symptoms are much less commonly encountered with food allergies as compared with environmental allergies such as pollen or animal dander allergies, where the allergens are airborne and inhaled directly into the respiratory tract. When respiratory reactions do occur in IgE-mediated food allergies, they are among the more severe and life-threatening symptoms. Those few food-allergic individuals who experience serious respiratory symptoms (asthma and laryngeal edema) in association are most likely to be at risk for life-threatening episodes (18). However, among the many symptoms involved in IgE-mediated food allergies, systemic anaphylaxis, also known as anaphylactic shock, is the most severe manifestation. Anaphylactic shock involves symptoms developing in rapid succession in multiple organ systems (gastrointestinal, respiratory, cutaneous, and cardiovascular). Death can occur from severe hypotension coupled with respiratory and cardiovascular complications. Anaphylactic shock is a common cause of death in the occasional fatalities associated with IgE-mediated food allergies (18,19). While the number of deaths occurring from IgE-mediated food allergies is not recorded in most countries, more than 100 deaths are thought to occur in the U.S. each year related to severe food allergies (20).

Perhaps the most common and mildest form of IgE-mediated food allergy is OAS (16). In OAS, symptoms, including pruritis, urticaria, and angioedema, occur primarily in the oropharyngeal area. OAS is usually associated with the ingestion of various fresh fruits and vegetables (16). Even though fresh fruits and vegetables are comparatively low in protein content, OAS is an IgE-mediated reaction to specific proteins present in these foods (16). Systemic symptoms are rarely encountered in OAS because the allergens in these fresh fruits and vegetables are apparently quite susceptible to digestive proteases in the gastrointestinal tract (21). The allergens involved in OAS are also apparently heat labile (21), since the heat-processed versions of these foods are not typically involved in elicitation of OAS. As noted earlier, individuals with OAS are initially sensitized to one or more pollens in the environment that cross-react with related proteins found in fresh fruits and vegetables (16,22). Birch and mugwort pollens are frequently implicated. With OAS, sensitization to the pollen increases the likelihood of sensitization to specific foods.

In some cases of IgE-mediated food allergy, exercise must have been done coincident with ingestion of the food for symptoms to occur (23). Several foods, including shellfish, wheat, celery, and peach, have been implicated in exercise-induced food allergies (3). The mechanism of this illness is not well understood, although the involvement of IgE antibodies is apparent.

19.2.3 Prevalence

The overall prevalence of IgE-mediated food allergies in the U.S. is estimated at 2 to 2.5% for all age groups combined (3). Only a few epidemiological investigations involving clinical confirmation of self-reported IgE-mediated food allergies have been conducted using representative, unselected groups of adults. For example, in a large-scale epidemiological investigation in the Netherlands (24), more than 10% of Dutch adults believed that they had adverse reactions to one or more specific foods, but food allergies were clinically confirmed in approximately 2%. The prevalence of IgE-mediated food allergies is higher among infants and young children than among adults (25). Epidemiological investigations among groups of unselected infants suggest that the prevalence of IgE-mediated food allergies is in the range of 4 to 8% in young infants (26,27).

The results of the epidemiological investigations are consistent with evidence obtained through consumer surveys. Surveys conducted in both the U.S. and England indicate that the self-perceived prevalence of peanut allergy is 0.5 to 0.6% among all age groups (28,29). In the U.S. survey, the self-reported prevalence of tree nut allergies combined with peanut allergies was 1.1% (28). Since peanut and tree nut allergies are often rather profound and easily diagnosed, these surveys are probably fairly reliable. If the prevalence of peanut and tree nut allergies alone exceeds 1% for all age groups and the prevalence of all food allergies among infants is 4 to 8%, then an overall estimate for IgE-mediated food allergies of 2 to 2.5% seems reasonable (3).

19.2.4 Persistence

Many food-allergic infants outgrow their food allergies within a few months to several years (30,31). Allergies to certain foods, such as cow's milk (30), are more commonly outgrown than others such as peanut allergy (31). The mechanisms involved in the loss of sensitivity to specific foods are not known, but the development of immunological tolerance is definitely involved (10).

19.2.5 Commonly Allergenic Foods

Eight foods or food groups, including cow's milk, eggs, fish, crustacea (e.g., shrimp, crab, lobster, and so on), peanuts, soybeans, tree nuts (e.g., walnuts, almonds, hazelnuts, and so on), and wheat are responsible for the vast majority of IgE-mediated food allergies on a worldwide basis (32). In 1999, the Codex Alimentarius Commission established an initial allergen list that incorporates these eight foods or food groups (sometimes referred to as the "Big 8"). Beyond these eight major foods or food groups, more than 160 other foods have been documented to cause IgE-mediated food allergies (33). Certain countries have identified additional commonly allergenic foods that are important in those geographic regions, including buckwheat in Japan and Korea, sesame seeds in Canada and probably soon in the European Union, and mustard seed and celery in certain European countries. Any food that contains protein is likely to cause allergic sensitization on at least rare occasions. The eight most commonly allergenic foods or food groups contain comparatively high amounts of protein and are commonly consumed in the diet in many countries of the world. However, several other commonly consumed foods with high protein contents such as beef, pork, chicken, and turkey are rarely allergenic, and so protein content alone cannot be used to identify foods that have the potential to be commonly allergenic (3).

Ingredients derived from the commonly allergenic foods will also be allergenic if they contain significant levels of residual protein from the source material. Several categories of food ingredients, including edible oils, protein hydrolyzates, lecithin, flavors, and gelatin can, on occasion, be derived from commonly allergenic sources. A thorough discussion of the potential allergenicity of ingredients derived from commonly allergenic sources is beyond the scope of this chapter, but other reviews can be consulted (3,34).

Edible oils will be used as an example. Edible oils do not typically cause allergic reactions. The refining process for edible oils removes virtually all of the protein from the source material when hot solvent extraction is used. Highly refined peanut, soybean, and sunflower seed oils are documented to be safe for ingestion by individuals allergic to the source material (35–37). Other edible oils from sources such as sesame seed and tree nuts may receive less processing and contain allergenic residues (38,39). Oils produced by other procedures such as cold-pressed oils may also contain allergenic residues (40).

Hypoallergenic foods can be developed from commonly allergenic foods by (a) removing the allergenic protein fraction (e.g., edible oil refining), (b) modifying the allergenic protein fraction through proteolysis, or (c) modifying the allergen through agricultural biotechnology. Subsequent sections in this chapter will describe various examples in more detail. However, proteolysis is the most important commercial process currently for the manufacturing of hypoallergenic foods. Hypoallergenic infant formulas are based upon extensively hydrolyzed casein or whey, commonly allergenic protein fractions from cow's milk (41).

19.2.6 Food Allergens

The majority of food allergens are naturally occurring proteins found in foods (4). However, only a small percentage of the many proteins occurring naturally in foods are known to be allergens (6). Several food allergens have been purified and characterized (4). Some foods contain principally one allergen (e.g., Brazil nut, codfish, and shrimp), while other foods contain multiple allergens (e.g., peanuts, eggs, cow's milk, and walnuts). Foods may contain both major and minor allergens. Major allergens are defined as those proteins that bind to serum IgE antibodies in more than 50% of patients with a specific food allergy. For example, cow's milk contains three major allergens: casein, β-lactoglobulin, and α-lactalbumin (42), which also happen to be the major proteins in cow's milk. Additionally, cow's milk contains several minor allergens such as bovine serum albumin (43). Peanuts contain at least three major allergens, namely, Ara h1, Ara h2, and Ara h3 (4). Peanuts also contain a large number of minor allergens (44). In contrast, codfish, Brazil nut, and shrimp contain primarily one major allergenic protein: Gad c1 (44), Ber e1 (45), and Pen a1 (46), respectively.

19.2.7 Diagnosis

Several diagnostic procedures are helpful in the clinical diagnosis of IgE-mediated food allergies. Self or parental diagnosis is often unreliable (26), and so the assistance of an allergy specialist is recommended. The patient may be asked to keep a food diary in an attempt to link adverse reactions to specific foods. This will aid the physician in determining if an adverse reaction is occurring following the ingestion of a particular food.

A particularly reliable diagnostic procedure is the double-blind, placebo-controlled food challenge (DBPCFC) (47). The DBPCFC will unequivocally link ingestion of a specific food to elicitation of a specific set of symptoms. It cannot be used when there is a history of life-threatening anaphylaxis to a suspected food (47). The DBPCFC is particularly useful when the role of a specific food or foods as the cause of the allergic reaction is nebulous.

The IgE-mediated allergic reactions are typically confirmed by the skin-prick test (SPT) or by the radioallergosorbent test (RAST) and similar immunoassays (48). The simplest procedure is the SPT (49). A food extract is applied to the skin, and the site is pricked with a needle to allow the extract (allergen) to enter. A wheal-and-flare reaction developing at the site demonstrates that IgE in the skin has reacted with some protein in the extract. The RAST test is an *in vitro* test utilizing a sample of the patient's blood serum (50). The RAST measures the binding of food-specific IgE antibodies in the patient's serum to a food extract bound to some solid matrix. The degree of binding is assessed with radiolabeled antihuman IgE. The results of the RAST are as reliable as the SPT, but these tests are considerably more expensive. However, this may be the test of choice for patients with extreme sensitivities because SPTs may be hazardous in such patients (52).

19.2.8 Treatment

The options for treating IgE-mediated food allergies are limited. The major approach to treatment is the specific avoidance diet (53). The patient must avoid the food or foods that cause the reaction; for example, if allergic to peanuts, simply avoid peanuts in all forms. Successful implementation of a specific avoidance diet can be daunting for the patient. Patients must have considerable knowledge of food composition. Dietitians may be helpful in teaching clients to interpret food labels to detect ingredients made from the offending food. Compliance with avoidance diets is improved if the number of foods eliminated is kept to a minimum.

Very few hypoallergenic foods are available for use by food-allergic patients. For example, most infants with cow's milk allergy can safely be fed alternative formulae such as casein hydrolyzate formulae (41). However, case reports have shown occasional adverse reactions to milk protein hydrolyzates (54–57).

Individuals with IgE-mediated food allergies are often sensitive to ingestion of rather small doses of the offending food (58,59). Allergic reactions can ensue, on occasion, following exposure to trace amounts of the offending food that might arise through various processing or preparation errors (60). Examples include the failure to adequately clean common equipment, the use of rework (a common practice in certain segments of the industry involving the incorporation of left-over or misformulated quantities of a food product into subsequent batches of related products with identical or related formulations), and the inadvertent addition of an ingredient that is not supposed to be in the formulation (61–64). The existence of unlabeled foods especially in restaurant or other food service settings is another concern for the development of safe and effective avoidance diets. Many adverse reactions have occurred in such settings (18,19).

For practical purposes, complete avoidance must be maintained for the implementation of a successful avoidance diet. However, there are threshold doses below which allergic individuals will not experience adverse reactions (58,59). Threshold doses seem to be in the low milligram range and vary among individuals allergic to any specific food (58,59). Also, the possibility exists that the threshold doses vary from one allergenic food to another (3).

In the implementation of specific avoidance diets, questions often arise regarding the need to avoid closely related foods. Cross-reactions do occur between closely related foods in the case of certain food groups. For example, cross-reactions are known to occur among the various crustacean species (shrimp, crab, lobster, and crawfish) (65), different species of avian eggs (66), and cow's milk and goat's milk (67). In contrast, the patterns of fish allergy are considerably variable from one individual to another (68). Furthermore, some peanut-allergic individuals are allergic to other legumes such as soybeans (69), but this is not a common occurrence (70). Clinical hypersensitivity to one legume, for example, peanuts or soybeans, does not warrant exclusion of the entire legume family from the diet unless allergy to each individual legume is confirmed by clinical challenge trials (70). Thus, it does not seem to be possible to offer uniform advice on the need to avoid foods that are related to the specific food implicated in the allergic sensitization.

Cross-reactions are also known to occur between certain types of pollens and foods especially with OAS (16). Examples include ragweed pollen and melons, mugwort pollen and celery, mugwort pollen and hazelnut, and birch pollen and various other foods including carrots, apples, hazelnuts, and potatoes (3). Cross-reactions are also known to occur between allergies to natural rubber latex and certain foods such as banana, chestnut, kiwi, and avocado (17).

The IgE-mediated allergic reactions to foods can be treated by various pharmacological approaches (71). Epinephrine (adrenaline) and antihistamines are particularly useful in treating the symptoms associated with allergic reactions to foods. Those patients with a history of life-threatening reactions to foods are advised to carry an epinephrine-filled syringe with them at all times (72).

19.2.9 Prevention

The prevention of allergic sensitization in infants requires early identification of high-risk infants, exclusion of commonly allergenic foods, including perhaps cow's milk, eggs, and peanuts, from the infant diet, breast-feeding for an extended period, possible use of hypoallergenic infant formulae, and the exclusion of commonly allergenic foods from the diet of the nursing mother (73,74). The IgE-mediated food allergies are most likely to develop in high-risk infants, defined as those infants born to parents with histories of allergic disease of any type (e.g., pollens, mold spores, animal danders, bee venoms, and food). Maternal dietary restrictions during pregnancy (excluding commonly allergenic foods such as peanuts) do not appear to be helpful in the prevention of allergy in the infant (73,74), suggesting that sensitization does not occur *in utero*. Breast-feeding delays, but may not prevent, the development of IgE-mediated food allergies (75). However, it appears as though some infants are sensitized to allergenic foods through exposure to the allergens in breast milk (76,77). For sensitization to occur in breast-fed infants, certain allergenic food proteins must resist digestion, be absorbed, at least to a small extent, from the small intestine, and be secreted in breast milk. The exclusion of specific allergenic foods such as peanuts from the diet of the lactating mother will help prevent sensitization through breast milk. While the maternal dietary elimination of peanuts is often recommended for women with high-risk infants, the elimination of milk and eggs is not because these foods are usually considered to be too important nutritionally to exclude from the diets of lactating women. The use of probiotics during lactation may also help lessen the likelihood of allergic sensitization (78).

Allergic sensitization may be prevented by the use of hypoallergenic infant formula in high-risk infants (79), although these formulae are more often used to prevent reactions after sensitization has already occurred. The use of partial whey hydrolyzate formula has been advocated for this purpose (80). High-risk infants may still develop food allergies once solid foods are introduced into the diet (75).

19.2.10 Effect of Processing on Allergenicity

Allergenic food proteins are typically quite stable to food processing conditions (21). Food allergens are considerably heat stable, and typical food processing conditions do not alter the allergenicity of the resulting products (21). Several exceptions exist to the heat stability of food allergens. The allergens in some fruits and vegetables are exceptions, being quite heat-sensitive (81). The allergens present in some fishes may be destroyed by canning processes, although other heating processes do not appear to affect these allergens (82).

Food allergens are typically resistant to proteolysis, and this allows these proteins to survive digestion (21). The resistance to proteolysis means that these allergens may survive, in whole or in part, the acid and enzymatic hydrolysis methods used to prepare protein hydrolyzates used as ingredients in various foods (21).

19.3 DELAYED HYPERSENSITIVITY (CELIAC DISEASE EXAMPLE)

Delayed hypersensitivities are cell-mediated reactions involving tissue-bound T lymphocytes that are sensitized to specific foodborne substances (83). Cell-mediated inflammatory reactions are often localized to certain sites within the body. With delayed hypersensitivity reactions, symptoms begin to appear 6 to 24 hours after consumption of the offending food.

Although the role of delayed hypersensitivity reactions in food allergies remains largely unknown, celiac disease can be cited as an example of a cell-mediated immune response associated with the ingestion of a specific group of foods. Celiac disease, also known as celiac sprue, nontropical sprue, or gluten-sensitive enteropathy, is a malabsorption syndrome occurring in sensitive individuals upon the consumption of wheat, rye, barley, triticale, spelt, and kamut (84). Following the ingestion of these grains or the protein-containing ingredients derived from these grains, the absorptive epithelium of the small intestine becomes damaged through a cell-mediated inflammatory response (85,86). The result is a decrease in the number of absorptive epithelial cells lining the small intestine. The mucosal enzymes necessary for digestion and absorption are also altered in the damaged cells. Thus, the absorptive cells are functionally compromised as a result of the cell-mediated inflammatory process.

19.3.1 Symptoms

The mucosal damage associated with the cell-mediated inflammatory reaction leads to nutrient malabsorption. The loss of absorptive function is characterized by diarrhea, bloating, weight loss, anemia, bone pain, chronic fatigue, weakness, muscle cramps, and, in children, growth retardation and failure to gain weight (84,87).

In addition, patients with celiac disease are at increased risk for certain chronic illnesses. Celiac sufferers are at increased risk for development of T-cell lymphoma (88). Celiac patients also seem to be more likely to have various autoimmune diseases, including dermatitis herpetiformis, thyroid diseases, Addison's disease, pernicious anemia, autoimmune thrombocytopenia, sarcoidosis, insulin-dependent diabetes mellitus, and IgA nephropathy (89).

19.3.2 Prevalence

The prevalence of celiac disease is a subject of some debate. The debate is fostered by the fact that different diagnostic approaches are used in different parts of the world. Celiac disease appears to be latent or asymptomatic in some individuals (90,91). The prevalence of celiac disease appears particularly high in certain European regions and in Australia (89,92). The estimated prevalence of celiac disease is as high as 1 in every 250 people in some European regions (89). In the U.S., the prevalence of celiac disease is, in contrast, estimated at about 1 of every 3000 individuals (93). Advanced diagnostic methods may be the reason for the lower prevalence rate in the U.S. (94). However, even within European populations, considerable variability is observed in the prevalence of celiac disease (89,92,95).

19.3.3 Causative Factor

The ingestion of prolamin protein fractions from wheat and related grains is associated with the development of celiac disease (85,93). The prolamin fraction of wheat is known as gluten, or gliadin, and so celiac disease is sometimes referred to as gluten-sensitive enteropathy. Apparently, a defect in the mucosal processing of gliadin in celiac patients provokes the generation of toxic peptides that contribute to the abnormal T-cell response and the subsequent inflammatory reaction (96). However, the mechanism involved in celiac disease and the exact role of gliadin remain to be determined.

19.3.4 Treatment

Celiac disease is treated with an avoidance diet (97). Celiac sufferers must avoid wheat, rye, barley, and related grains such as kamut, triticale, and spelt (97). The need to avoid ingredients that do not contain protein from the implicated grains is somewhat debatable, but widely practiced by many celiac sufferers. Celiac sufferers often choose to also avoid oats, although the role of oats in the elicitation of celiac disease has recently been refuted (98). However, commercially, oats are often contaminated by wheat, and so some caution may still be necessary with respect to the ingestion of oats by celiac sufferers. The degree of tolerance for wheat, rye, barley, and related grains among celiac sufferers is unknown, but ingestion of small amounts may trigger symptoms. Most celiac sufferers attempt to practice complete avoidance. An enzyme-linked immunosorbent assay (ELISA) has been developed for the detection of gluten in foods (99). The lower limit of detection of wheat gluten in this ELISA is 0.016%. The availability of this ELISA has led some to conclude that foods containing gluten below 200 ppm are not hazardous for celiac patients. While this has not been conclusively proven, a few isolated studies have concluded that levels of 10 mg of gliadin per day will be tolerated by most patients with active celiac disease (100). Clearly, the threshold dose may vary from one individual to another, since, in latent forms of celiac disease, normal dietary quantities of the offending grains seem to cause little problem.

19.4 CONCLUSIONS

Food allergies affect only a small percentage of the population. However, food-allergic reactions can be quite severe and even life threatening in some of these individuals. The avoidance of the offending food(s) is the key to management of food allergies. However, the avoidance of a specific food can be difficult, and complete success in avoidance is unlikely.

The IgE-mediated food allergies are well understood and relatively easy to diagnose, even though management can be difficult. Although many food allergens have been identified, further progress is needed in the identification and characterization of food allergens, the methods for their detection in foods, the removal or inactivation of these allergens to provide hypoallergenic foods, and the development of improved treatment modalities.

The existence of other types of allergic mechanisms in food allergies remains to be established. The existence of celiac disease serves as an example of an important illness that involves such mechanisms.

REFERENCES

1. Y.A. Mekori. Introduction to allergic disease. *Crit. Rev. Food Sci. Nutr.*, 36: S1–S18, 1996.
2. S.C. Bischoff, G. Sellge. Immune mechanisms in food-induced disease. In: D.D. Metcalfe, H.A. Sampson, R.A. Simon (Eds.) *Food Allergy: Adverse Reactions to*

Foods and Food Additives, 3rd edition. Malden, MA: Blackwell Publishing, 2003, 14–37.
3. S.L. Taylor, S.L. Hefle. Allergic reactions and food intolerances. In: F.N. Kotsonis, M. Mackey (Eds.) *Nutritional Toxicology*, 2nd edition. London: Taylor & Francis, 2002, 93–121.
4. R.K. Bush, S.L. Hefle. Food allergens. *Crit. Rev. Food Sci. Nutr.*, 36: S119–S163, 1996.
5. S.L. Taylor, S.L. Hefle. Allergylike intoxications from foods. In: M. Frieri, B. Kettelhut (Eds.) *Food Hypersensitivity and Adverse Reactions — A Practical Guide for Diagnosis and Management.* New York: Marcel Dekker, 1999, 141–153.
6. S.L. Taylor. Genetically-engineered foods: commercial potential, safety, and allergenicity. *Food Aust.*, 48: 308–311, 1996.
7. R.K. Bush, S.L. Taylor. Histamine. In: R. Macrae, R.K. Robinson, M. Sadler (Eds.) *Encyclopedia of Food Science, Food Technology and Nutriton*, Volume 4. London: Academic Press, 1993, 2367–2371.
8. S.L. Johnston, S.T. Holgate. Cellular and chemical mediators: their roles in allergic diseases. *Curr. Opin. Immunol.*, 2: 513–24, 1990.
9. S.H. Sicherer, H.A. Sampson. Cows' milk protein-specific IgE concentrations in two age groups of milk-allergic children and in children achieving clinical tolerance. *Clin. Exp. Allergy*, 29: 507–512, 1999.
10. S. Strobel. Oral tolerance: immune responses to food antigens. In: D.D. Metcalfe, H.A. Sampson, R.A. Simon (Eds.) *Food Allergy — Adverse Reactions to Foods and Food Additives*, 2nd edition. Boston: Blackwell Science, 1997, 107–135.
11. R.K. Chandra. Food allergy: setting the theme. In: R.K. Chandra (Ed.) *Food Allergy*. St. John's, Newfoundland: Nutrition Research Education Foundation, 1987, 3–5.
12. G. Lack, D.E.S. Fox, J. Golding. The role of the uterine environment in the pathogenesis of peanut allergy. *J. Allergy Clin. Immunol.*, 103: S95, 1999.
13. S.H. Sicherer, T.J. Furlong, B.D. Gelb, R.J. Desnick, H.A. Sampson. Peanut allergy in twins. *J. Allergy Clin. Immunol.*, 105: S181, 2000.
14. M. Aresery, A. Cartier, L. Wild, S.B. Lehrer. Occupational reactions to food allergens. In: D.D. Metcalfe, H.A. Sampson, R.A. Simon (Eds.) *Food Allergy: Adverse Reactions to Foods and Food Additives*, 3rd edition. Malden MA: Blackwell Publishing, 2003, 270–295.
15. D.J. Hendrick, R.J. Davies, J. Pepys. Bakers' asthma. *Clin. Allergy*, 6: 241–250, 1976.
16. E. Pastorello, C. Ortolani. Oral allergy syndrome. In: D.D. Metcalfe, H.A. Sampson, R.A. Simon (Eds.) *Food Allergy: Adverse Reactions to Foods and Food Additives*, 3rd edition. Malden MA: Blackwell Publishing, 2003, 169–182.
17. C. Blanco, T. Carrillo, R. Castillo, J. Quiralte, M. Cuevas. Latex allergy: clinical features and cross-reactivity with fruits. *Ann. Allergy*, 73: 309–314, 1994.
18. H.A. Sampson, L. Mendelson, J. Rosen. Fatal and near-fatal anaphylactic reactions to foods in children and adolescents. *N. Engl. J. Med.*, 327: 380–384, 1992.
19. J.W. Yunginger, K.G. Sweeney, W.Q. Sturner, L.A. Giannandrea, J.D. Teigland, M. Bray, P.A. Benson, J. York, L. Biedrzycki, D. Squillace, R. Helm. Fatal food-induced anaphylaxis. *J. Am. Med. Assoc.*, 260: 1450–1452, 1988.
20. R. Miller, A. Ghatek, P. Rothman, A. Neugut. Anaphylaxis in the United States: an investigation into its epidemiology. *J. Allergy Clin. Immunol.*, 105: S349, 2000.
21. S.L. Taylor, S.B. Lehrer. Principles and characteristics of food allergens. *Crit. Rev. Food Sci. Nutr.*, 36: S91–S118, 1996.

22. P.G. Calkhoven, M. Aalbers, V.L. Koshte, O. Pos, H.D. Oei, R.C. Aalberse. Cross-reactivity among birch pollen, vegetables and fruits as detected by IgE antibodies is due to at least three distinct cross-reactive structures. *Allergy*, 42: 382–390, 1987.
23. M.E. O'Connor, A.L. Schocket. Exercise- and pressure-induced syndromes. In: D.D. Metcalfe, H.A. Sampson, R.A. Simon (Eds.) *Food Allergy: Adverse Reactions to Foods and Food Additives*, 3rd edition. Malden MA: Blackwell Publishing, 2003, 262–269.
24. J.J. Neistijl Jansen, A.F.M. Kardinaal, G. Huijbers, B.J. Vlieg-Boestra, B.P.M. Martens, T. Ockhuizen. Prevalence of food allergy and intolerance in the adult Dutch population. *J. Allergy Clin. Immunol.*, 93: 446–456, 1994.
25. H.A. Sampson. Food allergy. *Curr. Opin. Immunol.*, 2: 542–547, 1990.
26. S.A. Bock, W.Y. Lee, L.K. Remigio, C.D. May. Studies of hyper-sensitivity reactions to foods in infants and children. *J. Allergy Clin. Immunol.*, 62: 327–334, 1978.
27. S.R. Halpern, W.A. Sellars, R.B. Johnson, D.W. Anderson, S. Saperstein, J.S. Reisch . Development of childhood allergy in infants fed breast, soy or cow milk. *J. Allergy Clin. Immunol.*, 51: 139–151, 1973.
28. S.H. Sicherer, A. Munoz-Furlong, A.W. Burks, H.A. Sampson. Prevalence of peanut and tree nut allergy in the U.S. determined by a random digit dial telephone survey. *J. Allergy Clin. Immunol.*, 103: 559–562, 1999.
29. S.E. Emmett, F.J. Angus, J.S. Fry, P.N. Lee. Perceived prevalence of peanut allergy in Great Britain and its association with other atopic conditions and with peanut allergy in other household members. *Allergy*, 54: 380–385, 1999.
30. D.J. Hill, C.S. Hosking. Patterns of clinical disease associated with cow milk allergy in childhood. *Nutr. Res.*, 12: 109–121, 1992.
31. S.A. Bock, F.M. Atkins. The natural history of peanut allergy. *J. Allergy Clin. Immunol.*, 83: 900–904, 1989.
32. Food and Agricultural Organization of the United Nations. *Report of the FAO Technical Consultation on Food Allergies*, Rome, Italy, November 13–14, 1995.
33. S.L. Hefle, J.A. Nordlee, S.L. Taylor. Allergenic foods. *Crit. Rev. Food Sci. Nutr.*, 36: S69–S89, 1996.
34. S.L. Taylor, S.L. Hefle. Ingredient issues associated with allergenic foods. *Curr. Allergy Clin. Immunol.*, 14: 12–18, 2001.
35. J.O'B Hourihane, S.J. Bedwani, T.P. Dean, J.O. Warner. Randomised, double-blind, crossover challenge study of allergenicity of peanut oils in subjects allergic to peanut. *Br. Med. J.*, 314: 1084–1088, 1997.
36. R.K. Bush, S.L. Taylor, J.A. Nordlee, W.W. Busse. Soybean oil is not allergenic to soybean-sensitive individuals. *J. Allergy Clin. Immunol.*, 76: 242–245, 1985.
37. A.B. Halsey, M.E. Martin, M.E. Ruff, F.O. Jacobs, R.L. Jacobs. Sunflower oil is not allergenic to sunflower seed-sensitive patients. *J. Allergy Clin. Immunol.*, 78: 408–410, 1986.
38. G. Kanny, C. De Hauteclocque, D.A. Moneret-Vautrin. Sesame seed and sesame seed oil contain masked allergens of growing importance. *Allergy*, 51: 952–957, 1996.
39. S.S. Teuber, R.L. Brown, L.A.D. Haapanen. Allergenicity of gourmet nut oils processed by different methods. *J. Allergy Clin. Immunol.*, 99: 502–507, 1997.
40. D.R. Hoffman, C. Collins-Williams. Cold-pressed peanut oils may contain peanut allergen. *J. Allergy Clin. Immunol.*, 93: 801–802, 1994.
41. D.J. Hill, C.S. Hosking. The management and prevention of food allergy. In: M. Frieri, B. Kettelhut (Eds.) Food hypersensitivity and adverse reactions — a

practical guide to diagnosis and management. New York: Marcel Dekker, 1999, 423–447.
42. S.L. Taylor. Immunologic and allergic properties of cows' milk proteins in humans. *J. Food Prot.*, 49: 239–250, 1986.
43. B.A. Baldo. Milk allergies. *Aust. J. Dairy Technol.*, 39: 120–128, 1984.
44. D. Barnett, B.A. Baldo, M.E.H. Howden. Multiplicity of allergens in peanuts. *J. Allergy Clin. Immunol.*, 72: 61–68, 1983.
45. S. Elsayed, H. Bennich. The primary structure of allergen M from cod. *Scand. J. Immunol.*, 4: 203–208, 1975.
46. J.A. Nordlee, S.L. Taylor, J.A. Townsend, L.A. Thomas, R.K. Bush. Identification of Brazil nut allergen in transgenic soybeans. *N. Engl. J. Med.*, 334: 688–692, 1996.
47. C.B. Daul, M. Slattery, G. Reese, S.B. Lehrer. Identification of the major brown shrimp (Penaeus aztecus) allergen as the muscle protein tropomyosin. *Int. Arch. Allergy Immunol.*, 105: 49–55, 1994.
48. S.A. Bock, H.A. Sampson, F.M. Atkins, R.S. Zeiger, S. Lehrer, M. Sachs, R.K. Bush, D.D. Metcalfe. Double-blind, placebo-controlled food challenge (DBPCFC) as an office procedure: a manual. *J. Allergy Clin. Immunol.*, 82: 986–997, 1988.
49. P.P. Van Arsdel, E.B. Larson. Diagnostic tests for patients with suspected allergic disease. Utility and limitations. *Ann. Int. Med.*, 110: 304–312, 1989.
50. S.H. Sicherer. In vivo diagnosis: skin testing and challenge procedures. In: D.D. Metcalfe, H.A. Sampson, R.A. Simon (Eds.) *Food Allergy: Adverse Reactions to Foods and Food Additives*, 3rd edition. Malden, MA: Blackwell Publishing, 2003, 104–117.
51. S. Ahlstedt, A. Kober, L. Soderstrom. In vitro diagnostic methods in the evaluation of food hypersensitivity. In: D.D. Metcalfe, H.A. Sampson, R.A. Simon (Eds.) *Food Allergy: Adverse Reactions to Foods and Food Additives*, 3rd edition. Malden, MA: Blackwell Publishing, 2003, 91–103.
52. D.D. Metcalfe. Diagnostic procedures for immunologically-mediated food sensitivity. *Nutr. Rev.*, 42: 92–97, 1984.
53. S.L. Taylor, R.K. Bush, W.W. Busse. Avoidance diets — how selective should we be? *N. Engl. Reg. Allergy Proc.*, 7: 527–532, 1986.
54. J.D. Saylor, S.L. Bahna. Anaphylaxis to casein hydrolyzate formula. *J. Pediatr.*, 118: 71–74, 1991.
55. E. Rosenthal, Y. Schlesinger, Y. Birnbaum, R. Goldstein, A. Benderly, S. Freier. Intolerance to casein hydrolyzate formula. *Acta Paediatr. Scand.*, 80: 958–960, 1991.
56. L. Businco, A. Cantani, M. Longhi, P.G. Giampietro. Anaphylactic reactions to a cow's milk whey protein hydrolyzate (Alfa-Re Nestle) in infants with cow's milk allergy. *Ann. Allergy*, 62: 333–335, 1989.
57. O. Tarim, V.M. Anderson, F. Lifshitz. Fatal anaphylaxis in a very young infant possibly due to a partially hydrolyzed whey formula. *Arch. Pediatr. Adolesc. Med.*, 148: 1224–1228, 1994.
58. S.L. Taylor, S.L. Hefle, C. Bindslev-Jensen, S.A. Bock, A.W. Burks, L. Christie, D.J. Hill, A. Host, J.O.'B Hourihane, G. Lack, D.D. Metcalfe, D.A. Moneret-Vautrin, P.A. Vadas, F. Rance, D.J. Skrypec, T.A. Trautman, I. Malmeheden Yman, R.S. Zeiger. Factors affecting the determination of threshold doses for allergenic foods: how much is too much? *J. Allergy Clin. Immunol.*, 109: 24–30, 2002.
59. J.O.B Hourihane, S.A. Kilburn, J.A. Nordlee, S.L. Hefle, S.L. Taylor, J.O. Warner. An evaluation of the sensitivity of subjects with peanut allergy to very low doses

of peanut: a randomized, double-blind, placebo-controlled food challenge study. *J. Allergy Clin. Immunol.*, 100: 596–600, 1997.

60. S.L. Taylor, J.A. Nordlee, J.H. Rupnow. Food allergies and sensitivities. In: S.L. Taylor, R.A. Scanlan (Eds.) *Food Toxicology — a Perspective on the Relative Risks*. New York: Marcel Dekker, 1989, 255–295.
61. J.E. Gern, E. Yang, H.M. Evrard, H.A. Sampson. Allergic reactions to milk-contaminated "non-dairy" products. *N. Engl. J. Med.*, 324: 976–979, 1991.
62. C. McKenna, K.C. Klontz. Systemic allergic reaction following the ingestion of undeclared peanut flour in a peanut-sensitive woman. *Ann. Allergy Asthma Immunol.*, 79: 234–236, 1997.
63. J.W. Yunginger, M.B. Gauerke, R.T. Jones, M.J.E. Dahlberg, S.J. Ackerman. Use of radioimmunoassay to determine the nature, quantity and source of allergenic contamination of sunflower butter. *J. Food Prot.*, 46: 625–628, 1983.
64. N. Laoprasert, N.D. Wallen, R.T. Jones, S.L. Hefle, S.L. Taylor, J.W. Yunginger. Anaphylaxis in a milk-allergic child following ingestion of lemon sorbet containing trace quantities of milk. *J. Food Prot.*, 61: 1522–1524, 1998.
65. C.B. Daul, J.E. Morgan, S.B. Lehrer. Hypersensitivity reactions to crustacea and mollusks. *Clin. Rev. Allergy*, 11: 201–222, 1993.
66. T. Langeland. A clinical and immunological study of allergy to hen's egg white. VI. Occurrence of proteins cross-reacting with allergens in hen's egg white as studied in egg white from turkey, duck, goose, seagull, and in hen egg yolk, and hen and chicken sera and flesh. *Allergy*, 38: 399–412, 1983.
67. H. Bernard, C. Creminon, L. Negroni, G. Peltre, J.M. Wal. IgE cross-reactivity with caseins from different species in humans allergic to cows' milk. *Food Agric. Immunol.*, 11: 101–111, 1999.
68. J. Bernhisel-Broadbent, S.M. Scanlon, H.A. Sampson. Fish hypersensitivity. I. *In vitro* and oral challenge results in fish allergic patients. *J. Allergy Clin. Immunol.*, 89: 730–737, 1992.
69. A.M. Herian, S.L. Taylor, R.K. Bush. Identification of soybean allergens by immunoblotting with sera from soy-allergic adults. *Int. Arch. Allergy Appl. Immunol.*, 92: 193–198, 1990.
70. J. Bernhisel-Broadbent, H.A. Sampson. Cross-allergenicity in the legume botanical family in children with food hypersensitivity. *J. Allergy Clin. Immunol.*, 83: 435–440, 1989.
71. C.T. Furukawa. Nondietary management of food allergy. In: L.T. Chiaramonte, A.T. Schneider, F. Lifshitz (Eds.) *Food Allergy — A Practical Approach to Diagnosis and Management*. New York: Marcel Dekker, 1988, 365–375.
72. F.M. Atkins. The basis of immediate hypersensitivity reactions to foods. *Nutr. Rev.*, 41: 229–234, 1983.
73. N.I. Kjellman, B. Bjorksten. Natural history and prevention of food hypersensitivity. In: D.D. Metcalfe, H.A. Sampson, R.A. Simon (Eds.) *Food Allergy — Adverse Reactions to Foods and Food Additives*, 2nd edition. Boston: Blackwell Scientific, 1997, 445–459.
74. R.S. Zeiger, S. Heller, M.H. Mellon, A.B. Forsythe, R.D. O'Connor, R.N. Hamburger, M. Schatz. Effect of combined maternal and infant food-allergen avoidance on development of atopy in early infancy: a randomized study. *J. Allergy Clin. Immunol.*, 84: 72–89, 1989.
75. R.S. Zeiger, S. Heller. The development and prediction of atopy in high-risk children: follow-up at seven years in a prospective randomized study of combined

maternal and infant food allergy avoidance. *J. Allergy Clin. Immunol.*, 95: 1179–1190, 1995.

76. P.P. Van Asperen, A.S. Kemp, C.M. Mellis. Immediate food hypersensitivity reactions on the first known exposure to food. *Arch. Dis. Child.*, 58: 253–256, 1983.
77. J.W. Gerrard. Allergy in breast fed babies to ingredients in breast milk. *Ann. Allergy*, 42: 69–72, 1979.
78. P.V. Kirjavainen, E. Apostolou, S.J. Salminen, E. Isolauri. New aspects of probiotics — a novel approach in the management of food allergy. *Allergy*, 54: 909–915, 1999.
79. L. Businco, S. Dreborg, R. Einarsson, P.G. Giampietro, A. Host, R.M. Keller, S. Strobel, U. Wahn. Hydrolyzed cow's milk formulae. Allergenicity and use in treatment and prevention. An ESPACI position paper. *Pediatr. Allergy Immunol.*, 4: 101–111, 1993.
80. Y. Vandenplas, B. Hauser, C. Van den Borre, C. Clybouw, T. Mahler, S. Hachimi-Idrissi, L. Deraeve, A. Malfroot, I. Dab. The long-term effect of a partial whey hydrolyzate formula on the prophylaxis of atopic disease. *Eur. J. Pediatr.*, 154: 488–494, 1995.
81. A. Jankiewicz, W. Baltes, K. Bogl, K. Werner, L.I. Dehne, A. Jamin, A. Hoffmann, D. Haustein, S. Vieths. Influence of food processing on the immunochemical stability of celery allergens. *J. Sci. Food Agric.*, 75: 357–370, 1997.
82. J. Bernhisel-Broadbent, D. Strause, H.A. Sampson. Fish hypersensitivity. II. Clinical relevance of altered fish allergenicity caused by various preparation methods. *J. Allergy Clin. Immunol.*, 90: 622–629, 1992.
83. W.T. Kniker. Delayed and non-IgE-mediated reactions. In: M. Frieri, B. Kettelhut (Eds.) *Food Hypersensitivity and Adverse Reactions — A Practical Guide to Diagnosis and Management*. New York: Marcel Dekker, 1999, 165–217.
84. J.A. Murray. Gluten-sensitive enteropathy. In: D.D. Metcalfe, H.A. Sampson, R.A. Simon (Eds.) *Food Allergy: Adverse Reactions to Foods and Food Additives*, 3rd edition. Malden, MA: Blackwell Publishing, 2003, 242–261.
85. J.H. Skerritt, J.M. Devery, A.S. Hill. Gluten intolerance: chemistry, celiac-toxicity, and detection of prolamins in foods. *Cereal Foods World*, 35: 638–644, 1990.
86. W. Strober. Gluten-sensitive enteropathy: a nonallergic immune hypersensitivity of the gastrointestinal tract. *J. Allergy Clin. Immunol.*, 78: 202–211, 1986.
87. R.F.A. Logan. Descriptive epidemiology of celiac disease. In: D. Branksi, P. Rozen, M.F. Kagnoff (Eds.). *Gluten-Sensitive Enteropathy*, *Frontiers in Gastrointestinal Research*, Volume 19. Basel: Karger, 1992, 1–14.
88. R.F.A. Logan, E.A. Rifkind, I.D. Turner, A. Ferguson. Mortality in celiac disease. *Gastroenterology*, 97: 265–271, 1989.
89. M.F. Kagnoff. Immunobiology of coeliac disease. In: W. Domschke, R. Stoll, T.A. Brasitus, M.F. Kagnoff (Eds.) *Intestinal Mucosa and Its Disease*. London: Kluwer Academic Publishers, 1998, 313–322.
90. J.M. Duggan. Recent developments in our understanding of adult coeliac disease. *Med. J. Aust.*, 166: 312–315, 1997.
91. R. Troncone. Latent coeliac disease in Italy. *Acta. Pediatr.*, 54: 1252–1257, 1995.
92. R. Troncone, L. Greco, S. Auricchio. Gluten-sensitive enteropathy. *Pediatr. Clin. North Am.*, 43: 355–373, 1996.
93. D.D. Kasarda. The relationship of wheat protein to celiac disease. *Cereal Foods World*, 23: 240–244,262, 1978.

94. A. Fasano. Where have all the American celiacs gone? *Acta Pediatr.*, (Suppl.). 412: 20–24, 1996.
95. E.K. George, M.L. Mearin, E.A. van der Velde, R.H.J. Houwen, J. Bouquet, C.F.M. Gijsbers, J.P. Vandenbroucke. Low incidence of childhood celiac disease in the Netherlands. *Pediatr. Res.*, 37: 213–218, 1995.
96. H.J. Cornell. Coeliac disease: a review of the causative agents and their possible mechanisms of action. *Amino Acids*, 10: 1–19, 1996.
97. E.I. Hartsook. Celiac sprue: sensitivity to gliadin. *Cereal Foods World*, 29: 157–158, 1984.
98. E.K. Janatuinen, P.H. Pikkarainen, T.A. Kemppainen, V.M. Kosma, M.K. Jarvinen, M.I.J. Uusitupa, R.J.K. Julkunen. A comparison of diets with and without oats in adults with celiac disease. *N. Engl. J. Med.*, 333: 1033–1037, 1995.
99. J.H. Skerritt, A.S. Hill. Enzyme immunoassay for determination of gluten in foods: collaborative study. *J. Assoc. Off. Anal. Chem.*, 74: 257–264, 1991.
100. W.T.J.M. Hekkens, M. van Twist de Graaf. What is gluten-free — levels and tolerances in the gluten-free diet. *Nahrung*, 34: 483–487, 1990.

20 The Production of Hypoallergenic Wheat Flour for Wheat-Allergic Patients

Soichi Tanabe and Jun Watanabe

CONTENTS

20.1 INTRODUCTION

Wheat is one of the world's most important food grains, and a variety of flour- and wheat-derived products are consumed throughout the world. However, hypersensitive responses to wheat have long been a public health problem, since in addition to cow's milk, eggs, and peanuts, wheat is one of the most common foods causing allergies. The adverse reactions to wheat flour may present different clinical outcomes, including enteropathy (diarrhea), asthma, atopic dermatitis, and exercise-induced anaphylaxis.

In this chapter, we describe the following: (a) symptoms of the adverse reactions to wheat flour, (b) identification of wheat allergens, (c) production of hypoallergenic wheat flour (HWF) and products, and (d) the suppression of the allergic reaction by consumption of this hypoallergenic flour.

20.2 SYMPTOMS OF ADVERSE REACTIONS TO WHEAT FLOUR AND THEIR RESPONSIBLE DETERMINANTS

There are at least four kinds of symptoms of the adverse reactions to wheat flour as summarized in Figure 20.1. In these cases, certain wheat proteins induce abnormal allergic reactions in sensitive individuals. Thus, for better understanding, it is helpful to know the following classification of wheat proteins according their solubilities (Table 20.1). Wheat seeds are composed of protein fractions including water-soluble "albumins," salt-soluble "globulins," ethanol-soluble "gliadins," and urea-, detergent-, or KOH-soluble "glutenins."

20.2.1 Celiac Disease

Gluten-sensitive enteropathy is called "celiac disease," which is caused by ingestion of the gliadin fraction (1). The reported prevalence of this disease is between

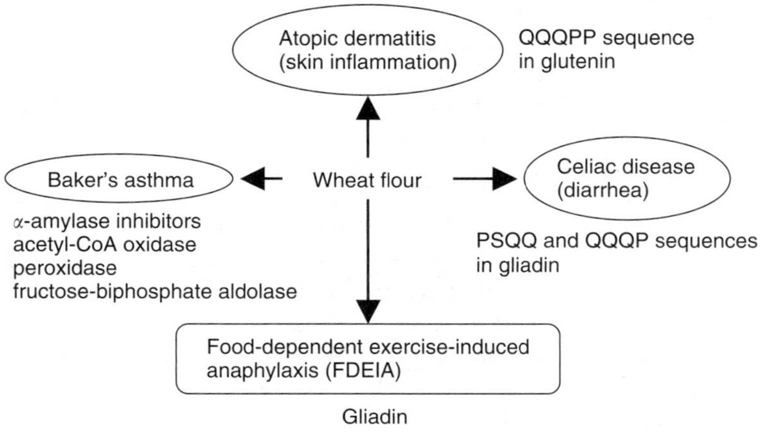

FIGURE 20.1 Wheat-sensitive disorders and their responsible determinants. Proteins or sequences responsible for the diseases.

TABLE 20.1
Weat Protein Classification.

Protein Fraction	Solubility
Albumin	Water-soluble
Globulin	Salt-soluble
Gliadin	Ethanol-soluble
Glutenin	Urea-, detergent-, or KOH-soluble

1:300 and 1:1000 in European countries (2). The amino acid sequences responsible for the disease have been characterized, and the minimum epitope structures were found to be Pro-Ser-Gln-Gln and Gln-Gln-Gln-Pro (1). Although the disease is mediated by T-lymphocyte-driven immunological activation in the gastrointestinal mucosa (2–4), the levels of total and wheat-specific immunoglobulin E (IgE) antibodies in celiac patients are usually not elevated (5–7). Therefore, in the general context of considering "allergy" synonymous with IgE-mediated hypersensitivity, celiac disease should not be classified as an allergic disorder (8). We recommend that readers who want the latest information about this disease consult publications such as *Gastroenterology* (9,10), *Gut* (11–13), and others. A hot debate about celiac disease has also appeared in *Science* (14).

20.2.2 Baker's Asthma

The inhalation of wheat flour often causes asthma (15), which is known as baker's asthma, a typical occupational allergic disease that has been known since ancient Roman times. Extensive studies have identified some proteins as allergens associated with asthma. Among them, alpha-amylase inhibitors (AIs) from the globulin fraction were identified as major allergens (15–17). The IgE-binding epitope structures of an AI (known as the 0.28 wheat AI) have already been determined (18). In addition, acyl-CoA oxidase (19), peroxidase (20), and fructose-bisphosphate aldolase (21) have been identified as other allergens.

20.2.3 Atopic Dermatitis

A third condition is skin inflammation — atopic dermatitis — that develops shortly after cereal-based products are ingested, usually resulting in eruption and itching. The wheat allergens associated with atopic dermatitis are so heterogeneous that many attempts have been made internationally to identify the allergens. However, little information was available on the molecular structures of the major allergens involved in atopic dermatitis when some groups, including ours, started to carry out systematic experiments (22–25). Later, several groups succeeded in identifying the wheat allergens (26–29). The proteins responsible for wheat-induced atopic dermatitis are described in Section 20.3.

20.2.4 Food-Dependent Exercise-Induced Anaphylaxis

Another allergic condition is food-dependent exercise-induced anaphylaxis (FDEIA) (30). FDEIA is a severe (life-threatening) form of allergy, in which the ingestion of a specific food before physical exercise induces symptoms of anaphylaxis. Although various foods (including shellfish, celery, hazelnuts, peanuts, soy, peas, and bananas) have been associated with FDEIA, the most frequently reported cause of these reactions seems to be wheat (31–33). Wheat-dependent, exercise-induced anaphylaxis is not as rare as previously thought but is a rather poorly recognized disorder. Although the mechanisms by which exercise provokes the reactions in food- or wheat-dependent, exercise-induced anaphylaxis remain unclear, the gliadin fraction is thought to be an allergen in FDEIA (31).

In the following sections, we will focus on wheat allergens that cause atopic dermatitis.

20.3 IDENTIFICATION OF WHEAT ALLERGENS AND THEIR EPITOPES

We evaluated the allergenicity of salt-soluble and salt-insoluble (gluten) fractions by means of enzyme-linked immunosorbent assay (ELISA) using serum from atopic patients allergic to wheat. As a result, most patients were found to be sensitive to gluten (34). Thus, we first tried to determine a major epitope structure in gluten responsible for atopic dermatitis.

20.3.1 Gluten Fraction

20.3.1.1 An Allergenic Peptide from Low-Molecular-Weight Glutenin

Gluten is insoluble in aqueous media. It was thus hydrolyzed with α-chymotrypsin to obtain soluble peptide fragments with allergenicity, since food allergens are often characterized by their high stability to digestive enzymes, with their epitope structures remaining unchanged (35). The resulting hydrolytic reaction product was centrifuged, and the supernatant was submitted to gel filtration and reversed-phase high-performance liquid chromatography (HPLC). The allergenicity of the fractionated eluate was evaluated by ELISA, and the primary structure of the peak with the highest allergenicity was determined.

The primary structure of the purified compound was a 30-mer peptide with the sequence (Ser-Gln-Gln-Gln-[Gln-]Pro-Pro-Phe)4 (Table 20.2A). This allergenic peptide shows high sequence similarities (almost 90%) to low-molecular-weight glutenin precursors (36,37). Therefore, we concluded that the peptide originated from low-molecular-weight glutenin. Similarities of about 70% were also obtained between the sequence of the allergenic peptide and those of low-molecular-weight glutenin precursors from durum wheat (38,39). Surprisingly, a high degree of similarity (53.6%) was also found between the allergenic peptide and a *Saccharomyces cerevisiae* protein (40). Thus, it should be noted that wheat allergic patients are also suspected to be sensitive to yeast using in bread making.

TABLE 20.2
IgE-Binding Abilities of Wheat Peptides

Peptide	Relative ELISA Value
(A)	
Ac-SQQQQPPF SQQQPPF SQQQQPPF SQQQPPF	1.0
Ac-SQQQQPPF SQQQPPF	1.1
Ac-SQQQQPPF	1.1
Ac-SQQQPPF	1.0
(B)	
Ac-**G**QQQPPF	1.1
Ac-S**G**QQPPF	nd*
Ac-SQ**G**QPPF	0.8
Ac-SQQ**G**PPF	1.0
Ac-SQQQ**G**PF	nd
Ac-SQQQP**G**F	nd
Ac-SQQQPP**G**	0.9
(C)	
Ac-QQQPP	0.9
Ac-**G**QQPP	nd
Ac-Q**G**QPP	0.7
Ac-QQ**G**PP	1.0
Ac-QQQ**G**P	nd
Ac-QQQP**G**	nd
QQQPP	0.6

*Not detectable

Source: S. Tanabe, S. Arai, Y. Yanagihara, H. Mita, K. Takahashi, M. Watanabe. A major wheat allergen has a Gln-Gln-Gln-Pro-Pro motif identified as an IgE-binding epitope. *Biochem. Biophys. Res. Commn.*, 219: 290–293, 1996.

Repeat sequences in allergenic peptides such as (Ser-Gln-Gln-Gln-[Gln-]Pro-Pro-Phe)$_4$, may be favorable for cross-linking IgE antibodies and triggering the release of chemical mediators from mast cells in the body. Indeed, there are 44 and 25 Gln-Gln-Gln-Pro-Pro sequences in high-molecular-weight glutenin subunits x and y, respectively (accession numbers CAC40686 and CAC40687). Like Gln-Gln-Gln-Pro-Pro repeat sequences, cod allergen (Gad c1, allergen M) is well known for containing three homologous IgE-binding tetrapeptides (41).

20.3.1.2 Gln-Gln-Gln-Pro-Pro as an IgE-binding Epitope

Next, the peptides listed in Table 20.2 were synthesized according to the solid-phase method. The N-terminal amino acid of each peptide was acetylated to mimic the conditions under which it exists in the intact form, and the allergenicity of each peptide was evaluated by ELISA (25). As shown in Table 20.2A, (Ser-Gln-Gln-Gln-[Gln-]Pro-Pro-Phe)$_4$, (Ser-Gln-Gln-Gln-[Gln-]Pro-Pro-Phe)$_2$, and

Ser-Gln-Gln-Gln-[Gln-]Pro-Pro-Phe bound to IgE almost equally. There was no difference between the relative ELISA values of Ser-$(Gln)_4$-Pro-Pro-Phe and Ser-$(Gln)_3$-Pro-Pro-Phe. These data suggest that the Ser-Gln-Gln-Gln-Pro-Pro-Phe motif is involved in binding to IgE antibodies.

To examine which amino acid residues in the motif are essential for binding to IgE, we attempted to replace each constituent amino acid residue in turn with Gly. When any of the asterisked amino acid residues in the sequence Ser-Gln*-Gln-Gln-Pro*-Pro*-Phe was replaced by Gly, the ELISA value dropped below the limit of detection (see Table 20.2B). These amino acid residues are therefore thought to be indispensable for binding to IgE. Tables 20.2B and 20.2C also show that Gln-Gln-Gln-Pro-Pro, which lacks the N- and C-terminals of Ser-Gln-Gln-Gln-Pro-Pro-Phe, gave an ELISA value equal to that obtained with the full peptide.

We further examined which amino acid residues in Gln-Gln-Gln-Pro-Pro are essential for binding to IgE (see Table 20.2C) and found that the N-terminal glutamine residue and the two proline residues are essential. It was thus concluded that the IgE-binding epitope of the allergenic peptide comprises Gln-Xaa-Xaa-Pro-Pro, where Xaa are replaceable amino acid residues. Indeed, the inhibition ELISA assay showed that Ac-Gln-Gln-Gln-Pro-Pro binds to wheat-specific IgE in the serum of patients. To analyze the binding between Ac-Gln-Gln-Gln-Pro-Pro and IgE antibody, Fukushi et al. (42) carried out the first nuclear magnetic resonance (NMR) analysis of Ac-Gln-Gln-Gln-Pro-Pro. Their data showed that the configurations of the amide bonds of the peptide backbone were all-*trans*.

As in Table 20.2C, the ELISA value obtained with Gln-Gly-Gln-Pro-Pro was lower by almost 30% than that obtained with Gln-Gln-Gln-Pro-Pro, and the value with nonacetylated Gln-Gln-Gln-Pro-Pro was almost half of that with acetylated Gln-Gln-Gln-Pro-Pro. From these data, the second glutamine residue in Gln-Gln-Gln-Pro-Pro and acetylation of the N-terminal amino group are both advantageous for binding to IgE.

Recombinant low-molecular-weight glutenins with many Gln-Gln-Gln-Pro-Pro motifs were expressed in *Escherichia coli* by a pET vector system. They were further confirmed to possess IgE-binding ability (43).

20.3.2 Albumin/Globulin Fractions, Especially Glycoproteins

Wheat alpha-amylase inhibitors (AIs) have been studied as wheat allergens for more than a quarter of a century (44) and are now recognized as a contributing allergen for both asthma and urticaria (15–17,44). Wheat AIs comprise a family of a number of differing monomeric, dimeric, and tetrameric proteins. As described above, the IgE-binding epitope structures of AI 0.28 have already been determined in the amino acid sequence (18). Moreover, a glycan moeity in one subunit of a tetrameric AI (CM-16) was also found to be allergenic in the case of baker's asthma (16). Although, other AIs, such as AI 0.19, AI 0.28, and AI 0.53, have also been reported as allergens, the involvement of glycans in allergic responses has not been fully proven.

In addition, Asn-linked glycochains have received recent attention in studies on cross-reactivity among pollen, insects, and food allergens (45,46). Garcia-Casado et al. (46) reported that the presence of a β 1-2 xylosyl residue attached to the β-linked mannose of the glycochain core in bromelain and peroxidase constitutes an IgE-reactive determinant. Indeed, we previously reported that patients sensitive to the salt-soluble fraction of wheat flour cross-react with bromelain (47). We, thus, tried to examine allergenic glycoproteins in wheat flour (Figure 20.2) and to clarify the existence of additional glycoproteins (48).

Wheat flour was extracted with 10 mM sodium dihydrogen phosphate, and ammonium sulfate was added to the extract to 50% saturation at pH 7. The precipitate was dialyzed against running water and then dissolved in 10 mM acetate buffer (pH 4.5) containing 0.5 M NaCl. The solution was submitted to CM (carboxymethyl)-cellulose and DEAE (diethylaminoethyl)-cellulose column chromatographies, and the IgE-binding crude fraction thus obtained was lyophilized. SDS-PAGE (sodium dodecyl sulfate–polyacrylamide gel electrophoresis) was carried out in a 7.5% gel, and the proteins in the gel were electrotransferred onto a PVDF (polyvinylidene difluoride) membrane. The same procedure was performed three times. One of the membranes was immunoassayed using serum from wheat-sensitive allergic patients (see Figure 20.2, lane A). Another membrane was

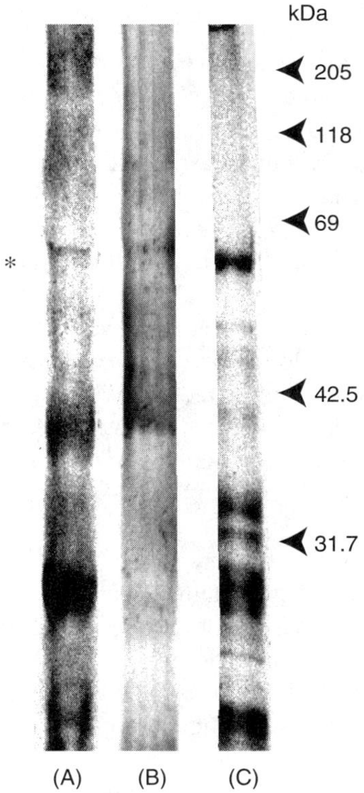

FIGURE 20.2 Western blot analysis of the 60-kDa glycoprotein. (A) Detected using patient sera. (B) Detected using the anti-peroxidase antibody. (C) Stained with Coomassie Brilliant Blue.

submitted to immunoassay with rabbit anti-HRP (horseradish peroxidase) (Sigma, P-7899) as a primary antibody that recognizes peroxidase type N-linked glycochains (45) (Figure 20.2, lane B). The third membrane was stained with CBB (Coomassie Brilliant Blue) R-250 (Figure 20.2, lane C). One unknown allergenic protein in all three lanes was detected at about 60 kDa (Figure 20.2, asterisked band). The protein reacted with the anti-HRP antibody (Figure 20.2, lane B), indicating that it possesses N-linked glycochain(s). Bands at about 40 kDa and 16 kDa are probably peroxidase (20) and AIs (20,44), respectively.

The N-terminal amino acid sequence of the asterisked band was analyzed, and the amino acid sequence similarity to other naturally occurring proteins was checked using a sequence database. No similarity was found between the sequence of the 60 kDa glycoprotein and any other proteins, including wheat allergens. Thus, the glycoprotein is confirmed to be newly found.

The involvement of glycochains in allergic symptoms remains unclear, however, it is certain that IgE antibodies against glycan moieties do exist. For example, Sánchez-Monge et al. (16) reported that glycochains are favorable for recognition by IgE antibodies.

20.3.3 Mannoglucan

In the meantime, it remains unclear whether a nonproteinaceous constituent in wheat also acts as an allergen. Unlike proteinaceous allergens, some nonproteinaceous substances would be more stable in the body, possibly acting as a remaining allergen to cause a longer-lasting allergic reaction. Thus, the existence of such nonproteinaceous allergens would explain why wheat allergy is difficult to treat. Next, we attempted to isolate a polysaccharide allergen from a water-soluble fraction of wheat flour and to clarify its chemical structure and immunological properties (49).

The water-soluble fraction of wheat flour was first subjected to DEAE-cellulose column chromatography to remove the proteinaceous substances. The unretained fraction was then subjected to ConA-agarose affinity column chromatography and gel filtration HPLC to obtain the fraction with allergenicity. The IgE-binding ability of wheat mannoglucan was confirmed by inhibition ELISA. The mean molecular weight was estimated to be approximately 50,000 Da.

ConA is a specific adsorbent with an affinity for mannose- and glucose-containing polysaccharides and glycoproteins. To clarify whether the allergenic compound consists of polysaccharide or glycoprotein, it was examined by IR spectrometry. The IR (infrared absorption) spectrum of the allergenic compound suggested the presence of –OH group, with no apparent characteristic absorption for amide groups. Therefore, the compound appears to consist mainly of polysaccharide. The polysaccharide was hydrolyzed, and the sugar composition of the hydrolyzate was analyzed by HPLC. The results revealed that the polysaccharide consists of glucose and mannose in a molar ratio of 4.4:1, with no other common sugars, such as xylose, galactose, fucose, *N*-acetyl glucosamine, or *N*-acetyl galactosamine, being detected. The polysaccharide allergen was converted to oligosaccharides by hydrolysis with cellulase, suggesting that the polysaccharide contains β-1,4-glycosidic linkages.

Judging from our detailed analysis, the polysaccharide is a novel allergen with linear β-1,4 linkages composed of glucose and mannose. While some studies have shown the presence of arabinoxylan and arabinogalactan in water extracts of wheat flour, our report (49) clearly demonstrated for the first time the occurrence of mannoglucan in wheat flour.

While orally administered mannoglucan allergen would be excreted because of its indigestible nature, it could be absorbed through the inhalation of wheat flour. In this case, it would not be degraded and would remain longer in the body as an allergen. This is the probable reason that patients sensitive to the water-soluble fraction of wheat flour are found to possess mannoglucan-specific IgE antibodies.

20.4 PRODUCTION OF HWF AND HYPOALLERGENIC PRODUCTS

20.4.1 Enzymatic Treatment of Wheat Flour to Produce Hypoallergenic Flour (HWF)

Based on the above described epitope structure — Gln-Xaa-Xaa-Pro-Pro — it was presumed that wheat flour would be made hypoallergenic by hydrolyzing the peptide bonds near the essential proline residues of the epitope. On the basis of this idea, we screened for a food-usable enzyme with high activity to hydrolyze peptide bonds near the proline residues. The enzyme chosen should also be characterized by its low amylase activity, which would be expected to minimize the development of sweetness, deterioration of starch functionality, and volume loss of the product. In addition, amylase inhibitors cannot be used because of their allergenic potencies (15–17,44). As a result of enzyme screening, bromelain and actinase showed high ability to hydrolyze peptide bonds near the proline residues with low amylase activities (50–53). However, some patients were also sensitive to bromelain itself (47). Thus, actinase was selected to produce HWF.

Based on our preliminary data, actinase-treated flour retained a small degree of allergenicity, indicating the contribution of glycoproteins (see Section 20.2.2) and mannoglucan (see Section 20.2.3). To decompose these carbohydrate allergens, we selected an enzyme with low amylase activity and the ability to reduce the remaining allergenicity. Cellulase was the best enzyme tested for this purpose and was used in the first step of the enzymatic process to avoid its inactivation by actinase. The reaction temperature was set at 50°C, at which no gelatinization of starches occurred. Favorably, the numbers of microorganisms decreased in this step. A reaction time of 1 hour and an enzyme/flour ratio of 0.5% were sufficient to decompose the carbohydrate allergens. The reaction product was cooled to 40°C, and then subjected to actinase treatment.

Actinase was used in the second step of the process to decompose proteinaceous allergens. Actinase was dissolved in water, and the solution was mixed with

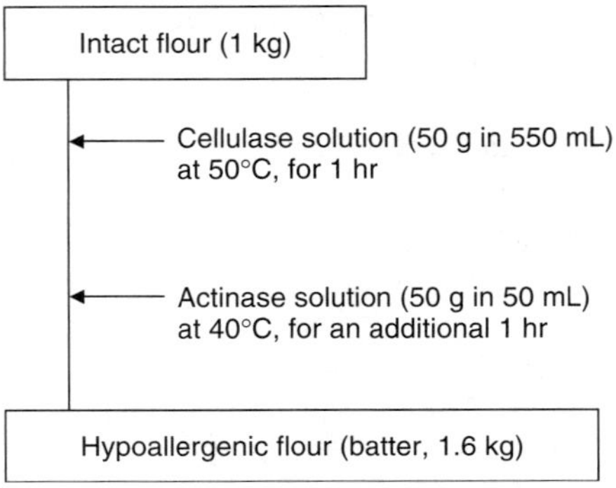

FIGURE 20.3 Method for the production of hypoallergenic wheat flour.

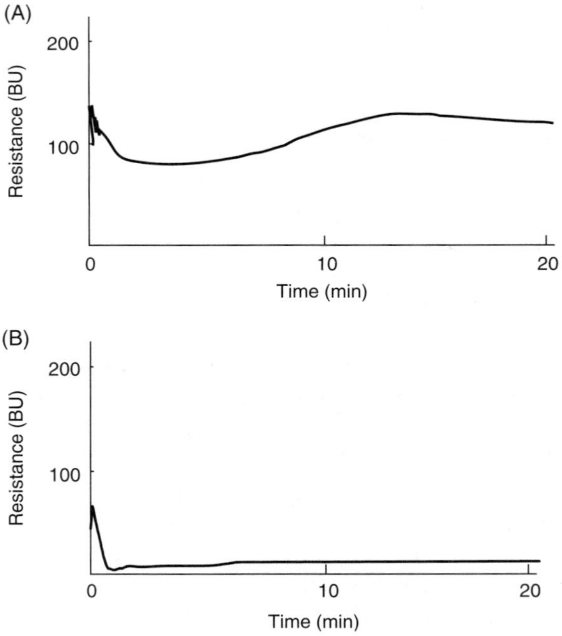

FIGURE 20.4 Resistograms of intact flour (A) and the hypoallergenic wheat flour (B). Hypoallergenic wheat flour (86g on a dry matter basis) was mixed with water (74ml) using a resistograph (Brabender) at 63 rpm and 30°C for 20 min to obtain a mixing time-resistance curve (B). A mixure of intact flour (100g) and water (60ml) was also resistographed in the same manner (A).

the cellulase-reaction mixture. The subsequent reaction was carried out at 40°C for 1 hour, yielding a product that was hypoallergenic in most cases. These processes are summarized in Figure 20.3.

20.4.2 Hypoallergenic Wheat Products

According to the resistographic data of the product (Figure 20.4B), soft flour is changed to the batter state by the proteolysis of gluten, and this characteristic limits its use as a material for food processing. To solve this problem, we gelatinized the starch present in the hypoallergenic batter prior to processing. The viscosity of the batter was increased by heating, and the resulting product was more easily handled. The partially and exhaustively gelatinized batter was suitable for making cupcakes, pizza, cookies, wafers (Figure 20.5), pasta-like noodles, and puffed items.

Clinical evaluation is being carried out to confirm the effectiveness of HWF, using cupcakes as models, and our preliminary results will be described in the next section.

20.4.3 Clinical Evaluation of HWF

We tried to clarify the safety and usefulness of hypoallergenic cupcakes in 15 patients (male/female: 8/7, 1.6 ± 0.9 years old, mean ± SD) with atopic

FIGURE 20.5 Hypoallergenic wheat products.

dermatitis and wheat allergy. All patients had a history of experiencing severe urticaria whenever they ingested cereal-based products. Among them, 13 patients showed no positive response after consuming hypoallergenic cupcakes. The results of the provocation of hypoallergenic cupcakes showed a positive immediate reaction (severe urticaria) in only two patients. Therefore, hypoallergenic cupcakes are safe for most patients with wheat allergy (54). Moreover, by taking the hypoallergenic cupcakes over a long period (more than 6 months), some patients were hyposensitized and developed the ability to eat normal wheat products (unpublished data). This suggests that HWF can act as an antiallergenic via allergen-specific immunotolerance. Thus, we next investigated the preventive and suppressive effects of HWF using an animal model.

20.5 PREVENTIVE AND SUPPRESSIVE EFFECTS OF HWF ON FOOD ALLERGY

It has been considered that food allergy is induced by the following mechanisms: absorption of an allergen from the intestinal tract → recognition of the allergen by T cells → production of IgE against the allergen by B cells → binding of the produced IgE to IgE receptors on mast cells → binding of the allergen to the IgE on mast cells and degranulation of the mast cells → allergic symptoms (Figure 20.6). The target for the allergy suppressive effect of HWF is outlined below.

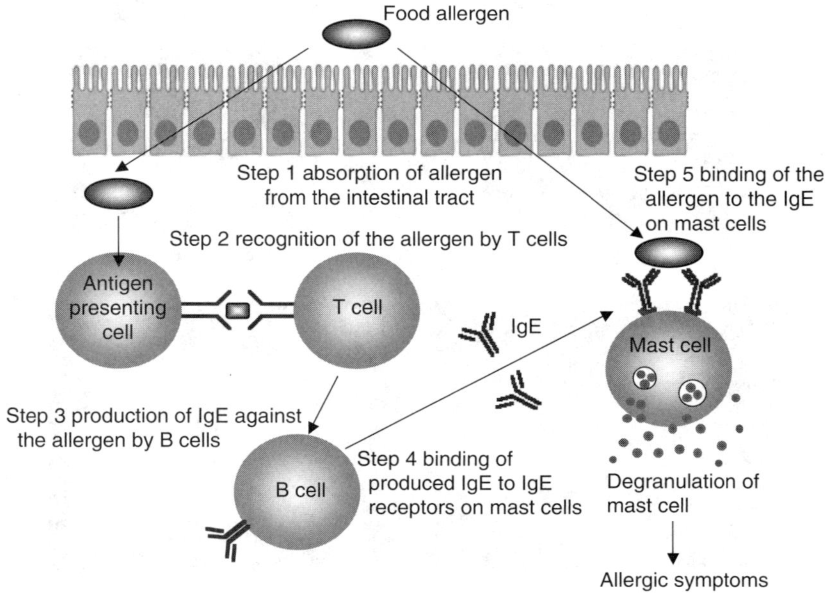

FIGURE 20.6 Mechanisms of food allergy.

20.5.1 Inhibition of Allergen Absorption from the Intestinal Tract by HWF

This target corresponds to step 1 as shown in Figure 20.6. The intestinal epithelium theoretically acts as a barrier restricting the permeation of macromolecules. However, Walker et al. (56) reported that even in a healthy person, a small amount of protein permeates the barrier and spreads in the circulation throughout the body. Moreover, it has been reported that antigen permeability through the intestinal tract is enhanced in allergic patients (57,58). In light of these facts, it is conceivable that enhancing the barrier function of the intestinal epithelium would help prevent food allergy. We established an *in vitro* assay system using a human colon carcinoma cell line, Caco-2, as a model for the intestinal tract and ovalbumin (OVA) as a model of an allergen (55,59). This model system was used to determine whether HWF inhibits OVA absorption. It was found that HWF possesses potent inhibitory activity against OVA absorption and that the active constituent derived from the cellulase preparation used to produce the flour. An active compound was isolated and identified as Trp-Ser-Asn-Ser-Gly-Asn-Phe-Val-Gly-Gly-Lys. According to a structure–activity relationship study, we also clarified that Trp residues without a free carboxyl group are required for the inhibitory activity.

20.5.2 Induction of Oral Tolerance by HWF

Induction of tolerance corresponds to steps 2 and 3 as shown in Figure 20.6. We think that this effect is the most important antiallergy effect of HWF.

Hyposensitization therapy consists of the repeated subcutaneous injection of an allergen into a specifically sensitized individual to induce the alleviation of allergic symptoms and has been applied to treat bronchial asthma (61) and bee venom allergy (62). Although hyposensitization therapy with whole bee venom is mostly effective, allergic side effects occur in up to 40% of patients (63). On the other hand, it has been reported that T-cell-determinant peptides are as effective as the whole allergen in experimental animals (64,65) and in clinical tests (66,67) (peptide immune therapy). Moreover Toda et al. (68) reported that T-cell-determinant peptides with a single amino acid substitution specifically suppress T-cell response.

HWF was intended to retain large amounts of peptide and thereby possess preventive and curative effects against wheat allergy. Peripheral blood mononuclear cells of wheat-allergic patients showed a proliferative response under stimulation with the water-soluble fraction of HWF, suggesting the existence of T-cell-determinant peptides (unpublished data). Moreover, peptides in HWF are absorbed from the intestinal tract in an *in vitro* model and in animal testing (69). We used test animals to determine whether HWF possesses preventive and curative effects against wheat allergy and whether its antiallergic effect arises from an induction of immune tolerance.

Despite the fact that gluten is a major wheat allergen, as mentioned above, few models in which to study the gluten-specific allergic reaction have been reported. A gluten-specific allergic inflammation model was first established to determine the allergy suppressive effect of HWF. Brown Norway rats are known as a high-IgE responder strain (70,71) and are used as a model of bronchial asthma because sensitized rats have been reported to show airway hyperresponsiveness and eosinophilic inflammation after antigen inhalation (72,73). For the allergen challenge, it was necessary to make the gluten soluble for its inhalation. Since we previously demonstrated that gluten solubilized by chymotryptic hydrolysis retains IgE-binding activity (24), a chymotryptic hydrolyzate of gluten was used as a "solubilized antigen" alternative to intact gluten. Significant eosinophilic infiltration in the bronchoalveolar lavage fluid (BALF) was observed when gluten-immunized rats inhaled solubilized gluten, suggesting the induction of gluten-specific airway inflammation.

The allergy suppressive effect of HWF was determined using the gluten-specific airway inflammation model. Rats were fed a diet containing HWF between immunization and inhalation of allergen. Rats in the control group were fed a control diet containing a mixture of amino acids (with an amino-acid composition the same as that of HWF) and wheat starch. The control diet corresponded to a diet without wheat flour. HWF inhibited the decrease in body weight after challenge, which may reflect the severity of allergic inflammation. More clearly, the results demonstrated that the consumption of HWF inhibited the increase in the numbers of eosinophils, lymphocytes, and neutrophils in BALF after allergen challenge (Figure 20.7). These results suggest that HWF prevents gluten-induced allergic inflammatory reactions. Wheat specific-IgG_1, IgG_{2a}, and IgE levels in the serum were significantly lower in rats fed HWF-containing diet than in those fed the control diet, suggesting an induction of oral tolerance by consumption of HWF. It is especially noteworthy that IgE levels, which contribute significantly to food allergy, decreased markedly. These results strongly suggest that HWF

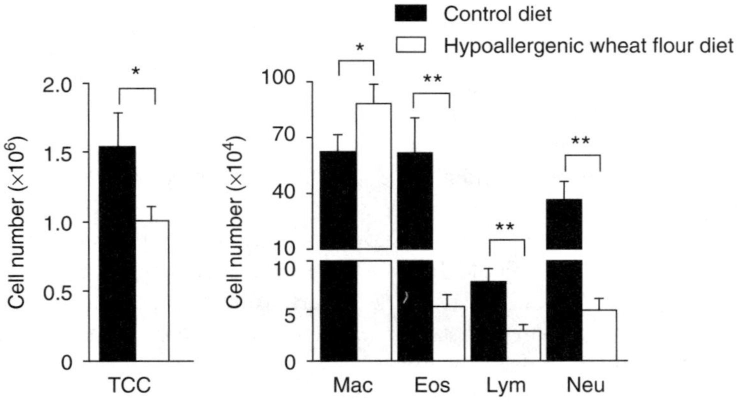

FIGURE 20.7 Comparsion of total cell count and cell profile in BALF between rats fed a control diet and those fed the hypoallergenic flour diet after allergen inhalation. Values are given as means ± SEM, $n = 9$. *, $p < 0.05$; **, $p < 0.01$. TCC, total cell count; Mac, alveolar macrophages; Eos, eosinophils; Lym, lymphocytes; Neu, Neutrophils.

actively suppresses the allergic reaction, probably by inducing oral tolerance. It is conceivable that the remission of wheat allergy in the clinical test was not due to spontaneous remission following allergen avoidance but to an "active" suppressive effect of HWF (see Section 20.3.3). A more sophisticated study is currently under way to clarify the precise mechanisms of the allergy-suppressive effect of HWF.

20.5.3 Haptenic Activity of HWF

Hepatic activity of HWF corresponds to step 5 as shown in Figure 20.6. It has been reported that monovalent hapten peptides can bind to IgE without inducing degranulation of mast cells or basophils (25). Such a monovalent hapten peptide in HWF is expected to inhibit degranulation by blocking the interaction between IgE molecules and the antigen. Indeed, we have reported that enzymatically treated gluten can inhibit histamine release from basophils in wheat allergic patients (74). A detailed analysis using an animal model is under way to clarify the haptenic activity of HWF.

REFERENCES

1. R. Sturgess, P. Day, H.J. Ellis, H.A. Gjertsen, M. Kontakou, P.J. Ciclitira. Wheat peptide challenge in coeliac disease. *Lancet*, 343: 758–761, 1994.
2. R. Troncone, N. Caputo, A. Zibella, G. Molitierno, L. Maiuri, S. Auricchio. Coeliac disease: a common food intolerance on an immunological basis. In: S. Auricchio, J.K. Visakorpi (Eds.). *Common Food Intolerances 1: Epidemiology of Coeliac Disease*. Basel: Karger, 1992, 1–11.
3. J.S. Trier. Celiac sprue. *N. Engl. J. Med.*, 325: 1709–1719, 1991.
4. M.N. Marsh. Gluten, major histocompatibility complex, and the small intestine. *Gastroenterology*, 102: 330–354, 1992.
5. C. Mietens, S.G.O. Johansson, H. Bennich. Serum concentration of immunoglobulins with special concentration of IgE in celiac disease. *Klin. Wochenschr.*, 49: 256–260, 1971.
6. H.J. Hodgson, R.J. Davies, A.E. Gent. Atopic disorders and adult coeliac disease. *Lancet*, 1(7951): 115–117, 1976.
7. S.L. Bahna, K. Tateno, D.C. Heiner. Elevated IgD antibodies to wheat in celiac disease. *Ann. Allergy*, 44: 146–151, 1980.
8. S.L. Bahna. Celiac disease: A food allergy? Contra! In: B. Wuthrich, C. Ortolani (Eds.). *Highlights in Food Allergy*. Basel: Karger, 1996, 211–215.
9. H. Arentz–Hansen, S.N. McAdam, Ø. Molberg, B. Fleckenstein, K.E.A. Lundin, J.D. Thomas, T.J.D. Jørgensen, G. Jung, P. Roepstorff, L.M. Sollid. Celiac lesion T cells recognize epitopes that cluster in regions of gliadins rich in proline residues. *Gastroenterology*, 123: 803–809, 2002.
10. P.H.R. Green, M. Barry, M. Matsutani. Serologic tests for celiac disease. *Gastroenterology*, 124: 585–586, 2003.
11. V.M. Salvati, T.T. MacDonald, M. Bajaj-Elliott, M. Borrelli, A. Staiano, S. Auricchio, R. Troncone, G. Monteleone. Interleukin 18 and associated markers of T helper cell type 1 activity in coeliac disease. *Gut*, 50: 186–190, 2002.

12. C. Esposito, F. Paparo, I. Caputo, M. Rossi, M. Maglio, D. Sblattero, T. Not, R. Porta, S. Auricchio, R. Marzari, R. Troncone. Anti-tissue transglutaminase antibodies from coeliac patients inhibit transglutaminase activity both *in vitro* and *in situ*. *Gut*, 51: 177–181, 2002.
13. G. Mazzarella, M. Maglio, F. Paparo, G. Nardone, R. Stefanile, L. Greco, Y. van de Wal, Y. Kooy, F. Koning, S. Auricchio, R. Troncone. An immunodominant DQ8 restricted gliadin peptide activates small intestinal immune response in *in vitro* cultured mucosa from HLA-DQ8 positive but not HLA-DQ8 negative coeliac patients. *Gut*, 52: 57–62, 2003.
14. F. Koning, W. Vader. Gluten peptides and celiac disease. *Science*, 299: 513–514, 2003.
15. L. Gómez, E. Martin, D. Hernandez, R. Sánchez-Monge, D. Barber, V. del Pozo, B. De Andres, A. Armentia, C. Lahoz, G. Salcedo, P. Palomino. Members of the alpha-amylase inhibitors family from wheat endosperm are major allergens associated with baker's asthma. *FEBS Lett.*, 261: 85–88, 1990.
16. R. Sánchez-Monge, L. Gómez, D. Barber, C. Lopez-Otin, A. Armentia, G. Salcedo. Wheat and barley allergens associated with baker's asthma. Glycosylated subunits of the alpha-amylase-inhibitor family have enhanced IgE-binding capacity. *Biochem. J.*, 281: 401–405, 1992.
17. M. Amano, H. Ogawa, K. Kojima, T. Kamidaira, S. Suetsugu, M. Yoshihama, T. Satoh, T. Samejima, I. Matsumoto. Identification of the major allergens in wheat flour responsible for baker's asthma. *Biochem. J.*, 330: 1229–1234, 1998.
18. B.J. Walsh, M.E. Howden. A method for the detection of IgE binding sequences of allergens based on a modification of epitope mapping. *J. Immunol. Meth.*, 121: 275–280, 1989.
19. W. Weiss, G. Huber, K.H. Engel, A. Pethran, M.J. Dunn, A.A. Gooley, A. Gorg. Identification and characterization of wheat grain albumin/globulin allergens. *Electrophoresis*, 18: 826–833, 1997.
20. R. Sánchez-Monge, G. Garcia-Casado, C. Lopez-Otin, A. Armentia, G. Salcedo. Wheat flour peroxidase is a prominent allergen associated with baker's asthma. *Clin. Exp. Allergy*, 27: 1130–1137, 1997.
21. A. Posch, W. Weiss, C. Wheeler, M.J. Dunn, A. Gorg. Sequence analysis of wheat grain allergens separated by two-dimensional electrophoresis with immobilized pH gradients. *Electrophoresis*, 16: 1115–1119, 1995.
22. E. Varjonen, J. Savolainen, L. Mattila, K. Kalimo. IgE-binding components of wheat, rye, barley and oats recognized by immunoblotting analysis with sera from adult atopic dermatitis patients. *Clin. Exp. Allergy*, 24: 481–489, 1994.
23. E. Varjonen, E. Vainio, K. Kalimo, K. Juntunen-Backman, J. Savolainen. Skin-prick test and RAST responses to cereals in children with atopic dermatitis. Characterization of IgE-binding components in wheat and oats by an immunoblotting method. *Clin. Exp. Allergy*, 25: 1100–1107, 1995.
24. M. Watanabe, S. Tanabe, T. Suzuki, Z. Ikezawa, S. Arai. Primary structure of an allergenic peptide occurring in the chymotryptic hydrolyzate of gluten. *Biosci. Biotechnol. Biochem.*, 59: 1596–1597, 1995.
25. S. Tanabe, S. Arai, Y. Yanagihara, H. Mita, K. Takahashi, M. Watanabe. A major wheat allergen has a Gln-Gln-Gln-Pro-Pro motif identified as an IgE-binding epitope. *Biochem. Biophys. Res. Commn.*, 219: 290–293, 1996.
26. C.P. Sandiford, A.S. Tatham, R. Fido, J.A. Welch, M.G. Jones, R.D. Tee, P.R. Shewry, A.J. Newman Taylor. Identification of the major water/salt insoluble wheat proteins involved in cereal hypersensitivity. *Clin. Exp. Allergy*, 27: 1120–1129, 1997.

27. M. Kusaba-Nakayama, M. Iwamoto, R. Shibata, M. Sato, K Imaizumi. CM3, one of the wheat alpha-amylase inhibitor subunits, and binding of IgE in sera from Japanese with atopic dermatitis related to wheat. *Food Chem. Toxicol.*, 38: 179–185, 2000.
28. I. Sander, A. Flagge, R. Merget, T.M. Halder, H.E. Meyer, X. Baur. Identification of wheat flour allergens by means of two-dimensional immunoblots. *J. Allergy Clin. Immunol.*, 107: 907–913, 2001.
29. T. Takizawa, H. Arakawa, K. Tokuyama, A Morikawa. Identification of allergen fractions of wheat flour responsible for anaphylactic reactions to wheat products in infants and young Children. *Int. Arch. Allergy Immunol.*, 125: 51–56, 2001.
30. A.L. Sheffer, N.A. Soter, E.R. McFadden, K.F. Austen. Exercise-induced anaphylaxis: a distinct form of physical allergy. *J. Allergy Clin. Immunol.*, 71: 311–316, 1983.
31. K. Palosuo, H. Alenius, E. Varjonen, M. Koivuluhta, J. Mikkola, H. Keskinen, N. Kalkkinen, T. Reunala. A novel wheat gliadin as a cause of exercise-induced anaphylaxis. *J. Allergy Clin. Immunol.*, 103: 912–917, 1999.
32. S. Harada, T. Horikawa, M. Icihashi. A study of food-dependent exercise-induced anaphylaxis by analyzing the Japanese cases reported in the literature. (in Japanese) *Jpn. J. Allergol.*, 49: 1066–1073, 2000.
33. M. Dohi, M. Suko, H. Sugiyama, N. Yamashita, K. Tadokoro, F. Juji, H. Okudaira, Y. Sano, K. Ito, T. Miyamoto. Food-dependent, exercise-induced anaphylaxis: a study on 11 Japanese cases. *J. Allergy Clin. Immunol.*, 87: 34–40, 1991.
34. S. Tanabe, M. Watanabe. Production of hypoallergenic wheat flour. *Food Sci. Technol. Res.*, 5: 317–322, 1999.
35. S.L. Taylor, R.F. Lemanske Jr., R.K. Bush, W.W. Busse. Food allergens: structure and immunologic properties. *Ann. Allergy*, 59: 93–99, 1987.
36. E.G. Pitts, J.A. Rafalski, C. Hedgcoth. Nucleotide sequence and encoded amino acid sequence of a genomic gene region for a low molecular weight glutenin. *Nucl. Acids Res.*, 16: 11376, 1988.
37. V. Colot, D. Bartles, R. Thompson, R. Flavell. Molecular characterization of an active wheat LMW glutenin gene and its relation to other wheat and barley prolamin genes. *Mol. Gen. Genet.*, 216: 81–90, 1989.
38. B.G. Cassidy, J. Dvorak. EMBL Data Library, S08683, 1990
39. R. D'Ovidio, O.A. Tanzarella, E. Porceddu. Nucleotide sequence of a low-molecular-weight glutenin from *Triticum durum*. *Plant Mol. Biol.*, 18: 781–784, 1992.
40. S.W. Rasmussen. Sequence of a 28.6 kb region of yeast chromosome XI includes the FBA1 and TOA2 genes, an open reading frame (ORF) similar to a translationally controlled tumour protein, one ORF containing motifs also found in plant storage proteins and 13 ORFs with weak or no homology to known proteins. *Yeast*, 10: S63–S68, 1994.
41. S. Elsayed, S. Sornes, J. Apold, H. Vik, E. Florvaag. The immunological reactivity of the three homologous repetitive tetrapeptides in the region 41–64 of allergen M from cod. *Scand. J. Immunol.*, 16: 77–82, 1982.
42. E. Fukushi, S. Tanabe, M. Watanabe, J. Kawabata. NMR analysis of a model pentapeptide, acetyl-Gln-Gln-Gln-Pro-Pro, as an epitope of wheat allergen. *Magn. Reson. Chem.*, 36: 741–746, 1998.
43. N. Maruyama, K. Ichise, T. Katsube, T. Kishimoto, S. Kawase, Y. Matsumura, Y. Takeuchi, T. Sawada, S. Utsumi. Identification of major wheat allergens by means of the *Escherichia coli* expression system. *Eur. J. Biochem.*, 255: 739–745, 1998.

44. J.M. James, J.P. Sixbey, R.M. Helm, G.A. Bannon, A.W. Burks. Wheat alpha-amylase inhibitor: a second route of allergic sensitization. *J. Allergy Clin. Immunol.*, 99: 239–244, 1997.
45. E. Batanero, M. Villalba, R.I. Monsalve, R. Rodriguez. Cross-reactivity between the major allergen from olive pollen and unrelated glycoproteins: evidence of an epitope in the glycan moiety of the allergen. *J. Allergy Clin. Immunol.*, 97: 1264–1271, 1996.
46. G. Garcia-Casado, R. Sanchez-Monge, M.J. Chrispeels, A. Armentia, G. Salcedo, L. Gomez. Role of complex asparagine-linked glycans in the allergenicity of plant glycoproteins. *Glycobiology*, 6: 471–477, 1996.
47. S. Tanabe, S. Tesaki, M. Watanabe, Y. Yanagihara. Cross-reactivity between bromelain and soluble fraction from wheat flour. (in Japanese) *Jpn. J. Allergol.*, 46: 1170–1173, 1997.
48. J. Watanabe, S. Tanabe, K. Sonoyama, M. Kuroda, M. Watanabe. An IgE reactive 60 kDa glycoprotein occurring in wheat flour. *Biosci. Biotechnol. Biochem.*, 65: 2102–2105, 2001.
49. S. Tanabe, J. Watanabe, K. Oyama, E. Fukushi, J. Kawabata, S. Arai, T. Nakajima, M. Watanabe. Isolation and characterization of a novel polysaccharide as a possible allergen occurring in wheat flour. *Biosci. Biotechnol. Biochem.*, 64: 1675–1680, 2000.
50. M. Watanabe, T. Suzuki, Z. Ikezawa, S. Arai. Controlled enzymatic treatment of wheat proteins for production of hypoallergenic flour. *Biosci. Biotechnol. Biochem.*, 58: 388–390, 1994.
51. M. Watanabe, Z. Ikezawa, S. Arai. Fabrication and quality evaluation of hypoallergenic wheat products. *Biosci. Biotechnol. Biochem.*, 58: 2061–2065, 1994.
52. S. Tanabe, S. Arai, M. Watanabe. Modification of wheat flour with bromelain and baking hypoallergenic bread with added ingredients. *Biosci. Biotechnol. Biochem.*, 60: 1269–1272, 1996.
53. M. Watanabe, J. Watanabe, K. Sonoyama, S. Tanabe. Novel method for producing hypoallergenic wheat flour by enzymatic fragmentation of constituent allergens and its application to food processing. *Biosci. Biotechnol. Biochem.*, 64: 2663–2667, 2000.
54. A. Yamamoto, S. Tanabe, T. Kojima, M. Sasai, Y. Hatano, M. Watanabe, Y. Kobayashi, S. Taniuchi. Hypoallergenic cupcake is a useful product for wheat-sensitive allergic patients. *J. Appl. Res.*, 4: 518–523, 2004.
55. S. Tesaki, J. Watanabe, S. Tanabe, K. Sonoyama, E. Fukushi, J. Kawabata, M. Watanabe. An active compound against allergen absorption in hypoallergenic wheat flour produced by enzymatic modification. *Biosci. Biotechnol. Biochem.*, 66: 1930–1935, 2002.
56. W.A. Walker, R. Cornell, L.M. Davenport, K.J. Isselbacher. Macromolecular absorption. Mechanism of horseradish peroxidase uptake and transport in adult and neonatal rat intestine. *J. Cell Biol.*, 54: 195–205, 1972.
57. H. Majamaa, E. Isolauri. Evaluation of the gut mucosal barrier: evidence for increased antigen transfer in children with atopic eczema. *J. Allergy Clin. Immunol.*, 97: 985–990, 1996.
58. T.W. Knutson, U. Bengtsson, A. Dannaeus, S. Ahlstedt, L. Knutson. Effects of luminal antigen on intestinal albumin and hyaluronan permeability and ion transport in atopic patients. *J. Allergy Clin. Immunol.*, 97: 1225–1232, 1996.
59. S. Tanabe, S. Tesaki, J. Watanabe, E. Fukushi, K. Sonoyama, J. Kawabata. Isolation and structural elucidation of a peptide derived from Edam cheese that inhibits β-lactoglobulin transport. *J. Dairy Sci.*, 86: 464–468, 2003.

60. J. Watanabe, S. Tanabe, M. Watanabe, T. Kasai, K. Sonoyama. Consumption of hypoallergenic flour prevents gluten-induced airway inflammation in Brown Norway rats. *Biosci. Biotechnol. Biochem.*, 65: 1729–1735, 2001.
61. J. Bousquet, F.B. Michel. Specific immunotherapy in asthma: Is it effective? *J. Allergy Clin. Immunol.*, 94: 1–11, 1994.
62. C.A. Akdis, T. Blesken, M. Akdis, B. Wüthrich, K. Blaser. Role of interleukin 10 in specific immunotherapy. *J. Clin. Invest.*, 102: 98–106, 1998.
63. U. Müller, A. Helbling, E. Berchtold. Immunotherapy with honeybee venom and yellow jacket venom is different regarding efficacy and safety. *J. Allergy Clin. Immunol.*, 89: 529–535, 1992.
64. T.J. Briner, M-C. Kuo, K.M. Keating, B.L. Rogers, J.L. Greenstein. Peripheral T cell tolerance induced in naive and primed mice by subcutaneous injection of peptide from the major cat allergen Fel d 1. *Proc. Natl. Acad. Sci. USA*, 90: 7608–7612, 1993.
65. G.F. Hoyne, R.E. O'Hehir, D.C. Wraith, W.R. Thomas, J.R. Lamb. Inhibition of T cell and antibody response to house dust mite allergen by inhalation of the dominant T cell epitope in naive and sensitized mice. *J. Exp. Med.*, 178: 1783–1788, 1993.
66. B.P. Wallner, M.L. Gefter. Immunotherapy with T cell-reactive peptides derived from allergens. *Allergy*, 49: 302–308, 1994.
67. P.S. Norman, J.L. Ohmann Jr, A.A. Long, P.S. Creticos, M.A. Gefter, Z. Shaked, R.A. Wood, P.A. Eggleston, K.B. Hafner, P. Rao, L.M. Lichtenstein, N.H. Jones, C.F. Nicodemus. Treatment of cat allergy with T-cell reactive peptide. *Am. J. Respir. Crit. Care Med.*, 154: 1623–1628, 1996.
68. M. Toda, M. Totsuka, S. Furukawa, K. Yokota, T. Yoshioka, A. Ametani, S. Kaminogawa. Down-regulation of antigen-specific antibody production by TCR antagonist peptides *in vivo*. Eur *J. Immunol.*, 30: 403–414, 2000.
69. M. Moriyama, C. Tokue, H. Ogiwara, H. Kimura, S. Arai. Chemical and nutritional properties of hypoallergenic wheat flour. *Biosci. Biotechnol. Biochem.*, 65: 706–709, 2001.
70. R. Pauwels, H. Bazin, B. Platteau, M. van der Straeten. The effect of age on IgE production in rats. *Immunology*, 36: 145–149, 1979.
71. A. Abadie, A. Prouvost-Danon. Specific and total IgE responses to antigenic stimuli in Brown-Norway, Lewis and Sprague-Dawley rats. *Immunology*, 39: 561–569, 1980.
72. W. Elwood, J.O. Lotvall, P.J. Barnes, K-F. Chung. Characterization of allergen-induced bronchial hyperresponsiveness and airway inflammation in actively sensitized brown-Norway rats. *J. Allergy Clin. Immunol.*, 88: 951–960, 1991.
73. A. Haczku, P. Macary, E.B. Haddad, T-J. Huang, D.M. Kemeny, R. Moqbel, K-F. Chung. Expression of Th-2 cytokines interleukin-4 and -5 and of Th-1 cytokine interferon-γ in ovalbumin-exposed sensitized Brown-Norway rats. *Immunology*, 88: 247–251, 1996.
74. S. Tanabe, S. Tesaki, Y. Yanagihara, H. Mita, K. Takahashi, S. Arai, M. Watanabe. Inhibition of basophil histamine release by a haptenic peptide mixture prepared by chymotryptic hydrolysis of wheat flour. *Biochem. Biophys. Res. Commn.*, 223: 492–495, 1996.

21 Milk Proteins

Koko Mizumachi and Jun-ichi Kurisaki

CONTENTS

21.1 INTRODUCTION

Cow's milk allergy is a serious health problem since milk is widely used for its nutritional values and useful properties in the manufacture of infant formula and a great number of processed foods. Infant formula is considered essential as an alternative to breast milk not only in the developed countries but also in the underdeveloped countries where there is rampant malnutrition, and therefore allergy to milk formula would cause serious growth retardation. As well, milk allergy in infancy would trigger hypersensitivity to other environmental and food allergies in atopic subjects. Due to the abundance of processed foods containing milk proteins in the market, there is a likelihood of unintentional intake of milk allergens. Therefore, it is quite important to develop hypoallergenic infant formulas and processed food products that contain milk components.

Milk contains about 3.5% protein that is classified into two categories: caseins (α_{s1}-, α_{s2}-, β-, and κ-caseins) and whey proteins (β-lactoglobulin, α-lactalbumin, serum albumin, immunoglobulins, proteose-peptones, and so on) (1); α_{s1}-casein and β-lactoglobulin are the major casein and whey proteins in bovine milk, respectively. Both of them are absent in human milk and are believed to be the potent allergens in cow's milk (2,3). Moreover, β-lactoglobulin is shown to be resistant to proteolytic digestion (4,5), and therefore it can easily reach and cross the intestinal mucosa without losing its allergenicity. Other than these proteins, α_{s2}-, β-, κ-caseins have also been reported as milk allergens (6,7).

In order to escape the immunological recognition of allergens or to diminish protein allergenicity, tremendous effort has been successfully made in the infant formula industry, and many hypoallergenic formulas with acceptable flavor have been developed by enzymatic hydrolysis of allergenic proteins. However, it should be borne in mind that the functional properties of milk proteins useful to the food industry, such as emulsifying, foaming, and coagulation activities, would be largely lost by such fragmentation. The ultimate goal is therefore to design functional foods that could make use of the beneficial properties of milk proteins and simultaneously suppress milk allergy.

In this chapter, the principles for the development of hypoallergenic milk products and the currently available products, mainly infant formulas, are introduced. In addition, a new concept of antiallergic food is proposed based on the recent immunological research (Figure 21.1).

21.2 CURRENTLY AVAILABLE HYPOALLERGENIC FORMULAS

The allergenicity of milk products can be reduced by technological procedures such as elimination, heat treatment, and enzymatic hydrolysis of antigen proteins.

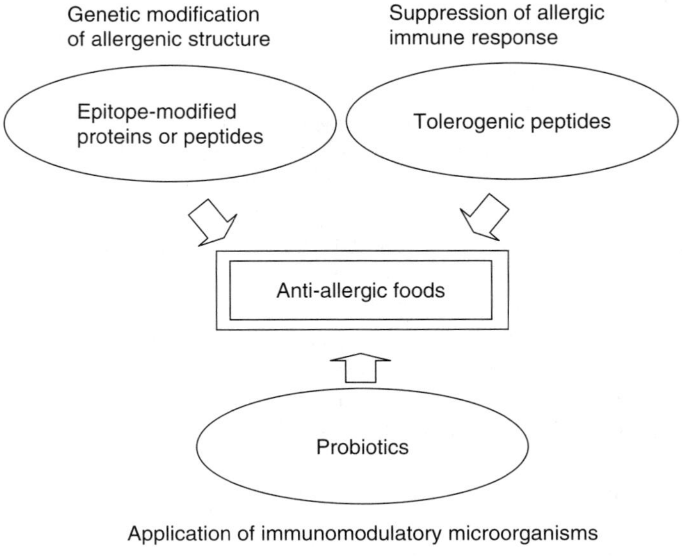

FIGURE 21.1 Approaches to antiallergic foods containing milk proteins.

TABLE 21.1
Types of Hypoallergenic Formulas and Their Properties

Type	Composition	Properties (Advantages or Drawbacks)
Alternatives		
Soy-based formulas	Soy protein	No cross-reactivity with cow's milk proteins
		Nutritional adequacy similar to that of cow's milk formulas
		Low price
		Aluminum and manganese toxicity
		Contain physiologically active substances (e.g., phytoestrogens)
		Allergenicity in soy-sensitive patients
Amino acid-based formulas	Elemental amino acids	No proteins
		More expensive than extensive hydrolyzates
		Unpalatable
Goat's milk formulas	Goat's milk	Low price
		Better taste
		High cross-reactivity with cow's milk protein
Hydrolyzates		
Extensive hydrolyzates	Casein or whey hydrolyzates	Expensive
		Bitter taste (casein hydrolyzates)
		Cross-reaction in highly sensitive patients
Partial hydrolyzates	Casein or whey hydrolyzates	Lower cost than extensive hydrolyzates
		Pleasant taste
		Residual allergenic activity

However, these treatments tend to deteriorate the nutritional value and the taste, or flavor, of the products, and adequate supplements are usually required. At present, a variety of hypoallergenic formulas have been produced with this in mind (Table 21.1) and are widely accepted by milk-allergic patients.

21.2.1 Alternatives or Elimination of Cow's Milk

The elimination of cow's milk from the diet, or the utilization of alternatives to cow's milk, is the first choice for the prevention of allergy (8). Several types of milk formulas made from vegetables such as soy or rice and milk from other animals such as goat, sheep, or mare have been proposed as substitutes for cow's milk (9). Soy-, amino acid-based formulas, and goat's milk formulas are commercially available in several countries.

21.2.1.1 Soy Protein-Based Formulas

Soy-based formulas are considered an effective alternative to cow's milk for allergic patients, since immunological cross-reactivity is not found between soy and cow's milk proteins (9–11). Infants fed a soy-based formula have shown normal growth and satisfactory bone mineralization (12,13). Therefore, the Committee on Nutrition of the American Academy of Pediatrics (AAP) concluded that soy-based formulas are well tolerated and safe in patients with immunoglobulin E (IgE)-mediated cow's milk allergy beyond 6 months of life (10). However, some of the infants with cow's milk allergy also exhibit allergic reactions to soy proteins, and there are still unsolved problems such as aluminum and manganese toxicity because of their high contents in soy-based formulas (14) and the exposure to phytoestrogens, which could cause hormone-like effects in infants (15).

21.2.1.2 Amino Acid-Based Formulas

For patients sensitive to multiple food allergens, amino acid-based formulas are recommended. The formulas currently available have been subjected to clinical testing and shown to meet the standards for hypoallergenicity (16); normal growth of children on these formulas has been confirmed by *in vivo* studies (17–19). However, the formulas are rather unpalatable and more expensive than the hydrolyzed formulas.

21.2.1.3 Goat's Milk

Goat's milk is commercially available and sometimes believed to be a good alternative to cow's milk for allergic children, but it is doubtful whether goat's milk is truly effective as a hypoallergenic milk. The composition of goat's milk is similar to that of cow's milk, except that goat's milk has lower contents of vitamin B_{12} and folic acid (9). Moreover, cow's milk and goat's milk both have a high degree of homology in proteins (20–23). Investigations on the allergenicity of goat's milk have shown that there is high cross-reactivity with cow's milk proteins in *in vitro* and *in vivo* studies (24,25). So far, there is no scientific evidence of the hypoallergenicity of goat's milk in allergic patients. Recently, variations caused by a genetic polymorphism were reported in goat α_{s1}-casein contents (26). The authors claimed that goat's milk with low α_{s1}-casein content might be useful for sensitive patients.

21.2.2 Destruction of Antigenic Structure

The allergic immune responses to milk proteins start from the recognition of the allergens by host immune cells such as T and B cells. Hence, the destruction of the antigenic structure recognized by these cells should inhibit the allergic response to milk proteins. The methods to disturb the antigenic structure are usually heat treatment and enzymatic hydrolysis.

21.2.2.1 Heat Denaturation

In many cases, heat treatment can reduce the antigenicity or allergenicity by destroying the conformational structure of a protein. The allergenicity of milk proteins may be also reduced by increasing the digestibility of proteins in the gut as a result of heat treatment. Heppell et al. (27) reported that extensively denatured whey proteins, heated at 100°C or 115°C for 30 minutes, failed to sensitize guinea pigs to anaphylaxis and significantly reduced antibody production, but caseins were resistant to heat treatment. Davis et al. (28) pointed out that heat denaturation did not always reduce the antigenicity but might present new antigenic sites. In fact, it has been reported that new sites are created by the interaction of β-lactoglobulin with α-lactalbumin under heat treatment (29). Therefore, at present, the combination of heat treatment and hydrolysis, with or without subsequent ultrafiltration, is used for reduction of the allergenicity of milk proteins.

21.2.2.2 Hydrolysis

Enzymatic hydrolysis is an efficient procedure to reduce the allergenicity of milk proteins, since the antibody-binding sites are often cleaved. Two kinds of formulas made of milk protein hydrolyzates have been developed: one is casein based and another whey based. Depending on the manufacturer's procedure, partially hydrolyzed formulas (PHF) and extensively hydrolyzed formulas (EHF) have been manufactured. PHF contain relatively long peptides having molecular weights between 8,000 and 20,000 Da and intact proteins. In contrast, EHF contain peptides with molecular weights below 5,000 Da (usually below 1,000 Da in Japan). Giampietro et al. (30) examined the allergenicity of EHF and PHF by clinical trial and found that 94 to 97% of patients tolerated EHF and 64% tolerated PHF after oral challenge in 32 children with cow's milk allergy.

The Committee on Nutrition of the AAP concluded that products claiming to be hypoallergenic formulas should not provoke reactions in 90% of patients with cow's milk allergy (95% confidence interval) in randomized, double-blind, placebo-controlled trials. The Committee at present recommends EHF and amino acid-based formulas, which can pass the above clinical tests, as a first choice of hypoallergenic formulas for most infants or children with cow's milk allergy (16). However, it should be borne in mind that some residual antigenic activities have been found even in all types of EHF, in both human and animal experiments (31–35).

It has been reported that dietary antigens are found in breast milk and cause allergy in breast-fed infants (36). Therefore, whey hydrolyzate formulas for lactating women have been developed, and their preventive effects have been described (37).

21.3 APPROACHES TO ANTIALLERGIC FOODS CONTAINING MILK PROTEINS

Hypoallergenic formulas are quite beneficial for the normal growth of milk-allergic infants and children. However, a variety of processed food products containing milk proteins as ingredients are widely available on the market, and thus cow's milk allergy may be provoked by an unintentional intake of milk proteins from such processed foods. In fact, Gern et al. (38) have reported serious allergic reactions from consumption of meat products containing hydrolyzed sodium caseinate.

Food products that are hypoallergenic to milk-allergic patients can be easily developed with hydrolyzates or without milk proteins (39). However, the opportunity to make use of the low-cost, performance, valuable functional properties, and nutritional values of the proteins would be lost. The development of a new type of product — antiallergic foods containing milk proteins — to regulate or inhibit allergic reactions are still awaited. In this section, we describe some possible approaches to prevent milk allergy, even in the presence of milk proteins, such as the modification of the allergenic structure of milk proteins, the suppression of allergic immune response by milk peptides, and the application of immunomodulatory microorganisms called probiotics.

21.3.1 Genetic Modification of Milk Proteins

Antibodies produced by B cells bind to extended sites on the surfaces of antigens, whereas T cells recognize antigen fragments, which are bound and presented by major histocompatibility complex (MHC) molecules on antigen presenting cells (APCs), with their T-cell receptors (TCRs). The sites on the antigen recognized by antibodies and T cells are termed B-cell epitopes (antigenic determinants) and T-cell epitopes, respectively.

In terms of cow's milk proteins, many studies on the B- and T-cell epitopes of major milk allergens such as β-lactoglobulin and caseins have been performed in mice or allergic patients using natural fragments of the protein or synthetic peptides covering the sequence of the molecule (Table 21.2) (40–60). Information on the B- and T-cell epitopes of milk proteins now facilitates the production of effectively mutated proteins with less antigenicity. The effects of the proteins with mutations at selected sites have been tested on immune responses. Cocco et al. (61) reported that the substitution of crucial amino acids of epitopes on α_{s1}-casein led to the reduction of binding to IgE antibodies from milk-allergic patients. We also demonstrated decreased antigenicity by genetic modification of the region aa 149–162 of β-lactoglobulin, which was one of the major B-cell epitopes in rabbit and mouse. The recombinant β-lactoglobulin (161His→Ala), prepared by the baculovirus–insect cell system (62), exhibited significantly reduced binding activity to region 149–162-specific antibodies (Mizumachi et al., unpublished data).

On recognition of the antigen fragments on the APCs, the T cells are activated to proliferate and differentiate into effector cells producing various kinds of cytokines. Recent studies have clearly indicated that T cells show different types of reactions in recognition of the altered peptides for TCRs; peptides with a single amino acid substitution at a TCR contact site on the T-cell epitope (63). Some kinds of altered peptide ligands (APLs) have been shown to induce partial T-cell activation with cytokine production in the absence of proliferation. The APLs, termed TCR antagonist peptides, inhibited both the proliferation of T cells and the production of cytokines (63). As T cells play a central role in the generation or regulation of an immune response, a method for the modification of T-cell responses by APLs such as TCR antagonist peptides can be effectively applied to the antigen-specific immunotherapy for allergies (64,65). In terms of milk proteins, Toda et al. (66) have demonstrated that TCR antagonist peptides, the amino acid-substituted analogue of a peptide corresponding to residues 119–133 of

TABLE 21.2
Epitope Mapping on Major Milk Proteins

Milk protein	Epitope	Species	Peptide Used for Mapping	Reference
α_{s1}-casein	B	Mouse	Proteolytic fragments	40
	B	Mouse	Proteolytic and synthetic peptides	41
	T, B	Mouse	Synthetic peptides	42,43
	B	Human	Synthetic peptides	44
	T, B	Human	Synthetic peptides and peptides by CNBr-cleavage	45
	B	Human	Synthetic peptides	46
β-casein	B	Rabbit	Proteolytic fragments and peptides by CNBr-cleavage	47–50
	B	Mouse, rabbit, goat	Synthetic peptides	51
	B	Human	Synthetic peptides	52
β-lactoglobulin	T	Mouse	Synthetic peptides	53
	T	Mouse	Synthetic peptides	54
	B	Human	Synthetic peptides	55
	B	Human	Peptides by CNBr-cleavage	56
	B	Rabbit and mouse	Synthetic peptide and phage-displayed peptides	57
	B	Human	Synthetic peptides	58
	B	Human	Tryptic peptides	59
	B	Mouse	Synthetic peptides	60

Abbreviations: B, B-cell epitope; T, T-cell epitope; CNBr, cyanogen bromide

β-lactoglobulin, could inhibit antibody response in mice. This implies that the modification of milk proteins to include antagonistic sequences in the molecules might be possible for the suppression of milk allergy.

The modification of milk protein composition by direct manipulation of the germline may also be useful for the development of hypoallergenic milk products. The genes that encode the major milk proteins have already been identified, and it is now possible to enhance the processing, nutritional, and biomedical properties of milk by the genetic modification of milk proteins (67). In murine studies, the modification of milk composition by deleting β-casein or α-lactalbumin (68,69), or replacing bovine α-lactalbumin with human α-lactalbumin (70), has been performed by germline manipulation techniques with the intent to prevent milk allergy.

Thus, forms of allergens modified by substitution of the amino acids of the epitopes and modification of milk components by germline manipulation techniques should have high potential in the development of hypoallergenic dairy foods by making good use of the functional properties of milk proteins. However, there are still difficulties to overcome for practical application in humans because tremendous time and effort would be required for the design of effective modified milk proteins for allergic patients. Moreover, public acceptance of the genetically modified products still needs to be won.

21.3.2 Oral Tolerance

Oral tolerance — oral administration of antigens, which induces specific immunological unresponsiveness to the antigens — has been attracting considerable attention as a potential therapy for autoimmune diseases and allergies (71,72).

There are several studies on oral tolerance to milk proteins such as α_{s1}-casein (73–75) and β-lactoglobulin in mice (76–79). It has been shown that in mice fed whey proteins or caseins, the anti-β-lactoglobulin or anti-α_{s1}-casein-specific antibody response was reduced (73,76). We also examined the immune response of mice to β-lactoglobulin after continuous feeding of a β-lactoglobulin solution or milk instead of drinking water and observed strong suppression of the anti-β-lactoglobulin antibody response and antigen-specific T-cell response (79).

The tolerance can be induced not only by milk proteins but also by milk peptides. There are several studies showing that the enzymatic hydrolyzates of milk proteins contain the tolerogenic peptides and are effective for the induction of antigen-specific tolerance (80–82). Hachimura et al. (80) have reported that oral administration of a tryptic digest of casein to mice induced suppression of the systemic immune response to casein. Pecquet et al. (81) reported that the peptides obtained by tryptic hydrolysis of β-lactoglobulin could induce oral tolerance to intact β-lactoglobulin at the humoral and cellular levels in mice. It was also shown that partially hydrolyzed formulas were able to induce oral tolerance to milk proteins in rats (82). Recently, several studies have shown that tolerance can be induced by oral administration of the dominant T-cell-epitope peptides of the antigen (83–85). We examined the immunosuppressive effect of peptides corresponding to T-cell-epitope peptides of β-lactoglobulin (53) and found that the peptide corresponding to residues 139–154 could effectively suppress β-lactoglobulin-specific antibody response in mice (Mizumachi et al., unpublished data).

All of these findings in animal models described above strongly suggest that milk peptides, if effectively chosen and designed not to include allergenic structures, could be applicable to milk-allergic patients in order to suppress an adverse immune response. However, detailed information about the allergenic epitopes and T-cell epitopes recognized by the patients should still be indispensable for practical application. Clinical trials have been already performed on an autoimmune disease by using the T-cell-epitope peptides of the antigen (86).

21.3.3 Probiotics

Probiotics represent a microbial supplement that beneficially influences the host by improving its intestinal microflora (87). For example, the prevention and management of gastrointestinal disorders, the modulation of local or systemic immune responses, the prevention and curing of tumors, hypotensive effects, and lowering of blood cholesterol by probiotics have been reported in animal and human studies (88).

In this context, the application of probiotics to the prevention of atopic disease has attracted considerable attention. Kalliomaki et al. (89) showed that *Lactobacillus rhamnosus* GG, isolated from human feces (Valio Ltd, Finland), was effective in the prevention or treatment of early atopic disease in children at higher risk. This is the first report on an antiallergic probiotic strain that prevents food allergy, including milk allergy.

We have also found a new probiotic strain, *Lactococcus lactis* subsp. *lactis* G50 (G50) separated from napier grass, had suppressive effects on IgE response to ovomucoid, a potent egg allergen, in BALB/c mice (90). Moreover, we have noticed that G50 feeding suppressed the IgE response to β-lactoglobulin while enhancing the IgA antibody response. Interestingly, oral tolerance to β-lactoglobulin was effectively induced with the simultaneous feeding of both β-lactoglobulin and G50 (Mizumachi et al., unpublished data). Thus, probiotics have the potential for the suppression of allergic reactions to milk proteins and could be useful in the development of antiallergic foods.

21.4 CONCLUSIONS

In this chapter, the currently available hypoallergenic formulas and possible approaches to suppress allergic reactions to cow's milk were described. Although the hydrolysis of milk proteins is the only practical method to reduce allergenicity at present, the epitope-modified milk proteins, tolerogenic milk peptides, and probiotics described in the text are good antiallergic candidates. Research into such immunomodulators would make it possible to develop new hypoallergenic foods containing milk proteins.

REFERENCES

1. H.E. Swaisgood. Chemistry of the caseins. In: P.F. Fox (Ed.). *Advanced Dairy Chemistry Volume1. Proteins.* London: Elsevier applied science, 1992, 63–110.
2. S.L. Taylor. Immunologic and allergic properties of cows' milk proteins in humans. *J. Food Prot.*, 49: 239–250, 1986.
3. J.W. Yunginger. Food antigens. In:D.D. Metcalfe, H.A. Sampson, R.A. Simon (Eds.). *Food Allergy: Adverse Reactions to Foods and Food Additives*, 2nd edition. Cambridge: Blackwell Science, 1996, 49–63.
4. G. Miranda, J.P. Pelissier. Kinetic studies of *in vivo* digestion of bovine unheated skim-milk proteins in the rat stomach. *J. Dairy Res.*, 50: 27–36, 1983.
5. I.M. Reddy, N.K. Kella, J.E. Kinsella, Structural and conformational basis of the resistance of β-lactoglobulin to peptic and chymotryptic digestion. *J. Agric. Food Chem.*, 36: 737–741, 1988.
6. H. Bernard, C. Creminon, M. Yvon, J.M. Wal. Specificity of the human IgE response to the different purified caseins in allergy to cow's milk proteins. *Int. Arch. Allergy Immunol.*, 115: 235–244, 1998.
7. K.M. Jarvinen, K. Beyer, L. Vila, P. Chatchatee, P.J. Busse, H.A. Sampson. B-cell epitopes as a screening instrument for persistent cow's milk allergy. *J. Allergy Clin Immunol.*, 110: 293–297, 2002.

8. R.S. Zeiger. Food allergen avoidance in the prevention of food allergy in infants and children. *Pediatrics*, 111: 1662–1671, 2003.
9. M.A. Muraro, P.G. Giampietro, E. Galli. Soy formulas and nonbovine milk. *Ann. Allergy Asthma Immunol.*, 89(Suppl): 97–101, 2002.
10. American Academy of Pediatrics. Committee on Nutrition. Soy protein-based formulas: recommendations for use in infant feeding. *Pediatrics*, 101(1 Pt. 1): 148–153, 1998.
11. L. Businco, G. Bruno, P.G. Giampietro. Soy protein for the prevention and treatment of children with cow-milk allergy. *Am. J. Clin. Nutr.*, 68(6 Suppl): 1447S–1452S, 1998.
12. J.J. Steichen, R.C. Tsang. Bone mineralization and growth in term infants fed soy-based or cow milk-based formula. *J. Pediatr.*, 110: 687–692, 1987.
13. L. Businco, B. G runo, P.G. Giampietro, A. Cantani. Allergenicity and nutritional adequacy of soy protein formulas. *J. Pediatr.*, 121(5 Pt 2): S21–S28, 1992.
14. G.E. Moro, A. Warm, S. Arslanoglu, V. Miniello. Management of bovine protein allergy: new perspectives and nutritional aspects. *Ann. Allergy Asthma Immunol.*, 89(Suppl): 91–96, 2002.
15. K.D. Setchell, L. Zimmer-Nechemias, J. Cai, J.E. Heubi. Exposure of infants to phytoestrogens from soy-based infant formula. *Lancet*, 350: 23–27, 1997.
16. American Academy of Pediatrics. Committee on Nutrition. Hypoallergenic infants formula. *Pediatrics*, 106: 346–349, 2000.
17. H.A. Sampson, J.M. James, J. Bernhisel-Broadbent. Safety of an amino acid-derived infant formula in children allergic to cow milk. *Pediatrics*, 90: 463–465, 1992.
18. E. Isolauri, Y. Sutas, S. Makinen-Kiljunen, S.S. Oja, R. Isosomppi, K. Turjanmaa. Efficacy and safety of hydrolyzed cow milk and amino acid-derived formulas in infants with cow milk allergy. *J. Pediatr.*, 127: 550–557, 1995.
19. S.H. Sicherer, S.A. Noone, C.B. Koerner, L. Christie, A.W. Burks, H.A. Sampson. Hypoallergenicity and efficacy of an amino acid-based formula in children with cow's milk and multiple food hypersensitivities. *J. Pediatr.*, 138: 688–693, 2001.
20. G. Preaux, G. Braunitzer, B. Schrank, A. Stangl. The amino acid sequence of goat β-lactoglobulin. *Hoppe Sylers Z. Physiol. Chem.*, 360: 1595–1604, 1979.
21. B. Roberts, P. DiTullio, J. Vitale, K. Hehir, K. Gordon. Cloning of the goat β-casein-encoding gene and expression in transgenic mice. *Gene*, 121: 255–262, 1992.
22. A. Coll, J.M. Folch, A. Sanchez. Nucleotide sequence of the goat κ-casein cDNA. *J. Anim. Sci.*, 71: 2833, 1993.
23. L.K. Rasmussen, E.S. Sorensen, T.E. Petersen, N.C. Nielsen, J.K. Thomsen. Characterization of phosphate sites in native ovine, caprine, and bovine casein micelles and their caseinomacropeptides: a solid-state phosphorus-31 nuclear magnetic resonance and sequence and mass spectrometric study. *J. Dairy Sci.*, 80: 607–614, 1997.
24. P. Spuergin, M. Walter, E. Schiltz, K. Deichmann, J. Forster, H. Mueller. Allergenicity of α-caseins from cow, sheep, and goat. *Allergy*, 52: 293–298, 1997.
25. B. Bellioni-Businco, R. Paganelli, P. Lucenti, P.G. Giampietro, H. Perborn, L. Businco. Allergenicity of goat's milk in children with cow's milk allergy. *J. Allergy Clin. Immunol.*, 103: 1191–1194, 1999.
26. C. Bevilacqua, P. Martin, C. Candalh, J. Fauquant, M. Piot, A.M. Roucayrol, F. Pilla, M. Heyman. Goats' milk of defective α_{s1}-casein genotype decreases intestinal and

systemic sensitization to β-lactoglobulin in guinea pigs. *J. Dairy Res.*, 68: 217–227, 2001.

27. L.M.J. Heppell, A.J. Cant, P.J. Kilshaw. Reduction in the antigenicity of whey proteins by heat treatment: a possible strategy for producing a hypoallergenic infant milk formula. *Br. J. Nutr.*, 51: 29–36, 1984.
28. P.J. Davis,S.C. Williams. Protein modification by thermal processing. *Allergy*, 53: 102–105, 1998.
29. A. Baer, M. Oroz, B. Blanc, Serological studies on heat induced interaction of α-lactalbumin and milk proteins. *J. Dairy Res.*, 43: 419–432, 1976.
30. P.G. Giampietro, N.I. Kjellman, G. Oldaeus, W. Wouters-Wesseling, L. Buinco. Hypoallergenicity of an extensively hydrolyzed whey formula. *Pediatr. Allergy Immunol.*, 12: 83–86, 2001.
31. A. Rosendal, V. Barkholt. Detection of potentially allergenic material in 12 hydrolyzed milk formulas. *J. Dairy Sci.*, 83: 2200–2210, 2000.
32. B. Niggemann, H. Nies, H. Renz, U. Herz, U. Wahn. Sensitizing capacity and residual allergnicity of hydrolyzed cow's milk formulae: results from a murine model. *Int. Arch. Allergy Immunol.*, 125: 316–321, 2001.
33. M.H. Ellis, J.A. Short, D.C. Heiner: Anaphylaxis after ingestion of a recently hydrolyzed whey protein formula. *J. Pediatr.*, 118: 74–77, 1991.
34. L. Businco, A. Cantani, M.A. Longhi, P.G. Giampietro. Anaphylactic reactions to a cow's milk whey protein hydrolyzate (Alfa-Re, Nestle) in infants with cow's milk allergy. *Ann. Allergy*, 62: 333–335, 1989.
35. J. Walker-Smith. Hypoallergenic formulas: are they really hypoallergenic? *Ann. Allergy Asthma Immunol.*, 90: 112–114, 2003.
36. J.W. Gerrard, M. Shenassa. Food allergy: two common types as seen in breast and formula fed babies. *Ann. Allergy*, 50: 375–379, 1983.
37. Y. Fukushima, K. Iwamoto, A. Takeuchi-Nakashima, N. Akamatsu, N. Fujino-Numata, M. Yoshikoshi, T. Onda, M. Kitagawa. Preventive effect of whey hydrolyzate formulas for mothers and infants against allergy development in infants for the first 2 years. *J. Nutr. Sci. Vitaminol. (Tokyo)*, 43: 397–411, 1997.
38. J.E. Gern, E. Yang, H.M. Evrard, H.A. Sampson. Allergic reactions to milk-contaminated "nondairy" products. *N. Engl. J. Med.*, 324: 976–979. 1991.
39. Y. Takahata, J. Kurisaki, K. Mizumachi, R. Shibata, T. Shigehisa, F. Morimatsu. IgE-antibody specificities of the patients allergic to meat products. *Anim. Sci. J.*, 71: 494–500, 2000.
40. A. Ametani, S. Kaminogawa, M. Shimizu, K. Yamauchi. Rapid screening of antigenically reactive fragments α_{s1}-casein using HPLC and ELISA. *J. Biochem.*, 102: 421–425, 1987.
41. A. Ametani, S.M. Kim, S. Kaminogawa, K. Yamauchi. Antibody response of three different strains of mice to α_{s1}-casein analyzed by using proteolytic and synthetic peptides. *Biochem. Biophys. Res. Commn.*, 154: 876–882, 1988.
42. A. Enomoto, D-H. Shon, Y. Aoki, K. Yamauchi, S. Kaminogawa. Antibodies raised against peptide fragments of bovine α_{s1}-casein cross-react both B and T cell determinants. *Mol. Immunol.*, 27: 581–586, 1990.
43. D-H. Shon, A. Enomoto, K. Yamauchi, S. Kaminogawa. Antibodies raised against peptide fragment of bovine α_{s1}-casein cross-react with the native protein, but recognize sites distinct from the determinant on the protein. *Eur. J. Immunol.*, 21: 1475–1480, 1991.

44. P. Spuergin, H. Mueller, M. Walter, E. Schiltz, J. Foster. Allergnic epitopes of bovine α_{s1}-casein recognized by human IgE and IgG. *Allergy*, 51: 306–312, 1996.
45. H. Nakajima-Adachi, S. Hachimura, W. Ise, K. Honma, S. Nishiwaki, M. Hirota, N. Shimojo, T. Katsuki, A. Ametani, Y. Kohno, A. Kaminogawa. Determinant analysis of IgE and IgG4 antibodies and T cells specific for bovine α_{s1}-casein from the same patients allergic to cow's milk: existence of α_{s1}-casein-specific B cells and T cells characteristic in cow's-milk allergy. *J. Allergy Clin. Immunol.*, 101: 660–671, 1998.
46. P. Chatchaatee, K-M. Järvinen, L. Bardina, K. Beyer, H.A. Sampson. Identification of IgE- and IgG-binding epitopes on α_{s1}-casein: differences in patients with persistent and transient cow's milk allergy. *J. Allergy Clin. Immunol.*, 107: 379–383, 2001
47. H. Otani, S. Iwasaki, F. Tokita. Studies on the antigenic structure of bovine β-casein. I. Antigenic reactivity of b-III with antiserum to β-casein. *Milchwissenschaft*, 39: 211–214, 1984.
48. H. Otani, S. Iwasaki, F. Tokita. Studies on the antigenic structure of bovine β-casein. III. Antigenic reactivity of peptide 1–93 with antiserum to β-casein. *Milchwissenschaft*, 39: 396–399, 1984.
49. H. Otani, S. Higashiyama, F. Tokita. Studies on the antigenic structure of bovine β-casein. IV. Antigenic activities of peptides obtained by cyanogens bromide cleavage of 94–209 region. *Milchwissenschaft*, 39: 469–472, 1984.
50. H. Otani, Y. Mine, F. Tokita. Studies on the antigenic structure of bovine β-casein. VI. Antigenic activities of peptides produced by tryptic and V8-proteolytic digestions of peptide 110–144. *Milchwissenschaft*, 43: 759–761, 1988
51. K. Mizumachi, J. Kurisaki, S. Kaminogawa. Localization of sequential antigenic determinants on bovine β-casein with synthetic peptides and antisera mouse, rabbit, and goat. *Biosci. Biotechnol. Biochem.*, 63: 911–916, 1999.
52. P. Chatchaatee, K-M. Jarvinen, L. Bardina, K. Beyer, H.A. Sampson. Identification of IgE and IgG binding on β- and κ-casein in cow's milk allergic patients. *Clin. Exp. Allergy*, 31: 1256–1262, 2001.
53. N.M. Tsuji, J. Kurisaki, K. Mizumachi, S. Kaminogawa. Localization of T-cell determinants on bovine β-lactoglobulin. *Immunol. Lett.*, 37(2-3): 215–221, 1993.
54. M. Totsuka, A. Ametani, S. Kaminogawa. Fine mapping of T-cell determinants of bovine β-lactoglobulin. *Cytotechnology*, 25: 101–113, 1997.
55. G. Ball, M.J. Shelton, B.J. Walsh, D.J. Hill, C.S. Hosking, M.E. Howden. A major continuous allergenic epitope of bovine β-lactoglobulin recognized by human IgE binding. *Clin. Exp. Allergy*, 24: 758–764,1994.
56. I. Selo, L. Negroni, C. Creminon, M. Yvon, G. Peltre, J. Wal. Allergy to bovine β-lactoglobulin: specificity of human IgE using cyanogen bromide-derived peptides. *Int. Arch. Allergy Immunol.*, 117: 20–28, 1998.
57. S.C. Williams, R.A. Badley, P.J. Davis, W.C. Puijk, R.H. Meloen. Identification of epitopes within β-lactoglobulin recognised by polyclonal antibodies using phage display and PEPSCAN. *J. Immunol. Meth.*, 213: 1–17, 1998.
58. A. Heinzmann, S. lattmann, P. Spuergin J. Forster, K.A. Deichmann. The recognition pattern of sequential B cell epitopes of β-lactoglobulin does not vary with the clinical manifestations of cow's milk allergy. *Int. Arch. Allergy Immunol.*, 120: 280–286, 1999.
59. I. Selo, G. Clement, H. Bernard, J. Chatel, C. Creminon, G. Peltre, J. Wal. Allergy to bovine β-lactoglobulin: specificity of human IgE to tryptic peptides. *Clin. Exp. Allergy*, 29: 1055–1063, 1999.

60. K. Kobayashi, A. Hirano, A. Ohta, T. Yoshida, K. Takahashi, M. Hattori. Reduced immunogenicity of β-lactoglobulin by conjugation with carboxymethyl dextran differing in molecular weight. *J. Agric. Food Chem.*, 49: 823–831, 2001.
61. R.R. Cocco, K-M. Jarvinen, H.A. Sampson, K. Beyer. Mutational analysis of major, sequential IgE-binding epitopes in α_{s1}-casein, a major cow's milk allergen. *J. Allergy Clin. Immunol.*, 112: 433–437, 2003.
62. K. Mizumachi, J. Kurisaki, N.M. Tsuji. High-level expression of recombinant bovine β-lactoglobulin in insect cells. *Proceeding of the 5th International meeting of the Japanese Association for Animal Cell Technology*, Omiya, 1992, 115–121.
63. J. Sloan-Lancaster, P.M. Allen. Altered peptide ligand-induced partial activation: molecular mechanisms and role in T cell biology. *Annu. Rev. Immunol.*, 14: 1–27, 1996.
64. M.D. Magistris, J. Alexander, M. Coggeshall, A. Altman, F.C. Gaeta, H.M. Grey, A. Sette, Antigen analog-major histocompatibility comples act as antagonists of the T cell receptor. *Cell*, 68: 625–634, 1992.
65. G.F. Hoyne, N.M. Kristensen, H. Yssel, J.R. Lamb. Peptide modulation of allergen-specific immune responses. *Curr. Opin. Immunol.*, 7: 757–761, 1995.
66. M. Toda, M. Totsuka, S. Furukawa, K. Yokota, T. Yoshida, A. Ametani, S. Kaminogawa. Sown-regulation of antige-specific antibody production by TCR antagonist peptides *in vivo*. *Eur. J. Immunol.*, 30: 403–414, 2000.
67. A.J. Clark, Genetic modification of milk proteins. *Am. J. Clin. Nutr.*, 63: 633S–638S, 1996.
68. S. Kumar, A.R. Clarke, M.L. Hooper, D.S. Horne, A.J. Law, J. Leaver, A. Springbett, E. Stevenson, J.P. Simons. Milk composition and lactation of β-casein-deficient mice. *Proc. Natl. Acad. Sci. USA*, 91: 6138–6142, 1994.
69. M.G. Stinnakre, J.L. Vilotte, S. Soulier, J.C. Mercier. Creation and phenotypic analysis of α-lactalbumin-deficient mice. *Proc. Natl. Acad. Sci. USA*, 91: 6544–6548, 1994.
70. A. Stacey, A. Schnieke, M. Kerr, A. Scott, C. McKee, I. Cottingham, B. Binas, C. Wilde, A. Colman.Lactation is disrupted by α-lactalbumin deficiency and can be restored by human α-lactalbumin gene replacement in mice. *Proc. Natl. Acad. Sci. USA*, 92: 2835–2839, 1995.
71. A.M. Mowat. The regulation of immune responses to dietary protein antigens. *Immunol. Today*, 8: 93–98, 1987.
72. H.L. Weiner. Oral tolerance: immune mechanisms and treatment of autoimmune diseases. *Immunol. Today*, 18: 335–343, 1997.
73. S.M. Kim, A. Enomoto, S. Hachimura, K. Yamauchi, S. Kaminogawa. Serum antibody response elicited by a casein diet is directed to only limited determinants of α_{s1}-casein. *Int. Arch. Allergy Immunol.*, 101: 260–265, 1993.
74. S. Hachimura, Y. Fujikawa, A. Enomoto, S.M. Kim, A. Ametani, S. Kaminogawa. Differential inhibition of T and B cell responses to individual antigenic determinants in orally tolerized mice. *Int. Immunol.*, 6: 1791–1797,1994.
75. T. Yoshida, S. Hachimura, S. Kaminogawa. The oral administration of low-dose antigen induces activation followed by tolerization, while high-dose antigen induces tolerance without activation. *Clin. Immunol. Immunopathol.*, 82: 207–215, 1997
76. A. Enomoto, M. Konishi, S. Hachimura, S. Kaminogawa. Milk whey protein fed as a constituent of the diet induced both oral tolerance and a systemic humoral response, while heat-denatured whey protein induced only oral tolerance. *Clin. Immunol. Immunopathol.*, 66: 136–142, 1993.

77. R. Fritsche, J.J. Pahud, S. Pecquet, A. Pfeifer. Induction of systemic immunologic tolerance to β-lactoglobulin by oral administration of a whey protein hydrolyzate. *J. Allergy Clin. Immunol.*, 100: 266–273, 1997.
78. S. Pecquet, A. Pfeifer, S. Gauldie, R. Fritsche. Immunoglobulin E suppression and cytokine modulation in mice orally tolerized to β-lactoglobulin. *Immunology*, 96: 278–285, 1999.
79. K Mizumach and J Kurisaki. Induction of oral tolerance in mice by continuous feeding with β-lactoglobulin and milk. *Biosci. Biotechnol. Biochem.*, 66:1287–1294, 2002.
80. S. Hachimura, Y. Takahashi, Y. Fujikawa, C. Tsumori, A. Enomoto, U. Yoshino, S. Kaminogawa. Suppression of the systemic immune response to casein by oral administration of tryptic digest of casein. *Biosci. Biotechnol. Biochem.*, 57: 1674–1677, 1993.
81. S. Pecquet, L. Bovetto, F. Maynard, R. Fritsche. Peptides obtained by tryptic hydrolysis of bovine β-lactoglobulin induce specific oral tolerance in mice. *J. Allergy Clin. Immunol.*, 105: 514–521, 2000.
82. R. Fritsche, J.J. Pahud, S. Pecquet, A. Pfeifer. Induction of systemic immunologic tolerance to β-lactoglobulin by oral administration of a whey protein hydrolyzate. *J. Allergy Clin. Immunol.*, 100: 266–273, 1997.
83. G.F. Hoyne, M.G. Callow, M.C. Kuo, W.R. Thomas. Inhibition of T-cell responses by feeding peptides containing major and cryptic epitopes: studies with the Der p I allergen. *Immunology*, 83: 190–195, 1994.
84. K. Hirahara, S. Saito, N. Serizawa, R. Sasaki, M. Sakaguchi, S. Inouye, Y. Taniguchi, S. Kaminogawa, A. Shiraishi. Oral administration of a dominant T-cell determinant peptide inhibits allergen-specific T_{H1} and T_{H2} cell responses in Cry j 2-primed mice. *J. Allergy Clin. Immunol.*, 102: 961–967, 1998.
85. F. Baggi, F. Andreetta, E. Caspani, M. Milani, L. Ronghi, R. Mantegazza, F. Cornelio, C. Antozzi. Oral administration of an immunodominant T-cell epitope downregulates Th1/Th2 cytokines and prevents experimental myasthenia gravis. *J. Clin. Invest.*, 104: 1287–1295, 1999.
86. S.R. Thurau, M. Diedrichs-Mohring, H. Fricke, C. Burchardi, G. Wildner. Oral tolerance with an HLA-peptide mimicking retinal autoantigen as a treatment of autoimmune uveitis. *Immunol. Lett.*, 68: 205–212, 1999.
87. A.C. Ouwehand, P.V. Kirjavaninen, C. Shortt, S. Salminen, Probiotics: mechanisms and established effects. *Int. Dairy J.*, 9: 43–52, 1999.
88. L.J. Fooks, G.R. Gibson. Probiotics as modulators of the gut flora. *Br. J. Nutr.*, 88 S: S39–S49, 2002.
89. M. Kalliomaki, S Salminen, H. Arvilommi, P. Kero, P. Koskinen, E. Isolauri. Probiotics in primary prevention of atopic disease: a randomised placebo-controlled trial. *Lancet*, 357: 1076–1079, 2001.
90. H. Kimoto, K. Mizumachi, T. Okamoto, J. Kurisaki, New *Lactococcus* strain with immunomodulatory activity: enhancement of Th1-type immune response. *Microbiol. Immunol.*, 48: 75–82, 2004.

22 Egg Proteins

Prithy Rupa and Yoshinori Mine

CONTENTS

22.1 INTRODUCTION

One of the most common causes of food allergy in infants and young children is eggs, although according to studies, most outgrow the allergy by the age of 5 years. Hen's egg is an essential ingredient in human as well as animal foods and is difficult to eliminate from the diet (1). Due to its high protein content, its introduction into the daily diet is necessary. Most people who are allergic react to the proteins in egg white, but some cannot tolerate the proteins in the yolk as well. Egg yolk has been considered less allergenic than egg white. Up to 24 different proteins have been isolated from egg white, but the antigenicity of most of these proteins is unknown. The prevalence of egg allergy is about 35% among food-allergic children and children with atopic dermatitis (2). It is more frequent in younger children who are younger than 3 years old than in older children and adults (3). It has been reported that the prevalence of egg allergy in children is maximum between 1.5 and 3 years of age. In the case of double-blind placebo-controlled food challenge (DBPCFC)-positive children with atopic dermatitis, the prevalence rate could be up to 45% (4). Seventy-six percent of sensitizations to egg proteins are found in children below 5 years, and 12% are found between the ages of 10 and 15 years.

22.2 PHYSICOCHEMICAL CHARACTERISTICS OF EGGS

Eggs should be stored at temperatures below 4°C and at a relative humidity of 70 to 80%. Eggs will age more in 1 day at room temperature than they will during 1 week under proper refrigeration. Eggs contain vitamins A, D, E, and K and the B-complex vitamins. Eggs are also a good source of the antioxidant carotenoids lutein and zeaxanthin.

The primary parts of an egg are the shell, yolk, and albumen. The shell, which is composed of calcium carbonate, is the outermost covering of the egg. The yolk constitutes just over one third of the egg and contains three fourths of the calories, most of the minerals and vitamins, and all of the lipids. The yolk also contains lecithin, the compound which is responsible for emulsification in products such as hollandaise sauce and mayonnaise. Egg yolk solidifies (coagulates) at temperatures between 65 and 70°C. The albumen, which is the clear portion of the egg, is often referred to as egg white. It constitutes about two thirds of the egg and contains more than half of the protein and riboflavin. There are about 40 different proteins in egg white. These distinct proteins play an important role during the developmental process and during the deterioration of the egg. The properties of egg white proteins with vital functional properties might be expected to undergo structural changes during the development process. Egg white constitutes the second line of defense against invading bacteria, next to the shell and shell membranes of the eggs.

Egg white proteins have been studied by a vast spectrum of scientists from different disciplines who have used them as a source of model (globular) proteins. Egg white proteins have different dynamic structures, and their amphiphilic nature possesses a range of physiological and biotechnological functions. The biological functions of egg white proteins are many, including prevention of the penetration of microorganisms to the yolk and provision of nutrients to the embryo during the late stages of development. Egg white coagulates, becoming firm and opaque, at temperatures between 62 and 65°C. Of the many different types of proteins found in egg white, most appear to possess antimicrobial properties or certain physiological functions to interfere with the growth and spread of the invading microorganisms. The multiple roles of this set of proteins are believed to prevent invading microorganisms from reaching the yolk and the embryo. The multiple conformational states of egg proteins are expected to have important consequences for the interpretation of their biological roles (1).

22.2.1 MAJOR EGG WHITE ALLERGENS

The allergenicity of egg white proteins has been the subject of many previously reported studies (5,6). The dominant egg white allergens are ovalbumin (OA), ovomucoid (OVM), ovotransferrin (OT), and lysozyme (LY) (7,8). The predominant egg white proteins have been well characterized and are very different with regard to their properties (9,10). Egg white proteins have been studied extensively, and most of them have been purified and their amino acid sequences determined. Some of the characteristics of the major egg white proteins are summarized in Table 22.1. Antigenic and allergenic properties of egg white proteins have been intensively

TABLE 22.1
Characteristics of Major Egg White Allergens

Proteins	% Dry Basis	pI	M.M (kDa)	Characteristics
Ovomucoid (*Gal d1*)	11	4.12	8.0	Inhibits trypsin
Ovalbumin (*Gal d2*)	54	4.5	44.5	Phosphoglycoprotein
Ovotransferrin (*Gal d3*)	12	6.1	76.0	Binds metalic ions
Lysozyme (*Gal d4*)	3.4	10.7	14.3	Lyses bacteria

investigated from the viewpoints of food science, immunology, and clinical allergology. OA was earlier reported as the major allergen in patients with hen's egg allergy (11,12) because of the use of commercial OA extracts contaminated with OVM. Recently, many groups have reported OVM to be the major allergen of hen's egg white (13–16). Some groups have reported OA, OVM, and OT to be major allergens and LY to be a weak allergen (17). The immunodominant epitopes for immunoglobulin E (IgE) response in animals from injection of IgE may differ from the IgE response in humans from ingestion of eggs.

22.2.2 Ovomucoid

Chicken OVM is a glycoprotein consisting of three tandem homologous domains, comprising 186 amino acids with a molecular weight of 28 kDa and a pI of 4.1 (18). It is a trypsin inhibitor and has a putative reaction site for the inhibition of serine proteinases. OVM constitutes 11% of the total protein weight in egg white. The molecule consists of three structurally independent tandem domains that are interconnected with peptide bonds, and each contains three intradomain disulfide bonds. Antibodies specific to OVM are detected frequently in the sera of patients suffering from hen's egg allergy (19). OVM is believed to play a significant role in the pathogenesis of allergic reactions to egg white compared with other egg white proteins (13–16). The titer of IgE antibodies to OVM was significantly greater in patients with positive results to oral egg challenge tests compared with that in patients with negative results. This is due to the relative stability of OVM to heat (20) and digestion with proteinases (21) and strong allergic reactivity (22) compared with other hen's-egg white proteins. Sensitivity of OVM to thermal inactivation is reported to be pH dependent (23). It has been shown that there is significantly more human IgG- and IgE-binding activity to the third domain than to the first and second domains of OVM (16). Using chemical modification of the third domain, it was found that the hydrophilic residues appeared to be more important for IgG binding and that hydrophobic residues were possibly more important for IgE binding (14). Each domain has common epitope(s) homology that is recognized by peripheral blood T cells from allergic patients (24). The first and the second domains contain four carbohydrate chains, and about half of all the third domain contains one large carbohydrate chain (the other half lacks the carbohydrate chain). It has been suggested that the carbohydrate chain in the third domain may play a role as an antigenic determinant of OVM against IgE (25,26). The disulfide bonds and tertiary structure of the third domain of OVM allow it to retain its allergenicity following proteolytic digestion (21). The reduction of OVM enhances its digestibility and reduces its allergenicity (27). OVM-specific T-cell lines (TCLs) and T-cell clones (TCCs) have been established recently (28). The significance of IgE-binding activity to enzymatic digests of OVM has been reported, and it has been concluded that subjects with high IgE-binding activity to pepsin-treated OVM are unlikely to outgrow egg white allergy (29).

The linear epitopes with IgE-binding capacity on OVM were determined by Cooke and Sampson (30) using overlapping peptides. Five IgE-binding epitopes and seven IgG-binding epitopes were identified using a pool of seven egg-allergic patients' sera. More recently, the fine mapping of the entire OVM gene and identification of B cell and T cell epitopes were reported (31). The epitopes were identified using pooled sera from patients with egg white allergy. Amino acids within each of the B cell- and T cell-binding epitope regions that were critical for immunoglobulin binding were identified. Three main IgE-binding epitopes were identified in the first domain, four in the second, and two major epitope regions in the third domain of the OVM sequence. A schematic representation of the location of the B cell and T cell epitopes on the entire OVM gene, reported by different groups, are summarized in Figure 22.1. A novel IgE epitope was reported by Holen et al. (32) recently, and they also reported T-cell epitope mapping of OVM using specific lymphocytes from egg-allergic patients. T-cell epitope mapping of the entire OVM sequence was reported using three different strains of mice (33). An undecane peptide attached to the N-terminal end of the recombinant third domain of OVM was found to suppress specific IgE- and IgG-binding activity (34).

22.2.3 Ovalbumin

Ovalbumin (OA) is a major globular component of egg white, and it belongs to the serpin superfamily. It is a phosphoglycoprotein with a molecular size of 45 kDa, consisting of 385 amino acids (35). The messenger ribonucleic acid (mRNA) of chicken OA has been sequenced (36). The IgE class of antibodies for OA have been shown to exist in sera from patients with egg white allergy (37). The molecular crystal structure of OA has been reported (38). A number of studies on its structural (39), functional (40), and immunological properties of egg OA (41) have been carried out. The treatment of OA with carboxymethylation, urea, or heat had no significant effect on its IgE-binding capacity to egg-allergic patient's sera (42). This would suggest that anti-OA IgE antibodies from egg-allergic patients' sera recognize mainly sequential epitopes. The specificities of IgE, IgG, and IgA class of antibodies for OA have previously been reported (43). Initial studies revealed that OA was the major allergen in hen's egg white. Proliferative responses of peripheral blood mononuclear cells to OA were significantly stronger in children with atopic dermatitis and in egg-sensitive children with immediate allergic symptoms (44).

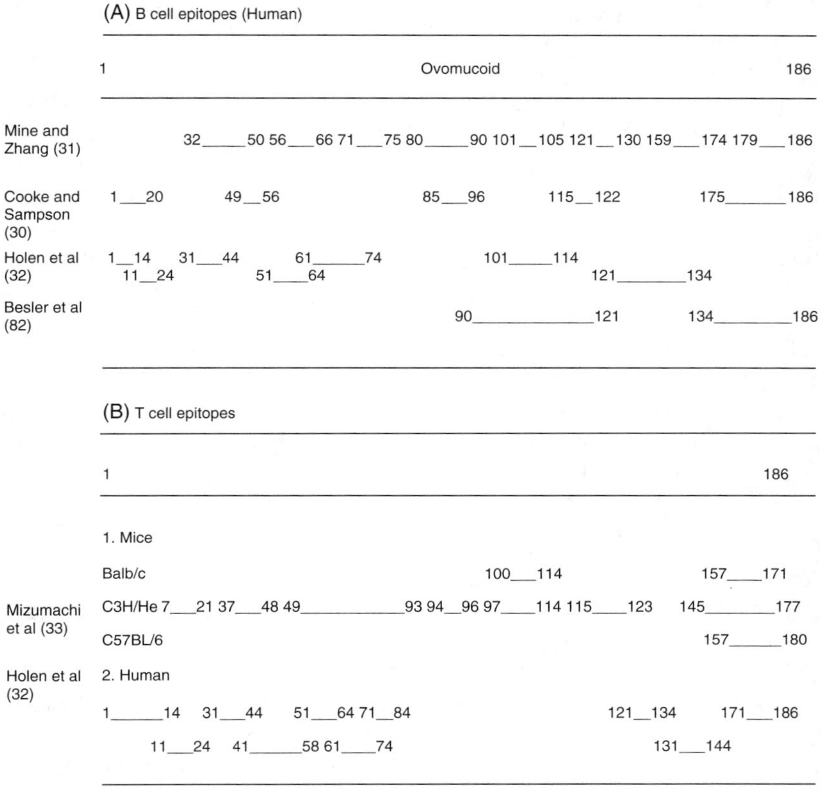

FIGURE 22.1 Schematic representation of the B-cell (A) and T-cell (B) epitope regions along the chicken ovomucoid molecule, as reported by different groups. (Ref. No. 82–85). The numbers represent amino acids, and the lines represent the length of the epitopes on the ovomucoid molecule.

Epitope mapping of OA was reported using enzyme-digested fragments and synthetic peptides (45,46). A synthetic decapeptide has been shown to encompass an Ig-binding haptenic epitope in OA (47), and antigenic and allergenic determinants of native OA have been identified using IgE antibodies from clinical patients (48). There seems to be a link between the allergen structure and IgE-binding epitopes, and it is believed that protein structure plays a crucial role in determining the immunodominant IgE B-cell epitopes (49,50). These reports have also been studied using Western blot and 2D-PAGE (two-dimentional polyacrylamide gel electrophoresis) quantitative precipitation inhibition assay. The T-cell proliferative response to OA in children with atopic dermatitis was observed (51). A synthetic peptide of OA that recognizes human IgE antibodies was also found to stimulate rabbit T cells (52). The same peptide also induced hypersensitivity reactions in Balb/c mice exposed via the respiratory route (53). T-cell clones for OA have been established earlier against patient's sera (49). Although a few B-cell epitopes have been identified, there are no obvious sequence motifs shared by these epitopes. Cyanogen bromide cleavage of a commercial preparation of OA was used by Kahlert et al. (54) to show IgE binding to peptide sequences 41–172 and 301–385. Fine mapping and structural analysis of the IgE B-cell epitopes in the entire chicken OA gene has been reported recently (55). Single amino acid substitutions within epitopes of OA can substantially enhance or abolish IgE antibody binding. Five distinct IgE-binding epitopic regions were identified in the primary sequence of chicken OA using pooled sera from 18 egg-allergic patients, and these immunodominant epitopes seemed to be distributed throughout the primary sequence of OA. The OA molecule consists of a five-stranded β-sheet running parallel to the long axis of the molecule and an α-helix protruding as a loop that forms the reactive center. Among the five IgE-binding epitopes identified, four of the epitopes are exposed to the protein surface. The major secondary structures of these four epitopes were β-sheet and β-turn structures (Figure 22.2). A schematic representation of the location of the B cell and T cell epitopes on the entire ovalbumin gene reported by different groups, are summarized in Figure 22.2.

22.2.4 Ovotransferrin

Ovotransferrin (OT) (Gald-3) is a single-chain glycoprotein with a molecular weight of about 80 kDa. It has a pI of 6, comprises 686 amino acids, and accounts for 13% of the protein in egg white (56). It consists of 12 disulfide bonds and has a 2.6% carbohydrate content. This protein has been demonstrated to be capable of delivering iron to cells and inhibiting bacterial multiplication. The nucleotide sequence of an almost double-stranded cDNA (copy deoxyribonucleic acid) of chicken OT (conalbumin) mRNA has been determined (57). The iron sequestering property of transferrins is also responsible for their bacteriostatic and bactericidal activities (58–60). OT can thus be used to produce iron-fortified food products and has bactericidal properties. OT is a covalent dimer protein, composed of N- and a C-terminal domains, each one binding to one atom-ion of transition metal. The C-terminal domain is glycosylated on Asn-473. Although the structure of OT has been evaluated in detail, epitope mapping and allergenic properties require more studies. Walsh et al. (61) found that OT was a minor allergen of egg white using radioallergosorbent test (RAST) studies with egg-allergic patient as subjects.

22.2.5 Lysozyme

Lysozyme (LY) makes up about 3.5% of egg white. It is a 14.3-kDa protein with a pI of 10.7. Holen and Elsayed (10) demonstrated that lysozyme could strongly bind IgE in all sera from egg-allergic individuals and that LY was a major allergen in hen's egg white. It is used in numerous foods as a bactericidal additive. LY is reported by several groups as one of the major allergens in egg white (62,63), although it is considered to be a relatively weak allergen when compared with the other major egg white allergenic proteins — OA, OVM, and OT. The three-dimensional structure of LY has been studied by several groups (64,65). The exact role of LY with regard to its allergenicity has been poorly studied and needs further investigation.

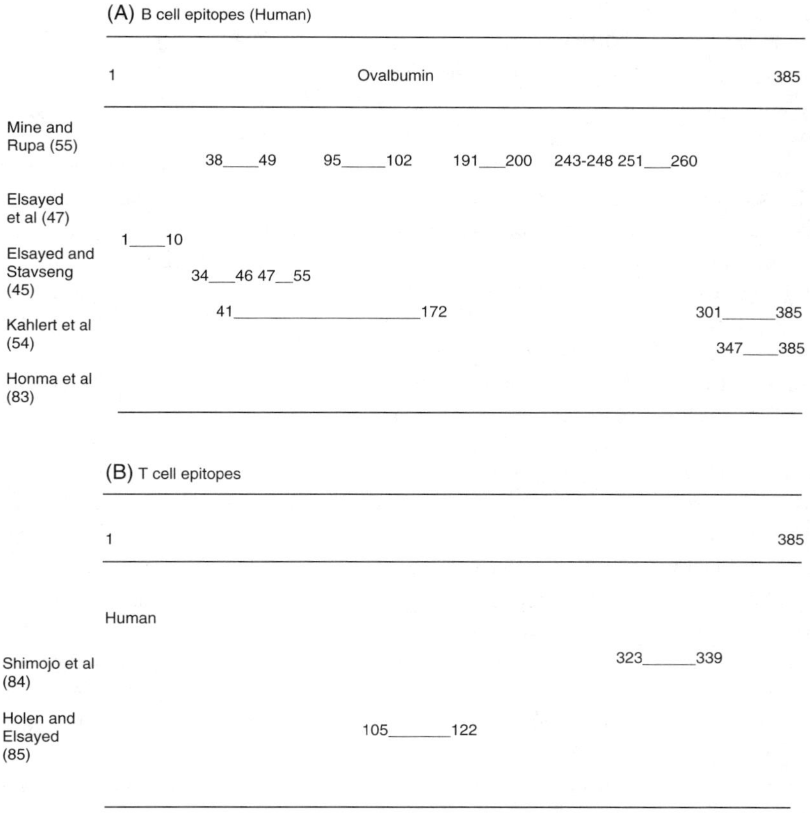

FIGURE 22.2 Schematic representation of the B cell (A) and T cell (B) epitope areas along the chicken ovalbumin molecule, as reported by different groups. (Ref. No. 82–85). The numbers represent the amino acid and the lines represent the length of the epitopes on the ovalbumin molecule.

22.3 PROCESSING OF EGGS AND ITS VALUE-ADDED HYPOALLERGENIC APPLICATION

Eggs are used as a multifunctional component for the commercial development of food products. Recent research has focused on finding nonfood uses for eggs and modifying egg chemistry to improve its nutritional value ("designer eggs"). By extracting fat and cholesterol from dried egg yolk powder using supercritical carbon dioxide, low-fat and low-cholesterol yolk forms have been generated. Fatty acid modification of yolk lipids and cholesterol-lowering eggs are being used on a large scale in the food industry; egg oil, a by-product of this process, has useful applications in the cosmetic and pharmaceutical industries. Choline and folate from egg yolk are added to enrich infant formulas. When cholesterol is added to infant formulas, care is taken to ensure that plasma cholesterol concentration and fatty acid composition exactly match the daily requirement of infants. Phosvitin acts as an emulsifier, and the combination of phosvitin and galactomannan yields a novel antioxidant with improved emulsifying property. Liposomes are used as a transport organ in the emulsification process, for example, in the cosmetic industry. Eggs are used as effective antibodies in humans as well as other animal models, and they do not cause any side effects. The antibodies are used as biochemical reagents due to their resistance to acid pH and digestive enzymes when removed and encapsulated. The composition of the whole egg has been engineered to enhance levels of omega fatty acids, folic acid, and vitamin E. The three parts of the egg are rich sources of nutrients: egg white being a source of proteins, yolk a source of lipids and proteins, and the egg shell a source of collagen and calcium after fractionation.

OVM has an important role in the pathogenesis of allergy when compared with other egg white proteins (66). Low-allergenic egg white products have been commercialized by depleting OVM through ethanol precipitation and are used as hypoallergenic formulas for patients sensitive to the OVM antigen (67). Heat-depleted OVM has been shown to be less allergenic than heated and freeze-dried preparations. OA is used as a master molecule and is a key reference in many biochemical and immunological research projects. LY is used as an antimicrobial agent, a preservative in various food components, and in the medical industry for therapeutic purposes. Ovomucin is used in cultured tumor cells as a cytotoxic agent. OT is known for its iron-binding property and is used to produce iron-fortified products. Avidin-biotin binding is used in molecular biology and immunological techniques and assays. The fatty acid DHA (docosahaxaenoic acid) can be easily incorporated into eggs to provide an alternative source other than fishes.

22.3.1 Heat Denaturation and Enzymatic Digestions

Heat treatment is a general food-processing method that induces the destruction of protein structures. Because of their unique properties such as gelation and foaming, egg white proteins have been used extensively in food processing. The functional properties of dried egg white can be significantly improved by heating the proteins at 80°C in a dry state. Egg white heated at 120°C becomes a completely new entity. Deamidation and heating at a controlled pH offers a new approach to improving the functionality of the egg protein (68). The thermal temperature required by bakers has been attributed to the denaturation and coagulation process of egg white. When heated, egg white undergoes visible changes, such as the appearance of a milky white color and the formation of a thermoirreversible gel coagulum. The thermal coagulation of egg white results in products that have stable textural properties. Pasteurization of liquid egg white is carried out at around 60°C for a few minutes. The gel strength, foam stability, and emulsion stability are increased to fourfold upon dry heating egg white for 10 days (68). The allergenic potential of egg white OVM is thought to depend on its stability to heat treatment and digestion. Little change in the digestibility of OVM by trypsin and chymotrypsin was observed following digestion with pepsin, indicating that pepsin-digested OVM retains its trypsin (protease)-inhibiting activity. The reduction of OVM, however, enhances its digestibility and lowers the allergenicity of the protein (27).

22.3.2 Suppression of Allergic Potential of Egg White by Genetic Modification

Recombinant DNA technology not only facilitates the characterization and analysis of the proteins that are allergenic but also provides the basis for producing allergens and their derivatives that are useful for diagnostic applications and specific immunotherapy (69). The advent of DNA technology has provided a way to develop and produce hypoallergenic variants using various strategies for gene mutation (70,71). Allergens that are genetically engineered need to meet the conformation, structure, and reactivity of the wild-type molecule. It is now possible to study allergens at the molecular level, determine their epitopes, and evaluate the mechanism of allergenic diseases. The therapeutic potential of several recombinant allergens has been reported (72,73). The key for specific immunotherapy is immune modulation of the T (helper) cells to prevent allergenicity (74,75). Recombinant allergens represent promising tools for the diagnosis and therapy of type 1 allergy. The cDNA coding for chicken OVM Gal d1 and Gal d2 was cloned in *Escherichia coli* under the control of the T-5 promoter fused with a six-his-tag at the amino terminal end (76,77). The third domain of OVM was cloned, and the IgE-binding properties were characterized and compared with the native third domain (78). These protein constructs have been purified to homogeneity, and a secondary structure has been confirmed using circular dichroism. It was found that the recombinant proteins have a less compact structure when compared with the native forms, and IgE-binding studies indicate that there is more binding in the case of recombinant forms than with the native protein. This reveals that the recombinant forms have similar sequential epitopes but may have more predominant conformational epitopes than their native analogues.

Allergen-specific IgE plays an important role in hypersensitivity reactions (79). Disruption of the three-dimensional structure and thereby destruction of the conformational epitopes in the recombinant molecule lead to low IgE-binding capacity but strong T-cell activation properties (80). Epitopes important for IgE binding were identified in chicken OVM and OA (31,55) genes using overlapping synthetic peptides and pooled sera from egg-allergic patients. Critical amino acids specific for IgE binding were evaluated. Substitution of these amino acids resulted in either loss or, in some cases, increase of IgE-binding activity. Five genetically modified mutations were made in the third domain of OVM using site-directed mutagenesis and characterized for their immunological properties (81). Two site mutations of glycine at position 32 and phenylalanine at position 37 substituted with methionine and alanine led to the complete loss of IgE-binding activity in human sera. They also interrupted the α-helical structure which comprised part of the IgE- and IgG-binding epitope regions. This clearly indicated that these amino acids play an important role in the

IgE-binding activity of the third domain of OVM as well as in the structural integrity of the protein. We are currently exploring the effect of this mutant and its immune response in a mouse model system. After detailed characterization of this mutant form in a mouse model, we will study the effect of this mutation in the whole OVM gene. Thus, this hypoallergenic molecule could be used effectively in immunotherapy for the suppression of allergic reactions to egg white. The structure–function relationship of the OVM gene need also be explored, thus leading to a solution to the problem of egg white allergy in humans.

REFERENCES

1. D.V. Vadehra, K.R. Nath. *Critical Reviews in Food Technology*. New York: CRC Press, 139–309, 1993.
2. J.F. Crespo, C. Pascual, A.W. Burks, R.M. Helm, M.M. Esteban. Frequency of food allergy in a pediatric population from Spain. *Pediatr. Allergy Immunol.*, 6: 39–43, 1995.
3. M. Eggesbo, G. Botten, R. Halvorsen, P. Magnus. The prevalence of allergy to egg: a population-based study in young children. *Allergy*, 56: 403–411, 2001.
4. S.A. Bock. The natural history of food sensitivity. *J. Allergy Clin. Immunol.*, 69: 173–177, 1982.
5. M.W. Yocum, M. Stoncham, A. Petruckevitch, J. Barton, R. Rona. A population study of food intolerance. *Lancet*, 343: 1127–1130, 1994.
6. H.A. Sampson, L. Mendelson, J.P. Rosen. Fatal and near fatal anaphylatic reactions to food in children and adolescents. *N. Engl. J. Med.*, 327: 380–384, 1992.
7. B. Aabin, L.K. Poulsen, K. Ebbehoj, A. Norgaard, H. Frokiaer, C. Bindslev-Jensen, V. Barkholt. Identification of IgE-binding egg white proteins: comparision of results obtained by different methods. *Int. Arch. Allergy Immunol.*, 109: 50–57, 1996.
8. J. Anet, J.F. Back, R.S. Baker, D. Barnett, R.W. Burley, M.E.H. Howden. Allergens in white and yolk of hen's egg: A study of IgE binding be egg proteins. *Int. Arch. Allergy Appl. Immunol.*, 77: 364–371, 1985.
9. R.W. Burley, D.V. Vedehra. *The Avian Egg: Chemistry and Biology*, New York: Wiley, 1989.
10. E. Holen, S. Elsayed. Characterization of four major allergens of hen egg white by IEF/SDS-PAGE combined with electrophoretic transfer and IgE-immunoautoradiography. *Int. Arch. Allergy Appl. Immunol.*, 91: 136–141, 1990.
11. H. Kahlert, A. Peterson, W.M. Becker, M. Schlaak. Epitope analysis of the allergen ovalbumin (Gal d2) with monoclonal antibodies and patient's IgE. *Mol. Immunol.*, 29: 1191–1201, 1992.
12. S. Elsayed. Four linear and two conformational epitopes on the major allergenic molecule of egg (Gal d1) ovalbumin. In: D. Kraft, A. Sehnon (Eds.). *Molecular Biology and Immnuology of Allergens*. Boca Raton, FL: CRC press, 1993, 287–290.
13. A. Urisu, H. Ando, Y. Morita, E. Wada, T. Yasaki, K. Yamada, K. Komada, S. Torii, M. Goto, T. Wakamatsu. Allergenic activity of heated and ovomucoid depleted egg white. *Int. Arch. Allergy Immunol.*, 100: 171–176, 1997.
14. J.W. Zhang, Y. Mine. Characterization of residues in human IgE and IgG binding site by chemical modification of ovomucoid third domain. *Biochem. Biophys. Res. Commn.*, 261: 610–613, 1999.

15. J. Bernhisel-Broadbent, H.M. Dintzis, R.Z. Dintziz, H.A. Sampson. Allergeneicity and antigenicity of chicken egg ovomucoid (Gal dIII) compared with ovalbumin (Gal dI) in children with egg allergy and in mice. *J. Allergy Clin. Immunol.*, 93: 1047–1059, 1994.
16. J.W. Zhang, Y. Mine. Characterization of IgE and IgG epitopes on ovomucoid using egg white allergic patient's sera. *Biochem. Biophys. Res. Commn.*, 253: 124–127, 1998.
17. D.R. Hoffman. Immunochemical identification of the allergens in egg white. *J. Allergy Clin. Immunol.*, 71: 481–486, 1983.
18. I. Kato, J. Schrode, W.J. Kohr, M. Laskowski, Jr. Chicken ovomucoid: determination of its amino acid sequence, determination of the trypsin reactive site, and preparation of all three of its domains. *Biochemistry*, 26: 193–201, 1987.
19. Y. Fujiwara, H. Tachibana, N. Eto, K. Yamada. Antigen binding of an ovomucoid-specific antibody is affected by a carbohydrate chain located on the light chain variable region. *Biosci. Biotechnol. Biochem.*, 64: 2298–2305, 2000.
20. K. Honma, M. Aoyagi, K. Saito, T. Nishimura, K. Sugimoto, H. Tsunoo, H. Niimi, Y. Kohno. Antigenic determinants on ovalbumin and ovomucoid: comparision of specificity of IgG and IgE antibodies. *Jpn. J. Allergol.*, 40: 1167–1175, 1991.
21. T. Matsuda, K. Watanabe, R. Nakamura. Immunochemical and physical properties of peptide-digested ovomucoid. *J. Agric. Food Chem.*, 31: 942–946, 1983.
22. E. Bleumink, E. Young. Studies on the atopic allergen in hen's egg. II. Further characterization of the skin reactive fraction in egg-white immunoelectrophoretic studies. *Int. Arch. Allergy Appl. Immunol.*, 40: 72–88, 1971.
23. Y. Konishi, J. Kurisaki, S. Kaminogawa, K. Yamauchi. Determination of antigenicity by radioimmunoassay and of trypsin inhibitory activities in heat or enzyme denatured ovomucoid. *J. Food Sci.*, 50: 1422–1426, 1985.
24. S.A. Bock, F.M. Atkins. Patterns of food hypersensitivity during sixteen years of double-blind, placebo-controlled food challenges. *J. Pediatr.*, 117: 561–567, 1990.
25. T. Matsuda, J. Gu, T. Kazuko, R. Nakamura. Immunoreactive glycopeptides separated from peptic hydrolyzate of chicken egg ovomucoid. *J. Food Sci.*, 50: 592–594, 1985.
26. T. Matsuda, R. Nakamura, I. Makashima, Y. Hasegawa, Y. Shimokata. Human IgE antibody to the carbohydrate-containing third domain of chicken ovomucoid. *Biochem. Biophys. Res. Commn.*, 129: 505–510, 1985.
27. J. Kovacs-Nolan, J.W. Zhang, S. Hayakawa, Y. Mine. Immunochemical and structural analysis of pepsin-digested egg white ovomucoid. *J. Agric. Food Chem.*, 48: 6261–6266, 2000.
28. K. Suzuki, R. Inoue, H. Sakaguchi, M. Aoki, Z. Kato, H. Kaneko, S. Matsushita, N. Kondo. The correlation between ovomucoid-derived peptides, human leucocyte antigen class II molecules and T cell receptor-complementarity determining region 3 compositions in patients with egg-white allergy. *Clin. Exp. Allergy*, 32: 1223–1230, 2002.
29. A. Urisu, K. Yamada, R. Tokuda, H. Ando, E. Wada, Y. Kondo, Y. Morita. Clinical significance of IgE-binding activity to enzymatic digests of ovomucoid in the diagnosis and the prediction of the outgrowing of egg white hypersensitivity. *Int. Arch. Allergy Immunol.*, 120: 192–198, 1999.
30. S.K. Cook, H.A. Sampson. Allergenic properties of ovomucoid in man. *J. Immunol.*, 159: 2026–2032, 1997.

31. Y. Mine, J.W. Zhang. Identification and fine mapping of IgG and IgE epitopes in ovomucoid. *Biochem. Biophys. Res. Commn.*, 292: 1070–1074, 2002.
32. E. Holen, B. Bolann, S. Elsayed. Novel B and T cell epitopes of chicken ovomucoid (Gal d1) induce T cell secretion of IL-6, IL-13 and INF-γ. *Clin. Exp. Allergy*, 31: 952–964, 2001.
33. K. Mizumachi, J. Kurisaki. Localization of T cell epitope regions of chicken ovomucoid recognized by mice. *Biosci. Biotechnol. Biochem.*, 67: 712–719, 2003.
34. Y. Mine, P. Rupa. Genetic attachment of undecane peptides to ovomucoid third domain can suppress the production of specific IgG and IgE antibodies. *Biochem. Biophys. Res. Commun.*, 311: 223–228, 2003.
35. A.D. Nisbet, R.I.I. Saundry, A.J.G. Moir, L.A. Fothergill, J.E. Fothergill. The complete amino acid sequence of hen ovalbumin. *Eur. J. Biochem.*, 115: 335–345, 1981.
36. L. McReynolds, B.W.O. Malley, A.D. Nisbet, J.E. Fothergill, D. Givol, S. Fields, M. Robertson, G.G. Brownlee. Sequence of chicken ovalbumin mRNA. *Nature*, 273: 723–728, 1978.
37. C.D. May, L. Rernigio, J. Feldman, S.A. Bock, R.I. Carr. A study of serum antibodies to isolated milk proteins and ovalbumin in infants and children. *Clin. Allergy*, 7: 583–595, 1977.
38. P.E. Stein, A.G.W. Leslie, J.T. Finch, W.G. Turnell, P.J. Mc Laughin, R.W. Carrell. Crystal structure of ovalbumin as a model for the reactive centre of serpins. *Nature*, 347: 99–102, 1990.
39. N. Takahashi, T. Koseki, E. Doi, M. Hirose. Role of an intrachain disufide bond in the conformation and stability of ovalbumin. *J. Biochem.*, 109: 846–851, 1991.
40. E. Doi, N. Kitabake. *Food Proteins and Their Applications. Structure and Functionality of Egg Proteins*. New York: Marcel Decker, 1997, 325–340.
41. M.J. Kim, J.W. Lee, H.S. Yook, S.Y. Lee, M.C. Kim, M.W. Byun. Changes in the antigenic and immunoglobulin E–binding properties of hen's egg albumin with the combinantion of heat and gamma irradiation treatment. *J. Food Sci.*, 65: 1192–1195, 2002.
42. Y. Mine, J.W. Zhang. Comparitive studies on antigenicity and allergenicity of native and denatured egg white proteins. *J. Agric. Food Chem.*, 50: 2679–2683, 2002.
43. K. Honma, Y. Kohmo, K. Saito, N. Shimojo, H. Tsunoo, H. Niimi. Specificities of IgE, IgG and IgA antibodies to ovalbumin. Comparison of binding activities to denatured ovalbumin or ovalbumin fragments of IgE antibodies with those of IgG or IgA antibodies. *Int. Arch. Allergy Appl. Immunol.*, 103: 28–35, 1994.
44. O. Fukutomi, N. Kondo, H. Agata, S. Shinoda, N. Kuwabara, M. Shinbara, T. Orii. Timing of onset of allergic symptoms as a response to a double-blind, placebo-controlled food challenge in patients with food allergy combined with a radioallergosorbent test and the evaluation of proliferative lymphocyte responses. *Int. Arch. Allergy Immunol.*, 104: 352–357, 1994.
45. S. Elsayed, L. Stavseng. Epitope mapping of region 11–70 of ovalbumin (Gal d1) using five synthetic peptides. *Int. Arch. Allergy Immunol.*, 104: 65–71, 1994.
46. T. Masuda, S.Y. Koseki, K. Yasumoto, N. Kitabatake. Characterization of anti-irradiation-denatured ovalbumin monoclonal antibodies. Immunochemical and structural analysis of irradiation denatured-ovalbumin. *J. Agric. Food Chem.*, 48: 2670–2674, 2000.
47. S. Elsayed, E. Holen, M.B. Haugstad. Antigenic and allergenic determinants of ovalbumin II. The reactivity of the NH2 terminal decapeptide. *Scand. J. Immunol.*, 27: 587–591,1988.

48. S. Elsayed, A.S. Hammer, M.B. Kalvenes, E. Florvaag, J. Apold, H. Vik. Antigenic and allergenic determinants of ovalbumin I. Peptide mapping, cleavage at the methionyl peptide bonds and enzymic hydrolysis of native and carboxymethyl OA. *Int. Arch. Allergy Appl. Immunol.*, 79: 101–107, 1986.
49. N. Shimojo, T. Katsuki, J.E. Coligan, Y. Nishimura, T. Sasazuki, H. Tsunoo, Y. Kohmo, H. Niimi. Identification of the disease related T cell epitope of ovalbumin and epitope-targeted T cell activation in egg allergy. *Int. Arch. Allergy Immunol.*, 105: 155–161, 1994.
50. M. Sen, R. Kopper, L. Pons, E.C. Abraham, A.W. Burks, G.A. Bannon. Protein structure plays a critical role in peanut allergen stability and may determine immunodominant IgE-binding epitopes. *J. Immunol.*, 169: 882–887, 2002.
51. S. Shinoda, N. Kondo, O. Fukutomi, H. Agata, Y. Suzuki, N. Shimozawa, S. Tomatsu, Y. Yamada, M. Takemura, A. Noma, T. Orii. Suppressive effects of elimination diets on T cell reponses to ovalbumin in hen's egg-sensitive atopic dermatitis patients. *Clin. Exp. Allergy*, 23: 689–695, 1993.
52. G. Johnsen, S. Elsayed. Antigenic and allergenic determinants of ovalbumin. III. MHC Ia-binding peptide (OA 323-339) interacts with human and rabbit specific antibodies. *Mol. Immunol.*, 27: 821–827, 1990.
53. H. Renz, K. Bradley, G. larsen, C. McCall, E.W. Gelfand. Comparision of the allergenicity of ovalbumin and ovalbumin peptide 323-339; differential expression of V β-expressing T cell populations. *J. Immunol.*, 151: 7206–7213, 1993.
54. H. Kahlert, A. Petersen, W.M. Becker, M. Schlaak. Epitope analysis of the allergen ovalbumin (Gal DII) with monoclonal antibodies and patient's IgE. *Mol. Immunol.*, 29: 1191–1201, 1992.
55. Y. Mine, P. Rupa. Fine mapping and structural analysis of immunodominant IgE allergenic epitopes in chicken egg ovalbumin. *Prot. Eng.*, 16: 747–752, 2003.
56. J. Williams, T.C. Elleman, I.B. Kingston, A.G. Wilkins, K.A. Kuhn. The primary structure of hen ovotransferrin. *Eur. J. Biochem.*, 122: 297–303, 1982.
57. J.M. Jeltsch, P. Chambon. The complete nucleotide sequence of the chicken ovotransferrin mRNA. *Eur. J. Biochem.*, 122: 291–295, 1982.
58. P. Valenti, G. Antonini, C. Von Hunolstein, P. Visca, N. Orsi, E. Antonini. Studies on the antimicrobial activity of OTF. *Int. J. Tissue React.*, 97: 105, 1983.
59. R.L. Jurado. Iron, infections, and anemia of inflammation. *Clin. Infect. Dis.*, 25: 888–895, 1997.
60. L.H. Vorland. Lactoferrin: a multifunctional glycoprotein. *APMIS*, 107: 971–981, 1999.
61. B.J. Walsh, C. Elliot, R.S. Baker, D. Barnett, R.W. Burley, D.J. Hill, M.E.H. Howden. Allergenic cross-reactivity of egg-white and egg-yolk proteins. An *in vitro* study. *Int. Arch. Allergy Appl. Immunol.*, 84: 228–232, 1987.
62. J.A. Bernstein, A. Kraut, D.I. Bernstein, R. Warrington, T. Bolen, C.P. Warren, I.L. Bernstein. Occupational asthma induced by inhaled egg lysozyme. *Chest*, 103: 532–535, 1993.
63. S. Fremont, G. Kanny, J.P. Nicolas, D.A. Moneret-Vautrin. Prevalence of lysozyme sensitization in an egg allergic population. *Allergy*, 52: 224–228, 1997.
64. P. Weber, I. Raynaud, L. Ettouati, M.C. Trescol-Biemont, P.A. Carrupt, J. Paris, C. Rabourdin-Combe, D. Gerlier, B. Testa. Molecular modeling of hen egg lysozyme HEL[52-61] peptide binding to I-Ak MHC class II molecule. *Int. Immunol.*, 12: 1753–1764, 1998.

65. D.H. Fremont, D. Monnaie, C.A. Nelson, W.A. Hendrickson, E.R. Unanue. Crystal structure of I-Ak in complex with a dominant epitope of lysozyme. *Immunity*, 8: 305–317, 1998.
66. A. Urisu, H. Ando, Y. Morita, E. Wada, T. Yasaki, K. Yamada, K. Komada, S. Torii, M. Goto, T. Wakamatsu. Allergenic activity of heated and ovomucoid-depleted egg white. *J. Allergy Clin. Immunol.*, 100: 171–176, 1997.
67. S. Tanabe, S. Tesaki, M. Watanabe. Producing a low ovomucoid egg white preparation by precipitation with aqueous ethanol. *Biosci. Biotech. Biochem.*, 64: 2005–2007, 2000.
68. Y. Mine. Effect of pH during the dry heating on the gelling properties of egg white proteins. *Food Res. Int.*, 29: 155–161, 1996.
69. R. Valenta. The future of antigen-specific immunotherapy for allergy. *Natl. Rev. Immunol.*, 2: 446–453, 2002.
70. M.B. Singh, N. De Weerd, P.L. Bhalla. Genetically engineered plant allergens with reduced anaphylactic activity. *Int. Arch. Allergy Immunol.*, 119: 75–85, 1999.
71. F. Ferriera, M. Wallner, H. Breiteneder, A. Hartl, J. Thalhamer, C. Ebner. Genetic engineering of allergens: Future therapeutic products. *Int. Arch. Allergy Immunol.*, 128: 171–178, 2002.
72. M.D. Chapman, A.M. Smith, L.D. Vailes, K.K. Arruda, V. Dhanaraj, J. Pones. Recombinant allergens for diagnosis and therapy of allergic disease. *J. Allergy Clin. Immunol.*, 106: 409–418, 2000.
73. H. Lowenstein, J.N. Larsen. Recombinant allergens/allergen standardization. *Curr. Allergy Asthma Rep.*, 5: 474–479, 2001.
74. H. Secrist, C.J. Chelen, Y. Wen, J.D. Marshall, D.T. Umetsu. Allergen immunotherapy decreases interleukin 4 production in CD4+ T cells from allergic individuals. *J. Exp. Med.*, 178: 2123–2130, 1993.
75. C. Ebner, U. Siemann, B. Bohle, M. Willheim, U. Wiedermann, S. Schenk, F. Klotz, H. Ebner, D. Kraft, O. Scheiner. Immunological changes during specific immunotherapy of grass pollen allergy: reduced lymphoproliferative responses to allergen and shift from TH2 to TH1 in T-cell clones specific for Phl p 1, a major grass pollen allergen. *Clin. Exp. Allergy*, 27: 1007–1015, 1997.
76. P. Rupa, Y. Mine. Structural and immunological characterization of recombinant ovomucoid expressed in *Escherichia coli*. *Biotechnol. Lett.,* 25: 427–433, 2003.
77. P. Rupa, Y. Mine. Immunological comparison of native and recombinant egg allergen, ovalbumin, expressed in *Escherichia coli*. *Biotechnol. Lett.*, 25: 1917–1934, 2003.
78. E. Sasaki, Y. Mine. IgE binding properties of the recombinant ovomucoid third domain expressed in *Escherichia coli*. *Biochem. Biophys. Res. commn.*, 282: 947–951, 2001.
79. B.J. Sutton, H.J. Gould. The human IgE network. *Nature*, 366: 412–418, 1993.
80. F. Ferriera, K. Hirtenlehner, A. Jilck, J. Godnickcvar, H. Breiteneder, R. Grimm, K. Hoffmann-Sommergruber, O. Scheiner, D. Kraft, M. Breitenbach, H.J. Rheinberger, C. Ebner. Dissection of immunoglobulin E and T lymphocyte reactivity of isoforms of the major birch pollen allergen Bet v 1: potential use of hypoallergenic isoforms for immunotherapy. *J. Exp. Med.*, 183: 599–609, 1996.
81. Y. Mine, E. Sasaki, J.W. Zhang. Reduction of antigenicity and allergenicity of genetically modified egg white allergen, ovomucoid third domain. *Biochem. Biophys. Res. Commn.*, 302: 133–137, 2003.

82. M. Besler, A. Petersen, H. Steinhart, A. Paschke. Identification of IgE-binding peptides derived from chemical and enzymatic cleavage of ovomucoid (Gal d1). [http://www.food-allergens.de] *Internet Symposium of Food Allergens*, 1(1): 1–12, 1999.
83. K. Honma, Y. Kohno, K. Saito, N. Shimojo, T. Horiuchi, H. Hayashi, N. Suzuki, T. Hosoya, H. Tsunoo, H. Niimi. Allergenic epitopes of ovalbumin (OVA) in patients with hen's egg allergy: inhibition of basophil histamine release by haptenic ovalbumin peptide. *Clin. Exp. Immunol.*, 103: 446–453, 1996.
84. N. Shimojo, T. Katsuki, J.E. Coligan, Y. Nishimura, T. Sasazuki, H. Tsunoo, T. Sakamaki, Y. Kohno, H. Niimi. Identification of the disease-related T cell epitope of ovalbumin and epitope-targeted T cell inactivation in egg allergy. *Int. Arch. Allergy Immunol.*, 105: 155–161,1990.
85. E. Holen, S. Elsayed. Specific T cell lines for ovalbumin, ovomucoid, lysozyme and two OA synthetic epitopes generated from egg allergic patient's PBMC. *Clin. Exp. Immunol.*, 26: 1080–1088, 1996.

23 Soybean

Tadashi Ogawa

CONTENTS

23.1 INTRODUCTION

Soybean and soybean products, which are important protein sources, are known as one of the major allergenic foodstuffs, in common with other protein-rich foods such as milk and eggs (1). In recent years, soybean protein isolate (SPI) prepared from defatted soybean meal has been increasingly employed in food processing as an ingredient in minced meat products, stuffing, pasta flour, yogurts, sauces, and flavor enhancers because of its high nutritional quality, processing functionality, and low cost. Accordingly, it has been difficult for patients allergic to soybean to get allergen-free products from commercially available processed foods; also, it has become necessary for these patients to have wide knowledge about the ingredients in the processed foods in which allergenic food residues are incorporated as processing aids. At present, strict elimination of offending foodstuffs is generally recommended as a conventional and effective treatment of food allergies. Long termavoidance of nutritionally fundamental or essential foods, however, may lead to malnutrition in young patients. There is, therefore, an urgent need to identify the protein components eliciting allergic manifestations for the reduction of the allergenicity of soybean products. The allergenicity of soybean is known to reside in the protein fractions, not in the soybean oil itself (2), whereas oxidized soybean oil has been shown to enhance the immunoglobulin E (IgE)-binding ability against soybean or other food proteins (3). In 1934, Duke (4) pointed out soybean as a possible important source of food allergy among people taking soymilk formula as milk substitutes. In 1980, a soybean allergen was first reported by Moroz and Yang (5); using the serum of a laboratory worker, who might have been sensitized through the airways, the allergen was isolated and identified as Kunitz type soybean trypsin inhibitor (KSTI). Shibasaki et al. (6) also reported that various allergenic protein components occurred in soybean protein fractions, and the IgE antibodies in the sera of the soybean allergic patients showed cross-reactivity among the 2S-, 7S-, and 11S-globulin fractions by radioallergosorbent test (RAST)-inhibition analyses. However, this method could not characterize the individual protein component responsible for cross-reactivity. They demonstrated that the most allergenic fraction was the 2S-globulin fraction, and then the 7S- and 11S-, in a decreasing order. Several investigators described the features of soybean allergens, but no detailed information was presented. Burks et al. (7) showed that allergenic proteins in soybean predominated in the 7S- or 11S-globulin fractions rather than in the 2S-globulin fraction, as a result of immunochemical analysis using the sera of soybean-sensitive patients with atopic dermatitis. Recently, Herian et al. (8) reported that the sera of patients sensitive to both peanuts and soybean bound to several protein components with molecular weights ranging from 50,000 to 60,000 Da (probably subunits of β-conglycinin) and also to the component with a molecular weight of about 20,000 Da, not identical with KSTI, which was strongly recognized by the IgE from patients allergic only to soybean. Rodrigo et al. (9) reported that the accidental inhalation of soybean dust caused asthma in individuals in Barcelona. Asthma patients inhaling soybean dust had specific IgE antibodies for the glycoproteins with molecular weights lower than 14,000 Da, which were assumed to be degradation products of β-conglycinin or unique protein species occurring in soybean hull, designated to be Gly m 1 and 2. Herian et al. (8) described that different soybean-allergic subjects were sensitive to quite different proteins that could be classified into three categories according to immunoblotting patterns. Results also showed that the IgE-binding proteins varied among patients, but the patients could not be classified into distinct groups according to their immunoblotting patterns (1). Recently, the occurrence of about 15 protein components binding with IgE antibodies in the sera of soybean-sensitive patients was demonstrated, three of which, named as Gly m Bd 30K, Gly m Bd 28K, and Gly m Bd 60K, were shown to be major allergenic proteins (1). Based on this information about allergenic proteins in soybean as the target to be removed, many approaches to reduce the allergenicity of soybean and soybean products have been proposed: (a) physicochemical procedures such as heat denaturation and precipitation, (b) destruction and modification of allergenic structures such as an introduction of polysaccharide moieties and enzymatic digestion, (c) breeding (selection of allergen-deficient varieties or induction of mutant), (d) genetic engineering, and (e) fabrication of nonallergenic constituents. Furthermore, more selective and sensitive methods for evaluation of the allergenicity of soybean products have concurrently been developed, which are applicable during the course of processing. The convenient methods to detect and determine the major allergens by immunoblotting and enzyme-linked immunosorbent assay (ELISA or sandwich ELISA) have been established using allergen-specific monoclonal antibodies. This chapter describes recent knowledge about soybean allergens and development of hypoallergenic products.

23.2 MAJOR ALLERGENS IN SOYBEAN

23.2.1 Gly m Bd 30K

The soybean allergenic protein, Gly m Bd 30K (1), which is most strongly and frequently recognized by the IgE antibodies in the sera of soybean-sensitive patients with atopic dermatitis, has been characterized to be a 34-kDa oil body-associated protein in soybean seed (10). This protein had been identified by Karinski et al. (11) from the fractionated soybean oil body membrane, whereas the cDNA (complementary deoxyribonucleic acid) was isolated and cloned as a vacuolar storage protein P34 with close homology to thiol proteases classified into the group of papain superfamily. The primary structure of Gly m Bd 30K was shown to have about 30% homology or 54% similarity with Der p1, a house dust mite allergen, which is thiol protease found in the feces of *Dermatophagoides preronyssius* (12). As shown in Figure 23.1, the mature P34 vacuolar protein consists of 257 amino acid residues and is derived by removal of a part of the 122 amino acid residues in the N-terminal of a precursor protein with a molecular weight of about 47,000 Da during maturation in a vacuole (11). The glycosylation site of Gly m Bd 30K was established to be located on the Asn170 residue of a mature protein (13), which consists of mannose, N-acetylglucosamine, xylose, and fucose in a mole ratio of 3:2:1:1, respectively, indicating one of typical plant aspargine N-linked high-mannose type glycans with xylose and fucose branches. The localization of Gly m Bd 30K (P34) in vacuoles of soybean cotyledons was confirmed by an electron microscopic immunostaining technique (14). In recent years, it was shown that the IgE-binding sites (B-cell epitopes) located on Gly m Bd 30K were located on the 3–12, 100–110, 229–238, 299–308, and 331–340 amino acid residues (15). Interestingly, all the epitope sites recognized by human IgE antibodies were shown to be quite different from those on the house dust mite allergen, Der p 1. Gly m Bd 30K was specifically associated with the proteins in the 7S-globulin fraction through the disulfide linkage. This property added important information to the strategy of development of hypoallergenic SPI. Furthermore, there are no soybean varieties lacking Gly m Bd 30K in the stock culture of soybean.

P34 (Gly m Bd 30K) deduced amino acid sequence

5 10 15 20 25 30

1 *M G F L V L L L F S L L G L S S S S S I S T H R S I L D L D*

31 *L T K F T T Q K Q V S S L F Q L W K S E H G R V Y H N H E E*

61 *E A K R L E I F K N N S N Y I R D M N A N R K S P H S H R L*

91 *G L N K F A D I T P Q E F S K K Y L Q A P K D V S Q Q I K M*

121 *A N K K M K K E Q Y S C D H P P A S W D W R K K G V I T Q V*

151 K Y Q G G C G R G W A F S A T G A I E A A H A I A T G D L V

181 S L S E Q E L V D C V E E S E G S Y N G W Q Y Q S F E W V L

211 E H G G I A T D D D Y P Y R A K E G R C K A N K I Q D K V T

241 I D G Y E T L I N S D E S T E S E T E Q A F L S A I L E Q P

271 I S V S I D A K D F H L Y T G G I Y D G E N C T S T Y G I N

301 H F V L L V G Y G S A D G V D Y W I A K N S W G F D W G E D

331 G Y I W I Q R N T G N L L G V C G M N Y F A S Y P T K E E S

361 E T L V S A R V K G H R R V D H S P L

(A)

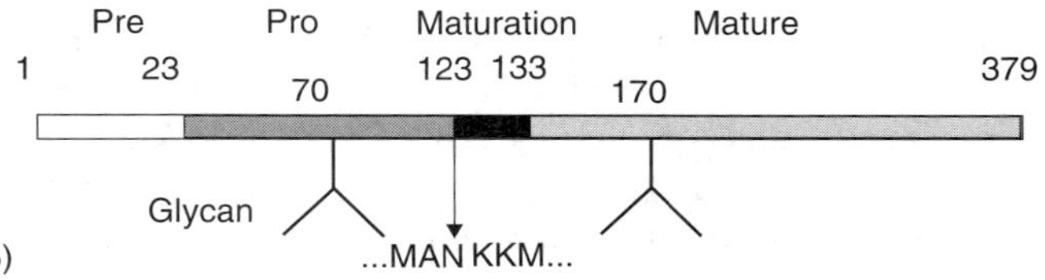

(B)

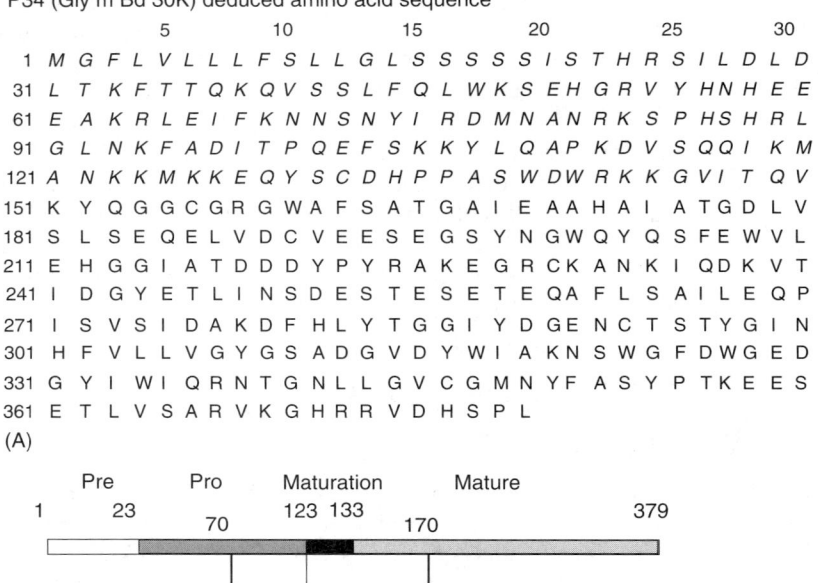

FIGURE 23.1 Molecular structure of P34 (Gly m Bd 30K). (A) Primary structure of P34 (pre- and prodomains indicated by italics); (B) Processing site, domains and glycosylation sites on the molecule of P34 precursor. (Adapted from A. Kalinski, J.M. Weisemann, B.F. Matthews, E.M. Herman. *J. Biol. Chem.*, 265, 13843–13848, 1990. With permission.)

The cDNA was cloned and the recombinant allergen without glycan moiety was prepared from E. coli, which was recognized by the sera of the soybean-sensitive patients, suggesting that rGly m Bd 30K could be applicable for diagnostic use as an allergen standard of RAST (16). In addition, the distribution of Gly m Bd 30K as the index of soybean allergenicity in soybean varieties and soybean products can be selectively determined by the use of monoclonal antibodies, F5 and H6 (17,18) (Figure 23.2). Furthermore, Okinaka et al. (19) showed that Gly m Bd 30K (P34) is a receptor of syringolide, which is an elicitor produced by *Pseudomonas syringe*, and the P34-elicitor complex probably induces a hypersensitive response through the inhibition of peroxisomal reduced nicotinamide-adenine dinucleotide (NADH)-dependent hydroxypyruvate reductase, suggesting that this allergenic protein might be associated with the defense system related to pathogenesis-related proteins.

23.2.2 Gly m Bd 28K

Gly m Bd 28K, a minor protein component in soybean (recognized by soybean-sensitive patients) has about 25% incidence as a major allergen. It was isolated

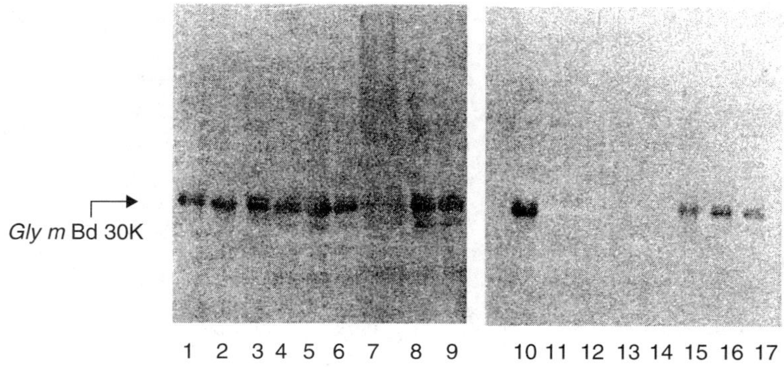

FIGURE 23.2 Immunoblotting analysis of Gly m Bd 30K in soybean products. Proteins in the extracts of soybean products were separated by SDS-PAGE and then blotted on nitrocellulose membranes. The allergen on the membrane was immunostained by a monoclonal antibody F5. Lanes: (1) isolated Gly m Bd 30K; (2) soybean grain; (3) soy milk; (4) *tofu* (soft type); (5) *tofu* (hard type); (6) *kori-tofu* (freeze-dried *tofu*); (7) *kinako* (baked soybean); (8) *abra-age* (fried *tofu*); (9) *yuba* (soy protein coagulant); (10) *miso* (soybean paste); (11) *syoyu* (soy sauce); (12) *natto* (fermented soybean); (13) meat ball; (14) beef croquette; (15) fried chicken; (16) fish sausage; and (17) hamburger. Products 1 to 12 are soybean products, and 13 to 17 are indicated to contain plant protein isolates as the ingredients. (Adapted from H. Tsuji, N. Bando, M. Kimoto, N. Okada, T. Ogawa. *J. Nutr. Sci. Vitaminol.*, 39, 389–397, 1993. With permission.)

and purified from 7S-globulin fraction prepared from defatted soybean flakes (products from Indiana, Ohio, and Michigan in the U.S.A.) (20). This purified allergen was shown to be a glycoprotein with a molecular mass of 26-kDa, and an isoelectric point of 6.1; an Asn-N-linked glycan moiety with the same sugar composition as that of Gly m Bd 30K was identified to be located on the Asn20 residue of Gly m Bd 28K. The N-terminal amino acid sequence analysis afforded FHDDEGGDKKSPKSLFMSDSTRVFK, and no homologous proteins (peptides) could be found in a database of proteins (21). In 2001, a cDNA clone encoding Gly m Bd 28K was isolated by Tsuji et al. (22). The open reading frame was demonstrated to encode a polypeptide composed of 473 amino acids, and the N-terminal region of this peptide contained the same amino acid sequence as that of the N-terminal peptide of Gly m Bd 28K with preceding 21 amino acids. It was shown that the polypeptide for the cDNA exhibits high homology with the MP-27/MP-32 proteins in pumpkin seeds and the carrot globulin-like protein, indicating that the protein deduced from this cDNA clone might be converted to Gly m Bd 28K and 23-kDa protein fragment during the development of soybean cotyledonous proteins. In addition, a nucleotide sequence deduced from the N-terminal amino acid sequence (21) coincided completely with a part of the sequence of unknown cDNA clone reported from Glycine max (GenBank accession no. AI416520), which was assumed to encode a vicilin-like protein similar to that reported from peanuts (23). When soybean varieties lacking this 28-kDa allergen were screened in the Japanese stock cultures and imported soybean seeds, about 80% of varieties examined were shown to lack the allergen, Gly m Bd 28K (M. Takahashi, personal communication) (Figure 23.3). The SPI prepared from defatted soybean flakes (IOM) was shown to contain this allergenic protein and the processed foods with SPI were also demonstrated to contain Gly m Bd 28K as well as Gly m Bd 30K (24).

23.2.3 Gly m Bd 60K (α-subunit of β-conglycinin)

The other major allergenic protein that was recognized by about 25% of the sera from soybean-sensitive patients with atopic dermatitis was identified as the α-subunit of β-conglycinin (25). The IgE antibodies recognizing the α-subunit showed no cross-reactivity against both of α'- and α-subunits of β-conglycinin, known to be highly homologous to the α-subunit. The α-subunit of β-conglycinin is a glycoprotein with a molecular weight of 57,000 Da and a pI of 4.90 (26). The amino acid sequence of the precursor deduced from the cDNA consisted of 543 amino acid residues (27). The epitope(s) of the IgE antibodies were shown to be located on the peptide of 232–383 residues from the N-terminal, which is highly homologous to the α-subunit and phaseolin, a storage protein of Phaseolus vulgalis (25) (Figure 23.4).

23.2.4 Other Allergenic Proteins

Soybean low-molecular-weight proteins identified by Rodrigo et al. (9) as allergens eliciting asthma in patients in Barcelona were identified as Gly m 1.0101 (Gly m 1A) and Gly m 1.0102 (Gly m 1B), which are isoforms with different

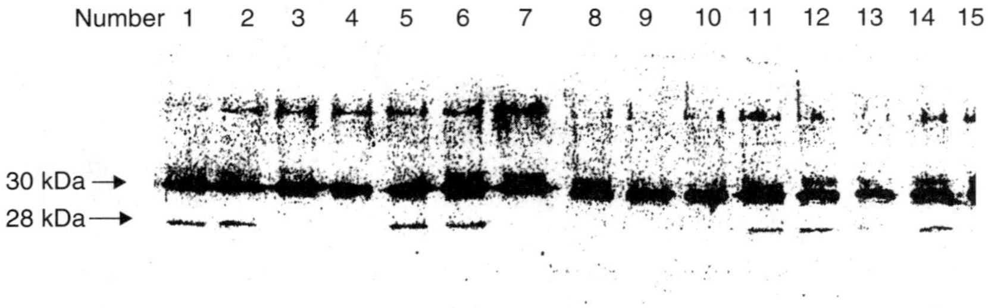

FIGURE 23.3 Distribution of Gly m Bd 30K and 28K in soybean varieties. Allergenic proteins were detected by immunoblotting technique using monoclonal antibodies F5 and H6. Lanes: (1) Toyomusume; (2) Suzumaru; (3) Fukuyutaka (Fukuoka); (4) Murayutaka; (5) Ryuuhou; (6) Toyokomachi; (7) Tachinagaha; (8) Enrei; (9) tachinagaha; (10) Fukuyutaka (Mie); (11) NK19-90 (China); (12) Binton (USA); (13) Burazil; (14) A2053 (USA), (15) Medium grain (USA). (Adapted from H. Tsuji, N. Bando, M. Kimoto, N. Okada, T. Ogawa. *J. Nutr. Sci. Vitaminol.*, 39, 389–397, 1993. With permission.)

A) YVVNPDNNENLRLITLAIPVNKPGRFESFFLSSTEAQQS
B) YVVNPDN**D**ENLR**M**ITLAPIVNKPGRFESFFLSST**Q**AQQS
C) Y**L**VNPD**PK**E**D**LRII**Q**LA**M**PVNNPQ-**IHE**FFLSSTE**K**QQS

YLQGFSRNILEASYDTKFEE–INKVLFSREEGQQQGRQR
YLQGFS**K**NILEASYDTKFEE–INKVLFS**G**REEGQQQGE**E**R
YLQ**E**FS**KH**ILEAS**F**N**S**KFEE**E**IN**R**VLF**E**–EEGQQ-----

LQESVIVEISKEQIRQLSKRAKSSSRKTISSEDKPFNLR
LQESVIVEISKKQIRELSK**H**AKSSSRKTISSEHKPFNLG
--E**G**VIV**N**I**DSE**QI**KE**LSK**H**AKSSSRK**SL**S**KQ**D**N**T**IGNE**

SRDPIYSNKLGKFFEITPEKNPQLRDLDIFLSIVDM 232–383
SRDPIYSNKLGK**L**FEIT–**QR**NPQLRDLD**V**FLS**V**VDM 268–417
FGNLTERTDN---------------**SLNVLISSIE**M 135–276

FIGURE 23.4 Comparison of the amino acid sequence of the epitope region on α subunit of β-conglycinin with the corresponding sites of α' subunit and phaseolin of *Phaseolus vulgaris*. (A) α subunit of β-conglycinin; (B) α' subunit of β-conglycinin; and (C) phaseolin. (Adapted from T. Ogawa, N. Bando, H. Tsuji, K. Nishikawa, K. Kitamura. *Biosci. Biotechnol. Biochem.*, 59, 831–833, 1995. With permission.)

molecular weights of 7,500 and 7,000 Da, respectively (28). Their amino acid sequences closely match a part of the hydrophobic protein first reported by Odani et al. (29), which is synthesized in endocarp on inner ovarian wall and is deposited on the seed surface during the development of the soybean (30). Patients in Barcelona with asthma have specific IgE antibodies for these unique glycoproteins distributed in soybean hull. Gly m 2 is demonstrated as an allergen in the hull and is also related to Barcelona asthma; it has a molecular weight of 8,000 Da and a pI of 6, which is homologous with a storage protein from cotyledones of *Vigna radiata* (cow pea) and with a disease response protein from *Pisum sativum* (green pea) (31). However, cross-reactivity between these proteins has not been elucidated. Soybean profilin, also identified as an allergen (Gly m 3) has a molecular weight of 14,000 Da and a pI of 4.4, is homologous to Bet v 2, a birch pollen allergen, with a sequence identity of 73% and other 11 plant profilins with 69 to 88% identity (32). These three allergens are recognized as inhalant allergens eliciting asthma in patients in Barcelona when cargo boats unload soybean grains because the allergenic proteins are located in a part of the hull of the grain surface. There are several reports on glycinin as an allergen, and these identify acidic subunits A1a, A1b, A2, A3, and A4 to be allergenic (33). The IgE epitope on the acidic chain of glycinin G1 is located between amino acid residues 192 and 306 (34). KSTI was first identified as a soybean allergen using the sera of laboratory workers with asthmawho had been sensitized through the airway while dealing with a fine powder of KSTI used as a reagent (5). KSTI was also identified as causing sensitization of occupationally exposed bakers (35). Examination of the sera of the patients showed that very few patients had IgE against KSTI (frequency of sensitization about 1.5% in the soybean-sensitive patients with atopic dermatitis (1).

23.3 DEVELOPMENT OF HYPOALLERGENIC SOYBEAN PRODUCTS

23.3.1 Determination of Allergenicity

There is a need to establish a convenient, selective and semiquantitative method for the determination or evaluation of the allergenicity of soybean and soybean products. The immunoblotting method (Western blot) is one of the useful techniques to detect very small amounts of allergens, but it is not a quantitative way to evaluate and compare allergenicity (amounts of allergens) during the course of reduction processing. The most suitable method for selective and quantitative analyses of allergens is considered to be enzyme-linked immunosorbent assay (ELISA) using allergen-specific antibodies. For measurement of the allergenicity of soybean, Gly m Bd 30K is the target allergen to be determined because about two thirds of soybean-sensitive patients examined have been shown to have IgE antibodies binding to Gly m Bd 30K. Furthermore, it is suitable to take Gly m Bd 30K as an index allergen molecule because of the wide distribution of this allergen in processed foods. Two monoclonal antibodies (mAb), F5 and H6, were prepared from the hybridoma of the spleen cells of BALB/c mice immunized with reductively carboxymethylated allergen (17), and a sandwich ELISA was established using these two mAbs (36). An immunoblot of Gly m Bd 30K by F5 mAb was carried out in various soybean products. As shown in Figure 23.2, all soybean products gave positive bands corresponding to the allergen. The allergen was detected in fermented products such as soybean paste (*Miso*), fermented soybean (*Natto*), and soy source (*Shoyu*). The positive bands were clearly observed in blots of meat balls, beef croquettes, and fried chicken, while no bands were seen in hamburgers and fish sausages, indicating that the former three products contained soybean protein isolate as an ingredient. These results confirm the facts obtained with sandwich ELISA. Gly m Bd 30K was shown to be a good index of soybean allergens for evaluating the allergenicity of soybean products; an additional support to this hypothesis was given by Yaklich et al. (37) as the results of the analysis of distribution of Gly m Bd 30K in the core collection of Glycine max accessions.

23.3.2 Molecular Breeding

The target molecular species to be removed are three major allergens, namely Gly m Bd 60K, 30K, and 28K. Among them, a new soybean line (Glycine max Tohoku 124) lacking both α- and α'-subunits of β-conglycinin (Gly m Bd 60K) was induced by irradiation with 20kR (1.0 kR h^{-1}) gamma-ray to Karikei 434 with a marked decrease in the level of the α-, α'-, and β-subunits of β-conglycinin; this was achieved by cross-hybridization as shown briefly in a pedigree chart (Figure 23.5) (38). The SDS-PAGE (sodium dodecyl sulfate–polyacrylamide gel electrophoresis) pattern of protein fraction of Tohoku 124 indicates that the seeds lack the α-subunit of β-conglycinin (Figure 23.6). Recently, it was confirmed that this mutant Tohoku 124 also lacks another major allergen, Gly m Bd 28K, together with the α-subunit of β-conglycinin from the results of immunoblotting analysis using monoclonal antibody C5 (24) specific to Gly m Bd 28K, as shown in Figure 23.7 (39). This fact indicates that the application of Tohoku 124 is beneficial for developing hypoallergenic soybean products because of the absence of the two major allergens Gly m Bd 28K and 60K. In 2000, Tohoku 124 was registered as a new variety named "Yumeminori" (which means "dream come true"). During screening studies of the three major allergenic proteins in soybean varieties, about 80% of edible varieties cultivated in Japan were shown to lack the Gly m Bd 28K allergen (see Figure 23.3). However, a mutant lacking Gly m Bd 30K could hardly be found, even by screening all of the soybean varieties and mutants available in the stock culture of the soybean breeding laboratory of the Tohoku National Agricultural Experiment Station (M. Takahashi and K. Kitamura, personal communication).

23.3.3 Reduction of Gly m Bd 30K by Genetic Modification

Herman et al. (40) achieved the removal of the most immunodominant soybean allergen Gly m Bd 30K, using a genetic modification technique. By eliminating the undesirable protein components, they made the use of soybean products

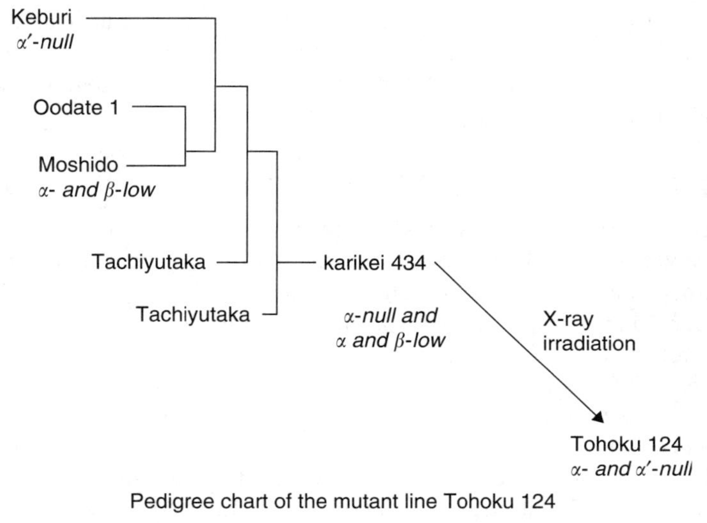

FIGURE 23.5 Pedigree chart for the preparation of Tohoku 124.

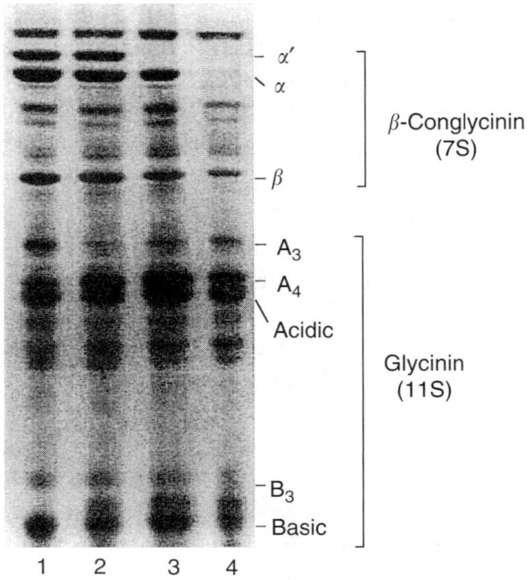

FIGURE 23.6 SDS-PAGE patterns of soybean proteins from various soybean varieties and Tohoku 124. Lanes: (1) Suzuyutaka; (2) Tachiyutaka; (3) Karikei 434; (4) Tohoku 124. (Adapted from X. Baur, M. Pau, A. Czuppon, G. Fruhmann. *Allergy*, 51, 326–330, 1996. With permission.)

possible for many sensitive patients allergic to soybean. They developed a genetically modified soybean line by transgene-induced gene silencing of the Gly m Bd 30K gene (40). The modified transgenic soybean seeds do not accumulate the Gly m Bd 30K allergenic protein but appear to be otherwise unchanged by the removal of this protein. The introduction of this transgenic technique to create the hypoallergenic soybean variety ("Yumeminori") may provide a powerful and safe hypoallergenic material for patients.

23.3.4 Physicochemical Reduction

Heat treatment is a common method of food processing and induces the destruction of protein structures. Epitopes, that is, the IgE-binding sites of allergens, are, however, assumed to be sequential structures on a peptide chain (about 10 amino acid residues in length) and so the reduction of allergenicity due to heat denaturation of proteins (unfolding of polypeptide chains) is not usually expected. It has been reported that the IgE-binding activity of Gly m Bd 30K is remarkably enhanced by an autoclave treatment of soybean grain during *natto* processing (41). As a unique technique of the hypoallergenic process, selective removal of Gly m Bd 30K from soymilk or defatted soymilk by centrifugation under a specified condition was achieved. Selective removal of Gly m Bd 30K was dependent on the solubility difference of the major storage proteins glycinin and β-conglycinin. In

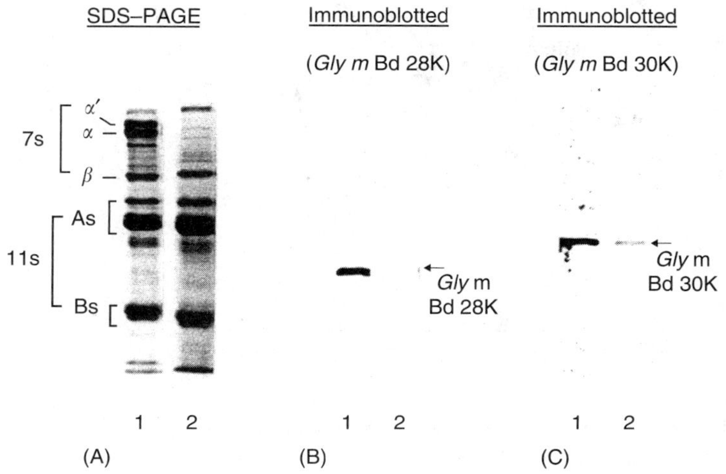

FIGURE 23.7 SDS-PAGE and immunoblotting patterns of defatted soy milk. Defatted soy milk with 1 M Na_2SO_4 without reducing agents was centrifuged to precipitate Gly m Bd 30K. The supernatant was treated with SDS-PAGE sample buffer and then run on a 12% gel for CBB staining (A); for immunostaining with a monoclonal antibody C5 specific to Gly m Bd 28K (B); and for immunostaining with a monoclonal antibody F5 specific to Gly m Bd 30K (C). Lanes: (1) IOM soybean; and (2) Tohoku 124. (Adapted from H. Tsuji, N. Okada, R. Yamanishi, N. Bando, M. Kimoto, T. Ogawa. *Biosci. Biotechnol. Biochem.*, 59, 150–151, 1995. With permission.)

the case of nondefatted soymilk, about 90% of Gly m Bd 30K could be removed into the oil pad layer formed by centrifugation in the presence of reducing agents (42). In the case of defatted soymilk, about 97% of Gly m Bd 30K could be removed as the precipitate in the presence of a reducing agent (10 mM sodium bisulfite) under specified conditions (1M Na_2SO_4 in an acidic pH of 4.5). The major storage soybean proteins, both glycinin and β-conglycinin, remained in the supernatant after centrifugation (Figure 23.8) (43). A small amount of Gly m Bd 30K, however, could not be removed from the supernatant in the absence of reducing reagents. It was indicated that a possible disulfide linkage between Gly m Bd 30K and the α- and α'- subunits of β-conglycinin were formed (44). This hypothesis was proved by using a mutant soybean Tohoku 124 lacking the α- and α'- subunits, the removal ratio of Gly m Bd 30K from defatted soymilk was improved (97 to 99.8%) without the addition of reducing agents for reductive cleavage of the disulfide linkage between Gly m Bd 30K and the α- or α'- subunits of β-conglycinin (44). As a result of the combination of the application of Tohoku 124 and a physicochemical procedure, substantially complete removal of the three major allergenic proteins (Gly m Bd 30K, α- subunit of β-conglycinin, and Gly m Bd 28K) from defatted soymilk was achieved (43). The average removal rate of the three allergens was increased to almost 99.9% based on the results of the

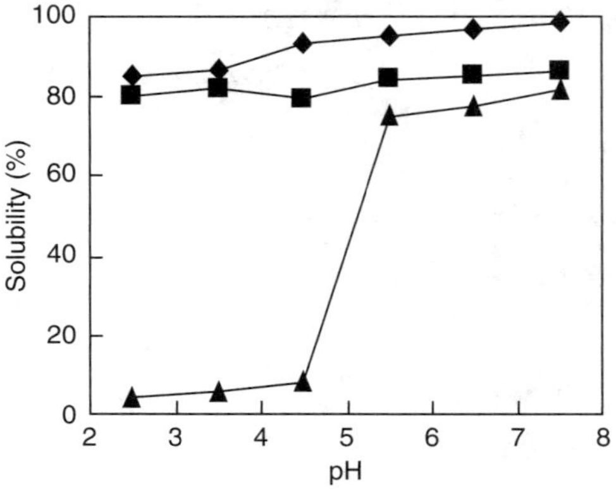

FIGURE 23.8 Solubility curves of soybean proteins. Defatted soy milk was treated with 1 M Na_2SO_4 under the reducing condition of 10 mM sodium bisulfite. Solubility was determined in given pH. (Adapted from M. Samoto, T. Akasaka, H. Mori, M. Manabe, T. Ookura, Y. Kawamura. *Biosci. Biotechnol. Biochem.*, 58, 2123–2125, 1994. With permission.)

densitometric measurement of ECL immunofluorescent intensity on X-ray film (see Figure 23.7). Since these procedures for the reduction of allergenicity do not include the methods of modifying protein structures and, especially, enzymatic digestion, the processing functionality of soybean storage proteins could be retained for making traditional soybean products, for example, *tofu* (soybean cake) and *ganmodoki* (cooked soybean cake).

23.3.5 Enzymatic Digestion

Enzymatic treatment of whole soybean seeds effectively reduces allergenicity. Autoclaved soybeans were treated with certain proteases from the *Bacillus* sp. at 37°C for 20 hours, which is the same condition as that for the fermentation procedure for *natto* with *Bacillus natto*, the Japanese traditional fermented soybean product (45). The product has only *natto*-like texture and no *natto*-like flavor and taste (plain). When the residual allergens were examined by immunoblot and ELISA (inhibition ELISA) tests, the product showed no binding activity against F5 mAb and patients' sera, and all of the proteins in enzyme-treated soybean grains were hydrolyzed into peptides with molecular weights of less than 10,000 Da (Figure 23.9). *Miso* (fermented soybean paste) also showed no residual immunoreactivity against patients' sera after fermentation for 3 months (Figure 23.10) (46). These facts indicate that fermented soybean products such as *natto*, *miso*, and *syoyu* (soy source) are candidates for naturally produced hypoallergenic soybean products (see Figure 23.2). Obata et al. (42) reported reduction of the allergenicity of *tofu* (soybean curd) by an enzymatic digestion method. *Tofu* was sliced into blocks of 2 cm thickness and wrapped with enzyme-treated cheese cloth. The enzyme from *Aspergillus soyae* was used. After treatment for 150 hours

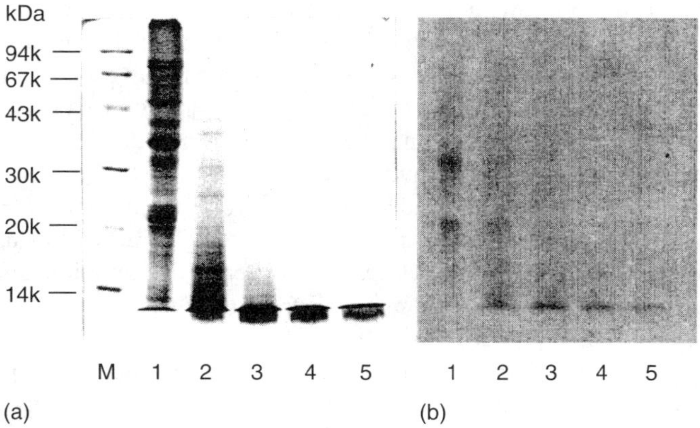

FIGURE 23.9 Hydrolysis of soybean proteins by protease N. Soybean grains soaked in water overnight were autoclaved at 121°C for 20 min. Ten milliliters of protease solution per gram of soybean (dry weight basis) was added to soybean grains, which was incubated for 20 h at 37°C with gentle shaking. Lanes: (1) control (0 unit); (2) 1 × 103 units; (3) 5 × 103 units; (4) 25 × 103 units; (5) 12.5 × 104 units. Abbreviation: M, molecular marker proteins. (Adapted from R. Yamanishi, H. Tsuji, N. Bando, Y. Yamada, Y. Nadaoka, T. Huang, K. Nishikawa, S. Emoto, T. Ogawa. *J. Nutr. Sci. Vitaminol.*, 42, 581–587, 1996. With permission.)

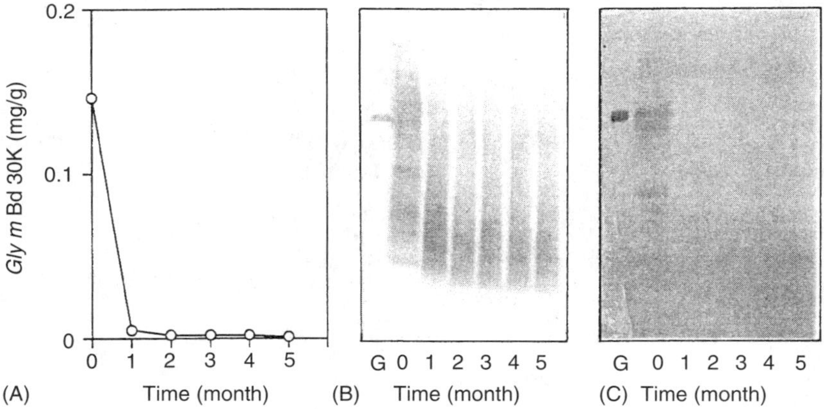

FIGURE 23.10 Fate of allergens in barley-koji *miso* during fermentation. Proteins in 0.6 mg of *miso* paste were applied on each lane. (A) Changed in the Gly m Bd 30k contents at various stages of fermentation; (B) proteins stained with CBB; (C) immunoblotting patterns of Gly m Bd 30K with a monoclonal antibody F5. Abbreviations: M, molecular marker proteins; G, isolated Gly m Bd 30k. (Adapted from H. Tsuji, N. Okada, R. Yamanishi, N. Bando, H. Ebine, T. Ogawa. *Food Sci. Technol. Int. Tokyo*, 3, 145–149, 1997. With permission.)

at 4°C, the reactivity of allergens against F5 mAb was found to be almost completely eliminated. The product had a soft texture like that of cheese but not of *tofu* and none of the bitter taste that generally appears along with protease digestions. Hypoallergenic food with *tofu*-like texture has also been produced using a coagulant (e.g., polysaccharide) and an enzyme-treated hypoallergenic soymilk (42).

Recently, a novel method of hydrolytic processing of soybean proteins was reported by Tsumura et al. (47). Under limited hydrolytic conditions (pH and temperature), selective digestion of *β*-conglycinin (but not glycinin) was attained. The key principle in selective digestion is based on different denaturation temperatures present in *β*-conglycinin and glycinin at neutral pH. The digestion of denatured soybean proteins could proceed more rapidly than that of native proteins with proteases at 70°C. Among *Bacillus* proteases used for the treatment, Proleather FG-F (Amano Pharmaceutical Co.) was found to be effective for selective hydrolysis of Gly m Bd 30K as well as *β*-conglycinin. The product obtained was proved to lose reactivity against the monoclonal antibody specific to these two allergens and the sera of the soybean-sensitive patients (Figure 23.11). As a result of the treatment with proteases under optimum conditions, the three major allergens could be digested. The product containing glycinin showed processing functionality that remained intact, such as gelation to produce *tofu* and emulsification activity (47).

23.3.6 Chemical Modification

An attempt to mask the allergenic sites (epitopes) of soybean proteins using the Maillard-type polysaccharide conjugation was examined. Acid-precipitated soybean proteins (APP or SPI) and galactomannan mixed in a weight ratio of 1:5

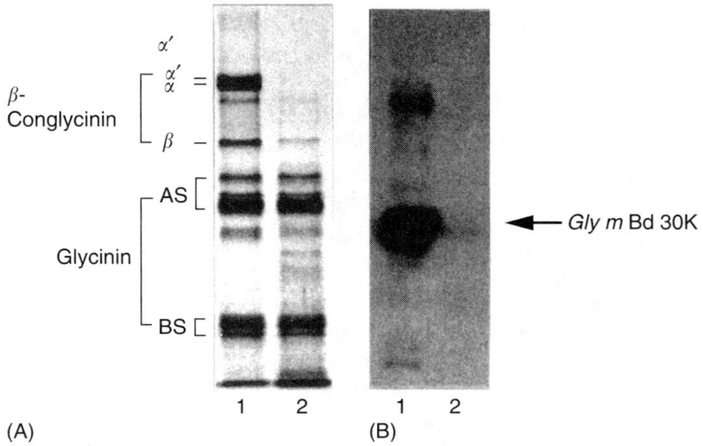

FIGURE 23.11 SDS-PAGE and immunoblotting patterns of the hydrolyzates prepared on a pilot scale. Five micrograms of protein was loaded on each lane for SDS-PAGE. (A) Staining with CBB; (B) Staining with F5 mAb. Lanes: (1) Undigested SPI; and (2) hydrolyzate with Proleather FG-F.

were dissolved in water at 10% (w/v) and freeze-dried. Maillard reaction was then induced at 60°C under 79% relative humidity (RH) in a desiccator for several days. The allergenic potential of APP was shown to be reduced by the conjugation of galactomannan molecules to the lysine residues of APP (48), which was confirmed by dot blotting of treated APP with F5 mAb and patients' sera and by inhibition ELISA.

23.3.7 Extrusion Cooking

Ohishi et al. (49) reported that the antigenicity of soybean meal against calves' sera was reduced to 0.1% of the original activity by extrusion cooking with screws containing kneading-disc elements and die-end temperatures exceeding 66°C. SDS-PAGE analysis of the cooked meal indicated that the reduction of antigenicity was due to the destruction or modification of protein molecules.

Frank et al. (50) reported that the production of texturized soy proteins with the processing steps including heating, mechanical pressure, and acid treatment (pH 4.5) leads to the disappearance of the major allergen Gly m Bd 30K, whereas a 38-kDa allergenic band corresponding to the glycinin fragment remained intact. However, further studies of the texturized soy protein could lead to the development of modified technologies using a complementary enzymatic treatment (50) in order to obtain more hypoallergenic products.

23.4 EVALUATION OF HYPOALLERGENIC SOYBEAN PRODUCTS AND PERSPECTIVES

The hypoallergenic soybean products have been developed and evaluated under the observation of physicians and dieticians. *In vitro* examination of IgE-binding activity has been carried out using immunoblot or ELISA techniques. In the case of *in vivo* examination of allergenicity, a single-blind food challenge test or open challenge test are practical in evaluating processed foods. In such tests, the products for challenge tests are served to patients who have a strict elimination diet of soybean and soybean products for about 3 weeks under the control of physicians. A challenge test is carried out after any of the symptoms disappear from the patients after elimination of the causative diet. As a standard case, for the first 5 days, the patients receive the hypoallergenic soybean products. If no symptomatic change is recognized in the patients during this period, they receive the control (allergen-containing) diets continuously for an additional 5 days. If some adverse reactions appear in the patients due to the presence of allergens, the challenge test is stopped. A preliminary challenge trial has confirmed that at least 80% of the soybean allergic patients can use the above-mentioned hypoallergenic products without any adverse reactions. Some of the products, evaluation of which have been completed, await distribution to soybean-sensitive patients through physicians. In addition, further information on soybean allergens, sensitization to soybean allergens, persistence and symptoms of soybean allergy, and diagnostic features is available as a database in an Internet Symposium on Food Allergens (51).

REFERENCES

1. T. Ogawa, N. Bando, H. Tsuji, H. Okajima, K. Nishikawa, K. Sasaoka. Investigation of the IgE-binding proteins in soybeans by immunoblotting with the sera of the soybean-sensitive patients with atopic dermatitis. *J. Nutr. Sci. Vitaminol.*, 37: 555–565, 1991.
2. R.K. Bush, S.L. Taylor. Soybean oil is not allergenic to soybean-sensitive individuals. *J. Allergy Clin. Immunol.*, 76: 242–245, 1985.
3. S. Doke, R. Nakamura, S. Torii. Allergenicity of food proteins interacted with oxidized lipids in soybean-sensitive individuals. *Agric. Biol. Chem.*, 53: 1231–1235, 1989.
4. W.W. Duke. Soybean as a possible important source of allergy. *J. Allergy*, 5: 300–305, 1934.
5. L.A. Morot, W.H. Yang. Kunitz soybean trypsin inhibitor. A specific allergen in food anaphylaxis. *N. Engl. J. Med.*, 302: 1126–1128, 1980.
6. M. Shibasaki, S. Suzuki, S. Tajima, H. Nemoto, T. Kuroume. Allergenicity of major component proteins of soybean. *Int. Arch. Allergy Appl. Immunol.*, 61: 441–448, 1980.
7. A.W. Burks, J.R. Brooks, H.A. Sampson. Allergenicity of major component proteins of soybean determined by enzyme linked immunosorbent assay (ELISA) and immunoblotting in children with atopic dermatitis and positive soy challenges. *J. Allergy Clin. Immunol.*, 81: 1135–1142, 1988.
8. A.M. Herian, S.L. Taylor, K.B. Robert. Identification of soybean allergens by immunoblotting with sera from soy-allergic adults. *Int. J. Allergy Appl. Immunol.*, 92: 193–198, 1990.
9. M.J. Rodrigo, F. Morell, R.M. Helm, M. Swanson, A. Greif, J.M. Antonio, J. Sunyer, C.E. Reed. Identification and partial characterization of the soybean-dust allergens involved in the Barcelona asthma epidemic. *J. Allergy Clin. Immunol.*, 85: 778–784, 1990.
10. T. Ogawa, N. Bando, H. Tsuji, K. Nishikawa, K. Kitamura. Identification of soybean allergenic protein, Gly m Bd 30K, with the soybean seed 34-kDa oil-body-assolciated protein. *Biosci. Biotechnol. Biochem.*, 57: 1030–1033, 1995.
11. A. Kalinski, J.M. Weisemann, B.F. Matthews, E.M. Herman. Molecular cloning of a protein associated with soybean seed oil bodies that is similar to thiol proteinases of the papain family. *J. Biol. Chem.*, 265: 13843–13848, 1990.
12. K.Y. Chua, G.A. Stewart, W.R. Thomas, R.J. Simpson, R.J. Dilworth, T.M. Plozza, K.J. Turner. Sequence analysis of cDNA coding of a major house dust mite allergen, Der p 1. Homology with cystein proteases. *J. Exp. Med.*, 157: 175–182, 1988.
13. N. Bando, H. Tsuji, R. Yamanishi, N. Nio, T. Ogawa. Identification of glycocylation site of a major soybean allergen, Gly m Bd 30k. *Biosci. Biotechnol. Biochem.*, 60: 347–348, 1996.
14. A. Kalinski, D.L. Merroy, R.S. Dwivedi, E.M. Herman. A soybean vacuolar protein (P34) related to thiol proteases is synthesized as a glycoprotein precursor during seed maturation. *J. Biol. Chem.*, 267: 12068–12067, 1992.
15. R.M. Helm, G. Cockrell, E. Herman, A.W. Burks, H.A. Sampson, G.A. Bannon. Cellular and molecular characterization of a major soybean allergen. *Int. Arch. Allergy Immunol.*, 117: 29–37, 1998.

16. M. Hosoyama, A. Obata, N. Bando, H. Tsuji, T. Ogawa. Epitope analysis of soybean major allergen Gly m Bd 30K recognized by the mouse monoclonal antibody using overlapping peptides. *Biosci. Biotechnol. Biochem.*, 60: 1181–1182, 2002.
17. E.E. Babiker, H. Azakami, T. Ogawa, A. Koto. Immunological characterization of recombinant soy protein allergen produced by *Esherichia coli* expression system. *J. Agric. Food Chem.*, 48: 571–575, 2000.
18. H. Tsuji, N. Bando, M. Kimoto, N. Okada, T. Ogawa. Preparation and application of monoclonal antibodies for a sandwich enzyme-linked immunosorbent assay for the major soybean allergen, Gly m Bd 30K. *J. Nutr. Sci. Vitaminol.*, 39: 389–397, 1993.
19. T.Okinaka, C.H. Yang, E. Herman, N.T. Keen. The P34 syringolide elicitor receptor interacts with a soybean photorespiration enzyme, NADH-dependent hydroxypyruvate reductase. *Mol. Plant Microbe Interact.*, 15:1213–1218, 2002.
20. H. Tsuji, N. Bando, M. Hiemori, R. Yamanishi, M. Kimoto, T. Ogawa. Purification and characterization of soybean allergen Gly m Bd 28K. *Biosci. Biotechnol. Biochem.*, 61: 942–947, 1997.
21. M. Hiemori, N. Bando, T. Ogawa, H. Shimada, H. Tsuji, R. Yamanishi, J. Terao. Occurrence of IgE antibody recobnizing N-linked glycan moiety of Gly m Bd 28K of soybean allergen. *Int. Arch. Allergy Immunol.*, 122, 238–245, 2000.
22. H. Tsuji, M. Hiemori, M. Kimoto, H. Yamashita, R. Kobatake, M. Adachi, T. Fukuda, N. Bando, M. Okita, S. Utsumi. Cloning of cDNA encoding a soybean allergen, Gly m bd 28K. *Biochim. Biophys. Acta*, 1518: 178–182, 2001.
23. A.W. Burks, D. Shin, G. Cockrell, J.S. Stanley, R.M. Helm, G.A. Bannon. Mapping and mutational analysis of the IgE-binding epitopes on Ara h 1, a legume vicilin protein and a major allergen in peanut hypersensitivity. *Eur. J. Biochem.*, 245: 334–339, 1997.
24. N. Bando, H. Tsuji, M. Hiemori, K. Yoshizumi, R. Yamanishi, M. Kimoto, T. Ogawa. Quantitative analysis of Gly m Bd 28K in soybean products by a sandwich enzyme-linked immunosorbent assay. *J. Nutr. Sci. Vitaminol.*, 44: 655–774, 1998.
25. T. Ogawa, N. Bando, H. Tsuji, K. Nishikawa, K. Kitamura. Alpha subunit of beta-conglycinin, an allergenic protein recognized by IgE antibodies of soybean-sensitive patients with atopic dermatitis. *Biosci. Biotechnol. Biochem.*, 59: 831–833, 1995.
26. V.H. Thanh, K. Shibasaki. Beta-conglycinin from soybean proteins isolation and immunological and physicochemical properties of the monomeric forms. *Biochim. Biophys. Acta*, 490: 370–384, 1977.
27. F.L. Sebastiani, E.S. Schmit, R.N. Beachy. Complete sequence of a cDNA of alpha subunit of beta-conglycinin. *Plant Mol. Biol.*, 15: 197–201, 1985.
28. R. Gonzalez, F. Polo, L. Zapatero, F. Caravaca, J. Carreira. Purification and characterization of major inhalant allergens from soybean hulls. *Clin. Exp. Allergy*, 22: 748–755, 1992.
29. S. Odani, T. Koide, T. Ono, Y. Seto, T. Tanaka. Soybean hydrophobic protein. Isolation, partial characterization and the complete primary structure. *Eur. J. Biochem.*, 162: 485–491, 1987.
30. M. Gijzen, S.S. Miller, K. Kufku, R.I. Buzzelul, B.L. Miki. Hydrophobic protein synthesized in the pod endcarp adheres to the seed surface. *Plant Physiol.*, 120: 951–959, 1999.
31. R. Codina, R.F. Lockey, E. Fernandez-Caldas, R. Rama. Purification and characterization of a soybean hull allergen responsible for the Barcelona asthma

outbreaks. II. Purification and sequencing of the Gly m 2 allergen. *Clin. Exp. Allergy*, 27: 424–430, 1997.

32. H.P. Rihs, Z. Chen, F. Rueff, A. Petersen, R. Royznek, H. Heimann, X. Baur. IgE binding of the recombinant allergen soybean profilin (rGly m 3) is mediated by conformational epitopes. *J. Allergy Clin. Immunol*, 104: 1293–1301, 1999.
33. R. Djurtoft, H.S. Pedersen, B. Aabin, V. Barkholt. Studies of food allergens: soybean and egg proteins. *Adv. Exp. Med. Biol.*, 289: 281–293, 1991.
34. M.G. Zeece, T.A. Beardslee, J.P. Markwell, G. Sarath. Identification of an IgE-binding region in soybean acidic Glycinin G1. *Food Agric. Immunol.*, 11: 83–90, 1999.
35. X. Baur, M. Pau, A. Czuppon, G. Fruhmann. Characerization of soybean allergens causing sensitization on occupationally exposed bakers. *Allergy*, 51: 326–330, 1996.
36. H. Tsuji, N. Okada, R. Yamanishi, N. Bando, M. Kimoto, T. Ogawa. Measurement of Gly m Bd 30k, a major soybean allergen, in soybean products by a sandwich enzyme-linked immunosorbent assay. *Biosci. Biotechnol. Biochem.*, 59: 150–151, 1995.
37. R.W. Yaklich, R.M. Helm, G. Cockrell, E.M. Herman. Analysis of the distribution of the major soybean seed allergens in a core collection of Glycine max Accessions. *Crop Sci.*, 39: 1444–1447, 1999.
38. K. Takahashi, H. Banba, A. Kikuchi, M. Ito, S. Nakamura. An induced mutant line lacking the alpha subunit of beta-conglycinin in soybean (Glycine max (L) Merrill). *Breeding Sci.*, 44: 65–66, 1994.
39. M. Samato, K. Takahashi, Y. Fukuda, S. Nakamura, Y. Kawamura. Substantially complete removal of the 34-kDa allergenic soybean protein, Gly m Bd 30K, form soy milk of a mutant lacking alpha and alpha′ subunits of conglycinin. *Biosci. Biotechnol. Biochem.*, 60: 1911–1913, 1996.
40. E.M. Herman, R.M. Helm, R. Jung, A.J. Kinney. Genetic modification removes an immunodominant allergen from soybean. *Plant Physiol.*, 132, 36–43, 2003.
41. R. Yamanishi, T. Huang, H. Tsuji, N. Bando, T. Ogawa. Reduction of the soybean allergenicity by the fermentation with *Bacillus natto*. *Food Sci. Technol. Int.*, 1: 14–17, 1995.
42. A. Obata, H. Hosoyama, T. Ogawa. Development of *Tofu*-like product from soy milk for soybean-sensitive patients. *Shokuhin Kogyo* (*in Japanese*), 41: 39–48, 1998.
43. M. Samoto, T. Akasaka, H. Mori, M. Manabe, T. Ookura, Y. Kawamura. Simple procedure for removing the 34-kDa allergenic protein, Gly m I, from defatted soy milk. *Biosci. Biotechnol. Biochem.*, 58: 2123–2125, 1994.
44. M. Samoto, C. Miyazaki, T. Akasaka, H. Mori, Y. Kawamura. Specific binding of allergenic soybean protein Gly m Bd 30K with alpha and alpha′ subunit of beta-conglycinin in soy milk. *Biosci. Biotechnol. Biochem.*, 60: 1006–1010, 1996.
45. R. Yamanishi, H. Tsuji, N. Bando, Y. Yamada, Y. Nadaoka, T. Huang, K. Nishikawa, S. Emoto, T. Ogawa. Reduction of the allergenicity of soybean by treatment with proteases. *J. Nutr. Sci. Vitaminol.*, 42: 581–587, 1996.
46. H. Tsuji, N. Okada, R. Yamanishi, N. Bando, H. Ebine, T. Ogawa. Fate of major soybean allergen, Gly m Bd 30K, in rice-, barley-, and soybean-koji *miso* (fermented soybean paste) during fermentation. *Food Sci. Technol. Int. Tokyo*, 3: 145–149, 1997.

47. K. Tumura, W. Kugimiya, N. Bando, M. Hiemori, T. Ogawa. Preparation of hypoallergenic soybean protein with processing functionality by selective enzymatic hydrolysis. *Food Sci. Technol. Res.*, 5: 171–175, 1999.
48. E.E. Babikwer, A. Hiroyuki, N. Matsudomi, H. Iwata, T. Ogawa, N. Bando, A. Kato. Effect of polysaccharide conjugation or transglutaminase treatment on the allergenicity and functional properties of soybean protein. *J. Agric. Food Chem.*, 46: 866–871, 1998.
49. A. Ohishi, K. Watanabe, M. Urushibata, K. Utsuno, K. Ikuta, K. Sugimoto, H. Harada. Detection of soybean antigenicity and reduction by twin-screw extrusion. *J. Am. Oil Chem.*, 71: 1391–1396, 1994.
50. P. Frank, D.A.M. Vautrin, B. Dousset, G. Kanny, P. Nabet, L. Guenard-Bilbaut, L. Parisot. The allergenicity of soybean-based products is modified by food technologies. *Int. Arch. Allergy Immunol.*, 128, 212–219, 2002.
51. M. Besler, R.M. Helm, T. Ogawa. Allergen Data Base Collection — (URL:http://www.food-allergens.de) update; Soybean (Glycine max). *Internet Symposium on Food Allergens*, 2 (Suppl.3): 1–37, 2000.

24 Meat Allergy

Soichi Tanabe and
Toshihide Nishimura

CONTENTS

24.1 INTRODUCTION

Meat and its products are important foods because of their high nutritional value and palatability. Especially in western diets, meat is a main source of protein, for example, Americans eat about 30 kg of beef per person per year. Generally, meat is less allergenic than common allergy-inducing foods such as cow's milk, hen's egg, wheat, peanuts, and so on. Thus, a quarter century ago, children with food allergies were advised to be placed on an elimination diet that included beef (1). However, there is increasing evidence that even meat can provoke allergic reactions in sensitizing patients. The prevalence of beef, pork, and chicken allergy was reported to be 73%, 58%, and 41%, respectively, among 57 subjects with suspected meat allergies in USA (2). In Japanese children with food allergies, the prevalence of chicken allergy was reported to be the highest (4.5%) with allergies to other meats following (3%), based on a questionnaire study (3). Although IgE binding to cooked meat is weaker than to raw meat, some patients are sensitive to well-cooked meat.

This brief chapter presents general information on meat allergy and its allergens, with the main focus on beef allergy. We will also discuss the stability of meat allergens, which may help readers in the design of hypoallergenic meat products.

24.2 BEEF ALLERGY AND ITS ALLERGENS

Beef allergy is reported to occur with an incidence between 3.3 and 6.5% among children with atopic dermatitis (1–4). This can increase to about 20% in children with cow's milk allergy (4). The major beef allergens are bovine serum albumin (BSA, *Bos d* 6, 66 to 67 kDa) (4–9) and bovine gamma globulin (BGG, *Bos d* 7, 160 kDa) (8–10). In addition, actin, myoglobin, and tropomyosin sometimes cause allergic reactions, while myosin is rarely allergenic (4,8). BSA and BGG are also present in milk, thus some children with cow's milk allergy are sensitive to beef as well. In this respect, as suggested by Ayuso et al. (10), sensitization to beef may be secondary to milk allergy (since milk is introduced first in the diet).

Serum albumin is one of the most widely studied and applied proteins in biochemistry. It is also the most abundant protein in the circulatory system, accounting for 60% of total serum protein, with a concentration of approximately 40 mg mL^{-1} (11–13). Serum albumins comprise about 580 amino acid residues and are characterized by a low content of tryptophan and a high content of cysteine and charged amino acids such as aspartic and glutamic acids, lysine, and arginine. The tertiary structure comprises three domains, I, II, and III (aa 1–190, 191–382, 383–581) that assemble to form a heart-shaped molecule. The fact that a major component of serum acts as an allergen is very surprising for the following reason: it is remarkable that albumins from animals, which are very similar in sequence, structure, and function to human serum albumin (HSA), are recognized by the human immune system as allergens instead of inducing tolerance.

Trials to identify the immunoglobulin E (IgE)-binding epitopes of BSA have been performed by several groups (5,6). For example, Beretta et al. (6) reported that the C-terminal region (aa 500–574) of BSA may be an epitopic area for patients based on analyses of tryptic hydrolyzates of BSA. In the meantime, we carried out a break-through study, in which the precise regions of IgE-binding as well as T-cell epitopes were identified (5).

Prior to our experiments, we hypothesized that BSA-specific antibodies and T cells react primarily with sequential epitopes in which the amino acid sequences differ greatly between BSA and HSA. To clarify this hypothesis, 16 peptides (Nos. 1–16) corresponding to such regions were synthesized as candidate epitopes (Table 24.1). Among them, at least two regions, aa 336–345 (No. 7) and aa 451–459 (No. 15), were found to be the major IgE-binding epitopes (Figure 24.1A). In inhibition ELISA (enzyme-linked immunosorbent assay), EYAV (aa 338–341) and LILNR (aa 453–457) bound to patient IgE antibodies and were found to be the cores of the IgE-binding epitopes.

Among eight IgE-binding epitopes, three were found to contain an EXXV motif (HPEYAVSVLL (No. 7), PVESKVT (No. 12), and VMENFVAF (No. 15)). The corresponding sequences in HSA are HPDYSVVLLL, PVSDRVT, and VMDDFAAF, respectively. Comparing two epitopic sequences (Nos. 7 and 15) in BSA with the corresponding sequences in HSA, it is likely that E residues (E338 and E547) are important for recognition by IgE antibodies, since the corresponding residues in HSA are D in both peptide Nos. 7 and 15. Therefore, two analogue

TABLE 24.1
Amino Acid Sequences Syntherized Peptides and Comparsion with the Corresponding Sequences in HSA

Peptide	Sequence Synthesized		Corresponding Sequences in HSA
No. 1	ESHAGCEKS	(57–65)	ESAENCDKS
No. 2	DDSPDLPKLKPDPNTLC	(107–123)	DDNPNLPRLVRPEVDVMC
No. 3	CDEFKADEKKFWGKY	(123–137)	CTAFHDNEETFLKKY
No. 4	LLYANKYNGVFQEC	(153–166)	LLFFAKRYKAAFTEC
No. 5	PKIETMREKVLTSS	(178–191)	PKLDELRDEGKASS
No. 6	EKDAIPEDLPPLTADFAEDK	(292–311)	ENDEMPADLPSLAADFVESK
No. 7	HPEYAVSVLL	(336–345)	HPDYSVVLLL
No. 8	PHACYTSVFDKLKHLVDEP	(364–382)	PHECYAKVFDEFKPLMEEP
No. 9	NCDQFEKLG	(389–400)	NCELFEQLG
No. 10	VGTRCCTKPESERM	(431–444)	VGSKCCHPEAKRM
No. 11	LSLILNRLC	(451–459)	LSVVLNQLC
No. 12	PVESKVT	(466–472)	PVSDRVT
No. 13	PKAFDEKLFT	(497–506)	PKEFNAETFT
No. 14	TLPDTEKQI	(513–521)	TLSEKERQI
No. 15	VMENFVAF	(545–552)	VMDDFAAF
No. 16	LVVSTQTAL	(573–581)	LVAASQAAL
	EYAV	(338–341)	
	LILNR	(453–457)	
E338D	HPDYAVSVLL	(336–345)	
E547D	VMDNFVAF	(545–552)	

Underlined residues represent amino acid differences between BSA and HSA. (Tanabe et al., 2002)

peptides, E338D (HPDYAVSVLL) and E547D (VMDNFVAF), with amino-acid substitutions from E to D were synthesized (see Table 24.1), and their IgE-binding abilities were characterized (14). As a result, the substitution of the glutamic acid in the EXXV sequence with aspartic acid led to a remarkable reduction in IgE-binding ability (Figure 24.1B). Thus, ^{338}E and ^{547}E in BSA were thought to be important for recognition by patient IgE antibodies. In other words, the difference between D (human type) and E (bovine type) at positions 338 and 547 seems to be a major cause of the allergenicity of BSA. According to the three-dimensional structure of HSA, these two D residues are located at the surface of the molecule (Figure 24.2). Although the three-dimensional structure of BSA is not available, the corresponding E residues at positions 338 and 547 in BSA are assumed to be located similarly to the D residues in HSA, since the tertiary structures of BSA and HSA are very similar to one another (12). Therefore, it is possible that IgE-antibodies in allergic patients easily recognize E residues at positions 338 and 547 on the surface of BSA, with the allergic reaction taking place subsequently.

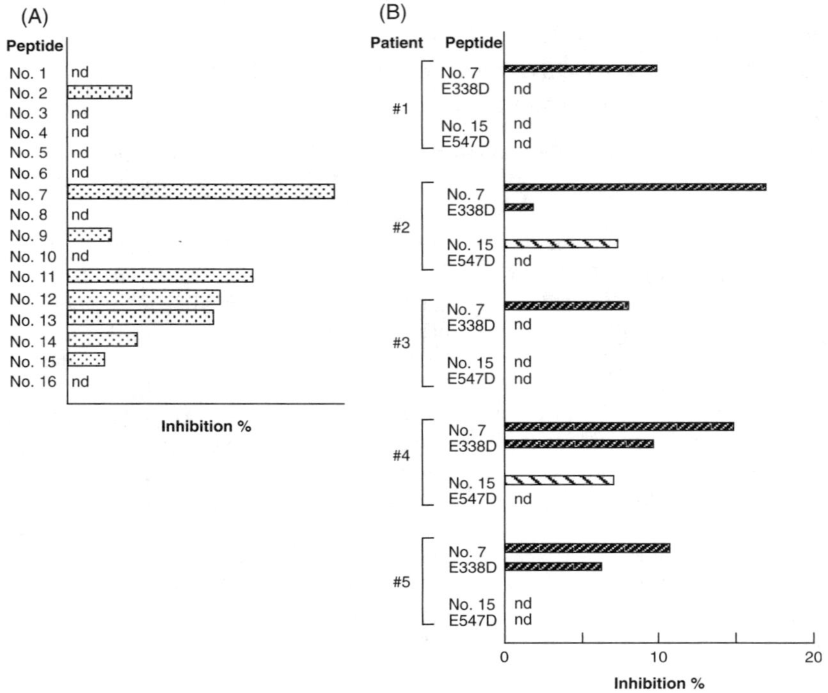

FIGURE 24.1 Inhibition ELISA of the synthetic peptides listed in Table 24.1. (A) Inhibition ELISA of 16 BSA peptides using pooled serum. From S. Tanabe, Y. Kobayashi, Y. Takahata, F. Morimatsu, R. Shibata, T. Nishimura. Some human B and T cell epitopes of bovine serum albumin, the major beef allergen. *Biochem. Biophys. Res. Commn.*, 293: 1348–1353, 2002, with modifications. With permission. (B) Inhibition ELISA of aa336–345 (No. 7), aa451–459 (No. 15), E338D and E547D using serum from 5 individual patients (#1–#5). From S. Tanabe, R. Shibata, T. Nishimura. Hypoallergenic and T cell reactive analogue peptides of bovine serum albumin, the major beef allergen. *Mol. Immunol.*, 41: 885–890, 2004. With permission.

As for the epitopes of bovine gamma globulin (BGG), Ayuso et al. (10) predicted conformational epitopes, since they found that IgE reactivity to BGG completely disappeared when beef extracts were treated under reducing conditions. In addition, they stressed BGG as the only major clinically relevant beef allergen (10).

24.3 PORK ALLERGY AND ITS ALLERGENS

Although pork allergy is relatively rare, it may manifest as atopic dermatitis and oral allergy syndrome (OAS), among other possibilities. The frequency of sensitization in the skin prick test (SPT) to pork was reported to be 2% in Germany (15).

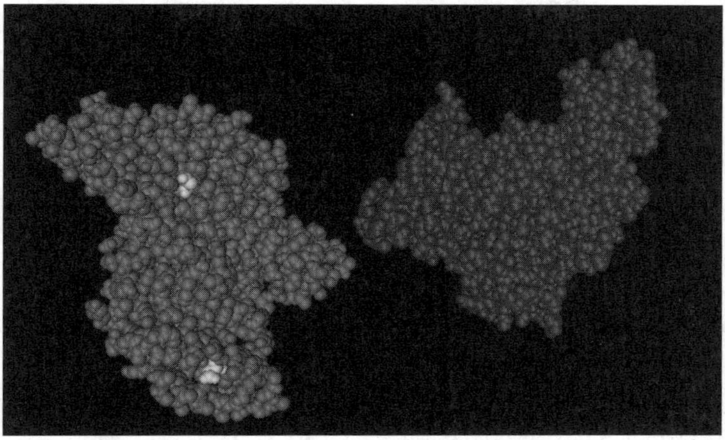

FIGURE 24.2 The molecular structure of HSA is shown in gray, and the two aspartic acid residues (upper, D338; lower, D547) are highlighted in white. The drawing was generated with software Cn 3 D version 4.1 (www.ncbi.nlm.nih.gov). From S. Tanabe, R. Shibata, T. Nishimura. Hypoallergenic and T cell reactive analogue peptides of bovine serum albumin, the major beef allergen. *Mol. Immunol.*, 41: 885–890, 2004. With permission.

The cross-reactivity of porcine meat and cat epithelia/dander has been reported and called "pork-cat syndrome" (16–18). Nearly all patients with IgE antibodies to pork also have IgE antibodies to cat epithelia/dander. However, among patients with IgE antibodies to cat epithelia/dander, only some (~20%) have IgE antibodies to pork (16).

Hilger et al. (18) performed immunoblotting and cross-inhibition assays and found that porcine serum albumin (PSA) and cat serum albumin (*Fel d* 2) are jointly recognized molecules. Inhibition assays have shown that the spectrum of IgE reactivity to cat serum albumin completely contains IgE reactivity to PSA, suggesting that sensitization to cats is the primary event (18).

The phenomenon of cross-sensitization between inhalants and food allergens is of increasing importance clinically. Allergic cross-reactions between pollen allergens and food allergens of plant origin are well known (e.g., birch pollen–hazelnut–apple and mugwort–celery allergic clusters).

Besides serum albumin, several other IgE-binding proteins (51-, 40-, and 28- to 30-kDa) have been detected in pork (19). It should be noted that even pork kidney and gut sometimes cause allergic reactions (19).

24.4 CHICKEN ALLERGY AND ITS ALLERGENS

Similar to the cross-reactivity between porcine meat and cat epithelia/dander (pork-cat syndrome) described earlier, there is cross-reactivity between chicken and hen's egg yolk, which is known as "bird-egg syndrome" (20–22). Egg yolk contains significant quantities of serum proteins. In bird-egg syndrome, the cross reacting allergen has been reported to be α-livetin in egg yolk (20), which is identical to chicken serum albumin (*Gal d* 5) (23). Chicken serum albumin can act as both an inhalant and a food allergen in patients with bird-egg syndrome (21). In this regard, sensitization to meat may be secondary to hair and dander sensitization (especially in veterinarians) as well as secondary to milk allergy, as described above.

Other than chicken serum albumin, several allergens have been reported by Ayuso et al. (2). Their immunoblotting data showed the existence of eight bands (150-, 66-, 45-, 31-, 28-, 24-, 20-, and 17-kDa) in raw chicken meat that react with patients' sera.

For the convenience of the reader, the major meat allergens are summarized in Table 24.2. The importance of serum albumin as a cross-reacting meat allergen needs to be reemphasized.

24.5 SERUM ALBUMIN AS A CROSS-REACTING MEAT ALLERGEN

Restani et al. (24) reported the cross-reactivity between serum albumins from different animal species. SPT and immunoblotting were performed for each of seven serum albumins (beef, sheep, pig, horse, rabbit, turkey, and chicken) (24). As a result, a clear relationship was found between the sequence homology of different serum albumins with BSA and the percentage of positive SPT and immunoblotting in the serum of beef-allergic children (Table 24.3). Therefore, the use of alternative meats in meat-allergic patients must be carefully evaluated on an individual basis (4).

As described earlier, cat serum albumin is one of the cat allergens. Also, dog serum albumin is known as a cross-reacting allergen (25); all patients allergic to cat serum albumin also showed high IgE titers to dog serum albumin, even when they had no contact with dogs. Thus, it can be said that serum albumin is an important inhalant allergen as well as food allergen.

TABLE 24.2
Major Meat Allergens (Summary)

Meat	Allergen	Cross-reactive allergen
Beef	Serum albumin (*Bos d* 6)	Other serum albumins - overlap with cow's milk allergy-
	Gamma globumin (*Bos d* 7)	Especially lamb and vension (but weakly to pork and chicken) - overlap with cow's milk allergy-
Pork	Serum albumin	Cat serum albumin (*Fel d* 2) - pork-cat syndrome-
Chicken	Serum albumin (*Gal d* 5)	Identical to α-livetin in egg yolk - bird-egg syndrome-

TABLE 24.3
Relationship Between Sequence Homology of Different Serum Albumins and Positive Responses Obtained in SPT and Immunoblitting in the Sera of Beef-Allergic Children (With Permission from the American College of Nutrition)

Species	% Homology with BSA	% of positive responses (n=6)	
		SPT	Immunoblotting
Beef	100.00	100	100
Sheep	92.26	100	66.7
Pig	78.84	50	50
Horse	74.14	50	0
Rabbit	-	33.3	0
Turkey	-	33.3	0
Chicken	44.98	0	0
Human	76.28	-	0

(Restani et al., 1997)

24.6 METHODS OF REDUCING THE ALLERGENICITY OF MEAT ALLERGENS

There are at least two treatments available that can reduce or abolish the allergenicity of meat allergens, heat and enzymatic treatment.

24.6.1 Heat Treatment

Fiocchi et al. (7) reported the effect of heating on the allergenicity of beef and BSA. In their study, 10 children with beef allergy and positive SPT to BSA were evaluated by SPT and DBPCFC (double-blind, placebo-controlled food challenge). Among them, 7 children were found to be sensitive to heated BSA (100°C for 5 minutes) by SPT, and 4 children were positive by DBPCFC. According to these results, heating appeared to partially reduce the allergenicity of BSA and beef but did not completely abolish allergenicity. On the other hand, Werfel et al. (26) reported that BSA and BGG were heat labile in the beef sample.

Pork allergens are also resistant to heat treatment, at least to some extent. Atanaskovic-Markovic et al. (27) reported that six pork allergens were detected even in cooked/roasted pork and concluded that cooked/roasted pork retains allergenic epitopes capable of inducing IgE-mediated food allergy.

A recent study by Quirce et al. (21) showed that heating reduced, but did not abolish, the allergenicity of chicken serum albumin (α-livetins). It should be noted that in some cases of chicken allergy, heating (140°C for 20 minutes in a conventional oven) may result in the formation of new allergenic moieties, as suggested by Ayuso et al. (2).

Of course, as shown using milk whey proteins, the thermal stability of proteins can be influenced by the pH (28), divalent cations (29,30), sugar content (30), and lipid concentration of the solution in which the proteins are dissolved. Thus, further detailed studies are needed to clarify the effectiveness of heat treatment.

Taken together, meat allergens are "partially" heat labile, and this fact can explain why some sensitized patients may tolerate well-cooked meat, but not raw meat. Patients reacting only to meat cooked to rare may not need to maintain a complete meat elimination diet. However, it should be noted that heating cannot "completely" abolish the allergenicity of meat allergens.

24.6.2 Enzymatic Treatment

Several protein hydrolyzate-based products such as whey-based formulas (31), hypoallergenic rice (32), and hypoallergenic flour (33), have been produced and considered suitable for use in the management of allergic patients. The merit of these products is that they possess the same nutritional value as the intact proteins. Therefore, if a hydrolytic reaction could decompose the structures of meat allergens, then the product (hypoallergenic meat) would be of great benefit for patients.

Fiocchi et al. (34) investigated the effect of peptic digestion on the allergenicity of BSA as well as ovine serum albumin (OSA) as measured by SPT, RAST (radioallergosorbent test) and SDS-PAGE (sodium dodecyl sulfate–polyacrylamide gel electrophoresis) analysis using 12 children with beef and lamb allergies (The reason OSA was included is that lamb is often suggested as a substitute for beef and milk in atopic children). All were SPT positive for intact BSA and OSA. The authors reported that peptic digestion decreased the allergic potential even after only 5 minutes of hydrolysis; 67% of the children became SPT negative for hydrolyzed BSA, and 75% became SPT negative for hydrolyzed OSA. After 2 hours of hydrolysis, only 17% reacted to hydrolyzed BSA and none to hydrolyzed OSA. However, 4-hour-hydrolyzed BSA still induced an SPT response in the same 7 17% of children who were positive to 2-hour-hydrolyzed BSA. From this data, it was concluded that peptic digestion can influence the allergenicity of serum albumins but not completely abolish allergenicity in severely allergic patients. The more rapid disappearance of SPT reactivity for OSA than for BSA during peptic digestion was also confirmed. This appears to support the clinical observation that ovine meat is generally less allergenic than bovine meat.

Although pepsin alone is incapable of abolishing the allergenicity of BSA completely, it remains possible that BSA allergenicity can be eliminated by using a combination of several enzymes. However, it should be noted that extensive hydrolysis of meat proteins usually leads to the loss of such physicochemical properties as gel-/network-forming activity.

24.6.3 Other Treatments

High-pressure treatment has recently been considered a useful technique for food processing, and the efficiency of this treatment in modifying meat properties has been reported (35). Changes in the allergenicity of BGG by high pressure treatment were investigated (36). However, high pressure treatment at 200 to 600 MPa failed to decrease the allergenicity of BGG.

It is of interest that industrially heat-treated (steaming) and sterilized (at 122°C for 40 minutes) homogenized beef evoked no clinical reaction in DBPCFC subjects (7,34). This treatment seems to be most effective for eliminating allergenicity.

24.7 SUMMARY

The major beef allergens are bovine serum albumin (*Bos d* 6) and bovine gamma globulin (*Bos d* 7). In addition, actin, myoglobin, and tropomyosin sometimes cause allergic reactions, while myosin rarely does.

Porcine serum albumin and cat serum albumin (*Fel d* 2) are jointly recognized as molecules in pork-cat syndrome, the cross-reactivity between pork meat and cat epithelia/dander. Besides serum albumin, several other IgE-binding proteins are detected in pork.

Chicken serum albumin (*Gal d* 5), which is identical to α-livetin in egg yolk, is responsible for bird-egg syndrome, the cross-reactivity between chicken and hen's egg yolk. In addition to serum albumin, several other IgE-binding proteins are detected in chicken meat.

Since serum albumin is a known dog and cat allergen, it is an important inhalant allergen as well as a food allergen.

Meat allergens are relatively heat labile and easily hydrolyzed by digestive enzymes; however, it should be noted that heating and digestive processes cannot "completely" eliminate the allergenicity in severe allergic patients.

REFERENCES

1. K.A. Ogle, J.D. Bullock. Children with allergic rhinitis and/or bronchial asthma treated with elimination diet: a five-year follow-up. *Ann. Allergy*, 44: 273–278, 1980.
2. R. Ayuso, S.B. Lehrer, L. Tanaka, M.D. Ibanez, C Pascual, A.W. Burks, G.L. Sussman, B. Goldberg, M. Lopez, G. Reese. IgE antibody response to vertebrate meat proteins including tropomyosin. *Ann. Allergy Asthma Immunol.*, 83: 399–405, 1999.
3. Y. Iikura, Y. Imai, T. Imai, A. Akasawa, K. Fujita, K. Hoshiyama, H. Nakura, Y. Kohno, K. Koike, H. Okudaira, E. Iwasaki. Frequency of immediate-type food allergy in children in Japan. *Int. Arch. Allergy Immunol.*, 118: 251–252, 1999.
4. A. Fiocchi, P. Restani, E. Riva. Beef allergy in chidren. *Nutrition*, 16: 454–457, 2000.
5. S. Tanabe, Y. Kobayashi, Y. Takahata, F. Morimatsu, R. Shibata, T. Nishimura. Some human B and T cell epitopes of bovine serum albumin, the major beef allergen. *Biochem. Biophys. Res. Commn.*, 293: 1348–1353, 2002.
6. B. Beretta, A. Conti, A. Fiocchi, A. Gaiaschi, C.L. Galli, M.G. Giuffrida, C. Ballabio, P. Restani. Antigenic determinants of bovine serum albumin. *Int. Arch. Allergy Immunol.*, 126: 188–195, 2001.

7. A. Fiocchi, P. Restani, E. Riva, G.P. Mirri, I. Santini, L. Bernardo, C.L. Galli. Heat treatment modifies the allergenicity of beef and bovine serum albumin. *Allergy*, 53: 798–802, 1998.
8. M. Besler, A. Fiocchi, P. Restani. Allergen data collection: beef. *Internet Symp. Food Allergens*, 3: 171–184, 2001.
9. G.D. Han, M. Matsuno, G. Ito, Y. Ikeuchi, A. Suzuki. Meat allergy: investigation of potential allergenic proteins in beef. *Biosci. Biotechnol. Biochem.*, 64: 1887–1895, 2000.
10. R. Ayuso, S.B. Lehrer, M. Lopez, G. Reese, M.D. Ibanez, M.M. Esteban, D.R. Ownby, H. Schwartz. Identification of bovine IgG as a major cross-reactive vertebrate meat allergen. *Allergy*, 55: 348–354, 2000.
11. X. Min He, D.C. Carter. Atomic structure and chemistry of human serum albumin. *Nature*, 358: 209–215, 1992.
12. T. Peters. Serum albumin. In: *Advances in Protein Chemistry*, Volume 37. New York: Academic Press, 1985, 161–245.
13. E.L. Gelamo, M. Tabak. Spectroscopic studies on the interaction of bovine (BSA) and human (HSA) serum albumins with ionic surfactants. *Spectrochim. Acta*, A56: 2255–2271, 2000.
14. S. Tanabe, R. Shibata, T. Nishimura. Hypoallergenic and T cell reactive analogue peptides of bovine serum albumin, the major beef allergen. *Mol. Immunol.*, 41: 885–890, 2004.
15. E. Böhler, T. Schäfer, S. Ruhdorfer, L. Weigl, D. Wessner, J. Heinrich, B. Filipiak, H.E. Wichmann, J. Ring. Epidemiology of food allergy in adults (in German). *Allergo. J.*, 10: 318–319, 2001.
16. M. Drouet, A. Sabbah. The pork/cat syndrome or crossed reactivity between cat epithelia and pork meat. In: B. Wüthrich, C. Ortolani (Eds.). *Highlights in Food Allergy.* Basel: Karger, 1996, 164–173.
17. A. Sabbah, C. Rousseau, M.G. Lauret, M. Drouet. The pork-cat syndrome: RAST inhibition test with *Fel d* 1 (in French). *Allerg. Immunol.*, 26: 259–260, 1994.
18. C. Hilger, M. Kohnen, F. Grigioni, C. Lehners, F. Hentges. Allergic cross-reactions between cat and pig serum albumin. *Allergy*, 52: 179–187, 1997.
19. R. Llatser, F. Polo, F. de la Hoz, B. Guillaumet. Alimentary allergy to pork. Crossreactivity among pork kidney and pork and lamb gut. *Clin. Exp. Allergy*, 28: 1021–1025, 1998.
20. Z. Szepfalusi, C. Ebner, R. Pandjaitan, F. Orlicek, O. Scheiner, G. Boltz-Nitulescu, D. Kraft, H. Ebner. Egg yolk alphalivetin (chicken serum albumin) is a cross-reactive allergen in the bird-egg syndrome. *J. Allergy Clin. Immunol.*, 93: 932–942, 1994.
21. S. Quirce, F. Maranon, A. Umpierrez, M. de las Heras, E. Fernandez-Caldas, J. Sastre. Chicken serum albumin (*Gal d* 5) is a partially heat-labile inhalant and food allergen implicated in the bird-egg syndrome. *Allergy*, 56: 754–762, 2001.
22. F. de Blay, C. Hoyet, E. Candolfi, R. Thierry, G. Pauli. Identification of alpha livetin as a cross reacting allergen in a bird-egg syndrome. *Allergy Proc.*, 15: 77–78, 1994.
23. J. Williams. Serum proteins and the livetins in hen's-egg yolk. *Biochem. J.*, 83: 346–355, 1962.
24. P. Restani, A. Fiocchi, B. Beretta, T. Velona, M. Giovannini, C.L. Galli. Meat allergy: III. Proteins involved and cross-reactivity between different animal species. *J. Am. Coll. Nutr.*, 16: 383–389, 1997.

25. B. Pandjaitan, I. Swoboda, F. Brandejsky-Pichler, H. Rumpold, R. Valenta, S. Spitzauer. *Escherichia coli* expression and purification of recombinant dog albumin, a cross-reactive animal allergen. *J. Allergy Clin. Immunol.*, 105: 279–285, 2000.
26. S.J. Werfel, S.K. Cooke, H.A. Sampson. Clinical reactivity to beef in children allergic to cow's milk. *J. Allergy Clin. Immunol.*, 99: 293–300, 1997.
27. M. Atanaskovic-Markovic, M. Gavrovic-Jankulovic, R.M. Jankov, O. Vuekovic, B. Nestorovic. Food allergy to pork meat. *Allergy*, 57: 960–961, 2002.
28. J.E. Kinsella, D.M. Whitehead. Proteins in whey: chemical, physical, and functional properties. *Adv. Food Nutr. Res.*, 33: 343–438, 1989.
29. E.A. Permyakov, L.A. Morozova, L.P. Kalinichenko, V.Y. Derezhklov. Interaction of alpha-lactalbumin with Cu^{2+}. *Biophys. Chem.*, 32: 37–42, 1988.
30. J.M. Garrett, R.A. Stairs, R.G. Annett. Thermal denaturation and coagulation of whey proteins: effect of sugars. *J. Dairy Sci.*, 71: 10–16, 1988.
31. P.G. Giampietro, N.I. Kjellman, G. Oldaeus, W. Wouters-Wesseling, L. Businco. Hypoallergenicity of an extensively hydrolyzed whey formula. *Pediatr. Allergy Immunol.*, 12: 83–86, 2001.
32. M. Watanabe. Hypoallergenic rice as a physiologically functional food. *Trends Food Sci. Technol.*, 4: 125–128, 1993.
33. M. Watanabe, J. Watanabe, K. Sonoyama, S. Tanabe. Novel method for producing hypoallergenic wheat flour by enzymatic fragmentation of the constituent allergens and its application to food processing. *Biosci. Biotechnol. Biochem.*, 64: 2663–2667, 2000.
34. A. Fiocchi, P. Restani, E. Riva, A.R. Restelli, G. Biasucci, C.L. Galli, M. Giovannini. Meat allergy: II. Effects of food processing and enzymatic digestion on the allergenicity of bovine and ovine meats. *J. Am. Coll. Nutr.*, 14: 245–250, 1995.
35. A. Suzuki, K. Kim, H. Tanji, Y. Ikeuchi. Effects of high hydrostatic pressure on postmortem muscle. *Recent Res. Devel. Agric. Biol. Chem.*, 2: 307–331, 1998.
36. G.D. Han, M. Matsuno, Y. Ikeuchi, A. Suzuki. Effects of heat and high-pressure treatments on antigenicity of beef extract. Biosci. *Biotechnol. Biochem.*, 66: 202–205, 2002.

25 Rice-Seed Allergenic Proteins and Hypoallergenic Rice

Tsukasa Matsuda, Masayuki Nakase, Angelina M. Alvarez, Hidehiko Izumi, Takeo Kato, and Yuichi Tada

CONTENTS

25.1 INTRODUCTION – ADVERSE ALLERGENIC REACTIONS TO RICE

Ingestion and inhalation of cereal grains and flours, including rice, sometimes provoke allergic disorders such as asthma, eczema and dermatitis, and gluten-sensitive enteropathy in a certain population (1,2). Clinical cases of allergic disorders caused by rice, which are less frequent than those of other cereals such as wheat have been reported in Japan and some European countries. In 1979, Shibasaki et al. (3) first reported in detail on rice allergy in Japanese patients, in which six patients showed immediate skin reaction and positive radioallergosorbent test (RAST) to soluble rice-seed extract, and five patients showed indurated skin reaction of delayed onset and lymphocyte proliferation reaction against the extract. Later, Ikezawa's group of Japanese dermatologists reported on the clinical effect of a replacement therapy for 43 patients with severe atopic dermatitis, who were suspected of having rice allergy (4). In this study, rice and wheat were replaced with hypoallergenic rice in their daily diet, and such a strict replacement therapy was shown to be useful or very useful in about 70% of the patients. In Europe, several clinical cases of rice allergy have been reported in Spain and Italy. A case study on hypersensitive reaction to rice (5) and another case of asthma and contact urticaria caused by handling rice and other cereals (6) were reported in Spain in 1992 and 1994, respectively. In Italy, clinical manifestations, including shock, vomiting, and diarrhea, in four infants with rice intolerance but without hypersensitivity to other foods were reported (7). In this case, double-blind oral challenges with rice were made and resulted in the induction of severe shock in three of four infants. Recently, there was report of a rare case of food allergy in the Netherlands in a 6-month-old girl, who had an anaphylactic reaction to rice flour, including a sudden onset of respiratory and gastrointestinal symptoms (8). There were also reports of clinical cases of allergic disorders induced by rice not only in infants but also in adults. A 43-year-old patient had a severe attack of bronchial asthma, and his serum immunoglobulin E (IgE) was RAST positive to rice (9). He had been eating hypoallergenic rice without experiencing any bronchial-asthma-induced attack. Although the incidences of rice-induced food allergy are not very frequent even in Asian countries, including Japan, where rice is consumed daily in large quantities, rice ingestion has sometimes caused severe allergic disorders such as sudden onset of skin, respiratory, and gastrointestinal symptoms and, in rarer cases, systemic anaphylaxis.

25.2 RICE-SEED PROTEINS AND ALLERGENS

25.2.1 Rice-Seed Proteins and Their Allergenic Potentials

From the viewpoint of food allergy caused by ingested rice grain or flour, in this chapter, rice-seed proteins mean proteins in the rice endosperm because rice is usually consumed as polished rice grains without the germ and the aleurone layer. Rice-seed proteins comprise about 8% (w/w) of the dried endosperm, and most of them accumulate in protein bodies as storage proteins for future use as nitrogen, sulfur, and carbon sources during the postgerminative period of development. These proteins are classified into four groups based on their solubility: water-soluble albumin, salt-soluble globulin, alcohol-soluble prolamin, and alkali-soluble glutelin (10). Rice endosperm contains 4 to 10% albumin plus globulin, 5 to 10% prolamin, and 80 to 90% glutelin, depending on the analytical methods, genotype, and environmental and cultivation conditions. Figure 25.1 shows a sodium dodecyl sulfate–polyacrylamide gel electrophoresis (SDS–PAGE) of fractionated rice seed proteins. Comparison of the total proteins with the proteins in each fraction demonstrates that glutelin is the major constituent of rice-endosperm proteins. The electrophoretogram also shows that the soluble proteins in the albumin and globulin fractions contain more diverse protein components than the insoluble proteins in prolamin and glutelin fractions.

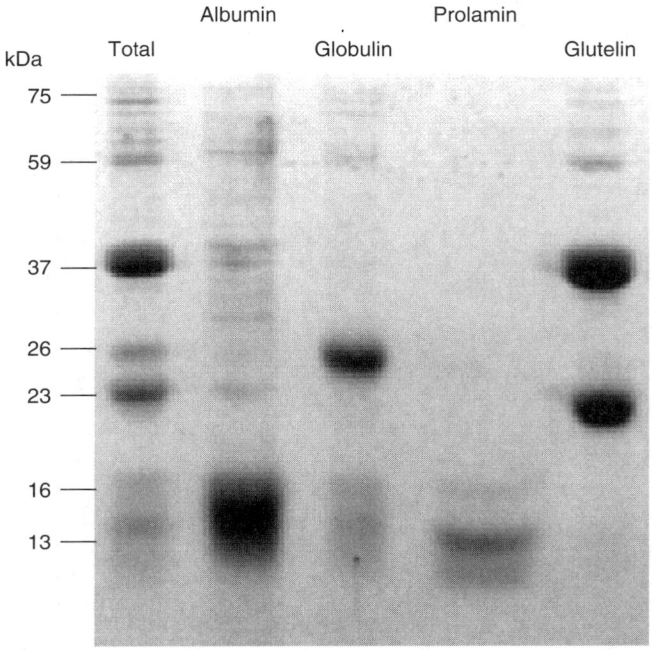

FIGURE 25.1 Solubility-dependent fractionation of rice-seed proteins. Rice-endosperm proteins were fractionated into water-soluble (albumin), salt-soluble (globulin), alcohol-soluble (prolamin), and alkali-soluble (glutelin) fractions. The component proteins in each fraction, as well as seed total proteins, were analyzed by SDS–PAGE (CBB-staining).

In an earlier study by Shibasaki et al. (3), a high degree of allergenic potential was suggested to be in the soluble protein fractions. In this regard, we extracted salt-soluble proteins (albumins plus globulins) from the rice endosperm, and they were further fractionated based on molecular size by size-exclusion chromatography. As shown in Figure 25.2A, the proteins were separated into two major peaks with two small shoulders based on ultraviolet (UV) absorbance measurement. The proteins in these peaks and shoulders were subjected to an IgE-binding analysis using enzyme-linked immunosorbent assay (ELISA) in patients' sera (Figure 25.2B). Most of the tested fractions clearly showed positive IgE-binding in the sera of four patients; especially, two peaks with IgE-binding were observed in fractions 25 and 29. Then, protein components in each fraction were analyzed by SDS–PAGE (Figure 25.2C), in which the major components of peaks 25 and 29 were affirmed to be ~33-kDa and ~14- to 16-kDa proteins. By employing immunoblot analysis using soluble rice proteins, Urisu et al. (11) screened proteins recognized by IgE antibodies from among 32 patients allergic to rice. This *in vitro* IgE-binding analysis indicated several proteins showing IgE-positivity in 25% or more of the 32 patients, specifically the ~15- to 16-kDa, 19-kDa, 25-kDa, ~33- to 35-kDa, 56-kDa, and 92-kDa proteins. Among them, the protein bands with strong IgE-binding

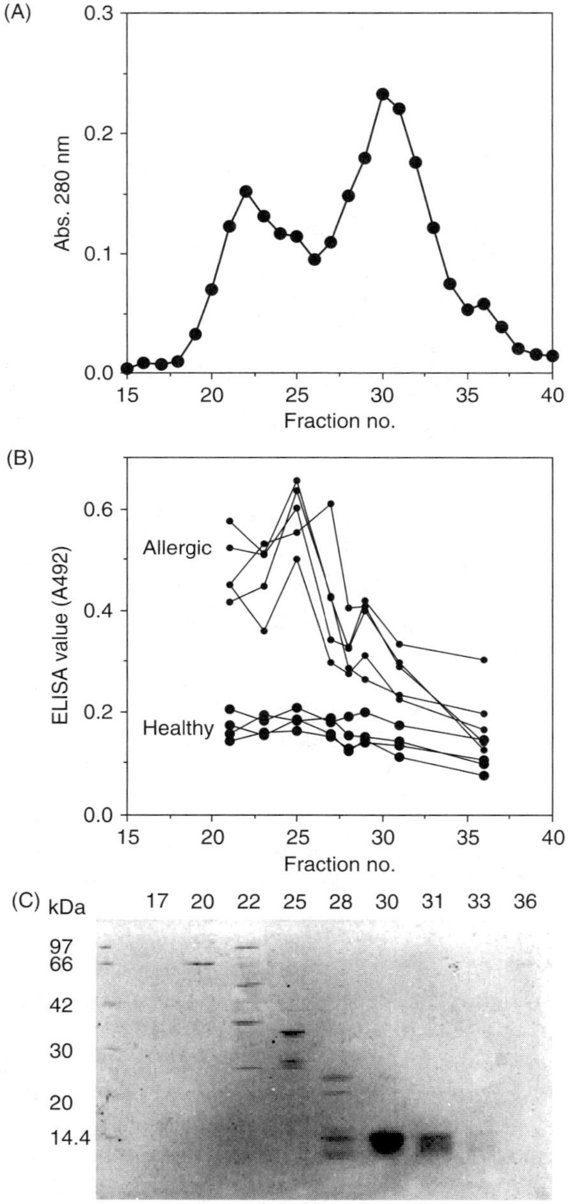

FIGURE 25.2 Molecular-size distribution of potential allergenic proteins in rice seeds. Salt-soluble rice proteins (albumin plus globulin) were fractionated by gel-filtration chromatography (Sephacryl S-200). The elution profile was monitored by measuring UV absorbance (280 nm) (panel A), and the IgE-binding activity of the fractions was assayed by ELISA, using allergic patients (dots) and healthy volunteers (closed circles) (panel B). The constituent proteins were analyzed by SDS–PAGE (panel C).

intensity were the ~33- to 35-kDa and the ~15- to 16-kDa proteins (11). Thus, several protein components with relatively lower molecular masses in the salt-soluble fraction are considered to be potential allergens causing rice allergy.

25.2.2 Rice-Seed Allergens

As indicated by specific binding with patients' IgE, several proteins have been purified and identified as potential allergens, and cDNA (complementary deoxyribonucleic acid) clones encoding these potential allergens have been obtained. Table 25.1 summarizes data on potential rice allergens identified and characterized so far. Three proteins with molecular masses of 18 kDa, 56 kDa, and 92 kDa, among those found in the screening by Urisu et al. (3), still remain to be investigated as potential rice allergens.

25.2.2.1 Allergens with MW of ~14- to ~16-kDa

A 16-kDa protein was first isolated from the salt-soluble fraction of rice-seed proteins, and its cDNA was cloned from a library of maturating seeds (12). The deduced amino acid sequence showed a limited but significant homology to the sequences of proteins belonging to the plant α-amylase-inhibitor/trypsin-inhibitor family (12) and to the lipid transfer-protein family (13,14). Interestingly, several members belonging to the amylase-/trypsin-inhibitor family have been reported to participate in allergic disorders such as baker's asthma caused by wheat (15) and barley (16) flours and food allergy caused by castor bean (17). Furthermore, lipid-transfer proteins have recently been suggested to be a panallergen commonly found in various vegetables and fruits (14). The overall sequence identities between the rice 16-kDa allergen and the inhibitor family members are only 20 to 40%, but all of the 10 cysteine residues are well conserved (12). Later, these cysteine residues were shown to form five intramolecular disulfide bonds, and each position of these disulfide bonds was estimated by characterization of disufide-containing peptides among the thermolytic digests of the 16-kDa protein (18). The speculated folding of the rice 16-kDa allergen is schematically shown in Figure 25.3. This folding pattern was identical to that reported for wheat α-amylase inhibitor (19), which was known to be the causative allergen in baker's asthma (15). These results suggest that the intramolecular disulfide linkages, aside from the position of 10 cysteine residues, have been well conserved within the α-amylase-/trypsin-inhibitor family. Allergenic proteins in food have been suggested to be stable against heat and proteases (20). The five disulfide bridges in a 135-residue polypeptide would stabilize the folding and the ordered structure, resulting in heat stability, resistance against proteolysis, and reversibility upon denaturation.

The monoclonal antibodies and polyclonal antiserum raised against the 16-kDa allergen cross-reacted with several other proteins having molecular masses of ~14 to 16 kDa and isoelectric points (pIs) between 6 to 9 (Figure 25.4). These results indicated the presence of some other proteins structurally similar and immunologically cross-reactive to the 16-kDa allergen. More than 10 cDNA clones and two genomic clones encoding these ~14- to 16-kDa proteins have been

TABLE 25.1
Rice Allergens Identified Based on the Binding to Patient's Serum IgE

Protein Size (by SDS–PAGE)	cDNA (Accession No.)		Amino Acids	Molecular Mass (mature form)	pI	Properties
~14- to 16-kDa						α-amylase inhibitor limited homology to lipid transfer protein
	RA17	(D11431)	140	14.8 kDa	6.45	
	RA14	(D11432)	139	15.2 kDa	7.00	N-glycosylation site
	RA16	(D42142)	130	14.6 kDa	7.72	
	RA5	(D11430)	131	14.4 kDa	8.28	
26-kDa	A3-12	(X63990)	164	18.9 kDa	6.74	pyroglutamic acid at the N-terminal
33-kDa	Glb33	(AB017042)	290	32.4 kDa	5.39	glyoxalase I, two homologous domains

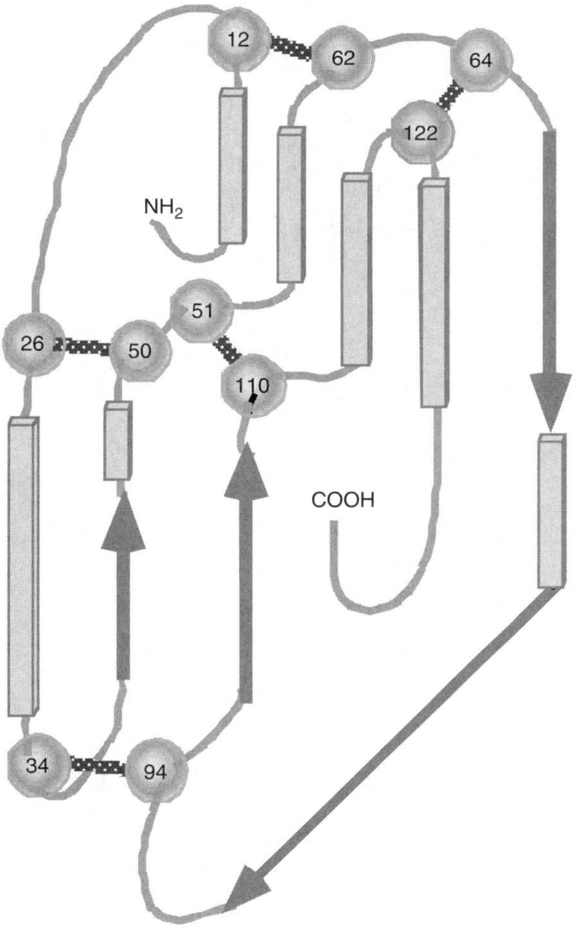

FIGURE 25.3 Speculated folding profile of rice ~14- to 16-kDa allergen. Helix and sheet structures as predicted by the Chou and Fasman method are drawn as arrows and vertical bars, respectively, and unordered structure is drawn as solid lines. Cysteine residues are shown as open circles with residue numbers, and disulfide bridges are represented by gray lines between two circles.

obtained so far (13,21), and sequence similarities in the amino acid levels among these clones range from 70 to 95%, indicating that the ~14- to 16-kDa proteins are products of a multigene family. Some of the deduced amino acid sequences had a potential N-glycosylation site near the C-terminus. On the other hand, the ~14- to 16-kDa proteins extracted from rice seeds were separated into five components by ion-exchange chromatography; and by analysis of their amino acid sequences, the protein components were found to have corresponding cloned cDNAs (22). The potential allergenicity of each of the ~14- to 16-kDa components was assessed by

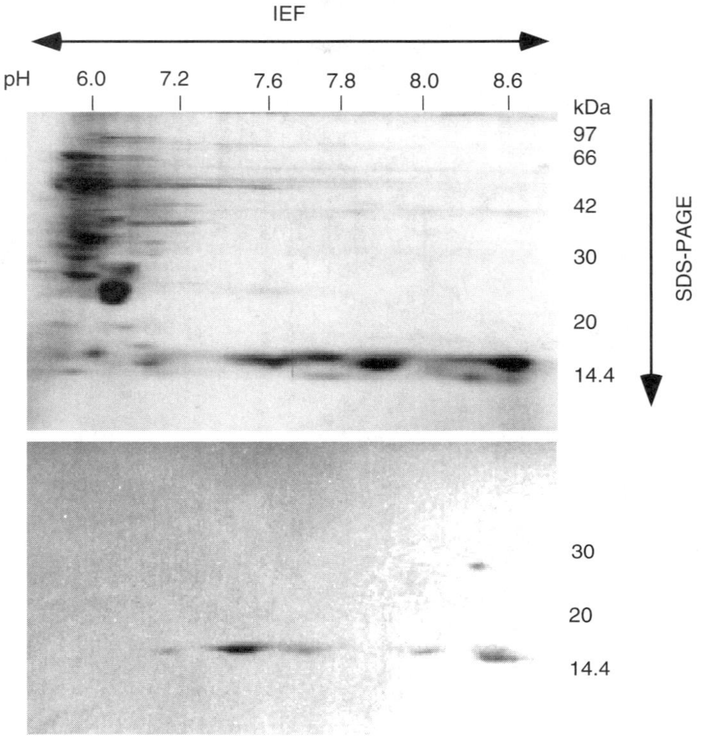

FIGURE 25.4 Heterogeneity in isoelectric point of ~14- to 16-kDa allergens. Salt-soluble rice-seed proteins (albumin plus globulin) were separated by IEF/SDS two-dimensional PAGE (2D-PAGE) and stained with CBB (upper panel). The ~14- to 16-kDa allergens were immunologically detected by using a monoclonal antibody specific for a 16-kDa allergen (lower panel).

measuring their IgE-binding in 17 patients' sera. The IgE antibodies from some patients reacted almost equally with all of the five proteins, while the rest reacted with one or two proteins dominantly. Thus, all five proteins are potentially allergenic. These ~14- to 16-kDa allergens might have multiple different epitopes recognized by patients' IgE, and some of the epitopes are common among these proteins and shared by some patients.

25.2.2.2 Allergens with MW of 26-kDa

The 26-kDa protein, which is the major component of the rice-endosperm globulin fraction (see Figure 25.1), is sometimes termed 28- or 25-kDa protein (based on SDS–PAGE), 19-kDa protein (based on the deduced sequence), or α-globulin in the literature. This protein migrates as 26-kDa in SDS–PAGE (23,24), but the molecular mass calculated from the amino acid sequence deduced from its cDNA sequence is 19 kDa (25). The reason for the slower migration of this 19-kDa/26-kDa protein by SDS–PAGE might be due to unique conformational properties of the protein in SDS-containing solution.

Limas et al. (26) first reported on the potential allergenicity of the 26-kDa protein as measured from the immunoblot of purified salt-soluble rice proteins for IgE from patients with suspected cereal allergy. In screening for rice allergens Urisu et al. had identified a 26-kDa protein as an allergen recognized by IgE from 12 patients, even though strong IgE-binding was observed only in a few patients (11).

The 26-kDa allergen is expressed at a higher level than are the other globulins in maturing seeds and consists of a large proportion of salt-soluble rice-endosperm proteins, even though this protein is encoded by a single gene (27). Nakase et al. (28) cloned a genomic DNA fragment containing the promoter and coding regions of the 26-kDa allergen and later isolated a cDNA clone encoding a novel b-ZIP protein (termed REB: Rice Endosperm b-ZIP) that specifically bound to the 26-kDa allergen gene promoter (29). Recently, by using transient gene expression analysis, REB was demonstrated by Yang et al. (30) to be a strong transcriptional activator of the 26-kDa allergen gene in maturing rice seeds. Such a strong promoter driven by REB might explain the higher expression of the 26-kDa allergen in rice endosperm.

25.2.2.3 Allergens with MW of 33-kDa

The 33-kDa protein was detected as a major component in the higher molecular mass fraction showing strong IgE binding (see Figure 25.2). Furthermore, the 33-kDa band was strongly stained with IgE antibodies from 12 patients in the screening of rice allergens (11). Usui et al. (31) purified the rice 33-kDa protein from a salt-soluble fraction and cloned its cDNA. The purified protein showed remarkable IgE-binding in more than 60% of patients' sera tested; also, such IgE binding (to the 33-kDa protein) correlated well with that of the total salt-soluble proteins (Figure 25.5). These results indicate that the 33-kDa protein is a major or dominant IgE-binding protein among the salt-soluble rice proteins and that it could be a major allergen. The 33-kDa allergen cDNA coded for a protein of 291 amino acids with two 120-amino acid residue repeats. Its amino acid sequence showed similarity to glyoxalase I from various organisms, including humans, plants, yeast, and bacteria. A comparison of sequence similarity and overall polypeptide structure of the rice 33-kDa protein with other proteins, including glyoxalase I, is presented schematically in Figure 25.6. Indeed, both the native 33-kDa allergen purified from rice seeds and the recombinant protein expressed in *Escherichia coli* had glyoxalase I activity that catalyzed condensation of methylglyoxal and glutathione into S-lactoylglutathione. Both the native and the recombinant 33-kDa proteins also reacted with the patients' IgE and showed a molecular mass of 33 kDa in SDS–PAGE, which agreed well with the size of the 33-kDa protein polypeptide (32.4 kDa) calculated from the amino acid sequence. These results, in addition to the absence of a potential N-glycosylation site in its sequence, suggest that the 33-kDa allergen has no antigenic/allergenic

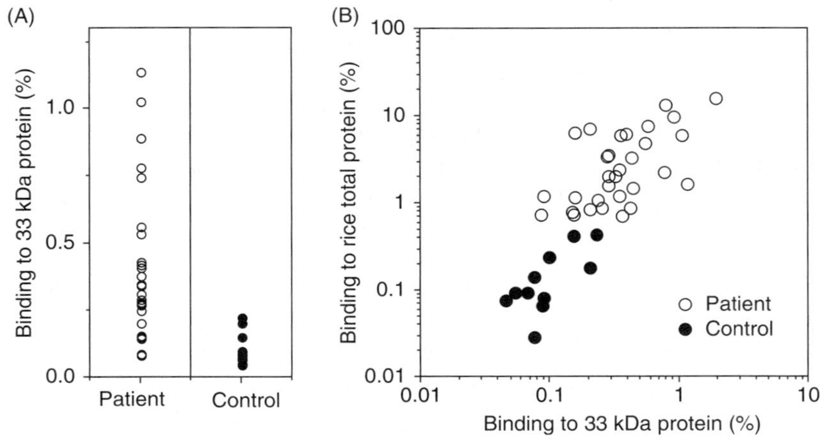

FIGURE 25.5 IgE-binding to rice 33-kDa allergen. Human IgE, which bound to purified 33-kDa allergen, was measured by radioallergosorbent assay (RAST) individually in patients' sera and control sera (panel A). A plot of IgE-binding to 33-kDa allergen against that of rice-seed total protein is shown in panel B.

N-linked sugar chain such as xylose- and fucose-containing sugar chains found in some plant and insect proteins (32). The antibody specific for the rice 33-kDa allergen recognized immunologically cross-reactive 33-kDa proteins from the seeds of some other rice species (Usui and Matsuda, unpublished) and cereals such as wheat, maize, and barley, indicating a wide presence of possible 33-kDa allergen homologues.

25.2.3 Other Allergens

Some potential allergens in rice seeds other than the ~14- to 16-kDa, 26-kDa, and 33-kDa allergens remain to be investigated. It has already been suggested that IgE antibodies from several patients also recognized the 56-kDa and 92-kDa bands in SDS–PAGE (11), but neither the proteins nor the genes have yet been isolated and characterized. A 60-kDa protein was reported to be a potential allergen with IgE binding, based on its N-terminal amino acid sequence, and identified as uridine diphosphate (UDP)-glucose starch glycosyltransferase (EC: 2.4.1.21) (33). A cDNA encoding a rice lipid transfer protein, which might be a potential rice allergen, has been cloned, but the protein has not yet been purified and characterized.

25.3 HYPOALLERGENIC RICE

25.3.1 Degradation and Removal by Enzymatic Hydrolysis

Hypoallergenic foods have been developed and industrially produced for replacement therapy, the prevention of sensitization, and the diagnosis of several food

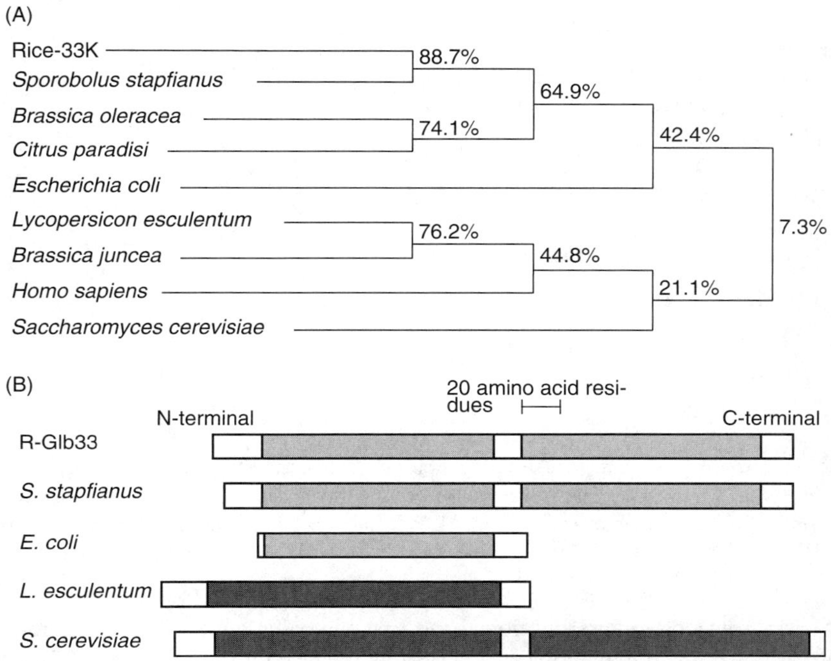

FIGURE 25.6 Structural comparison of 33-kDa allergen with glyoxalase I and homologous proteins. Phylogenetic analysis of the 33-kDa allergen and some homologous proteins based on their amino acid sequences (panel A). Schematic presentation of rice 33-kDa allergen and some other glyoxalase I proteins (panel B). The regions with sequence homology are shown by gray and dark gray squares.

allergies. In most cases, proteins, including allergens in foods, are proteolytically degraded and removed, resulting in the loss or reduction of allergenicity. Rice grains seem to have some advantages for the enzymatic reduction or removal of allergenicity: (a) rice is usually consumed as cooked grains, not flour, and the component playing important roles in texture and taste is starch, not proteins. This is not the case with wheat, in which it is the proteins (gluten) that are indispensable for the functional properties of wheat flour used for making bread, pasta, and noodles; (b) most of the potential allergenic proteins in rice are salt soluble. This physicochemical property would make it possible for the allergens to be digested enzymatically and to be extracted or washed out; and (c) the major storage proteins, glutelin and prolamin, which occupy 90% or more of total rice-seed proteins, are not allergenic generally and accumulated in specialized protein bodies. These major storage proteins in the protein bodies escape proteolytic attack in the grains, and remain intact even after the hypoallergenic treatments, resulting in the maintenance of the nutritional value of rice as a major protein source in many Asian countries, including Japan.

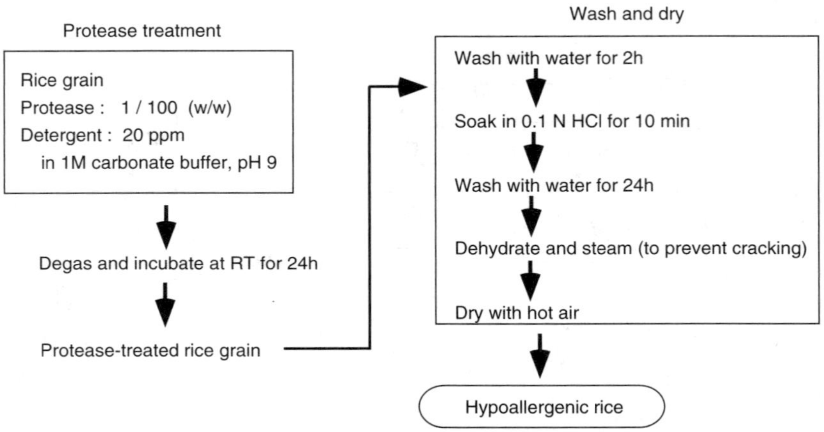

FIGURE 25.7 A procedure for the production of a dehydrated type of hypoallergenic rice by enzymatic treatment. The proteolytic digestion, wash, and dry steps are summarized according to the reported method. (From M. Watanabe, J. Miyakawa, Z. Ikezawa, Y. Suzuki, T. Hirano, T. Yoshizawa, S. Arai. *J. Food Sci.*, 55, 781–783, 1990. With permission.)

Arai and co-workers developed the first hypoallergenic rice preparation in 1990 (34). In the hypoallergenic process that they established (Figure 25.7), polished rice grains were treated with a protease in an alkali solution containing a detergent and then washed with water to remove the hydrolyzed proteins and the detergent. Finally, the rice grains were dried and then supplied to patients. Most of the (95% or more) albumin/globulin proteins, including the allergens, were eliminated from the grains by this enzymatic and washing treatment (34,35). After this hypoallergenic rice was administered to rice-allergic patients suffering from severe and refractory dermatitis, there was a marked alleviation of the allergic symptoms (4).

Another variety of hypoallergenic rice has been developed and produced by a Japanese food company. Nakajo and co-workers found that the albumin/globulin proteins were effectively extracted from polished rice grains in hot salt-solution and utilized this property of the rice proteins for selective removal of allergens (36). They also treated the rice grains with proteases after the saline treatment to degrade the remaining albumin/globulin proteins, including the allergens. Then, the hypoallergenic rice grains were steamed, packed into retort pouches, sterilized, and supplied as "cooked-type" hypoallergenic rice. This cooked-type hypoallergenic rice has been produced commercially since 1997, and about 100,000 packages per year are now sold on the market.

In such hypoallergenic rice products, most of the salt-soluble proteins are washed out or proteolytically degrade, but the major storage proteins remain in the seeds. Typical SDS–PAGE patterns of proteins remaining in these hypoallergenic rice seeds are shown in Figure 25.8. When the total proteins were extracted and analyzed, major bands found in the lane of nontreated control seeds were also detected

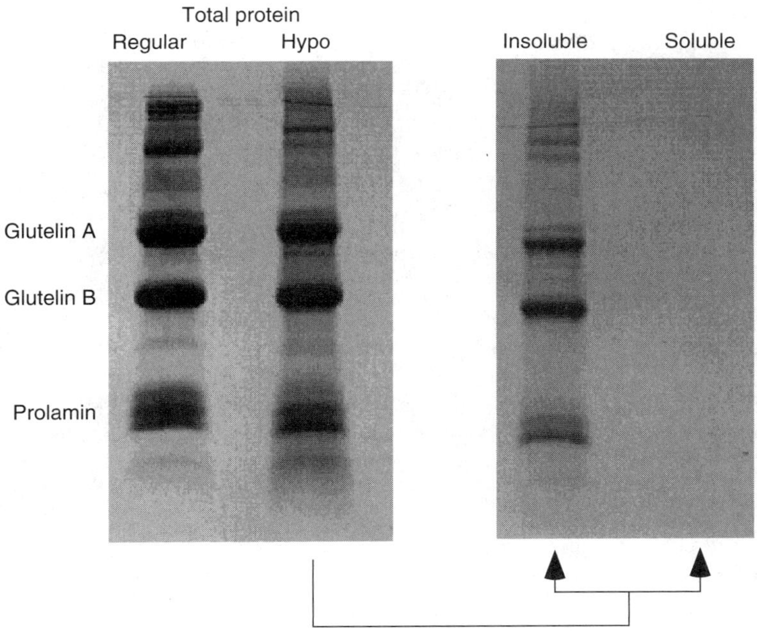

FIGURE 25.8 Proteins remaining in cooked-type hypoallergenic rice. Total proteins and salt-soluble proteins are extracted from the cooked-type hypoallergenic rice and analyzed by SDS–PAGE.

in the lane of the hypoallergenic seeds. On the other hand, salt-soluble proteins remaining in the hypoallergenic seeds were undetectable. Thus, by taking advantage of the differences in the distribution and properties of rice-seed proteins, the salt-soluble proteins, which include most allergenic proteins, are selectively removed from the rice grains in these enzymatically processed hypoallergenic rice products.

A unique enzymatic method similar to the two described above is one that utilizes a fermented soybean seasoning called "*miso*" for reduction of the potential allergenicity of rice grains (37). When polished rice grains were incubated with a 10% *miso* solution, but not with heat-treated *miso*, the 26-kDa and ~14- to 16-kDa allergens in the grains were decreased from 60 to 15% during the incubation. *Miso* is well known to possess koji-derived proteolytic activity, which degrades soybean proteins, including a soybean allergen, Gly m Bd 30K, during fermentation (38). The *miso* proteases also degraded proteins in rice grains used for koji preparation (39). Therefore, some koji-derived proteases remaining in *miso* could be expected to degrade rice-seed proteins, including the allergens.

25.3.2 Release and Removal by High Hydrostatic Pressure

Numerous studies have been done on the effects of high hydrostatic pressure on food processing (40). High-pressure technology was also applied in the reduction

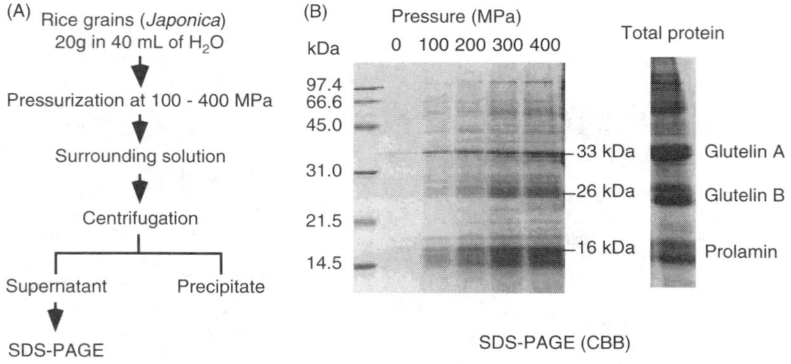

FIGURE 25.9 Release of several salt-soluble proteins from rice grains by pressurization. Rice grains were pressurized, as summarized in (A), and the released proteins were analyzed by SDS–PAGE (B).

of the allergenic potential in rice grains (41). When polished rice grains immersed in water or salt solutions were pressurized at 100 to 400 MPa for ~15 to 120 minutes, the amount of proteins extracted from the grains was ~three to five times as much as that of the control without pressurization. A typical pressurization procedure and the SDS–PAGE pattern of extracted proteins are shown in Figure 25.9. The major protein components effectively extracted by pressurization were identified to be the 33-kDa, 26-kDa, and ~14- to 16-kDa allergens based on their N-terminal amino acid sequences and recognition by specific antibodies. The combination of pressurization and protease treatment was suggested to be effective in the removal of allergenic proteins from rice grains.

25.3.3 Suppression of Biosynthesis by Genetic Modification

Some genetic methods using antisense ribonucleic acids (RNAs) or double-stranded RNAs to suppress a target gene specifically by initiating "gene silencing" have been developed (42,43). We attempted to develop genetically modified hypoallergenic rice by introducing antisense allergen genes into rice and suppressing the ~14- to 16-kDa allergen gene expression in maturing rice seeds (44,45). The antisense gene, which was derived from the 16-kDa allergen cDNA (RA17) and driven by promoters of several seed-specific genes — prolamin, glutelin, and a starch branching enzyme (BE) — was introduced into rice cell protoplasts by the electroporation method. "Selfed" seeds obtained from the transgenic rice, which had an expected size of the antisense gene, were analyzed for their allergen content by immunoblotting. Figure 25.10 shows a genomic Southern blot analysis of transgenic lines and SDS–PAGE and immunoblot analyses of their seeds. The seeds from some lines having the transgenic antisense gene(s) in their genome showed a weak or almost no band stained by the allergen-specific antibody. The introduced antisense gene was stably transmitted to

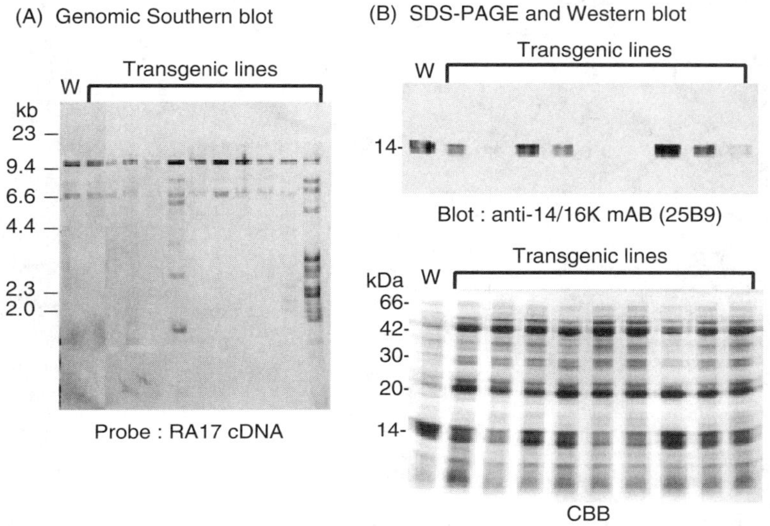

FIGURE 25.10 Suppression of ~14- to 16-kDa allergen biosynthesis in the antisense-transgenic rice plant. Introduction of the antisense gene was confirmed by genomic Southern analysis (A) using RA17 cDNA as a probe. Seeds of transgenic rice plants were analyzed by SDS–PAGE (lower panel of B) and Western blot analysis (upper panel of B). "W" represents wild-type rice seeds.

the progeny plants, and the phenotype of low allergen content was stably inherited by the second-generation lines. The ~14- to 16-kDa allergen contents of nontransgenic and transgenic rice seeds were estimated by ELISA to be about 300 μg and 60 to 70 μg per seed, respectively.

Thus, the content of ~14- to 16-kDa allergens in some transgenic lines was reduced by not more than 20% relative to the nontransgenic controls. However, no transgenic line was found to be free from these allergens. To know the reason for this incomplete suppression, the ~14- to 16-kDa allergens remaining in seeds of several transgenic lines were analyzed by two-dimensional (2D) gel electrophoresis (46). Figure 25.11 shows typical 2D-PAGE and immunoblots of salt-soluble seed proteins. Expression of each of the ~14- to 16-kDa allergens with pIs ranging between 5 and 9 was not evenly suppressed in a single transgenic rice seed. Among the ~14- to 16-kDa allergens, those remaining in the transgenic seed were mostly basic components with a pI of about 8, whereas those reduced successfully were the acidic components. Such selective or preferential suppression was explained by the facts that the cDNA used for the antisense gene construct encoded the most acidic component RA17 and that its cDNA sequence homology was higher for the acidic and lower for the basic components (45,46). Thus, the effectiveness of antisense RNA would strictly depend on the homology between the sequence used in the antisense gene and the target sequence, even though a single antisense RNA could suppress the expression of several homologous genes in a multigene family.

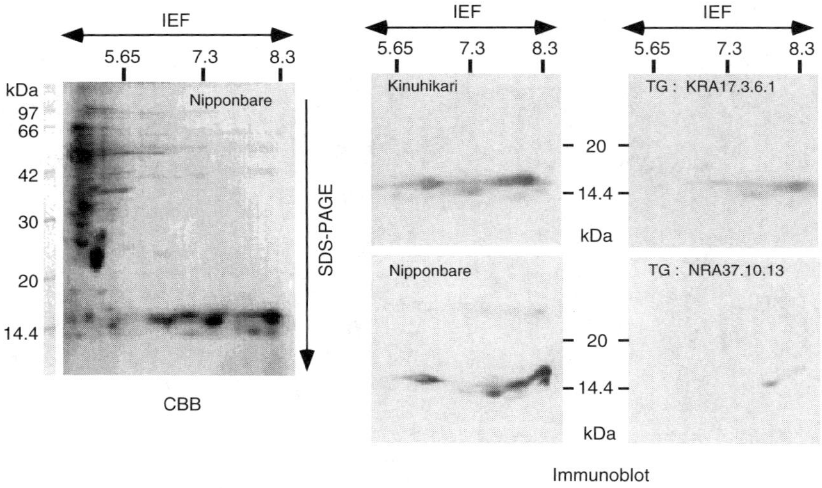

FIGURE 25.11 Selective suppression of ~14 to 16-kDa allergen in the transgenic rice plant. Salt-soluble rice-seed protein of the wild-type (Nipponbare and Kinuhikari) and transgenic rice seeds (TG) was separated by IEF/SDS two-dimensional PAGE and stained with CBB or immunologically detected with an antibody specific for the 16-kDa allergen.

It is still uncertain whether such hypoallergenic rice seeds obtained from transgenic rice plants can be tolerated by patients allergic to rice because even a small amount of residual allergens could elicit allergic reactions in these patients. Furthermore, most patients have IgE antibodies specific to some allergens other than the ~14- to 16-kDa allergen (11). Toward the development of genetically modified hypoallergenic rice, further basic studies would be needed for the identification of the other allergens and their genes and the elucidation of detailed mechanisms regulating allergen gene expression in rice seeds.

REFERENCES

1. B.A. Baldo, C.W. Wrigley. Allergies to cereals. In: Y. Pomeranz (Ed.). *Advances in Cereal Science and Technology*, Volume VI. St. Paul: American Association of Cereal Chemists, 1984, 289–356.
2. R.P. Sturgess, H.J. Ellis, P.J. Ciclitira. Cereal chemistry, molecular biology, and toxicity in coeliac disease. *Gut*, 32: 1055–1060, 1991.
3. M. Shibasaki, S. Suzuki, H. Nemoto, T. Kuroume. Allergenicity and lymphocyte-stimulating property of rice protein. *J. Allergy Clin. Immunol.*, 64: 259–265, 1979.
4. Z. Ikezawa, K. Miyakawa, H. Komatsu, C. Suga, J. Miyakawa, A. Sugiyama, T. Sasaki, H. Nakajima, Y. Hirai, Y. Suzuki. A probable involvement of rice allergy in severe type of atopic dermatitis in Japan. *Acta, Dermatol. Venereol.(Stockh)*, Suppl. 176: 103–107, 1992.
5. C. Granel, A. Olive, L. Randazzo, G. Tapi, A. Martinez, M. Valencia. Hypersensitivity reaction to rice. A case report. *Allergol. Immunopathol. (Madr.)*, 20: 171–172, 1992.

6. A. Lezaun, J.M. Igea, S. Quirce, M. Cuevas, F. Parra, M.D. Alonso, J.A. Martin, M.S. Cano. Asthma and contact urticaria caused by rice in a housewife. *Allergy*, 49: 92–95, 1994.
7. F. Cavataio, A. Carroccio, G. Montalto, G. Iacono. Isolated rice intolerance: clinical and immunologic characteristics in four infants. *J. Pediatr.*, 128: 558–560, 1996.
8. S.K. Klein, E.M. Kremers, W.B. Vreede. Six-month old girl with an anaphylactic reaction to rice flour, a rare food allergy. *Ned. Tijdschr. Geneeskd.*, 145: 1471–1473, 2001.
9. T. Arai, T. Takaya, Y. Ito, K. Hayakawa, S. Toshima, C. Shibuya, M. Nomura, N. Yoshimi, M. Shibayama, Y. Yasuda. Bronchial asthma induced by rice. *Intern. Med.*, 37: 98–101, 1998.
10. B.O. Juliano. The rice grain and its gross composition. In: DF. Houston (Ed.). *Rice Chemistry and Technology*. St. Paul: American Association of Cereal Chemistry, 1972, 16–74.
11. A. Urisu, K. Yamada, S. Masuda, H. Komada, E. Wada, Y. Kondo, F. Horiba, M. Tsuruta, T. Yasaki, M. Yamada, S. Torii, R. Nakamura. 16-kilodalton rice protein is one of the major allergens in rice grain extract and responsible for cross-allergenicity between cereal grains in the Poaceae family. *Int. Arch. Allergy Appl. Immunol.*, 96: 244–252, 1991.
12. H. Izumi, T. Adachi, N. Fujii, T. Matsuda, R. Nakamura, K. Tanaka, A. Urisu, Y. Kurosawa. Nucleotide sequence of a cDNA clone encoding a major allergenic protein in rice seeds. Homology of the deduced amino acid sequence with members of alpha-amylase/trypsin inhibitor family. *FEBS Lett.*, 302: 213–216, 1992.
13. A.M. Alvarez, T. Adachi, M. Nakase, N. Aoki, R. Nakamura, T. Matsuda. Classification of rice allergenic protein cDNAs belonging to the α-amylase/trypsin inhibitor gene family. *Biochim. Biophys. Acta*, 1251: 201–204, 1995.
14. R. Sanchez-Monge, M. Lombardero, F.J. Garcia-Selles, D. Barber, G. Salcedo. Lipid-transfer proteins are relevant allergens in fruit allergy. *J. Allergy Clin. Immunol.*, 103: 514–519, 1999.
15. J.M. James, J.P. Sixbey, R.M. Helm, G.A. Bannon, A.W. Burks. Wheat alpha-amylase inhibitor: a second route of allergic sensitization. *J. Allergy Clin. Immunol.*, 99: 239–244, 1997.
16. D. Barber, R. Sanchez-Monge, L. Gomez, J. Carpizo, A. Armentia, C. Lopez-Otin, F. Juan, G. Salcedo. A barley flour inhibitor of insect alpha-amylase is a major allergen associated with baker's asthma disease. *FEBS Lett.*, 248: 119–122, 1989.
17. F.S. Sharief, S.S.L. Li. Amino acid sequence of small and large subunits of seed storage protein from *Ricinus communis*. *J. Biol. Chem.*, 257: 14753–14759, 1982.
18. H. Izumi, M. Sugiyama, T. Matsuda, R. Nakamura. Structural characterization of the 16-kDa allergen, RA17, in rice seeds. Prediction of the secondary structure and identification of intramolecular disulfide bridges. *Biosci. Biotechnol. Biochem.*, 63: 2059–2063, 1999.
19. E. Poerio, C. Caporale, L. Carrano, P. Pucci, V. Buonocore. Assignment of the five disulfide bridges in an alpha-amylase inhibitor from wheat kernel by fast-atom-bombardment mass spectrometry and Edman degradation. *Eur. J. Biochem.*, 199: 595–600, 1991.
20. J.D. Astwood, J.N. Leach, R.L. Fuchs. Stability of food allergens to digestion *in vitro*. *Natl. Biotechnol.*, 14: 1269–1273, 1996.
21. T. Adachi, H. Izumi, T. Yamada, K. Tanaka, S. Takeuchi, R. Nakamura, T. Matsuda. Gene structure and expression of rice seed allergenic proteins belonging to the alpha-amylase/trypsin inhibitor family. *Plant Mol. Biol.*, 21: 239–248, 1993.

22. M. Nakase, T. Adachi, A. Urisu, T. Miyashita, A.M. Alvarez, S. Nagasaka, N. Aoki, R. Nakamura, T. Matsuda. Rice (*Oryza sativa* L.) α-amylase inhibitors of 14-16 kDa are potential allergens and products of a multigene family. *J. Agric. Food Chem.*, 44: 2624–2628, 1996.
23. S. Komatsu, H. Hirano. Rice seed globulin: a protein similar to wheat seed glutenin. *Phytochemistry*, 31: 3455–3459, 1992.
24. M. Nakase, A.M. Alvarez, T. Adachi, N. Aoki, R. Nakamura, T. Matsuda. Immunochemical and biochemical identification of the rice seed protein encoded by cDNA clone A3-12. *Biochem. Biosci. Biotech.*, 60: 1031–1032, 1996.
25. H.B. Krishnan, S.G. Pueppke. Nucleotide sequence of an abundant rice seed globulin: homology with the high molecular weight glutelins of wheat, rye and triticale. *Biochem. Biophys. Res. Commn.*, 193: 460–466, 1993.
26. G.G. Limas, M. Salinas, I. Moneo, S. Fischer, B. Wittmann-Liebold, E. Mendez. Purification and characterization of ten new rice NaCl-soluble proteins: identification of four protein-synthesis inhibitors and two immunoglobulin-binding proteins. *Planta*, 181: 1–9, 1990.
27. B.S. Shorrosh, L. Wen, K.C. Zen, J.K. Huang, J.S. Pan, MA. Hermodson, K. Tanaka, S. Muthukrishnan, G.R. Reeck. A novel cereal storage protein: molecular genetics of the 19 kDa globulin of rice. *Plant Mol. Biol.*, 18: 151–154, 1992.
28. M. Nakase, H. Hotta, T. Adachi, A.M. Alvarez, N. Aoki, R. Nakamura, T. Masumura, K. Tanaka, T. Matsuda. Cloning of the rice seed α-globulin-encoding gene: sequence similarity of the 5′-flanking region of the those of the genes encoding wheat high-molecular-weight glutenin and barley D hordein. *Gene*, 170: 223–226, 1996.
29. M. Nakase, N. Aoki, T. Matsuda, T. Adachi. Characterization of a novel rice bZIP protein which binds to the α-globulin promoter. *Plant Mol. Biol.*, 33: 513–522, 1997.
30. D. Yang, L. Wu, Y.S. Hwang, L. Chen, N. Huang. Expression of the REB transcriptional activator in rice grains improves the yield of recombinant proteins whose genes are controlled by a *Reb*-responsive promoter. *Proc. Natl. Acad. Sci. USA*, 98: 1438–11443, 2001.
31. Y. Usui, M. Nakase, H. Hotta, A. Urisu, N. Aoki, K. Kitajima, T. Matsuda. A 33 kDa allergen from rice (*Oryza sativa* L. *Japonica*): cDNA cloning, expression and identification as a novel glyoxalase I. *J. Biol. Chem.*, 276: 11376–11381, 2001.
32. R. van Ree, M. Cabanes-Macheteau, J. Akkerdaas, J.P. Milazzo, C. Loutelier-Bourhis, C. Rayon, M. Villalba, S. Koppelman, R. Aalberse, R. Rodriguez, L. Faye, P. Lerouge. (1,2)-Xylose and (1,3)-fucose residues have a strong contribution in IgE binding to plant glycoallergens. *J. Biol. Chem.*, 275: 11451–11458, 2000.
33. Z. Ikezawa, K. Tsubaki, H. Osuna, T. Shimada, K. Moteki, H. Sugiyama, K. Katumata, H. Anzai, S. Amano. Usefulness of hypoallergenic rice (AFT-R 1) and analysis of the salt insoluble rice allergen molecule. *Arerugi*, 48: 40–49, 1999.
34. M. Watanabe, J. Miyakawa, Z. Ikezawa, Y. Suzuki, T. Hirano, T. Yoshizawa, S. Arai. Production of hypoallergenic rice by enzymatic decomposition of constituent proteins. *J. Food Sci.*, 55**:** 781–783, 1990.
35. M. Watanabe, T. Yoshizawa, J. Miyakawa, Z. Ikezawa, K. Abe, T. Yanagisawa, S. Arai. Quality improvement and evaluation of hypoallergenic rice grains. *J. Food Sci.*, 55: 1105–1107, 1990.
36. M. Nakajo, S. Nakano, Y. Sato. Japanese Patent No. 391060812; or M. Nakajo, S. Nakano. Hypoallergenic cooked rice "Care Rice". *Arerugi no Rinsho (in Japanese)*, 18: 536–538, 1998.

37. H. Izumi, S. Kondo, H. Kasho, T. Matsuda, R. Nakamura. Decrease in rice allergenic proteins of polished rice grains by incubating with a miso solution. *Biosci. Biotechnol. Biochem.*, 64: 2248–2251, 2000.
38. H. Tsuji, N. Okada, R. Yamanishi, N. Bando, H. Ebine, T. Ogawa. Fate of major soybean allergen, Gly m Bd 30K, in rice-, barley-, and soybean- koji miso (fermented soybean paste) during fermentation. *Food Sci. Technol. Int.* (*Tokyo*), 3: 145–149, 1997.
39. S. Nikkuni, T. Ishiyama, C. Suzuki. Changes in rice proteins during miso fermentaion. *Nippon Shokuhin Kogyo Gakkaishi (in Japanese)*, 38: 316–322, 1991.
40. R. Hayashi. Progress of high-pressure use. In: R. Hayashi (Ed.). *High-Pressure Bioscience and Food Science*. Kyoto: San-Ei Publications, 1993, 1–17.
41. T. Kato, A. Katayama, S. Matsubara, Y. Oumi, T. Matsuda. Release of allergenic proteins from rice grains induced by high hydrostatic pressure. *J. Agric. Food Chem.*, 48: 3124–3129, 2000.
42. C.J.S. Smith, C.F. Watson, C. Ray, C.R. Bird, P.C. Morris, W. Schuch, D. Grierson. Antisense RNA inhibition of polygalacturonase gene expression in transgenic tomatoes. *Nature*, 334: 724–726, 1998.
43. N.A. Smith, S.P. Singh, M.B. Wang, P.A. Stoutjesdijk, A.G. Green, P.M. Waterhouse. Gene expression: total silencing by intron-spliced hairpin RNAs. *Nature*, 407: 319–320, 2000.
44. Y. Tada, M. Nakase, T. Adachi, R. Nakamura, H. Shimada, M. Takahashi, T. Fujimura, T. Matsuda. Reduction of 14-16kDa allergenic proteins in transgenic rice plants by antisense gene. *FEBS Lett.*, 391: 341–345, 1996.
45. R. Nakamura, T. Matsuda. Rice allergenic protein and molecular-genetic approach for hypoallergenic rice. *Biosci. Biotechnol. Biochem.*, 60: 1215–1221, 1996.
46. Y. Tada, H. Akagi, T. Fujimura, T. Matsuda. Effect on an antisense sequence on rice allergen genes comprising a multigene family. *Breeding Sci.*, 53: 61–67, 2003.

26 Buckwheat Allergy

Chein-Soo Hong and Kyu-Earn Kim

CONTENTS

26.1 INTRODUCTION

Buckwheat (*Fagopyrum esclentum* Moench) is an herbaceous plant of the Polygonaceae family, which includes reum (rhubarb) and rumex (sorrel). It does not have any affinity for the Gramineae family. Generally, there are many species of buckwheat in the world, and mainly nine species have agricultural significance. Generally, *Fagopyrum* has two groups of species: annual and multiennal species. The annual species (A) are *F. esculentum* Moench, *F. tataricum*, L., and *F. giganteum* Krotov. The multiennual species (B) include *F. cymosum* Meissn, *F. suffruticosum* Fr. Schmidt, and *F. ciliatum* Jaegt. Among these species, currently, the most commonly grown buckwheat species is *F. esculentum* Moench (common buckwheat, or sweet buckwheat), while *F. tartaricum* is also available in some mountainous regions.

Buckwheat is an important crop in some areas of the world and has been cultivated in China for more than a thousand years (1). It is grown in both the East and the West: in Asian countries such as China, Korea, Tibet, Nepal, and Bhutan as well as in Russia, Slovenia, Italy, France, Germany, Canada, Poland, and the United States, as well as in Brazil. Russia is now the biggest producer of buckwheat, and China ranks second.

Due to its unusually short growing season, it can be cultivated twice a year. Buckwheat requires less water and less nutrients from the soil than do other main crops. It can be grown under less ideal conditions than rice, at high altitudes, and both the grain and the leaves can be used as food (2). Buckwheat bears triangular seeds, with black hulls covering the light green to white kernel. The flower of buckwheat is white, pink, or yellow. In fact, buckwheat is the last available honey-producing plant before the arrival of winter.

26.2 GENERAL CHEMICAL AND NUTRITIONAL PROPERTIES OF BUCKWHEAT SEEDS

Buckwheat is rich in nutrients. Buckwheat seeds contain 100 to 125 mg of proteins per gram of seeds, 650 to 750 mg of starch, 20 to 25 mg of fat, and 20 to 25 mg of mineral. The protein content in buckwheat flour is less only than oat flour but significantly higher than those in rice, wheat, millet, sorghum, maise, and zan-ba (a traditional food in Tibet). The amino acids in buckwheat protein are well balanced and are rich in lysine (generally the first limiting amino acid in other plant proteins) and arginine. Buckwheat proteins can show a strong supplemental effect with other vegetable proteins to improve the dietary amino acid balance and, as a result, increase the protein biological value (BV). Besides its high-quality proteins, buckwheat is also rich in many rare components with healing effects for some chronic diseases. Among these are flavones, flavonoids, phytosterols, fagopyrines, and the thiamin-binding proteins that are found in buckwheat seed. Many nutraceutical compounds exist in the seeds and other tissues of buckwheat (3).

Buckwheat proteins are rich in lysine, as noted earlier, and contain less glutamic acid than do other cereal proteins. Glutamic acid, a nonessential amino acid, is the major component of reserve proteins in cereals, and lysine is the limiting essential amino acid of several cereal species. Thus, buckwheat proteins may serve as excellent supplement of the amino acid pattern of several cereal proteins. Buckwheat flour is superior to wheat flour in magnesium, copper, and iron content. Furthermore, buckwheat flour does not present hemagglutinin activity, and its tannin content is negligible. Rheological assays indicate that buckwheat flour does not contain gluten and has high starch content that is not degraded by mechanical action (4). Buckwheat tissues (seeds, flowers, leaves, and stems) can serve as very useful sources for high-quality flavonoids, though their flavonoid content varies with the growing phase and is significantly influenced by the contents of phenylalanine and tyrosine, by the activity of kinetin in the tissues, and by the different existing forms of nitrogen in the soil.

Rutin, a major flavonoid, was identified in buckwheat seed (5). Rutin and other flavonoids can be separated from the buckwheat seeds, flower, and shrub by simple chemical methods. Six flavonoids have been isolated and identified in buckwheat grain. All six of them — rutin, orientin, vitexin, quercetin, isovitexin, and isoorentin — were found in buckwheat hulls.

The plant sterols (or the so-called phytosterols) found in buckwheat seed, although at a low level, have shown positive effects in lowering blood cholesterol level as well as pharmaceutical effects, including antiviral, antitumor, and cholesterol lowering effects. The fagopyrines in buckwheat seeds have unique properties and can be used for the treatment of type II diabetes. Ingestion of the entire plant, either green or dried, can cause serious photosensitization in exposed or light-colored skin, including those of cattle, goats, sheep, swine, and turkeys.

Thiamin-binding protein (TBP) exists ubiquitously in Spermatophyta seeds and functions for thiamin transportation and storage in the plant. The TBP of buckwheat seed is an oligomer. On sodium dodecyl sulfate–polyacrylamide gel electrophoresis (SDS–PAGE), buckwheat TBP migrates as a single band corresponding to a molecular weight of 42 to 45 kDa, depending on whether a reducing agent has been applied, and its isoelectric point (Ip) is about 5.3. The TBP isolated from buckwheat consists of polypeptides linked by disulfide bonds.

From the results mentioned above, it is clear that buckwheat is rich in nutrients and its amino acid composition is more ideal than those of many other grains. Its flavonoids and fagopyritols may have many powerful medical properties suitable for the development and production of functional foods, but the mechanisms involved need to be further elucidated. The advantages of buckwheat may make it a crop of greater importance in the future. There is also a growing interest in buckwheat products as health food and as a substitute for wheat flour for gluten-allergic persons.

26.3 USES OF BUCKWHEAT

Buckwheat has been used extensively to manufacture foods for both human and domestic animals, pillow contents, and ornaments, among others. Buckwheat flour is used for making griddle cakes, waffles, black bread, biscuits, crepes, buckwheat wine, buckwheat sauce, buckwheat cakes, and buckwheat-lotus confectionery and is also used to make noodles. Noodles made from buckwheat flour-water dough have been popular in countries such as Japan, Korea, China, and Italy. Buckwheat has been widely used in Korea and Japan to make food items [*soba*, *manjuu*, *sobagaki*, *memill muk* (a kind of jelly), and so on]. Jewish people make a pure buckwheat soup called *kasha*. Buckwheat bread made with milk, pure buckwheat flour, and yeast has been recommended by allergy specialists for patients sensitive to wheat and rye. Roasted buckwheat is served with meat for extra flavor. Steamed buckwheat groats served with the juice of steak poured over it is a tasty dish. The outer hulls of buckwheat removed in milling are used as fuel, as packing material for bottled goods and bulbs, and for milling into stock feeds. The husks of buckwheat are used as a pillow stuffing. Roasted and ground hulls are sometimes mixed with spices, particularly black pepper. The middlings consist of the portion of the grain just beneath the hull and are used as fertilizer and as feed. The whole buckwheat grain is used to feed poultry. The straw is chiefly used for manure but may also be used for feed. Buckwheat flowers provide an excellent source of honey for bees.

Buckwheat was introduced to Scandinavia during the Middle Ages, but during later centuries, it totally disappeared from the Scandinavian countries and was replaced by other crops. During recent years, buckwheat has experienced a revival as an ingredient in health foods (6). Buckwheat proteins have unique amino acid compositions with special biological activities, including lowering cholesterol, having antihypertensive effects, and, similar to dietary fiber, improving constipation and obesity by interrupting the *in vivo* metabolisms. The trypsin inhibitors isolated from buckwheat seeds are heat stable and can cause poor digestion if not suitably cooked before consumption (3).

Buckwheat products are increasing in popularity as health food because of their high content of protein, vitamins B_1 and P, dietary fiber and low fat content. Buckwheat is a major dietary source of rutin, a phenolic flavonoid found to possess antitumor, antimutagenic, anticarcinogenic, and anti-inflammatory activities (7). Buckwheat tea has been reported to have preventive properties against leg edema (8), and buckwheat food products can reduce high blood pressure and serum cholesterol and possibly contribute to lowering the incidence of cardiovascular disease in an ethnic minority in China (9). Since buckwheat can be used to produce a gluten-free flour, it can be recommended for those with gluten-sensitive enteropathy (celiac disease) (10). Pomeranz and Robbins (11) reported that buckwheat seeds have a well-balanced amino acid composition and are rich in arginine and lysine. Thus, buckwheat seed can be used as a complement to many other cereal grains that are deficient in lysine.

In 1978, Ikeda and Kusano (12) first reported on a number of antinutritional factors, including trypsin inhibitors, in common buckwheat seeds (*F. esculentum*) that cause low digestibility; this led to a low utilization rate of buckwheat food products. *F. tataricum* also contains trypsin inhibitors, mainly buckwheat trypsin inhibitor (BTI)-I, -II, and -III. The trypsin inhibitors in buckwheat seeds are heat resistant and very stable at high temperatures, particularly in acidic environments. Early in 1975, Carroll and Hamilton (13) suggested that plant proteins have cholesterol-lowering effects compared with animal proteins such as casein. Kayashita (14) suggested that buckwheat proteins have strong cholesterol-lowering effects compared with proteins from other plants. Protein extracts from buckwheat can be used as potential functional food ingredients for treating hypertension, obesity, alcoholism, and constipation. A patent for the production of a peptide with high hypotension potency that repressed the activity of angiotensin-converting enzyme (ACE) was approved by the Japanese Commission in 1993 (15). The enzymatic hydrolyzates of buckwheat have strong ACE-inhibitory activities. Subtilisin and thermolysin are the two enzymes used for the production of hydrolyzates from buckwheat with the strongest ACE-inhibitory effects.

The cholesterol-lowering effect of buckwheat proteins is due to their low digestibility and other dietary-fiber-like components. An increase in neutral sterols in cholesterol-fed rats was observed during animal-feeding experiments using buckwheat (16).

Buckwheat flavonoids are very effective in inhibiting the oxidation of lipids in foods and can be used as natural antioxidants without the drawbacks of traditional antioxidants such as butylated hydroxyanisole (BHA), butylated hydroxytoluene (BHT), and propyl gallate (PG) used in the food industry. The flower of buckwheat contains flavonoids up to 13.8%. Waterman (17) reviewed the medicinal effects of plant flavonoids. Flavonoids are known for their effectiveness in reducing cholesterol levels in blood, keeping capillaries and arteries strong and flexible, and providing a preventative action against high blood pressure. Flavonoids demonstrate strong antioxidant effects and thus can protect human lymphocyte deoxyribonucleic acid (DNA) from oxidative damage (18). Some experiments suggest that flavonoids are important, even essential, in the maintenance of the brain and the nervous system (19).

26.4 ADVERSE EFFECTS OF BUCKWHEAT

Ingestion of buckwheat can produce fagopyrism and allergic reactions. Fagopyrism is a toxic reaction from fagopyrin, a polyaromatic derivative of naphthodianthrone. It causes a toxic photosensitization in animals such as sheep and cattle fed buckwheat and exposed to sunlight. The clinical symptoms of fagopyrism are itching, erythema, constipation, and digestive disturbances of mild to serious degree (2). Allergy to buckwheat was first reported in the scientific literature in 1909 by Smith (20). The pathogenetic mechanism of this allergy is type I hypersensitivity reaction. The allergic reaction can follow ingestion of buckwheat, occupational exposure, or domestic exposure through sleeping on a pillow stuffed with buckwheat husk (21). In 1926, Peshkin (22), in a discussion of the etiology of bronchial asthma in children, reported the positive skin reaction for buckwheat in 100 consecutive cases. In 1920, Highman and Michael (23) reported a case of urticaria following buckwheat ingestion, which was eventually cured by avoidance of buckwheat. In 1931, Rowe (24) reported 27 positive skin reactions to buckwheat in 500 consecutive cases of allergy. He estimated that 1% of allergic patients are clinically sensitive to buckwheat. In 1935, Blumstein (25), in a review of 500 consecutive patients, found 19 who showed positive skin reactions to buckwheat. Of these, 8 were clinically sensitive, six had bronchial asthma, 1 had allergic rhinitis, and one had gastrointestinal allergy.

In 1964, Matsumura et al. (26) from Japan reported six cases of buckwheat asthma induced by inhalation of buckwheat flour attached to buckwheat chaff used as stuffing in pillows. Kang and Min (27) from Korea reported three cases of buckwheat allergy. Their chief complaints were wheezing and dyspnea for 1 year in all cases. One case was accompanied by gastrointestinal symptoms, and another had suffered anaphylactic shock, but all showed skin reactivity to buckwheat flour. Two patients who had used buckwheat husk stuffed pillows also showed skin reactivity to buckwheat husk.

26.5 PATHOGENETIC MECHANISM OF BUCKWHEAT ALLERGY

Food hypersensitivities develop in genetically predisposed individuals when oral tolerance fails to develop normally or breaks down. Immunoglobulin E (IgE)-mediated reactions develop when food-specific IgE antibodies residing on mast cells and basophils bind circulating ingested food allergens and activate the cells to release a number of potent mediators and cytokines. A number of non–IgE-mediated food hypersensitivity disorders have been described. There is little evidence to implicate antigen-antibody complex–mediated hypersensitivity in food-related disorders. Cell-mediated hypersensitivity reactions probably contribute to a number of gastrointestinal disorders (28).

Buckwheat allergic reactions are mostly pure IgE-mediated reactions, and suspected cases can easily be verified by skin prick tests or determination of specific IgE. Buckwheat flour is a food allergen, an occupational allergen, and a hidden domestic allergen. Buckwheat protein works as a very strong allergen because severe systemic allergic reaction can be caused by very small amounts of the buckwheat allergen.

Sensitization to food allergens may occur in the gastrointestinal tract after the ingestion of a food item is considered a traditional or class 1 food allergy; that occurring after inhalation of an airborne allergen that cross-reacts with a specific food is considered a class 2 food allergy (29). The sensitization of buckwheat is suggested to occur by two routes: one is through the respiratory tract by inhalation of buckwheat flour and by inhalation of the dust of buckwheat husk, and the other is through the gastrointestinal tract by ingestion of foods containing buckwheat flour. Cases of occupational buckwheat asthma have been reported to have originated from noodle shops in Japan and Korea, a creperie in Spain, a pancake restaurant in France, and a health-food shop in Switzerland. Therefore, inhalation of buckwheat flour is an important mode of sensitization in buckwheat allergy. In the United States, a woman suffered anaphylaxis after eating buckwheat crepes. She had been sensitized 4 years earlier when working in a factory that made buckwheat husk pillows (21). This report showed that a person sensitized by inhalation of buckwheat flour could experience allergic symptoms upon ingestion of buckwheat foods.

In 1964, Matsumura et al. (26) reported six cases of Japanese children with buckwheat asthma, probably induced by buckwheat flour in buckwheat-chaff pillows (BCP). Lee et al. (30) reported a case of a young boy with chronic intractable asthma who had been treated for nonatopic asthma and who had been using a BCP for almost 5 years. The boy's first asthmatic symptom had occurred 6 months after his exposure to BCP. Eventually, Lee et al. (30) concluded that the patient had buckwheat flour allergy caused by long-term exposure to BCP and that he had never eaten food containing buckwheat flour. They also noted that one third of the asthmatic children and their family members in the Kyung-Gi province in South Korea had used BCP due to health beliefs (30). In 1987, Hong et al. (31) reported that buckwheat chaff, rice chaff, cotton, sponge, and feather were used as pillow-stuffing materials in South Korea, of which buckwheat chaff was the most common; up to 30.5% of allergic asthmatic adults in Korea had used BCP in the belief that it would improve their health. However, Hong et al. (31) did not consider buckwheat flour as a significant sensitizing inhalant allergen in its users because all patients showed stronger and more prevalent positive reactions to house dust mites than to buckwheat flour. He suggested that BCP would also be a place where house dust mites would thrive (31).

Matsumura et al. (26) and Lee et al. (30) demonstrated certain allergenic substances coming from BCP that acted as sensitizing agents causing buckwheat flour allergy in sensitized patients, especially in children. These children had no specific history of exposure to inhalation of buckwheat flour or ingestion of buckwheat foods; they had only used BCP for long periods prior to the development of asthma. Matsumura et al. (26) suggested that sensitization occurred due to inhalation of buckwheat flour attached to the buckwheat chaff. However, among many children who had used BCP, only a few were sensitized. Before being stuffed into pillows, buckwheat chaff is generally washed aggressively. Therefore, it is less likely that the flour remains in BCP. However, using BCP for long periods of time may sensitize the user, especially during childhood, due to dust containing allergenic components being released from buckwheat husk.

26.6 PREVALENCE OF BUCKWHEAT ALLERGY

Food allergy is one of the common allergic diseases, and the incidence and principal allergy-causing foods are different in various countries. Horesh (32) conducted skin tests for buckwheat sensitivity over a period of 45 years in his private office and in the University Hospitals of Cleveland's allergy clinic in all children who were suspected of being allergic. Tests were done by the scratch technique. He reviewed 500 consecutive allergic patients who ranged in age from 6 months to 18 years. Thirty-six (7%) had positive skin test reactions to buckwheat. Twenty-nine patients showed 1+ reaction. Two patients showed 2+ reaction, and six patients showed 4+ reaction. Only the six who had 4+ reactions were considered clinically sensitive to buckwheat (1.2%). All six of these had bronchial asthma. Three had allergic rhinitis, and two had urticaria and angioedema in addition to asthma. These figures closely agree with those of Peshkin (22), Rowe (24), and Blumstein (25),and hence it can be estimated that about 1% of allergic patients in the United States are sensitive to buckwheat. The extent to which sensitivity to buckwheat occurs in infants and children probably varies widely by area, depending on whether buckwheat is readily available and used extensively for food or for other purposes. One would expect a higher incidence in rural areas and in some foreign countries such as Japan where the chaff is widely used for stuffing pillows. Horesh (32) reported brief case histories of buckwheat asthmatic children: a 7-year-old Caucasian girl developed asthma through inhalation of buckwheat flour after preparing buckwheat griddle cakes for her parents; a 16-year-old Caucasian girl visiting friend's home had an asthma attack caused by inhalation of incredibly small amounts of buckwheat flour from lace curtains starched with a solution of buckwheat flour — a procedure which was customarily followed in Poland — a 10-year-old boy with known buckwheat food allergy suffered a severe systemic allergic reaction after ingestion of buckwheat honey; an 8-year-old Caucasian boy with buckwheat food allergy had an asthma attack when the wind blew from the direction of a buckwheat mill at some distance away; and two children (6 and 8 years old) had asthma attacks after they had eaten buckwheat griddle cakes. Because of the widespread use of buckwheat in Japan, buckwheat allergy is more common in that country than in others (33). The prevalence of buckwheat allergy in children in Yokohama in Japan was evaluated. Data on the children allergic to buckwheat flour were collected by sending questionnaires to 341 nurses in elementary schools. Among the total subjects of this investigation, 92,680 children, the incidence of buckwheat allergy was 0.22% (140 boys and 54 girls). In a large majority, the clinical symptoms of buckwheat allergy were urticaria (37.3%), skin itching (33.3%), and wheezing (26.5%). Four children (3.9%) experienced anaphylactic shock requiring emergency medical treatment. The incidence of anaphylactic shock caused by buckwheat ingestion was higher than those due to egg and milk allergies. In fact, seven childlren developed an allergic reaction from eating buckwheat noodles served at school lunch and one from eating a picnic meal. Thus, allergy to buckwheat is not rare in school children, and therefore it is important to withdraw buckwheat food products from school lunches and picnic meals (34).

Food and food-additive hypersensitivities in Japanese adult asthmatics were evaluated by skin scratch test and food challenge test. Of the 3,102 subjects, 625 (20.1%) had a positive test to one or more food allergens. The most common food allergens were shrimp (27.7%), crab (27.7%), yeast (23.8%), and buckwheat (15.8%). Positive food challenge responses occurred in 30 of 60 subjects (50%). The foods that most provoked a reaction were buckwheat, shrimp, crab and bread (35).

In Korea, Kim et al. (36) reviewed the clinical histories of 3,320 children with atopic asthma and performed skin tests with food allergens. (a) Out of the 3,320 asthmatic children (3 to 15 years old), 379 (11.4%) had clinical histories of food allergies (Figure 26.1). From the questionnaire, these 379 patients reported 554 allergic reactions to 58 foods (1.5 allergic reactions per patient). (b) The 10 most common foods implicated in allergic reactions were egg (27.7%), pork (14.8%), peach (14%), mackerel (4.4%), chicken (11.1%), milk (10%), buckwheat (7.4%), crab (6.3%), wheat (4.7%), and tomato (4.4%), in the order of frequency (Figure 26.2). (c) The rate of positive skin tests corresponding to the clinical history was collectively very low (22.6%), but it was the highest in the case of buckwheat (92.9%).

Lee (37) reviewed the results of allergy skin tests (scratch tests) in 5,003 consecutive allergic patients who ranged in age from 4 to 15 years. The rates of positive skin tests in the case of inhalant allergens were as follows; *Dermatophagoides farinae* (87.9%), *Dermatophagoides pyteronyssinus* (80.1%), cockroach (12.5%), *Alternaria* (10%), and cat hair and dander (7.4%). The food allergens were crab (4.3%), buckwheat (4%), shrimp (3.7%), chicken (3.1%), milk (2.6%), and egg (2.1%), among others (Table 26.1).

Kim et al. (38) carried out allergy skin tests with food allergens (62 items) in 1,425 adult Korean patients who visited allergy clinics with various allergic

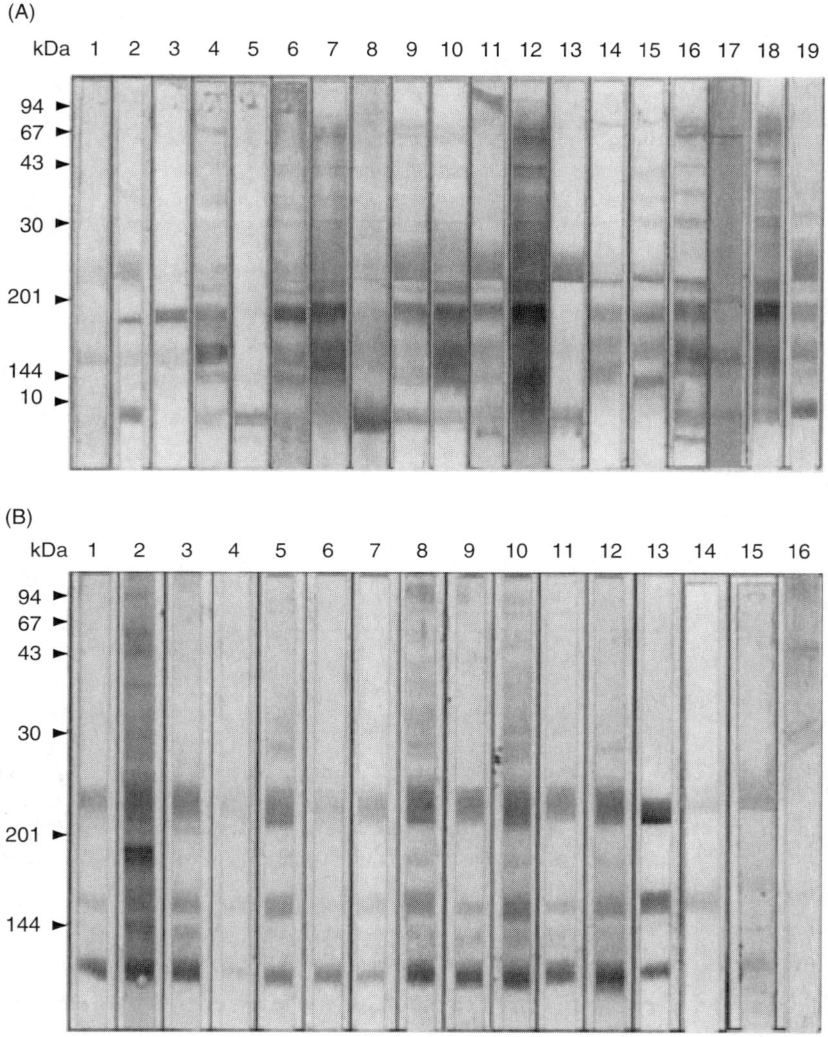

FIGURE 26.1 (A) IgE immunoblotting features of 19 BW-allergic subjects; (B) asymptomatic control subjects (lanes 1~15) and nonatopic control (lane 16).

symptoms. The food allergens to which sensitization rates were above 1% were the pupa of a silkworm (9.4%), shrimp (5.8%), chestnut (5%), curry (2.8%), potato (2.5%), soybean (1.8%), rice (1.8%), buckwheat (1.5%), cabbage (1.5%), mackerel (1.5%), abalone (1.4%), lobster (1.3%), turban shell (1.1%), and arrowroot (1%) in decreasing order. The sensitization rates of food allergens were generally higher in males and young adults than in females and older individuals. Atopic patients with allergies from inhalant allergens showed higher sensitization rates of food allergens than nonatopic patients (see Table 26.1).

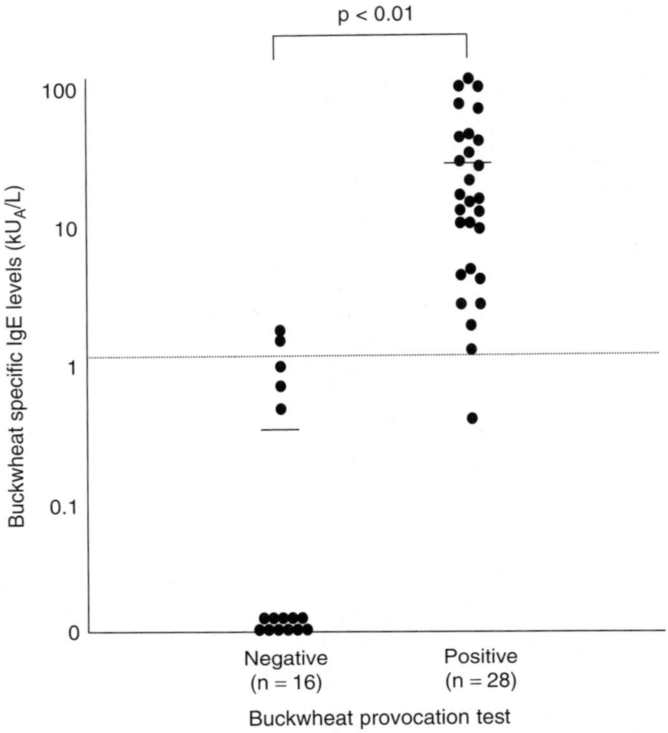

FIGURE 26.2 The serum levels of BW-specific IgE antibodies in BW-alletgic patients. The optimal cutoff level (1.26 kU$_A$/L is drawn as a broken line. Horizontal lines represent geometric means. As shown in figure, the serum levels of BW-specific IgE were significantly higher in the BW-provocation test positive group (29.23 ± 6.21 kU$_A$/L) than in negative group (0.35 ± kU$_A$/L, $p < 0.01$).

26.7 ALLERGIC REACTION AND SYMPTOMS

Although buckwheat allergy is not common, buckwheat is considered to be a very potent allergen. Small amounts of this allergen can provoke serious symptoms. Allergy to buckwheat was first reported by Smith in 1909 (20). Buckwheat allergy manifests several symptoms on the skin and in the respiratory and gastrointestinal tracts and also produces systemic symptoms, but 60% of affected individuals reported symptoms caused by exposure through the airways. The most common symptoms were asthma attacks following buckwheat exposure; 139 patients (82%) suffered from buckwheat asthma. Less common were nasal symptoms (23%), ocular symptoms (13%), urticaria (45%), and gastrointestinal symptoms (33%). In 18% of the patients, anaphylactic shock occurred from exposure to buckwheat (2). In Japan, since buckwheat is widely used to make a noodle called *soba* and buckwheat husk is used as pillow stuffing, there have been many reports of buckwheat asthma since 1937 (39).

TABLE 26.1
Skin Test Results of Food Allergens in Child Atopic Diseases and Adult Respiratory Diseases

Allergens	Children[a] % (n = 5003)	Adults[b] % (n = 1425)
Yeast, baker's	4.6	ND
Crab	4.3	ND
Buckwheat	4.0	1.5
Shrimp	3.7	5.8
Milk	2.6	0.4
Tomato	2.1	0.5
Egg	2.1	0.5
Chicken meat	2.0	0.6
Pork	1.8	0.1
Soybean	1.4	1.8
Soy sauce	1.4	ND
Wheat flour	1.4	0.7
Tuna	1.3	0.3
Rice	1.3	1.8
Cheese	1.1	ND
Soybean paste	1.1	ND
Pupa, silkworm	ND	9.4
Curry	ND	2.8
Potato	0.5	2.5
Cabbage	ND	1.5
Mackerel	0.7	1.5
Abalone	ND	1.4
Turban shell	ND	1.1
Arrowroot	ND	1.0

[a] n = 5,003 children with atopic diseases. From: K.Y. Lee. *Pediatr. Allergy Respir. Dis. (Korea)*, 8 (Suppl. 2), S43–S48, 1998. With permission.

[b] n = 1,425 adult patients with respiratory allergic diseases. From S.H. Kim, H.R. Kang, K.M. Kim, T.B. Kim, S.S. Kim, Y.S. Chang, C.W. Kim, J.W. Bahn, Y.K. Kim, S.H. Cho, H.S. Park, J.M. Lee, K.U. Min, C.S. Hong, N.S. Kim, Y.Y. Kim. *J. Asthma Allergy Clin. Immunol.*, 23, 502–514, 2003. With permission.

Nakamura and Murohisa (40) first described the term "buckwheat allergose" in relation to hypersensitivity reactions to buckwheat. Nakamura et al. (41) also reported nine cases of buckwheat allergose, and the most distinguishing traits of buckwheat allergose were as follows: (a) it is difficult to draw the boundary lines among hypersensitive manifestations of buckwheat allergy such as asthma attacks, urticaria eruption, gastrointestinal disorders, nasal symptoms, and congestion of conjunctiva; (b) the manifestation develops when the allergenic substance invades the body through the mouth as well as through the airways; (c) judging from the results of various kinds of allergy examinations, hypersensitive manifestations can be attributed to the allergic mechanism based on antigen–antibody reaction, in the strictest sense of the term. The buckwheat allergose is considered to be the model of type 1 allergy (that is immediate-type allergy) proposed by Coombs and Gell; (d) the antigenicity of buckwheat is extremely strong, and therefore hyposensitization treatment with buckwheat must not be automatically employed as there is the risk of a severe and dangerous reaction due to the injection of the buckwheat extract. In Japan, Nakamura and Yamaguchi (41) reported the results and recommendations of a nationwide clinical investigation of 169 cases with the buckwheat allergose. The results derived from the questionnaires were as follows: (a) males were predominant in number (male:female = 1.64:1); (b) initial occurrence of buckwheat hypersensitivity was (n = 115) mostly in children under 10 years of age, followed by teenagers (n = 20) and individuals in their twenties (n = 16) and thirties (n = 5). Only one patient was over the age of 40. At the first incidence, in most cases, the hypersensitive patients noticed manifestations after consuming buckwheat or using a pillow stuffed with buckwheat husk; (c) of the 131 patients whose family histories were clearly documented, buckwheat allergose was found in 13 families (0.10%); (d) in nine cases, the development of hypersensitive manifestations of buckwheat allergy was related to occupational exposure to buckwheat; (e) among symptoms of buckwheat allergy, allergose asthma attacks were most frequently seen, followed by urticaria and gastrointestinal disorders. The combination of these symptoms varied case by case, and it was difficult to distiguish among them. In addition to this, it is worth mentioning that anaphylactic shock caused by buckwheat allergen was manifested in 18 cases, including our case; (f) in 126 cases (74.6% of all), the manifestation of allergose developed when the allergenic substance entered the body through the mouth and in 102 cases (60.4%) through airways. In 62 cases, the manifestations developed through both modes of entrance; and (g) the clinical response to hyposensitization treatment with buckwheat was favorable in 8 out of a total of 26 cases, effective to some degree in 7 cases, ineffective in 4 cases, and suitably reserved to estimate in 7 cases. However, it is worth noting that in almost all cases of buckwheat allergose, avoidance of buckwheat was quite effective (42). Buckwheat allergose is generalized as multisystem allergic reactions , which can be classified as anaphylaxis.

In South Korea, the first case of buckwheat hypersensitivity to be reported was in 1984 by Kang and Min (27), followed by two cases caused by ingestion of noodles made from buckwheat flour (*soba*) (43) and one case with positive challenge to the inhalation of dust from buckwheat husk pillow and the extract of house dust (44). A 32-year-old woman with a 1-year history of intermittent asthma had been using a BCP for 2 years; she showed strong skin reactivity to house dust mites and the dust extract of buckwheat husk from BCP. On inhalation challenge test, she showed an early bronchoconstriction response (% fall of FEV_1 and FEF 25 to 75; 21.8% and 57%) to house dust extract (1:50 w/v) and late bronchoconstriction response (% fall of FEV_1 and FEF 25 to 75; 20.2% and 38.6%) to the dust extraction of buckwheat husk (1:10 w/v). At late response of buckwheat husk challenge, she complained of respiratory difficulty and wheezing in her chest but recovered by inhalation of isoproterenol (5 mg mL^{-1}). She stopped using BCP, and her condition improved dramatically (44).

In China, the occurrence of buckwheat allergy is low, even though consumption and occupational exposure to both common and tartary buckwheat is common in the Shanxi province in northern China. However, a study showed indications of nonspecific intolerance to buckwheat food products in some subjects (45).

In Switzerland, allergic reactions to buckwheat are very rare and mostly due to ingested buckwheat in the form of *pozoccheri* and bread. Schmacher et al. (46) reported six cases of allergic reactions to buckwheat. There were six cases of urticaria, four of quincke edema, and five of asthma. Four patients reacted to ingested buckwheat, and two patients were allergic to inhaled buckwheat from occupational exposure. Sensitizations were demonstrated in all patients by positive skin tests and specific IgE (radioimmunosorbent assay, RAST) (46). There was a report of anaphylaxis caused by eating a buckwheat burger unknowingly(47).

Several cases of asthma attacks caused by buckwheat flour were reported in children in Japan (nine cases) (48) and Korea (three cases) (30). These patients suffered asthma attacks immediately after going to bed at night. All the patients had been using BCP for 1 to 2 years prior to the onset of asthma. They showed strong skin reactivity to buckwheat allergen, high levels of specific IgE to buckwheat, and positive challenge test with inhalation buckwheat flour; their asthma symptoms stopped within a few days after avoidance of BCP. In fact, it is unusual for young children in South Korea to eat foods containing buckwheat flour; however, traditionally, many Korean parents use BCP for their children to improve their health and intelligence, particularly during infancy. For this reason, there is a possibility of sensitization to dust allergens from buckwheat chaff Lee et al. (30) reported a clinical immunological study on three cases of childhood nocturnal asthma, mainly due to buckwheat flour sensitization by long-term exposure to BCP. The children were 5 to 7 years old and had nocturnal asthma for 2 to 3 years due to prior exposure to BCP for 2 to 6 years. All showed a strong reaction on skin prick test with buckwheat extract and also demonstrated high titers of specific IgE to buckwheat by the AlaSTAT radioimmunoassay method. All three patients showed dramatic improvement of nocturnal symptoms within 3 days after BCP elimination. With buckwheat flour bronchial provocation test, two showed dual asthmatic responses, and one experienced an immediate asthmatic response.

Davidson et al. (49) reported a case of anaphylaxis secondary to buckwheat ingestion. A 38-year-old woman developed itching and pain in her throat within 30 minutes after eating a meal comprising three buckwheat crepes with strawberry sauce, coffee with cream, and orange juice. Her symptoms then progressed to diffuse urticaria, chest tightness, and decreasing blood pressure. Her past medical history was significant for mild asthma controlled with intermittent use of inhaled albuterol. The patient had a history of occupational exposure to buckwheat hulls and kapok fibers in the manufacture of pillows and cushions. The patient had left this job 4 years earlier due to severe work-related asthmatic reactions that were diagnosed as kapok allergy after inhalation challenge with kapok. Three weeks before the episode of anaphylaxis, the patient had ingested one bite of a buckwheat crepe without adverse effects. The patient denied any other previous ingestion of buckwheat. Skin testing revealed a 3+ reaction to buckwheat extract. A skin test to a battery of foods, including wheat, milk, egg white, and strawberry, were uniformly negative. RAST for antibuckwheat IgE was strongly positive. Aspirin challenge was performed with single doses of up to 650 mg and a cumulative daily dose of 1,075 mg. There were no adverse reactions, and the challenge was considered negative. Therefore, the study concluded that the possible cause of this patient's anaphylaxis was ingestion of buckwheat (49).

There was a case report of fatal buckwheat-induced anaphylaxis. This involved an 8-year-old girl with a history of atopic dermatitis and asthma. She had consumed one half of a package of buckwheat noodles called "*zaru soba*" that had been purchased at a local convenience store and immediately thereafter swam vigorously. Approximately 30 minutes later, she developed severe systemic allergic reactions and experienced cardiorespiratory arrest. She never regained consciousness and died after another cardiorespiratory arrest 13 days later. The analysis of buckwheat-specific IgE antibody by immunoblotting showed seven bands, of which four bands, with molecular weights of 16, 20, 24, and 58 kDa, were specific to the patient as compared with subjects not allergic to buckwheat (50).

Asthma may develop from indirect exposure to buckwheat flour (51). A 26-year-old housewife who had had buckwheat allergy since age 14 noticed, over a period of seven years, the development of rhinitis and asthma symptoms 30 minutes after her husband came home from working at a buckwheat flour mill. She had strong positive reaction to buckwheat extract by skin prick test and also developed bronchoconstriction after inhalation challenge test with buckwheat allergen (51).

The first reported case in the United States was a person who developed asthma and worsening allergic rhinitis after exposure to a buckwheat-stuffed pillow. The positive skin prick and ImmunoCAP test to buckwheat along with the positive clinical response to buckwheat pillow elimination supported an IgE-mediated mechanism in explaining this patient's asthma and allergic rhinitis induced by buckwheat pillow (52).

In the Netherlands, there was a report of a case of anaphylactic reaction in an 18-year-old man after eating "*poffertjes*" (small Dutch pancakes). Buckwheat is a principal ingredient of *poffertjes.* It is highly likely that the patient had been sensitized by sleeping on a BCP (53). There was also a case of anaphylaxis caused by buckwheat present as an additive in pepper (54).

26.8 OCCUPATIONAL LUNG DISEASE

Occupational exposure to buckwheat has been implicated as a cause of rhinitis, conjunctivitis, contact urticaria, and occupational asthma. In one study, 13 of 28 persons employed in the manufacture and packaging of buckwheat flour developed work-related symptoms (6). There was a case of a housewife with buckwheat-induced asthma whose husband had been working in a buckwheat flour mill (50). In France, a 29-year-old man working in a pancake restaurant was confirmed to have asthma caused by buckwheat flour by specific challenge test using a computerized device that generated particles. A very small amount of buckwheat flour (10 μg) induced an immediate fall of the FEV_1 to 56% of the initial value. No bronchial reaction was observed with lactose or with wheat flour (55).

In Spain, there was a case report of occupational asthma, contact urticaria, and gastrointestinal symptoms related to exposure to buckwheat flour (56). A 29-year-old woman, an ex-smoker, with a family history of atopy had been working for 6 years in a creperie, where buckwheat flour was used to prepare crepes. Shortly after she started on this job, she began to notice sneezing, rhinorrhea, and nasal itching. Four years later, she developed dyspnea, wheezing, and contact urticaria whenever exposed to buckwheat flour. In the last few years, she experienced nausea, vomiting, epigastric pain, and occasional urticaria a few minutes after ingestion of crepes prepared with buckwheat flour. She noted a strong reaction of skin testing with buckwheat flour extract. Rub testing was positive only with buckwheat flour, producing a wheal-and-flare reaction on the contact area. Precipitins to buckwheat were not found in the patient's serum. Histamine release test was positive with all concentrations of buckwheat flour extract. Buckwheat-specific IgE antibodies were confirmed by reversed enzyme immunoassay (REIA) and REIA-inhibition assays. With inhalation provocation test, a dual response was noted; 10 minutes after oral provocation with 10 mg of buckwheat flour, the patient experienced nausea, vomiting, and epigastric pain. The symptoms disappeared a few minutes after she vomited. The spirometric values remained normal. When she avoided exposure to buckwheat flour, the respiratory and cutaneous symptoms markedly decreased, until she became totally asymptomatic (56).

In Sweden, 13 out of 28 persons employed in a company that imported, prepared, and distributed plant products used in spices and in so-called health foods, developed work-related symptoms in the form of rhinitis, conjunctivitis, asthma, itching, or urticaria. The symptoms occurred in connection with specific work operations, especially in the grinding and packaging of buckwheat. Out of 25, a total of 7 (28%) had at least one positive allergy test (skin prick test or RAST) with buckwheat. Furthermore one person had positive RAST with castor-oil bean extract. The correlation between positive allergy tests and work-related symptoms was significant. The levels of airborne dust in the breathing zones of the workers when they performed dust-forming work were around or below 5 mg m^{-3}. When buckwheat flour was packaged, the airborne dust levels were about 1 to 2 mg m^{-3}. Thus, exposure to comparatively low levels of buckwheat dust may induce a definite risk of rapidly ensuing allergy (6).

In Korea, Park and Nahm (57) reported a case of occupational asthma and rhinitis caused by buckwheat flour, which was confirmed by bronchoprovocation tests, in a patient running a buckwheat flour noodle shop. This 26-year-old nonsmoking woman had been making noodles with buckwheat and wheat flours for 7 years. Her job included mixing buckwheat flour. She had been experiencing profuse rhinorrhea and sneezing for 3 years and began to feel shortness of breath 2 months before admission for evaluation. She reported that these symptoms were aggravated after handling buckwheat flour. A skin prick test showed immediate positive responses to buckwheat and wheat flour extracts along with the extracts of some tree and grass pollens. Specific IgE was detected to allergens of cockroach (RAST class 3), ragweed (class 2), blue grass (class 2), rye grain (class 3), and wheat grain (class 3). Ingestion of foods containing buckwheat and wheat flours provoked no symptoms. She showed early asthmatic reaction after inhalation of 10 mg of buckwheat flour in a capsule and also exhibited an early asthmatic reaction after inhalation of 10 mg of wheat flour. The test detected specific IgE and IgG_4 of buckwheat in the patient's serum (57).

26.9 HYPERSENSITIVITY REACTION TO BUCKWHEAT OTHER THAN ALLERGY

Pulmonary hemosiderosis related to nonimmediate buckwheat protein hypersensitivity was reported by Agata et al. (58). An 8-year-old girl with a history of mild, recurrent asthma for the past 5 years presented with nausea, bloody stool, and hemoptysis. She had experienced wheezing 2 hours after she had eaten buckwheat noodles for lunch. She had eaten buckwheat noodles again for supper, and nausea and hemoptysis developed 2 hours after ingestion. Continuous hemoptysis and bloody stools resulted in hospitalization the next day. A chest roentgenogram revealed fluffy pulmonary infiltrates and atelectasis of the right lower lobe. Results of tests on early-morning gastric aspirate showed numerous macrophages that were hemosiderin laden. She was diagnosed with pulmonary hemosiderosis and treated with transfusion and oral prednisolone. During the following year, she twice experienced hemoptysis, again after eating buckwheat noodles; however, pulmonary hemorrhage did not recur when dietary buckwheat was fastidiously avoided. The skin prick test for buckwheat allergen was negative. Skin patch testing showed a 3+ reaction to buckwheat but a negative reaction to milk and egg. RAST results for buckwheat protein were also negative. The proliferative responses of peripheral blood mononuclear cell (PBMC) to buckwheat protein were higher than those to ovalbumin, bovine serum albumin, β-lactoglobulin, and casein. The levels of specific IgG antibodies to buckwheat protein in the patient's serum were higher (OD492; 0.572) than those in the sera of three egg-sensitive patients with atopic dermatitis (OD492: 0.313 $\pm$0.027) and those in the sera of four healthy controls (OD492; 0.228 $\pm$0.059). With the above clinical history and results of laboratory tests, the authors concluded that the patient had pulmonary hemosiderosis related to nonimmediate buckwheat protein hypersensitivity (58).

26.10 ALLERGEN CHARACTERIZATION

The major food allergens identified as class 1 allergens are water-soluble glycoproteins that have a molecular mass of 10 to 70 kDa and are stable to treatment with heat, acid, and proteases. Most class 2 allergens are plant-derived proteins that are highly heat labile and difficult to isolate, often making standardized extracts for diagnostic purposes unsatisfactory. A limited number of the class 1 and class 2 food allergens have been identified, cloned, sequenced, and expressed as recombinant proteins. Many plant-related allergens are homologous with pathogen-related proteins, seed storage proteins, profilins, peroxidases, or protease inhibitors common to many plants (28).

Yoshimasu et al. (59) reported seven IgE-binding bands between the 10- and 20-kDa regions on SDS–PAGE-reduced gels in the crude extracts of buckwheat flour. They showed strong IgE activity around the 18- and 14-kDa regions, which were suggested to be the major allergens, and they evaluated N-terminal amino acid sequences of the 14- and 18-kDa allergens of buckwheat. The 18-kDa sequence (RDEGFDLGETQMSSKCMR) showed homology (80% identity) against a 16.7-kDa protein (*Oryza sativa* dehydrin, messenger ribonucleic acid [mRNA]). The N-terminal sequences of the 14-kDa protein (KYEGALKRIEGEGCK) showed some homology with a 76.8-kDa rice protein (phenylalanine ammonia lyase) when the first nine amino acids were analyzed with three substitutions (70% identity). When the water-soluble protein fraction was heated, exposed to acidic and alkaline conditions, and fully denatured, IgE-binding activity was reduced. When the faction was partially denatured through urea, IgE-binding activity increased. Furthermore, IgG-binding activity was detected with proteins only above the 20-kDa region. They concluded that proteins with molecular masses around 14 and 18 kDa were the major allergenic proteins in buckwheat-allergic patients' sera (59).

Park et al. (60) described the allergenic components of buckwheat-binding sera from buckwheat-allergic patients and symptomatic controls. The prevalence of IgE binding to 24-kDa (pI 8.3), 16-kDa (pI 5.6), and 9-kDa (pI 5.0/6/0) allergens was higher than 50% in buckwheat-allergic and asymptomatic subjects. However, the specific IgE to split the 19-kDa (pI 6.5/7.0) allergens was more specially found in buckwheat-allergic patients than in asymptomatic subjects (78% versus 7%). N-terminal amino acid sequences of the 19-kDa and 16-kDa allergens showed moderate and weak homology to the 19-kDa globulin protein of rice and α-amylase/trypsin inhibitor of millet, respectively (19-kDa [pI 6.5] : Gly-Asp-Tyr-Pro-Leu-Glu-X-Cys-Arg-Gln-Lys-Ile-Glu-His; and 16 kDa [pI 5.6]: Arg-Asp-Glu-Gly-Phe-Asp-Leu-Gly-Glu-Thr-Gln-Met-Ser-Ser-Lys-Cys-Met-Arg-Gln-Val-Lys-Met-Asn-Glu-Pro). The N-terminus of the 9-kDa isoallergens (9 kDa [pI 5.0/6.0] : Ser-Asp-Lys-Pro-Gln-Gln-Leu-Leu-Glu-Glu-CysArg-Tyr-Leu-X-Arg-Iso) were not different from each other and were identified as the reported trypsin inhibitors of buckwheat. Attenuation of the IgE binding to the 9-kDa allergen was found with periodate oxidation. Therefore, the allergens of 24, 19, 16, and 9 kDa are strong candidates to be major allergens, and the 19-kDa allergen was relatively specific to buckwheat-allergic patients (60) (see Figure 26.1).

Lee et al. (30) reported several distinct IgE-binding bands with masses of 59, 53, 50, 47, 41, 36, 34, 6, and 4 kDa, detected by immunoblotting using sera from children with BCP-induced buckwheat allergy. These patients' sera did not show significant binding to the 24-kDa components that had been identified previously as a major band of buckwheat flour allergen.

Tanaka et al. (61) reported that the 24-kDa protein that had previously been reported to be a major allergen reacted to IgE antibodies present in sera from almost all subjects (19 of 20 cases), regardless of symptoms. On the other hand, the 16- and 19-kDa proteins bound to the IgE antibodies present in sera from 9 of the 10 patients with immediate hypersensitivity reactions. After pepsin treatment, the 16-kDa protein, but not the 19- and 24-kDa proteins, remained undigested and preserved the capacity for IgE binding. The 16-kDa buckwheat protein was resistant to pepsin digestion and appeared to be responsible for immediate hypersensitivity reactions, including anaphylaxis, while the pepsin-sensitive 24-kDa protein was responsible for CAP test but not immediate hypersensitivity reactions (61).

26.11 CROSS ALLERGENICITY

Unger (62) reported a case of cross-reactivity between rhubarb and buckwheat as evidenced by feeding tests and skin tests. Also, there was cross-reactivity between IgE antibodies against buckwheat and rice. The IgE antibodies from the buckwheat-tolerant subjects with high levels of IgE antibodies against buckwheat might recognize the epitopes on buckwheat antigens that cross-react with rice antigens, whereas IgE antibodies from buckwheat-sensitive subjects might bind to buckwheat-specific epitopes. Rice is a staple for Asian people and is consumed abundantly starting from early infancy. Oral ingestion of boiled rice grains rarely induces overt symptoms. Cereal antigens such as rice might easily induce tolerance to IgE-mediated immediate reactions. Buckwheat determinants with cross-allergenicity with rice antigens might therefore not induce immediate hypersensitivity reactions, and buckwheat-specific determinants might produce the reactions (63,64). Lee et al. (65) produced polyclonal rabbit antisera to rice, wheat, barley, and buckwheat. They showed 18.4% inhibition on rice-IgG enzyme-linked immunosorbent assay (ELISA) by buckwheat antigen and 37.3% inhibition on buckwheat-IgG ELISA by rice antigen. Park et al. (60) showed the N-terminal amino acid sequences of two major allergens, the 19-kDa and 16-kDa proteins of buckwheat flour, had moderate and weak homology to the 19-kDa globulin protein of rice and α-amylase/trypsin inhibitor of millet, respectively.

Some reports also demonstrated cross-allergenicity between buckwheat and wheat flours. Varjonen (66) revealed that buckwheat skin tests were reactive in 13 cases of 33 wheat-positive atopic dermatitis (AD) children on skin prick tests.

Park and Nahm (57) reported the case of a 26-year-old woman working at a buckwheat noodle shop who had occupational asthma and rhinitis provoked by buckwheat flour and wheat flour, as confirmed by bronchoprovocation tests. A skin prick test showed immediate positive responses to buckwheat and wheat flour extracts along with the extracts of some tree and grass pollens. Specific IgE was detected against the allergens of cockroach (RAST class 3), ragweed (class 2), blue grass (class 2), rye grain (class 3), and wheat grain (class 3). They detected specific IgE against buckwheat in the patient's serum. A buckwheat IgE inhibition ELISA showed significant inhibition with serially diluted addition of rye grass, wheat flour, and ragweed antigens, as well as buckwheat flour extract. This study suggested the possibility of cross-allergenicity between buckwheat and wheat flours and between buckwheat flour and pollens such as rye grass and ragweed (57).

Recently, a number of reports describing anaphylactic reactions to food items in patients with latex allergy have appeared. The cases of three patients who developed anaphylactic reactions to both latex and food items were presented, and the importance of the association of latex and cross-reactivity with food items was stressed. The food items that led to anaphylactic reactions were banana and avocado; banana, avocado, and buckwheat; and banana, avocado, and tomato. The first case in the cross-reactivity of latex to buckwheat and tomato was reported by Abeck et al. (67).

Case histories of two patients with latex allergy who experienced a severe anaphylactic reaction after consumption of small pancakes (Dutch: "*poffertjes*") prepared with buckwheat flour have been presented (68). One of these two female patients, a goalkeeper on a soccer team, had been exposed to latex by wearing gloves covered with a layer of natural rubber. The other had been exposed to natural rubber latex in a fish factory, where all personnel were required to wear latex gloves while cleaning fish. The first woman had noticed rhinoconjunctivitis and urticaria on her face while playing soccer, especially if the gloves were new. She experienced pruritus oris and generalized urticaria after eating banana and kiwi. The other patient noticed itching and contact urticaria on the skin of her hands when wearing latex gloves and rhinoconjunctivitis and asthma in the working environment. She also had oral allergy symptoms after eating banana and kiwi. On latex inhibition RAST, Maat-Bleeker and Stapel (68) showed no inhibition by preincubation of serum 1 with different amounts of buckwheat protein. On the other hand, a strong inhibition of buckwheat RAST could be obtained with latex extract. In serum 2, very small amounts (less than 1 μg) of latex extract induced more than 95% inhibition of the buckwheat RAST. Thus, these cases showed that the clinical history, as well as the results of RAST-inhibition experiments, strongly indicates cross-reactivity between latex and buckwheat (68).

A nurse with known allergy to natural rubber latex (NRL) developed anaphylaxis 10 minutes after eating a few pieces of a muesli bar (69). On prick-to-prick test with the 20 ingredients of the muesli bar, only buckwheat showed an immediate 2+ skin test reaction. Specific serum IgE antibodies were found to NRL, buckwheat, and fig. Preparation of NRL and fig strongly inhibited the binding of the patient's IgE antibodies to the buckwheat CAP test, whereas buckwheat or fig extracts inhibited IgE binding to the NRL CAP test to a lesser degree. As the patient was able to tolerate the ingestion of fig, it was concluded that buckwheat was probably the elicitor of the patient's anaphylactic reaction. These results confirm the presence of antibodies cross-reaction to NRL, buckwheat, and fig in the same patient. The author concluded that development of buckwheat allergy due to cross-sensitization with NRL is highly probable in their patient, since she was not heavily exposed to buckwheat products. Patients with NRL allergy should be aware of possible associated food hypersensitivities to buckwheat or fig (69).

26.12 DIAGNOSIS OF BUCKWHEAT ALLERGY

Food allergy is clinically classified into two types, immediate and nonimmediate. Immediate-type food allergies are diagnosed by a patient's history of adverse reaction to foods, positive skin test, and determination of specific IgE to foods. However, these methods will not help diagnose nonimmediate type of food allergic reactions. For confirmative diagnosis of immediate-type food allergy double-blind placebo-controlled food challenge (DBPCFC) must be applied. However, when definite adverse reactions to food, positive skin test, and determination of specific IgE strongly support specific food reactivity, open food challenge may be done, except in patients with a previous life-threatening reaction (70).

Histories of allergic reactions after contact with buckwheat flour are most important in the diagnosis of buckwheat allergy. Clinically symptomatic patients with buckwheat allergy usually show strong skin reactivity and high levels of specific IgE to buckwheat. Most allergic symptoms to buckwheat are objective manifestations such as urticaria, asthma symptoms, and hypotension, among others. Thus, open food challenge tests have been applied for confirmative diagnosis of buckwheat allergy in a few studies.

Lee et al. (71) performed an open food challenge test (OFCT) in 276 cases using 33 kinds of suspected foods based on history or skin test results. Increasing doses of natural foods were given every 20 minutes, and the clinical manifestations elicited by challenge test were observed for 1 hour. (a) Of the 276 cases, 128 were positive for OFCT, and the provocation rate was 46.4%. (b) The provocation rate of individual foods was highest in buckwheat flour (72.3%), followed by crab (60%), shrimp (45.8%), milk (41.6%), and egg (41.4%). (c) Urticaria was the most common symptom (84.4%) elicited by OFCT. There were various allergic manifestations such as asthma, rhinitis, conjunctivitis, and burning sensation in the throat on oral challenge test. Anaphylactic shock was observed in three cases with buckwheat flour allergy. (d) In most cases (51.5%), symptoms occurred after ingestion of a small amount (1 to 100 g) of the offending food. However, the symptoms in 27 cases were elicited when the food was just rubbed on the lips or licked (71) (Table 26.2).

Park et al. (72) also reviewed 55 cases where an open food challenge test with buckwheat had been performed and reported the following results: (a) Of the 55 cases, 40 (72.7%) showed positive buckwheat oral provocation test. (b) The most common clinical finding after the oral provocation test was urticaria; and 60% showed severe allergic reactions such as asthma attack or anaphylactic shock. (c) The rate of correlation among past history, positive skin test, and open food challenge test was very high (86.1%). The predictive value of the allergy skin test and past history for positive buckwheat allergy challenge results was very high. Therefore, troublesome oral challenge tests may not be needed in all patients with suspicious buckwheat allergy if definite past history of buckwheat allergy and strong skin reactivity are present (72). Park et al. (60) reported that Pharmacia CAP-specific IgE to buckwheat showed 100% sensitivity (100%), 53% specificity, 100% negative predictive value, and 73% positive predictive value for the diagnosis of

TABLE 26.2
Clinical Manifestations in Open Food Challenge Test

Foods	Cases	U	A,R,C	BS	AP,V	AN
Buckwheat flour	34	29	21	7	8	3
Crab	21	18	6	4	2	
Egg	17	13	8	2	3	
Shrimp	11	8	7	3	1	
Milk	10	7	5		2	
Peach	8	8	4		1	
Yeast	7	5	3			
Tomato	4	4	1			
Others	16	16	9	7	1	
Total No.	128	108	64	23	18	3
%		84.4	50.0	18.0	14.1	2.3

U: urticaria, A: asthma, R: rhinitis, C: conjunctivitis, BS: burning sensation in throat, AP: abdominal pain, V: vomiting, AN: anaphylaxis

buckwheat allergy. Sohn et al. (73) investigated the positive and negative predictive values of buckwheat-specific IgE antibodies (Pharmacia CAP System) in subjects with buckwheat allergy aiming to reduce the need for buckwheat challenge, which can be more risky than other challenges of food allergies. Twenty-eight buckwheat-allergic subjects with symptoms after open food challenge with buckwheat and 16 asymptomatic control subjects with positive skin test to buckwheat were recruited. The optimal cutoff level of buckwheat-specific IgE was 1.26 kUA L^{-1}, and the sensitivity, specificity, positive and negative predictive values were 93.10, 93.33, 79.75, and 97.96%, respectively (73) (see Figure 26.2).

The oral Prausnitz-Kustner (P-K) reaction was introduced for the diagnosis of food allergy and applied in 41 children suspected of IgE-mediated food allergy (74). The recipients of the children's sera were their mothers. Blood from the food-allergic children was taken in sterile conditions, and the sera were separated aseptically. One part of the sera was incubated for 30 minutes at 56°C. Each 0.1 mL of heat-treated and nontreated sera was injected intradermally at different sites on the forearms of the children's mothers. Forty-eight hours after receiving the sera, the mothers consumed the suspected allergenic foods. After ingestion of the foods, the sera injection sites were monitored for 2 hours for allergic reactions. A positive reaction was observed in nine children when using hen's eggs, cow's milk, chicken meat, buckwheat, red-bean, salmon meat, or common-dolphin meat. Among them, seven had anaphylactic skin reactions at the injection sites (systemic urticaria or angioedema). According to the authors' suggestion, the oral P-K test may be useful for the diagnosis and prediction of food-induced anaphylaxis. The sera that showed the positive oral P-K reaction to buckwheat or chicken meat were zero in the RAST scores. These results suggest that a soluble substance might be involved in the oral P-K reaction and that the RAST technique does not always detect the

IgE antibody that recognizes the food antigens degenerated during the absorption *in vivo* (74). This test may be useful to diagnose food allergies that produce severe systemic reactions. However, the clinical usefulness of oral P-K reaction was not established precisely.

A sensitive *in vitro* serological method for the detection of offending allergens in nonimmediate types of food allergy has not been established yet. The proliferative responses of PBMCs to each offending food antigen in patients with nonimmediate types of food allergy were significantly higher than those of healthy controls and patients with immediate types of food allergy, respectively (75). Moreover, in each case of nonimmediate type allergy, proliferative responses to food antigens other than the offending food were not detected. When PBMCs were stimulated twice with the offending food antigen, the same results were obtained. These results indicate that the proliferative response of PBMCs to food antigens is specific to each offending food antigen in the nonimmediate types of food allergy. Therefore, proliferative responses of PBMCs to each food antigen are useful for detection of allergens in nonimmediate types of food allergy (75).

26.13 TREATMENT AND PREVENTION

Treatment of buckwheat allergy involves complete elimination of buckwheat-containing foods and of articles containing buckwheat materials from the daily living environment. Buckwheat-allergic patients have to be aware of foodstuffs containing buckwheat flour, and complete avoidance of contact with such foods is an essential for buckwheat allergic patients.

A 52-year-old Japanese female nurse with a past history of immediate-type buckwheat allergy developed an anaphylactic reaction (asthma attack, dyspnea, heat sensation on her face and neck, generalized urticaria, congestion of her conjunctiva, and swelling of her ears, lips, and hands) after eating *pyongyang* cold noodles at a restaurant in Japan. The specific serum IgE to buckwheat by RAST was 3+. The skin prick testing (prick-to-prick) with the *pyongyang* cold noodles that she had ingested showed 4+, whereas distilled water showed no reaction. *Pyongyang* cold noodles (*reimen*) are popular in Japan and are usually on the menu at Korean restaurants and at Korean barbecue restaurants in Japan. There are many kinds of these noodles, most of which are made from buckwheat, starch, and flour. The buckwheat content in *pyongyang* cold noodles is generally higher than that of starch or flour. Japanese people generally are not aware that buckwheat is one of the components of these noodles (33).

Some foodstuffs can act as aeroallergens, especially in children with coexistent food allergy and allergic asthma. Roberts et al. (76) demonstrated that food allergens may have a significant role as aeroallergens in the etiology of asthmatic symptoms in children with food allergies. Fish allergens may cause symptoms via the inhaled route; three out of four children in this series reacted on challenge to aerosolized fish allergen. It seems that fish allergens can become aerosolized in sufficient quantities even without cooking the fish. They also reported that inhalation tests of chickpea and buckwheat allergens also showed positive reactions among their cases. There is now considerable concern that peanut allergens may be able to behave in a similar way. This highlights the importance of considering foods as aeroallergens in children with coexistent food allergy and asthma. Even with rigorous dietary avoidance, continued environmental exposure may contribute to persistent asthma symptoms in children. In such a situation, for these children, dietary avoidance alone may not be sufficient; further environmental measures may be required to limit exposure to aerosolized food allergens during cooking (boiling, frying, or steaming) (76).

In Japan, hyposensitization treatment of buckwheat allergens has been tried when the effects of avoidance of buckwheat allergens were not favorable or could not be expected because of their ubiquitous presence (42). However, the hyposensitization treatment was tried in only a few cases in Japan. Out of 26 cases,

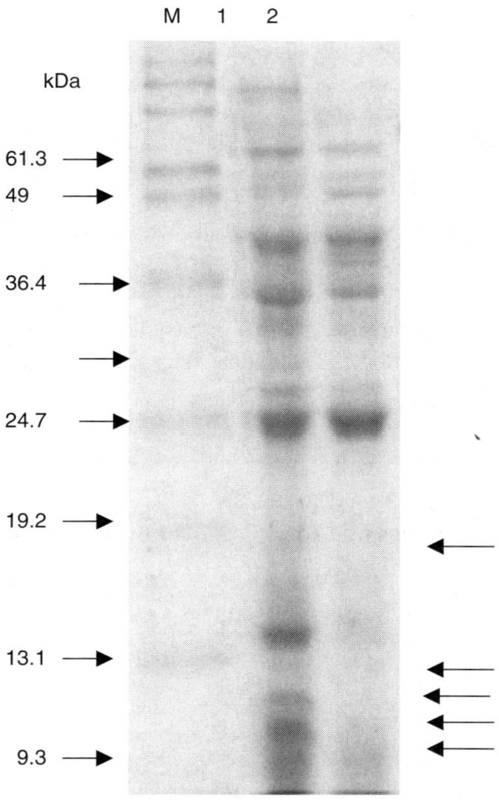

FIGURE 26.3 SDS-PAGE of raw buckwheat and hypoallergenic buckwheat. Lane M: molecular weight, Lane 1: raw extract, Lane2: hypoallergenic extract (treated with 0.5M NaCl). Arrows on the right indicate disappeared or weakened protein bands of hypoallergenic extract compared with those of crude extract.

hyposensitization treatment had favorable responses in 8 cases, was effective to some degree in 7, was ineffective in 4, and was suitably reserved to estimate in 7. Nonetheless, the buckwheat antigen is so strong that the hyposensitization procedure may not be easily applied for fear of severe and dangerous reactions (42).

Several trials for reducing allergenic components in foods are being carried out, for instance, by thermal treatment, irradiation, and hyperpressured or enzymatic (actinase and surfactant, proteinase) hydrolysis of major allergenic foods. Also, more recently, allergenic proteins of cereals and legumes were reduced by a transgenic approach that genetically modified the organism. Lee et al. (77) prepared a hypoallergenic buckwheat flour (HABF) by salt extraction (0.5 M NaCl) of raw buckwheat flour. The protein concentration of the extracts of HABF was four times lower than that of the raw buckwheat flour. Protein analysis of the HABF by SDS–PAGE revealed almost total disappearance of 12.5-kDa, 36-kDa, 43-kDa, and other high-molecular-weight bands (50-kDa, 60-kDa, and so on) (Figure 26.3). In six patients with buckwheat flour allergy they carried out IgE immunoblotting with raw buckwheat antigen, and all the sera showed IgE binding with the 23.5-, 30-, and 50-kDa bands, which frequently reacted with the IgE antibodies of children with buckwheat-flour allergy. On IgE blotting with HABF, all six cases showed no IgE binding with the 23.5-, 30-, and 50-kDa bands, but one who had high RAST class showed IgE binding with the 60- and 16-kDa bands. In oral challenge tests with HABF, they observed only mild allergic symptoms in two patients (Table 26.3). They concluded that raw buckwheat flour and HABF had a significant difference in allergenicity. The HABF was found to be useful in the elimination diet of many children with atopic buckwheat flour allergy (77). For primary prevention of buckwheat allergy, the buckwheat chaffs should no longer be used as pillow stuffing because there are many probable cases of sensitization to buckwheat flour following the long-term use of BCP in children and adults.

TABLE 26.3
Laboratory Characteristics and Oral Provocation Test in Patients

Cases	TEC	IgE	Skin Reaction		RAST	Oral Provocation Test	
	(mm^{-3})	(IU mL^{-1})	RBF	HABF	Class	RBF	HABF
1	920	507	SP	SP	3+	P	P
2	670	896	SP	SP	3+	P	N
3	2620	140	SP	SP	4+	P	N
4	60	183	P	N	2+	P	N
5	670	238	SP	SP	0	P	P
6	750	398	SP	SP	1+	P	N

U: urticaria, A: asthma, R: rhinitis, C: conjunctivitis, BS: burning sensation in throat, AP: abdominal pain, V: vomiting, AN: anaphylaxis

26.14 PROGNOSIS OF BUCKWHEAT ALLERGY

There were no reports on the long-term prognosis of buckwheat allergy. In one publication, allergic reaction was observed in a patient upon ingestion of even a very small amount of buckwheat, even after she had been away from the sensitizing environment of her workplace for several years (4 to 5 years) (49).

REFERENCES

1. C.G. Campbell. Buckwheat. In: N.W.S. Simmond (Ed.). *Evolution of Crop Plants*. London: Longman, 1976, 235–237.
2. G. Wieslander. Review of buckwheat allergy. *Allergy*, 51: 661–665, 1996.
3. S.Q. Li, Q.H. Zhang. Advances in the development of functional foods from buckwheat. *Crit. Rev. Food Sci. Nutr.*, 41: 451–464, 2001.
4. M.L.P. De Francischi, J.M. Salgado, R.F.F. Leitao. Chemical, nutritional and technological characteristics of buckwheat and non-prolamine buckwheat flours in comparison of wheat flour. *Plant Food. Hum. Nutr.*, 46: 323–329, 1994.
5. D. Gabrovska, V. Fiedlerova, M. Holasova, E. Maskova, H. Smrcinov, J. Rysova, R. Winterova, A. Michalova, M. Hutar. The nutritional evaluation of underutilized cereals and buckwheat. *Food Nutr. Bull.*, 23 (3 Suppl.): 246–249, 2002.
6. C.-J. Goehte, G. Wieslander, K. Ancker, M. Forsbeck. Buckwheat allergy: health food, an inhalation health risk. *Allergy*, 38: 155–159, 1983.
7. B.D. Oomah, G. Mazza. Flavonoids and antioxidative activities in buckwheat. *J. Agric. Food Chem.*, 44: 1746–1750, 1996.
8. N. Ihme, H. Kiesewetter, F. Jung, K.H. Hoffmann, A. Birk, A. Muller, K.I. Grutzner. Leg oedema protection from a buckwheat herb tea in patients with chronic venous insufficiency: a single-centre, randomized, double-blind, placebo-controlled clinical trial. *Eur. J. Clin. Pharmacol.*, 50: 443–447, 1996.
9. J. He, M.J. Klag, P.K. Whelton, J.P. Mo, J.Y. Chen, M.C. Qian, P.S. Mo, G.Q. He. Oats and buckwheat intakes and cardiovascular disease risk factors in an ethnic minority in China. *Am. J. Clin. Nutr.*, 61: 366–172, 1995.
10. G. Wieslander, D. Norback. Buckwheat allergy among gluten sensitive persons in Sweden. In: C. Campbell, R. Przybylski (Eds.). *Advances in Buckwheat Research. Section IV: Allergenic Properties*. Winnipeg, Canada: Organizing Committee on the 7th International Symposium on Buckwheat, 1998, 26–30.
11. Y. Pomeranz, G.S. Robbins. Amino acid composition of buckwheat. *J. Agric. Food Chem.*, 20: 270–274, 1972.
12. K. Ikeda, T. Kusano. Isolation and some properties of a trypsin inhibitor from buckwheat grain. *Agric. Biol. Chem.*, 42: 309–314, 1978.
13. K.K. Carroll, R.M.G. Hamilton. Symposium: nutritional perspectives and atherosclerosis. Effects of dietary protein and carbohydrate on plasma cholesterol levels in relation to atherosclerosis. *J. Food Sci.*, 40: 18–23, 1975.
14. J. Kayashita. Hypocholesterolemic effect of buckwheat protein extract in rats fed cholesterol-enriched diets. *Nutr. Res.*, 15: 691–698, 1995.
15. F. Koyama, Y. Nakamura . Antihypertensive and healthy foods containing tripeptide. [Jpn Kokai Tokkyo Koho JP 0597798, 1993; cited by SQ Li, QH Zhang.] *Crit. Rev. Food Sci. Nutr.*, 41: 451–464, 2001.

16. J. Kayashita, I. Shimaoka, M. Nakajoh, M. Yamazaki, N. Kato. Consumption of buckwheat protein lowers plasma cholesterol and raise fecal neutral sterols in cholesterol-fed rats because of its low digestibility. *J. Nutr.*, 127: 1395–1400, 1997.
17. P.G. Waterman. Plant flavonoids in biology and medicine; biochemical, pharmacological and structure-activity relationship. *Phytochemistry*, 25: 2698, 1986.
18. M. Noroozi. Effects of flavonoids and vitamin C on oxidative DNA damage to human lymphocytes. *Am. J. Clin. Nutr.*, 67: 1210–1218, 1998.
19. J.P. Schroder-van der Elst. Synthetic flavonoids cross the placenta in the rat and are found in fetal brain. *Am. J. Physiol.*, 274: E253–E256, 1998.
20 H.L. Smith. Buckwheat poisoning with report a case in man. *Arch. Intern. Med.*, 3: 350–359, 1909.
21. G. Wieslander, D. Norback. Buckwheat allergy. *Allergy*, 56: 703–704, 2001.
22. M.M. Peshkin. Asthma in children etiology. *Am. J. Dis. Child.*, 31: 763, 1926.
23. W.J. Highman, J.C. Michael. Protein sensitization in skin diseases. *Arch. Derm. Symp.*, 2: 544, 1920.
24. A.H. Rowe. *Food Allergy*. Philadelphia: Lea and Febiger, 1931.
25. G.J. Blumstein. Buckwheat sensitivity. *J. Allergy*, 7: 74–79, 1935.
26. T. Matsumura, K. Tateno, S. Ugami, T. Kuroume. Six cases of buckwheat asthma. *J. Asthma Res.*, 1: 219–227, 1964.
27. S.-Y. Kang, K.U. Min. Three cases of buckwheat allergy. *J. Korean Med. Assoc.*, 27: 765–768, 1984.
28. H.A. Sampson. Food allergy. *J. Allergy Asthma Clin. Immunol.*, 111: S540–S547, 2003.
29. H. Breiteneder, C. Ebner. Molecular and biochemical classification of plant-derived food allergens. *J. Allergy Clin. Immunol.*, 106: 27–36, 2000.
30. S.Y. Lee, K.S. Lee, C.H. Hong, K.Y. Lee. Three cases of childhood nocturnal asthma due to buckwheat allergy. *Allergy*, 56: 763–766, 2001.
31. C.S. Hong, H.S. Park, S.H. Oh. *Dermatophagoides farinae*, an important allergenic substance in buckwheat-husk pillows. *Yonsei Med. J.*, 28: 272–281, 1987.
32. A.J. Horesh. Buckwheat sensitivity in children. *Ann. Allergy*, 30: 685–689, 1972.
33. K. Sugiura, M. Sugiura, R. Shiraki, S. Hayakawa, Y. Kato, M. Shamoto, J. Uchida, T. Ihara, T. Ueno, A. Itoh. A case of anaphylactic reactions with generalized urticaria due to Pyongyang cold noodles. *ACI Int.*, 15: 85–87, 2003.
34. Y. Takahashi, S.I. Ichikawa, Y. Aihara, S. Yokota. Buckwheat allergy in 90,000 school children in Yokohama. *Arerugi*, 47: 26–33, 1998.
35. Y. Arai, Y. Sano, K. Ito, E. Iwasaki, T. Mukouyama, M. Baba. Food and food additives hypersensitivity in adult asthmatics. 1 Skin scratch test with food allergens and food challenge in adult asthmatics. *Arerugi*, 47: 658–666, 1998.
36. K.E. Kim, B.J. Jeong, K.Y. Lee. The incidence and principal foods of food allergy in children with asthma. *Pediatr. Allergy Respir. Dis. (Korea)*, 5: 96–105, 1995.
37. K.Y. Lee. Allergens detected by allergy skin test in children with atopic diseases. *Pediatr. Allergy Respir. Dis. (Korea)*, 8 (Suppl. 2): S43–S48, 1998.
38. S.H. Kim, H.R. Kang, K.M. Kim, T.B. Kim, S.S. Kim, Y.S. Chang, C.W. Kim, J.W. Bahn, Y.K. Kim, S.H. Cho, H.S. Park, J.M. Lee, K.U. Min, C.S. Hong, N.S. Kim, Y.Y. Kim. The sensitization rates of food allergens in Korean people who visited allergy clinics: a multi-center study (in Korean). *J. Asthma Allergy Clin. Immunol.*, 23: 502–514, 2003.
39. T. Misawa. Allergic diseases. *J. Jap. Soc. Intern. Med.*, 25: 133–262, 1937.
40. S. Nakamura, B. Murohisa. Studies on asthma bronchiale. Report 5. On the buckwheat allergose. *Arerugi*, 19: 702–717, 1970.

41. S. Nakamura, M. Yamaguchi, M. Oishi, T. Hayama. Studies of the buckwheat allergose. Report 1: On the cases with the buckwheat allergose. *Allerg. Immunol.*, 20/21: 449–456, 1974/1975.
42. S. Nakamura, M. Yamaguchi. Studies on the buckwheat allergose. Report 2: Clinical investigation on 169 cases with the buckwheat allergose gathered from the whole country of Japan. *Allerg. Immunol.*, 20/21: 457–465, 1974/1975.
43. Y.B. Kwon, K.Y. Lee. Two cases of children with buckwheat allergy confirmed by oral challenge test. *J. Korean Acad. Pediatr.*, 28: 82–87, 1985.
44. Y.S. Jang, C.S. Hong, S.K. Kim. A case of bronchial asthma induced by house dust and buckwheat husk. *J. Korean Soc. Allergol.*, 5: 48–52, 1985.
45. G. Wieslander, D. Norback, Z. Wang, Z. Zhang, Y. Mi, R. Lin. Buckwheat allergy and reports on asthma and atopic disorders in Taiyuan city, Northern China. *Asian Pacific J. Allerg. Immunol.*, 18: 147–152, 2000.
46. E. Schumacher, P. Schmid, B. Wuerthrich. Zur Pizokel-Allergie: ein Beitrag ueber die buckweizenallergie. *Schweiz Med Wochenschr*, 123: 1559–1562, 1993.
47. B. Wuthrich, A. Trojan. Wheatburger anaphylaxis due to hidden buckwheat. *Clin. Exp. Allergy*, 25: 1263, 1995.
48. T. Matsumura, K. Tateno, S. Yugami, H. Fuzii, T. Kimura, M. Todokoro, T. Okabe. Studies on bronchial asthma in childhood. 1. Bronchial asthma induced by buckwheat flour attached to buckwheat chaff in pillows. *Arerugi*, 18: 902–911, 1969.
49. A.E. Davidson, M.A. Passer, G.A. Settipane. Buckwheat-induced anaphylaxis: a case report. *Ann. Allergy*, 69: 439–440, 1992.
50. T. Noma, I. Yoshizawa, N. Ogawa, M. Ito, K. Aoki, Y. Kawano. Fatal buckwheat dependent exercised-induced anaphylaxis. *Asian Pacific J. Allergy Immunol.*, 19: 283–286, 2001.
51. E. Okumura. A case of bronchial asthma from buckwheat in a housewife whose husband is working in a buckwheat flour mill. *Jpn. J. Ind. Health*, 22: 382–383, 1980.
52. S.B. Fritz, B.L. Gold. Buckwheat pillow-induced asthma and allergic rhinitis. *Ann. Allergy Asthma Immunol.*, 90: 355–358, 2003.
53. C.J. van Ginkel. Sensitisation to 'poffertjes' as a result of sleeping on a pillow containing buckwheat. *Neth. Tijdschr. Geneeskd.*, 146: 624–625, 2002.
54. M. Yuge, Y. Niimi, S. Kawana. A case of anaphylaxis caused by buck-wheat as an addition contained in pepper. *Arerugi*, 50: 555–557, 2001.
55. D. Choudat, C. Villette, J.-F. Dessanges, M.-F. Combalot, J.-F. Fabries, A. Lockhart, J. Dall'Ava, F. Conso. Occupational asthma induced by buckwheat flour. *Rev. Mal. Respir.*, 14: 319–321, 1997.
56. R. Valdivieso, I. Moneo, J. Pola, T. Munoz, C. Zapata, M. Hinojosa, E. Losada. Occupational asthma and contact urticaria caused by buckwheat flour. *Ann. Allergy*, 63: 149–152, 1989.
57. H.-S. Park, D.-H. Nahm. Buckwheat flour hypersensitivity: an occupational asthma in a noodle maker. *Clin. Exp. Allergy*, 26: 423–427, 1996.
58. H. Agata, N. Kondo, O. Fukutomi, M. Takemura, H. Tashita, Y. Kobayashi, S. Shinoda, T. Nishida, M. Shinbara, T. Orii. Pulmonary hemosiderosis with hypersensitivity to buckwheat. *Ann. Allergy Asthma Immunol.*, 78: 233–237, 1997.
59. M.A. Yoshimasu, Z.W. Zhang, S. Hayakawa, Y. Mine. Electrophoretic and immunochemical characterization of allergenic proteins in buckwheat. *Int. Arch. Allergy Immunol.*, 123: 130–136, 2000.
60. J.W. Park, D.B. Kang, C.W. Kim, S.H. Ko, H.Y. Yum, K.E. Kim, C.-S. Hong, K.Y. Lee. Identification and characterization of the major allergens of buckwheat. *Allergy*, 55: 1035–1041, 2000.

61. K. Tanaka, K. Matsumoto, A. Akasawa, T. Nakajima, T. Nagasu, Y. Iikura, H. Saito. Pepsin-resistant 16-kD buckwheat protein is associated with immediate hypersensitivity reaction in patients with buckwheat allergy. *Int. Arch. Allergy Immunol.*, 129: 49–56, 2002.
62. L. Unger. *Bronchial Asthma*. Springfield: Charles C Thomas, 1945, 159. [Cited in AJ Horesh,1972]
63. K. Yamada, A. Urisu, Y. Kondou, E. Wada, H. Komada, Y. Inagaki, M. Yamada, S. Torii. Cross-allergenicity between rice and buckwheat antigens and immediate hypersensitive reactions induced by buckwheat ingestion. *Arerugi*, 42: 1600–1609, 1993.
64. K. Yamada, A. Urisu, Y. Morita, Y. Kondo, E. Wada, H. Komada, M. Yamada, Y. Inagaki, S. Torii. Immediate hypersensitive reactions to buckwheat ingstion and cross allergenicity between buckwheat and rice antigens in subjects with high levels of IgE antibodies to buckwheat. *Ann. Allergy Asthma Immunol.*, 75: 56–61, 1995.
65. K.Y. Lee, S.Y. Lee, B.J. Jeoung, K.U. Kim, D.S. Kim. Cross reactivity among four cereals: rice, barley, wheat flour and buckwheat flour. *J. Korean Soc. Allergol.*, 13: 1–10, 1993.
66. E. Varjonen, E. Vainio, K. Kalimo, K. Juntunen-Backman, J. Savolainen. Skin-prick test and RAST responses to cereals in children with atopic dermatitis. Characterization of IgE-binding components in wheat and oats by an immunoblotting method. *Clin. Exp. Allergy*, 25: 1100–1107, 1995.
67. D. Abeck, M. Borries, C. Kuwert, V. Steinkraus, D. Vieluf, J. Ring. Food-induced anaphylaxis in latex allergy. *Hautarzt*, 45: 364–367, 1994.
68. F.D.E. Maat-Bleeker, S.O. Stapel. Cross-reactivity between buckwheat and latex. *Allergy*, 53: 538–539, 1998.
69. R. Schiffner, B. Przybilla, T. Burgdorff, M. Landthaler, W. Stalz. Anaphylaxis to buckwheat. *Allergy*, 56: 1020–1021, 2001.
70. Y. Arai, Y. Sano, K. Ito, E. Iwasaki, T. Mukouyama, M. Baba. Food and food additives hypersensitivity in adult asthmatics. 1. Skin sratch test with food allergens and food challenge in adult asthmatics. *Arerugi*, 47: 658–666, 1998.
71. K.Y. Lee, K.E. Kim, B.J. Jeong. Immediate type reaction of food allergy confirmed by open food challenge test: diagnostic value of history and skin test in food allergy. *Pediatr. Allergy Respir. Dis.* (*Korea*), 7: 173–186, 1997.
72. K.H. Park, S.M. Park, H.H. Lee, H.Y. Kim, B.J. Jeong, K.E. Kim, K.Y. Lee. A clinical study on oral buckwheat provocation test. *Pediatr. Allergy Respir. Dis.* (*Korea*), 8: 30–36, 1998.
73. M.H. Sohn, S.Y. Lee, K.E. Kim. Prediction of buckwheat allergy using specific IgE concentrations in children. *Allergy*, 58: 1308–1310, 2003.
74. T. Matsumoto, M. Murakami. The oral Prausnitz-Kustner reaction in food allergies. *Arerugi*, 40: 620–625, 1991.
75. N. Kondo, O. Fukutomi, H. Agata, Y. Yokoyama. Proliferative reponses of lymphocytes to food antigens are useful for detection of allergens in nonimmediate types of food allergy. *J. Invest. Allergol. Clin. Immunol.*, 7: 122–126, 1997.
76. G. Roberts, N. Golder, G. Lack. Bronchial challenges with aerosolized food in asthmatic, food-allergic children. *Allergy*, 57: 713–717, 2002.
77. K.Y. Lee, H.H. Lee, K.E. Kim, KH. Park, Y.K. Kim, B.J. Jeong. A comparative study on allergenicity of raw and hypoallergenic buckwheat flour. *Pediatr. Allergy Respir. Dis.* (*Korea*), 7: 13–22, 1997.

Section VI

Modern Approaches to Bioactive Proteins and Peptides

27 Database of Protein and Bioactive Peptide Sequences

Jerzy Dziuba and Anna Iwaniak

CONTENTS

27.1 INTRODUCTION

Until now, the nutritional value of proteins was determined according to the following criteria: the source of amino acids necessary to regulate the body functions, influence of protein intake on body mass (1,2), type and content of antinutritional compounds occurring with proteins (3), allergenic properties of proteins (4,5), and the maximum yield of bioactive peptides obtained from 1 g of protein (6). At present, an additional criterion of protein value determination is applied, that is, the possibility of bioactive peptide release from proteins (7,8). Such peptides can interact with appropriate organism's receptors and thus regulate its physiological functions. Currently, the use of bioactive peptides in therapy, especially in the treatment of cancer, viral infections, and immune, cardiac, and neurological system disorders, is of interest (9). Biologically active peptides are also recommended as functional food components (i.e., designed food with special desired biological and functional properties) (10,11). Particular attention is paid to antihypertensive peptides, including phosphopeptides and immunomodulating peptides, as physiological food components or to nutraceuticals for formon production (formon = food hormone) (12). Many bioactive peptides have been isolated from bacteria, fungi, plants, and animals and then characterized in detail. These peptides can be dipeptides, complex linear peptides, cyclic oligopeptides, or polypeptides, often modified by glycosylation, phosphorylation, and amino acid residue acylation (7).

Mass development of techniques applied in molecular biology or biochemistry contributed to the formation of a new scientific discipline called bioinformatics, which is based on the use of virtual methods oriented at data storage, availability, and analysis (13). Computer methods can support or, in many cases, substitute experiments (14), and their application is consistent with the paradigm of contemporary genomics assuming theoretical research from sequence to function (15,16). Growing interest in the application of computer techniques resulted in the creation of many structure and sequence databases (17) and programs deducing the protein function from its sequence, predicting the secondary structure of proteins and peptides (18), or designing proteolytic processes (19). Computer databases are applied to biomacromolecule classification taking into account the similarity between the sequence motifs, the identification of proteins based on sequences, or the determination of their fragments by the use of mass spectrometry (20). Such databases enable searching for bioactive fragments in the protein chain (21), and hierarchical classification of proteins (22). The information obtained can be useful while modeling the biomacromolecule structure (23,24).

There are many databases of protein sequences and structures, and their number is still growing. The main advantages resulting from their use are easy access to information and reproducibility of results (17). The databases discussed can be divided as follows:

1. PDB: Protein Data Bank (25)
2. CATH: database of hierarchical classification of protein domain structures (25)
3. SWISS-PROT database (13)
4. PROSITE: database of patterns and profiles necessary to classify proteins according to the similarity of protein domains (26)
5. InterPro: database of domains and structural motifs (27)
6. SCOP: Structural Classification of Protein database (28)
7. CluStr: database of protein clusters (29)

There is also the Complex Carbohydrate Structure Database (CCSD) designed for quick carbohydrate structure searching. The CSSD contains information about the structures of all naturally occurring carbohydrates, starting from trisaccharides. Such a database may be applied to create other databases such as those containing information about nuclear magnetic resonance (NMR) spectroscopy and three-dimensional structures obtained by roentgen crystallography (30).

27.2 AN OVERVIEW OF COMPUTER METHODS APPLIED IN PROTEIN SEQUENCE ANALYSIS

27.2.1 PROTEIN DATA BANK

The Protein Data Bank (PDB) was established at Brookhaven National Laboratories (BNL) in 1971 as an archive for biological macromolecular crystal structures. The main purpose of its establishment was to help scientists understand the structure–function relationships in biological systems. Most of the three-dimensional macromolecular structure data present in the PDB (http://www. rcsb.org/pdb) (31) was obtained by means of X-ray crystallography, NMR, or theoretical modeling. X-ray crystal diffraction usually cannot resolve the position of hydrogen atoms or reliably distinguish nitrogen from oxygen or carbon. Thus, the chemical identity of the terminal side-chain atoms is uncertain for Asp, Gln, and Thr. All of the above-mentioned methods have some limitations. NMR determines molecules not much greater than 30 kDa. The structures obtained by theoretical modeling tend to be less precise in comparison with those obtained experimentally. One type of modeling, called homology modeling, involves fitting a known sequence to the experimentally determined three-dimensional structure of a sequence-similar molecule. Such results are more likely to be reliable than *ab initio* modeling results (32).

All data collected in the PDB contain general information about structures as well as additional information collected for those structures, determined by X-ray methods. The following information about an individual structure can be obtained from the PDB source (i.e., genus), species, variant, full sequence of the macromolecular components, chemical structure of cofactors and prosthetic groups, names of all the components in the structure with qualitative descriptions, three-dimensional coordinates, and literature citation (33).

27.2.2 CATH

The CATH (http://www.biochem.ucl.ac.uk./bsm/cath) (34) database was created in the Brookhaven Protein Databank and is a hierarchical domain classification of protein structures. Classification of these structures is based on four major levels in hierarchy, namely class, architecture, topology (fold family), and homologous superfamily (25). Class is described as the secondary structure composition and packing within the structure. It can be assigned automatically to over 90% of the known structures. There are three major classes recognized as mainly-alpha, mainly-beta, and alpha-beta. The last class (alpha-beta) includes both alpha/beta and alpha + beta structures.

A-level (architecture) concerns the overall shape of the domain structure. It is determined by the orientation of the secondary structures excluding connectivity between them. At this level, sandwich, barrel, or trefoil architectures can be found, and it is possible to distinguish among the structures of the same class, of different architecture but not of different topology. Structures grouped into families at the topology (fold families) level characterize both the overall shape and connectivity of the secondary structures. They can belong to the same C-, A-, or T-level but not directly to the same homologous superfamily. Fold families are highly populated within mainly beta two-layer sandwich architectures and the alpha-beta three-layer sandwich architectures. Homologous superfamily levels classify protein domains according to high-functional and evolutionary similarities. Proteins belonging to this family are considered evolutionarily related. Structures within each H-level are clustered on sequence (S-level) identity, which indicates similar structures and functions. Proteins derived from different species can be found there (22).

27.2.3 SWISS-PROT Database

The SWISS-PROT database (http://www.expasy.org) (35) was established in 1986 at the University of Geneva's Department of Medical Biochemistry. At present, this database is maintained at the Swiss Bioinformatics Institute (SIB) and the European Bioinformatics Institute (EBI) in Hinxton (United Kingdom).

The SWISS-PROT database consists of sequence entries, each of them corresponding to a single continuous sequence as contributed to the bank or reported in the literature. There are two types of data in this protein database — core data and annotation data. In core data, the user can find sequence data, references, and taxonomy. Annotation data include the functions of proteins, posttranslational modifications, domains and sites, secondary and quaternary structures, similarities to other proteins, diseases associated with any number of protein deficiencies, and sequence variants.

Currently, SWISS-PROT is integrated with about 45 other different databases. Cross-references are provided in the form of pointers to information related to SWISS-PROT entries and found in data collections other than SWISS-PROT (36).

27.2.4 Prosite

PROSITE (http://www.expasy.org/prosite) (37) is a database of patterns and profiles. Patterns are formed from alignments of related sequences derived from different sources such as literature, well-characterized protein families, sequence databases, and sequence clustering. To start with, the alignments are checked for conserved regions, especially for proteins that can be experimentally assigned to catalytic or binding substrate activity. The basis of the pattern is the form of regular expressions that states the position of each amino acid. To complement the pattern, the profiles have also been elaborated. The database uses a symbol-comparison table in the conversion of frequency distribution into weights, resulting in a table of position-specific weights. A symbol-comparison table comprises values describing comparisons between amino acid pairs. This table provides scores characterizing the probability of the replacement of one amino acid by another at a particular position within the sequence alignment. These numbers are applied for the calculation of the similarity score of alignment between the profile and sequences in SWISS-PROT. The alignment with the similarity score equal to or greater than a given cut-off value constitutes a true hit. The profile is refined until only the intended set of protein sequences scores above the threshold for the profile (27).

27.2.5 InterPro

InterPro (http://www.ebi.ac.uk/interpro) (38) is the source of comprehensive information about protein families, domains, and repeats. It consists of 11,972 entries possessing accession numbers and names related to relevant families, repeats, or posttranslational modifications. Each entry includes a minimum one signature from individual databases describing the relevant group of proteins, annotation about proteins matching the entry, and a list of precomputed matches against the whole of SWISS-PROT/trEMBL databases. The match lists can be viewed in a tabular or graphical form. All of these forms include hits of all signatures from the same and other InterPro entries. The proteins can also be viewed graphically in a condensed form that computes the consensus domain boundaries from all proteins within each entry and splits the protein sequence into different lines for every InterPro entry matched. The results give an indication of the family an unknown protein belongs to, including the domain composition. Further information concerning related proteins and a protein family is available in the InterPro annotations and hits (22).

27.2.6 SCOP — A Database of Structural Classification of Proteins

SCOP (http://scop.mrc-lmb.cam.ac.uk/scop) (39) provides comprehensive information about evolutionary and structural relationships among proteins of known structure. Protein classification is based on hierarchical levels such as family, superfamily, common folds, and class.

Proteins belonging to the same family have to match the following criteria: (a) residue identities of 30% and greater; and (b) proteins with lower sequence identities, but very similar functions and structures with the sequence identity of 15%. Families whose proteins have low sequence identities but whose functional features suggest a common evolutionary origin are grouped into superfamilies. Both families and superfamilies have a common fold if their proteins have the same major secondary structures in the same arrangement with the same topological connections. Different folds can be put together into classes. Most of the folds are assigned to one of the five structural classes based on their secondary structures. They are: (a) all alpha (the structure is essentially composed of alpha-helices); (b) all beta (the structure is essentially composed of beta-sheets); (c) alpha and beta (proteins with alpha-helices and beta-strands are largely interspersed); (d) alpha

+ beta (for those proteins in which alpha-helices and beta-strands are largely segregated; and (e) multidomain (for those with domain of different fold and for which no homologues are known at present) (40).

27.2.7 CluStr Database

The CluStr database (http://www.ebi.ac.uk/clustr) (41) is the database of Clusters of SWISS-PROT and trEMBL proteins, built on the basis of sequence similarity. This database can be applied to the prediction of the functions of individual proteins and protein sets, automatic annotation of newly sequenced proteins, removal of redundancy from protein databases, searching for new protein families, proteome analysis, and provision of data for phylogenetic analysis. Clusters for mammalian proteins, plant proteins, and three complete eukaryote genomes (*Caenorhabditis elegans, Saccharomyces cerevisiae, Drosophila melanogaster*) have been built. The cluster approach is based on two steps. First, the matrix of "all-against-all" protein sequences is built and then computed using the Smith-Waterman algorithm. Second, clusters are built using single-linkage algorithms for different levels of protein similarity (29).

27.2.8 HOMOLOGY and PREDICT 7

Apart from databases used for protein classification based on different criteria, there are many computer programs that analyze proteins. One of them is HOMOLOGY. It was created at the Institute of Biochemistry, Wrocław University, Poland. This program checks the locations of homologous and semihomologous fragments of proteins and finds the insertion and deletion sites in these fragments. HOMOLOGY found a practical use in the analysis of repartition processes — analysis of evolutionary affinity of proteinase inhibitors derived from pumpkin seeds *(Cucurbita maxima)* — as well as in the location of antigen determinants in spectrins, proteins that are composed of 2000 amino acid residues (42).

Since it has become known that protein conformation is defined by amino acid sequence, many studies concerning the prediction of secondary structures have been carried out (43,44). One such program is PREDICT 7. This program analyzes protein sequences focusing on the following algorithms: location of potential sites of N-glycosylation, analysis of hydropathy indices, location of fragments exposed on the molecule surface, and location of antigenic determinants. PREDICT 7 analyzes protein sequences composed of 1800 amino acid residues. Results are obtained as text files that can be easily transferred and read in other computer applications (18).

Other programs used in protein sequence and structure analyses are WinGen 1.0 and WinPep 1.2. They are designed for an analysis of nucleotide and amino acid sequences. WinGen makes a translation of nucleotide sequences into amino acid sequences and transfers them to WinPep. WinPep detects the length of the sequence, isoelectric point, molecular mass, and hydrophobicity and molar absorption coefficients in polypeptides. The sequence can also be searched taking into consideration the presence of any sequence motifs (45).

27.3 THE DATABASE OF PROTEIN AND BIOACTIVE PEPTIDE SEQUENCES — BIOPEP

27.3.1 DATABASE DESCRIPTION

Biochemistry-related sciences constitute an area characterized by the rapid development and improvement of computer techniques. This encouraged us to create the BIOPEP database. BIOPEP (http://www.uwm.edu.pl/biochemia) (46) was established in 1999 by the Chair of Food Biochemistry, University of Warmia and Mazury in Olsztyn. It consists of two major databases, namely proteins and bioactive peptides. Each of them provides some information about the sequences, number of amino acid residues, molecular and monoisotopic masses, and additional information such as the role of a protein in a biological system, the accession number of a given protein in the SWISS-PROT or trEMBL database, the activity of a given peptide, and references. The number of amino acid residues in a protein or peptide, as well as masses (molecular and monoisotopic), is calculated automatically after sequence insertion. It is possible to insert short di- or tripeptides and peptides with posttranslational modifications into both databases. Currently, in our database, there are 569 protein sequences and 1572 peptide sequences showing 44 different kinds of activity. In the protein database, there are sequences of proteins of the same specimen differing in genetic variants as well as sequences of the same type of proteins derived from different species. Thus, among the lysozymes, those derived from the moth (*Hydrophora cecropia*) can be found. It should be noted that there are no Greek characters in our database because sometimes they are not readable to all World Wide Web users.

An example of protein data in BIOPEP is shown in Figure 27.1. The user can search for information in two different ways: (a) by viewing both databases, or

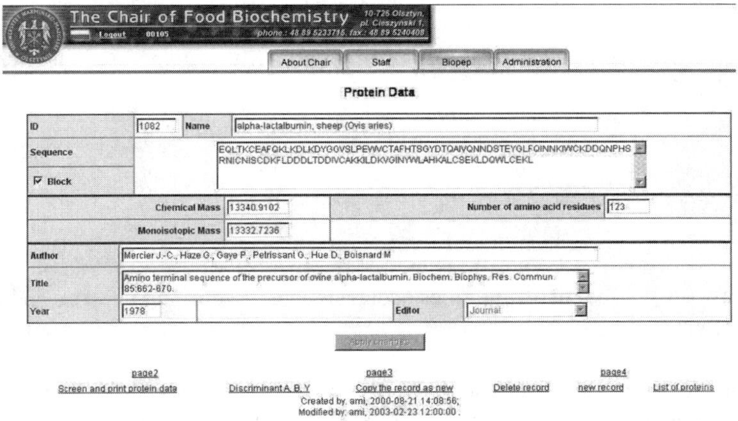

FIGURE 27.1 An example of information about α-lactalbumin from sheep (*Ovis aries*) in BIOPEP.

(b) by pasting the sequence of a protein/bioactive peptide and checking if the sequence inserted matches the hits.

27.3.2 New Criteria of Protein Evaluation in BIOPEP

Apart from the above-mentioned databases, BIOPEP includes a form called "Record Operations." On this form, the user can list all the proteins and peptides and make an evaluation of proteins according to the following quantitative and qualitative discriminants (47).

1. The profile of potential biological activity of a given protein, defined as the kind and location of a bioactive fragment in the protein chain
2. The occurrence frequency of fragments with a given kind of activity in the protein chain (*A*):

$$A = \frac{a}{N}$$

   where a = number of fragments with a given kind of activity in the protein chain, and N = number of amino acid residues.

3. Potential activity of protein fragments (*B*) [mM^{-1}]:

$$B = \frac{\sum_{i=1}^{k} \frac{a_i}{EC_{50i}}}{N}$$

   where a_I = number of repetitions of i-th bioactive fragment in the protein chain, EC_{50i} = the concentration of i-th bioactive peptide corresponding to its half maximum activity [mM], and k = number of fragments with a given kind of activity (46).

4. Relative occurrence frequency of fragments with a given kind of activity (Y_j) [%]:

$$Y_j = \frac{A_j}{\sum_{j=1}^{l} A_j} \times 100\%$$

   where A_j = occurrence frequency of fragments with a given kind of activity, l = number of all activities occurring in a protein, and j = a given kind of activity.

The above-mentioned discriminants can be determined for both protein sequences accumulated in BIOPEP and those inserted by the user.

BIOPEP also has an option enabling the potential prediction of bioactive peptide release from proteins. This is possible because of an additional database of endopeptidases with known specificity (about 25) and a recognition sequence. The recognition sequence is defined as the fragment of protein recognized by an enzyme [21].

27.3.3 Analysis of Protein in BIOPEP According to the New Criteria

An example of a profile of potential biological activity of α-lactalbumin from sheep (*Ovis aries*) is presented in Figure 27.2.

The sequence of this protein (ID number 1082) includes 123 amino acid residues. Twenty-nine bioactive fragments were found within the whole protein chain. Each of them has peptide ID, name (if known), activity, number of peptides with such activity found in the protein sequence, peptide sequence, and location in the protein chain. Bioactive fragments found in sheep α-lactalbumin are characterized by the following activities: antihypertensive (11), antibacterial (4), opioid (3), immunomodulating (3), regulating the phosphoinositole mechanism (1), regulating the ions flow (1), ligand of bacterial permease (1), dipeptidyl peptidase

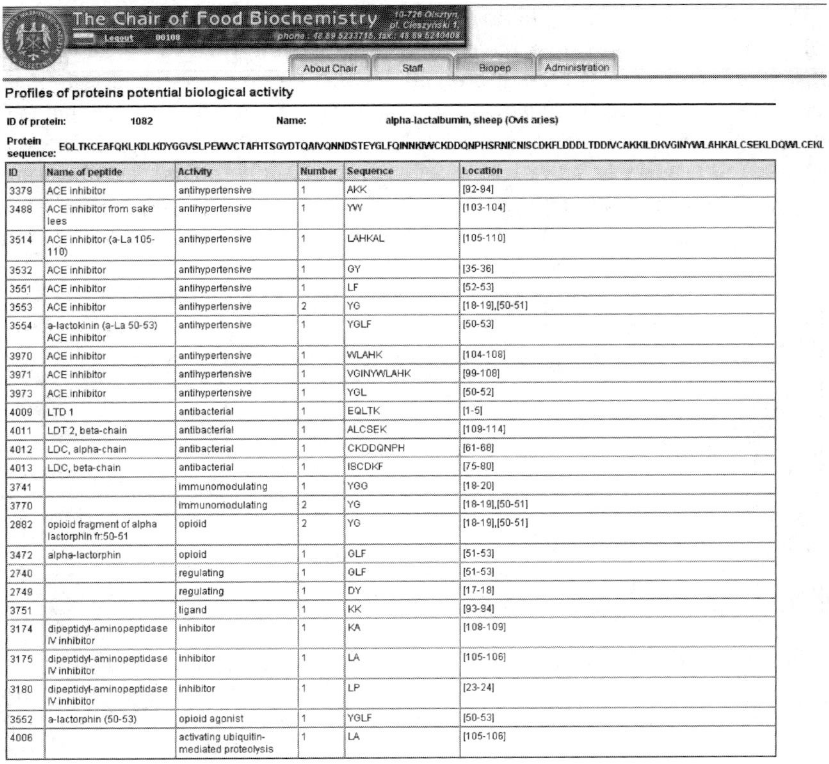

ID	Name of peptide	Activity	Number	Sequence	Location
3379	ACE inhibitor	antihypertensive	1	AKK	[92-94]
3488	ACE inhibitor from sake lees	antihypertensive	1	YW	[103-104]
3514	ACE inhibitor (a-La 105-110)	antihypertensive	1	LAHKAL	[105-110]
3532	ACE inhibitor	antihypertensive	1	GY	[35-36]
3551	ACE inhibitor	antihypertensive	1	LF	[52-53]
3553	ACE inhibitor	antihypertensive	2	YG	[18-19],[50-51]
3554	a-lactokinin (a-La 50-53) ACE inhibitor	antihypertensive	1	YGLF	[50-53]
3970	ACE inhibitor	antihypertensive	1	WLAHK	[104-108]
3971	ACE inhibitor	antihypertensive	1	VGINYWLAHK	[99-108]
3973	ACE inhibitor	antihypertensive	1	YGL	[50-52]
4009	LTD 1	antibacterial	1	EQLTK	[1-5]
4011	LDT 2, beta-chain	antibacterial	1	ALCSEK	[109-114]
4012	LDC, alpha-chain	antibacterial	1	CKDDQNPH	[61-68]
4013	LDC, beta-chain	antibacterial	1	ISCDKF	[75-80]
3741		immunomodulating	1	YGG	[18-20]
3770		immunomodulating	2	YG	[18-19],[50-51]
2882	opioid fragment of alpha lactorphin fr.50-51	opioid	2	YG	[18-19],[50-51]
3472	alpha-lactorphin	opioid	1	GLF	[51-53]
2740		regulating	1	GLF	[51-53]
2749		regulating	1	DY	[17-18]
3751		ligand	1	KK	[93-94]
3174	dipeptidyl-aminopeptidase IV inhibitor	inhibitor	1	KA	[108-109]
3175	dipeptidyl-aminopeptidase IV inhibitor	inhibitor	1	LA	[105-106]
3180	dipeptidyl-aminopeptidase IV inhibitor	inhibitor	1	LP	[23-24]
3552	a-lactorphin (50-53)	opioid agonist	1	YGLF	[50-53]
4006		activating ubiquitin-mediated proteolysis	1	LA	[105-106]

FIGURE 27.2 An example of a profile of potential biological activity of α -lactalbumin from sheep (*Ovis aries*) in BIOPEP.

IV inhibitors (3), opioid agonist (1), and activating ubiquitin-mediated proteolysis (1). The numbers given in parentheses denote the number of bioactive fragments found in the sheep α-lactalbumin sequence. These bioactive fragments consisted mostly of three amino acid residues, which can be important if peptide absorption from the digestive tract to blood is taken into account. Di- and tripeptides are more easily absorbed than free amino acids (48).

Antihypertensive, immunomodulating, and opioid fragments were noted in the majority of the profiles of potential biological activity of all protein sequences present in BIOPEP. The best-known group is that of antihypertensive peptides. We found that the best source of bioactive peptides was bovine milk proteins and the worst ones faba bean and rape proteins. One of the milk proteins, bovine β-casein, had not been known as the precursor of dipeptidyl peptidase IV inhibitors — one of the enzymes participating in the digestion processes (49). This protein had also not been known as a source of antioxidative fragments. Both dipeptidyl peptidase IV inhibitors and antioxidative peptides, as fragments occurring in β-casein, were not taken into consideration in the research results published recently (50). It means that supplementation of the database with new sequences of bioactive peptides offers the opportunity to find useful information about well-known proteins.

The profile of potential biological activity of proteins allows a quantitative evaluation of protein as the source of bioactive peptides. Apart from that criterion, BIOPEP allows the evaluation of proteins according to quantitative discriminants such as the occurrence frequency of bioactive fragments in the protein chain (A), potential biological activity of protein fragments (B), and the relative occurrence frequency of fragments with a given kind of activity (Y). Examples of the above-mentioned discriminant determination for sheep α-lactalbumin are shown in Figure 27.3. Discriminants A and Y were calculated for all activities found in this

A, B, Y calculation

Number of peptides in database: 1382

ID	NAME	Activity	A	B	Y
1082	alpha-lactalbumin, sheep (Ovis aries)	activating ubiquitin-mediated proteolysis	0.008130081300813		3.448275862069
1082	alpha-lactalbumin, sheep (Ovis aries)	antibacterial	0.032520325203252		13.793103448276
1082	alpha-lactalbumin, sheep (Ovis aries)	antihypertensive	0.089430894308943	0.00086383893241128	37.931034482759
1082	alpha-lactalbumin, sheep (Ovis aries)	immunomodulating	0.024390243902439		10.344827586207
1082	alpha-lactalbumin, sheep (Ovis aries)	inhibitor	0.024390243902439		10.344827586207
1082	alpha-lactalbumin, sheep (Ovis aries)	ligand	0.008130081300813		3.448275862069
1082	alpha-lactalbumin, sheep (Ovis aries)	opioid	0.024390243902439		10.344827586207
1082	alpha-lactalbumin, sheep (Ovis aries)	opioid agonist	0.008130081300813	2.710027100271E-05	3.448275862069
1082	alpha-lactalbumin, sheep (Ovis aries)	regulating	0.016260162601626		6.8965517241379

FIGURE 27.3 Calculation of qualitative discriminants A, B, Y for cheep α-lactalbumin (*Ovis aries*) in BIOPEP database. A - the occurrence frequency of fragments with given kind of activity in the protein chain; B - potential activity of protein fragments; Y - relative occurrence frequency of fragments with given kind of activity.

protein. Discriminant *B* was estimated only for two activities: antihypertensive and opioid agonist. In order to determine this parameter, we have to know the value of EC_{50} (i.e., the concentration of i-th bioactive fragment corresponding to half-maximum activity). BIOPEP cannot calculate parameter *B* if EC_{50} is unknown.

The highest value of discriminant *A* ($A = 0.0894$) for sheep α-lactalbumin was obtained for antihypertensive activity. It was also confirmed by the highest value of parameter *Y* (37.7931%), as well as by the profile of potential biological activity of that protein — among its 29 bioactive fragments, 11 were antihypertensive. In some cases, the value of discriminant *A* was equal to zero, for example, the occurrence frequency of fragments with antiamnestic activity for the myosin light chain from chicken β protein. It does not mean that myosin is a "bad" source of bioactive peptides but implies that it can be a better source of peptides with other kinds of activity.

Taking into account all protein sequences from the BIOPEP database, we found that the highest relative values of discriminants *A* were characteristic of bovine β-casein (genetic variant A^2) and bovine β-lactoglobulin and concerned antihypertensive activity. The values were 0.19 and 0.0092, respectively. The presence of peptides with antihypertensive activity was also confirmed experimentally by Meisel (51) and Yamamoto (52). High-frequency occurrence of fragments with antihypertensive activity (*A*) for chicken connectin can suggest that connectin — apart from its basic function as a component of connective tissue — can also be the precursor of endogenic angiotensin-converting enzyme (ACE) (EC 3.4.15.1) inhibitors with an antihypertensive effect. This is consistent with the hypothesis proposed by Karelin (53), according to which all proteins present in an organism can play the role of peptide reserves regulating its functions. The fact that the highest values of parameter *A* are characteristic of antihypertensive activity is related to the largest number of antihypertensive peptides in BIOPEP. It provides a higher probability of their occurrence in protein sequences (50).

Potential biological activity of protein fragments (*B*) is a more accurate measure of the protein value than the occurrence frequency of fragments with a given kind of activity (*A*) and the relative occurrence frequency of fragments with a given kind of activity (*Y*). One of the limitations while using parameter *B* in protein evaluation is the necessity to know the EC_{50} value, which is mostly measured for antihypertensive peptides. However, the algorithms presented enable easy distinction in the value of proteins as the source of bioactive peptides. Such criteria as discriminants *A* and *Y* can be a more universal, widely applied measure of the protein value.

Taking into consideration all protein sequences from BIOPEP, as well as the values of parameter *A* for all of them, we found that the majority of proteins can be the source of antihypertensive, dipeptidyl peptidase IV inhibitors, immunomodulating, antioxidative, neuroactive, and antithrombotic peptides. Table 27.1 shows the average values *A* and *B* for antihypertensive activity in selected groups of proteins.

In BIOPEP, there was only one chicken connectin, and thus all chicken meat proteins from our database were taken into consideration. In the case of β-caseins (bovine), different genetic variants of the same specimen were considered. Low values of standard deviations indicate that mutations concerning changes in single amino acid residues cause small changes in the occurrence frequency of

TABLE 27.1
Average Values of the Occurrence Frequency of Antihypertensive Fragments (*A*) and Potential Antihypertensive Activity (*B*) of Selected Protein Groups

Proteins	Number of Proteins in a Given Group	Average Value ± Standard Deviation (sd)	
		$A \pm$ sd	$B \pm$ sd
Chicken Meat Proteins:			
Troponins	10	0.0451 ± 0.0077	0.00154 ± 0.0011
Myosins			
Connectins			
Tropomyosins			
β-lactoglobulins:			
Bovine (gen. var. A)	3	0.088 ± 0.0035	0.0015 ± 0.0005
Goat			
Sheep			
α-caseins (bovine):			
Genetic variants:	6	0.1945 ± 0.0018	0.0042 ± 0.0034
A^1, A^2, A^3, B, C, E			
Soybean globulins	5	0.0690 ± 0.0347	0.0070 ± 0.0058
α/β-wheat gliadins	10	0.0715 ± 0.0180	0.0057 ± 0.0017

antihypertensive activity and potential biological activity of these proteins. Among α/β-wheat gliadins, the differences were significant. One explanation is that there are larger differences in gliadins than in caseins. In β-lactoglobulins derived from different species, we obtained much larger differences than in lactoglobulins from the same specimen.

In comparison with other widely used criteria of protein evaluation, additional ones like the profile of potential biological activity of the protein, the occurrence frequency of fragments with a given kind of activity in the protein chain (*A*), potential biological activity of protein fragments (*B*), and relative occurrence frequency of fragments with a given kind of activity (*Y*) are consistent with the information concerning the amino acid content of proteins (1). The percentage of amino acids does not provide complete information about the effects of protein intake on the organism. Similarly, the values *A*, *B*, and *Y* or the profile of potential biological activity of a protein also do not allow one to determine precisely if bioactive fragments are released by proteolytic enzymes and then absorbed from the digestive tract or under what conditions this occurs.

The discriminants of protein evaluation available in BIOPEP can be a measure of protein selection applied in nutraceutical production (i.e., physiologically active functional food compounds — foods with specific therapeutic properties) [54,11]. Table 27.2 shows the percentages of determinants characterizing the structure of selected proteins from BIOPEP. The above-mentioned parameters such as the content of proline, basic amino acids, hydrophobic fragments, α-helix, random coil, β-sheet, and β-turn were calculated by means of PREDICT 7 (18). These

TABLE 27.2
Parameters Determining the Structure of Selected Proteins Calculated Theoretically by PREDICT 7

Protein	Pro	Basic Amino Acids	Hydro-phobic Fragments	α-Helix	Random Coil	β-Turn	β-sheet
Myosin, light chain, chicken *(Gallus gallus)*	6.84	13.68	41.57	63.7	21.1	6.3	8.9
Connectin, chicken *(Gallus gallus)*	5.79	18.37	29.22	45.37	23.18	18.51	12.94
β-Lactoglobulin, gen. var. A, bovine *(Bos taurus)*	4.9	12.2	56.2	64.8	17.3	3.1	14.8
β-Casein, gen. var. A^2, bovine *(Bos taurus)*	16.7	9.5	63.1	23.9	41.6	16.3	18.2
Basic 7S soybean subunit precursor, *(Glycine max)*	5.1	7.3	56.2	9.5	32.1	38.7	33.57
α/β-wheat gliadin, clone PW 1215 *(Triticum aestivum)*	13.76	4.34	22.1	5.79	32.1	42.02	14.88

calculations were performed especially to this chapter. Results have not been published so far. The basis for their calculation were conformational coefficients. They are measures of the ability of individual amino acid residues to stabilize secondary structures (43,44). The percentages of structural parameters for β-lactoglobulin-the member of lipocalin family (55) were consistent with the experimental results (56). The results obtained for β-casein were different from the experimental results. The content of α-helix is higher (57), which is related to the presence of phosphate residues in casein, ignored by the computer program.

The results presented in Table 27.2 were the basis for estimating the correlations between parameters determining the protein structure and the occurrence frequency of antihypertensive fragments (A) and the potential antihypertensive activity (B) of proteins. These correlations are given in Table 27.3.

We found low correlations between the parameters determining the protein structure and discriminants A or B. The highest correlations were observed between the proline content and discriminant A and between random coils and A. In the case of the random coil content and log A the situation was similar. The correlation coefficients were 0.304, 0.317, and 0.315, respectively, and the correlation was significant at $p < 0.01$ and $p < 0.001$. Proline is a common C-terminal amino acid occurring in antihypertensive peptides (21). According to Meisel (57) and Yamamoto (52), the presence of such amino acids as lysine, tyrosine, and arginine in antihypertensive peptides favors the formation of random coil conformation and destabilizes

TABLE 27.3
Correlations between the Parameters Characterizing the Protein Structure and Quantitative Discriminants Determining the Protein Value as Precursors of Antihypertensive Peptides

	Values of Correlation Coefficients between Quantitative Discriminants of the Protein Value and Percentages of:						
	Pro	Basic Amino Acids	Hydrophobic Fragments	α-Helix	Random Coil	β-Turn	β-Sheet
A	0.304*	−0.229*	—	−0.227*	0.317***	—	—
Log A	0.281*	−0.276*	—	−0.335***	0.315***	—	—
B	0.230*	−0.287**	—	−0.245*	—	—	0.286**
Log B	0.233*	−0.265*	0.286**	−0.470***	—	—	—

Correlation significant at: $*p < 0.05$; $**p < 0.01$; $***p < 0.001$.

α-helices. It was confirmed by the negative values of the correlation coefficient between the α-helix content and the logarithmic values of *A* and *B* discriminants.

Negative correlations were obtained for the basic amino acid content and discriminant *B*, as well as for the logarithms of values *A* and *B* and the basic amino acid content. There was also a significant correlation ($p < 0.01$) between the percentage of β-sheet and log, as well as between the hydrophobic fragment content and log*B*. A less significant correlation was observed between the proline content and potential antihypertensive activity (*B*). The correlations between linear or logarithmic dependencies and parameters determining the protein structure indicate that logarithmic dependencies better reflect the relationships between the structure of a protein and its value as the source of antihypertensive peptides.

The occurrence frequency of bioactive fragments in the protein chain (*A*) was the basis for dividing the proteins from BIOPEP into families. The protein families were distinguished on the basis of six most common activities in proteins, that is, antihypertensive, dipeptidyl IV peptidase inhibitors, immunomodulating, opioid, neuropeptide, and antioxidative. The average values of discriminant *A* and standard deviations were determined for each family, and a *t*-test was performed. The result of this analysis is shown in Table 27.4. It was possible to distinguish at least four families for each of the above activities. For two activities — antihypertensive and antioxidative — five families were found. Within the range of a given kind of activity, there were statistically significant differences in the occurrence frequency of bioactive fragments in the protein chain. Significant differences were observed in antioxidative activity between families I and II and between II and IV at $p < 0.05$. For antihypertensive, dipeptidyl peptidase IV inhibitors, immunomodulating, and neuropeptide activities, the differences were significant at $p < 0.001$. No statistically significant differences were observed for proteins grouped according to the occurrence frequency of opioid fragments. Table 27.5 presents examples of proteins belonging to families grouped by the occurrence frequency of antihypertensive fragments.

TABLE 27.4
Statistical Analysis of Protein Families

Activity	Family	Ranges of A	n[a]	Mean value	sd[b]	t[c]
Antihypertensive	I	0–0.006	3	0.002	0.003464	43.39[e]
	II	0.0061–0.039	32	0.0251	0.009302	
	III	0.0391–0.1016	53	0.065638	0.017195	10.187[e]
	IV	0.1016–0.1567	12	0.1335	0.016533	12.63[e]
	V	0.15671–0.25	9	0.1988	0.018418	12.25[e]
Dipeptidyl peptidase IV inhibitors	I	0–0.0015	9	0	0	7.25[e]
	II	0.00151–0.0215	29	0.012714	0.005054	
	III	0.02151–0.0636	51	0.041976	0.011365	13.93[e]
	IV	0.06361–0.21	20	0.11507	0.038387	27.95[e]
Immuno-modulating	I	0–0.006	61	0.001018	0.001916	3.85[e]
	II	0.0061–0.012	21	0.008908	0.001615	
	III	0.0121–0.019	12	0.015675	0.001894	1.409
	IV	0.0191–0.03	15	0.022294	0.001936	9.166[e]
Opioid	I	0–0.001	72	5.41×10^{-5}	0.000225	2.383
	II	0.0011–0.012	8	0.007344	0.003448	
	III	0.0121–0.024	27	0.021273	0.00371	1.967
	IV	0.0241–0.036	2	0.03055	0.003606	3.52
Neuropeptide	I	0–0.004	65	0.000381	0.001092	33.478[e]
	II	0.0041–0.0104	35	0.0074	0.001299	
	III	0.01041–0.013	8	0.011963	0.000739	19.013[e]
	IV	0.0131–0.17	1	0.1612	—	187.70[e]
Antioxidative	I	0–0.001	70	1.42×10^{-5}	0.00012	34.13[e]
	II	0.0011–0.006	19	0.003906	0.001059	
	III	0.0051–0.009	7	0.007755	0.001009	2.26[d]
	IV	0.0091–0.0208	10	0.011589	0.003621	
	V	0.02081–0.028	3	0.024	0	2.39[d]
						6.16[e]

[a]n = number of proteins in family; [b]sd = standard deviation; [c]statistical significance of differences; [d] = difference significant at $p < 0.005$; [e] = difference significant at $p < 0.001$.

An analysis of the above data shows that division is not consistent with the protein classification by primary structure. Thus, proteins belonging to, for example, rice prolamins can occur in a few families grouped according to the occurrence frequency of fragments with a given kind of activity. The above rice prolamins were present in the second, third, and fifth families.

The procedure applied to protein classification is quite flexible. An increasing number of bioactive peptides in BIOPEP affects changes in the values of discriminants, and thus the proteins analyzed can be found in other families. Different techniques applied to, for example, protein structure classification in bioinformatics laboratories also show certain limitations. For example, an analysis of protein

TABLE 27.5
Characteriztics of Protein Families Grouped by the Occurrence Frequency of Antihypertensive Fragments (A)

Family	Types of proteins
I	Wheat purothionin A-1, barley γ-hordothionin, chicken skeletal muscle troponin
II	Rice 10 KD prolamin; pumpkin 11S globulin; basic 7S soybean globulin, sunflower 11 S globulin; wheat purothionins: α-, γ-; barley α-hordohtionin; berry monellin; wheat γ-gliadin; rabbit α-lactalbumin; bovine, pigeon, chicken lysozymes; Che Y; chicken light chain myosin; chicken, porcine troponins; chicken tropomyosins α-, β-; bovine elastin, bovine, chicken α-collagens, rat epidermal retinol binding protein
III	Sorghum kafirins; rice prolamin; cotton legumins, oat globulin; α/β-wheat gliadins; barley γ-hordein; storage cocoa protein, wheat glutenin; fava bean legumin-like chain; fycocyanine; sheep, camel, goat, mare, human, guinea pig, rat, kangaroo, rabbit, bovine, α-lactalbumins; bovine, goat, sheep β-lactoglobulins; moth, human, dog lysozymes; flavodoxin; goat, human, bovine κ-caseins; bovine α-casein; human lactoferrin; chicken connectin; bilin-binding protein; bovine odorant-binding protein; bovine retinol-binding protein
IV	Rice globulin; sorghum kafirin; rice prolamin; bery monellin, gingko biloba, pumpkin, soybean, green pea legumin like chains; bovine αs_1-caseins genetic variants: A, B, C, D
V	Rice prolamins; bovine β-caseins genetic variants: A^3, B, C, E; bovine β-casein genetic variant F

functions on the basis of protein similarities can be applied in two cases: (a) only for proteins with known standard homology sequences; and (b) when it is difficult to determine common protein functions directly, especially when homologies between sequences are too distant (58).

27.3.4 Proteolytic Process Design in BIOPEP

There are different ways of producing biologically active peptides from proteins. For instance, peptides derived from milk can be released in three ways: (a) enzymatic hydrolysis with digestive enzymes, (b) fermentation of milk with proteolytic starter cultures, and (c) the action of enzymes derived from microorganisms (59). Apart from the different industrial methods of obtaining bioactive peptides, there are many programs for computer simulation of proteolysis, for example, PROTEOLYSIS (19).

Another part of the "Record Operations" form in BIOPEP is the option called "Enzyme Action." This option is suitable for proteolytic process design in the aspect of obtaining bioactive peptides (46). This option includes a database containing 25 endopeptidases with known specificity. Users can select a protein sequence from BIOPEP or enter their own sequence and then select one or a maximum of three enzymes from our database. Results will show potential peptides that may be released after the action of the selected enzymes, including the location of peptides in the protein chain. By clicking on the command "Search for active fragments," the user can find which of the peptides released shows biological activity.

An example of bioactive peptide release from sheep α-lactalbumin is presented in Figure 27.4, which shows the data selected from BIOPEP to check whether the enzymes chosen will release any bioactive peptides. The results of proteolysis simulation in the above protein are shown in Figures 27.5 and 27.6.

Among the many peptides released from sheep α-lactalbumin, one showed biological activity (peptide DY — regulating the ions flow).

Information concerning liberated bioactive peptides can be important from the perspective of the large-scale production and application of bioactive peptides. At present, many experiments are being carried on to determine if peptides derived from proteins can be used as nutraceuticals. Food supplementation with bioactive

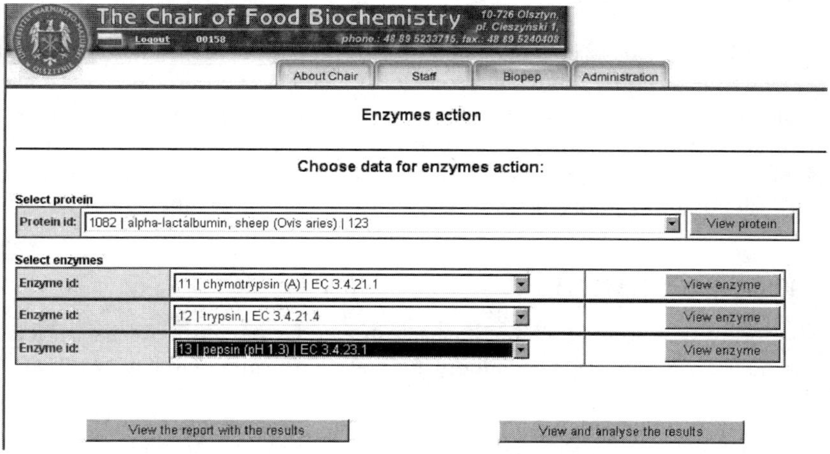

FIGURE 27.4 Proteolysis process design for sheep α-lactalbumin (*Ovis aries*) in BIOPEP. Selected enzymes: chymotrypsin (BIOPEP ID - 11), trypsin (BIOPEP ID - 12), pepsin (BIOPEP ID - 13).

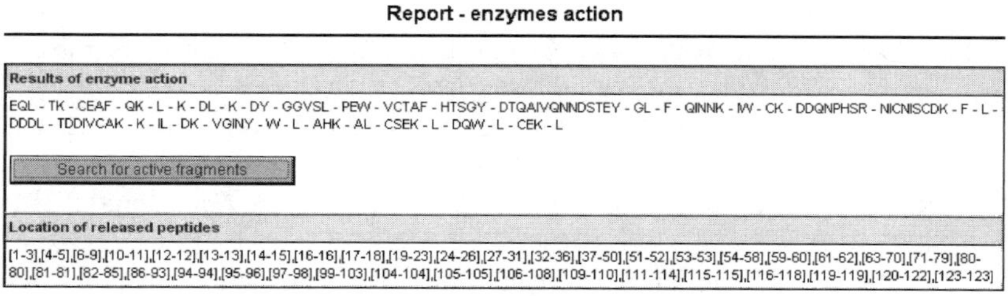

FIGURE 27.5 Peptides released from sheep α-lactalbumin (*Ovis aries*).

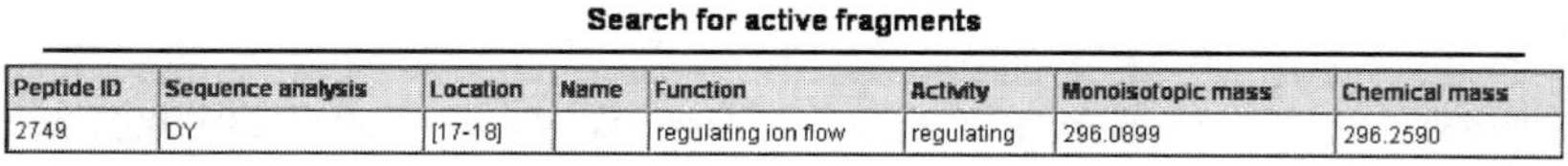
Search for active fragments

Peptide ID	Sequence analysis	Location	Name	Function	Activity	Monoisotopic mass	Chemical mass
2749	DY	[17-18]		regulating ion flow	regulating	296.0899	296.2590

FIGURE 27.6 An example of bioactive peptide found in sheep α-lactalbumin (*Ovis aries*).

peptides requires the use of specific enzymes or genetically modified organisms (GMOs). Thus, proteolytic process design can be a challenge in many branches of the food industry.

However, it should be emphasized that the physicochemical conditions of enzymatic hydrolysis are important in bioactive peptide release. Selectivity of enzyme is also the reason for release of peptides with long sequences. Shorter fragments showing biological activity can be blocked in the longer chains released. Bonds susceptible to proteolysis may be located in a bioactive fragment, which can result in obtaining a different, nonbioactive product. Therefore, an analysis of the possibilities of bioactive peptide release will require additional, experimental confirmation.

27.4 CONCLUSIONS

The discriminants described above can be a valuable source of information supplementing the knowledge about proteins and their properties. The advantage of the methods applied is simplicity of data saving and storage. Automated methods provide information about biomacromolecules and their functions, but only biochemists or biologists can interpret the results in the appropriate way, and in some cases, such interpretations have to be confirmed experimentally.

ACKNOWLEDGMENTS

This work was supported by the State Committee for Scientific Research (KBN), grant no. 021/P06/99/08. The authors are grateful to Marta Niklewicz, Ph.D., for help and support in proteolytic process design.

REFERENCES

1. M. Friedman. Nutritional value of proteins from different food sources. A review. *J. Agric. Food Chem.*, 44: 6–29, 1996.
2. S. Fukudome, M. Yoshikawa. Opioid peptides from wheat gluten: their isolation and characterisation. *FEBS Lett.*, 296: 107–111, 1992.
3. K. Anantharaman, P.A. Finot. Nutritional aspects of food proteins in relation to technology. *Food Rev. Int.*, 9: 629–655, 1993.
4. R.K. Bush, S.L. Hefle. Food allergens. *Crit. Rev. Food Sci. Nutr.*, 36: S119–S163, 1996.
5. T. Matsuda, R. Nakamura. Molecular structure and immunological properties of food allergens. *Trends Food Sci. Technol.*, 4: 289–293, 1993.
6. H. Meisel. Casokinins as bioactive peptides in the primary structure of casein. In: V.Ch. Weinheim (Ed.). *Food Protein, Structure and Functionality*. John Wiley and Sons Ltd., 1993, 223–227.
7. E. Schlimme, H. Meisel. Bioactive peptides derived from milk proteins. Structural, physiological and analytical aspects. *Nahrung*, 39: 1–20, 1995.
8. H. Meisel. Biochemical properties of regulatory peptides derived from milk proteins. *Biopolymers*, 43: 119–128, 1997.

9. P.W. Latham. Therapeutic peptides revisited. *Nat. Biotechnol.*, 17: 755–758, 1999.
10. S. Arai. Studies on functional foods in Japan — state of the art. *Biosci. Biotechnol. Biochem.,* 60: 9–15, 1996.
11. H. Korhonen, A. Pihlanto-Leppälä, P. Rantamäki, T. Tupasela. Impact of processing on bioactive proteins and peptides. *Trends Food Sci. Technol.*, 9: 307–319, 1998.
12. M. Gobetti, P. Ferranti, E. Smacchi, F. Goffredi, F. Addeo. Production of angiotensin-I-converting-enzyme-inhibitory peptides in fermented milks started by *Lactobacillus delbrueckii* ssp. *bulgaricus* SS1 and *Lactococcus lactis* ssp. *cremoris* FT4. *Appl. Environ. Microbiol.*, 66: 3898–3904, 2000.
13. H.W. Mewes, R. Doelz, D.G. George. Sequence databases: an indispensible source for biotechnological research. *J. Biotechnol.*, 35: 239–256, 1994.
14. M.S. Boguski. Bioinformatics — a new era. *Trends Biotechnol.*, 16S: 1–2, 1998.
15. J. Skolnick, J.S. Fetrow. From genes to protein, structure and function: novel applications of computational approaches in the genomic era. *Trends Biotechnol.*, 18: 34–39, 2000.
16. J. Skolnick, J.S. Fetrow, A. Kolinski. Structural genomics and its importance for gene function analysis. *Nat. Biotechnol.*, 18: 283–287, 2000.
17. I.D. Campbell, A.K. Downing. Building protein structure and function from modular units. *Biotechnology*, 12: 168–172, 1994.
18. R.S. Carménes, J.P. Freije, M.M. Molina, J.M. Martin. PREDICT 7, a program for protein structure prediction. *Biochem. Biophys. Res. Commn.*, 2: 687–693, 1989.
19. M.M. Vorob'ev, I.A. Goncharova. Computer simulation of proteolysis. Peptic hydrolysis of partially demasked β-lactoglobulin. *Nahrung*, 42: 61–67, 1998.
20. M. Kanehisa. Database of biological information. *Trends Biotechnol.*, 16: 24–26, 1998.
21. J. Dziuba, P. Minkiewicz, D. Nałęcz, A. Iwaniak. Database of biologically active peptide sequences. *Nahrung*, 43 (3): 190–195, 1999.
22. C.A. Orengo, A.D. Michie, S. Jones, D.T. Jones, M.B. Swindells, J.M. Thornton. Cath — a hierarchic classification of protein domain structures. *Structure*, 5: 1093–1108, 1997.
23. M.S. Johnson, N. Srinivasan, R. Sowdhamini, T.L. Blundell. Knowledge-based protein modelling. *Crit. Rev. Biochem. Mol. Biol.*, 29: 1–68, 1994.
24. A.F. Neuwald, J.S. Liu, D.J. Lipman, C.E. Lawrence. Extracting protein alignments models from the sequence database. *Nucl. Acids Res.*, 25: 1665–1677, 1997.
25. J.E. Bray, A.E. Todd, F.M.G. Pearl, J.M. Thornton, C.A. Orengo. The CATH dictionary of homologous superfamilies (DHS): a concensus approach for identifying distant structural homologues. *Protein Eng.*, 13, 3: 153–165, 2000.
26. K. Hofmann, P. Bucher, P. Falquet, A. Bairoch. The PROSITE database, its status in 1999. *Nucl. Acid Res.*, 27(1): 215–219, 1999.
27. N.J. Mulder, R. Apweiler. Tools and resources for identifying families, domains and motifs. *Genome Biol.*, 3: 1–8, 2001.
28. A.G. Murzin, S.E. Brenner, T. Hubbard, C. Chothia. SCOP: a structural classification of proteins database for the investigation of sequences and structures. *J. Mol. Biol.*, 247: 536–540, 1995.
29. E.V. Kriventseva, W. Fleischmann, E.M. Zdobnov, R. Apweiler. CluStr: a database of clusters of SWISS-PROT+TrEMBL proteins. *Nucl. Acids Res.*, 29: 33–36, 2001.
30. A.J. Van Kuik, J.F.G. Vliegenthart. Database of complex carbohydrates. *Trends Food Sci. Technol.*, 4: 73–77, 1993.
31. PDB database [http://www.rcsb.org/pdb].

32. PDB database [http://www.rcsb.org/pdb/experimental_methods].
33. H.M. Berman, J. Westbrook, Z. Feng, G. Gilliland, T.N. Bhat, H. Weissig, I.N. Shindyalov, P.E. Bourne.The protein bank. *Nucl. Acids Res.*, 28: 235–242, 2000.
34. CATH database [http://www.biochem.ucl.ac.uk./bsm/cath].
35. SWISS-PROT database [http://www.expasy.ch].
36. SWISS-PROT database [http://www.expasy.ch/sprot/userman].
37. PROSITE database [http://www.expasy.ch/prosite].
38. InterPro database [http://www.ebi.ac.uk/interpro].
39. SCOP database [http://www.scop-mrc-lmb.cam.ac.uk/scop].
40. A.G. Murzin. Structure classification-based assessment of CASP3 predictions for the fold recognition targets. *Prot. Struct. Funct. Genet.*, S3: 88–103, 1999.
41. CluStr database [http://ebi.ac.uk/clustr].
42. J. Leluk. Analiza struktury pierwszorzędowej białek z zastosowaniem programu HOMOLOGY. *Proceedings of XXIXth Meeting of Polish Biochemistry Society*, Wrocław, 1999, 348–350.
43. P.Y. Chou, G.D. Fasman. Conformational parameters for amino acids in helical, β-sheet and random coil regions calculated from proteins. *Biochemistry*, 13: 211–221, 1974.
44. P.Y. Chou, G.D. Fasman. Prediction of protein conformation. *Biochemistry*, 13: 222–245, 1974.
45. L. Henning. WinGene/WinPep: User-friendly software for the analysis of amino acid sequence. *BioTechniques*, 26: 1170–1172, 1999.
46. J. Dziuba, A. Iwaniak, M. Niklewicz. Database of protein and bioactive peptide sequences — BIOPEP [http://www.uwm.edu.pl/biochemia], 2003.
47. J. Dziuba, A. Iwaniak, P. Minkiewicz. Computer-aided characteristics of proteins as potential precursors of bioactive peptides. *Polimetry*, 48 (1): 50–53, 2003.
48. A.D. Siemensma, W.J. Weijer, H.J. Bak. The importance of peptide lengths in hypoallergenic infant formulae. *Trends Food Sci. Technol.*, 4: 16–21, 1993.
49. D.F. Cunningham, B. O'Connor. Proline specific peptidases. *Biophys. Acta*, 1343: 160–186, 1997.
50. J. Dziuba, P. Minkiewicz. Influence of glycosylation on micelle-stabilizing ability on biological properties of C-terminal fragments of cow's κ-casein. *Int. Dairy J.*, 6: 1017–1044, 1996.
51. H. Meisel. Overview on milk protein derived peptides. *Int. Dairy J.*, 8: 363–373, 1998.
52. N. Yamamoto. Antihypertensive peptides derived from food proteins. *Biopolymers*, 43: 129–134, 1997.
53. A.A. Karelin, E.Y. Blischenko, V.T. Ivanov. A novel system of peptidergic regulation. *FEBS Lett.*, 428: 7–11, 1998.
54. P. Jelen, P. Lutz. *Functional Foods. Biochemical and Processing Aspects*. G. Mazza (Ed.). Lancaster, Basel: Technomic Publishing Co. Inc., 1998, 357–380.
55. S. Brownlow, J.H.M. Cabral, R. Cooper, D.R. Flower, S.J. Yewdall, I. Polikarpov, A.C.T. North, L. Sawyer. Bovine β-lactoglobulin at 1.8 Å resolution — still an enigmatic lipocalin. *Structure*, 5: 481–495, 1997.
56. M. Darewicz, J. Dziuba, P.W.J.R. Caessens, H. Gruppen. The secondary structure of dephosphorylated β-casein and its plasmin derived peptides after adsorption at teflon/water interfaces. *Proceedings of Xth Symposium, Euro Food Che*, Budapest, 1999, 270–282.

57. H. Meisel. Overview on milk protein derived peptides. *Int. Dairy J.*, 8: 363–373, 1998.
58. A. Šali. Functional links between proteins. *Nature*, 402: 23–25, 1999.
59. H. Korhonen, A. Pihlanto-Leppälä. Milk protein-derived bioactive peptides — novel opportunities for health promotion. *Dairy Nutr. Healthy Future Bull. IDF*, 363: 17–26, 2001.

28 Rational Designing of Bioactive Peptides

Shuryo Nakai and Nooshin Alizadeh-Pasdar

CONTENTS

28.1 INTRODUCTION

Since the early 1980s, the advances in molecular biology, protein crystallography, and computational chemistry have greatly promoted the "rational drug design." The concept of rational drug design was made more implicit based on the similarities between the bioactivity of different compounds. The general drug-target scheme suggests the possibility of creating structure-based rational drug design. It is reasonable to assume that drugs that produce the same biological response interact with the same, but unknown, biological target. They must, therefore, have some common set of the required structural features in order to evoke the above biological responses. This common set of structural features is the so-called "pharmacophore." According to Parrill and Reddy (1), this assumed similarity of drugs with similar effects suggests an alternative set of tasks that can accomplish rational drug design. Three basic tasks have to be fulfilled: first, the appropriate protein target for a given therapeutic need must be identified; second, the structure of the target protein must be distinguished; and, finally, the structure of a drug must be designed to interact with the target protein.

Optimization is the eternal aim of research. To accomplish this aim, the quantitative structure-activity relationship (QSAR) is a prerequisite. A variety of QSAR technologies in conjunction with combinatorial chemistry has frequently been used for rational drug design (1). As a mathematical approach, a variety of regression analyses, both linear and nonlinear, have been utilized to correlate independent variables (predictors) with responses (functions). Hydrophobicity and variables of electronic and structural properties have been most frequently employed as predictors (Hansch analysis) (2). Meanwhile, random search is another approach to circumvent mathematical complications; however, a large number of iterations are usually required to reach the global optimum, or even with a risk that they may never reach it. We have circumvented this problem by using (a) a regulated random design to avoid the localization of search spaces, (b) the centroid search to best utilize the result of the preceding random search, and (c) the mapping of response surfaces to speculate search direction (3). This random-centroid optimization (RCO) has enabled us to find the global optimums within 50 iterations for most of the nonlinear mathematical models used in our study (3). Currently, the most advanced global optimization techniques, for example, simulated annealing and genetic algorithm, require thousands of iterations using computers. We have been successful in applying the RCO for genetics (RCG) to site-directed single-site (4) as well as double-site mutagenesis simultaneously (5) in peptide sequences to improve their biological functions.

As in the two cases of rational drug design, we have developed two computer-aided methods, namely, QSAR-based methods and RCG optimization, regardless of whether any structural information is available or not, respectively. In addition, we have also developed a software package for computer-aided peptide sequence analysis, including (a) principal component similarity (PCS) for classification (6), (b) homology similarity analysis (HSA) for QSAR (7), and (c) homology similarity search (HSS) for identifying functional motifs in the lysozyme and cystatine sequences (8, 9). Rational combinations of all these computer programs would expedite the optimization of peptide designs.

In the post-genome era, the pharmaceutical industry has embraced the means to identify new biological targets derived from the human genome for target-based drug discovery approaches. Genomics is driving a substantial effort in protein structure determination or prediction of unknown structures. These structures will provide the opportunity to undertake structure-guided discovery programs, such as high-throughput screening. In addition, these structures could provide structure-based criteria to prioritize certain targets for focused drug discovery programs (10). However, the necessity of rational design for food proteins is not always the same as that of rational drug design. Drastic modification in protein sequences is not acceptable from the aspect of human health, despite potential outstanding improvement in the functionality. Furthermore, as far as the extremely complicated structures of macromolecules such as proteins are concerned, both techniques — QSAR and combinatorial chemistry — have not readily been applicable to food research even by using the technologies of the most modern rational drug designs.

The objective of this chapter is to compare the need for the study on rational bioactive peptide design in food research and development with that for rational drug design. As a post-genome approach, the importance of the analysis of DNA (deoxyribonucleic acid) arrays and protein sequences should be emphasized. The chapter is divided into four parts: (a) QSAR, (b) functional sequence analysis, (c) optimization, and (d) rational designing strategy. As detailed discussions of the first three parts have recently been included in a chapter of the book entitled *Proteins in Food Processing* (11), the post-genomic approach for rational peptide design, including the recent technology of bioinformatics, will be specifically emphasized in this chapter.

28.2 QSAR

The pioneer of the QSAR concept, Dr. Corwin Hansch, recently wrote a review article (12). As the area of study of his group is organic chemistry, their QSAR has been mostly in chemical and pharmaceutical research. Major descriptors used were hydrophobic interactions, electronic effects, and steric parameters, which were called three-dimensional (3D) QSAR. The QSAR in toxicology was recently thoroughly reviewed by Schultz et al. (13). For the selection of predictor variables for the three different categories, the readers are recommended to refer to Hansch et al. (14). As a hydrophobic descriptor, log P has been most frequently used, where P is the n-octanol/water partition coefficient; or more generally, lipophilicity log P, where P is the oil/water partition coefficient in this case, is accepted. For the electronic parameter, the Hammett free-energy-related parameter σ, which was the log ratio of the ionization constants of substituted and unsubstituted compounds, was the first choice. For steric parameters, Taft's E_s constant was used to demonstrate the ortho-/meta- or para-substitution of aromatic compounds. Other representations of steric effects frequently used included molar volume, van der Waals radius, interatomic distance, and many others. Molar refractivity (MR), which is defined on the basis of molecular weights, refractive index, and density, is also a useful expression as a steric parameter (12). However, there have been no simple solutions to the problem of understanding chemical-biological interactions with enzymes, organs, tissues, and organisms.

It is relatively easy to apply 3D QSAR to amino acids alone; however, a difficulty with peptides or proteins, and not amino acids *per se*, is their functional properties in biology. This situation is especially true when peptides or proteins are in biological cells. An excellent multivariate QSAR approach was proposed for peptides by Hellberg et al. (15). Three principal components, z_1, z_2, and z_3, were derived for each amino acid from 29 physicochemical property indices of the 20 amino acids, using principal components analysis (PCA). This 3z method was applied to four peptides, namely, (a) oxytocin analogues with nine amino acid residues, (b) pseudopeptides, (c) pepstatin analogues with five residues, and (d) bradykinin-potentiating pentapeptides. Partial least squares (PLS) regression was employed for the observed/calculated activity correlation study. Although PCA software usually normalizes each variable prior to computation to avoid the effects of scale differences, the potentially important conformational effects of peptides, which may appear as minor factors affecting the target biological activity, may be ignored. This is especially true when a variable includes the components of other property indices.

Using 19 derivatives of the 15-residue murine LFcin (lactoferricin) and bovine LFcin (LFB or bLFcin), Strøm et al. (16) employed two parameters for α-helices measured for micelle affinity, three parameters describing α-helix propensities, two parameters related to charge, four hydrophobicity parameters, and one parameter relating to the molecular surface (Figure 28.1). This excellent prediction explicitly supports the importance of the helix, charge, and hydrophobicity in the antimicrobial activity of peptides. However, these parameters were used for the quantification of the physicochemical properties of entire peptide sequences without finding roles of segments or motifs within the same sequences. Although PCA or PCS may probably screen important factors to be used in regression analysis for QSAR study, it may be difficult to identify and isolate the truly influential factors included in the underlying mechanism of the minimum inhibitory concentration (MIC) of LFcins.

We have analyzed QSAR of a total 71 LFcin derivatives reported in the literature, including Strøm's peptides, by using artificial neural networks (ANN) for regression analysis, of which the input variables were screened by using PCS and HSA. It was found that positions 4 to 9 were the most important in determining the MIC, followed by cationic charge pattern at positions 4 to 9 as well as 1 to 3, in this order. The reference peptide used in this study was ^{17}FKCRRWQWR MKKLGA^{31} of bLFcin. Thus, it was concluded that pattern similarity analysis of segments within the sequences was more useful than that of the property variables of entire sequences (14,15) in screening the important distribution attributes of amino acid residue properties to the QSAR. In our study, it was found that using

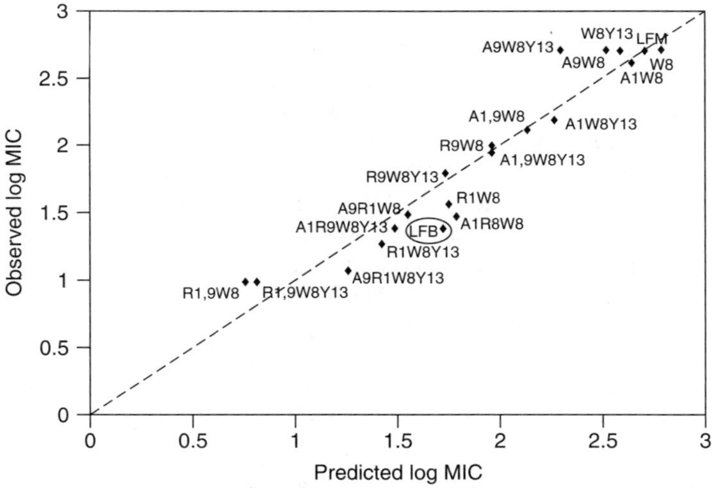

FIGURE 28.1 Observed vs. predicted antibacterial activity against *Eschrichia coli* of all of the 19 different murine lactoferricin derivatives and bovine lactoferricin. The predicted values were derived from the QSAR study using PLS for regression by Strøm et al. (From MB Strøm, Ø Rekdal, W Stensen, JS Svendsen. *J. Pept. Res.*, 57, 127–139, 2001. With permission.)

the currently best scale alone for each side-chain property was adequate to conduct a correct QSAR study (Figure 28.2) (7). The critical differences between Figures 28.1 and 28.2 are the differences in the reliability of the observed values of MIC. In the case of a smaller number of observations made in the same laboratory, as seen in Figure 28.1, it is easy to get the higher correlation computed. The high computed correlation for many more observed MIC data reported from different laboratories, as shown in Figure 28.2, should be attributable mainly to the efficient screening of highly influential predictor variables, which were measured for functionally important segments within sequences.

In a discussion of the convenience in the use of PLS in QSAR, the importance of cross-validation in avoiding overfitting was emphasized (17). Because of the substitution of polynomial variables with hidden variables, the PLS is regarded as a linear algorithm. However, ANN is more flexible with many more layers of hidden variables, thus being more suitable for nonlinear fitting than PLS (18,19). Cronin and Schultz (20) compared multiple linear regression (MLR) analysis, PLS regression, and regression ANN for the same toxicological data and found further support for the superiority of ANN over PLS as well as MLR.

28.3 FUNCTIONAL SEQUENCE ANALYSIS

In the recent review on protein sequence analysis in silico, Michalovich et al. (21) described the methodology for the transfer of the functional annotation of known

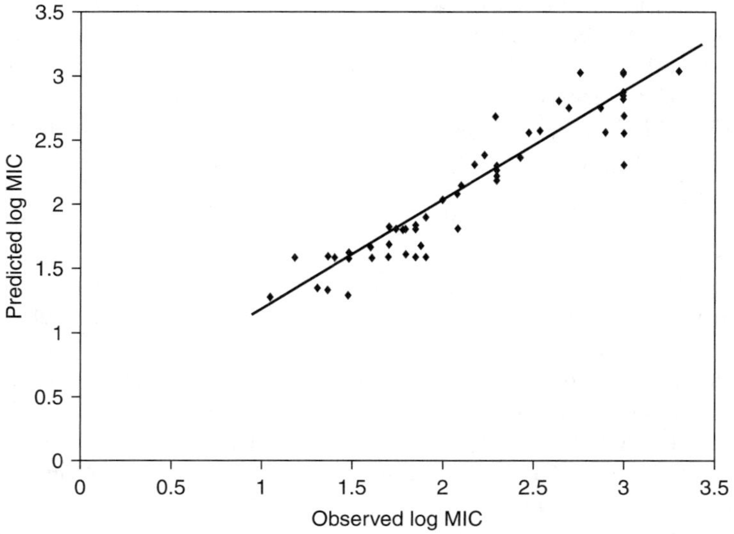

FIGURE 28.2 Predicted vs. observed log MIC values against *E. coli* of 65 different lactoferricin derivatives as regressed by using ANN (From S. Nakai, J.C.K. Chan, E.C.Y. Li-Chan, J. Dou, M. Ogawa. *J. Agric. Food Chem.*, 51, 1215–1223, 2003. With permission.).

proteins to a novel protein. Computer-assisted technology searches a well-maintained and annotated source database and assigns similarity. Sequence-based search and profile-based search are conducted using BLAST and PSI-BLAST (www.ncbi.nlm.nih.gov/BLAST/), respectively; the "hidden Markov model," that is, HMM (for instance, in PFAM, see www.sanger.ac.uk/Software/Pfam/), is more powerful for searching for a family more distant than the family acceptable to the BLAST. Structure-based annotation is conducted using a combination of PSI-BLAST and GenThreader (21). As a result, hypotheses to implement biological functions become available for many previously uncharacterized genes where the biological or molecular function is still unaccountable.

In contrast to this approach, which uses peptide sequence analysis based on the principle of evolutionary conservation, a direct approach such as peptide QSAR has been used. A crucial difference between the two approaches is the necessity of information on the 3D structure in the case of the former; this information is helpful, but not indispensable, in the case of peptide QSAR by substituting the 3D structure with simpler steric parameters to account for the functional mechanism. Because of our discovery of the importance of functional motifs within peptide sequences (7), we have attempted a mechanism study of the emulsifying capacity of peptides as a function, based on hydrophobicity distribution, as suggested by Giuliani et al. (22). In addition to homology similarity analysis (HSA), new software of homology similarity search (HSS) was introduced as part of the HSA software (23). This HSS algorithm is similar to the HSA algorithm as far as the basic element for computation is concerned, which is the computation of (a) the pattern similarity constant, and (b) the average value of the physicochemical properties of side chains. The pattern similarity is computed against functional segments within the reference sequence. The HSS program searches the distribution of functional motifs within sequences.

As the reference of HSS search for the emulsifying ability of peptides with the 10 to 32 residues listed in Table 28.1, synthetic peptide S with sequence ELELELELELELELEL of Saito et al. (24) was used because among the total 15 surface-active peptides used for computation, it showed the highest emulsifying activity index (EAI) value of 61. A difficulty in screening predictor variables from databases in QSAR work is the lack of objectively comparable functional response variables; in this case, it is the emulsifying ability data. Although the most popular EAI was chosen in this study for conducting a regression analysis, the effects of the concentration of protein in solution used for oil emulsification (0.1 to 2%) and pH could not be ignored. Excellent emulsification data of β-lactoglobulin (25) did not report an EAI, but a comparable absorption rate was computed from surface tension measurement, which should be closely related to emulsification activity. Although there is a close relationship between the surface hydrophobicity of proteins and their surface absorption, it may not immediately provide reliable EAI values for comparison with other peptides.

As a search unit, ELE (positions 1 to 3 of #1 in Table 28.1) was employed, thereby yielding polar/apolar periodicity demonstrated on the HSS patterns (Figure 28.3). As shown in A1 (#1) and A2 (#2) in the pattern similarity curves of Figure 28.3, a regularity in polar/apolar alternate distribution is required for good emulsification.

TABLE 28.1
Short-Chain Peptides with 10 to 32 Residues

Peptide	Sequence	Emulsification	Absorption	Reference
1. Peptide S	E L E L E L E L E L E L E L E L	EAI 61 (0.1%)	Abs(500).372	[8]
2. Peptide H	L E E L L E E L L E E L L E E L	" " " 27 (0.1%)	" " " .157	[8]
3. Peptide R	L E L L E E E L L E E E L L E L	" " " 12 (0.1%)	" " " .071	[8]
4. (O-4632 SIGMA)	Y Q E A F R R F F G P V	" " " 3.5 (0.1%)	Abs(500).020	This study
5. (F-9145 SIGMA)	H H L G G A K Q A G N V	" " " 7.5 (0.1%)	" " " .043	This study
6. (S-7152 SIGMA)	S F L L R N P N N K Y E P F	" " " 10 (0.1%)	" " " .055	This study
7. Salmine	P R R R R S S S R P I R R R R P R R A S R R R R R R G G R R R R	" " " 7 (0.1%)	" " " .04	This study
8. Bovine lactoferricin	F K C R R WQ W R MK K L G A P S I T C V R R A F	" " " 14 (0.1%)	" " " .08	This study
9. beta-Lg (21–40)	S L A M A A S D I S L L D A N S A P L R		AbsRate ~.8	[16]
10. (41–60)	V Y V E E L K P T P E G D L D I L L Q K		" " " 1.27	[16]
11. (61–70)	WE N G E C A N K K		" " " <.1	[16]
12. (149–162)	L S F N P T Q L E E N C H I		" " " <.1	[16]
13. alpha$_{s1}$CN(1–23)	R P K H P I K H Q G L P Q E V L N E N L L R F		" " " 10 (1.0%)	[15]
14. beta-CN(1–25)	R E I E E I N V P G E I V E S L S S S E E S I T R		" " " 10 (2.0%)	[14]
15. beta-CN(193–209)	Y Q Q P V I G P V R G P F P I I V		" " " 8 (2.0%)	[14]

Abbreviations: EAI, emulsification activity index; AbsRate, absorption rate computed from surface tension
(From S. Nakai, N. Alizadeh-Pasdar, J. Dou, R. Buttimor, D. Rousseau, A. Paulson (23). With permission.)

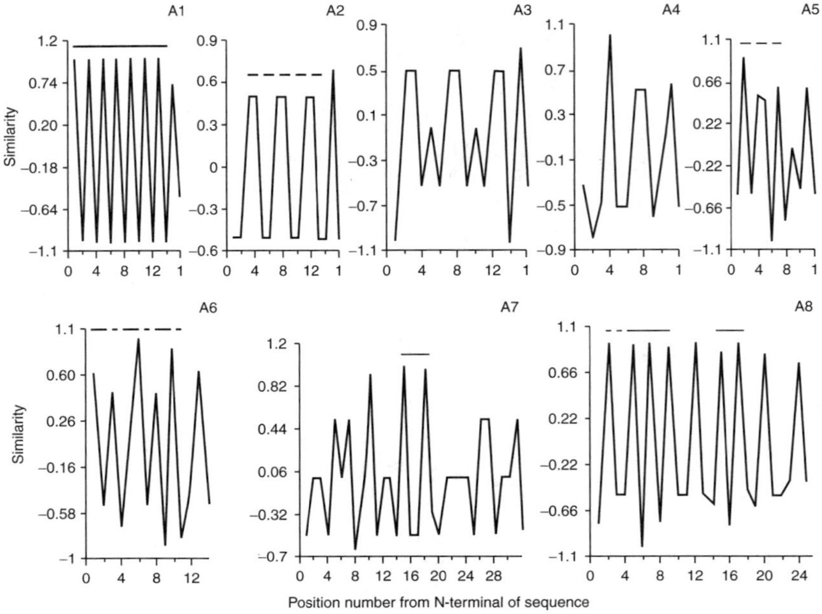

FIGURE 28.3 HSS patterns of peptides emulsification using hydrophobicity index. A1: peptide S (#1 in Table 28.1), A2: peptide H (#2), A3: peptide R (#3), A4: Sigma O-4632 (#4), A5: Sigma F-9145 (#5), A6: Sigma S-7152 (#6), A7 salmine (#7), A8: bovine lactoferricin (#8). Horizontal line at the top of each pattern is to show the hydrophobicity similarity density (periodicity). (From S. Nakai, N. Alizadeh-Pasdar, J. Dou, R. Buttimor, D. Rousseau, A. Paulson (23). With permission.)

The lines drawn at the top of the hydrophobicity similarity patterns show the location of high (solid line) and medium (broken line) hydrophobic periodicity estimated by visual comparison. Peptides A3 and A4 in Figure 28.3 show too broad intervals between the main similarity peaks in peptide R (#3) and Sigma peptide O-4632 (#4). In Figure 28.3, A7 and A8 compare two cationic antimicrobial peptides (CAP), that is, salmine (#7) and bovine lactoferricin (#8), respectively. Salmine suffers from irregular similarity compared with bovine lactoferricin. It has been hypothesized that the highly periodic polar/apolar segments alone can neatly predict the emulsifying capacity of peptides without taking other properties into account (Figure 28.4), except for minor roles being played by compressibility and charge (23). Good relationships between the estimated crude hydrophobic periodicity and the emulsifying ability of three Sigma peptides (#4-#6) and four peptides selected from the hydrolyzates of β-lactoglobulin (#9 to #12 not shown in Figure 28.4 due to unavailability of EAI values) were also separately observed (23). Under the circumstances of this study, we had to have some reservations regarding the reliability of the correlation result shown in Figure 28.4. This is due to inadequate confidence in the quantitative estimates of hydrophobic periodicity values representing a "similarity

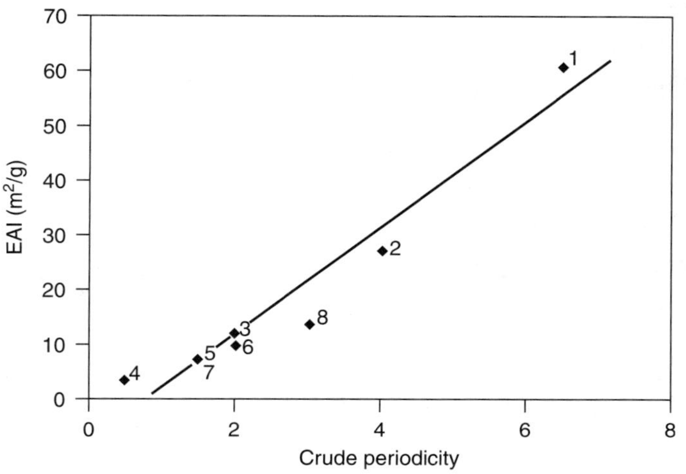

FIGURE 28.4 Correlation of EAI value with visual estimate of periodicity of hydrophobicity distribution of peptides as shown in Figure 28.3. (From S. Nakai, N. Alizadeh-Pasdar, J. Dou, R. Buttimor, D. Rousseau, A. Paulson (23). With permission.)

density" (similarity constants/residue) of a segment with uniformity in high similarity constants as well as comparable EAI values. The similarity density may not be extremely difficult to define quantitatively. Also, with the advent of easier, low-cost peptide synthesis that has become available recently, larger-scale experiments to support or deny the hypothesis proposed in this study are required soon.

This finding of the critical role of regularity in the polar/apolar cycle in the peptide emulsification may be important, since it may suggest that the location, distribution, or sometimes periodicity of active motifs is essential in the mechanism of a function in general. It may be more realistic when the same HSS software was found useful for the identification of active and binding sites in unknown peptide sequences. Examples were lysozymes and cystatins (8, 9). Using positions 54–57 (34–37) and 80–83 (51–54) of hen lysozyme (Table 28.2) as reference points, the search for the active sites was conducted. The position numbers in parentheses are the positions in individual enzyme sequences, whereas the numbers preceding those digits are position numbers of the multiple sequence alignment (see Table 28.2). So many algorithms have been listed on the World Wide Web for the multiple sequence alignment that selection of the the best algorithm for a specific study is extremely difficult. The greatest problem is whether ungapped (Table 28.1) or gapped (Table 28.2) alignment should be used. Due to the least involvement of evolutionary conservation, ungapped sequences were intentionally preferred for our peptide emulsification study (23) as shown in Table 28.1. Another crucial point is the distance between sequence families, with difficult homology comparison of remotely separated families, which may immediately affect the accuracy of functionality prediction (Table 28.2). The PCS scattergram illustrates CH-lysozyme family departed from c-, g- and v- lysozyme families (9).

TABLE 28.2
Homology Patterns of Lysozymes Belonging to Four Families

```
                              K V F G R C E L A A A M K R H G L           D N Y
                              K V F E R C E L A R T L K R L G M           D G Y
            R T D C Y G N V N R I D T T G A S C K T A K P E G L S Y C G V S A S
M N I F E M L R I D E G L R L K I Y K D T E G Y Y T I G I G H L         L T K S
          D T S G V Q G I D V S H W Q G S I N W S S V K S A G M S F A Y I K A T
                  |                   |                   |                   |

R G Y S L G N W V C A A K F E S N F N T                 Q A T N R N T   D G S T
R G I S L A N W M C L A K W E S G Y N T                 R A T N Y N A G D R S T
K K I A E R D L Q A M D R Y K T I I K K V G E K L C V E P A V I A G I I S R E S
P S L N A A K S E L D K A I G R N C N G           V I T K D E A E K L F N Q D V
E G T N Y K D D R F S A N Y T N A Y N A G   I I R G A Y H F A R P N A S S G T A
                  |                   |                   |                   |
                 50

D Y G I L Q I N S R W W C N D G R T   P G S R N L C N I P         C S A L L S S
D Y G I F Q I N S R Y W C N D G K T   P G A V N A C H L S         C S A L L Q D
H A G K V L K N G W G D R G N G F G L M Q V D K R S H K P Q G T W N G E V H I T
D A A V R G I L R N A K L K P V Y D S L D A V R R C A L I N     M V F Q M G E T
Q A D Y F A S N G G G W S R D N R T L P G V L D I E H N P S G A M C Y G L S T T
                  |                   |                   |                   |
                                     100

D I T A S V N           C A K K I V S D
N I A D A V A           C A K R V V R D
Q G T T I L I N         F I K T I Q K K F P S W T K         D Q
G V A G F T N             S L R M L Q Q                     K R
Q M R T W I N D F H A R Y K A R T T R D V V I Y T T A S W W N T C T G S W N G M
                  |                   |                   |                   |
                                                         150

                                                   G  N  G  M  N  A  W  V  A  W  R  N  R  C  K  G  T  D
                                                   P  Q  G  I  R  A  W  V  A  W  R  N  R  C  Q  N  R  D
                        Q                       L  K  G  G  I  S  A  Y  N  A  G  A  G  N  V  R  S  Y  A  R  M  DI
                                          W  D  E  A  A  V  N  L  A  K  S  R  W  Y  N  Q  T  P  N  R  A  K  R
AA K  S  P  F  W  V  A  H  W  G  V  S  A  P  T  V  P  S  G  F  P  T  W  T  F  W  Q  Y  S  A  T  G  R  V  G  G  VS
                        |                             |                             |                          |
                                                                                                             200

      V Q A  W I R G C  R                           Hen 129
      V R Q  Y V Q G C  G                           Human 130
G T T H D D  Y A N D V  V A R  A Q Y  Y K Q  H G Y  Goose 185
    V I T T  F R T G T  W D A  Y K N  L             phargeT4 164
G D V D R N  K F N G S  A A R  L L A  L A N  N T A  Streptomyces coelicolor 164
                |                                |
```

Highlighted: active site
Underlined: binding site

From S. Nakai, E.C.Y. Li-Chan, J. Dou (9). In preparation with permission.

Since the peptidoglycan-lysing activity is in the glutamic 55(E^{35}) and aspartic 81(D^{52}) side chains in hen lysozyme, the segments flanking these positions in the sequences were the focus of homology similarity search (HSS). The HSS using charge and turn (loop) indices was successful in finding the binding sites in the lysozymes of hen, human (Figure 28.5A), and goose and in T4 as well as CH type lysozymes (*Streptomyces coelicolor* 217) (see Figure 28.5B). The potential substrate binding sites should have high similarity constant and average value of hydrogen-bonding strength, as shown with arrow signs in the HSS pattern. This new CH-type lysozyme had a unique β/α-barrel fold unlike the common helix/strand double domain fold of other lysozyme families (26); therefore, a self-search using its own sequence, different from other lysozymes, was used. The reader may notice

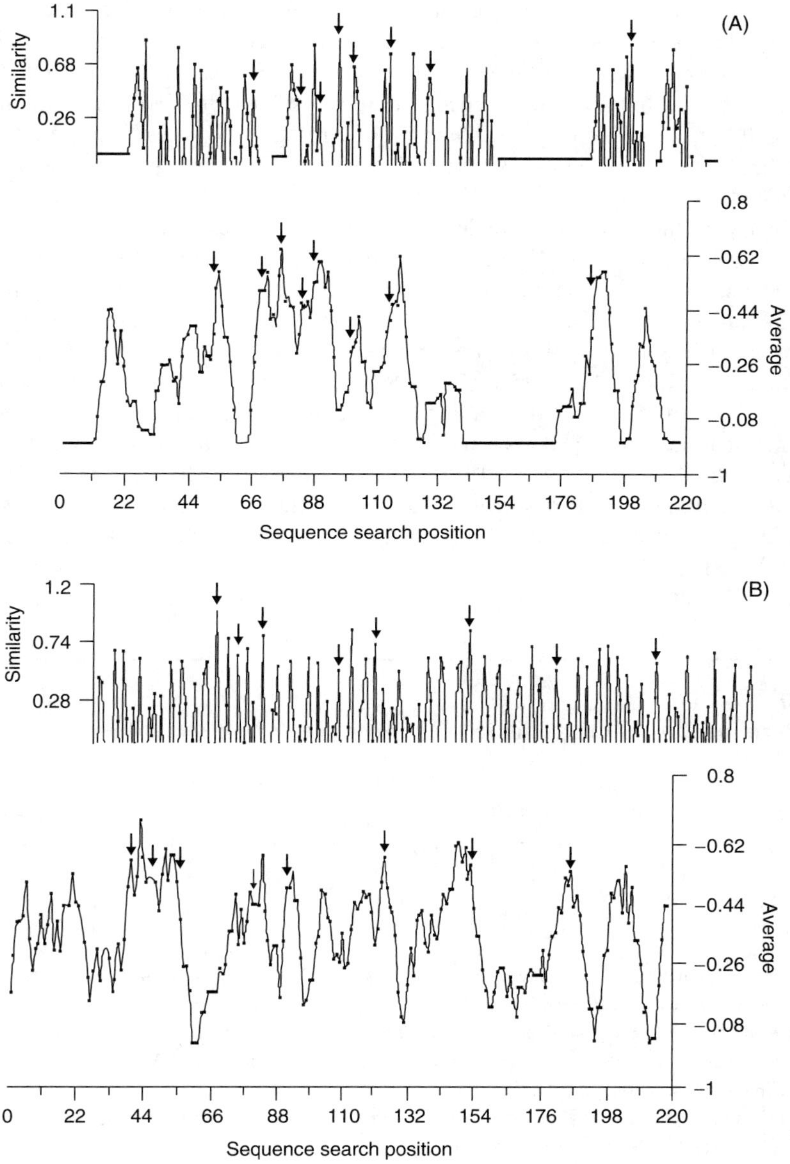

FIGURE 28.5 HSS patterns of substrate-binding sites. A: Human lysozyme against hen lysozyme using the segment at alignment positions 84–89 (see Table 28.2) as the reference. B: *Streptomyces* lysozyme using the segment at alignment positions 40–45 (see Table 28.2) as the reference (self-search). Arrows show the alignment positions where similarity constant and average hydrogen-bonding strength are both high, thus being most likely to be the binding sites. Abscissa scale is the alignment positions. (From S. Nakai, E.C.Y. Li-Chan, J. Dou (9). With permission.)

that the HSS patterns of the substrate binding site are intentionally selected for Figure 28.5 and not those of the active sites. This is because of the higher similarity in active site segments, which appear in the HSS patterns, compared with loosely defined binding site segments, despite the fact that both are extremely important in the functional mechanism of bioactive peptides. The results of searches for the active sites are shown in Table 28.3. The two active sites of a known sequence of hen lysozyme, ^{34}F**E**SN^{37} (54–57) and ^{51}T**D**YG^{54} (80–83), were used as references in the search (see Table 28.2). The active sites identified by HSS for human, goose, and T4 lysozymes are exactly the same as reported in the literature.

The sequence of a new lysozyme fold of Rau et al. (26) was used assuming that it was unknown. Due to a new fold, the references used for other lysozymes were not efficient in identifying the active sites of this new lysozyme. Therefore, alignment positions 40 to 45 (^{35}T**E**GTNY^{40}), instead of hen's 84–89 (I^{55}–S^{60}), were used as the reference for binding-site search. This selection was made due to the fact that the binding site is found near the active site, that is, 41E^{36} (26). Four other positions — 47, 55, 92, and 124 — were also used, as shown in Figure 28.5B (only the best examples are shown) in addition to three less potential sites for self-searching to accurately identify other sites in good agreement with the reported sites (26). In reality, if this sequence was totally unknown, the search would be repeated by altering the reference segment until the most reasonable site would be

TABLE 28.3
Determination of Active Sites in Sequences of Lysozymes in Different Families

	Hen		Human		Goose	
Position	55(35)E	81(52)D	55(35)E	81(53)D	79(73)E	103(97)D
Charge 1	1.0/5.0	0.98/5.2	1.0/5.0	0.98/5.2	0.67/6.9	0.83/7.3
Charge 2	0.98/5.0	1/0/5.2	0.98/5.0	1.0/5.2	0.78/6.9	0.83/7.3
Turn 1	1.0/1.12		0.98/1.17		0.06/1.03	
Turn 2		1.0/1.21		1.0/1.22		0.62/.92

The first position number is the alignment position shown in Table 1. Position numbers in brackets are those in the sequence of each lysozyme.
Active sites 1 and 2 : Sites 54-57 and 80-83 (alignment position numbers in Table 1) of hen lysozyme were used as references. These position numbers correspond to 34-37 and 50-54, respectively in the sequence of hen lysozyme.
The first digit in each cell is similarity constant and the second digit is average values of side chain charge index and propensity scale value for turn.

	T4		CH-		
Position	11(11)E	20(20)D	14(9)D	104(98)D	106(100)E
Charge 1	0.61/4.5	0.55/5.4	0.98/5.2	0.74/4.5	0.95/5.6
Charge 2	0.55/4.5	0.55/5.4	1.0/5.2	0.65/4.5	0.86/5.6
Turn 1	0.76/1.16				
Turn 2		0.53/1.07	0.95/.95	0.81/.92	0.87/.97

(From S. Nakai, E.C.Y. Li-Chan, J. Dou (9). In preparation. With permission.)

found, thus, this HSS search could not be totally useless as it could be used for finding candidate sites in the unknown lysozyme. However, it may be worth noting that although the active sites have been frequently reported, literally no publication exists on the binding sites for most enzymes. The author hope that a simple computer-technique, despite its potential for lack of reliability, may be useful for identifying the functions of unknown proteins. This could be a search technique of some sort, which would be a start in the right direction, especially considering that no information on the subject is available at present.

The HSS technique was applied also to cysteine proteinases (human cathepsins B, S, K, L, and H and papain) and their inhibitors, namely, cystatins (human A, B, and C and hen egg white cystatins) using stefin B-papain complex as a reference (9). Six binding sites in cathepsin sequences and three zones — an N-terminal inhibitory site and two binding loops — of cystatin sequences were used for computing pattern similarity constants. Structural differences of cathepsin H, physiologically important cathepsin B from other cathepsins, and inhibitory activity of human stefins A and B against cathepsins B and H, better than those of other cystatins, could be found by using this new approach. Principal component similarity and homology similarity analyses also explained the mechanism of reduced amyloidosis of mutant human cystatin C, which was due to a decline in strand propensity with a concomitant increase in helix propensity.

28.4 OPTIMIZATION

In addition to the QSAR approach, optimization could be a powerful tool for functional proteomics. We have been successful in applying our random-centroid optimization (RCO) to site-directed mutagenesis (3–5). Single-site mutations were conducted in the 16 amino acid residues of the active-site helix (G139-Y154) of *Bacillus stearothermophilus* neutral protease. After 13 mutations, it was found that the substitution V143E substantially improved the thermostability by 6.5°C of half-life temperature along with simultaneous 32% activity increase of the recombinant enzyme (4). Meanwhile, a simultaneous double-site mutation of human cystatin C sequence with 120 residues improved the papain inhibitory activity fivefold in G12W/J86V by avoiding insolubilization during expression by *Pichia pastoris* as well as the purification process caused probably by amyloidosis (5). The greatest advantage of RCG (RCO for genetics) is finding the best mutants with minimum efforts in terms of number of mutants to be produced and time for laborious, and thus costly, mutation.

RCO has already been successfully applied in food formulation (27). Recently, during a study of the activity of whey proteins against the invasion of HeLa cells by *Salmonella typhymurium*, the effects of mixing α-lactalbumin (La), β-lactoglobulin (Lg) and lactoferrin (Lf) in cow's milk were investigated (28). It was found that mixing the former two whey proteins, which were proven to be anti-invasive independently, interfered with the antiinvasive activity of lactoferrin (Figure 28.6). It is worth noting that the trend curves of La and Lg are negative when mixed with Lf but are positive when used alone. These RCO maps show that there are almost no synergistic effects that can be expected when La and Lg are mixed with Lf. This

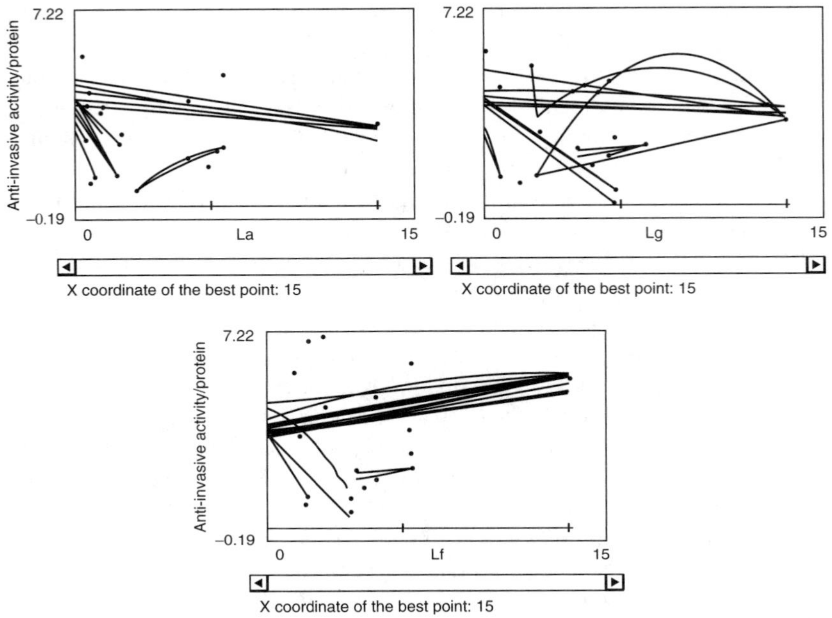

FIGURE 28.6 RCO maps of antiinvasive constant (antiinvasive activity mg^{-1} protein) vs. protein concentration ($mg\ mL^{-1}$). La: α-lactalbumin, Lg: β-lactoblobulin, Lf: lactoferrin. Disrupting HeLa cell membrane released the internalized bacteria, and the bacterial colony forming units (CFU) were enumerated and compared with the negative control (without sample proteins). The results were recorded as percentage invasion: Antiinvasive activity % = [100 − 100(CFU of internalized bacteria in protein/Control CHU)]. (From A. Madadi. Anti-invasive activity of bovine colostrums and whey proteins against *Salmonella typhymurium.* [Ph.D. thesis] Vancouver: University of British Columbia, 2003. With permission.)

result was derived from 27 reports of mixing as well as from other available data. RCO has a broad application in many different fields of scientific and even nonscientific studies, although it is rather empirical and not strongly theory based. A difficulty in the broader application of software such as RCO is the fact that most of the scientists, especially researchers, do not realize its substantially improved efficiency until they have applied the technology in their projects. In the case of random designs, there is a great chance of encountering novel, revolutionary discoveries. However, the randomization should be regulated, as in RCO, to avoid wasting effort and time. Figure 28.6 is just a recent example of many cases where an optimization plan was not employed.

28.5 RATIONAL DESIGNING STRATEGY

Excellent review articles on protein therapeutics (29) and on drug-binding sites in proteins (30) have been published. Because of longer intake and longer-lasting effects of foods, safety regulations imposed on food processing are much stricter than those for drug production. Modifications of food proteins, whether they are chemical or genetic, should be restricted to the minimum. All techniques discussed in this chapter are based on the utilization of endogenous proteins in plant and animal food, with or without the slightest modification, hopefully with only physical treatments that do not alter the protein sequences. Thus, the adverse side effects of many drugs can be minimized. The readers may recognize that the software described here are employed in only the smallest modification required for native protein sequences or are used only for the elucidation of functional mechanisms. These software are available for downloading from the website (ftp://ftp.agsci.ubc.ca/foodsci/), along with computing services through online requests at (shuryo.nakai@ubc.ca), if so desired.

Since RCO has had successful application in food formulation, it is possible to also use it for other mixture design purposes (27). However, the most important aspect of a rational design strategy in the post-genome era is matching the progress in genomics with the quick developments in mutation technology. A recent revolutionary discovery is the utilization of human peptides and proteins of human origin and their mutants in food formulation, mainly for prophylactic purposes. Recombinant human lactoferrin (rhLF) as well as human cystatin C [5] are good examples. The greatest advantage of this approach is the production costs; according to our calculations, the price of 1 liter of culture media for genetic fermentation is less than a dollar. rhLF with the least allergic reaction can be produced in transgenic plants, animals, and microbes at reasonable costs (31); furthermore, rhLF is multifunctional, that is, it is cytoproliferative, immunostimulatory, antiangiogenic, antiviral (31), antiallergic, and even probiotic (32). However, its major effect is the upregulation and release of interleukin (IL)-18 in the gut (33; also www.agennix.com). As lactoferrin is a multifunctional protein, functioning as a key component of mammalian host defense, there is a great chance of developing a new food based on the mechanism of functions of this proteins (34). One of the most interesting topics with regard to lactoferrin is the difference in functional mechanisms between bovine LF and human LF. The greatest puzzle is why rhLF is more functional than bLF when it has been widely publicized that the antimicrobial effects of bLFcin are stronger than those of hLFcin (35). Apparently, one key role is the difference in cytokine stimulation, which could be elucidated using functional genomics, including our new homology similarity approach, in the near future. In addition to the aforementioned rhLF and human cystatin C, human collagen is also being investigated for use in bone regeneration (36). It was reported that the introduction of protease inhibitory drugs drastically decreased the mortality and morbidity associated with acquired immune deficiency syndrome (AIDS) (37). It should be emphasized that these proteins are not "genetically modified proteins," to which consumers have a strong objection; instead, they are genetically produced natural or least modified human proteins, potentially least likely to be rejected by the human body. Furthermore, increasing numbers of clinical reports are appearing in the literature to demonstrate the cost-advantage of natural peptides and proteins over expensive drugs; in addition, the natural peptides and proteins have fewer side effects less likelihood of developing resistance. Therefore, the prevention of diseases with prophylactic foods may become a realty in the years to come.

The approaches — based on given 3D structure — to identify binding pockets, including geometric analyses of molecular surfaces, comparisons of molecular structures, similarity searches in databases of protein cavities, and docking scans to reveal areas of high-ligand complementarity, have been reported (30). However, even when using the most modern computing system, docking two proteins may take an hour of computation (29). Our approach has a simpler in algorithm and faster computation on a PC, and thus, computation can be completed in minutes, despite the fact that this approach is rather empirical.

28.6 CONCLUSIONS

A new protocol — multiple sequence alignment → PCS → HSA → (RCG) → ANN — is effective and useful in the rational design of food peptides. This protocol is a revised version of the previously published QSAR protocol [6] and incorporates a new computer program HSA (file name of the program), including HSS (file name) approach. After computing the sequence alignments, the PCS classifies them, thereby facilitating screening of functionally important samples, and the HSA segregates segments in the sequences; this is followed by regression ANN analysis to compute the structure–activity relationships. The RCO, independently or as a summary of the above protocol, can be used for optimizing the design plan of bioactive peptides.

ACKNOWLEDGMENT

This work was financially supported by a Multidisciplinary Network Grant (MNG) Program entitled "Structure-Function of Food Macromolecules" (Dr. Rickey Y. Yada of University of Guelph as the principal investigator) from the Natural Sciences and Engineering Research Council of Canada.

REFERENCES

1. A.L. Parrill, M.R. Reddy (Eds.). *Rational Drug Design: Novel Methodology and Practical Application*. ACS Symposium Series 719. Washington, DC: American Chemical Society, 1999.
2. A.J. Stuper, W.E. Brugger, P.C. Jurs (Eds.). *Computer Assisted Studies of Chemical Structure and Biological Function.* New York: John Wiley & Sons, 1979.
3. S. Nakai, J. Dou, K.V. Lo, C.H. Scaman. Optimization of site-directed mutagenesis I. New random-centroid optimization for Windows useful in research and development. *J. Agric. Food Chem.*, 46: 1642–1654, 1998.
4. S. Nakai, J. Dou, S. Nakamura, C.H. Scaman. Optimization of site-directed mutagenesis. II. Application of random-centroid optimization to one-site mutation of

B. stearothermophilus neutral protease to improve thermostability. *J. Agric. Food Chem.*, 46: 1655–1661, 1998.

5. M. Ogawa, S. Nakamura, C.H. Scaman, H. Jing, D. Kitts, J. Dou, S. Nakai. Enhancement of proteinase inhibitory activity of recombinant human cystatin C using random-centroid optimization. *Biochim. Biophys. Acta*, 1599: 115–124, 2002.
6. S. Nakai, M. Ogawa, S. Nakamura, J. Dou, K. Funane. A computer-aided strategy for structure-function relation study of food protein using unsupervised data mining. *Int. J. Food Prop.*, 6: 25–47, 2003.
7. S. Nakai, J.C.K. Chan, E.C.Y. Li-Chan, J. Dou, M. Ogawa. Homology similarity analysis of sequence of lactoferricin and its derivatives. *J. Agric. Food Chem.*, 51: 1215–1223, 2003.
8. Pattern similarity analysis based on side-chain properties in searching for functional motifs in protein sequences. I, Lysozyme families and II. Cystatine proteinase inhibiton by cystatins. These two articles were combined and published as shown in Reference 9.
9. S. Nakai, E.C.Y. Li-Chan, J. Dou. Pattern similarity study of functional sites in protein sequences: lysozymes and cystatins. *BMC Biochem.,* 9: pp. 1–17 at www.biomedcentral.com/1471-2091/6/9.
10. M.D. Schmidt. Structural proteomics: the potential of high-throughput structure determination. *Trends Microbiol.*, 10: S27–S31, 2002.
11. S. Nakai. Modelling protein behaviour. In: R. Yada (Ed). *Protein in Food Processing.* Cambridge, UK: Woodhead Publishing, 2004, chapter 11. 245–269.
12. C. Hansch, D. Hoekman, A. Leo, D. Weininger, D. Selassie. Chem-Bioinformatics: comparative QSAR at the interface between chemistry and biology. *Chem. Rev.*, 102: 783–812, 2002.
13. T.W. Schultz, M.T.D. Cronin, J.D. Walker, A.O. Aptula. QSARs in toxicology: a historical perspective. *J. Mol. Struct.* (*Theochem.*), 622: 1–22, 2003.
14. C. Hansch, A. Leo, E. Hoekman. *Exploring QSAR. Hydrophobic, Electronic, and Steric Constants.* Washington, DC: American Chemical Society, 1995. Oxford University Press USA, New York, ISBN 0-8412-2991-0.
15. S. Hellberg, M. Sjöström, B. Skagerberg, S. Rold. Peptide quantitative structure-activity relationships, a multivariate approach. *J. Med. Chem.*, 30: 1126–1135, 1987.
16. M.B. Strøm, Ø. Rekdal, W. Stensen, J.S. Svendsen. Increased antibacterial activity of 15-residue murine lactoferricin derivatives. *J. Pept. Res.*, 57: 127–139, 2001.
17. S. Wold, M. Sjöström, L. Eriksson. PLS-regression: a basic of chemometrics. *Chemom. Intel. Syst.*, 58: 109–130, 2001.
18. F.R. Burden, D.A. Winkler. A quantitative structure-activity relationships model for the active toxicity of substituted benzenes to *Tetrahymena pyriformis* using Bayesian–regulated neural networks. *Chem. Res. Toxicol.*, 13: 436–440, 2000.
19. G.A. Bakken, P.C. Jurs. QSARs for 6-azasteroids as inhibitors of human type 1 5α-reductase: prediction of binding affinity and selectivity relative to 3-BHSD. *J. Chem. Inf. Comput.*, 41: 1255–1265, 2001.
20. M.T.D. Cronin, T.W. Schultz. Development of quantitative structure-activity relationships for the toxicity of aromatic compounds to *Tetrahymena pyriformis*: comparative assessment of the methodologies. *Chem. Res. Toxicol.*, 14: 1284–1295, 2001.
21. D. Michalovich, J. Overington, R. Fagan. Protein sequence analysis *in silico*: application of structure-based bioinformatics to genomic initiatives. *Curr. Opin. Pharmacol.*, 2: 574–580, 2002.

22. A. Giuliani, R. Benigni, J.P. Zbilut, C.L. Webber Jr, P. Sirabella, A. Colosimo. Nonlinear signal analysis methods in the elucidation of protein sequence-structure relationships. *Chem. Rev.*, 102: 1471–1291, 2002.
23. R. Buttimor, D. Rousseau, A. Paulson. Pattern similarity analysis of amino acid sequences for peptide emulsification. *J. Agric. Food Chem.,* 52: 927–934, 2004.
24. M. Saito, M. Ogasawara, K. Chikuni, M. Simizu. Synthesis of a peptide emulsifier with an amphiphilic structure. *Biosci. Biotechnol. Biochem.*, 59: 388–392, 1995.
25. S. Turgeon, S.F. Gauthier, D. Molleé, J. Léonil. Interfacial properties of tryptic peptides of β-lactoglobulin. *J. Agric. Food Chem.*, 40: 669–675, 1992.
26. A. Rau, R. Hogg, R. Marquardt, R. Hulgenfeld. A new lysozyme fold: crystal structure of the muramidase from *Streptomyces coelicolor* at 1.65? *J. Biol. Chem.*, 276: 31994–31999, 2001.
27. J. Dou, S. Toma, S. Nakai. Random-centroid optimization for food formulation. *Food Res. Int.*, 26: 27–37, 1993.
28. A. Madadi. Anti-invasive activity of bovine colostrums and whey proteins against *Salmonella typhymurium.* [Ph.D. thesis] Vancouver: University of British Columbia, 2003.
29. Z. Weng, C. DeLisi. Protein therapeutics; promises and challenges for the 21st century. *Trends Biotechnol.*, 20: 29–35, 2002.
30. C. Sotriffer, G. Klee. Identification and mapping of small-molecule binding sites in proteins: Computational tools for structure-based design. *Il Farmaco* 57: 243–251, 2002.
31. Recombinant human lactoferrin. *Immunotherapy Weekly*, Nov.20, 2002 (www.newsrx.com/News/2002-11-20LW.html).
32. B.W.A. van der Strate, L. Beljaars, G. Molema, M.C. Harmsen, D.K.F. Meijer. Antiviral activities of lactoferrin. *Antivir. Res.*, 52: 225–239, 2001.
33. D. Tomé, H. Debabbi. Physiological effects of milk protein components. *Int. Dairy J.*, 8: 383–392, 1998.
34. P.P. Ward, S. Uribe-Luna, O.M. Conneely. Lactoferrin and host defense. *Biochem. Cell Biol.*, 80: 95–102, 2002.
35. A.S. Naidu. Lactoferrin, In: A.S. Naidu (Ed.). *Natural Food Antimicrobial Systems.* Boca Raton, FL: CRC Press, 2000, 17–102.
36. P.F. Wang, D.L. Kaplan. Genetic engineering of fibrous proteins: spider dragline silk and collagen. *Adv. Drug Deliv. Rev.*, 54: 1131–1143, 2002.
37. A. Wlodawer. Rational approach to AIDS drug design through structural biology. *Annu. Rev. Med.*, 52: 595–614, 2002.

29 Engineering Hen Egg-White Lysozyme

Akio Kato

CONTENTS

29.1 INTRODUCTION

Lysozyme is widely distributed in both the animal and plant kingdoms. Since lysozyme exists abundantly in chicken egg white and can be easily purified by crystallization, it has been investigated extensively and was the first enzyme for which the amino acid sequence was determined and the tertiary structure was elucidated by X-ray crystallographic analysis (1). The recent development of recombinant techniques has enabled the elucidation of the molecular mechanism of the structural and functional properties of lysozyme. In early studies, recombinant lysozyme was investigated in the *Escherichia coli* (*E. coli*) expression system. However, since the prokaryotic cells have a different secretion system from eukaryotic cells, the correctly folded lysozyme could not be expressed. The correct folding of disulfide-rich proteins such as lysozyme is difficult in *E. coli* because of the absence of endoplasmic reticulum (ER), in which proteins are post-translationally folded. Kumagai et al. (2) found that hen egg-white lysozyme was correctly processed and folded in *Saccharomyces cerevisiae* (*S. cerevisiae*). The

N-terminus of recombinant lysozyme is lysine, and its stability is the same as that of the native lysozyme. Since *S. cerevisiae* is a typical eukaryotic cell with an ER system, lysozyme is correctly processed and folded in the yeast. Thus, the relationship between the functional and structural properties of lysozyme has been investigated at a molecular level using protein engineering. Although the catalytic sites of lysozyme were confirmed by chemical modifications, these sites were easily reconfirmed by site-directed mutagenesis of responsible amino acids. The mutation of Glu_{35} to Asp or Asp_{52} to Glu in hen (3) or human (4) lysozyme resulted in inactivation of the enzyme. These observations indicate that the catalytic groups are strictly defined and that the replacement of even one methylene group is not allowed. Taniyama et al. (5) reported the folding mechanism of disulfide bond-deficient human lysozyme C77/95A mutant secreted in *S. cerevisiae*. Although the mutant that was deficient in other disulfide bonds could not be secreted, the only C77/95A mutant was secreted eightfold greater than wild-type in *S. cerevisiae*. The stability of the C77/95A mutant greatly decreased despite the correct folding.

These observations show that the disulfide bond Cys77–95 contributes to the stabilization of the folded form of human lysozyme. Thus, the relationship of the structural and functional properties of lysozyme has been clearly elucidated by genetic engineering. However, the ultimate goal of protein engineering is the molecular design and construction of a protein with a novel function. For this purpose, we used lysozyme as a model protein for designing a functionally novel protein. Our strategy was the enhancement of heat stability and bactericidal action. The former was performed with the hyperglycosylation in the yeast (*S. cerevisiae*) expression system, and the latter was achieved by genetic fusion of a hydrophobic peptide with the C-terminus of lysozyme. In addition to these dramatic changes in the functional properties of lysozyme by genetic modification, this chapter describes the molecular diseases derived from lysozyme such as allergy and amyloidosis and the genetic engineering of lysozyme that may contribute to the elucidation of the molecular mechanism for these diseases.

29.2 HYPERGLYCOSYLATION OF LYSOZYME BY GENETIC ENGINEERING

We have reported that a Maillard-type lysozyme polysaccharide conjugate showed dramatic improvement of heat stability and emulsifying property (6,7). Therefore, the dramatic improvement of the functional properties of protein by glycosylation to remodel proteins using genetic engineering is highly fascinating. Attempts were made to construct glycosylated lysozymes in a yeast expression system using genetic engineering (8,9). In yeast cells, the proteins having Asn-X-thr/Ser sequence were N-glycosylated in the ER, and the attached oligosaccharide chain could be elongated with further extension of a large polymannose chain in the Golgi apparatus, as shown in Figure 29.1. For this reason, we attempted to construct the yeast expression plasmid carrying

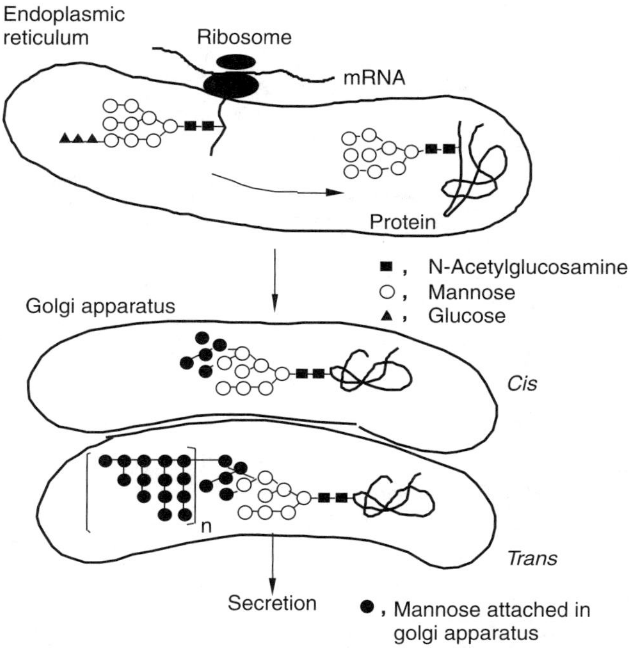

FIGURE 29.1 Glycosylation processing of protein in ER and Golgi apparatus in yeast. Glc3-Man9-(GlcNAc)2 oligosaccharide unit attaches to asparagines in protein. After carbohydrate processing, polymannosylation occurs in Golgi apparatus by polyammosyl transferase.

lysozyme cDNA (complementary deoxyribonucleic acid). The mutant cDNAs of hen egg-white lysozyme having N-glycosylation signal sequence (Asn-X-Thr/Ser) at positions 49, 67, 70, and 103 were inserted into the *Sal*-1 site of pYG-100, located downstream of the GPD promoter (Figure 29.2). The expression vectors were introduced into *S. cerevisiae* AH-22. The yeast *S. cerevisiae* carrying mutant lysozyme cDNAs was cultivated at 30°C for 5 days, and then glycosyl mutant lysozymes secreted in the medium were isolated by CM Toyopearl, Sephadex G-50, and concanavalin A-Sepharose column chromatography. Glycosylated lysozyme was obtained only from G49N mutant lysozyme, while it was not obtained from the mutants G67N, P70N, and M105T, as shown in Figure 29.3. It is probable that the N-linked glycosylation seems to be inhibited by the steric hindrance around positions 67, 70, and 103. The SDS–PAGE (sodium dodecyl sulfate–polyacrylamide gel electrophoresis) patterns indicated that a large molecular size of polymannosyl lysozyme was secreted and stained strongly with carbohydrate reagent (see right panel, Figure 29.3). A large molecular size of N-glycosylated lysozyme with a polymannose chain was predominantly expressed in the yeast carrying the G49N lysozyme expression plasmid. The secreted amount of polymannosyl lysozyme was much

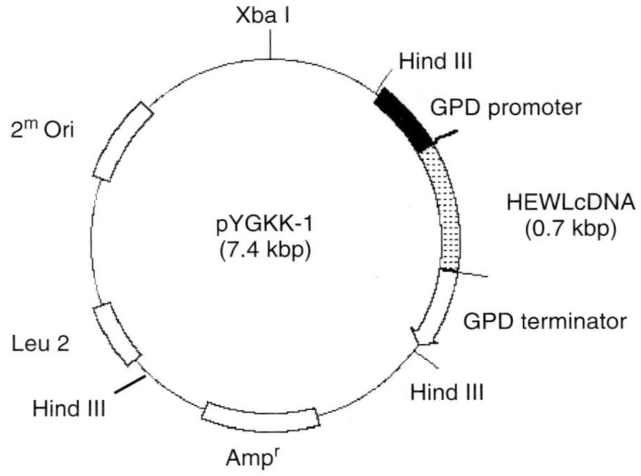

FIGURE 29.2 Insertion of lysozyme cDNA into pYG100 (pYGKK-1). Lysozyme cDNA is inserted downstream of GPD promoter.

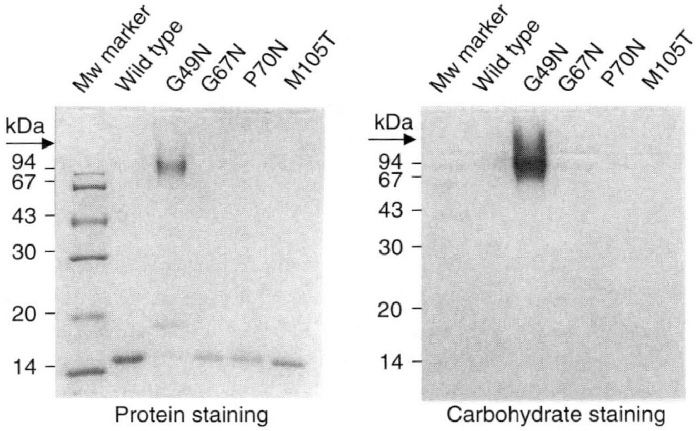

FIGURE 29.3 SDS-PAGE patterns of wild-type and mutant lysozymes. Left panel, protein staining; Right panel, carbohydrate staining. G49N, Glycine 49 was substituted with asparagines; G67N, glycine 67 was substituted with asparagines; P70N, proline 70 was substituted with asparagines; M105T, methionine 105 was substituted with threonine.

higher than that of the oligomannosyl lysozyme. The mutant G49N lysozyme was secreted in the two types of glycosylated forms, a small oligomannose chain ($Man_{18}GlcNAc_2$)-linked form and a large polymannose chain ($Man_{310}GlcNAc_2$)-linked form, according to carbohydrate analysis. Both types of glycosylated lysozymes were susceptible to endo β-N-acetylglucosaminidase cleavage of their carbohydrate chains. The average molecular weights of oligomannosyl and polymannosyl lysozymes were 18 and 71 kDa, respectively. The length of the polymannose chain was found to be 200 to 350 residues per molecule of lysozyme according to the estimation of molecular mass distribution by low-angle laser light scattering measurement. The protein conformation, estimated by circular dichroism (CD) analysis, was completely conserved in these glycosylated lysozymes. The enzymatic activities of oligomannosyl and polymannosyl lysozymes were 100% and 91%, respectively, of wild-type protein when glycol chitin was used as a substrate. Further, we were successful in constructing another polymannosyl lysozyme R21T, in which arginine 21 was substituted with threonine (9).

In addition to these mutants having a single-glycosylation site, a mutant (R21T/G49N) having a double-glycosylation site was also constructed. The oligomannosyl and polymannosyl lysozymes were secreted in the yeast carrying the cDNA of R21T, G49N, and R21T/G49N mutants. The carbohydrate contents of these glycosylated lysozymes are shown in Table 29.1. The carbohydrate content in double-glycosylated mutant R21T/G49N was almost the same as that in the single-glycosylated mutants G49N and R21T. This suggests that further extension of mannose residues may be suppressed in yeast cells. Although many polymannosyl proteins were known to secrete in *S. cerevisiae*, the number of polymannosyl residues was usually 50 to 150. Why was the length of polymannose residues of lysozyme much longer than that of general mannoproteins? The glycosylated lysozyme may be retained in the Golgi apparatus longer than other proteins, resulting in the hyperglycosylation. The hyperglycosylated lysozyme was an interesting novel protein constructed by genetic engineering. We could elucidate the role of carbohydrate chains in the functional properties of proteins using single polymannosyl and oligomannosyl lysozymes and double-glycosylated lysozyme.

TABLE 29.1
Carbohydrate Composition of Glycosylated Lysozymes

Mutant Lysozymes	Contents (moles/mole lysozyme)	
	N-acetylglucosamine	Mannose
G49N		
Polymannosyl	2	310
Oligomannosyl	2	18
R21T		
Polymannosyl	2	338
Oligomannosyl	2	14
R21T/G49N		
Polymannosyl	4	290

29.3 HEAT STABILITY OF OLIGOMANNOSYL AND POLYMANNOSYL LYSOZYMES

The enzymatic activity of glycosylated lysozymes during heating was measured as an indication of the apparent heat stability (Figure 29.4). As expected, the thermal stability of polymannosyl lysozyme was much higher than that of oligomannosyl lysozyme. No coagulation was observed in the polymannosyl lysozyme on heating up to 95°C, whereas the wild-type lysozyme completely coagulated at temperatures above 85°C. In addition, about 60% of the residual enzymatic activities were retained in the polymannosyl lysozyme after heating up to 95°C, while about 15% of residual enzymatic activities were retained in oligomannosyl lysozyme. This result suggests that the attachment of polysaccharide is much more effective for improvement of the thermal stability of proteins than that of oligosaccharide. It is probable that polysaccharide attached to lysozyme may confer the ability to stabilize the aqueous phase around the protein molecule and to protect the intermolecular interaction with unfolded molecules, thereby causing stabilization of the protein structure against heating.

To understand the role of polyglycosylation in protein stability, the thermodynamic changes in the denaturation of various polymannosyl lysozyme mutants (R21T, G49N, and R21T/G49N) constructed by genetic modification were analyzed using differential scanning calorimetry (DSC) (10). As shown in Figure 29.5, the denaturation temperature (Td, peak) and the enthalpy change (ΔH, peak area) for the unfolding of the lysozymes were reduced with an increase in the length of the polymannose chain and the number in the binding site to a protein; however, the polymannosyl lysozymes revealed apparent heat stability in that no aggregation was observed and the enzymatic activity

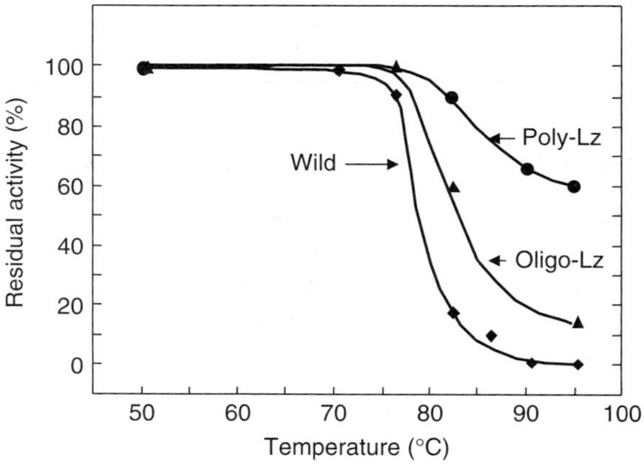

FIGURE 29.4 Thermal stability of glycosylated lysozyme (G49N) during heating. ◆, Wild type lysozyme; ▲, oligomannosyl lysozyme; ●, polymannosyl lysozyme.

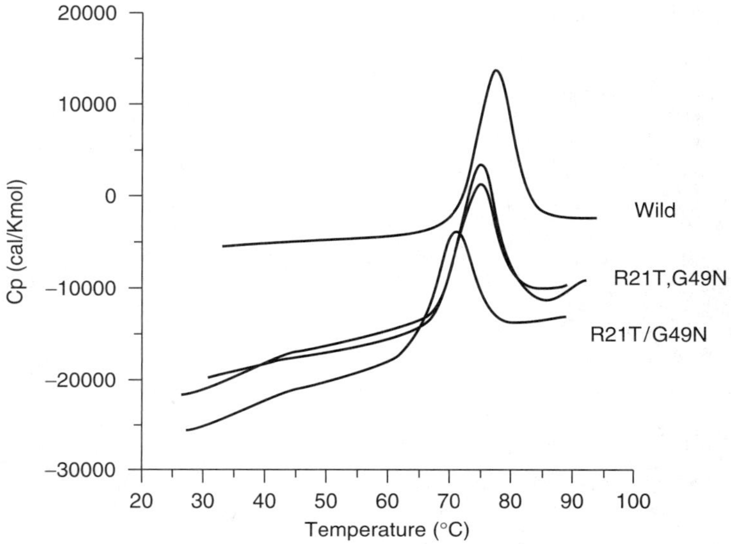

FIGURE 29.5 Typical excess heat capacity curves of polymannosyl lysozymes at pH4. Wild, wild-type lysozyme; R21T, mutant polymannosyl lysozyme in which arginine 21 was substituted with threonine; G49N, mutant polymannosyl lysozyme in which glycine 49 was substituted with asparagines; R21T/G49N, mutant double polymannosyl lysozyme in which arginine 21 and glycine 49 were substituted with threonine and asparagines, respectively.

was conserved under conditions in which the wild-type lysozyme coagulated. Despite destabilization of the lysozyme by attachment of the carbohydrate chain, the polymannosyl lysozymes revealed apparent heat stability. In order to elucidate the contradiction, the reversibility of the denaturation was investigated by DSC analysis, which compared the DSC curves of the unheated polymannosyl lysozymes with those of the heat denatured ones. The DSC measurement was again carried out using the sample solution cooled immediately after attaining or surpassing the Td. The DSC curves were drawn at pH 4, which was the critical point to maintain the solution without aggregation during the denaturation. As shown in Figure 29.6, the peak areas of the DSC curve for the wild-type lysozyme was significantly reduced by reheating after denaturation (see Figure 29.6, top). On the other hand, the DSC curves of the polymannosyl lysozyme (G49N) overlapped by reheating after attaining the Td (see Figure 29.6, middle), and that of the double mannosyl lysozyme (R21T/G49N) also overlapped by reheating after denaturation (see Figure 29.6, bottom). This result indicates that during the process of refolding, the polymannosyl chain contributes to the proper folding and enhances the

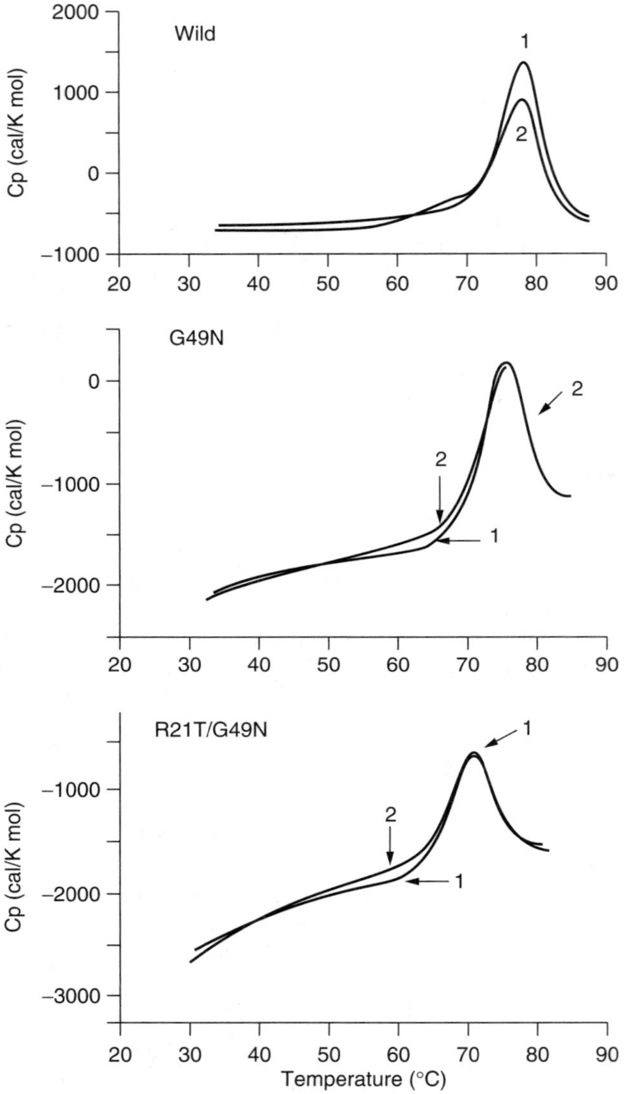

FIGURE 29.6 Reversibility of the denaturation of wild-type (top), G49N (middle), and R21T/G49N (bottom) lysozymes. DSC measurement was again carried out using the sample solution cooled immediately after attaining (middle) or surpassing (top and bottom) the Td. 1, first measurement of DSC; 2, second measurement of DSC after cooling the same solution of first measurement.

reversibility of refolding. Based on these results, the polymannosyl lysozyme seems to easily refold due to the excellent reversibility of denaturation, despite the decreases in the enthalpic stabilization caused by the strain in the protein molecule from the introduction of a polysaccharide chain.

29.4 EMULSIFYING PROPERTY OF OLIGOMANNOSYL AND POLYMANNOSYL LYSOZYMES

The emulsifying properties were greatly increased by the attachment of polymannose chains (Figure 29.7), as expected from the data of the Maillard-type lysozyme–polysaccharide conjugate mentioned above. The polymannosyl lysozyme constructed by genetic engineering showed much higher emulsifying properties than the oligomannosyl lysozyme (11,12). This suggests that the length of polysaccharide is very important and critical for the emulsifying property of protein–polysaccharide conjugates. It is interesting that the polymannosyl lysozyme shows better emulsifying properties than do commercial emulsifiers. The role of polysaccharide in the stabilization of emulsion is considered as follows. The hydrophobic residues in a protein molecule are anchored in the oil droplets during emulsion formation. The polysaccharide moiety orients in the aqueous outer layer of oil–water emulsion acting as hindrance layer as well as decelerator of migration of oil droplets for their coalescence due to high surface viscosity. Thus, it was evidenced, on a molecular basis, that the dramatic improvements of the thermal stability and emulsifying property of lysozyme are brought about by the attachment of polysaccharide, not of oligosaccharide.

The effect of the number of glycosylation sites in proteins on the functional properties of polymannosyl lysozyme is another subject of interest. This might be elucidated only by genetic protein glycosylation because the chemical glycosylation

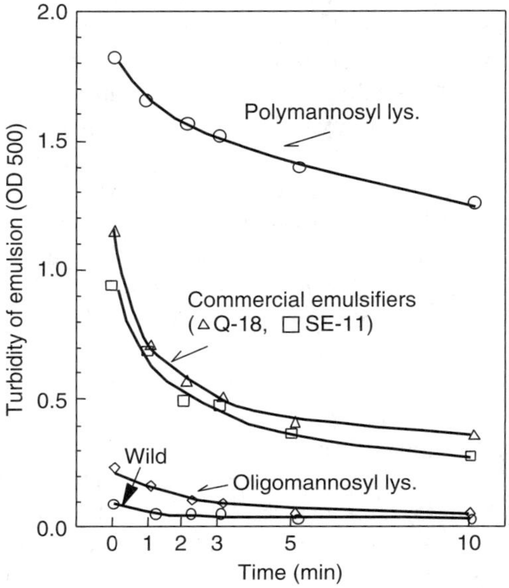

FIGURE 29.7 Emulsifying properties of oligomannosyl and polymannosyl lysozyme (G49N). The turbidity of emulsion homogenized a mixture of 1 mL corn oil and 3 mL water (0.1 % glycosyl lysozyme or emulsifier) at 12,000 rpm for 1min was followed for 10 min.

of proteins, including Maillard-type glycosylation, cannot control the number of glycosylation sites. In order to solve this problem, the single-polymannosyl mutants (R21T, G49N) and the double-polymannosyl mutant (R21T/G49N) were successfully constructed by genetic engineering. The double-polymannosyl lysozyme revealed much higher emulsifying properties than did the single-polymannosyl lysozyme (12). This suggests that the formation of a thick steric stabilizing adsorbed layer around the emulsion is further enhanced and the coalescence of oil droplets is further effectively inhibited when the number of glycosylation sites are increased.

Some novel and promising approaches to improve the functional properties such as heat stability and emulsifying properties of hen egg-white lysozyme by genetic engineering were described above. In order to understand the molecular mechanism of the dramatic improvement of the functional properties of lysozyme, oligomannosyl and polymannosyl lysozymes were constructed in a yeast expression system using genetic engineering. The importance of the size of the saccharide chain in the effective improvement of functional properties was proven by the genetic modification of lysozyme.

29.5 MODIFICATION OF ANTIMICROBIAL ACTION OF LYSOZYME

The bactericidal action of lysozyme is stronger against Gram-positive bacteria than against Gram-negative bacteria because in Gram-negative bacteria, the cell envelope consisting of lipopolysaccharide obstructs the access of lysozyme to the peptidoglycan layer. The covalent attachment of fatty acids (myristic and palmitic acids) to lyszoyme broadens the bactericidal action against Gram-negative bacteria (13). The lipophilized lysozyme facilitates the access to, and invasion of, the outer membrane to hydrolyze the peptidoglycan, thereby killing the Gram-negative bacteria. Based on this observation, to extend the action against Gram-negative bacteria, the molecular design of lysozyme to attach a hydrophobic pentapeptide for anchoring lysozyme in the outer membrane was attempted. First, attempts were made to attach the hydrophobic pentapeptide (Phe-Phe-Val-Ala-Pro) in milk casein to the C-terminus of lysozyme. The peptide readily forms β-strands of the same length as myristic acid. The ribbon model structure of the hydrophobic pentapeptide-fused lysozyme (H5) is shown in Figure 29.8. The H5 lysozyme revealed strong bactericidal action against *E. coli* (14). Different lengths of hydrophobic peptides were attached to the C-terminus of the hen egg-white lysozyme to determine the most effective length of the hydrophobic bactericidal peptides (15). The oligonucleotides encoding Phe-Val-Pro (H3), Phe-Phe-Val-Ala-Pro (H5), and Phe-Phe-Val-Ala-Ile-Ile-Pro (H7) were fused to the C-terminus Leu-129 of lysozyme cDNA. The reconstructed cDNAs were inserted into the yeast expression vector. The hydrophobic peptide-fused lysozymes were secreted in yeast carrying the reconstructed cDNA. The hydrophobic peptide-fused lysozymes retained 75 to 80% lytic activity of the wild-type protein. The bactericidal action of the tripeptide-fused lysozyme

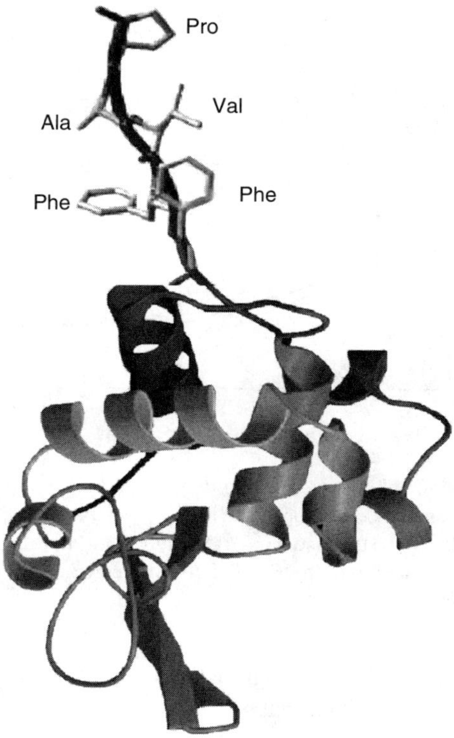

FIGURE 29.8 Ribbon model structure of hydrophobic pentapeptide-fused lysozyme. The hydrophobic pentapeptide, Phe-Phe-Val-Ala-Pro, was attached to the C-terminus of lysozyme. The peptide easily forms β-strand.

against *E. coli* increased slightly, compared with that of the wild-type lysozyme, while that of the pentapeptide- and heptapeptide-fused lysozymes greatly increased (Figure 29.9). These results suggest that the length of the pentapeptide is sufficient to kill Gram-negative bacteria. The hydrophobic pentapeptide attached to the C-terminus of lysozyme may contribute to the penetration into the outer membrane of the bacteria. Ibrahim et al. (14), using model experiments with *E. coli* phospholipid liposome, indicated that the enhanced bactericidal action of H5 lysozyme against *E. coli* is due to the disruption of the electrochemical potential of the inner membrane in cooperation with the inherent function of lysozyme in the outer membrane and peptide glycan. Thus, the bactericidal action of lysozyme was exerted against both Gram-negative and Gram-positive bacteria. The bactericidal lysozyme can be used for industrial applications. However, the secretion amount of H5 lysozyme was found to be very low in *S. cerevisiae*. In order to overcome the low secretion, we succeeded in increasing 400 times the secretion amount of H5 lysozyme in the *Pichia pastoris* expression system (16). Thus, it has been shown that lysozyme can be

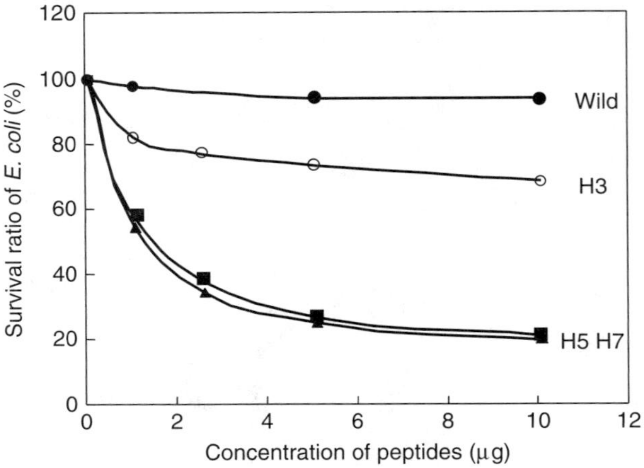

FIGURE 29.9 Bactericidal action of wild and hydrophobic peptide-fused lysozymes to *E.coli* cells. The different concentrations of wild lysozyme (●), Phe-Val-Pro(H)- lysozyme (○), Phe-Phe-Val-Ala-Pro (H5)-lysozyme (■), and Phe-Phe-Val-Ala-Ile-Ile-Pro (H7)-lysozyme (▲)were incubated with the *E.coli* cells (105 cells/ml) at 37°C for 30 min in 50 mM acetate buffer, pH5.5.

converted by genetic modification to have bactericidal action against both Gram-positive and Gram-negative bacteria.

Genetic modifications provide a very powerful method for improving functional properties such as heat stability, emulsifying properties, and antimicrobial action that might be of interest for industrial applications. Advances in yeast expression systems have enabled us to design and create new functional proteins. We have previously used the *S. cerevisiae* expression system to secrete various proteins, but the yield of the target proteins was very low. Recently, the *P. pastoris* expression system has been developed to secrete large amounts of heterologous proteins to allow investigation of the characteristics of new functional proteins.

29.5.1 Reduction of Antigenic Structure of Lysozyme by Genetic Modifications

The antigenic determinants of hen egg-white lysozyme were identified as Arg-45 and Arg-68 that were spacially adjacent on the surface of the groove of Fab in the antibody (17), as shown in Figure 29.10. The Arg-45 and Arg-68 in hen egg-white lysozyme are Tyr and Lys in the human lysozyme, respectively. This is the reason why lysozyme is one of the allergenic proteins in hen egg-white. The importance of Arg-68 in the antigenicity of hen egg-white lysozyme was confirmed by many investigators. We reported that the monoclonal antibody (mAb) of hen egg-white lysozyme could sensitively detect conformational changes (18). The binding of mAb with lysozyme was decreased both by denaturation with heat and guanidine-hydrochloride (HCl), corresponding to the denaturation curves of

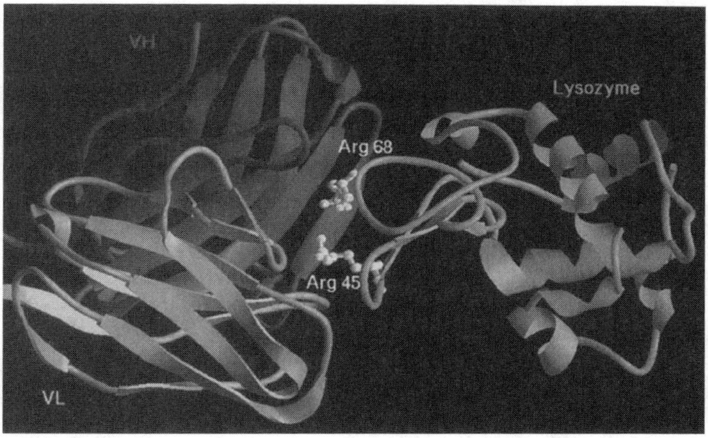

FIGURE 29.10 HEL-antibody complex interface region. HEL is zigzag (right), the Fab heavy chain in helix (left upper), and light chain in helix (left bottom). Arg.68 and Arg45 of HEL, in white, can be seen extending into the groove of the Fab. Adapted from Chacko et al.(17). With permission.

lysozyme. This suggests that the epitope region of lysozyme is specific to the tertiary structure around two loops containing Arg-45 and Arg-68, as shown in Figure 29.10. In order to reduce the antigenic structure of lysozyme, we constructed various mutant lysozymes. As shown in Figure 29.11, unstable mutant lysozymes such as helix-destabilized mutant (K13D), disulfide bond-deletion mutant (C94/76A) considerably decreased the binding with mAbs, 4G5 and TL-G1100. The mutant R68Q, in which Arg-68 was substituted with glutamine, also decreased the binding with mAbs. However, the reduction was only partial, and the reactivity still remained with mAbs. On the other hand, the oligomannosyl and polymannosyl lysozymes (G49N) greatly decreased the binding capacity with mAbs, suggesting that the extended carbohydrate chain may result in steric hindrance around the epitope region near Arg-45. On the basis of the observation that polymannosyl lysozyme masked the allergenic structure of lyszoyme, we attempted the model experiment of immune tolerance. The polymannosyl lysozyme G49N and R21T were intraperitoneally injected into mice and the production of immunoglobulin G (IgG) and IgE measured. As shown in Figure 29.12, the production of IgG and IgE by lysozyme were suppressed by the intraperitoneal injection of polymannosyl lysozymes (G49N, R21T) and the lysozyme–galactomannan conjugate prepared through the Maillard reaction (19). The polymannosyl lysozyme G49N was more effective in reducing the production of IgE than R21T because G49N completely masked the epitope region. On the other hand, the injection of the galactomannan–lysozyme conjugate increased the production of IgG but decreased the production of IgE. These observations suggest that masking the epitope of antigenic proteins by polysaccharide attachment may be effective in reducing antigenicity and can be used to induce immune tolerance.

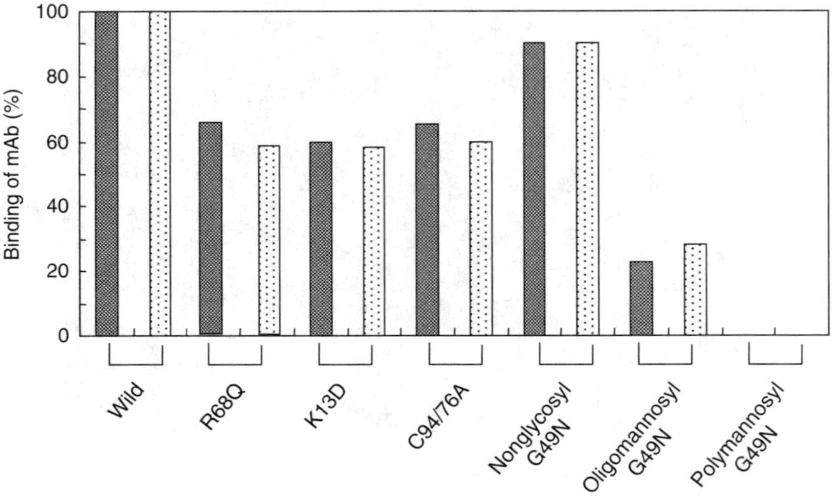

FIGURE 29.11 Binding of various mutant lysozymes with mAbs. Two monoclonal antibodies (TLG1100, ▩ and 4G5, ⬚) were used. The binding of mAb was expressed as a percentage of the wild-type lysozyme.

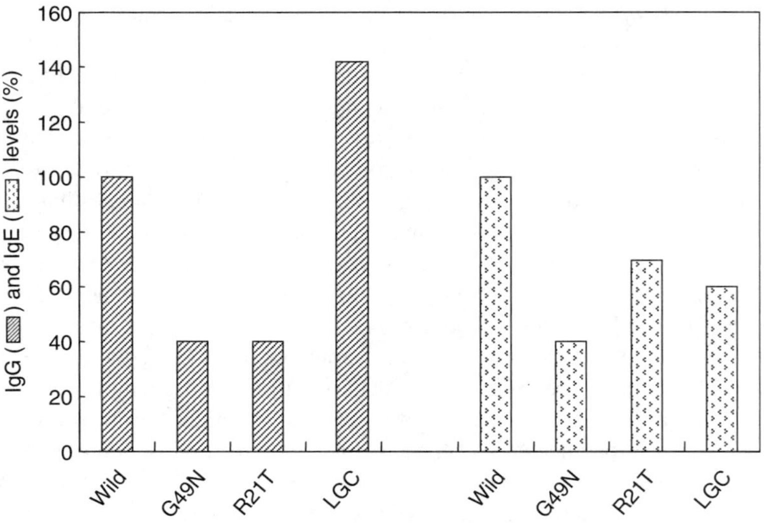

FIGURE 29.12 Effects of polysaccharide attachment to lysozyme on the production of HEL- specific IgG(▨) and (⬚) IgE. The production level was expressed as a percentage of the wild-type lysozyme.

29.5.2 Amyloidogenic Mutant Lysozymes

The two naturally occurring human lysozyme mutants have been found to form amyloidosis that causes fatal disease (20). This is significant especially in the context of an increase in the number of amyloidogenic diseases such as Alzheimer's disease and transmissible spongiform encephalopathy. At present, there are more than 20 known amyloidogenic diseases. Although the diverse human proteins that can form amyloid fibrils *in vivo* have unrelated sequences and tertiary folds, they can polymerize into fibrils with similar ultrastructural appearances and identical tinctorial properties. The core structure of all amyloid fibrils consists of β-sheets with the strands perpendicular to the long axis of fiber. The two known natural mutations in the human lysozyme gene both cause autosomal-dominant hereditary amyloidosis. Affected individuals are heterozygous for single-base changes that encode the nonconservative amino acid substituions Ile56Thr and Asp67His, respectively. These mutants become less stable than the native form and reduce the cooperativity of unfolding process, although they are all enzymatically active. Booth et al. (20) proposed a model for lysozyme amyloidosis, in which association of the partly folded forms of the variants occurs through the unstable β-domain (Figure 29.13). The FTIR data indicate that fibrillogenesis involves an increase in β-structure; conversion of α-helix to β-structure will be easier in the molten globule state (amyloidogenic mutants) than in the native state because of the much lower cooperativity of the unfolding process. The self-association through the β-domain causes the initiation of fibril formation. This provides the template for further deposition of protein and for the development of the stable, mainly β-sheet, core structure of the fibril. The Asp-67 and Ile-56 are pivotal residues to stabilize the β-domain in human lysozyme. The importance of these residues is supported by the fact that they are conserved as Asp-66 and Ile-55, respectively, in hen egg-white lysozyme. We constructed the amyloidogenic mutants Asp66His and Ile55Thr using genetic engineering and secreted them in a yeast expression system. As shown in Figure 29.14, the thermal denaturation curves of amyloidogenic mutant lysozymes are consistent in both human and hen lysozymes, although the amyloidogenic mutants of hen egg-white lysozyme are less stable than those of human lysozyme. The dramatic decrease in the stability of amyloidogenic mutants lead to the molten globule-like state that lacks global cooperativity to convert from the soluble to fibrillar form.

In addition, we reported the further characterization of the corresponding amyloidogenic mutants of hen egg-white lysozyme (21). The amyloidogenic mutants (Ile55Thr and Asp66His) of hen egg white lysozyme were remarkably less soluble than the wild-type lysozyme. To enhance the secretion of these mutants, we constructed the glycosylated amyloidogenic mutant lysozymes (I55T/G49N and D66H/G49N) with the N-glycosylation signal sequence (Asn-X-Thr) by substitution of glycine with asparagine at position 49. The secretion of these glycosylated mutant lysozymes was greatly increased in *S. cerevisiae*, compared with the nonglycosylated type. Both glycosylated amyloidogenic lysozymes retained about 40% lytic activity when incubated at pH 7.4 for 1 hour at the physiological temperature

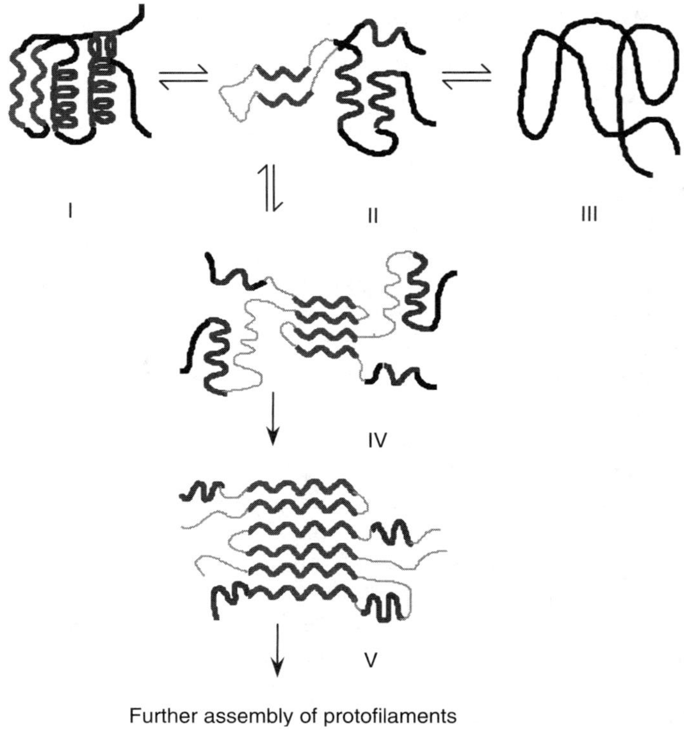

FIGURE 29.13 Proposed model for lysozyme amyloid fibril formation. Zigzag, β-sheet structure; helix, helical structure. A partly folded, molten globule-like form of protein (II), distinct from the native (I) and denatured (III) states, self-associates through the β-domain (IV) to initiate fibril formation. Adapted from Booth et al. (20). With permission.

of 37°C, whereas the nonglycosylated amyloidogenic lysozymes eventually lost all enzymatic activity under these conditions. These results suggest that the glycosylated chains could mask the β-strand of amyloidogenic lysozymes from the intermolecular cross-β-sheet association, thus blocking the amyloid formation of mutant lysozymes. As described here, the genetic insertion of the carbohydrate chain could suppress the formation of amyloid fibrils. Further studies are necessary to determine whether the carbohydrate chain attachment is effective for the suppression of *in vivo* amyloidosis in higher eukaryotes. Such investigation would point to effective means to treat amyloidosis.

29.5.3 Quality Control of Abnormal Proteins by Molecular Chaperones

The mechanisms of the molecular diseases such as amyloidosis that develop from the accumulation of unfolded proteins are being investigated in our

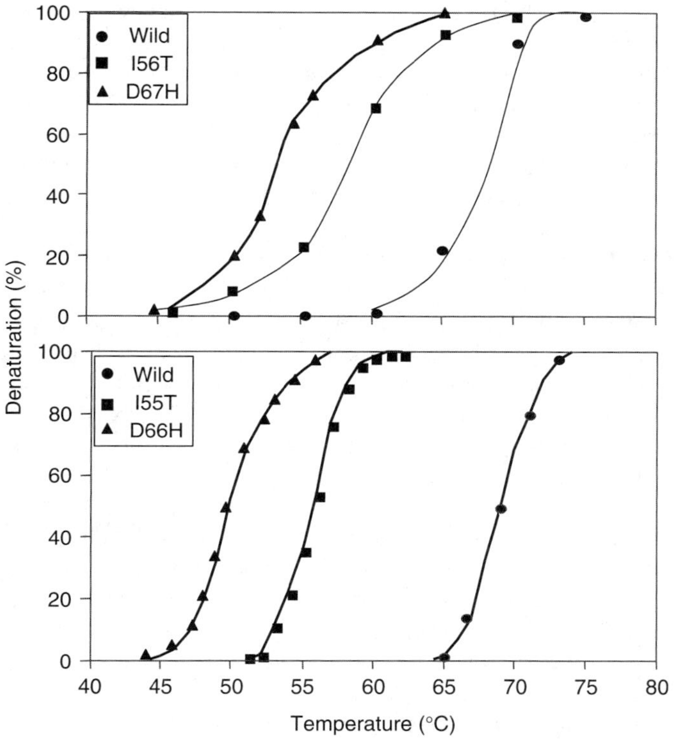

FIGURE 29.14 Denaturation curves of human (upper) and hen (lower) amyloidogenic lysozymes. Denaturation was determined by changes in the ellipticity at 222 nm during heating 40°C to 80°C.

studies as introduced here. The endoplasmic reticulum (ER) quality control system is extremely important to the eukaryotic organism that retains misfolded intermediates in the ER until these substrates fold correctly or until the misfolded proteins are degraded. It seems likely that the stable conformation of the nascent protein leads it to the secretion pathway and the unstable conformation of the nascent proteins leads it to the degradation pathway. The quality control of abnormal proteins is carried out by the cooperative work of chaperone and proteasome in the ER. Therefore, the lowering of the function of the ER quality control leads the organism to experience overexpression of misfolded proteins that may result in serious diseases such as amyloidosis. The molecular chaperone calnexin is a component of the ER quality control system through its oligosaccharide moieties for glycoproteins. Interestingly, disruption of the calnexin gene leads to the deletion of ER quality control for glycoproteins but does not affect the growth of *S. cerevisiae*. Thus, the evaluation system for quality control of various unstable proteins using calnexin-disrupted *S. cerevisiae* was established (22). The amyloidogenic lysozymes were glycosylated to evaluate the quality control in yeast. The glycosylated amyloidogenic lysozymes I55T/G49N and D66H/G49N were both expressed in wild-type and calnexin-disrupted *S. cerevisiae*. As shown in Figure 29.15, the secretion of D66H/G49N was greatly increased in calnexin-disrupted *S. cerevisiae*, while the secretion was very low in wild-type *S. cerevisiae*, although the secretion of another mutant, I55T, was not changed much. It is probable that mutant I55T may escape the quality control because the crystal structure of human I56T is similar to that of wild-type lysozyme (23). In parallel, the induction level of the other molecular chaperones Bip (Hsp-70) and PDI (protein disulfide isomerase) located in the ER was investigated when these amyloidogenic lysozymes were expressed in wild-type and calnexin-disrupted *S. cerevisiae.* The mRNA (messenger ribonucleic acid) concentrations of Bip and PDI were increased when the amyloidogenic lysozymes were secreted in calnexin-disrupted *S. cerevisiae.* This observation indicates that calnexin is important for quality control of abnormal proteins and that the decrease in its function brings about secretion of amyloidogenic lysozyme in cooperation with overexpression of other chaperones due to unfold protein response (UPR). Therefore, the molecular chaperones related to the quality control of proteins are extremely important for suppressing the secretion of abnormal proteins. This is also supported from the data for other unfolded mutants such as K13D/G49N and C76A/G49N (21). These observations suggest that the lowering of normal quality control by chaperone causes molecular diseases such as unfolding diseases (e.g., amyloidosis).

As shown in this chapter, the genetic engineering of lysozyme is very powerful for elucidating the molecular mechanism of structural and functional properties. Novel functional lysozymes can be designed and constructed by using posttranslational modification in yeast expression systems. This technique should contribute to the elucidation of the mechanisms of molecular diseases such as allergy and amyloidosis and to the development of preventive measures against them.

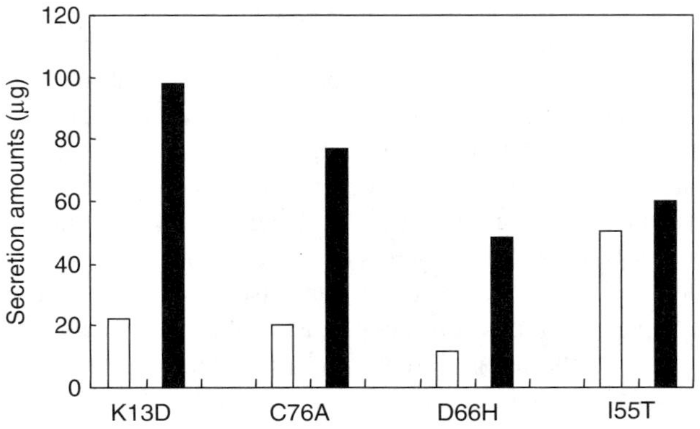

FIGURE 29.15 Secretion amounts of various mutant lysozymes in *Saccharomyces cerevisiae*. Secretion amounts were indicated as micrograms per 1 liter of medium.

ACKNOWLEDGMENT

Appreciation is extended to Professor Soichiro Nakamura, Shimane University, Japan; Professor Hisham Radwan Ibrahim, Kagoshima University, Japan; and Yutao Song, National Institute of Health, United States, for their excellent postdoctoral work in Yamaguchi University. This review was mainly written on the basis of the PhD works of Drs. S. Nakamura, H.R. Ibrahim and Y. Song.

REFERENCES

1. C.C.F. Blake, D.F. Koenig, G.A. Mair, A.C.T. North, D.C. Phillips, V.R. Sarma. Structure of hen egg-white lysozyme. A three-dimensional Fourier synthesis at 2 A resolution. *Nature*, 206: 757–761, 1965.
2. I. Kumagai, S. Kojima, E. Tamaki, K. Miura. Conversion of Trp62 of hen egg white lysozyme to Tyr by site-directed mutagenesis. *J. Biochem.*, 102: 733–740, 1987.
3. T. Imoto. *Protein Engineering*. M. Ikehara (Ed.). *Strategy to Examine the Function of Hen Lysozyme by Protein Engineering*. Japan Scientific Societies Press & Springer-Verlag, Tokyo. 1990, 153–158.
4. M. Muraki, K. Harata, Y. Hayashi, M. Machida, Y. Jigami. The importance of precise positioning of negatively charged carboxylate in the catalytic action of human lysozyme *Biochim. Biophys. Acta*, 1079: 229–237, 1991.
5. Y. Taniyama, K. Ogasawara, K. Yutani, M. Kikuchi. Folding mechanism of mutant human lysozyme C77/95A with increased secretion efficiency in yeast. *J. Biol. Chem.*, 267: 4619–4629, 1992.
6. A. Kato, Y. Sasaki, R. Furuta, K. Kobayashi. Functional protein-polysaccharide conjugate prepared by controlled dry-heating of ovalbumin-dextran mixtures. *Agric. Biol. Chem.*, 54: 107–112, 1990.
7. A. Kato, K. Kobayashi. Excellent emulsifying properties of protein-dextran conjugates. In: M. El-Nokaly, D. Cornell (Eds.). *Microemulsions and Emulsions in Foods*. ACS Symposium Series 448, 1991, 213–229.
8. S. Nakamura, H. Takasaki, K. Kobayashi, A. Kato. Hyperglycosylation of hen egg white lysozyme in yeast. *J. Biol. Chem.*, 268: 12706–12712, 1993.
9. A. Kato, H. Takasaki, M. Ban. Polymannosylation to asparagines-19 in hen egg white lysozyme in yeast. *FEBS Lett.*, 355: 76–80, 1994.
10. A. Kato, S. Nakamura, M. Ban, H. Azakami, K. Yutani. Enthalpic destabilization of glycosylated lysozymes constructed by genetic modification. *Biochim. Biophys. Acta.*, 1481: 88–96, 2000.
11. S. Nakamura, K. Kobayashi, A. Kato. Novel surface functional properties of polymannosyl lysozyme constructed by genetic modification. *FEBS Lett.*, 328: 259–262, 1993.
12. A. Kato, S. Nakamura, H. Takasaki, S. Maki. Novel functional properties of glycosylated lysozymes constructed by chemical and genetic modifications. In: N. Parris, A. Kato, L.K. Creamer, J. Pearce (Eds.). *Macromolecular Interactions in Food Technology*. ACS Symposium Series 650, 1996, 243–256.
13. H.R. Ibrahim, A. Kato, K. Kobayashi. Antimicrobial effects of lysozyme against Gram-negative bacteria due to covalent binding of palmitic acid. *J. Agric. Food Chem.*, 39: 2077–2082, 1991.

14. H.R. Ibrahim, M. Yamada, K. Matsusita, K. Kobayashi, A. Kato. Enhanced bactericidal action of lysozyme to *Escherichia coli* by inserting a hydrophobic pentapeptide into its C terminus. *J. Biol. Chem.*, 269: 5059–5063, 1994.
15. H. Arima, H.R. Ibrahim, T. Kinoshita, A. Kato. Bactericidal action of lysozymes attached with various sizes of hydrophobic peptides to the C-terminal using genetic modification. *FEBS Lett.*, 415: 114–118, 1997.
16. S.T. Liu, A. Saito, H. Azakami, A. Kato. Expression, purification, and characterization of an unstable lysozyme mutant in *Pichia pastoris*. *Protein Expr. Purif.*, 27: 304–312, 2003.
17. S. Chacko, E. Silverton, L. Kam-Morgan, S. Smith-Gill, G. Cohen, D. Davies. Structure of an antibody-lysozyme complex. Unexpected effect of a conservative mutation. *J. Mol. Biol.*, 245: 261–274, 1995.
18. A. Kato, T. Shimizu, S. Saga. Conformational changes in mutant lysozymes detected with monoclonal antibody. *FEBS Lett.*, 371: 17–20, 1995.
19. K. Arita, E.E. Babiker, H. Azakami, A. Kato. Effect of chemical and genetic attachment of polysaccharides to proteins on the production of IgG and IgE. *J. Agric. Food Chem.*, 49: 2030–2036, 2001.
20. D.R. Booth, M. Sunde, V. Bellotti, C.V. Robinson, W.L. Hutchinson, P.E. Fraser, P.N. Hawkins, C.M. Dobson, S.E. Radford, C.C.F. Blake, M.B. Pepys. Instability, unfolding and aggregation of human lysozyme variants underlying amyloid fibrillogenesis. *Nature*, 385: 787–793, 1997.
21. Y.T. Song, J. Sata, A. Saito, M. Usui, H. Azakami, A. Kato. Effects of calnexin deletion in *Saccharomyces cerevisiae* on the secretion of two glycosylated lysozymes. *J. Biochem.*, 130: 757–764, 2002.
22. Y.T. Song, H. Azakami, S. Begum, J.W. He, A. Kato. Different effects of calnexin deletion in *Saccharomyces cerevisiae* on the secretion of two glycosylated amyloidogenic lysozymes. *FEBS Lett.*, 512: 213–217, 2002.
23. J. Funahashi, K. Takano, K. Ogasawara, Y. Yamagata, K. Yutani. The structure, stability, and folding process of amyloidogenic mutant human lysozyme. *J. Biochem.*, 120: 1216–1223, 1996.

30 New Methodologies for the Synthesis of Oligopeptides and Conformation-Constrained Peptidomimetics

Kazuhiro Chiba

CONTENTS

30.1 INTRODUCTION

Systematic synthesis of peptides and peptidomimetics plays an important role in their structure elucidation and estimation of active conformations in target receptors. In order to approach the final goal of structure analysis of peptides, synthetic methodology based on combinatorial, high-throughput peptide synthesis may be employed to explore adequate peptide sequences among the large number of oligopeptides at hand. Furthermore, systematic synthesis of conformation-constrained peptidomimetics provides important structural information on the native and biologically active peptides. This chapter provides information about (a) a new methodology for liquid-phase peptide synthesis using thermomorphic biphasic organic solutions, and (b) a new anodic method for the introduction of functional groups for the conformation control of proline moieties that would enable regulation of secondary structures of several kinds of peptides. In the first part of the chapter, convenient synthetic methods for oligopeptides are introduced. Combinations of typical organic solvents composed of cyclohexane and qualified aprotic polar organic solvents were found to realize an effective, biphasic thermomorphic system in arbitrary ratios of upper and lower phases that enable a practical application of a liquid-phase peptide synthesis. In the second part, a new synthetic method of bicyclic dipeptidomimetics is introduced to obtain several kinds of conformation-constrained synthetic peptidomimetics.

30.2 NEW LIQUID-PHASE PEPEPTIDE SYNTHESIS (LPPS) METHOD BASED ON THERMOMORPHIC BIPHASIC ORGANIC SOLUTIONS

Much of the current research has used solid-phase as a platform for organic reactions (1–3). While solid-phase resins offer many advantages with regard to compound isolation and ease of handling, the insoluble nature of the resins complicates the characterization of compounds attached to them and may lead to reagent accessibility problems. One successful merging of solid- and liquid-phase chemistry is the use of soluble polymer platforms that selectively remove excess reagents and by-products (4–11). On the other hand, use of a liquid–liquid biphasic thermomorphic process, in which a reagent or a catalyst is designed as a residue in one of the liquid phases and as a product in the other liquid phase, can be an enabling approach for the commercial application of chemical reactions with high selectivity, efficiency, and ease of handling for separation of solutes (12, 13). Based on the immiscibility of perfluorinated hydrocarbons with both organic and inorganic solvents, a novel "fluorous biphasic system" for catalysis was proposed (14–19). Fluorous biphasic systems consist of a fluorous phase containing a preferentially fluorous soluble reagent and a catalyst with limited solubility in this phase (20–27). The thermomorphic liquid–liquid separation system composed of organic solvents of different polarity might allow efficient reactions, for example, by association of apolar products and polar substrates or catalysts in a one-phase solution, which could be spatially separated after the completion of reactions. It should further open the door for construction of liquid-phase combinatorial and industrial chemistry due to the ease of separation of products, catalysts, reagents, electrolytes, and soluble platforms for organic synthesis. This provided us the incentive to explore a novel liquid-phase peptide synthesis in the thermomorphic system using typical organic solvents. Hence, a new methodology for peptide synthesis based on combination of cyclohexane and some typical organic solvents was designed in order to realize the aimed miscible biphasic system in arbitrary ratios of upper and lower phases. By using the cyclohexane-based biphasic thermomorphic system, the synthesis of liquid-phase peptide was efficiently accomplished.

It is well known that solid-phase synthesis plays a very important role in the preparation of peptides. However, their sequential reactions generally require excess amino acids and reagents at each step (Figure 30.1). It is also difficult to confirm the completion of each step of the reaction because the products are combined on the insoluble resins. Furthermore, after separation of the resins from excess reagent solutions, the products should be chemically cleaved from the

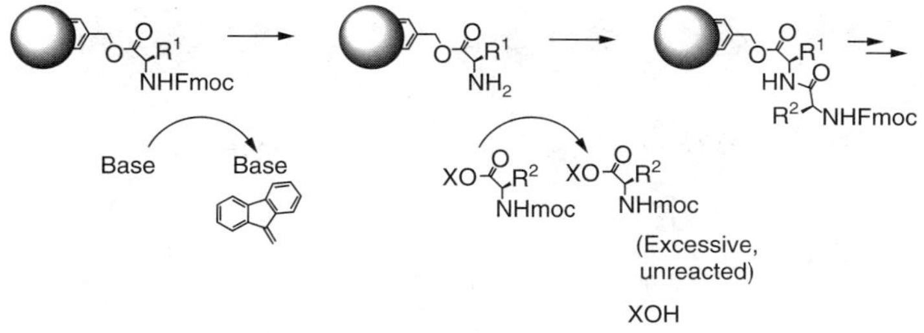

FIGURE 30.1 A protocol for typical solid-phase peptide synthesis.

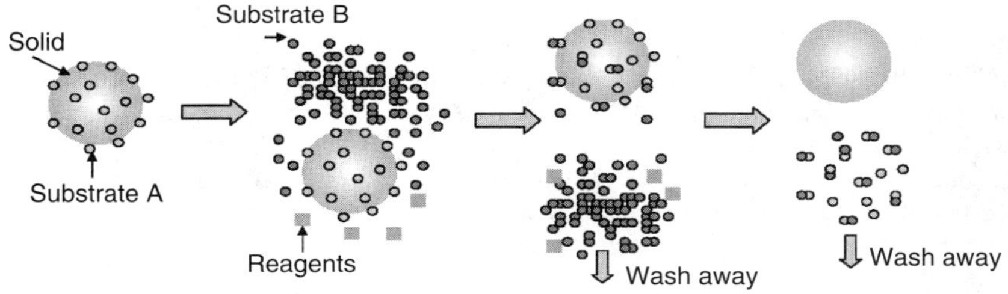

FIGURE 30.2 An example of the solid-phase synthesis.

resins followed by their purification (Figure 30.2). In this case, if the reaction is performed in homogeneous solutions that enable facile separation of the products and unused reagents, it might overcome the complexity of the procedures of general peptide synthesis. In order to achieve the desired property, solvent systems should be composed of a typical organic solution — a combination of less-polar and polar solvents that cause thermomorphic changes in homogeneous and biphasic solutions at ambient temperature.

Suitable solvent constructions for thermomorphic solutions that could be reversibly mixed at "desired" temperatures in "arbitrary ratios" of upper- and lower-layer volumes were explored. Among numerous compositions of organic solvents, an appropriate mixture of cyclohexane (CH) and nitroalkane (NA) was initially found to have the desired function, that is, a 75:25 (v/v) mixture of CH and NA (a 20:80 (v/v) mixture of nitromethane [NM] and nitroethane [NE]). At 25°C, this solvent mixture was separated into an upper CH(main) phase and a lower NA(main) phase (Figure 30.3, colored by methylene blue, MB). After heating to 45°C, it was soon mixed to form a stable, homogeneous phase. Furthermore, by cooling back to 25°C, the solution was immediately separated into the two initial phases, completely recovering the homogeneously dissolved MB into the lower NA layer. This clear partition of a solute suggested the ability of the thermocontrolled solubilization and partition of the designed solutes in this solvent system. This prompted the seeking of the promising property of the biphasic organic liquid system by expanding the solvent combinations and ratios. Figure 30.4 shows the correlations between

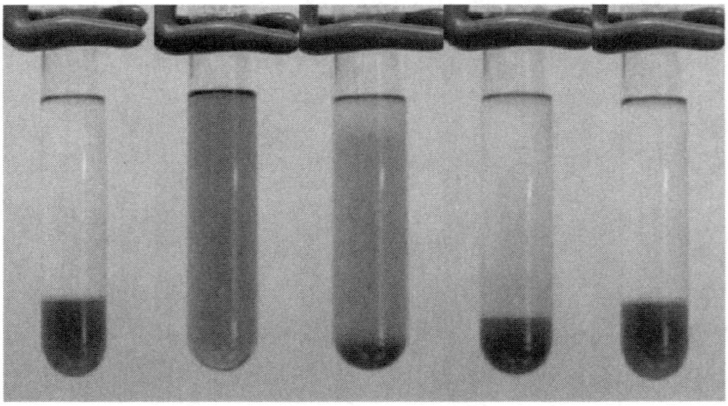

FIGURE 30.3 (A) An organic solvent mixture composed of CH-NE-NM (75:20:5 v/v/v) formed the biphasic system at 25°C (a lower NA layer was colored by methylene blue). (B) The solvent mixture was immixed at 45°C. (C, D) After cooling back to 25°C, it immediately began to exclude the CH phase. (E) Finally, it formed the initial biphasic solution to recover methylene blue in the lower layer.

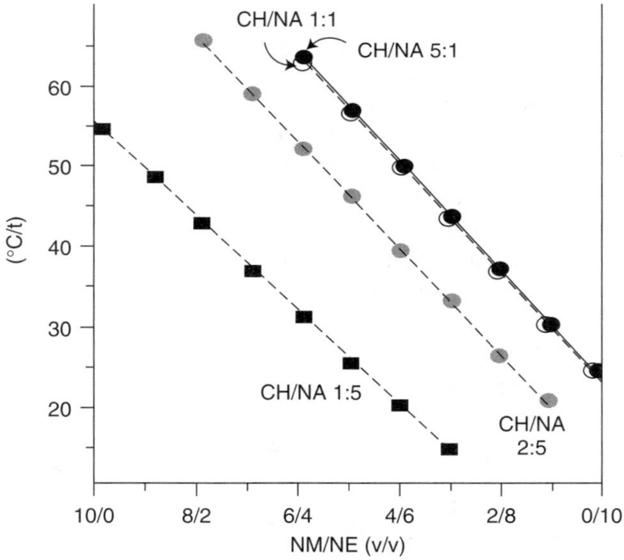

FIGURE 30.4 Correlations between immixing temperatures and solvent compositions based on cyclohexane and nitroalkanes (CH: cyclohexane, NA: nitroalkane, NM: nitromethane, NE: nitroethane). Effect of the solvent compositions (CH:NA ratios and NM:NE ratios) on the miscible temperatures. Each biphasic solvent mixture (10 mL in an 18 × 200 mm test tube, 10°C) was gradually heated (*ca.* 3°C min^{-1}) with stirring to form a homogeneous solution at the temperature plotted above.

the miscible temperature and the ratio of NE and NM. By increasing the content of NE, the miscible temperature was linearly lowered. For example, a biphasic mixture of CH-NE-NM (CH:NA = 17:83, NE:NM = 70:30) formed a homogeneous phase at 14°C and higher temperatures. On the other hand, the miscible temperature increased with an increase in the NM ratio. The miscible temperature was also affected by the CH ratio , requiring a higher temperature with an increase in the CH ratio to complete the one-phase formation. The miscible temperature, however, was not affected when the ratio of CH:NA (v/v) was more than 50:50 (v/v). This shows that the miscible temperature and the ratio of CH:NA can be arbitrarily selected by controlling the ratio of NE:NM. Furthermore, a similar property was also found in the mixture of CH-dimethylacetamide (DMA)-dimethylformamide (DMF) and of CH–acetonitrile (AN)–propionitrile (PN) (Figure 30.5). In the mixture of CH–DMF–DMA, the miscible temperature varied between 18°C and 47°C with change in the ratio of DMF:DMA. A clear phase separation was also observed for the mixture of CH–AN–PN between 33°C and 61°C.

In these thermomorphic organic solutions, partition ratios of typical solutes were determined. Quaternary ammoniums salts, Lewis acids, and polar photosensitizers

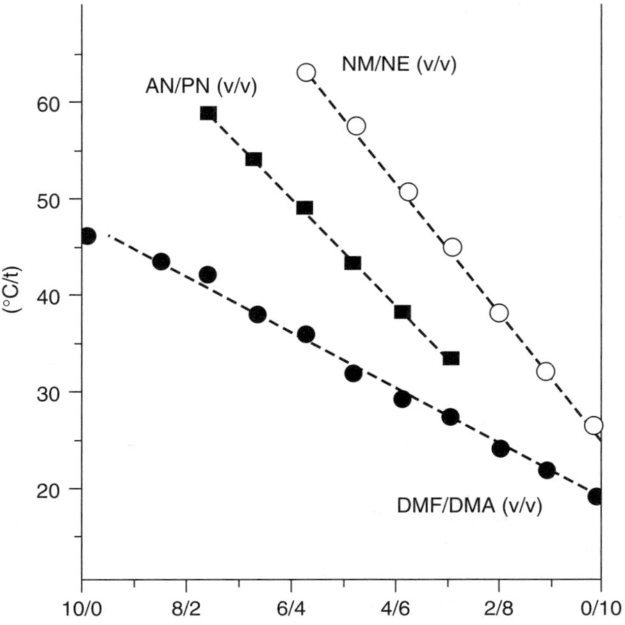

FIGURE 30.5 Correlations between immixing temperatures and solvent compositions based on cyclohexane and polar organic solvents (DMF: dimethylformamide, DMA: dimethylacetamide, AN: acetonitrile, PN: propionitrile). Effect of solvent composition on the miscible temperature. The solvent mixtures were composed of 50:50 (v/v) mixture of CH–nitriles or CH–amides with varying ratios of AN:PN or DMF:DMA, respectively.

TABLE 30.1
Partition Ratios of Typical Solutes

Solute	(mM)	Lower layer (%)	Upper layer (%)	Solvents[b]
TEAP[a]	(25)	98	2	A
	(100)	>99	ND	B
	(100)	>99	ND	C
TBABF[c]	(15)	>99	ND	A
LPC[d]	(30)	>99	ND	A
MB[e]	(6)	>99	ND	A
BF_3/Et_2O	(120)	>99	ND	A
$AlCl_3$	(32)	>99	ND	A
$Yb(OTf)_3$	(6)	>99	ND	A
$C_{16}H_{33}SH$	(10)	ND	>99	A
$C_{21}H_{43}COOH$	(10)	ND	>99	A
$C_{18}H_{37}OH$	(100)	ND	>99	A
$C_{18}H_{37}NH_2$	(20)	ND	>99	A
	(20)	ND	>99	B
$(C_8H_{17})_3N$	(20)	ND	>99	A
	(20)	ND	>99	B
1	(10)	ND	>99	A
	(130)	3	97	A
	(130)	ND	>99	B

[a] TEAP: Tetraethyl-ammonium perchlorate
[b] Solvent constitutions: (A) CH/NM/NE(2/1/1); (B) CH/DMF/DMA(2/1/1); (C) CH/AN/PN(2/1/1);
[c] Tetrabutylammonium tetraflouroborane
[d] LPC: Lithium perchlorate.
[e] MB: Methylene blue.

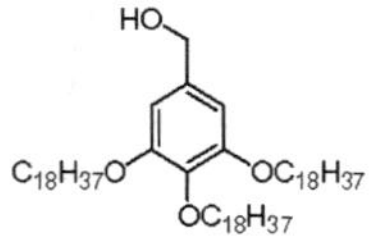

1

were mainly partitioned to polar organic layers (Table 30.1). On the other hand, less polar compounds possessing thiol, carboxyl, alcohol, and amino groups were selectively partitioned in the cyclohexane layer.

In order to apply for practical peptide synthesis, all of the products must be separated from the excess reagents by removing the products into the separated cyclohexane solution. Therefore, a platform (cyclohexane-soluble-platform [CHSP]) on which products were combined for partitioning in the cyclohexane phase after biphasic separation of the thermomorphic solutions was introduced. These should also be synthesized easily to perform selective partition with combined chemicals. Furthermore, after completion of the reactions and processings, the resultant chemicals must be recovered from the mixture by moderate cleavage reactions. In the thermomorphic system, a platform, (3,4,5-trioctadecyloxy phenyl)methan-1-ol (28) for example, was found to work effectively in the sequential peptide synthesis. On the other hand, as DMF is widely used for current solid-phase Fmoc-peptide synthesis, a thermomorphic solution composed of CH–DMF–DMA (50:25:25) was introduced for new liquid-phase peptide synthesis. A clearly separated biphasic solution was formed at <33°C. After warming up (higher than 33°C), it changed to a homogeneous solution (Figure 30.6). In this homogeneous organic solution (at 35°C), the platform was dissolved. Furthermore, the platform was selectivity partitioned in the CH layer after cooling, followed by the formation of the biphasic solution (at 25°C) which allowed sequential peptide synthesis in the liquid phase. In the homogenized reaction mixture at 35°C, peptide-chain elongation was accomplished by treatment with only 3 mol equivalents of activated Fmoc amino acids. To a solution of Fmoc-valine (170 mg) in CH_2Cl_2 (3 mL), 125 mg of DCC (dicyclohexylcarbodiimide) was added (Figure 30.7). The solution was stirred for 15 minutes at ambient temperature. After filtration, the filtrate was concentrated *in vacuo* to dryness. The residue was then dissolved in 3 mL of DMF–DMA 1:1, v/v and the solution was combined with the cyclohexane-soluble platform [CHSP] **1** in CH (50 mg 3 mL^{-1}) at room temperature. To the biphasic solution, 6.6 mg of DMAP (4-dimethylaminopyridine) was added and then heated to 35°C to form a homogenized solution. After it was allowed to stand for 60 minutes, the solution was cooled to 5°C to form the biphasic solution (from the upper layer, [CHSP]-Val- Fmoc **2** [see Scheme 30.1], which was then isolated in 99% yield). The upper CH-layer was separated, followed by the addition of 10% diethylamine–DMF–DMA (50:50, 3 mL). The solution was heated to 35°C and allowed to stand for 60 minutes; the CH layer was then separated after cooling. From the CH solution, [CHSP]-Val-NH_2 **3** was isolated in 95% yield. Fmoc-Gly (57 mg), HOBt (55 mg), and DIPCD (diisopropylcarbodiimide, 25 mg) were dissolved in 2 mL of DMF–DMA 1:1, v/v and

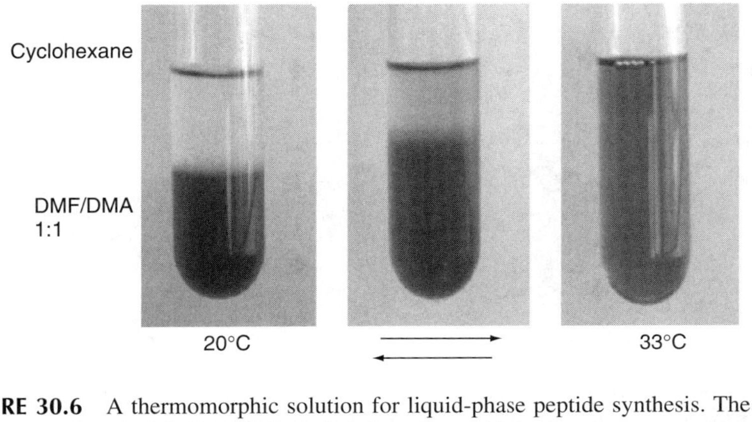

FIGURE 30.6 A thermomorphic solution for liquid-phase peptide synthesis. The solution was composed of CH:DMF:DMA (2:1:1) to form a biphasic solution at ambient temperature and a homogeneous solution at 33°C and higher. Lower amide layer was colored by methylene blue.

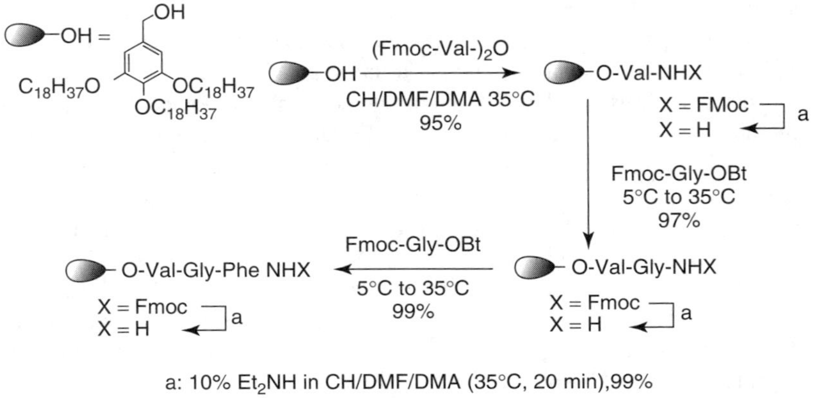

FIGURE 30.7 An example of the liquid-phase peptide synthesis using the biphasic organic solvent mixture. (A) (Fmoc-Val-)$_2$O in CH:DMF:DMA of 50:25:25 (35°C), 95%; (B) 10% Et$_2$NH in DMF:DMA of 50:50 (35°C), 99%; (C) Fmoc-Gly-OBt, DIPCD in DMF:DMA of 50:50 (5 to 35°C), 99%; (D) 10% Et$_2$NH in DMF:DMA of 50:50 (35°C), 99%; (E) Fmoc-Phe-OBt, DIPCD (5 to 35°C), 97%; (F) 10% Et$_2$NH in DMF:DMA of 50:50 (35°C), 99%

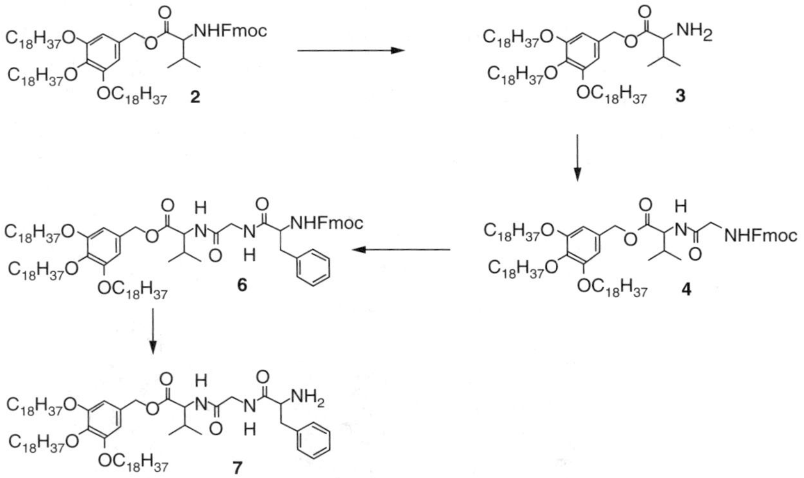

SCHEME 30.1 Structures of synthesized peptides in the thermomorphic biphasic system.

allowed to stand for 150 minutes at room temperature. After cooling to 5°C, 2 mL of the [CHSC]-Val-NH$_2$ **3** (60 mg) in CH solution was added to form the biphasic solution. The reaction mixture was gradually heated to 35°C (for 60 minutes) and allowed to stand for an additional 60 minutes at 35°C. The solution was cooled back (5°C) to afford [CHSC]-Val-Gly-Fmoc **4** in the separated upper layer in 99% yield. The Fmoc group of **3** in CH (3 mL) was removed by the addition of 10% diethylamine–DMF–DMA (50:50, 3 mL) in the mixed solution at 35°C (for 30 minutes), followed by the separation of the cooled CH layer, in which deprotected product [CHSP]-Val-Gly-NH_2 **5** was obtained in 99% yield. Fmoc-Phe (63 mg), HOBt (63 mg), and DIPCD (25 mg) were dissolved in 2 mL of DMF and allowed to stand for 150 minutes at room temperature. After cooling to 5°C, 2 mL of the [CHSP]-Val-Gly-NH_2 (60 mg) in CH solution were added to form the biphasic solution. The reaction mixture was gradually heated to 35°C (for 60 minutes) and allowed to stand for an additional 60 minutes at 35°C. The solution was cooled back (to 5°C) to afford [CHSP]-Val-Gly-Phe-Fmoc **6** in the separated upper layer in 99% yield. The Fmoc group of **6** in CH (3 mL) was removed by the addition of 10% diethylamine–DMF–DMA (50:50, 3 mL) in the mixed solution at 35°C (heated for 30 minutes), followed by the separation of the cooled CH layer in which the deprotected product [CHSP]-Val-Gly-Phe-NH_2 **7** was obtained in 99% yield. In these sequential reactions, a large excess of activated Fmoc-amino acid was not required, and the peptide intermediates (e.g., **3** to **7**) were easily isolated in excellent yields and were characterized by nuclear magnetic resonance (NMR) or time-of-flight mass spectrometer (TOF-MS) (29).

In summary, the biphasic, thermomorphic liquid–liquid separation systems were constructed for liquid-phase peptide synthesis by using a CH-soluble platform in CH and typical organic solvents. It is noteworthy that the regulation of the separation and mixing of solutes can be achieved by a moderate thermocontrol in the practical range of 15 to 65°C. Furthermore, products can be easily isolated from the CH layer to complement existing solid-phase peptide synthesis, offering advantages with regard to compound isolation, characterization, and reagent accessibility in reaction mixtures. Since polar media — NM, DMF, and AN — have been widely used for organic reactions with varied catalysts and reagents, their extensive chemical applications can be also ensured with flexibly controllable temperatures and upper- or lower-layer ratios.

30.3 ELECTROLYTIC CYCLIZATION OF DIPEPTIDES INVOLVING PROLINE-MOIETY FOR CONSTRUCTION OF CONFORMATION-CONSTRAINED PEPTIDOMIMETICS

There is considerable interest in the synthesis of unnatural and conformation-constrained amino acids, peptidomimetics, and small peptide fragments encompassing these residues. It is well known that proline plays a specific role in peptide secondary structure formation. Furthermore, the importance of substituted prolines and peptidomimetics involving proline in the design of new catalysts or in the chemical synthesis of pharmacologically or biologically interesting molecules is recognized. From a synthetic standpoint, electrochemical means are among the most useful methods for the modification of proline derivatives (see Scheme 30.2) (30–33). For example, new routes have been developed to construct functionalized, conformation-constrained peptidomimetics. This has been

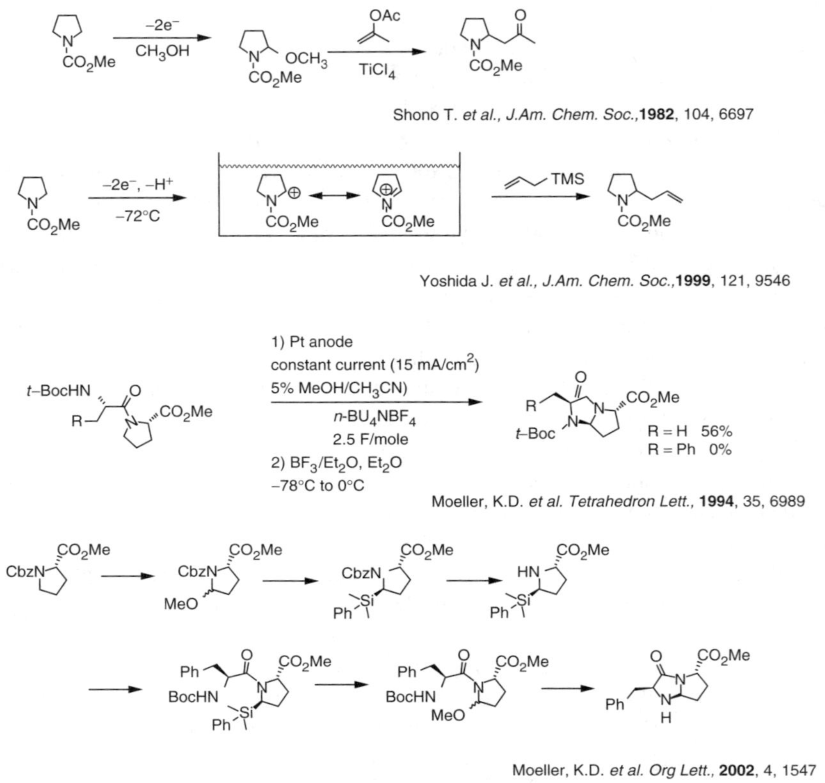

SCHEME 30.2 Examples of electrochemical modification of pyrrolidine and proline moieties.

accomplished by introducing silyl groups as electron-auxiliaries to lower the oxidation potentials. These potentials enable anodic substitution of the silyl group with a methoxy group that works as a trigger for following Lewis acid-catalyzed cyclization reactions via *N*-acyliminium cation for the construction of peptidomimetics. On the other hand, Yoshida, et al. (30–33) reported a "cation pool" method that can electrochemically generate and stabilize the iminium cations of carbamates in divided electrolytic cells at low temperatures (lower than −47°C); this is followed by a reaction with nucleophiles, in the absence of an electrolytic current, to introduce functional groups on the α-carbon to nitrogen.

In anodic oxidation systems, a divided cell is often introduced to avoid cathodic re-reduction of electrogenerated products or their undesired reactions with cathodic products. However, application of higher electrolytic potentials is generally required, and there is an accumulation of electrogenerated acid accompanying the generation of the products. This provided the incentive to develop an extended, simple electrochemical method that would enable anodic generation and accumulation of unstable *N*-acyliminium cations of prolines in an undivided system for their diverse functionalization. In this case, an electrolytic medium would play an important role in avoiding the re-reduction and decomposition of the reactive intermediates. Furthermore, if one can also introduce an electron auxiliary that converts to the corresponding *N*-acyliminium cation under neutral, lower oxidation potential conditions, it should further open the door for the introduction of varied functional groups in the proline residues of peptides. Thus, a new method was introduced for the introduction of nucleophiles on the carbon α' to nitrogen of proline derivatives, including phenylsulfanylated prolines, as a precursor of the iminium cations and for the intramolecular cyclization of proline-based peptides to construct conformation-constrained peptidomimetics involving proline moiety (see Scheme 30.3).

Initially, modification in the α'-position to the nitrogen of proline derivatives was investigated (Figure 30.8). Electrolysis of Moc-Pro-OMe **1** and Ac-Pro-OMe **2** was performed in a 1 M lithium perchlorate–nitromethane electrolyte solution (34–43) in the presence of 50 mM AcOH, using an undivided cell, a glassy carbon anode, and a platinum cathode. Because of the higher oxidation potential of proline derivatives (E_{ox} 1.9 V versus Ag–AgCl), compared with that of thiols, electrooxidation of proline derivatives in the presence of thiophenols did not give the desired product. On the other hand, electrolysis of compound **8** or **9** conducted under constant current conditions (2 mA, 2.2 F mol^{-1}) at 0°C followed by the addition of thiophenol **10** (three equivalents) afforded the desired product **11** or **12** in 86% yield (see Scheme 30.4). On the other hand, the product was scarcely obtained under other typical electrolytic conditions, such as tetraalkylammonium tosylate in acetonitrile or in nitromethane. This result suggested that the intermediate generated by anodic oxidation of proline derivatives was highly stabilized in the reaction medium, which also assisted the progress of the following C–S bond-

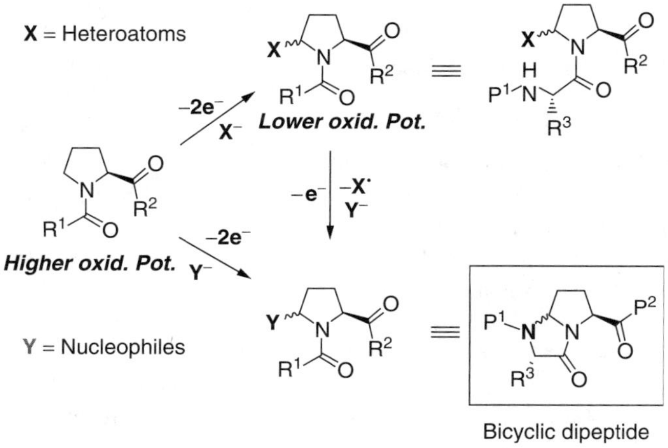

SCHEME 30.3 Synthetic plan for conformation-constrained proline residues and bicyclic dipeptidomimetics using electrochemical modification of proline moiety.

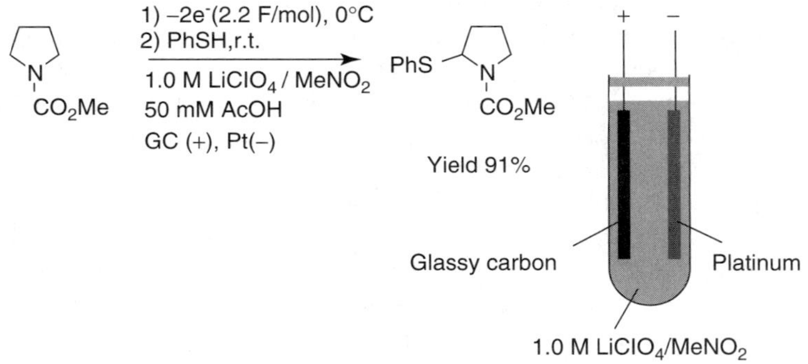

FIGURE 30.8 Anodic stepwise modification of a pyrolidine derivative. In a lithium perchlorate–nitromethane electrolyte system, anodically activated pyrrolidine formed stable cationic intermediate that can be trapped by thiols dissolved after the completion of electrolysis. The phenylsulfanylated pyrrolidine can be easily converted to varied derivatives.

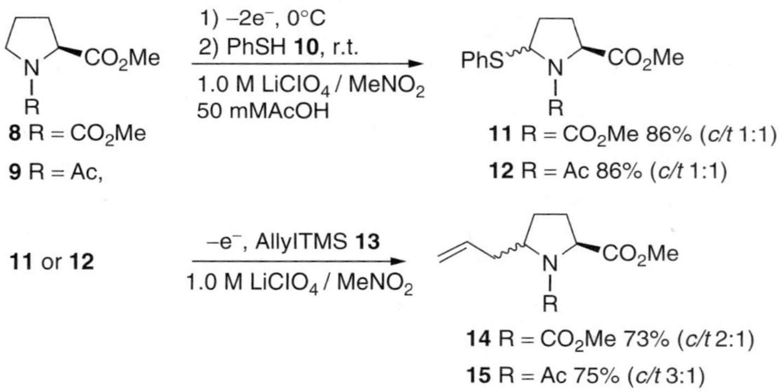

SCHEME 30.4 Stepwise anodic modification of proline derivatives.

forming reaction. The oxidation potential of α'-phenylsulfanyl proline derivatives was lowered to E_{ox} 1.2 V versus Ag–AgCl, which enabled us to try the oxidative cleavage of the C–S bond in the presence of electron-rich olefins. The anodic oxidation of compound **11** or **12** allowed the preparation of the alkylated products. Oxidative C–C bond formation of **11** or **12** with allyltrimethylsilane **13** (1.2 equivalents) in lithium perchlorate–nitromethane resulted in a 73 to 75% yield of allylated product **14** or **15** at room temperature. Although it is generally difficult to generate cationic intermediates of *N*-acylprolines followed by C–C bond formation that is triggered by anodic C–S bond cleavage, the desired reaction successfully occurred in this medium. It was presumed that the lowered oxidation potential based on the introduction of thiophenol to iminium cations and the electrolytic reaction medium with moderate Lewis acidity led to the desired oxidative C–S bond fission followed by a nucleophilic attack of the carbon nucleophile under mild conditions.

Furthermore, the application of a direct C–C bond formation of iminium cations with **13** was also investigated. After completion of anodic oxidation of **8** or **9** in lithium perchlorate–nitromethane in the presence of AcOH (2.2 F mol^{-1}) at 0°C followed by the addition of allyltrimethylsilane at room temperature, the desired products **14** or **15** were produced in good yield (see Scheme 30.5).

Finally, synthesis of the bicyclic conformation-constrained dipeptides using the sequential electrolytic reactions of dipeptides involving proline-moiety was attempted (see Scheme 30.6). Thus, a phenylsulfanyl group was electrochemically introduced at the α'-position of the proline-moiety of Boc-Gly-Pro-OMe **16** in order to generate the corresponding iminium cation intermediates at a lowered oxidation potential. The following anodic oxidation of the sulfide **17** afforded a cyclized dipeptide **18** in good yield in lithium perchlorate–nitromethane using undivided cell at room temperature. The cyclized dipeptides possess amino and carboxyl groups after deprotection, and they can be inserted into the synthetic peptide chains to regulate peptide conformations.

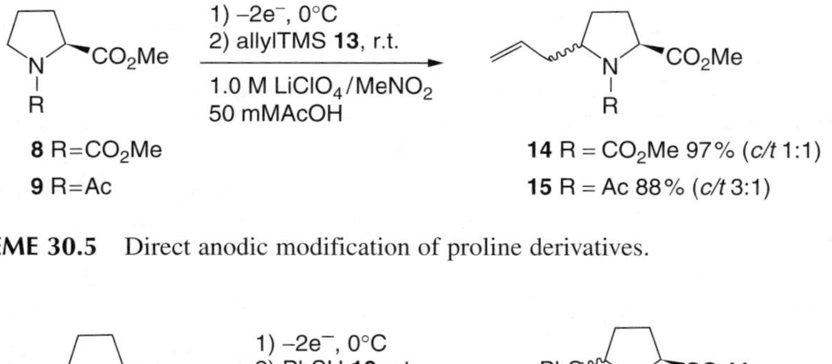

SCHEME 30.5 Direct anodic modification of proline derivatives.

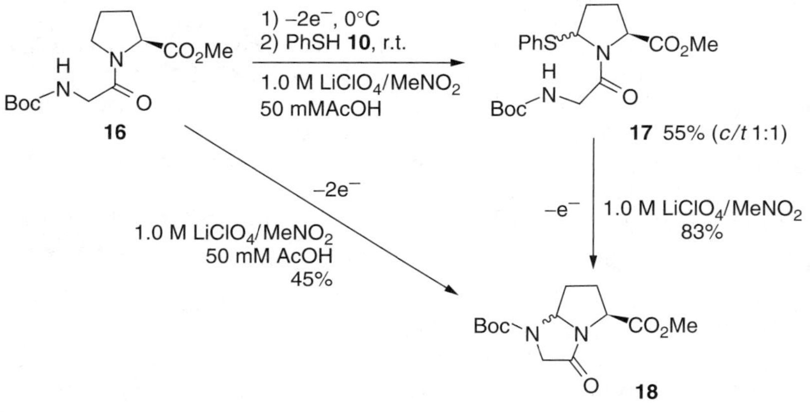

SCHEME 30.6 Synthesis of bicyclic diterpene moiety involving proline skeleton via anodic modification.

In conclusion, a practical new pathway for the synthesis of substituted proline derivatives or conformationally constrained dipeptides was developed. Amides and carbamates of proline derivatives were efficiently generated and trapped by nucleophiles in a lithium perchlorate–nitromethane solution. The introduction of thiophenol as a nucleophile lowered the oxidation potential of proline derivatives, leading to the oxidative fission of C–S bond; this was followed by intermolecular C–C bond-forming reaction with a nucleophilic attack of carbon nucleophiles under mild conditions. In addition, the intramolecular carbamate nitrogen atom of the dipeptide attached to the prolyl iminium cation under mild, electrooxidative conditions. In the sequential electrolytic reaction system, the reaction medium should work to stabilize iminium cation to assist the generation and accumulation of the intermediates via direct electrooxidation of proline moiety or anodic cleavage of the C–S bond.

REFERENCES

1. L.A. Thompson, J.A. Ellman. Synthesis and applications of small molecule libraries. *Chem. Rev.*, 96: 555–600, 1996.
2. J.A. Ellman. Design, synthesis, and evaluation of small-molecule libraries. *Acc. Chem. Res.*, 29: 132–143, 1996.
3. R.W. Armstrong, A.P. Combs, P.A. Tempest, S.D. Brown, T.A. Keating. Multiple-component condensation strategies for combinatorial library synthesis. *Acc. Chem. Res.*, 29: 123–131, 1996.
4. H. Han, M.M. Wolfe, S. Brenner, K.D. Janda. Liquid-phase combinatorial synthesis. *Proc. Natl. Acad. Sci., U.S.A.* 92: 6419–6423, 1995.
5. H. Han, K.D. Janda. Azatides: Solution and liquid phase syntheses of a new peptidomimetic. *J. Am. Chem. Soc.*, 118, 2539–2544, 1996.
6. H. Han, K.D. Janda. Soluble polymer-bound ligand-accelerated catalysis: asymmetric dihydroxylation. *J. Am. Chem. Soc.*, 118: 7632–7633, 1996.
7. D.J. Gravert, K.D. Janda. Organic synthesis on soluble polymer supports: liquid-phase methodologies. *Chem. Rev.*, 97: 489 1997.
8. C.W. Harwig, D.J. Gravert, K.D. Janda. Soluble polymers: new options in both traditional and combinatorial synthesis. *Chemtracts.*, 12: 1–26, 1999.
9. R.J. Booth, J.C. Hodges. Solid-supported reagent strategies for rapid purification of combinatorial synthesis products. *Acc. Chem. Res.*, 32: 18–26, 1999.
10. D.L. Flynn. Phase-trafficking reagents and phase-switching strategies for parallel synthesis. *Med. Res. Rev.*, 19: 408–431, 1999.
11. J.J. Parlow, R.V. Devraj, M.S. South. Solution-phase chemical library synthesis using polymer-assisted purification techniques. *Curr. Opin. Chem. Biol.*, 3: 320–336, 1999.
12. D.E. Bergbreiter. Palladium-catalyzed C–C coupling under thermomorphic conditions *J. Am. Chem. Soc.*, 122: 9058–9064, 2000.
13. D.E. Bergbreiter. P.L. Osburn, J.D. Frels. Nonpolar polymers for metal sequestration and ligand and catalyst recovery in thermomorphic systems. *J. Am. Chem. Soc.*, 123: 11105–11106, 2001.
14. I.T. Horvath, J. Rabai. Facile catalyst separation without water: fluorous biphase hydroformylation of olefins. *Science*, 266: 72–75, 1994.

15. J.A. Gladysz. Are teflon "Ponytails" the coming fashion for catalysts? *Science*, 266, 55–56, 1994.
16. B. Cornils. Fluorous biphase systems: the new phase-separation and immobilization technique. *Angew. Chem. Int. Ed. Engl.*, 36: 2057–2059, 1997.
17. I.T Horvath. Fluorous biphase chemistry. Accounts of chemical research. *Acc. Chem. Res.*, 31: 641–650, 1998.
18. R.H. Fish. Fluorous biphasic catalysis: a new paradigm for the separation of homogeneous catalysts from their reaction substrates and products. *Chem. Eur. J.*, 5: 1677–1680, 1999.
19. P.Jr Wentworth, K.D. Janda. Liquid-phase chemistry: recent advances in soluble polymer-supported catalysts, reagents and synthesis. *Chem. Commun.*, 19: 1917–1924, 1999.
20. D.P. Curran, S. Hadida. Tris(2-(perfluorohexyl)ethyl)tin hydride: a new fluorous reagent for use in traditional organic synthesis and liquid phase combinatorial Synthesis. *J. Am. Chem. Soc.*, 118: 2531–2532, 1996.
21. G. Pozzi, S. Banfi, A. Manfredi, F. Montanari, S. Quici. Towards epoxidation catalysts for fluorous biphase systems: synthesis and properties of two Mn(III)-tetraarylporphyrins bearing perfluoroalkylamido tails. *Tetrahedron,* 52: 11879–11888, 1996.
22. A. Studer, S. Hadida, R. Ferritto, S.Y. Kim, P. Jeger, P. Wipf, D.P. Curran. Fluorous synthesis: a fluorous-phase strategy for improving separation efficiency in organic synthesis. *Science*, 275: 823–826, 1997.
23. J.J.J. Juliette, I.T. Horvath, J.A. Gladysz. Transition metal catalysis in fluorous media: practical application of a new immobilization principle to rhodium-catalyzed hydroboration. *Angew. Chem. Int. Ed. Engl.*, 36: 1610–1612, 1997.
24. S. Kainz, D. Koch, W. Baumann, W. Leitner. Perfluoroalkyl-substituted arylphosphines as ligands for homogeneous catalysis in supercritical carbon dioxide. *Angew. Chem. Int. Ed. Engl.*, 36: 1628–1630, 1997.
25. B. Cornils. Fluorous biphase systems: the new phase-separation and immobilization technique. *Angew. Chem. Int. Ed. Engl.*, 36: 2057–2059, 1997
26. D. Rutherford, J.J.J. Juliette, C. Rocaboy, I.T. Horvath, J.A. Gladysz. Transition metal catalysis in fluorous media application of a new immobilization principle to rhodium-catalyzed hydrogenation of alkenes. *Catal. Today*, 42: 381–388, 1998.
27. I.T. Horvath, G. Kiss, R.A. Cook, J.E. Bond, P.A. Stevens, J. Rabai, E.J. Mozeleski. Molecular engineering in homogeneous catalysis: one-phase catalysis coupled with biphase catalyst separation. The fluorous-soluble $HRh(CO)\{P[CH_2CH_2(CF_2)_5CF_3]_3\}_3$ hydroformylation system. *J. Am. Chem. Soc.*, 120: 3133–3143, 1998.
28. H. Tamaki, T. Obata, Y. Azefu, K. Toma. A novel protecting group for constructing combinatorial peptide libraries. *Bull. Chem. Soc. Jpn.*, 74: 733–738, 2001.
29. K. Chiba, Y. Kono, S. Kim, K. Nishimoto, Y. Kitano, M. Tada. A. Liquid-phase peptide synthesis in cyclohexane-based biphasic thermomorphic systems, *Chem. Commun.*, 16: 1766–1767, 2002.
30. T. Shono, Y. Matsumura, K. Tsubata. Electroorganic chemistry. 46. A new carbon-carbon bond forming reaction at the .alpha.-position of amines utilizing anodic oxidation as a key step. *J. Am. Chem. Soc.*, 103: 1172–1176, 1981.
31. R.D. Long, K.D. Moeller. Conformationally constrained peptide mimetics: The use of a small lactam ring as an HIV-1 antigen constraint. *J. Am. Chem. Soc.*, 119: 12394–12395, 1997.

32. J. Yoshida, M. Itoh, S. Isoe. Electrooxidative coupling of α-heteroatom-substituted organostannanes and organosilanes. *J. Chem. Soc. Chem. Commun.*, 6: 547–549, 1993.
33. J. Yoshida, S. Suga, S. Suzuki, N. Kinomura, A. Yamamoto, K. Fujiwara. Direct oxidative carbon-carbon bond formation using the "cation pool" method. 1. Generation of iminium cation pools and their reaction with carbon nucleophiles. *J. Am. Chem. Soc.*, 121: 9546–9549, 1999.
34. K. Chiba, M. Tada. Diels-Alder Reaction of Quinones generated *in situ* by electrochemical oxidation in lithium perchlorate-nitromethane. *J. Chem. Soc. Chem. Commun.*, 21: 2485–2486, 1994.
35. K. Chiba, M. Jinno, R. Kuramoto, M. Tada. Stereoselective Diels-Alder reaction of electrogenerated quinones using PTFE-fiber coated electrode. *Tetrahedron Lett.*, 39: 5527–5530, 1998.
36. K. Chiba, M. Fukuda, S. Kim, Y. Kitano, M. Tada. Dihydrobenzofuran synthesis by an anodic [3+2] cycloaddition of phenols and unactivated alkenes. *J. Org. Chem.*, 64: 7654–7656, 1999.
37. K. Chiba, R. Uchiyama, S. Kim, Y. Kitano, M. Tada. Benzylic intermolecular carbon-carbon bond formation by selective anodic oxidation of dithioacetals. *Org. Lett.*, 3: 1245–1248, 2001.
38. M. Jinno, Y. Kitano, M. Tada, K. Chiba. Electrochemical generation and reaction of *o*-quinodimethanes from {[[2-(2,2-Dibutyl-2-stannahexyl)phenyl]methyl]thio} benzenes. *Org. Lett.*, 1: 435–438, 1999.
39. S. Kim, Y. Kitano, M. Tada, K. Chiba. Alkylindan synthesis *via* an Intermolecular [3+2] Cycloaddition between unactivated alkenes and *in situ* generated *p*-quinomethanes. *Tetrahedron Lett.*, 41: 7079–7083, 2000.
40. S. Kim, Y. Kitano, M. Tada, K. Chiba. Benzylic Nitroalkylation by paired electrolysis of benzyl sulfides in nitroalkanes *J. Electroanal. Chem.*, 507: 152–156, 2001.
41. K. Chiba, T. Miura, S. Kim, Y. Kitano, M. Tada. Electrocatalytic intermolecular olefin cross-coupling by anodically induced formal [2+2] cycloaddition between enol ethers and alkenes. *J. Am. Chem. Soc.*, 123: 11314–11315, 2001.
42. S. Kim, K. Hayashi, Y. Kitano, M. Tada, K. Chiba. Anodic modification of proline derivatives using a lithium perchlorate/nitromethane electrolyte solution. *Org. Lett.*, 4: 3735–3737, 2002.
43. K. Chiba, T. Miura, S. Kim, Y. Kitano, M. Tada. Anodic modification of proline derivatives using a lithium perchlorate/nitromethane electrolyte solution. *J. Am. Chem. Soc.*, 123: 11134, 2001.

31 Lacticin 3147

Paul D. Cotter, Colin Hill, and R. Paul Ross

CONTENTS

31.1 INTRODUCTION

Bacteriocins are antimicrobial peptides/proteins produced by bacteria. A number of bacteriocins are produced by lactic acid bacteria, and these are usually assigned to one of three classes. Class I, the lantibiotics, are a group of ribosomally synthesized peptides that undergo extensive posttranslational modification resulting in the formation of lanthionine bridges and other unusual amino acids. Class II are small, heat stable, non-lanthionine-containing, membrane-active peptides; and Class III comprises large, heat-labile proteins (1). Lacticin 3147, the subject of this chapter, is a two-peptide lantibiotic containing seven lanthionine groups in total. Lanthionines are formed when hydroxy amino acids (serine and threonine) are selectively dehydrated to form dehydroalanine (Dha) and dehydrobutyrine (Dhb), respectively (2). The resultant α,β-unsaturated residues may then undergo Michel addition reactions with the thiol group of specific cysteine residues to form the characteristic lanthionine (lan) and β-methyl-lanthionine (meLan) residues that have been shown to contribute to tolerance of high temperatures (3), oxidizing conditions (4), low pH (5), and proteolysis (6).

Nisin is the prototypical lantibiotic. It was discovered in 1928 (7), was added to the positive list of food additives by the European Union as additive number E234 in 1983, and was approved by the U.S. Food and Drug Administration (FDA) for use as a biopreservative in 1988 (8). It remains the only lantibiotic that is commercially produced and sold in more than 40 countries. While other bacteriocins produced by GRAS (generally regarded as safe) organisms such as lactic acid bacteria (LAB) have been the subject of intensive research over the past 20 years (5), only a relatively small number have demonstrated significant potential for food and biomedical applications. One of the most obvious exceptions is the lantibiotic lacticin 3147. In fact, since its original discovery in 1996 the study of lacticin 3147 has provided us with an excellent model of how modern approaches to bioactive proteins and peptides can be utilized, both independently and in conjunction with classic techniques, to study a novel bacteriocin from both fundamental and applied perspectives.

Lacticin 3147 produced by a strain of *Lactococcus lactis*, subsp. *lactis* (DPC3147), is isolated from an Irish kefir grain. Kefir grains (or buttermilk plants) have been used in Irish households for many generations for the souring of milk for soda bread preparation and are colonized by heterogenous flora consisting of lactococci (10^9 g^{-1}), leuconostoc (10^8 g^{-1}), lactobacilli (10^6 g^{-1}), acetic acid bacteria (10^5 g^{-1}), and yeast (10^6 g^{-1}) (9). Kefir grains were screened for bacteria-producing antimicrobial peptides on the basis that no contamination of kefir grains with undesirable microorganisms has been reported despite the grains not being treated aseptically in homes. DPC3147 was one of six strains, isolated from four separate grains that had been sourced from different locations in Ireland, that were found to produce this lantibiotic. Another lacticin 3147 producer, *L. lactis* IFPL105, has been isolated from goat's milk in Spain (10,11). In both these cases, lacticin 3147 producers were isolated using classic screening strategies involving standard direct plating. In addition, producers of this bacteriocin were frequently reisolated from kefir grains using a novel method for the isolation of microbially derived inhibitory substances (12). This method involves a rapid throughput multiwell plate assay and, though designed for the screening of kefir grains, has the potential to be modified for the isolation of antimicrobial peptides from other sources. Kefir grains were incubated in 10% reconstituted skim milk for 20 hours at 32°C, the fermentates were aliquoted into multiwell plates, and a known number of indicator cells were added to each well. The fermentates were incubated for a further 20 hours, and counts were carried out to determine whether a reduction in indicator cell numbers had occurred. A reduction in cell-forming units indicated the presence of an inhibitory substance, and these fermentates were selected for further investigation (12).

The antimicrobial activity of lacticin 3147 was assayed initially using an agar well diffusion assay (13). Briefly, molten agar at 48°C was seeded with an indicator strain (50 μL of an overnight culture per 20 mL agar), dispensed into sterile Petri dishes, and allowed to solidify. Wells of approximately 4.6 mm in diameter were made, and 50 μL aliquots of a twofold serial dilution of the bacteriocin preparation were dispensed into the wells. It was found that growth was inhibited in all Gram-positive bacteria tested (including *Bacillus* sp., *Clostridum* sp., *Enterococcus* sp., *Lactobacillus* sp., *Lactococcus* sp., *Leuconostoc* sp., *Listeria* sp., *Pediococcus* sp., *Staphylococcus* sp., and *Streptococcus* sp.) (10,14). A more sensitive assay which demonstrated that lacticin 3147 displays activity in buffer against Gram-positive bacterial pathogens that appear insensitive in standard plate assays has also been

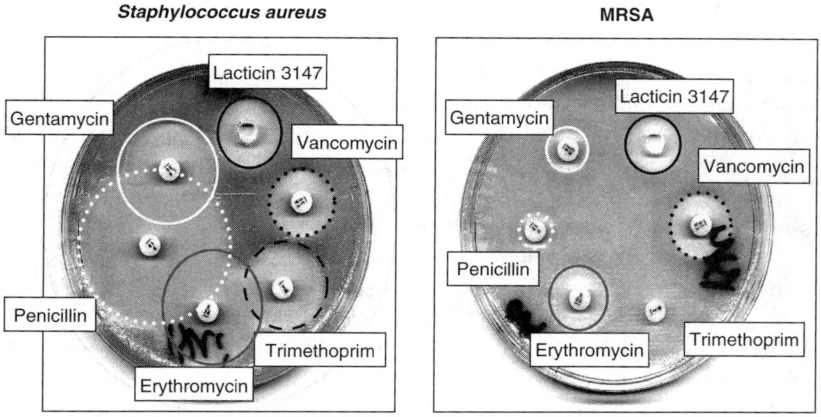

FIGURE 31.1 Lacticin 3147 is active against methicillin resistant *S. aureus* (MRSA).

developed. This assay was used to determine that lacticin 3147 is effective against methicillin-resistant *Staphylococcus aureus* (Figure 31.1), vancomycin-resistant *Enterococcus faecalis*, penicillin-resistant *Pneumococcus*, *Propionibacterium acne*, and *Streptococcus mutans* (15).

Lacticin 3147 is produced during the exponential phase, with a peak during early stationary phase. It is active at both acidic and neutral pHs and is heat stable, particularly at an acidic pH. The production of lacticin 3147 is associated with a 60,232 bp conjugative plasmid pMRC01, while production by IFPL105 requires a 46-kb nonconjugative plasmid pBAC105. The molecular characteristics of the genetic determinants and the bacteriocin will be discussed in a later section; the potential applications for lacticin and lacticin producing cultures will be discussed in the following section.

31.2 APPLICATIONS FOR LACTICIN 3147 IN FOOD AND IN MEDICINE

31.2.1 Inhibition of Nonstarter Lactic Bacteria and the Regulation of Flavor and Quality in Cheese

Flavor and quality are the key characteristics of hard cheeses such as cheddar, and these factors govern their value and market destination. During cheese manufacture, parameters such as processing conditions, ripening temperatures, and packaging are controlled, but factors such as milk composition and adventitious nonstarter flora, for example, nonstarter lactic acid bacteria (NSLAB), can be highly variable. Cheddar manufacture relies upon the growth of a few deliberately introduced strains of *L. lactis*, termed the starter culture. During a 6-month cheese ripening period, NSLAB can reach 10^8 colony-forming units per gram (cfu g^{-1}) of cheese. While the precise contribution of various NSLAB to the quality of cheddar cheese is unclear, they are thought to mediate proteolysis and influence flavor development during ripening, often positively (16–18). However they are also associated with a major defect in cheese, namely, the development of calcium-D-lactate crystals produced as a consequence of their ability to racemize L-lactate to D-lactate (19), which can result in major economic losses. NSLAB populations have also been associated with "slit defects" in cheddar, while the variable nature of NSLAB populations probably contributes to the inconsistencies often associated with cheddar flavor and quality. It was postulated that the use of bacteriocins as biological tools could offer the cheeseaker greater control and predictability with respect to these adventitious cheese flora.

The potential application of lacticin 3147 in cheddar cheese manufacture was first tested using a combination of three natural kefir isolates as starters. Importantly, it was found that acid production by these cultures was similar to that observed in commercial starters. This is in contrast to nisin-producing starters, which generally perform poorly as cheese starters as a consequence of their slow acid production, reduced proteolytic activity, and phage-sensitivity (20,21). Initial results were very encouraging in that the use of lacticin 3147 producers as starters resulted in the complete absence of NSLAB from cheese after 6 months, compared with a figure of $10^{7.5}$ CFU g^{-1} after 4 months in a cheese made with a regular starter. A strain DPC4275 was subsequently created by conjugally transferring pMRC01 into a commercial cheddar cheese starter *L. lactis* DPC4268. DPC4275 was found to be suitable for use as a cheese starter in that acid was produced at only a slightly lower rate and the gross composition of the cheese was unaffected. Although less bacteriocin was produced by the transconjugant, a significant 100-fold reduction in NSLAB was observed relative to that observed in a control cheese containing DPC4268 (14).

As a result of these initial trials, pMRC01 was transferred to a variety of lactococcal starter cultures used for cheese and lactic butter production. The transconjugants successfully maintained the plasmid and were observed to produce acid, acetolactate, and diacetyl at rates approaching those of the corresponding parent strain (22). These transconjugants all have the ability to limit the growth of NSLAB. However, while NSLAB may be a major source of inconsistency in the quality of some cheeses, it is desirable not to eliminate the potentially positive effects that lactobacilli and other starter adjuncts can have on cheese flavor (16). Consequently, a system to develop adjunct strains that are less sensitive to lacticin 3147 has been created. While resistance to lacticin 3147 has not been observed among sensitive organisms, increased tolerance can be obtained on repeated exposure to increasing concentrations of the bacteriocin. Thus, *Lactobacillus paracasei* subsp. *paracasei*, DPC5336, isolated from a commercial cheddar cheese, was exposed to increasing concentrations of lacticin 3147, which resulted in the creation of a stable, more tolerant variant DPC5337. A trial comparing the ability of the two adjunct strains to grow in cheddar cheese with DPC3147 as a starter demonstrated that the resistant strain was present in 10- to 100-fold greater numbers than the sensitive parent after 6 months. A similar trend was observed when DPC4275 was used as the starter. It was confirmed by genetic fingerprinting that the tolerant *Lactobacillus* became the predominant microflora in the cheese, which was also found to be of higher quality by commercial graders than a control cheese (23).

In addition to controlling quality and flavor, lacticin 3147 has also been used to accelerate ripening through the controlled lysis of LAB resulting in the release of intracellular peptidase. The first indications of this potential were observed when lacticin 3147 was shown to induce lysis of two LAB commonly used in the manufacture of goat's milk cheese, resulting in elevated release of the intracellular marker enzyme postproline dipeptidyl aminopeptidease (Pep X) (24). The practical application of this observation was first tested by introducing *L. lactis* IFPL105 into cheese curd slurries. As predicted, this resulted in the lysis of starter adjuncts with high peptidase activity. By optimizing the conditions, it was possible to avoid late acidification by the starter and to promote early cell lysis with the subsequent release of intracellular peptidases and the acceleration of ripening without altering the cheese-making process (25). As with the cheddar cheese studies, it was found that this procedure was improved upon by the transfer of the relevant lacticin-producing plasmid (pBAC105) to a commonly used starter culture, *L. lactis* IFPL359, to generate *L. lactis* IFPL3593. Again, accelerated cheese ripening due to early cell lysis of adjuncts was observed (24). In addition to the enhanced ripening observed in the presence of IFPL3593, an improvement in flavor also resulted. This development of enhanced flavor may be due to enhanced branched chain amino acid transamination, which is involved in cheese flavor development. When IFPL359 was treated with lacticin 3147, an increase in branched-chain amino acid transamination was observed. This occurred as a consequence of cell permeabilization, permitting free diffusion of amino acids into the cell and cell lysis, both of which result in the increased accessibility of the enzymes to their substrate (26).

31.2.2 Inhibition of Pathogens in Food

The greatest potential for broad-host-range bacteriocins may lie in their ability to inhibit the growth of Gram-positive pathogens. In the following section, a variety of mechanisms for incorporating lacticin into a food system are described; however, in all cases, the common theme is the inhibition of pathogens.

As mentioned before, the production of lacticin 3147 by a starter culture is frequently the method of choice. The lacticin-producing transconjugant DPC4275 has been used to determine the success of lacticin 3147 in inhibiting the growth of *L. monocytogenes* in cottage cheese. Cottage cheese was chosen as a test food because postproduction contamination by *L. monocytogenes* could be a problem in its manufacture. The inhibition of this pathogen is especially desirable as it is responsible for 27.6% of the total deaths caused by food pathogens (27) and for 71% of all recalls of food products due to bacterial contamination between 1993 and 1998 in the U.S. (28). One of the more lacticin-tolerant *L. monocytogenes* strains, ScottA, was introduced into a cream dressing of cottage cheese (pH 5.2) at a level of 10^4 cells g^{-1}. At 4°C, a 99.9% reduction was observed within 5 days when DPC4275 was used as a starter, while no decline was observed when utilizing a starter that did not produce lacticin 3147 (29). While the inhibitory effect of a nisin concentrate has been demonstrated in cottage cheese (30), the utilization of a starter producing the bacteriocin may be more widely accepted as it would not be considered an additive and thus would not require special labelling. DPC4275 has also been used as a protective culture to inhibit *Listeria* on the surface of mold-ripened cheese. The pH on the surface of such cheeses can range from 6.5 to 8, and such conditions are favorable for the growth of *L. monocytogenes*. It was observed that the presence of the lacticin 3147 producer on the cheese surface reduced the numbers of deliberately inoculated *L. monocytogenes* Scott A 1,000-fold, while in a parallel experiment, a nisin producer caused no reduction of *Listeria* (31).

When *L. lactis* DPC4275 was used as an alternative to *Lactobacillus sake* and *Staphylococcus carnosus* as a starter in salami manufacture, it was found to be favorable for preservation and hygienic stability because the pH values dropped below 5.1 and an Aw of 0.9 was achieved. In addition, the salami thus produced demonstrated good sensory and colorimetric qualities (32). However, one of the limitations associated with the use of a bacteriocinogenic *L. lactis* meat starter culture is the inhibition of growth and bacteriocin production by sodium nitrite and sodium chloride. However, it was found that when DPC4275 was treated with 2.5 ppm manganese (Mn) and 250 ppm magnesium (Mg), which have previously been used as cryoprotective agents in the lyophilization of LAB (33,34), the culture reached significantly higher levels in beaker sausage by day 10 and demonstrated improved acid and bacteriocin production (35). Significantly, reductions in *L. innocua* and *S. aureus* populations were observed in spiked beaker sausage samples in the presence of DPC4275 relative to a nonbacteriocinogenic control strain (35).

Another mechanism for incorporating bacteriocin involves the development of a lacticin 3147–enriched whey powder. Such a powder was generated through a DPC3147 fermentation of a 10% solution of demineralized whey powder followed by spray drying. It was found that 10 and 15% solutions of this powder were effective against *L. monocytogenes* and *S. aureus*, respectively, and this powder was also effective when incorporated into an infant milk formula. Again, these results emphasize the advantage of lacticin 3147 over nisin, which is reported to be most effective in foods due to its acidic pH (<6) and low protein/fat content and thus would not be suitable for infant milk formula (36). The incorporation of this powder also resulted in an 85% and 99.9% reduction in *L. monocytogenes* in cottage cheese and natural yogurt within 2 hours, while the *B. cereus* numbers were reduced by 80% within 3 hours. in soups containing lacticin powder. Reconstituted powder was most stable at –20°C, retained full activity over a 9-month period of study, and lost no activity when subjected to temperatures up to 70°C or even higher temperatures. It is anticipated that this lacticin powder could be used to replace the dairy component of some foods with a view to increased safety and longer shelf life (37).

It has also been shown that it is possible to benefit from the antimicrobial activity of lacticin 3147 by incorporating it into a cellulose-based food packaging material. Bioactive packaging allows industry to combine the preservative functions of antimicrobials with the protective functions of preexisting packaging concepts. In this form, activity remained stable over a 3-month trial (38).

With a view toward commercial-scale production, tests were carried out to determine whether lacticin 3147 production could be adapted to allow continuous production by cells immobilized in calcium alginate. The feasibility of continuous production was demonstrated over a 180-hour period (39). The advantages of this, in addition to the development of a more stable and long-term means of producing lacticin 3147, include improved process control, simplified cell recovery procedures, decreased cell washout, and the ability to apply high dilution rates in continuous processes (40).

The inhibition of pathogens by lacticin 3147 can often be augmented by other antimicrobial treatments. When hydrostatic pressure was used with lacticin 3147 for the treatment of milk and whey to improve the quality of minimally processed dairy foods, it was found that treatments had a synergistic effect (41). In addition, treatment of lacticin 3147 preparations at pressures >400 MPa resulted in a doubling of activity, possibly as a consequence of the individual lacticin peptides coming together to form active bacteriocin (41). When lacticin 3147 was combined with other bacteriocins, varying degrees of inhibition resulted. A synergistic effect was observed also when lacticin was combined with nisin, lactococcins A, B, and M, or enterocin A. The effects were additive when used with lacticin 481, and no interaction occurred with lacticin 115, pediocin PA-1, or lacticin 104. When assayed at pH 7 and pH 5, it was found that inhibition of *L. monocytogenes* and *S. aureus* was greater at pH 7, and thus, no advantage is gained from combining lacticin 3147 with low pH stress (M. Galvin, S.M. Morgan, R.P. Ross, T. Beresford, and C. Hill, personal communication, 2001).

31.2.3 Mastitis Prevention in Cattle

Bovine mastitis is one of the most persistent and costly diseases in dairy cows. It is usually treated with intramammary antibiotics. There are a number of disadvantages to this, one of which is the potential presence of residues in the milk of treated cows, resulting in milk being withheld. Nisin has been shown to be effective against a number of mastitic pathogens (42); its use does not compromise downstream processing, and it is readily degraded by the digestive system. However, the poor solubility of nisin at physiological pH is again an issue. Despite this, nisin has been used successfully in combination with lysostaphin (43) and is also the active agent in two commercially available teat wipes.

Lacticin 3147 has also been shown to be active against mastitic pathogens when incorporated into a teat seal product. The teat seal is an oil-based formulation that forms a physical barrier against infection in the area of the teat canal and sinus. These seals are usually introduced after antibiotic treatment and a drying-off period. In the case of teat seal containing lacticin 3147, it was found that bacteriocin activity improved upon addition of Tween 80 to the bacteriocin prior to mixing with the teat seal (44). Lacticin did not suffer a loss of activity at physiological pH, did not cause teat irritancy, and retained activity 8 days after its introduction into an adult cow's teat (45).

This technology was tested to determine its efficacy *in vivo* using 68 uninfected quarters of 18 cows. Following infusion with either teat seal alone or teat seal containing lacticin 3147, *S. dysgalactiae* was introduced into each teat sinus at a level predicted to cause a 50% infection rate in the teats with the seal alone. Over the following 8 days, 61% of control quarters (seal) and only 6% of treated quarters (seal and lacticin) either developed clinical mastitis or were shedding the challenge organism (45). The success of this treatment is probably due to the fact that in addition to the barrier protection of the seal, the seal also localizes the inhibitor in the teat sinus, with the result that less bacteriocin is required to prevent infection. A similar trend was again observed when *S. aureus* was introduced to simulate a natural mastitis infection. Teat seal with lacticin 3147 reduced the number of teats shedding viable cells when an inoculum of either 1.7×10^3 or 6.8×10^3 was used. In teats that continued to shed *S. aureus*, the numbers present were also reduced relative to the control experiment involving teat seal without lacticin 3147. Given the success of this trial, and given that *S. aureus* cells are the least lacticin sensitive of mastitis-causing pathogens, it is likely that this strategy would also be effective in the treatment of mastitis caused by *S. uberis*, *S. agalactiae*, and *S. bovis* (46).

31.2.4 Tolerance, Resistance, and Immunity

With the growing use of bacteriocins as antimicrobial agents, a new field of study characterizing the emergence of bacteriocin-resistant bacteria and the associated underlying mechanisms has developed. This is important because the two factors that are most likely to limit the use of bacteriocins are (a) a limited spectrum of activity (although, as outlined earlier, lacticin 3147 inhibits all Gram-positive bacteria tested), and (b) the emergence of spontaneously resistant variants, which is discussed below. These studies are likely to be critical when one considers the frequency with which antibiotic-resistant pathogens have emerged.

For clarity, the following definitions, as applied in this chapter, are included to differentiate between *tolerant* and *resistant* mutants. A *resistant* mutant is one that becomes insensitive to any level of the bacteriocin, while a *tolerant* strain has the ability to withstand a certain level of the bacteriocin. The first analysis into the emergence of lacticin 3147–resistant cultures stemmed from the use of lacticin to screen for tranconjugants into which pMRC01 had been introduced. Of the 11 *L. lactis* strains tested, only DPC 220 produced mutants tolerant to 400 arbitrary units (AU) per milliliter, though at a frequency of less than 10^{-8} per donor (22). Subsequent studies showed the corresponding figure for *S. dysgalactiae* M was <1 in 10^{-6}, while no tolerant *S. aureus* 10 were observed. Importantly, these tolerant derivatives were sensitive to increased concentrations of lacticin and are thus not termed resistant under our definition (44). Even among cells that are more likely to tolerate bacteriocin, such as those that have been acid adapted to induce a general cross-protective stress response, it was found that tolerance to lacticin

3147 increased only moderately; however, in contrast, such cells were significantly more nisin tolerant than their nonadapted counterparts (47).

This study was extended to determine if it was possible to isolate resistant variants of *L. monocytogenes* or *L. lactis*. While no resistant isolates were recovered, it was again found that strains could tolerate relatively low levels of lacticin 3147. These tolerant spontaneous mutants were classified as being either R1 or R2 mutants. R1 mutants are those that tolerate a certain level of lacticin 3147 without previous exposure, while R2 mutants are generated from R1 strains that demonstrate enhanced tolerance as a consequence of being exposed to incremently higher levels of lacticin. Tolerant R1 spontaneous mutants emerge at a very low frequency in general (in the majority of cases between 10^{-6} and 10^{-7}) (48,49). It was found to be possible to generate R2 mutants by increasing the tolerance of strains in incremental steps up to 13,000 AU (MG1363), 8,000 AU (ScottA), and 11,000 AU (LO28) AU per milliliter (49). This tolerance was found to be stable. When sensitive and tolerant strains were subsequently compared, it was found that (a) the bacteriocin adsorbed equally well to both strains, (b) the growth rates of the strains were similar at optimum growth temperatures but the tolerant strain grew more slowly at lower temperatures, and (c) the tolerant strain was found to be more hydrophobic (48). Thus, while tolerant strains can be found at low frequencies, this is only observed in relatively low levels of lacticin, and tolerance can only be increased with small concentrations of lacticin in a stepwise manner. Given that these situations are unlikely to occur in a food system, their appearance and stability should not be of concern to the food industry (48). It would thus seem that the application of lacticin 3147 is not restricted by the potential problems outlined above, as it demonstrates activity against all Gram-positive bacteria tested; while bacteria can be preconditioned to increase their tolerance, spontaneous resistance to lacticin 3147 has not been observed. In addition to allowing us to predict the likelihood that mutants resistant to a particular bacteriocin will emerge, these studies also identify ways in which it may be possible to generate tolerant strains that may be of practical use. It has already been outlined how the lacticin-tolerant variant *Lactobacillus paracasei* subsp. *paracasei* of DPC5336 (DPC5337) was used as an adjunct in the preparation of cheddar cheese. A study of DPC5337 showed that it did not flocculate as well as DPC5336 did, suggesting that cell surface changes existed. This correlates with the observation that DPC5337 cells adsorb less bacteriocin (23). This study demonstrates that while lacticin 3147 has an obvious role in the inhibition of undesirable bacteria, it can also be used to select for particular strains in a complex microbial background.

In contrast to resistance, *immunity* is defined as the self-protection mechanism possessed by bacteriocin-producing strains in order to prevent cell death from the action of their own bacteriocin. Following the same logic as that used in the application of a tolerant adjunct, it has been suggested that the lacticin immunity gene *ltnI* (see the section that follows) could also be used to permit strain selection by lacticin 3147 (50). Furthermore due to the restrictions associated with the use of antibiotic markers in food applications, there is a need for alternative types of selection from a biotechnology point of view. With this in mind, a number of food-grade plasmids that use bacteriocin resistance as a selectable marker have been engineered (51,52). Given the broad range of activity of lacticin 3147 and the low frequency of tolerant variants, it is likely that the *ltnI* gene can be successfully applied in a similar fashion.

31.3 MOLECULAR ANALYSIS OF LACTICIN 3147 PRODUCTION AND IMMUNITY

The antimicrobial activity of lacticin 3147 results from the combined action of two posttranslationally modified peptides, LtnA1 and A2. Both peptides are required for optimal activity and combine to act on the target cell membrane to elicit the leakage of potassium ions (53). pMRC01, the plasmid associated with bacteriocin production and immunity from DPC3147, has been sequenced (60,232 bp). The plasmid can be divided into three ~20-kb regions encoding (a) conjugative transfer functions, (b) phage resistance and plasmid replication, and (c) lacticin production and immunity. A GC (guanine cytosine) content of 30%, rather than the 36 to 38% typical of *L. lactis* chromosomal DNA (deoxyribonucleic acid) (54), suggests horizontal transfer from an organism with a lower GC (guanine cytosine) composition. A combination of bioinformatics and experimental studies has unveiled the function of a number of genes. Of the 64 identified open reading frames (ORFs), 10 have been associated with the production of and immunity to lacticin 3147. These 10 (12.6 kb in length) are located within a 20-kb region flanked by two IS*946* insertion elements. The six-gene operon associated with production and modification consist of *ltnA1*, *A2*, *M1*, *T*, *M2*, and *D* (Figure 31.2), with *ltnA1* and *ltnA2* encoding the structural peptides. While it has been shown that LtnA1 has some independent activity when present at high concentrations (45), both LtnA1 and LtnA2 must be produced to detect activity in cell-free supernatant (53). Both peptides contain unusual thioether amino acids, termed lanthionines. Lacticin 3147 is one of four known lantibiotic two-peptide bacteriocins, the others being cytolysin ($CylL_L$ and $CylL_S$) (55) plantaricin W (Plwα and Plwβ) (56), and staphylococcin C55 (C55a and C55b) (57). In addition, the genome sequence of *Bacillus halodurans* (58) has identified two potential two-component lantibiotics designated BhaA1 and BhaA2. The primary unmodified amino acid sequence of LtnA1 shares 78% similarity with C55a (58) and 40% with Plwα (59), while LtnA2 shares 63% similarity with C55b and 26% with Plwβ (see Figure 31.2). These results indicate that the strains C55 and DPC3147 produce closely related two-component lantibiotic systems. Indeed, downstream of *sac* αA and *sac* βB, the C55a and -b determinants, homologues of *ltnM1*, *ltnT*, *ltnM2*, and *ltnD* are located (60). Twomey et al. (61) classified lantibiotic peptides into six categories: the nisin, lacticin 481, mersacidin, LtnA2, cytolysin, and lactocin S groups. LtnA1, C55a, Plwα, BhaA1, RumB (EMBL AF320327), mersacidin, and actagardine belong to the mersacidin group. The best-studied members of the mersacidin subgroup are mersacidin (62) and actagardine (63), both of which have globular rigid structures with overlapping bridging structures. While a number of peptides from two-component lantibiotics belong to the mersacidin group, the four members of the

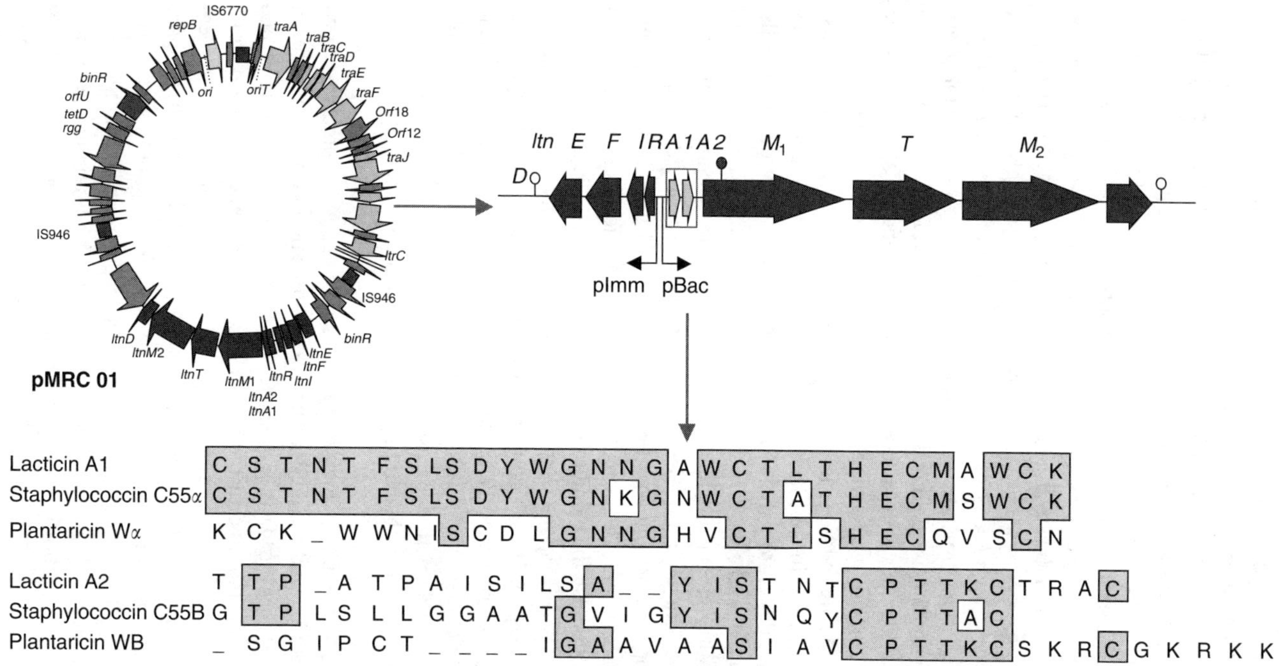

FIGURE 31.2 The lacticin genes are present on the 60.2-kbp plasmid pMRC01.[a] Two of these genes, *ltnA1* and *ltnA2*, encode the structural peptides, the unmodified amino acid sequence of which closely resembles those of other two-component lantibiotics, that is, Staphylococcin[b] (57) and Plantaricin W.[c] (From [a]M.P. Ryan, M.C. Rea, C. Hill, R.P. Ross. *Appl. Environ. Microbiol.*, 62, 612–619, 1996; and B.A. Dougherty, C. Hill, J.F. Weidman, D.R. Richardson, J.C. Venter, R.P. Ross. *Mol. Microbiol.*, 29, 1029–1038, 1998; [b] M.A. Navaratna, H.G. Sahl, J.R. Tagg. *Appl. Environ. Microbiol.*, 64, 4803–4808, 1998; [c] H. Holo, Z. Jeknic, M. Daeschel, S. Stevanovic, I.F. Nes. Plantaricin W from *Lactobacillus plantarum* belongs to a new family of two-peptide lantibiotics. *Microbiology*, 147, 643–651, 2001. With permission.)

two-component systems are more closely related to each other than they are to other members of the group. LtnA2, C55b, Plwβ, and BhaA2 are the only four members of the LtnA2 group. These hydrophobic molecules are thought to have a general linear structure with a rigid central region. In general, the two peptides of the two-component lantibiotics act synergistically to mediate their effect. LtnA1 and C55a, but not LtnA2 and C55b, have intrinsic antimicrobial properties (45,58,64), while both the plantaricin W peptides possess activity (56). The similarity between LtnA1 and mersacidin suggests that this peptide may interact with a peptidoglycan component of the cell and disrupt synthesis of the cell wall. Mersacidin blocks transglycosylation — the enzymatic reaction by which the sugar chains of peptidoglycan are polymerized (65) — by forming a tight complex with the membrane-bound peptidoglycan precursor lipid II. LtnA1 may serve as the docking molecule, whereas LtnA2, which has a proposed elongated linear-type structure, may facilitate pore formation. Chiral phase analysis of the two peptides has demonstrated that both peptides possess D-alanine residues and are thus, together with lactococcin S, the only ribosomally synthesized prokaryote peptides possessing D-amino acids (64). These residues are proposed to arise from the stereospecific hydrogenation of Dha, which is formed from the dehydration of a gene-encoded serine residue (64). When LtnA1 and LtnA2 combine, they demonstrate a bactericidal activity that is augmented by energizing of the target cells. It is thus likely that the presence of a proton motive force promotes the interaction of the bacteriocin with the cytoplasmic membrane, resulting in the formation of pores. These pores have been shown to be selective for certain small molecules such as K^+ ions and inorganic phosphate. This loss leads to dissipation of the membrane potential and hydrolysis of internal adenosine triphosphate (ATP), leading to an eventual collapse of the pH gradient at the membrane and ultimately to cell death (66). Another more distantly related two-component lantibiotic, cytolysin, has been found to exhibit hemolytic activity. To compare the biological activity of lacticin 3147 with cytolysin, the lacticin-encoding determinants were expressed in *E. faecalis* FA2-2 to facilitate comparison with an *E. faecalis* FA2-2 containing pAD1, a plasmid that encodes cytolysin production. While both demonstrated a broad spectrum of activity, it was significant that lacticin 3147, unlike cytolysin, did not exhibit a hemolytic activity against equine blood cells.

ltnM1 and *ltnM2* showed homology to the genes involved in the production and export of lantibiotics (67) and were confirmed as modification proteins, with each having a specific role (53). LtnM1 is responsible for the modification of LtnA1, and LtnM2 is responsible for the modification of LtnA2 (53). A triple-stem loop structure that is thought to restrict transcription of the downstream genes has been identified within *ltnM1* (68). LtnT is homologous to an ATP-binding transporter implicated in the secretion of cytolysin. This transporter contains a proteolytic domain that is probably involved in the cleavage of the leader peptides during export.

Another set of genes, *ltnRIFE*, have been associated with immunity to lacticin 3147. LtnE and F show homology to multicomponent ABC transporters identified as having a role in immunity to staphylococcal (69) and lactococcal lantibiotics, although such a role for LtnE and F has yet to be definitively proven. Although LtnI bears no homology to other bacteriocin immunity proteins, it was found that disruption of *ltnI* resulted in a complete loss of immunity. When cloned on an expression vector, it was found that the resultant plasmid was capable of conferring levels of immunity comparable with pMRC01. Furthermore, when the levels of expression of *ltnI* were regulated, it was found that immunity was proportional to the transcription of the gene (50), making it an excellent candidate as a food-grade selectable marker as described above. LtnR is homologous to a number of transcriptional regulators of the repressor type (50). It was found to regulate immunity to lacticin by binding to a region that overlaps the Pimm promoter but does not seem to play a role in lacticin biosynthesis (68). Mutation of *ltnR* results in hyperimmunity, while its overexpression causes sensitivity to lacticin, indicating that it is a negative regulator of the immunity genes.

Many of the studies performed to test the functionality of the various *ltn* genes were carried out with subclones. The entire 12.6-kb lacticin-coding region of pMRC01 was cloned into pCI372 to generate pOM02. In a similar way, other subclones that lack one or more of the *ltn* genes were generated (50). This powerful technology has been superseded by the generation of a novel technique that facilitates modification of pMRC01 itself in a food-grade manner. This system uses splicing by overlap extension (SOE) polymerase chain reaction (PCR) (70) followed by a double homologous cross-over event that is facilitated through the use of the $RepA^+$ temperature-sensitive helper plasmid (pVE6007) and a $RepA^-$ cloning vector (pORI280). The net result is the exchange of an existing region of DNA on pMRC01 with modified DNA. This strategy has the potential to allow the removal or replacement of particular genes or allow the introduction of point mutations to determine the importance of individual nucleotides and amino acids (71). It also overcomes the problems that can be associated with subcloning, that is, difficulties in amplifying and cloning large fragments of DNA and the increased risk of the misincorporation of bases in such large fragments (71).

Sequencing of the 46-kb plasmid pBAC105 has demonstrated that it possesses a region of DNA identical to the bacteriocin-encoding region of pMRC01. Like pMRC01, this region is again flanked by IS*946* elements. However, outside of this region, no homology exists between the two plasmids (10,11).

Analysis of the lacticin 3147 producers at the nucleotide and amino acid levels is continuing. Significant targets include the determination of the three-dimensional structure of LtnA1 and A2, a better understanding of the role of individual amino acids (particularly the modified residues), and an analysis of the immunity and biosynthesis processes. These advances will be necessary to permit the next generation of modern approaches for the application of these bioactive peptides, for example, the overproduction of bacteriocin, the introduction and alteration of bridging patterns and amino acids to increase activity, and the creation of new hybrid bacteriocins.

31.4 DISCUSSION

This chapter has outlined the identification, characterization, and application of a novel bacteriocin, lacticin 3147, through the use of classic and molecular techniques. The approaches used should be widely applicable in the search for additional bacteriocins with desirable traits. The isolation of novel bacteriocins would greatly benefit from the application of high-throughput strategies such as that used to screen kefir strains. When lacticin 3147 was isolated from a strategy designed to screen kefir grains for a bacteriocin producer with potential uses in dairy fermentations, it transpired that it could also be applied for additional purposes. There are reports of successful screening strategies to identify probiotic peptide producers in gut microflora (72) and an inhibitor of cariogenic bacteria in the oral cavity (73,74). Following the same general principle, it may be possible to isolate, for example, extremely aciduric peptides from acidic springs or barotolerant peptides from the ocean bed. However, like lacticin 3147, once peptides are located, it becomes necessary to characterize host range and to determine activity under different physiological conditions. On the basis of these findings, one can choose what applications a particular bacteriocin is best suited for. Lacticin 3147 possesses a number of desirable attributes. It is heat stable, active at neutral and acidic pHs, is active against all Gram-positive bacteria, and does not readily generate resistant colonies. The next hurdle may involve the production of the bacteriocin by a different host. The options available depend on whether the relevant genes are located on the plasmid or the chromosome and whether they are associated with a mobilizable element. While a number of bacteriocins are associated with conjugative plasmids or transposable elements, this is not always the case. In other instances, plasmids have been transferred by isolation and electroporation, by coconjugation following the introduction of a conjugative plasmid or by subcloning into a suitable vector. The lacticin 3147 producer pMRC01 benefits from possessing an origin of transfer (*oriT*) that has facilitated its transfer to more than 30 lactococcal strains (14,22,75). In fact ,so successful is its transfer that *oriT* has been demonstrated as being suitable for the development of novel mobilizable food-grade vectors for the genetic enhancement of strains (76). pOM02 (a recombinant plasmid carrying the lacticin genes) has also been used to heterologously express lacticin 3147 in *Enterococcus*. This achievement suggests that it may be possible to construct lacticin-producing derivatives of a number of different genera such as probiotic strains that would then have the potential to modify the gut flora (77).

Lacticin has been applied in a number of modern approaches, including its use to control NSLAB and flavor, the incorporation of a bacteriocin into a teat seal or food packaging, production as a skim-milk powder, and the development of continuous production systems. However, it is likely that the majority of modern approaches that will be utilized in the coming years will facilitate fundamental investigations to understand the mode of action, determine the structure of the peptide, appreciate the role of individual peptides and amino acids, characterize tolerant mutants, and identify the factors that control biosynthesis. A greater appreciation of all of these factors will make apparent the most appropriate use for individual bacteriocins, determine how frequently tolerant or resistant strains will emerge, whether the combined use of different bacteriocins could be synergistic, and what alterations might need to be made to, for example, increase the heat resistance of a peptide. The use of bioinformatics through the comparison of closely related bacteriocins will also be of great value. In doing so, it may be possible to identify common motifs that may be responsible for peptide interaction, receptor binding, target site, and so on. Conversely, it may be possible, through the analysis of unconserved regions, to explain variations in antimicrobial spectra or to even identify strains likely to produce bacteriocin on the basis of their genome sequence. Thus, it is hoped that through a combination of classic techniques, some modern approaches outlined here, and the development of additional novel procedures, we can maximize the undoubted potential that exists for the application of bacteriocins.

ACKNOWLEDGMENTS

Funding from the Irish Government under the National Development Plan 2000–2006 is acknowledged.

REFERENCES

1. T.R. Klaenhammer. Genetics of bacteriocins produced by lactic acid bacteria. *FEMS Microbiol. Rev.*, 12: 39–85, 1993.
2. E. Gross, J.L. Morrell. The structure of nisin. *J. Am. Chem. Soc.*, 93: 4634–4635, 1971.
3. A. Hurst. Nisin. In: D. Perlman, A.I. Laskin (Eds.). *Advances in Applied Microbiology*, Volume 27. New York: Academic Press, 1981, 85–123.
4. H.G. Sahl, R.W. Jack, G. Bierbaum. Biosynthesis and biological activities of lantibiotics with unique post-translational modifications. *Eur. J. Biochem.*, 230: 827–853, 1995.
5. L. De Vuyst, E. Vandamme. Nisin, a lantibiotic produced by *Lactococcus lactis* subsp. *lactis*: Properties, biosynthesis, fermentation and applications. In: L. De Vuyst, E.J. Vandamme (eds.). *Bacteriocins of Lactic Acid Bacteria: Microbiology, Genetics and Applications*. London: Blackie Academic and Professional, 1985, 151–221.
6. G. Bierbaum, C. Szekat, M. Josten, C. Heidrich, C. Kempter, G. Jung, H.G. Sahl. Engineering of a novel thioether bridge and role of modified residues in the lantibiotic Pep5. *Appl. Environ. Microbiol.*, 62: 385–392, 1996.
7. L.A. Rogers. The inhibiting effect of *Streptococcus lactis* on *Lactobacillus bulgaricus*. *J. Bacteriol.*, 16: 321–325, 1928.
8. J. Delves-Broughton. Nisin and its use as a food preservative. *Food Tech.*, 44: 100–117, 1990.
9. M.C. Rea, T. Lennartsson, P. Dillon, F.D. Drinan, W.J. Reville, M. Heapes, T.M. Cogan. Irish kefir-like grains: their structure, microbial composition and fermentation kinetics. *J. Appl. Bacteriol.*, 81: 83–94, 1996.
10. D. Casla, T. Requena, R. Gomez . Antimicrobial activity of lactic acid bacteria isolated from goat's milk and artisanal cheeses: characteristics of a bacteriocin produced by *Lactobacillus* curvatus IFPL 105. *J. Appl. Bacteriol.*, 81: 35–41, 1996.

11. M.C. Martinez-Cuesta, G. Buist, J. Kok, H.H. Hauge, J. Nissen-Meyer, C. Pelaez, T. Requena. Biological and molecular characterization of a two-peptide lantibiotic produced by *Lactococcus lactis* IFPL105. *J. Appl. Microbiol.*, 89: 249–260, 2000.
12. S.M. Morgan, R. Hickey, R.P. Ross, C. Hill. Efficient method for the detection of microbially-produced antibacterial substances from food systems. *J. Appl. Microbiol.*, 89: 56–62, 2000.
13. E. Parente, C. Hill. A comparison of factors affecting the production of two bacteriocins from lactic acid bacteria. *J. Appl. Bacteriol.*, 73: 290–298, 1992.
14. M.P. Ryan, M.C. Rea, C. Hill, R.P. Ross. An application in cheddar cheese manufacture for a strain of *Lactococcus lactis* producing a novel broad-spectrum bacteriocin, lacticin 3147. *Appl. Environ. Microbiol.*, 62: 612–619, 1996.
15. M. Galvin, C. Hill, R.P. Ross. Lacticin 3147 displays activity in buffer against gram-positive bacterial pathogens which appear insensitive in standard plate assays. *Lett. Appl. Microbiol.*, 28: 355–358, 1999.
16. P.F. Fox, P.L.H. McSweeney, C.M. Lynch. 1998. Significance of non-starter lactic acid bacteria in cheddar cheese. *Aust. J. Dairy Technol.*, 53: 83–89, 1998.
17. P.L.H. McSweeney, C.M. Lynch, E.M. Walsh, P.F. Fox, K.N. Jordan, T.M. Cogan, F.D. Drinan. Role of non-starter lactic acid bacteria in cheddar cheese ripening. In: T. Cogan, P.F. Fox, R.P. Ross (Eds.). *4th Cheese Symposium*. Fermoy, Ireland: Teagasc., 1995, 32–45.
18. P.L.H. McSweeney, E.M. Walsh, P.F. Fox, T.M. Cogan, F.D. Drinan, M. Castelo-Gonzalez. A procedure for the manufacture of cheddar cheese under controlled bacteriological conditions and the effect of adjunct lactobacilli on cheese quality. *Ir. J. Agric. Food Res.*, 33: 183–192, 1994.
19. T.D. Thomas, V.L. Crow. Mechanism of D(-)-lactic acid formation in cheddar cheese. *NZ. J. Dairy Sci. Technol.*, 18: 131–141, 1983.
20. E. Lipsanka. Use of nisin-producing lactic streptococci in cheesemaking. *Bull. Int. Dairy Fed.*, 73: 1–24, 1973.
21. E. Lipsanka. Nisin and its application, In M. Woodbine (Ed.). *Antimicrobials and Antibiosis in Agriculture*. London, United Kingdom: Butterworths, 1977, 103–130.
22. M. Coakley, G.F. Fitzgerald, R.P. Ross. Application and evaluation of the phage resistance- and bacteriocin-encoding plasmid pMRC01 for the improvement of dairy starter cultures. *Appl. Environ. Microbiol.*, 63: 1434–1440, 1997.
23. M.P. Ryan, R.P. Ross, C. Hill. Strategy for manipulation of cheese flora using combinations of lacticin 3147-producing and -resistant cultures. *Appl. Environ. Microbiol.*, 67: 2699–2704, 2001.
24. M.C. Martinez-Cuesta, T. Requena, C. Pelaez. Use of a bacteriocin-producing transconjugant as starter in acceleration of cheese ripening. *Int. J. Food Microbiol.*, 70: 79–88, 2001.
25. M.C. Martinez-Cuesta, P. Fernandez de Palencia, T. Requena, C. Palaez. Enhancement of proteolysis by a *Lactococcus lactis* bacteriocin producer in a cheese model system. *J. Agric. Food Chem.*, 46: 3863–3867, 1998.
26. M.C. Martinez-Cuesta, T. Requena, C. Pelaez. Effect of bacteriocin-induced cell damage on the branched-chain amino acid transamination by *Lactococcus lactis*. *FEMS Microbiol. Lett.*, 217: 109–113, 2002.
27. P.L. Mead, L. Slutsker, V. Dietz, L.F. McCaig, J.S. Bresee, C. Chapiro, P.M. Griffin, R.V. Tauxe. Food-related illness and death in the United States. *Emerg. Infect. Dis.*, 5: 607–625, 1999.

28. S. Wong, D. Street, S.I. Delgado, K.C. Klontz. Recalls of foods and cosmetics due to microbial contamination reported to the U.S. Food and Drug administration. *J Food Prot.*, 63: 1113–1116, 2000.
29. O. McAuliffe, C Hill, R.P. Ross. Inhibition of *Listeria monocytogenes* in cottage cheese manufactured with a lacticin 3147-producing starter culture. *J. Appl. Microbiol.*, 86: 251–256, 1999.
30. N. Benkerroum, W.E. Sandine. Inhibitory action of nisin against *Listeria monocytogenes. J. Dairy Sci.*, 71: 3237–3245, 1988.
31. R.P. Ross, C. Stanton, C. Hill, G.F. Fitzgerald, A. Coffey. Novel cultures for cheese improvement. *Trends Food Sci. Technol.*, 11: 96–104, 2000.
32. A. Coffey, M. Ryan, R.P. Ross, C. Hill, E. Arendt, G. Schwarz. Use of a broad-host-range bacteriocin-producing *Lactococcus lactis* transconjugant as an alternative starter for salami manufacture. *Int. J. Food Microbiol.*, 43: 231–235, 1998.
33. G.F. De Valdez, H. Diekmann. Freeze-drying conditions for starter cultures for sourdoughs. *Cryobiology*, 30: 185–190, 1993.
34. S. Desmons, S.S. Zgoulli, J. Destain, P. Thonart. Freeze-drying of a lactic acid bacteria: *Lactobacillus acidophilus*. In: *Proceedings of the 11th forum for Applied Biotechnology*, Volume 62. Belgium: University of Gent, 1997, 1713–1716.
35. A. Scannell, G. Schwarz, C. Hill, R.P. Ross, E.K. Arendt. Pre-inoculation enrichment procedure enhances the performance of bacteriocinogenic *Lactococcus lactis* meat starter culture. *Int. J. Food Microbiol.*, 64: 151–159, 2001.
36. S.M. Morgan, M. Galvin, J. Kelly, R.P. Ross, C. Hill. Development of a lacticin 3147-enriched whey powder with inhibitory activity against foodborne pathogens. *J. Food Prot.*, 62: 1011–1016, 1999.
37. S.M. Morgan, M. Galvin, R.P. Ross, C. Hill. Evaluation of a spray-dried lacticin 3147 powder for the control of *Listeria monocytogenes* and *Bacillus cereus* in a range of food systems. *Lett. Appl. Microbiol.*, 33: 387–391, 2001.
38. A.G. Scannell, C. Hill, R.P. Ross, S. Marx, W. Hartmeier, E.K. Arendt. Development of bioactive food packaging materials using immobilised bacteriocins lacticin 3147 and nisaplin. *Int. J. Food Microbiol.*, 60: 241–249, 2000.
39. A.G. Scannell, C. Hill, R.P. Ross, S. Marx, W. Hartmeier, E.K. Arendt. Continuous production of lacticin 3147 and nisin using cells immobilized in calcium alginate. *J. Appl. Microbiol.*, 89: 573–579, 2000.
40. J. Huang, C. Lacroix, H. Daba, R.E. Simard. Pediocin 5 production and plasmid stability during continuous free and immobilized cell cultures of *Pediococcus acidilactici* UL5. *J. Appl. Bacteriol.*, 80: 635–644, 1996.
41. S.M. Morgan, R.P. Ross, T. Beresford, C. Hill. Combination of hydrostatic pressure and lacticin 3147 causes increased killing of *Staphylococcus* and *Listeria. J. Appl. Microbiol.*, 88: 414–420, 2000.
42. J.R. Broadbent, Y.C. Chou, K. Gillies, J.K. Kondo. Nisin inhibits several gram-positive, mastitis-causing pathogens. *J. Dairy Sci.*, 72: 3342–3345, 1989.
43. E.R. Oldham, M.J. Daley. Lysostaphin: use of a recombinant bactericidal enzyme as a mastitis therapeutic. *J. Dairy Sci.*, 74: 4175–4182, 1991.
44. M.P. Ryan, W.J. Meaney, R.P. Ross, C. Hill. Evaluation of lacticin 3147 and a teat seal containing this bacteriocin for inhibition of mastitis pathogens. *Appl. Environ. Microbiol.*, 64: 2287–2290, 1998.
45. M.P. Ryan, J. Flynn, C. Hill, R.P. Ross, W.J. Meaney. The natural food grade inhibitor, lacticin 3147, reduced the incidence of mastitis after experimental challenge with *Streptococcus dysgalactiae* in nonlactating dairy cows. *J. Dairy Sci.*, 82: 2625–2631, 1999.

46. D.P. Twomey, A.I. Wheelock, J. Flynn, W.J. Meaney, C. Hill, R.P. Ross. Protection against *Staphylococcus aureus* mastitis in dairy cows using a bismuth-based teat seal containing the bacteriocin, lacticin 3147. *J. Dairy Sci.*, 83: 1981–1988, 2000.
47. W. van Schaik, C.G. Gahan, C. Hill. Acid-adapted *Listeria monocytogenes* displays enhanced tolerance against the lantibiotics nisin and lacticin 3147. *J. Food Prot.*, 62: 536–539, 1999.
48. N.S. Dodd. Sensitivity and tolerance of *Lactococcus* sp. and *Listeria monocytogenes* ScottA to lacticin 3147. M.Sc. thesis. Cork, Ireland: University College Cork, 1996.
49. A.W.P.E. Klijn. Spontaneous resistance against lacticin 3147 in *Lactococcus lactis* and *Listeria monocytogenes*. M.Sc. thesis. Cork, Ireland: University College Cork, 2001.
50. O. McAuliffe, C. Hill, R.P. Ross. Identification and overexpression of *ltnl*, a novel gene which confers immunity to the two-component lantibiotic lacticin 3147. *Microbiology*, 146: 129–138, 2000.
51. G.E. Allison, T.R. Klaenhammer. Functional analysis of the gene encoding immunity to lactacin F, lafI, and its use as a *Lactobacillus*-specific, food-grade genetic marker. *Appl. Environ. Microbiol.*, 62: 4450–4460, 1996.
52. A. von Wright, S. Wessels, S. Tynkkynen, M. Saarela. Isolation of a replication region of a large lactococcal plasmid and use in cloning of a nisin resistance determinant. *Appl. Environ. Microbiol.*, 56: 2029–2035, 1990.
53. O. McAuliffe, C. Hill, R.P. Ross. Each peptide of the two-component lantibiotic lacticin 3147 requires a separate modification enzyme for activity. *Microbiology*, 146: 2147–2154, 1998.
54. R. Kilpper-Balz, G. Fischer, K.H. Schleifer. Nucleic acid hybridisation of group N and group D streptococci. *Curr. Microbiol.*, 7: 245–250, 1982.
55. M.C. Booth, C.P. Bogie, H.G. Sahl, R.J. Siezen, K.L. Hatter, M.S. Gilmore. Structural analysis and proteolytic activation of *Enterococcus faecalis* cytolysin, a novel lantibiotic. *Mol. Microbiol.*, 21: 1175–1184, 1996.
56. H. Holo, Z. Jeknic, M. Daeschel, S. Stevanovic, I.F. Nes. Plantaricin W from *Lactobacillus plantarum* belongs to a new family of two-peptide lantibiotics. *Microbiology*, 147: 643–651, 2001.
57. M.A. Navaratna, H.G. Sahl, J.R. Tagg. Two-component anti-*Staphylococcus aureus* lantibiotic activity produced by *Staphylococcus aureus* C55. *Appl. Environ. Microbiol.*, 64: 4803–4808, 1998.
58. H. Takami, K. Nakasone, Y. Takaki, G. Maeno, R. Sasaki, N. Masui, F. Fuji, C. Hirama, Y. Nakamura, N. Ogasawara, S. Kuhara, K. Horikoshi. Complete genome sequence of the alkaliphilic bacterium *Bacillus halodurans* and genomic sequence comparison with *Bacillus subtilis*. *Nucleic Acids Res.*, 28: 4317–4331, 2000.
59. S. Garneau, N.I. Martin, J.C. Vederas. Two-peptide bacteriocins produced by lactic acid bacteria. *Biochimie*, 84: 577–592, 2002.
60. T. Yamaguchi, T. Hayashi, H. Takami, M. Ohnishi, T. Murata, K. Nakayama, K. Asakawa, M. Ohara, H. Komatsuzawa, M. Sugai. Complete nucleotide sequence of a *Staphylococcus aureus* exfoliative toxin B plasmid and identification of a novel ADP-ribosyltransferase, EDIN-C. *Infect. Immun.*, 69: 7760–7771, 2001.
61. D. Twomey, R.P. Ross, M. Ryan, B. Meaney, C. Hill. Lantibiotics produced by lactic acid bacteria: structure, function and applications. *Antonie Van Leeuwenhoek*, 82: 165–185, 2002.
62. T. Prasch, T. Naumann, R.L. Markert, M. Sattler, W. Schubert, S. Schaal, M. Bauch, H. Kogler, C. Griesinger. Constitution and solution conformation of the antibiotic

mersacidin determined by NMR and molecular dynamics. *Eur. J. Biochem.*, 244: 501–512, 1997.

63. N. Zimmermann, G. Jung. The three-dimensional solution structure of the lantibiotic murein-biosynthesis-inhibitor actagardine determined by NMR. *Eur. J. Biochem.*, 246: 809–819, 1997.
64. M.P. Ryan, R.W. Jack, M. Josten, H.G. Sahl, G. Jung, R.P. Ross, C. Hill. Extensive post-translational modification, including serine to D-alanine conversion, in the two-component lantibiotic, lacticin 3147. *J. Biol. Chem.*, 274: 37544–37550, 1999.
65. H. Brotz, G. Bierbaum, P.E. Reynolds, H.G. Sahl. The lantibiotic mersacidin inhibits peptidoglycan biosynthesis at the level of transglycosylation. *Eur. J. Biochem.*, 246: 193–199, 1997.
66. O. McAuliffe, M.P. Ryan, R.P. Ross, C. Hill, P. Breeuwer, T. Abee.1998. Lacticin 3147, a broad-spectrum bacteriocin which selectively dissipates the membrane potential. *Appl. Environ. Microbiol.*, 64: 439–445, 1998.
67. B.A. Dougherty, C. Hill, J.F. Weidman, D.R. Richardson, J.C. Venter, R.P. Ross. Sequence and analysis of the 60 kb conjugative, bacteriocin-producing plasmid pMRC01 from *Lactococcus lactis* DPC3147. *Mol. Microbiol.*, 29: 1029–1038, 1998.
68. O. McAuliffe, T. O'Keeffe, C. Hill, R.P. Ross. Regulation of immunity to the two-component lantibiotic, lacticin 3147, by the transcriptional repressor LtnR. *Mol. Microbiol.*, 39: 982–993, 2001.
69. J.I. Ross, E.S. Eady, J.H. Cove, S. Baumberg. Identification of a chromosomally encoded ABC-transport system with which the staphylococcal erythromycin exporter MsrA may interact. *Gene*, 153: 93–98, 1995.
70. R.M. Horton, Z.L. Cai, S.N. Ho, L.R. Pease. Gene splicing by overlap extension: tailor-made genes using the polymerase chain reaction. *Biotechniques*, 8: 528–535, 1990
71. P.D. Cotter, C. Hill, R.P. Ross. A food-grade approach for functional analysis and modification of native plasmids in *Lactococcus lactis*. *Appl. Environ. Microbiol.*, 69: 702–706, 2003.
72. S. Flynn, D. van Sinderen, G.M. Thornton, H. Holo, I.F. Nes, J.K. Collins. Characterization of the genetic locus responsible for the production of ABP-118, a novel bacteriocin produced by the probiotic bacterium *Lactobacillus salivarius* subsp. *salivarius* UCC118. *Microbiology*, 148: 973–984, 2002.
73. J.D. Hillman, B.I. Yaphe, K.P. Johnson. Colonization of the human oral cavity by a strain of *Streptococcus mutans*. *J. Dent. Res.*, 64: 1272–1274, 1985.
74. J.D. Hillman, A.L. Dzuback, S.W. Andrews. Colonization of the human oral cavity by a *Streptococcus mutans* mutant producing increased bacteriocin. *J. Dent Res.*, 66: 1092–1094, 1987.
75. D. O'Sullivan, A. Coffey, G.F. Fitzgerald, C. Hill, R.P. Ross. Design of a phage-insensitive lactococcal dairy starter via sequential transfer of naturally occurring conjugative plasmids *Appl. Environ. Microbiol.*, 64: 4618–4622, 1998.
76. R.M. Hickey, D.P. Twomey, R.P. Ross, C. Hill. Exploitation of plasmid pMRC01 to direct transfer of mobilizable plasmids into commercial lactococcal starter strains. *Appl. Environ. Microbiol.*, 67: 2853–2858, 2001.
77. M.P. Ryan, O. McAuliffe, R.P. Ross, C. Hill. Heterologous expression of lacticin 3147 in *Enterococcus faecalis*: comparison of biological activity with cytolysin. Heterologous expression of lacticin 3147 in *Enterococcus faecalis*: comparison of biological activity with cytolysin. *Lett. Appl. Microbiol.*, 32: 71–77, 2001.

32 Membrane-Based Fractionation and Purification Strategies for Bioactive Peptides

Yves Pouliot, Sylvie F. Gauthier, and Paule Emilie Groleau

CONTENTS

32.1 INTRODUCTION

Bioactive peptides can be obtained by *in vitro* hydrolysis of protein substrate using appropriate enzymes or by proteolysis from fermentation by specific bacteria. For example, a number of peptides produced by the enzymatic hydrolysis of caseins and whey proteins have been shown to have various biological activities with high potential for nutraceutical applications. Several review papers report on the biologically active sequences found in dairy proteins (1–5). Bioactive peptides are also found in fermented milk produced with lactic acid bacteria (6,7) or in fermented whey prepared with different bacteria (8).

The separation of bioactive peptides from enzymatic hydrolyzates or fermented milk presents a promising area for creating new value-added ingredients. In both cases, bioactive peptides are produced in low concentrations, along with other peptidic sequences. Fractionation is often needed to concentrate bioactivity up to a level where the active dose of a given peptide will be delivered in a small amount of finished product (e.g., <5 g). This makes the finished product suitable for food formulations such as bars, drinks, or even capsules (if <1 g). The design of fractionation processes for bioactive peptide mixtures, however, requires detailed characterization of the mixtures to be fractionated, an appropriate understanding of the structures and functions of bioactive peptide, and knowledge of the main fractionation techniques available. Fractionation and purification methods are chosen based on the nature of the product to be separated and on specific requirements dictated by the final application of the molecule. The added value of the separated product must outweigh processing costs, and in general, highly expensive methods such as chromatography will have to be used exclusively when the purity of the product is essential for its commercialization.

Membrane separation processes have been widely used in the concentration of bioactive peptides. Although they are especially adapted to industrial-scale processing, they often need to be combined with other techniques to achieve the separation of a specific peptide. This chapter aims at providing tools in the development of membrane-based fractionation processes for bioactive peptides. Pressure-driven membrane-based separation will be introduced in the next section, and more general strategies for separating peptides will be presented thereafter.

32.2 PRESSURE-DRIVEN MEMBRANE-BASED SEPARATION

Pressure-driven membrane-based separation processes are techniques in which a liquid is separated from a mixture as the latter is forced through a porous membrane by means of an applied pressure.

32.2.1 SEPARATION DOMAINS

Pressure-driven membrane-based separation processes comprise microfiltration (MF), ultrafiltration (UF), nanofiltration (NF), and reverse osmosis (RO). The differences in the separation domain of MF, UF, NF, and RO arise from the separation properties of the membrane used. Table 32.1 summarizes the main

TABLE 32.1
Comparative Features of Some Pressure-Driven Membrane Processes

	Microfiltration (MF)	Ultrafiltration (UF)	Nanofiltration (NF)	Reverse Osmosis (RO)
Pore size	>0.1 μm	1–500 nm	0.1–1 nm	<0.1 nm
Sieving mechanism	Size	Size (and charge)	Size & charge	Size
Separation domain	Particles, globules (>0.1 μm)	Proteins/peptides (1–300 kDa)	Salts, solutes, amino acids	Monovalents (Na^+, K^+, Cl^-)
Material	Polymeric, mineral			
Configuration	Spiral wound, tubular, flat sheets, hollow fibers			
Operating pressure	0.1–2 bars	1–10 bars	15–30 bars	30–50 bars

differences among these processes. Pore size varies from <0.1 nm (RO) to >0.1 μm (MF), and the separation domain varies accordingly from monovalent salts to particles. Size represents the main sieving mechanism in MF and RO, whereas charge is also involved in the retention mechanism in UF and NF.

A wide variety of membrane materials and configurations are available for all membrane-based processes. Polymeric materials such as polysulfone (UF) and polyamide (NF, RO) are widely used because of their low cost and chemical stability. The spiral-wound configuration is well adapted for industrial conditions and is largely used in UF, NF, and RO. The tubular configuration is mainly used in MF with ceramic membranes. Pressure-driven membrane-based processes are operated at pressures that vary inversely with their pore size (e.g., MF uses much lower pressures [0.1 to 2 bars] than RO [30 to 50 bars]).

The differences in the physicochemical properties of mixed peptides (hydrolyzates or fermentation media) are often small. A separation technique that can discriminate small differences in charge, size, and species hydrophobicity is often needed. Given the low molecular weight of peptides (200 up to 5,000 Da), peptide fractionation is best achieved using UF and NF, which are processes that separate solutes based on both size and charge.

32.2.2 Tangential Filtration

Tangential flow filtration refers to the direction of the feed flow at the membrane's surface. Understanding the phenomena occurring during tangential filtration is critical for optimal use of membrane-based processes in the fractionation of peptide mixtures. Figure 32.1 illustrates the various phenomena occurring during tangential filtration in pressure-driven membrane-based processes. Concentration polarization (CP) occurs when rejected solutes accumulate on the membrane surface as a result of the convective flow. Convection is induced by the transport of solutes close to the membrane, and its intensity is proportional to the transmembrane pressure. The higher solute concentration then gives rise to a back-diffusion phenomenon that restores the osmotic pressure and decreases electrostatic repulsions. In some cases,

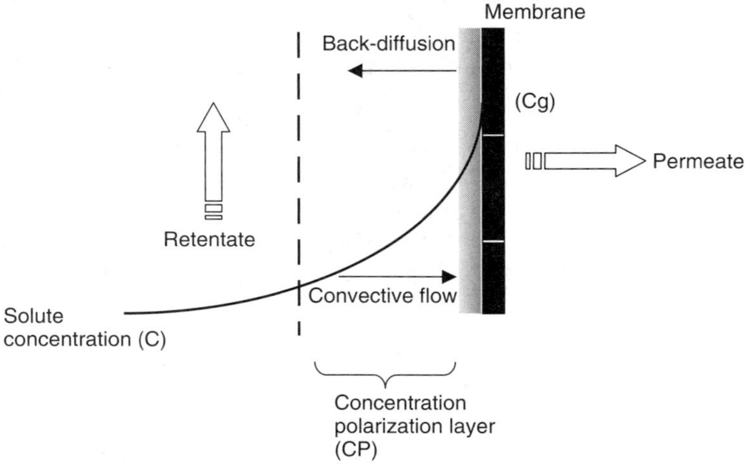

FIGURE 32.1 Formation of a concentration polarization layer (CP) during tangential filtration.

the higher solute concentration can lead to the formation of a gel layer that can irreversibly affect membrane flux and selectivity. The negative effects of CP can be minimized by operating at low concentration factors or by adjusting recirculation velocity (shear rate), transmembrane pressure, and permeation rate. The formation of CP essentially explains why membranes often show higher selectivity than their molecular weight cutoff suggests.

Although UF has been used to separate proteins, protein hydrolyzates contain smaller molecules (peptides) that can only be fractionated by tighter membranes. Nanofiltration (NF) is an intermediate technique between UF and RO, in which transport phenomena are governed by both convection and diffusion (9). The charge is known to be an important discrimination factor in the transport of solutes through a NF membrane. The charge of the membrane determines, to a large extent, the properties of the polarized layer and, hence, the selectivity of the membrane. The charge of a particle influences its selective adsorption on the membrane, the establishment of interactions in the polarized layer, and the electronegative potential of the system, known as the Donnan potential. This Donnan effect is an important separation parameter in NF. It results from an electrostatic equilibrium between two solutions separated by a membrane.

The charge effects resulting from the establishment of a CP layer may become a separation factor, which is particularly important in NF membranes. Mechanisms underlying separation selectivity are far more complex than just the sieving effect. Membrane filtrations have the advantage of being easily adaptable to large-scale processes, but their downside lies in the fact that the selectivity is variable and that their behavior is difficult to predict. Membrane separation processes will often be used in combination with other technologies to increase their separation potential.

32.3 STRATEGIES FOR PEPTIDE FRACTIONATION

As a prerequisite to defining a strategy for fractionation, the process generating the peptide mixture must be well standardized and the co-products (or contaminants) of the peptides must be known. Table 32.2 lists a number of compositional differences between peptide mixtures derived from enzymatic hydrolysis and bioprocesses. Enzymatic hydrolyzates generally offer simpler feed solutions, which are made up of peptides, the nonhydrolyzed protein fraction, and the enzyme. Peptide mixtures issued from bioprocesses are generally more diluted and contain bacteria as well as a number of their metabolites.

32.3.1 Background Data

Obtaining background data on the bioactive peptide to be fractionated from a complex mixture is critical. As a starting point, the design of an efficient fractionation process for peptide mixtures requires the following information:

1. Characterization of the bioactive peptide (amino acid sequence, mass, pI, hydrophobicity)
2. An accurate methodology for the quantification of the bioactive peptide
3. A relevant and rapid *in vitro* methodology to evaluate the bioactivity of the peptide
4. Feed composition and sensitivity to processing conditions
5. Experimental data on the high-performance liquid chromatography (HPLC) conditions for the separation of peptides at the analytical scale

In addition, the target or final bioactive peptide concentration must be clearly defined.

TABLE 32.2
Comparative Characteristics of Peptide Mixtures Obtained from Enzymatic Hydrolysis and Bioprocesses

Enzymatic Hydrolysis	Bioprocess
In vitro enzymatic hydrolysis	Fermentation (extracellular proteinases)
Prepared from purified protein source	Prepared from a foodstuff (e.g., milk)
Prepared using enzymes of various sources (animal, bacterial, fungal)	Prepared using a wide variety of microorganisms
High protein content (20–25%)	Low protein content (1–5%)
May contain active enzymes	May contain active bacteria and enzymes
Peptide profile dependent upon the hydrolysis conditions used	Complex peptide profile (short)
Contaminants dependent upon the purity of the protein source	Numerous contaminants (membrane foulants)

32.3.2 Overview of the Options

Figure 32.2 provides an overview of the steps involved in the fractionation of bioactive peptides; these steps range from peptide production by enzymatic hydrolysis/fermentation to peptide purification. Pretreatments may be used to remove suspended material or molecules that could interfere with subsequent purification steps. Membrane-based separation can be used as a preliminary step for the removal of enzymes, nonhydrolyzed proteins, and bacteria as well as a step to further fractionate the peptide mixture. Purification techniques may finally be used to produce fractions particularly rich in certain selected peptides.

As shown in Figure 32.3, the design of the separation or purification steps will be largely dependent on peptide properties. If the mixture is characterized by peptides of wide-ranging molecular weights, a size exclusion technique can be used. The amino acid composition is also critical for purification purposes. For

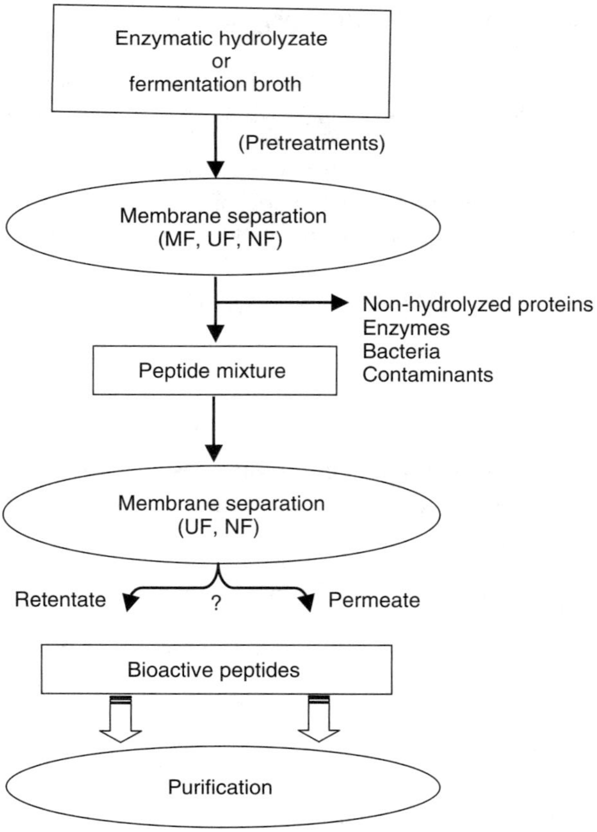

FIGURE 32.2 Potential processing steps from peptide production to peptide purification.

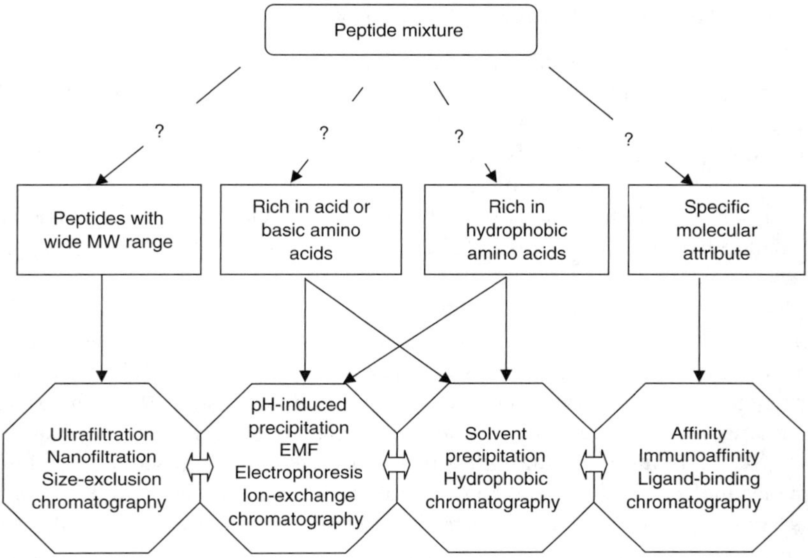

FIGURE 32.3 Options for purification techniques as a function of molecular properties of peptides.

example, a peptide mixture rich in acidic (Asp, Glu) or basic (Arg, Lys) amino acids will be suitable for purification techniques based on charge, whereas solvent extraction will be applied to a peptide mixture rich in hydrophobic peptides. Specific molecular attributes (e.g., metal-binding residues) are increasingly used to design affinity and immunoaffinity separation techniques.

The following sections of this chapter will illustrate how these options are used for separating various peptide mixtures.

32.3.3 Pretreatments

Pretreatments are usually designed to improve the selectivity and yield of the downstream processing steps. The selection of an appropriate pretreatment is essentially based on economical considerations, but other factors such as the introduction of contaminants or the modification of properties of the peptide mixture must also be considered. Typical pretreatments include prefiltration, adsorbents, and selective precipitation (with salts or solvents).

32.3.3.1 Adsorbents

Activated carbon, charcoal, and diatomaceous earths can be used as adsorbents for the removal of some organic molecules, free amino acids, or volatile compounds. However, the poor selectivity of these media and the loss of peptidic material often limit their use.

32.3.3.2 Salt- or pH-Induced Precipitation

Precipitation of proteins can be induced by changes in solvent type, pH, temperature, and ionic strength. Precipitation through pH variations is an easy and efficient strategy well suited for certain proteins. Most proteins are less soluble at the pH value corresponding to their isoelectric point (pI), since they carry no charge to interact with the solvent molecules. For example, milk caseins precipitate at their pI (i.e., at pH ~4.6). However, some whey proteins like undenatured β-lactoglobulin are soluble even at their pI (at pH ~5.2). Isoelectric precipitation can also be applied to separate peptides.

Ionic strength can also induce precipitation, as ions generally tend to screen charges on molecules and hence induce hydrophobic interactions. Ions are also known to induce structural changes in water molecules. Ions like NH_4^+ and SO_4^{2-} disrupt the structure of water, increase its surface tension, and decrease the solubility of nonpolar molecules; this phenomenon is known as salting out. Ions like Ca^{2+}, I^- and SCN^-, however, stabilize the structure of water and increase the solubility of nonpolar molecules, which is known as salting in. Ions like Na^+ and Cl^- have no direct effect on the structure of water but can screen charges on molecules (10). Hence, different ions can be used to induce interactions between peptides and trigger their precipitation.

Léonil et al. (11) studied the effect of salts and pH on the solubility of peptides derived from tryptic casein hydrolyzate to correlate their solubility behavior with their physicochemical properties. An optimum precipitation was obtained at pH 3.5 and with a salt (NaCl) concentration of 0.25 M. Nine major peptides were found in the precipitate, all having a marked hydrophobicity. One peptide, β-CN 114–169, is a bioactive peptide that affects gastric emptying and was partially separated by this acid precipitation, along with some other peptides.

In another study by Groleau et al. (12), an acid precipitation was used to fractionate peptides from a tryptic hydrolyzate of β-lactoglobulin (β-LG). A bioactive peptide known for its angiotensin-converting enzyme (ACE)-inhibitory and bactericidal activity, β-LG 15–20 (13), was found within the hydrophobic peptides separated from the total mixture by a drop in pH. However, no work has been carried out to confirm the yield and purity of this peptide in the precipitate. This method, although simple and easily applicable, is not selective.

32.3.3.3 Solvent Precipitation

Solvents of different polarities are used to induce protein or peptide precipitation. Solvent precipitation is not very suitable for separating bioactive peptides because solvent residues are considered undesirable in the food and nutraceutical industries. However, ethanol may be used, as it is a volatile solvent and is easily removed by evaporation. Adamson et al. (14) successfully used a combination of acid and ethanol precipitation to separate casein phosphopeptides (CPP) from a tryptic casein hydrolyzate. The precipitate consisted of phosphopeptides accounting for approximately 13% of the initial protein. Bacteriocins (antimicrobial peptides) were also purified by solvent extraction (15), where the surface-active peptides were concentrated at the interface until they formed reverse submicelles that aggregated into reverse micelles and gelled.

32.3.4 Membrane-Based Separation

As highlighted in section 32.2, pressure-driven membrane-based processes such as ultrafiltration (UF) and nanofiltration (NF) have been used to fractionate proteins, peptides, and amino acids in whey products.

Turgeon and Gauthier (16) fractionated peptides in enzymatically hydrolyzed whey protein concentrate using a two-step ultrafiltration (UF) process schematized in Figure 32.4. The hollow fiber polysulfone UF membranes had a molecular weight cutoff (MWCO) of 30 kDa and 1 kDa, respectively. The two-step UF process of whey protein hydrolyzates resulted in a mixture of polypeptides (MP) with improved overall composition, in comparison with the ratio of commercial whey protein concentrate to higher protein content and the lower lactose and lipid content. Furthermore, MP consisted of peptides smaller than 5,000 Da, which could serve as emulsifiers in food systems and cosmetic products (17–19).

Several researchers studied the use of NF membranes in peptide fractionation in model solutions of amino acids and peptides. Tsuru et al. (20) investigated the

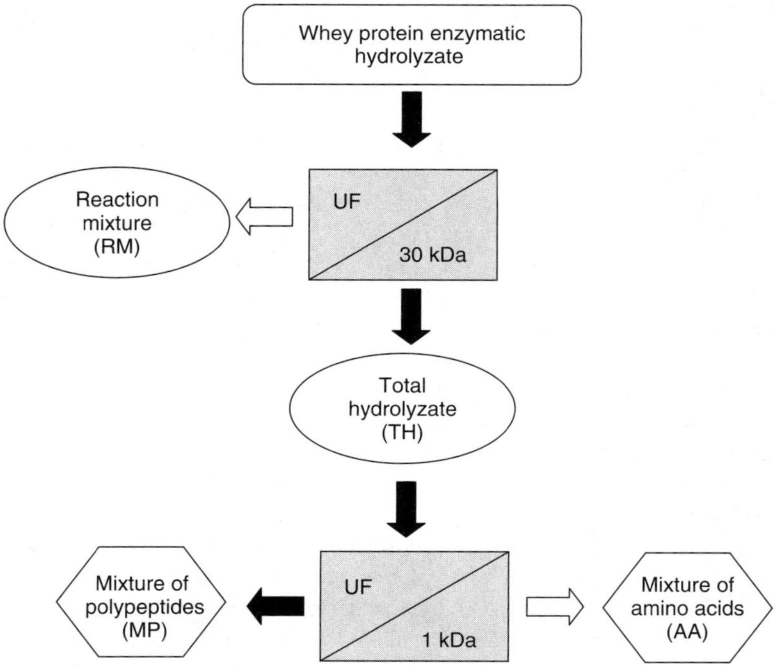

FIGURE 32.4 Two-steps UF-process leading to a mixture of polypeptides (MP), a fraction containing emulsifying peptides. (Adapted from S.L. Turgeon, S.F. Gauthier. *J. Food Sci.*, 55: 106–110, 1990. With permission.)

separation of amino acids and small peptides (two and three amino acids) with NF membranes. Membranes having a MWCO below 300 Da were shown to be ineffective in separating amino acids, while membranes with a MWCO between 2,000 Da and 3,000 Da could efficiently separate mixtures of amino acids and peptides. The predominance of charge effects on the rejection of the various species by the membranes was illustrated by the impact of pH adjustments of the feed before NF. Garem et al. (21) studied the separation of a complex mixture of 15 amino acids using two commercial membranes composed of polyamide (MWCO of 500 Da) and cellulose polyether polyamide (MWCO of 1,000 Da), both layered onto a polysulfone support. Both membranes behaved differently. The lower-MWCO membrane showed high steric rejection for all the amino acids, while the separation obtained with the higher-MWCO membrane was strongly dependent on the pH and, to a lesser extent, on the ionic strength and the transmembrane pressure. The maximum selectivity was obtained at pH 10, low ionic strength, and low transmembrane pressure. Kimura and Tamano (22) used charged UF membranes to separate amino acids with a molecular weight value between 75 and 200 Da. The membranes were composed of sulfonated polysulfone with a charge density of 1.6 meq g^{-1} and a MWCO of 10,000 Da. These authors also found that the separation of amino acids was based on differences in charges and that the rejection can be influenced by pH variations. Another study on the fractionation of model peptide solution by Martin-Orue et al. (23) proposed that the number of charges on peptides, and not global net charge, should be considered in predicting peptide transmission.

NF membranes have also been used to fractionate complex protein hydrolyzates that contain bioactive peptides. Nau et al. (24) used inorganic membranes to separate β-casein peptides and demonstrated that peptide separation is regulated by two mechanisms: size exclusion and electrostatic repulsions. They observed that the formation of the polarized layer generally increased peptide transmission, but the operating parameters that increased the buildup of the polarized layer also increased fouling, at the expense of peptide transmission. Optimal operating parameters are thus often the result of a compromise. They also emphasized that high ionic strength favored steric rejection by screening peptide charges, and low ionic strength promoted electrostatic repulsion (rejection). The effect of ionic strength and pH on the fractionation of β-LG tryptic hydrolyzates was also investigated by Pouliot et al. (25,26). It was found that maximum membrane selectivity was obtained at pH 9, while NaCl addition (0.5 M) increased peptide permeability.

Garem et al. (27) used NF to fractionate β-casein peptides and managed to partially separate acidic from basic peptides at pH 8. They proposed that selectivity is affected not only by the charge of the membrane but also by the presence of high-molecular-weight negatively charged components in the polarized layer. A later study by Lapointe et al. (28) demonstrated the importance of the polarized layer at the membrane surface on the transmission of tryptic peptides from β-LG. The authors observed that the presence of high-molecular-weight compounds in the polarized layer could not only increase charge density, as proposed previously

(27), but also attract basic peptides in CP and enhance their transmission. Hydrophobic residues contained in the sequence of high-molecular-weight compounds or peptides can, however, modify the charge effect as a result of hydrophobic interactions.

A great advantage of filtration techniques, such as UF or NF, is that membranes can be added to the production process of bioactive peptides (by hydrolysis or fermentation), and the product can be separated continuously in the so-called bioreactors. Bordenave et al. (29) used this technology to continuously separate α-lactorphin, an opioid peptide produced by the peptic hydrolysis of α-lactalbumin.

32.3.5 Purification Techniques

Purification techniques are designed to isolate a specific compound from a complex mixture. At the laboratory scale, highest purity is often targeted, and techniques that offer the most powerful resolution are generally favored. For industrial applications, efficiency of the process becomes important, and the purification technique must be a balance between purity and efficiency.

32.3.5.1 Chromatographic Methods

The chromatographic techniques used in a preparative mode rely on ion exchange, hydrophobic interactions, size exclusion, or affinity.

32.3.5.1.1 Ion-Exchange Chromatography

Ion-exchange chromatography (IEC) is used to selectively separate peptides with different charge characteristics. Three major groups of materials are used in the manufacture of ion exchangers: polystyrene, cellulose, and acrylamide. The functional groups for IEC are amine derivatives, whereas cation-exchange chromatography uses sulfopropyl and carboxyl groups (30). The matrix, along with washing and elution buffers, provides selective peptide separation. The retention of the charged molecule on the matrix is induced by electrostatic interactions. The elution of the so-fixed molecule involves its displacement by a new counter-ion with a greater affinity for the fixed charges, which becomes the new nonfixed ion (31).

The type of ion-exchange resin used is determined by the properties of the molecules to be separated. For proteins or peptides, the isoelectric point is the first parameter to consider. The complexity of the mixture to be separated may turn out to be a limiting factor because many different peptides can generate equal charges, which makes them difficult to separate from one another.

32.3.5.1.2 Hydrophobic Interaction Chromatography

Hydrophobic interaction chromatography (HIC) uses a matrix made of hydrophobic material and the proteins/peptides to separate are bound in the presence of high concentration of a lyotropic salt such as ammonium sulfate (32). However, HIC is not commonly used because hydrophobic interactions are not very specific and are often too weak to allow tight binding. At the laboratory scale, reverse-phase chromatography (RP-HPLC) is also based on hydrophobicity. With this technique, peptides are allowed to bind on a silica column and are eluted as the concentration of hydrophobic solvent increases. RP-HPLC is, however, most commonly used as an analytical tool and is hardly applicable to large-scale separation processes.

32.3.5.1.3 Size-Exclusion Chromatography

Size-exclusion chromatography (SEC) or gel filtration chromatography separates molecules based on their size. The matrix is composed of a gel made from spherical beads with pores of a specific size distribution. In general, small molecules diffuse better into pores than do larger molecules, and so their retention times are higher. Consequently, molecules are eluted in the order of decreasing molecular weight. A review on SEC applied to peptides has been published by Irvine (33). SEC can be used to separate relatively large peptides, as the gel matrix (composed of agarose, dextran, or polyacrylamide) is not suitable for peptides of <1 kDa. A few peptide-designed SEC columns can differentiate peptides with molecular weights below 1 kDa, but such procedures are more appropriate for analytical techniques.

32.3.5.1.4 Affinity Chromatography

Affinity chromatography is a highly selective separation technique based on molecular recognition. Antibody–antigen, receptor—antagonist, and enzyme–substrate are examples of molecules interacting naturally. The choice of the matrix is dependent on the type of ligand used. Ligands can be used to bind specific functional groups on any molecule such as sulfhydryl or ammonia groups or to bind more specifically one single molecule such as avidin-biotin or antisense peptides. In general, ions can be used as ligands for specific binding in affinity chromatography or as chelating agents to cross-link molecules and cause them to precipitate. The affinity technology is not a recent technique and several review papers have been published (34,35). Applied on an industrial scale, purification of bioactive peptides by affinity chromatography may simplify the process, increase yield, and downsize capital equipment that would generally improve process economics (36). However, this method may suffer from irregularity and decrease in effectiveness with time.

Peptides have also been used as ligands for purification of different proteins. Peptides, being smaller than most ligands conventionally used, have several advantages. They are nontoxic, have a stable conformation, and often bind molecules with weak affinity, which requires no harsh conditions to elute (37). A tetrapeptide was used to bind fibrinogen proteins under mild conditions (38). S-protein was successfully purified from a ribonuclease hydrolyzate with specific peptides (37). The reverse was also performed, using immobilized S-protein to find interacting peptides from a peptide library (39). Gurgel et al. (40) also identified peptides that could specifically bind α-lactalbumin. Affinity for peptide purification using peptide ligands was demonstrated by Zhang et al. (41). They showed that peptides isolated from peptide libraries could recognize target peptidic sequences with excellent specificity. These affinities were modest compared with those of monoclonal antibody–peptide epitope interactions, but they are sufficient to support affinity purification strategies.

32.3.5.2 Other Techniques

32.3.5.2.1 Electromembrane Filtration (EMF)

The migration of charged peptides under an electric field (electrophoresis) has been combined with pressure-driven membrane-based processes to generate electromembrane filtration (EMF). A review of the theory related to EMF was published by Huotari et al. (42). Applications have essentially been used to enhance the permeation flux or improve membrane selectivity. Lentsh et al. (43) separated bovine serum albumin (BSA) from polyethylene glycol molecules using EMF. Daufin et al. (44) used EMF to separate peptides from a model solution of amino acids and peptides prepared from a tryptic hydrolyzate of casein. Moreover, it was demonstrated that positive peptides were actively transmitted or retained because of their high electrophoretic mobility, resulting in permeate or retentate enrichment according to the direction of the electric field. The selectivity, however, seemed to be impaired by peptide–membrane interactions. Nau et al. (45) demonstrated that many phenomena could regulate the selectivity of separation and that EMF needed to be more extensively studied. Ionic and hydrophobic interactions among the peptides and with the membrane complicated the interpretation of data resulting from the applied electric field. In later work, Bargeman et al. (46) selectively enhanced the separation of a positively charged amino acid, lysine, and applied this technique to separate the antibacterial cationic peptide lactoferricin. EMF has recently been used to separate bioactive peptides from a casein hydrolyzate (46). Fragment α_{S2}-casein 183–207 is a cationic antibacterial peptide that could be enriched from 7.5% of the total protein to 25% in the permeate. EMF can be considered a suitable technique for separating the cationic antibacterial peptide from a casein hydrolyzate, as the final product purity and the efficiency of the process are comparable with chromatographic techniques but may be more cost-effective.

32.3.5.2.2 Isoelectric Focusing

Isoelectric focusing (IEF) is an electrophoresis technique that uses the electromigration potential of charged species to migrate in a pH gradient created by carrier ampholytes between two electrodes. IEF was conventionally used as an analytical technique, but the development of matrix-free units made possible the use of IEF as a preparative technique. Separation of peptides can also be achieved with this technique. Since peptides have a greater diffusion coefficient than do proteins, they can be used to generate their own pH gradient, without additional ampholytes. Yata et al. (47) used matrix-free preparative IEF to fractionate peptides from protein digests, without additional ampholytes. Because of the amphoteric nature of peptides, no buffers or electrolytes were added, and the collected fractions were in aqueous form and free of chemicals. This technique, referred to as autofocusing, was subsequently used by Groleau et al. (48) to fractionate a tryptic hydrolyzate from β-LG. The Rotofor cell (Bio-Rad) used in both these studies can only fractionate small volumes (40 mL) of diluted solutions (12). Productivity is therefore low, and the technique needs to be upscaled for practical applications.

A larger-scale system was developed by Righetti et al. (49) and by Akahoshi et al. (50). Larger volumes were used (400 mL), less precipitation was achieved, and higher recovery yield was obtained. According to the authors, the separation potential of this technique has to be taken into consideration, since it might have a great influence on the preparation of bioactive peptide-based functional foods.

32.4 EXAMPLES AND APPLICATIONS

Examples have been mentioned throughout the description of the methods, but combined techniques are also often used to isolate bioactive peptides. The following sections will provide examples designed for casein phosphopeptides (CPP), lactoferricin, and nisin (a bacteriocin).

32.4.1 Separation of Casein Phosphopeptides (CPP)

Phosphopeptides from casein, released by tryptic digestion, are recognized for their calcium-binding properties, which play an important role in calcium bioavailability in the small intestine as well as in bone and teeth mineralization. CPP have been separated by different methods from casein hydrolyzates. Ultrafiltration (51) and selective precipitation (14) yielded low-purity product. Methods such as immobilized metal ion chromatography (IMAC) in which phosphorylated residues of CPP are bound to a copper ion matrix have been developed. CPP were shown to interact with ions at low pH values in the presence of sodium chrloride (NaCl) and were then eluted with phosphate buffer. Two sequences were selectively separated, one peptide from casein α_{s1} (fragment 106–119) and one from casein β (fragment 1–25). This method is fairly simple but is limited to the separation of those two peptides; other CPP out of a total of nine CPP produced in a tryptic hydrolyzate of caseins are not separated.

Ellegard et al. (52) used a more complex, albeit more efficient, combination of large-scale methods to separate CPP from a tryptic hydrolyzate of sodium caseinate solution. As shown in Figure 32.5, the product of the hydrolysis reaction was subjected to selective precipitation at pH 4.6 to eliminate hydrophobic peptides. The soluble fraction was purified by ultrafiltration and diafiltration. Phosphorylated peptides were further separated from nonphosphorylated peptides by ion-exchange chromatography, on an agarose anion-exchange resin. Nonphosphorylated peptides were repelled from the resin by phosphorylated peptides of higher negative charges. sodium hydroxide (NaOH) was used to elute bound phosphorylated peptides. At the end of this process, 16% of the material was recovered as CPP. As mentioned by the authors, the economy of the process is strongly dependent on this low yield, which is mainly a result of the parameters used for chromatography.

32.4.2 Production of Lactoferricin

Ion-exchange chromatography is used to isolate lactoferricin, a bioactive peptide derived from lactoferrin, an iron-binding milk protein. After peptic hydrolysis of lactoferrin, lactoferricin is released and can be isolated on a cation exchanger with

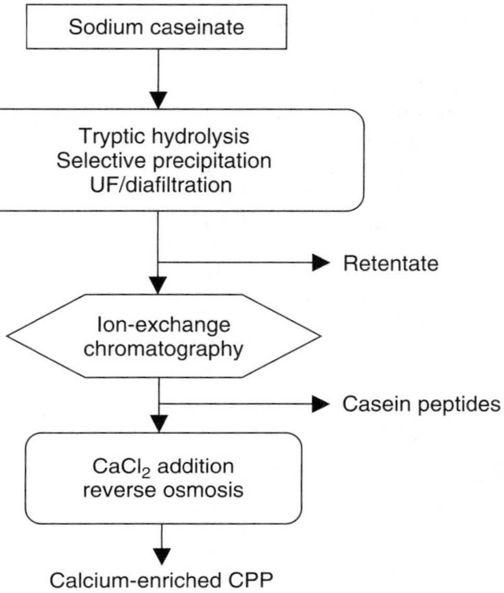

FIGURE 32.5 Preparation of CPP. (From K.H. Ellegard, C. Gammelgard-Larsen, E.S. Sorensen, S. Fedesov. *Int. Dairy J.*, 9: 639–652, 1999. With permission.)

an ammonium buffer to bind this cationic peptide. Lactoferricin remains bound to the matrix until elution is performed with 2 M NaCl (53,54). Lactoferricin can, however, be produced *in situ* by filtering cheese whey through a cation-exchange membrane, which selectively binds lactoferrin and can be directly hydrolyzed with pepsin (Figure 32.6). Inactive lactoferrin fragments are then washed off the membrane with ammonia, and lactoferricin can be eluted again with 2 M NaCl. The elution product can be freeze-dried as a powder with over 95% purity (55).

32.4.3 Recovery of Nisin from Fermentation Broths

Bacteriocins are defined as bacteria-derived proteinaceous compounds having a relatively narrow spectrum of bactericidal activity. The cationic and hydrophobic nature of bacteriocins is used for their recovery from complex fermentation broths (56). Carolissen-Mackay et al. (57) reviewed purification protocols for bacteriocins from lactic acid bacteria. Most laboratory purification protocols (ammonium sulfate precipitation, ion-exchange, immunoaffinity, or hydrophobic chromatography) provide high purity products but can be difficult to adapt to large-scale purification. Methods based on adsorption/desorption are more commonly used for large-scale recovery of bacteriocins.

Nisin, an antimicrobial peptide produced by certain *Lactococcus* strains, has practical application in the food industry, as it is licensed for food use in more than 40 countries (56). Nisin A, a food-grade bacteriocin, was purified by a single-step immunoaffinity method (58). An immunoadsorption matrix was developed by

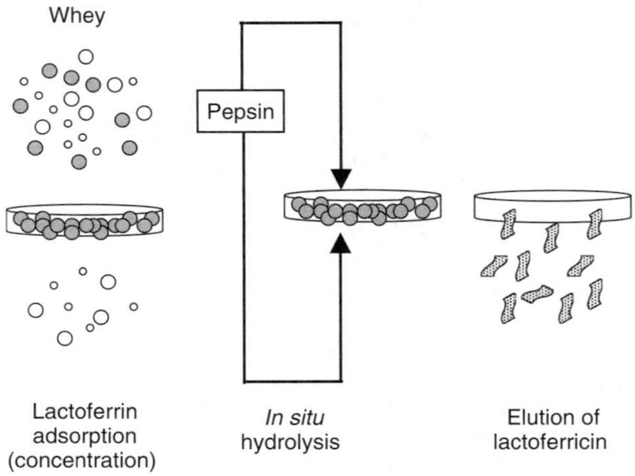

FIGURE 32.6 Purification technique developed for lactoferricin. (From Recio, 1999. With permission)

direct binding of anti-nisin A monoclonal antibodies to activated sepharose. The fermentation broth containing nisin was passed on a nisin-binding matrix. Nisin was then eluted with a glycine-HCl (hydrochloride) buffer. This one-step purification method provides a final yield that is 30-fold higher than those of conventional methods and could easily be scaled up, according to the authors.

Combinations of techniques have also been used to recover bacteriocins. A two-step separation process (adsorption on hydrophobic interaction chromatography resin followed by a cation-exchange resin) was developed by Stoffels et al. (59) and provided 100% nisin recovery and a 60-fold concentration, compared with that of the initial media. Boyaval et al. (60) developed another two-step purification based on phase partitioning in a detergent and adsorption/desorption on a cation-exchange resin. In a process developed by Uteng et al. (61), bacteriocins were bound on a sepharose cation-exchanger column and eluted with NaCl. In the second step, the NaCl fraction from the cation exchanger was applied at a high flow rate on a low-pressure reverse-phase column. Bacteriocins bound to the column and were detected as a single peak that could be collected. A 90% pure bacteriocin fraction was obtained in a shorter time than with the conventional method.

32.5 CONCLUSIONS

Membrane-based separation techniques, as unit operations in a general fractionation/purification process, are powerful tools for peptide separation. Although some processes have been applied on a commercial scale, a larger number of processes are too complex and expensive to be industrialized. High-resolution purification techniques that could perform peptide separation in a minimum of processing steps are still needed. The development of new research areas such as proteomics may prove useful in defining new tools (i.e., proteins or ligands) that could specifically bind bioactive peptides in one single step.

REFERENCES

1. D.A. Clare, H.E. Swaisgood. Bioactive peptides: a prospectus. *J. Dairy Sci.*, 83: 1187–1195, 2000.
2. N.P. Shah. Effects of milk-derived bioactives: an overview. *Br. J. Nutr.*, 84: 3–10, 2000.
3. D. Tomé. In: *Milk and Health. Proceedings of the 25th International Dairy Congress 21–24, September 1998. Aarhus, Denmark.* The Danish National committee of the IDF, 1999, pp. 163–181.
4. E. Smacchi, M. Gobbetti. Bioactive peptides in dairy products: synthesis and interaction with proteolytic enzymes. *Food Microbiol.*, 17: 129–141, 2000.
5. I. Gill, R. Lopez-Fandino, X. Jorba, E.N. Vulfson. Biologically active peptides and enzymatic approaches to their production. *Enzyme Microb. Technol.*, 18: 162–183, 1996.
6. P.L. Leclerc, S.F. Gauthier, H. Bachelard, M. Santure, D. Roy. Antihypertensive activity of casein-enriched milk fermented by lactobacillus helveticus. *Int. Dairy J.*, 12: 995–1004, 2002.
7. N. Yamamoto, M. Maeno, T. Takano. Purification and characterization of an antihypertensive peptide from a yogurt-like product fermented by lactobacillus helveticus CPN4. *J. Dairy Sci.*, 82: 1388–1393, 1999.
8. M.A.F. Belem, B.F. Gibbs, B.H. Lee. Proposing sequences for peptides derived from whey fermentation with potential bioactive sites. *J. Dairy Sci.*, 82: 486–493, 1999.
9. P.Y. Pontalier, A. Ismail, M. Ghoul. Mechanism for the selective rejection of solutes in nanofiltration membranes. *Sep. Purif. Technol.*, 12: 175–181, 1997.
10. T.E. Creighton. *Proteins: Structures and Molecular Properties*, 2 edition, W.H. Freeman and Company, New York, USA, 2002.
11. J. Léonil, D. Mollé, S. Bouhallab, G. Henry. Precipitation of hydrophobic peptides from casein hydrolyzate by salt and pH. *Enzyme Microb. Technol.*, 16: 591–595, 1994.
12. P.E. Groleau, P. Morin, S.F. Gauthier, Y. Pouliot. Effect of physicochemical conditions on peptide-peptide interactions in a tryptic hydrolyzate of β-lactoglobulin and identification of aggregating peptides. *J. Agric. Food Chem.*, 51: 4370–4375, 2003.
13. A. Pihlanto-Leppala, T. Rokka, H. Korhonen. Angiotensin I converting enzyme inhibitory peptides derived from bovine milk proteins. *Int. Dairy J.*, 8: 325–331, 1998.
14. N.J. Adamson, E.C. Reynolds. Characterization of tryptic casein phosphopeptides prepared under industrially relevant conditions. *Biotechnol. Bioeng.*, 45: 196, 1995.
15. N.A. Kelly, B.G. Reuben, J. Rhoades, S. Roller. Solvent extraction of bacteriocins from model solutions and fermentation broths. *J. Chem. Technol. Biotechnol.*, 75: 777–784, 2000.
16. S.L. Turgeon, S.F. Gauthier. Whey peptide fractions obtained with a two-step ultrafiltration process: production and characterization. *J. Food Sci.*, 55: 106–110, 1990.

17. V.L. Johnson. Proteins in cosmetic and toiletries. *Drug and Cosmetic Industry*, 126: 36, 1980.
18. J.M. Chobert, C. Bertrand-Harb, M.G. Nicolas. Solubility and emulsifying properties of caseins and whey proteins modified enzymatically by trypsin. *J. Agric. Food Chem.*, 36: 883–892, 1988.
19. S.W. Lee, M. Shimizu, S. Kaminogawa, K. Yamauchi. Emulsifying properties of peptides obtained from the hydrolyzate of β-caseins. *Agric. Biol. Chem.*, 51: 161, 1987.
20. T. Tsuru, T. Shutou, S.-I. Nakao, S. Kimura. Peptide and amino acid separation with nanofiltration membranes. *Sep. Sci. Technol.*, 29: 971–984, 1994.
21. A. Garem, J. Léonil, G. Daufin, J.L. Maubois. Nanofiltration d'acides aminés sur membranes organiques : influence des paramètres physico-chimiques et de la pression transmembranaire sur la sélectivité. *Lait*, 76: 267–281, 1996.
22. S. Kimura, A. Tamano. Separation of amino acids by charged ultrafiltration membranes. In: E. Droli, N. Nakagaki (Eds.). *Membranes and Membrane Processes*. New York: Plenum Press, 1986.
23. C. Martin-Orue, S. Bouhallab, A. Garem. Nanofiltration of amino acid and peptide solutions: mechanisms of separation. *J. Memb. Sci.*, 142: 225–233, 1998.
24. F. Nau, F.L. Kerhervé, J. Léonil, G. Daufin, P. Aimar. Separation of β-casein peptides through UF inorganic membranes. *Bioseparation*, 3: 205–215, 1993.
25. Y. Pouliot, M.C. Wijers, S.F. Gauthier, L. Nadeau. Fractionation of whey protein hydrolyzates using charged UF/NF membranes. *J. Memb. Sci.*, 158: 105–114, 1999.
26. Y. Pouliot, S.F. Gauthier, J. L'Heureux. Effect of peptide distribution on the fractionation of whey protein hydrolyzates by nanofiltration membranes. *Lait*, 80: 113–122, 2000.
27. A. Garem, G. Daufin, J.L. Maubois, B. Chaufer, J. Léonil. Ionic interactions in nanofiltration of β-casein peptides. *Biotechnol. Bioeng.*, 57: 109–117, 1998.
28. J.F. Lapointe, S.F. Gauthier, Y. Pouliot, C. Bouchard. Effect of hydrodynamic conditions on fractionation of β-lactoglobulin tryptic peptides using nanofiltration membranes. *J. Memb. Sci.*, 212: 55–67, 2002.
29. S. Bordenave, F. Sannier, G. Ricart, J.M. Piot. Continuous hydrolysis of goat whey in an ultrafiltration reactor: generation of alpha-lactorphin. *Prep. Biochem. Biotechnol.*, 29: 189–202, 1999.
30. C.K. Larive, S.M. Lunte, M. Zhong, M.D. Perkins, G.S. Wilson, G. Gokulrangan, T. Williams, F. Afroz, C. Schoneich, T.S. Derrick, C.R. Middaugh, S. Bogdanowich-Knipp. Separation and analysis of peptides and proteins. *Anal. Chem.*, 71: 389–423, 1999.
31. E.F. Rossomando. Ion-Exchange chromatography. In: M.P. Deutscher (Ed.). *Guide to Protein Purification*. New York: Academic Press, Inc., 309–316, 1990.
32. S.C. Burton, D.R.K. Harding. Hydrophobic charge induction chromatography: salt independent protein adsorption and facile elution with aqueous buffers. *J. Chromatogr.*, A 814: 71–81, 1998.
33. G.B. Irvine. Size-exclusion high-performance liquid chromatography of peptides: a review. *Anal. Chim. Acta*, 352: 387–397, 1997.
34. V. Gaberc-Poreka, V. Menart. Perspectives of immobilized-metal affinity chromatography. *J. Biochem. Biophys. Meth.*, 49: 335–360, 2001.
35. N. Labrou, Y.D. Clonis. The affinity technology in downstream processing. *J. Biotechnol.*, 36: 95–119, 1994.

36. C.R. Lowe, A.R. Lowe, G. Gupta. New developments in affinity chromatography with potential application in the production of biopharmaceutical. *J. Biochem. Biophys. Meth.*, 49: 561–574, 2001.
37. P.Y. Huang, R.G. Carbonell. Affinity purification of proteins using ligands derived from peptide libraries. *Biotechnol. Bioeng.*, 47: 288–297, 1995.
38. A. Pingali, B. McGuinness, H. Keshishian, J. Fei-Wu, L. Varaday, F. Regnier. Peptides as affinity surfaces for protein purification. *J. Mol. Recognit.*, 9: 426–432, 1996.
39. P.Y. Huang, R.G. Carbonell. Affinity chromatographic screening of soluble combinatorial peptide libraries. *Biotechnol. Bioeng.*, 63: 633–641, 1999.
40. P.V. Gurgel, R.G. Carbonell, H.E. Swaisgood. Studies of the binding of α-lactalbumin to immobilized peptide ligands. *J. Agric. Food Chem.*, 49: 5765–5770, 2001.
41. Z. Zhang, W. Zhu, T. Kodadek. Selection and application of peptide-binding peptides. *Nature Biotechnol.*, 18: 71–74, 2000.
42. H.M. Huotari, G. Tragadardh, I.H. Huisman. Crossflow membrane filtration enhanced by an external DC electric field: a review. *Trans. IChemE.*, 77: 461–468, 1999.
43. S. Lentsch, P. Aimar, J.L. Orozco. Enhanced separation of albumin-poly(ethylene glycol) by combination of ultrafiltration and electrophoresis. *J. Membr. Sci.*, 80: 221–232, 1993.
44. G. Daufin, F. Kerhervé, P. Aimar, D. Mollé, J. Léonil, F. Nau. Electrofiltration of solutions of amino acids or peptides. *Lait*, 75: 105–115, 1995.
45. F. Nau, F. Kerhervé, J. Léonil, G. Daufin. Selective separation of tryptic b-casein peptides through ultrafiltration membranes: influence of ionic interactions. *Biotechnol. Bioeng.*, 46: 246–253, 1995.
46. G. Bargeman, M. Dohmen-Speelmans, I. Recio, M. Timmer, C. van der Horst. Selective isolation of cationic amino acids and peptides by electro-membrane filtration. *Lait*, 80: 175–185, 2000.
47. M. Yata, K. Sato, K. Ohtsuki, M. Kawabata. Fractionation of peptides in protease digest of proteins by preparative isoelectric focusing in the absence of added ampholytes: a biocompatible and low-cost approach referred to autofocusing. *J. Agric. Food Chem.*, 44: 76–79, 1996.
48. P.E. Groleau, R. Jimenez-Flores, S.F. Gauthier, Y. Pouliot. Fractionation of β-lactoglobulin tryptic peptides by ampholyte-free isoelectric focusing. *J. Agric. Food Chem.*, 50: 578–583, 2002.
49. P.G. Righetti, F. Nembri, A. Bossi, M. Mortarino. Continuous enzymatic hydrolysis of β-casein and isoelectric collection of some of the biologically active peptides in an electric field. *Biotechnol. Prog.*, 13: 258–264, 1997.
50. A. Akahoshi, K. Sato, Y. Nawa, Y. Nakamura, K. Ohtsuki. Novel approach for large-scale, biocompatible, and low-cost fractionation of peptides in proteolytic digest of food protein based on the amphoteric nature of peptides. *J. Agric. Food Chem.*, 48: 1955–1959, 2000.
51. V. Gagnaire, A. Pierre, D. Mollé, J. Léonil. Phosphopeptides interacting with colloidal calcium phosphate isolated by tryptic hydrolysis of bovine micelles. *J. Dairy Res.*, 63: 405–422, 1996.
52. K.H. Ellegard. C. Gammelgard-Larsen, E.S. Sorensen, S. Fedesov. Process scale chromatographic isolation, characterization and identification of tryptic bioactive casein phosphopeptides. *Int. Dairy J.*, 9: 639–652, 1999.

53. D.A. Dyonisius, J.M. Milne. Antibacterial peptides of bovine lactoferrin: purification and characterization. *J. Dairy Sci.*, 80: 667–674, 1997.
54. I. Recio, S. Visser. Two-ion-exchange chromatographic methods for the isolation of antibacterial peptides from lactoferrin in situ enzymatic hydrolysis on an ion-exchange membrane. *J. Chromatogr.*, A 831: 191–201, 1999.
55. M. Tomita, H. Wakabaya, K. Yamauchi, S. Teraguchi, H. Hayasawa. Bovine lactoferrin and lactoferricin derived from milk: production and applications. *Biochem. Cell Biol.*, 80: 109–112, 2002.
56. E. Parente, A. Ricciardi. Production, recovery and purification of bacteriocins from lactic acid bacteria. *Appl. Microbiol. Biotechnol.*, 52: 628–638, 1999.
57. V. Carolissen-Mackay, G. Arendse, J.W. Hastings. Purification of bacteriocins of lactic acid bacteria: problems and pointers. *Int. J. Food Microbiol.*, 34: 1–16, 1997.
58. A.M. Suarez, J.I. Azcona, J.M. Rodriguez, B. Sanz, P.E. Hernandez. One-step purification of nisin A by immunoaffinity chromatography. *Appl. Environ. Microbiol.*, 63: 4990–4992, 1997.
59. G. Stoffels, H.G. Sahl, A. Gudmundsdottir. Carnocin U149, a potential biopreservative produced by carnobacterium piscicola: large scale purification and activity against various Gram-positive bacteria including *Listeria* sp. *Int. J. Food Microbiol.*, 20: 199–210, 1993.
60. P. Boyaval, P. Bhugallo-Vial, F. Duffes, A. Metivier, X. Dousset, D. Marion. Réacteurs à hautes densités cellulaires pour la production de solutions concentrées de bactériocines. *Lait*, 78: 129–133, 1998.
61. M. Uteng, H.H. Hauge, I. Brondz, J. Nissen-Meyer, G. Fimland. Rapid two-step procedure for large-scale purification of pediocin-like bacteriocins and other cationic antimicrobial peptides from complex culture medium. *Appl. Environ. Microbiol.*, 68: 952–956, 2002.

Index